DATE DUE

~~NO 3 98~~		
~~FE 18 '99~~		
~~FE 10 00~~		
~~AG 3 '00~~		
~~OCT 7 00~~		
~~NO 4 00~~		
~~NO 22 00~~		
~~DE 9 00~~		
~~AP 7 01~~		
~~DE 21 02~~		
~~NO 5 03~~		

DEMCO 38-296

Auto Fuel and Emission Control Systems
TECHNOLOGY

by

James E. Duffy
Automotive Writer

Howard Bud Smith
Executive Editor-Technology
Goodheart-Willcox Co.

South Holland, Illinois
THE GOODHEART-WILLCOX COMPANY, INC.
Publishers

Library of Congress Catalog Card Number 91-38813
International Standard Book Number 0-87006-932-2

3 4 5 6 7 8 9 10   92   96 95

Library of Congress Cataloging in Publication Data

Duffy, James E.
    Auto fuel and emissions systems / by James E.
Duffy, Howard Bud Smith.
        p.   cm.
    Rev. ed. of: Auto fuel systems. c1987
    Includes index.
    ISBN 0-87006-932-2
    1. Automobiles—Fuel systems. 2. Auto-
mobiles—Fuel Systems—Maintenance and repair.
3. Automobiles—Motors—Exhaust systems.
4. Automobiles—Motors—Exhaust systems—
Maintenance and repair.
I. Smith, Howard Bud. II. Duffy, James E.
Auto fuel systems. III. Title.
TL214.F8D83 1992
629.25'3—--dc20   91-38813
                                                          CIP

INTRODUCTION

Auto Fuel and Emission Control Systems Technology is designed to help prepare you to properly diagnose and repair all types of automotive fuel systems. It provides full coverage of both the theory and the service procedures that you will need to be a fuel system technician.

A few years ago, carburetors and simple mechanical fuel pumps were used on almost all passenger cars. This made fuel system service a comparatively simple job. Today, late model cars come equipped with electronically controlled carburetors, electronic fuel injection—both multi and single point, as well as diesel injection. Various emission control devices, engine sensors, and the computer also add to fuel system complexity. As a result, there is a demand for technicians capable of repairing fuel systems and related components.

The book is divided into five sections and 28 chapters. The first section is a quick review of fundamentals: engine operation, safety, special service tools and equipment, measurements, formulas, fuels, and combustion. The other sections cover the operation, construction, and service of fuel emission system components.

Usually, one chapter will cover the theory and construction of a part or subsystem. Then, the following chapter will summarize the diagnosis, testing, and repair of the same system. As a result, a buildup of knowledge occurs as you progress through the book.

Each chapter opens with learning objectives so you will know what you are expected to learn in the chapter. The end of each chapter has "Know These Terms" and "Review Questions" so that you can check how well you understand what you have studied. Both conventional and ASE certification type questions are used.

Auto Fuel and Emission Control Systems Technology will prepare you to pass three of the ASE certification tests—Engine Performance, Engine Repair, and Electrical Systems. Potential questions for these tests relate to fuel system service. You will also need additional work experience and study in other fields.

To help you learn the "language" of a fuel system technician, all technical terms are printed in italics and defined immediatley after their first use. This highlights the word and reminds you it is a word or term you must understand.

Auto Fuel and Emission Control Systems Technology is a valuable guide to anyone wanting to learn about today's automotive fuel systems. It will help students enter the trade more easily, as well as help the experienced mechanic become more informed of new advances in technology. It is an excellent reference for the car owner who wants to know more about the operation and service of fuel systems.

James E. Duffy
Howard Bud Smith

CONTENTS

SECTION 4—FUEL-RELATED SYSTEMS

SECTION 5—ON-THE-JOB SKILLS

IMPORTANT SAFETY NOTICE

Proper service and repair methods are critical to the safe, reliable operation of automobiles. The procedures described in this book are designed to help you use a manufacturer's service manual. A service manual will give the how-to details and specifications needed to do competent work.

This book contains safety precautions which must be followed. Personal injury or part damage can result when basic safety rules are not followed. Also, remember that these cautions are general and do not cover some specialized hazards. Refer to a service manual when in doubt about any service operation!

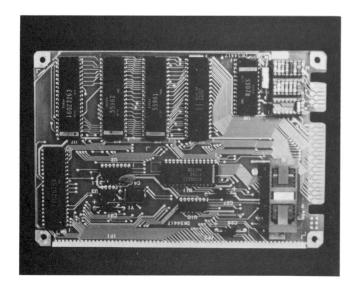

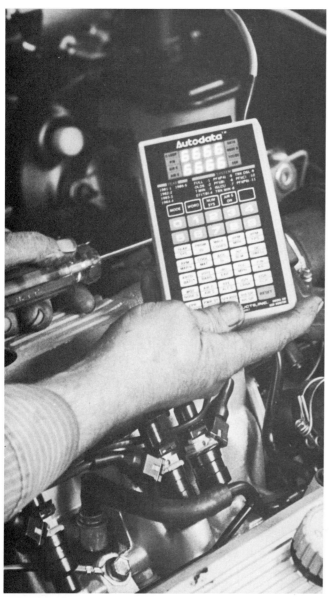

This book will help you become a competent fuel system technician by summarizing the most important information relating to fuel system operation and repair. Shown are a few of the components and tests that you will learn about when studying this textbook. (Chevrolet, Buick, Microtron Products, and Peerless Instruments)

1 REVIEW OF ENGINES AND ENGINE SYSTEMS

After studying this chapter, you will be able to:
- *List major parts of an automotive engine.*
- *Describe the operation of an automotive engine.*
- *List the major systems of an engine.*
- *Describe the operation of major engine systems.*
- *Explain how modern automotive engine systems interact.*
- *Relate engine operation to fuel system operation.*

With today's "high-tech" cars, no one system of an automotive engine can operate independently of other systems. A malfunction in one system can cause faulty operation of an entirely different system. To find a problem and solve it, you have to check several of the engine's systems or even the mechanical condition of the engine itself.

To cite a few examples, a faulty electrical part, such as an engine sensor, can affect the delivery of fuel to the engine. Similarly, poor mechanical condition of the engine (low compression in one cylinder) may create symptoms suggesting faulty fuel delivery by a gasoline injector.

It is wasteful and frustrating to replace parts by trial and error. A general understanding of ALL engine systems enables the mechanic to test and eliminate possible causes in an orderly and logical sequence.

For this reason, this text chapter reviews basic engine operation. This will help you more fully understand later chapters detailing fuel system construction, operation, diagnosis, and repair.

BASIC ENGINE OPERATION

The sole function of an auto's engine is to provide power. A car engine is capable of producing heat. It then harnesses the energy in the heat to propel the vehicle and energize all of the vehicle's systems.

Initially, power for starting is provided by the battery

Fig. 1-1. Engine provides energy for all systems. Modern automobile has many accessories requiring extra power from engine. (Saab)

which supplies electrical power to an electric starting motor. Once it has "fired up," the engine takes over and operates various systems which are necessary to keep the engine running.

A gasoline engine, for example, powers the ignition system which supplies voltage for producing an electrical spark to start combustion in the cylinders. It supplies turning force to the water pump that circulates coolant to carry away the excess heat of combustion.

Piston action creates a vacuum for fuel induction (intake), vacuum for power brakes, and vacuum for use by other systems. Hydraulic pumps, running off the crankshaft of the engine, provide "muscle" for steering and oil for engine protection. The list goes on with every additional accessory taking its energy from one source—the engine. See Fig. 1-1.

Converting heat energy

The *engine,* also called a *power plant,* is designed to ignite a combustible mixture of vaporized fuel. Usually, the fuel is gasoline or diesel fuel. The burning vapors produce heat. Shown in Fig. 1-2, heat from the ignited fuel mixture causes expansion of gases in the engine combustion chamber. This creates great pressure.

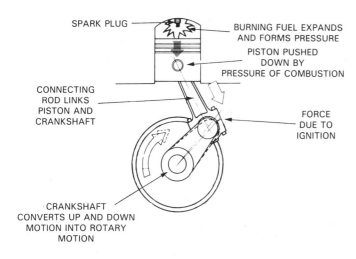

Fig. 1-2. Pressure of burning, expanding gases pushes piston down, turning crankshaft at same time.

Certain parts of the engine convert this pressure into motion. These parts include the piston, connecting rod, and crankshaft. They transform the gaseous pressure into a spinning motion. This provides power for moving the car and operating its various systems.

Because the burning, expanding fuel vapors have nowhere else to go in the combustion chamber, they push on the piston. The piston is forced to move downward. The connecting rod moves down too. It forces the crank to rotate with a powerful spinning motion. The crank's rotary motion can be used to drive gears, chains, belts, and other devices in the automobile. Finally, engine power is utilized to rotate the drive wheels to propel the car over the road, Fig. 1-3.

Understanding piston travel

Piston travel in the cylinder is controlled by the crankshaft. As shown in Fig. 1-4, the points at which

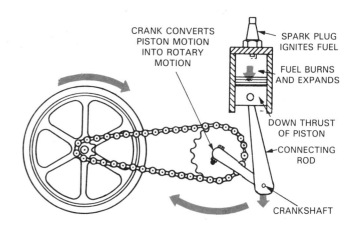

Fig. 1-3. Simple diagram showing how reciprocating motion of engine piston is changed to rotary motion suitable for turning drive wheel.

it changes direction are called TDC (top dead center) and BDC (bottom dead center). TDC occurs when the piston is at the HIGHEST point in the cylinder. BDC is when the piston is at the LOWEST point in the cylinder.

FOUR-STROKE CYCLE

Most engines today operate on a four-stroke cycle. One *stroke* of the piston is its movement from TDC to BDC. At the same time, the crankshaft will rotate 180 degrees or a half turn.

One *cycle* in a four-stroke cycle engine consists of four strokes of the piston and two complete revolutions of the crankshaft. The engine produces power on every fourth stroke. Fig. 1-5 illustrates one complete cycle. A cycle is also known as a COMPLETE SERIES OF EVENTS.

The *intake stroke* of a four-stroke cycle gasoline engine draws fuel into the cylinder as the piston moves from TDC to BDC. As shown in Fig. 1-5A, the intake valve is open to allow fuel entry. The exhaust valve is closed. As the piston travels, it causes a *vacuum* (low pressure area) in the cylinder. Higher atmospheric pressure (out-

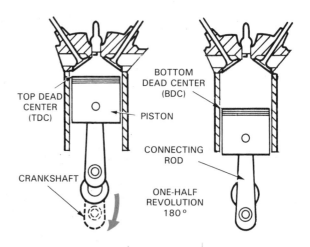

Fig. 1-4. Left. When piston has traveled as far upward as it can go, it is at top dead center (TDC) and about to begin downward travel. Right. Piston shown at bottom dead center (BDC). It is about to move upward again.

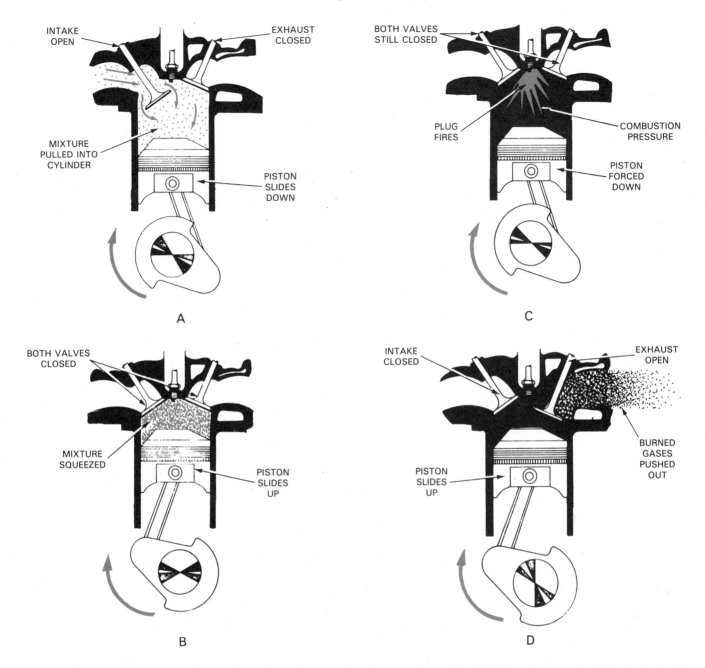

Fig. 1-5. Four-stroke cycle. A—On intake stroke, piston begins downward travel and intake valve opens. Air-fuel mixture is drawn into combustion chamber. B—Compression stroke begins as intake valve closes and piston begins upward travel. C—Power stroke begins as air-fuel mixture burns and expands. Piston is forced down rapidly to spin crankshaft. D—Exhaust stroke occurs with exhaust valve open so spent (burned) gases can enter exhaust manifold. (TRW)

side air pressure) outside the cylinder pushes the air-fuel vapor into the cylinder.

At BDC, the crank rotation causes the piston to reverse direction. It travels back up the cylinder, producing the *compression stroke.* This piston movement squeezes or compresses the air-fuel mixture prior to combustion. During this stroke both valves are closed, Fig. 1-5B.

The fuel-air mixture is under heavy compression as the piston nears TDC. The mixture becomes highly *combustible* (will burn easily) because of the pressure.

The *power stroke* begins as the spark plug "fires" and the pressurized air-fuel mixture ignites. The expanding, hot gases from the burning fuel build up pressure rapidly. Since both valves are still closed, the piston is pushed downward, rapidly spinning the crankshaft, as shown in Fig. 1-5C.

This is the only stroke that does NOT consume power. The fuel charge may be ignited by a SPARK (gasoline engine) or by high compression stroke pressure and HOT AIR (diesel engine).

At the bottom of the power stroke, just as the piston changes direction of travel, the exhaust valve opens. The *exhaust stroke* pushes burned gases out of the cylinder. The gases move out through the exhaust port into the exhaust system, Fig. 1-5D.

In an operating engine, these series of events happen over and over again in very rapid succession. The engine is designed so that valve opening and closing, piston travel, fuel injection, and ignition are timed for high efficiency. This requires careful engineering.

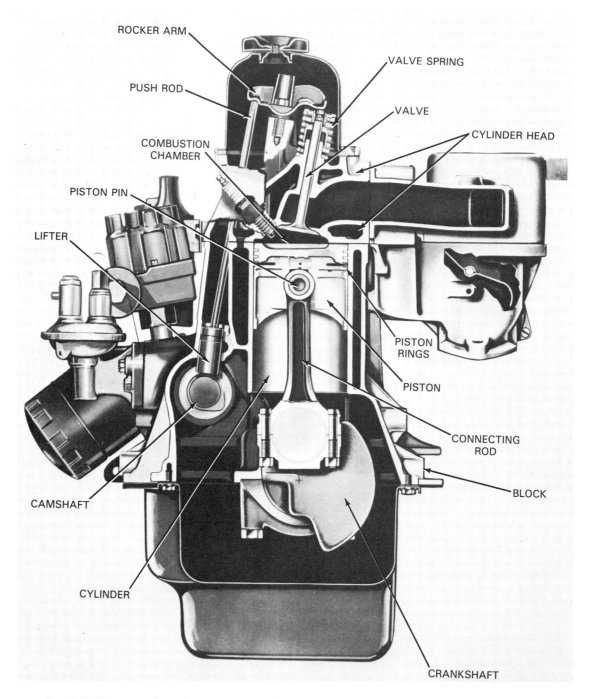

Fig. 1-6. Cutaway view of complete engine. Note names of various parts. (American Motors)

BASIC ENGINE PARTS

Large or small, all piston engines have the same basic parts. Refer to Fig. 1-6 as each part is described.
1. BLOCK—This is a heavy, machined casting that holds all other parts of the engine. It may be made of cast iron or aluminum.
2. CYLINDERS—These are round holes or sleeves bored or pressed into the block. They provide a guide for the pistons and they control piston reciprocating movement.
3. PISTONS—These are basically plugs open at one end to receive the connecting rod. The other end is closed to prevent the escape of burning gases to

the bottom of the block. The closed end also provides a stout surface to withstand the violent pushing action of the hot combustion gases. The pistons transfer energy to the connecting rod.
4. RINGS—They are designed to seal the small space between the piston and cylinder. Rings are strips of tempered steel or cast iron shaped in a circle. They fit into machined grooves in the piston. Rings prevent combustion pressure and oil from leaking past the narrow space between the piston and cylinder wall.
5. CONNECTING ROD—Heavy shaft with machined bores (holes) at each end. It links the piston to the crankshaft.

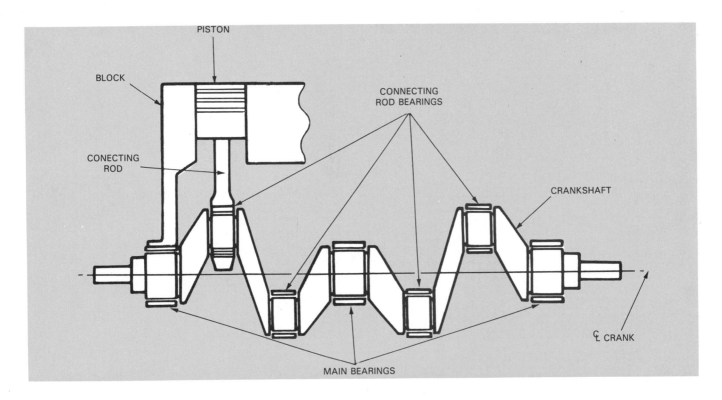

Fig. 1-7. Simple drawing of crankshaft. Crank is held in block by main bearings. (Ford)

6. CRANKSHAFT—A shaft with offsets where rods are connected. It changes the reciprocating motion of the pistons to rotary motion. See Fig. 1-7.
7. CYLINDER HEAD—This is a heavy "cap" that bolts to the top of the block to seal the tops of the cylinders. In most engines, it also contains the valves, valve springs, and rocker arm assemblies. Sometimes, it also contains the camshaft.
8. COMBUSTION CHAMBER—This is the space between the top of the piston and the cylinder head. The air-fuel mixture is introduced to this space and burned.
9. FLYWHEEL—This is a heavy metal wheel which attaches to the rear of the crankshaft. It provides momentum to keep the crankshaft turning smoothly between power strokes.
10. VALVES—Long-stemmed plugs, located in the cylinder head, that open and close to perform several jobs:
    a. Seal the combustion chamber during the power stroke.
    b. Control entrance of the air-fuel mixture into the combustion chamber.
    c. Provide for escape of exhaust gases (spent, burned mixture) from the combustion chamber.
11. CAMSHAFT—Linked with the crankshaft, this shaft rotates whenever the engine is running. Along its length, Fig. 1-8, are egg-shaped bumps called *lobes.* During rotation, these lobes open and close the engine valves in a carefully timed sequence. Linkage to the crankshaft is made with a system of gears, or sprockets and a chain, or sprockets and a belt.
12. VALVE TRAIN—Parts which are used to open and close the engine valves.

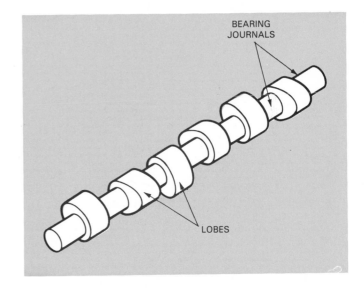

Fig. 1-8. Camshaft changes rotary motion into up and down motion for the valve train by using a series of lobes. (Ford Motor Co.)

VALVE TRAIN ARRANGEMENT

The *valve train,* Fig. 1-9, transfers the lifting motion of the cam lobes to the engine valves. It is a system of parts consisting of lifters, push rods, rocker arms, and valve spring assemblies. Because of design differences, not every engine has all of these parts in its drive train.

Lifters or tappets, as they are sometimes called, rest on top of the camshaft lobes. Since the lobes are off-center, they cause the lifters to move up and down. The purpose of the lifter is to transfer this motion to the rest

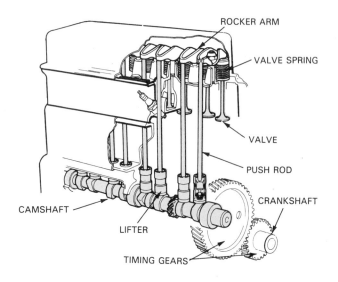

Fig. 1-9. Valve train includes valves and all mechanisms which open and close them (note colored portion). Refer to this illustration as you review operation of valve train. (Federal Mogul)

Valve train operation

Since the valve train is mechanically connected to the crankshaft, it is put in motion as the crankshaft and pistons move. As the piston travels through its four-stroke cycle, the valve train is also moving. This motion is perfectly timed to open one of the valves during the intake stroke and another one at the exhaust stroke.

Engine valves

Engine valves open and close the ports through which fuel and air (gasoline engine) or air (diesel engine) enters the cylinder and through which exhaust leaves. Normally there are two valves, Fig. 1-11, for each cylinder.

The larger is an *intake valve.* It controls the port to the intake manifold and, thus, controls flow into the cylinder.

The smaller *exhaust valve* controls the flow of exhaust gases out of the cylinder. It covers the port leading from the combustion chamber to the exhaust manifold.

An engine valve has several basic parts. Refer again to Fig. 1-11.

of the valve train. A hydraulic (oil filled) lifter is shown in Fig. 1-10.

Push rods are long, hollow steel tubes located between the lifters and the rocker arms. Their job is to continue the up-down motion started by the lobes on the camshaft. They sometimes serve to lubricate each rocker arm and valve assembly. Engine oil, under pressure, is directed up through the hollow push rods where it can drip over the rocker arms and valve stems.

The *rocker arms* pivot on a shaft or half-ball socket to transfer and reverse the motion of the push rods. When moved by a push rod, they will open the engine valves.

The *valve spring assembly* includes a spring, keepers, and a seal. The assembly fits over the valve stem and keeps the valve closed when it is not being forced open by the valve train parts. The *valve seals* keep too much oil from being pulled into the valve guide.

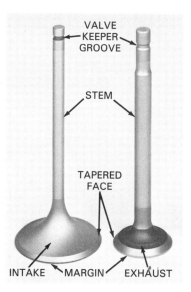

Fig. 1-11. Typical engine valves. Each cylinder has at least two—one for intake and one for exhaust. (Dana Corp.)

The *valve head* is the disc-shaped end which fits over the port. Its inner surface, called the *valve face,* is carefully machined at an angle to provide a leakproof mating surface with the valve seat on the port.

The *valve margin* is a part of the outer edge of the valve head. It forms a narrow, flat edge between the face and the head of the valve. The margin allows the valve to stand up under high combustion temperatures. Without the margin, the rim of the valve would be too thin and the metal would melt and burn off.

The valve head is attached to a long machined and polished shaft called the *valve stem.* The stem extends through a *valve guide* which is a hole machined through the cylinder head. Refer to Fig. 1-12. The stem transfers valve train motion from the rocker arm to the valve head.

Machined into the valve stem near its end are *keeper grooves* or *lock grooves.* They provide a place to fasten small keepers or locks which hold the valve spring onto the stem under tension.

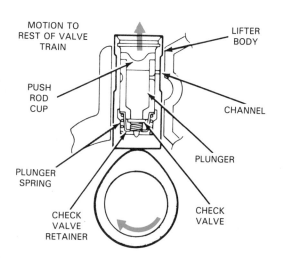

Fig. 1-10. Hydraulic lifter can expand to take up slack in valve train and quiet valve action. It is located between camshaft and pushrods or rocker arms. (Ford Motor Co.)

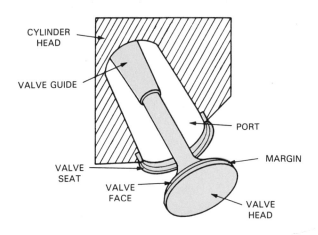

Fig. 1-12. Valve and valve seat. Cylinder head has been cut away to reveal valve guide.

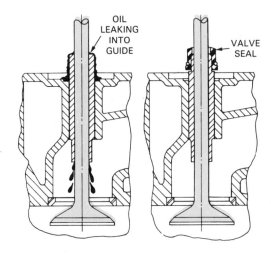

Fig. 1-13. Valve seal fits around valve stem. It deflects oil away from valve stem. (American Hammered Piston Rings)

Valve seals

Without *valve seals,* oil could leak down between the valve stem and valve guide and into the port. Once there, it would be burned, adding to oil consumption. The seal is a rubber-like ring that fits over the valve stem and acts like an umbrella to direct oil away from the stem and guide. It keeps the oil from running down into the valve stem. Refer to Fig. 1-13.

ENGINE FRONT END

The *engine front end,* Fig. 1-14, includes a drive mechanism (gears, sprocket and belt, or sprocket and chain) which use crankshaft power to operate the camshaft. It sometimes drives the oil pump, distributor, and diesel injection pump. The engine front end also includes a front cover, oil seal, and a crankshaft damper.

The *engine front cover* bolts over the crankshaft snout and holds the seal that surrounds the front end of the crankshaft. Its main job is to protect the camshaft drive and seal in the engine oil which lubricates the chain or gear drive mechanism.

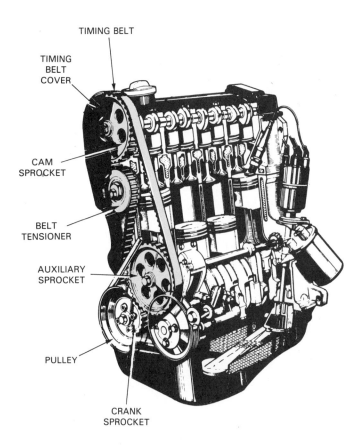

Fig. 1-14. Cutaway of modern four cylinder engine. Note drive mechanism for camshaft. (Chrysler)

The *crank damper,* shown in Fig. 1-15, is a heavy wheel on the crankshaft snout. Mounted in rubber, it helps prevent crankshaft vibration and damage. Other names for it are *harmonic balancer* and *vibration damper.*

Valve-camshaft arrangement

There are two basic valve-camshaft arrangements in use on modern automotive engines. They are known as the overhead valve system and the overhead camshaft system.

In the *overhead valve system* (OHV), the intake and exhaust valves are mounted in the cylinder head. The

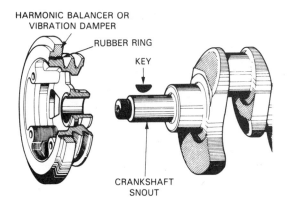

Fig. 1-15. Vibration damper uses its weight and a cushion of rubber to smooth out crankshaft twisting vibrations from pressure of pistons on crankshaft. (Peugeot)

camshaft is located in the cylinder block. This arrangement, shown in Fig. 1-16, requires the use of valve lifters, push rods, and rocker arms to transfer the camshaft motion to the valves.

The **overhead cam engine** (OHC) also has the intake and exhaust valves in the cylinder head. However, unlike the OHV arrangement, the camshaft is also located in the head. The valves, thus, are operated directly by the camshaft without the use of push rods, Fig. 1-17.

Camshaft drive operation

To see how the camshaft drive works, see Fig. 1-18. Suppose that the piston is at TDC and is starting the intake (downward) stroke. The lobe for the intake valve pushes on the lifter. Pushing action travels up through

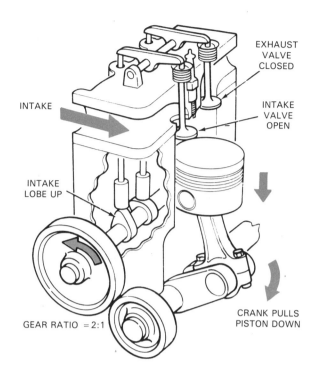

INTAKE STROKE

Fig. 1-18. Simplified drawing of overhead valve engine as intake stroke is underway. Note cam and valve action. Crankshaft rotates camshaft at half crank speed. Intake lobe pushes on lifter, opening intake valve. (Ford)

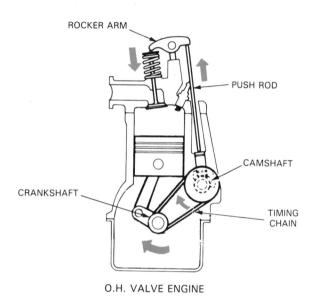

O.H. VALVE ENGINE

Fig. 1-16. Overhead valve engine. In this type, valves are located in head and camshaft is in block. Push rods transfer motion from camshaft to rocker arms. (Ford)

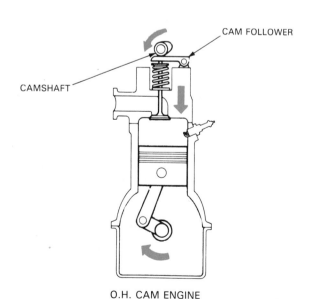

O.H. CAM ENGINE

Fig. 1-17. Overhead cam engine. Cam is located in head close to valves.

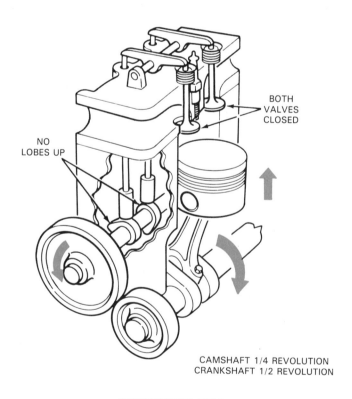

CAMSHAFT 1/4 REVOLUTION
CRANKSHAFT 1/2 REVOLUTION

COMPRESSION STROKE

Fig. 1-19. No cam lobes are up and piston slides up. Thus, both valves are closed so air-fuel mixture, unable to escape, is compressed.

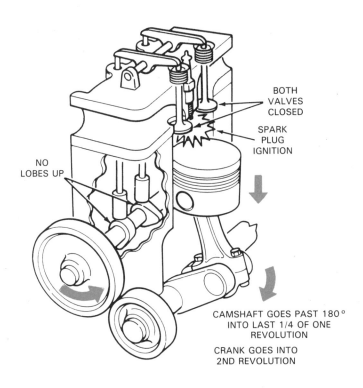

POWER STROKE

Fig. 1-20. Both valves closed (no cam lobes up) and crankshaft has turned 360 degrees while camshaft has turned 180 degrees.

the push rod, to the rocker arm, and to the valve stem. The valve opens just as the piston starts the intake stroke.

By the time the piston is at BDC, the cam has turned far enough so that the lobe is past the lifter and the lifter, under pressure from the valve spring, drops back. The valve is allowed to close.

Then, on the compression stroke both valves are closed because NO cam lobes are pushing up on lifters, Fig. 1-19. The air-fuel mixture is compressed because it cannot escape from the combustion chamber.

At about TDC, the air-fuel mixture ignites, burns, and forces the piston down because no lobes are up and both valves remain closed. This is the power stroke, as shown in Fig. 1-20.

In the fourth stroke of the piston, a cam is causing the exhaust valve to open. The piston pushes the spent gases out the exhaust port. Refer to Fig. 1-21.

ENGINE DESIGNS

While single cylinder engines are common on lawnmowers, motorcycles, boats, chain saws, and garden tillers, automotive engines have a number of cylinders. Modern cars normally have four, six, or eight cylinders.

Having many cylinders makes the engine run smoother since there is less time lapse between each power stroke. Also, such engines can develop more power than single cylinder engines. Fig. 1-22 has simplified drawings of blocks for common automotive engine configurations.

EXHAUST STROKE

Fig. 1-21. Exhaust valve is pushed open so exhaust gases can flow out exhaust port and into exhaust manifold. (Ford Motor Co.)

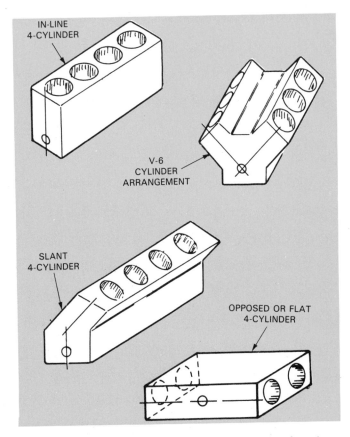

Fig. 1-22. Basic shapes or cylinder configurations of modern automotive engines.

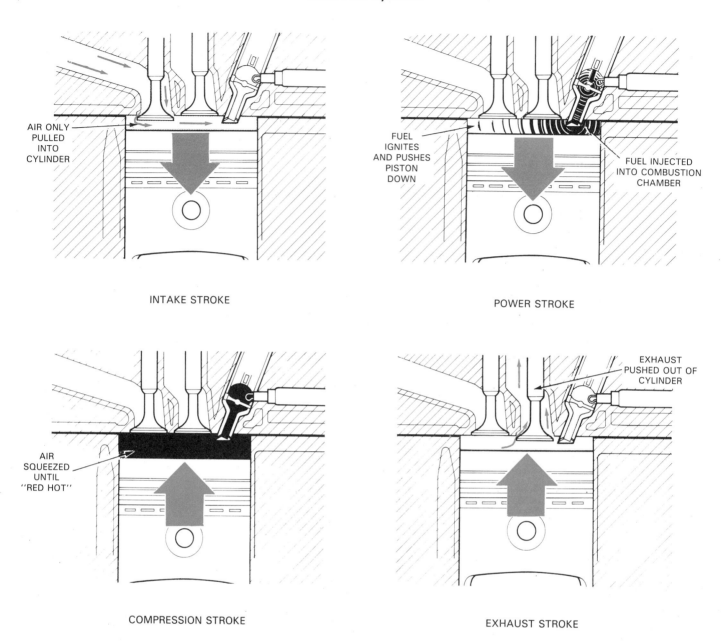

INTAKE STROKE

AIR ONLY PULLED INTO CYLINDER

POWER STROKE

FUEL IGNITES AND PUSHES PISTON DOWN

FUEL INJECTED INTO COMBUSTION CHAMBER

COMPRESSION STROKE

AIR SQUEEZED UNTIL "RED HOT"

EXHAUST STROKE

EXHAUST PUSHED OUT OF CYLINDER

Fig. 1-23. Study operation of diesel engine. (Mercedes Benz)

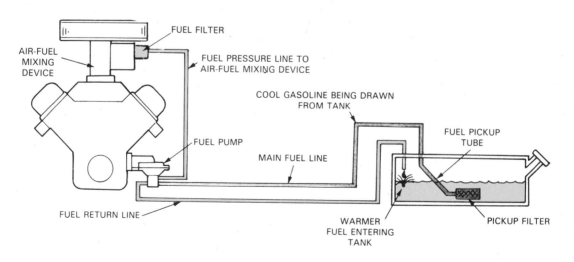

FUEL FILTER

AIR-FUEL MIXING DEVICE

FUEL PRESSURE LINE TO AIR-FUEL MIXING DEVICE

COOL GASOLINE BEING DRAWN FROM TANK

FUEL PICKUP TUBE

FUEL PUMP

MAIN FUEL LINE

FUEL RETURN LINE

WARMER FUEL ENTERING TANK

PICKUP FILTER

Fig. 1-24. Essential parts of gasoline type fuel system include tank, fuel pickup, fuel lines, fuel pump, air cleaner and intake and an air-fuel mixing device (carburetor or injectors).

Diesel engines

The *diesel engine,* having no spark ignition, uses high compression to cause combustion. Look at Fig. 1-23.

During the intake stroke, AIR ALONE is drawn into the cylinder. Compression reduces the volume to 1/20 of original. Combustion chamber pressure reaches about 500° psi (3 445 kPa) at the end of the stroke. This heats the air to about 1000°F or about 538°C.

Toward the end of the compression stroke, the fuel is injected or sprayed into the combustion chamber. The air is hot enough to ignite the fuel.

FUEL SYSTEMS

The job of the *fuel system* is simply to deliver a mixture of air and fuel to the engine's combustion chambers, Fig. 1-24. It must add just the right amount of fuel to the air entering or already in the cylinder. If the mixture is incorrect, it will not be very volatile (burnable) and the engine may not start or run efficiently.

The fuel system must also be able to change the air-fuel ratio (percentages of air in relation to fuel) as engine operating conditions change. Engine temperature, speed, and load alter the fuel mixture requirements of the engine.

FUEL SYSTEM TYPES

Three basic types of fuel systems are used in modern autos: carburetion, gasoline injection, and diesel injection.

Fuel may be mixed with the incoming air before being introduced to the cylinder (carburetor or gasoline injection system), or fuel and air can be introduced separately (diesel injection system). Though older cars commonly use carburetors, gas and diesel injection is now more popular in late model cars. As you will learn, one type of injection system is designed for gasoline fuels, another for diesel fuels.

Carburetor system

A *carburetor fuel system* relies upon engine vacuum (suction) to draw fuel into the engine. Airflow through the carburetor controls the amount of fuel used. Air-fuel ratio is, thus, controlled automatically, Fig. 1-25.

A fuel pump pulls fuel out of the fuel tank and sends it to the carburetor. The intake strokes cause a vacuum inside the combustion chamber. Atmospheric pressure causes airflow through the carburetor. The air mixes with fuel before being drawn into the engine cylinders through the intake manifold.

The amount of air-fuel taken into the engine is controlled by a throttle valve (air valve) inside the carburetor. It is connected to the driver's gas pedal. As the pedal is depressed, the throttle valve opens. More air flows through the carburetor, drawing in more fuel for more power output.

Gasoline injection system

In the *gasoline injection system,* fuel is sprayed into incoming air, just as the air is about to pass the intake valve and enter the combustion chamber. See Fig. 1-26.

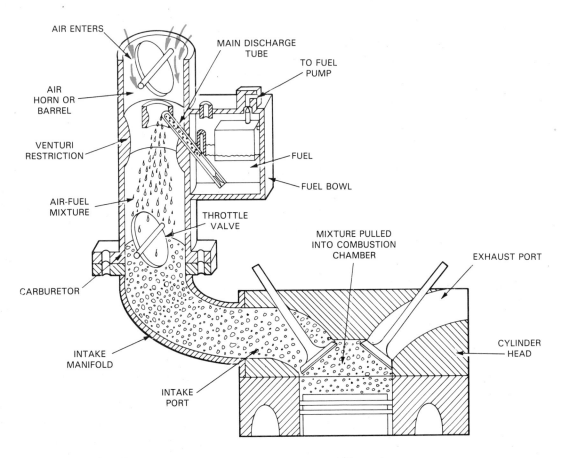

Fig. 1-25. Cutaway view of carburetor fuel system shows how system operates.

GASOLINE INJECTION SYSTEM

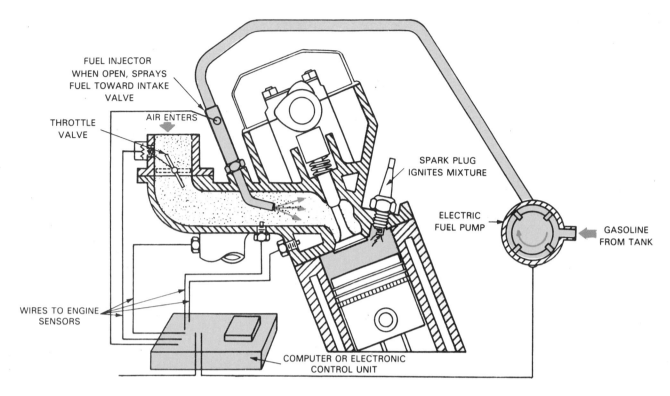

Fig. 1-26. Gasoline injection system. Study cutaway drawing to see how system operates.

Modern gasoline injection systems use a computer, engine sensors, and electrically operated injectors (also called fuel valves) to meter fuel into the engine. An electric pump supplies fuel to the injectors at a constant pressure.

As with the carburetor, a throttle valve controls airflow, engine speed, and power. When the throttle is open, the computer holds the injector open longer, allowing more fuel to enter the air stream into the cylinder. With the throttle closed, the computer closes the injec-

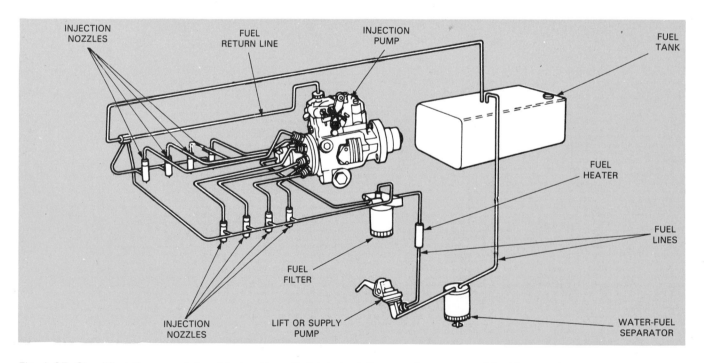

Fig. 1-27. Simplified diagram of diesel fuel system. Its job is to deliver precise amount of fuel into each engine cylinder at the right time. (Ford)

tors sooner, reducing the amount of fuel injected into the engine.

Diesel injection

A *diesel fuel system,* Fig. 1-27, is primarily mechanical. Fuel is forced directly into the combustion chamber, as shown in Fig. 1-28. This occurs near top dead center with the air already compressed. The vaporized fuel hits the hot, compressed air and burns violently.

The diesel injection pump controls fuel injection quantity and engine power output. No throttle valve (air control valve) is needed.

COOLING SYSTEM

The *cooling system* keeps the engine at a constant temperature. It speeds warmup of the engine and removes excess heat to prevent engine damage.

The cooling system's main parts are the water pump, radiator, and thermostat, Fig. 1-29. The *water pump* forces the coolant to circulate through the engine water jackets (passages). The *coolant* (antifreeze-water mixture) absorbs heat from the cylinder walls and other parts of the block and head. Heated coolant is forced into the radiator. As it flows through the *radiator,* the coolant loses its heat to the cooler air flowing through the radiator. Again, refer to Fig. 1-29.

The *thermostat* acts like an automatic temperature control valve. When the coolant is cold, the thermostat prevents coolant circulation to the radiator. As the

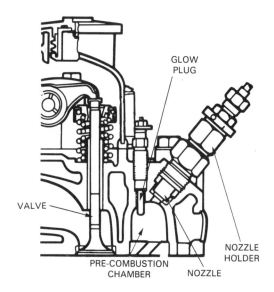

Fig. 1-28. Note location of diesel injector. Injectors deliver fuel directly into cylinder combustion chamber right after air is introduced and compressed.

coolant becomes hot, the thermostat opens, allowing circulation.

The cooling systems of modern automobiles are designed to operate at temperatures near the boiling point of water. A *pressure cap* maintains a constant pressure in the system usually 15 to 17 psi (103-117 kPa). This raises the boiling temperature and keeps the coolant from steaming out of the system.

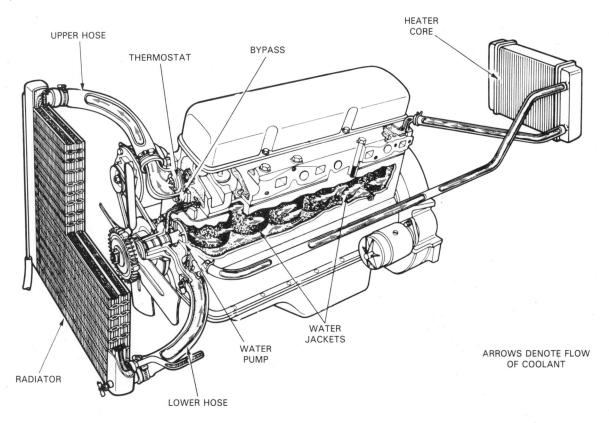

Fig. 1-29. Typical automotive cooling system. Water pump circulates coolant. It rises through upper hose to radiator where it loses heat to outside air. Coolant returns through lower hose and recirculates through water jackets, picking up heat from the cylinders. Heater core receives heated coolant and distributes heat to the passenger compartment. (Chrysler)

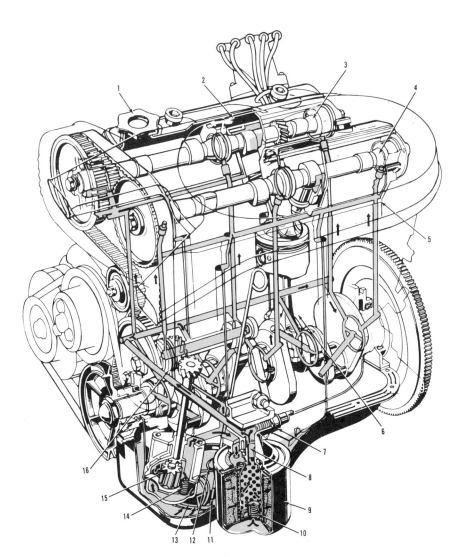

1 — Oil filler cap
2 — Oil spurt for camshaft and tappet
3 — Exhaust camshaft
4 — Intake camshaft
5 — Oil return from tappets
6 — Oil spurt for cylinder walls
7 — Low oil sending unit
8 — Oil delivery-filter to engine components
9 — Oil filter
10 — Filter by-pass valve
11 — Dipstick
12 — Oil sump drain plug
13 — Oil pump suction intake
14 — Relief valve
15 — Oil pump
16 — Duct for oil pump drive gears spurt

Fig. 1-30. Typical engine lubrication system. Note names of parts. Oil passages, also called oil galleries, are full of oil under pressure whenever the engine is operating.

LUBRICATION SYSTEM

Using a pump and a system of oil passages throughout the engine, the *lubrication system* supplies lubricating oil to high friction points in the engine. The cutaway view in Fig. 1-30 shows the entire system. Note how the oil is picked up from the oil pan by the pump and circulated to various moving parts. Small amounts of oil wash over the other high friction points and eventually drain back into the oil pan to be used again.

ELECTRICAL SYSTEMS

A car's *electrical system* includes the ignition system, starting system, charging system, computer control system, and other systems.

Ignition system

Gasoline engines must have an *ignition system* to ignite the air-fuel mixture. This system produces extremely high voltages which cause current to arc across the tip of the spark plug. This event starts combustion and is perfectly timed at each cylinder.

There are three basic types of ignition systems: the older breaker point ignition system, the electronic ignition system, and the computer-coil (no distributor)

ELECTRONIC IGNITION

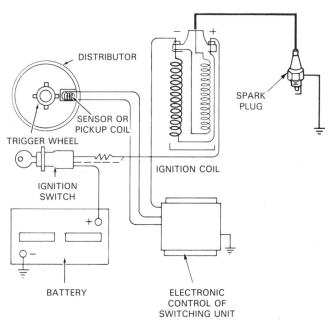

Fig. 1-31. Simple illustration of ignition system. Electronic circuit switches on and off to "fire" coil and spark plug.

20

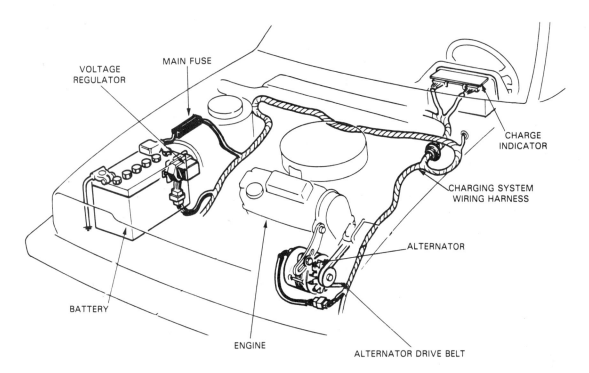

Fig. 1-32. Heart of charging system is alternator. Powered by engine, it produces current to operate car's electrical systems and to recharge battery. (Honda)

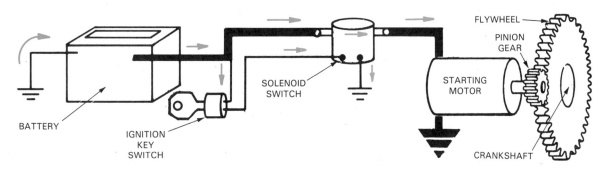

Fig. 1-33. Starting system. When switch is turned to ''start'' mode, solenoid energizes starting motor which engages and spins engine flywheel to start engine. (GM Corp.)

ignition system. Modern cars use either the electronic, Fig. 1-31, or the computer-coil system.

Operation of electronic ignition

The electronic ignition has basically the same parts as the older point ignition. Included are:
1. Primary circuit.
   a. Battery.
   b. Ignition switch.
   c. Primary coil windings.
   d. Pick-up or trigger assembly.
   e. An electronic switching unit.
2. Secondary circuit.
   a. Secondary coil winding.
   b. Distributor cap.
   c. Distributor rotor.
   d. High tension leads.
   e. Spark plugs.

When the engine is running (ignition on), the distributor produces tiny electrical signals for the amplifier or electronic control unit (ECU). One signal is produced for each power stroke. The ECU amplifies (increases) these pulses into on/off current signals for the ignition coil. The coil reacts to this on/off condition by producing the high voltage needed to ''fire'' the spark plugs.

When the ignition key is turned off, the ignition coil stops functioning, the plugs stop firing, and the engine stops running.

CHARGING SYSTEM

The *charging system* replaces the electrical energy drawn from the battery when starting the engine. To re-energize the battery, the charging system sends current back into the battery. See Fig. 1-32. A belt, running off a crankshaft pulley, spins the *alternator* to create the current. A *voltage regulator* controls the output of the alternator.

STARTING SYSTEM

The *starting system* consists of an electric motor and its controls. The motor turns the engine crankshaft until

21

the engine starts and is able to run on its own power. Refer to Fig. 1-33.

The controls include contacts in the ignition switch and a solenoid. The *solenoid* is a high current, magnetic switch that connects the motor directly to the battery.

When the ignition key is moved to the "start" position, current flows to parts of the starting system. A gear on the starting motor engages a gear on the engine flywheel and spins the crankshaft. When the engine starts, the ignition key is released to the "run" position. This cuts off electrical current to the starting system and it shuts off.

EXHAUST SYSTEM

The *exhaust system* muffles (quiets) the engine combustion noise and carries exhaust gases from the cylinder head exhaust ports to the rear of the car. Refer to Fig. 1-34 for system parts. Can you trace flow of exhaust through the system?

EMISSION CONTROL SYSTEMS

A number of *emission control* devices are used to reduce the amount of air pollutants released to the atmosphere. One system traps fuel vapors. A second takes care of harmful crankcase vapors. A third removes toxic chemicals from the engine exhaust. These are discussed in later chapters.

COMPUTER CONTROL SYSTEMS

On-board *computers* are employed in modern autos, not only to meet federal emission control standards, but to improve fuel economy without sacrificing perfor-

mance. Most modern cars have electronic controls which sense, correct, and maintain the engine's air-fuel mixture.

The "brains" of the system is a central control, called an electronic control module (ECM), electronic control unit (ECU), or just computer. It receives *inputs* (signals) from sensors located in various places (such as intake and exhaust manifolds) where they monitor engine operating conditions.

Typically, these signals will trigger *commands* (electrical impulses) from the computer to control the:

1. Fuel injectors, carburetor controls, or electric fuel pump.
2. Idle speed of motor.
3. Electronic spark timing.
4. Automatic transmission torque converter clutch.
5. Emission control systems.
6. Other engine actuators (solenoids or servomechanisms).

Many computers or electronic control units are also programmed to test the performance of the systems they control. When a malfunction of a system occurs, a digital display or indicator will show that a system is malfunctioning.

Electronically controlled systems, and their operation will be fully covered in later chapters. Fig. 1-35 shows a schematic of a computer control system.

SUMMARY

All automotive systems are dependent on one another for proper operation. It would be useless to undertake the study of automotive fuel systems without first having a fundamental understanding of how other systems function and interact with the fuel system.

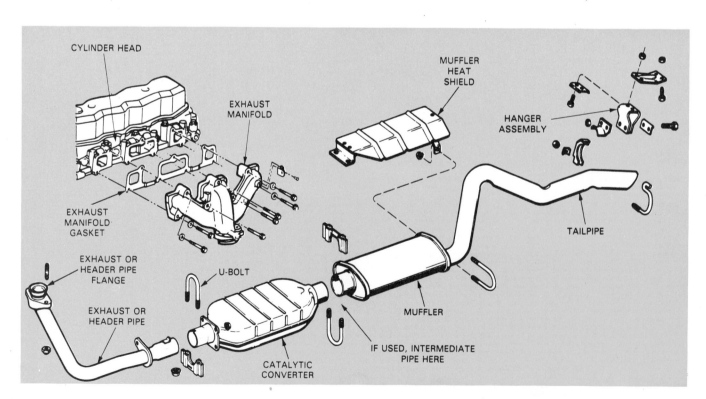

Fig. 1-34. Exploded view of complete exhaust system. Study names of parts. (AMC)

The engine produces the power for operating all systems and propelling the car. The engine block contains or supports all the other parts and systems. Operating principle of the engine is based on the four-stroke cycle: intake, compression, power, and exhaust. A crankshaft converts the reciprocal movement of pistons to rotary motion. This motion is transferred to the drive wheels.

Proper timing of the various events that support the four-stroke cycle depends on the camshaft and valve train. Driven by the crankshaft, these parts open and close the valve ports at the proper time in relation to the piston strokes.

The common engine designs are based on the shape of the block and arrangement of the cylinders. Modern cars utilize the inline engine and the V-type engine in four, six, or eight cylinders. The engines are either of the gasoline or diesel type.

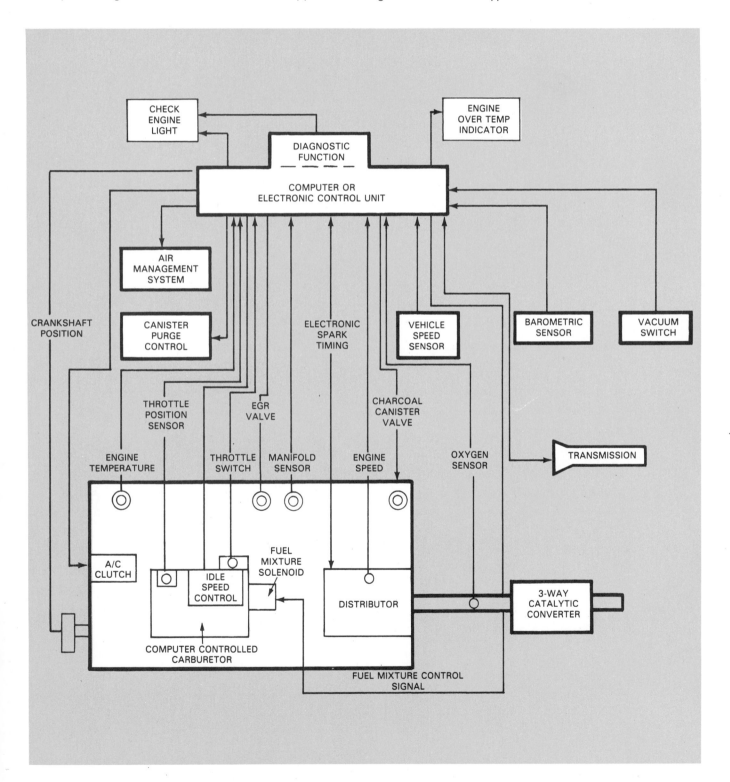

Fig. 1-35. Schematic of computer system which controls spark timing, air-fuel mixture, and exhaust emissions. (General Motors)

Fuel systems include a fuel tank, lines, a pump, and a mechanism for mixing air and fuel. There are three basic fuel systems: carburetion type, gasoline injection type, and diesel injection type.

Cooling systems are designed to keep the engine at a constant temperature. Parts include a water pump, radiator, and a thermostat. The system contains coolant which circulates within the engine and through the radiator where it is cooled.

An electrical system supplies current for all of the car's needs. It has subsystems including an ignition system, a starting system, and a charging system. The ignition system supplies sparks to ignite the fuel charge in each of the cylinders. The starting system spins the flywheel to start the engine. The charging system generates electrical current to operate the other electrical subsystems and to recharge the battery.

The exhaust system muffles the noise of combustion and carries the gases away from the engine.

Electronics and computers are becoming increasingly important in modern automobiles. They control the fuel system, ignition spark advance, idle speed, automatic transmission lockup, and parts of the emission control systems. An electronic control unit (computer) is the "brain" which receives messages from sensors and sends commands that adjust these systems for efficient operation.

KNOW THESE TERMS

Engine sensor, Injector, Crankshaft, TDC, BDC, Cycle, Intake stroke, Compression stroke, Power stroke, Exhaust stroke, Block, Cylinders, Pistons, Rings, Connecting rod, Cylinder head, Combustion chamber, Flywheel, Valve train, Valves, Camshaft, Lifters, Push rods, Rocker arms, Valve spring assembly, Valve face, Valve margin, Keepers, Valve guides, Crankshaft damper, OHV engine, OHC engine, Gasoline injection system, Diesel injection system, Thermostat, Electronic ignition system, Computer-coil ignition, Magnetic pick-up assembly, ECU, Voltage regulator, Emission control systems.

REVIEW QUESTIONS—CHAPTER 1

1. _____ and _____ are the points at which the piston changes direction in its travel up and down in the cylinder.
2. List the four strokes in the four-stroke cycle engine and describe what happens during each stroke.

MATCHING TEST: On a separate sheet of paper, match the terms with the appropriate statements.

3. ____ Strips of tempered steel that seal opening between piston and cylinder.
4. ____ Component linking piston to crankshaft.
5. ____ A shaft with offsets where connecting rods are fastened.
6. ____ Heavy "cap" or "lid" that seals off top of cylinders.
7. ____ Space between tops of pistons and cylinder head where fuel is burned.
8. ____ Long-stemmed plugs that seal off ports to intake and exhaust manifolds.
9. ____ Heavy metal wheel attached to rear of crankshaft.
10. ____ Heavy machined casting that holds all other parts of the engine.
11. ____ Holes cast and bored into block for pistons.
12. ____ Basically, a plug open at one end which moves up and down in cylinder.
13. ____ A shaft with lobes for opening and closing valves.
14. ____ Parts used to open and close engine valves.

a. Block.
b. Piston.
c. Cylinders.
d. Rings.
e. Crankshaft.
f. Combustion chamber.
g. Cylinder head.
h. Flywheel.
i. Valve train.
j. Valves.
k. Camshaft.
l. Connecting rod.

15. In an overhead valve engine, lifters are located between the push rods and rocker arms. True or false?
16. Without _____ _____ the valves could easily be burned by the gases of combustion.
17. An _____ type engine has its camshaft located in the cylinder head.
18. Name the three basic types of fuel systems for automotive engines.
19. Explain the operation of an ignition system.
20. List five things that a car's computer can control.

2 SAFETY, SPECIAL SERVICE TOOLS AND EQUIPMENT

After studying this chapter, you will be able to:
- *Identify common tools and equipment used during fuel system service.*
- *Explain the basic purpose of fuel system service tools and equipment.*
- *Use safety related tools and equipment.*
- *List special safety rules for fuel system service.*
- *List general shop safety rules.*

This chapter reviews safety rules and introduces fuel system related service tools and equipment. This is an important chapter. It will give you essential information for becoming a safe and competent fuel system technician. Study carefully!

REVIEW OF SAFETY

An auto shop can be a very safe and enjoyable place to work. However, if basic safety rules are NOT followed, it can be a dangerous place.

Every year, thousands of mechanics are injured or killed on the job. Most of these accidents occurred because a safety rule was broken.

The injured mechanics learned to respect safety rules after experiencing a painful injury. You must learn to respect safety rules. Study and follow those given in this textbook.

TYPES OF ACCIDENTS

Basically, you should be aware of and try to prevent six kinds of accidents.
1. FIRES.
2. EXPLOSIONS.
3. ASPHYXIATION (airborne poisons).
4. CHEMICAL BURNS.
5. ELECTRIC SHOCK.
6. OTHER PHYSICAL INJURIES.

IMPORTANT! If an accident or injury occurs in the shop, notify your instructor immediately. Use common sense on deciding whether to get a fire extinguisher or to take other action.

FIRES

Fires are serious accidents capable of causing severe burns. Burns suffered over large areas of the body may lead to permanent disfigurement or even death. Numerous combustible substances (gasoline, oily rags, paints, and thinners) are found in an auto shop. Any of these can produce a fire.

Gasoline is, by far, the most dangerous material in an auto shop. Just a cupful can instantly engulf a car in flames. However, safe use of gasoline is possible by following a few rules:

1. Store gasoline and other flammables in approved, sealed containers, Fig. 2-1.

Fig. 2-1. Approved container is only safe way to store fuels. Gasoline container should be clearly labeled so it is not mistaken for another substance. (Chrysler Corp.)

2. Bleed off fuel pressure before disconnecting fuel system components.
3. When disconnecting a car's fuel line or hose, wrap a shop rag around the fitting to keep fuel from spilling.
4. Disconnect the car battery before working on a fuel system.
5. Wipe up fuel spills immediately. Do NOT place ''quick dry'' (oil absorbent) on gasoline or diesel oil because the absorbent will become highly flammable.
6. Keep sources of heat away from fuel systems.
7. Never use gasoline or diesel oil as a cleaning solvent.
8. Do not prime an engine with fuel. Pouring fuel into the air inlet (carburetor or throttle body) is dangerous. A backfire could cause serious burns!
9. Never look into a carburetor or throttle body when cranking or running the engine. Do not cover the air inlet with your hand. If the engine were to backfire, it could cause a serious burn.

Oily rags, can also start fires. Store them in an approved *safety can* (special can with lid).

Paints, thinners, and other combustible materials should be stored in a fire cabinet. Also, never place flammables near a source of sparks (grinder, welder, or water heater), or heat (furnace, for example).

Note the location of all fire extinguishers in your shop. A few seconds time can be a ''lifetime'' during a fire! Fig. 2-2 illustrates various fire extinguisher types. Study them!

Electrical fires can be started when a ''hot wire'' (wire carrying current to component) touches ground (vehicle frame or body). The wire can begin to heat up, melt the insulation, and burn. Then, other wires can do the same. Dozens of wires could burn up in a matter of seconds.

To prevent electrical fires, always disconnect the battery when told to do so in a service manual.

CLASS A: USE ON WOOD, PAPER, ETC.
CLASS B: USE ON OIL, GAS, GREASE, ETC.
CLASS C: USE ON ELECTRICAL EQUIPMENT.

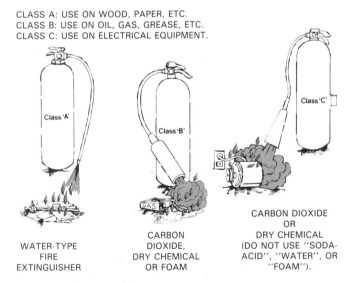

WATER-TYPE FIRE EXTINGUISHER

CARBON DIOXIDE, DRY CHEMICAL OR FOAM

CARBON DIOXIDE OR DRY CHEMICAL (DO NOT USE ''SODA-ACID'', ''WATER'', OR ''FOAM'').

Fig. 2-2. Fire extinguishers are rated for type of fire for which they are designed. Note types. (Cummins)

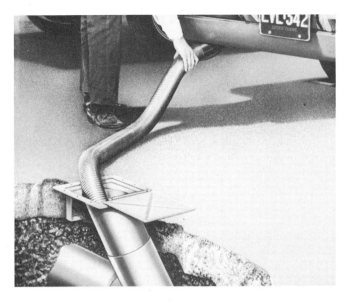

Fig. 2-3. Never operate an auto engine in closed space without routing exhaust into shop exhaust system.

EXPLOSIONS

Several types of explosions are possible in an auto shop. You should be aware of these sources so you can avoid practices that can cause injury or death.

Car batteries can explode! Hydrogen gas can surround car batteries being charged or discharged (used). This gas is highly explosive. The slightest spark or flame can cause the battery to explode. Chunks of battery case and acid can strike your eyes and face. Blindness, facial cuts, acid burns, and scars can result.

Fuel tanks can explode, even when they seem to be empty! A drained fuel tank can still contain gasoline, fumes, and varnish. When gum heats and melts, it will give off vapors that can ignite. Keep sparks and heat away from fuel tanks.

When a fuel tank explodes, one side will usually blow out. Then, the tank can be propelled across the shop with great force. You or other workers could be seriously injured or killed.

Various other sources can cause shop explosions: for example, special sodium-filled engine valves, welding tanks, propane-filled bottles, and engine starting fluids. All can explode if mishandled. These hazards will be discussed in later chapters.

ASPHYXIATION

Asphyxiation is caused by breathing toxic or poisonous substances in the air. Mild cases will cause dizziness, headaches, and vomiting. Severe asphyxiation can cause death.

The most dangerous source of asphyxiation in an auto shop is an automobile engine. An engine's EXHAUST GASES ARE DEADLY POISON. As shown in Fig. 2-3, connect a shop exhaust system suction hose to the tailpipe of any car being operated in the shop. Also, make sure the exhaust system is turned ON.

CHEMICAL BURNS

Various solvents (parts cleaners), battery acid, and a few other shop substances can cause *chemical burns* to the skin. Always read the directions and warnings on solvents and chemicals.

Carburetor cleaner (decarburizing type cleaner), for example, is very caustic. It can severely burn your hands in a matter of seconds. Always wear rubber gloves. If a skin burn occurs, follow label directions immediately. Wear safety glasses to protect against splashed solvent.

ELECTRIC SHOCK

Electric shock can occur when using improperly grounded electric power tools or test equipment. Never use an electric tool unless it has a functional GROUND PRONG (third, round prong on plug socket). This prevents current from accidentally passing through your body. Also, never use an electric tool when standing on a wet shop floor.

OTHER PHYSICAL INJURIES

Other physical injuries, such as cuts, broken bones, and strained backs, are the outcome of hundreds of different accidents. As a mechanic, you must constantly plan and evaluate every repair technique. Decide whether every operation is safe and take action as required.

REVIEW OF GENERAL SAFETY RULES

These *general safety rules* should be remembered. They will help you avoid injury.

1. WEAR EYE PROTECTION, Fig. 2-4, during operations that endanger your eyes. This includes operating power tools, working around a running car engine, and carrying batteries.
2. NO GOOFING OFF! Avoid anyone who does NOT take shop work seriously. A joker is "an accident waiting to happen."
3. KEEP YOUR SHOP ORGANIZED! Return all tools and equipment to their proper storage areas. Never leave tools, creepers, or parts on the floor.
4. DRESS LIKE A MECHANIC! Remove rings, bracelets, necklaces, watches, and other jewelry. They can get caught in engine fans, belts, driveshafts. Also, roll up long sleeves and secure long hair, they too can get caught in revolving parts.
5. NEVER CARRY SHARP TOOLS or PARTS in your pockets! They can easily puncture skin.
6. WEAR FULL FACE PROTECTION when grinding or welding, and during other operations where severe hazards are present!
7. WORK LIKE A PROFESSIONAL! When learning auto mechanics, it is easy to get excited about your work. However, avoid working too fast. You could overlook a repair procedure or safety rule.
8. USE THE RIGHT TOOL FOR THE JOB! There is usually a "best tool" for each repair task. Determine whether a different tool will work better than another, especially when you run into difficulty.
9. KEEP GUARDS OR SHIELDS IN PLACE! If a power tool has a safety guard, use it!
10. LIFT WITH YOUR LEGS, not with your back! When lifting heavy parts, keep your knees bent while your back is kept straight.
11. USE ADEQUATE LIGHTING! A portable shop light not only increases working safety, it increases working speed and precision.
12. VENTILATE AREA WHEN NEEDED! Turn on the shop ventilation fan any time there are fumes in the shop.
13. JACK UP A CAR SLOWLY AND SAFELY. A car can weigh between one and two tons. It is NOT safe to work under a car held only by a floor jack; use jack stands. See Fig. 2-5.

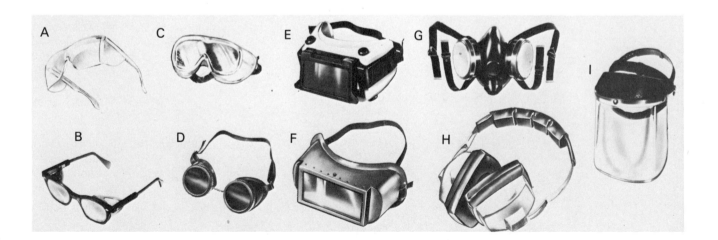

Fig. 2-4. Different types of eye protection are designed for many different types of work. The safe automotive shop will be properly equipped for every type of work performed. A—Safety glasses. B—Safety glasses with side shields. C—Goggles. D, E, and F—Welding goggles. G—Respirator. H—Noise mufflers. I—Face shield. (Snap-On Tools)

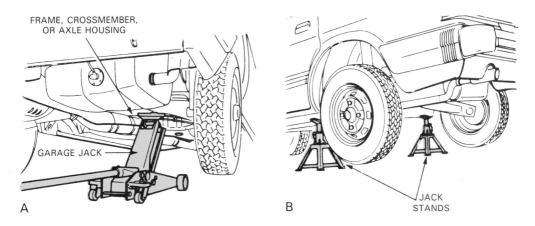

FRAME, CROSSMEMBER,
OR AXLE HOUSING

GARAGE JACK

A

JACK
STANDS

B

Fig. 2-5. Lifting car can be dangerous unless done properly. Left. Be sure to check manual for proper lift point before using floor jack. Right. Jack stands are a necessary safety precaution when working under car. (Subaru)

14. DRIVE SLOWLY WHEN IN THE SHOP AREA! With all of the other students and cars in the shop, it is very easy to have an accident.
15. REPORT UNSAFE CONDITIONS TO YOUR IN-STRUCTOR! If you notice any hazards, let your instructor know about them.
16. STAY AWAY FROM ENGINE FANS! The fan on a car engine can cut like a KNIFE. It can inflict serious injuries. Also, if a part or tool is dropped into the fan, it may fly out and hurt someone.
17. NO SMOKING! No one should smoke in an auto shop. Smoking is a serious fire hazard, considering that a spark could land on fuel lines, cleaning solvents, paints, and other flammables.
18. GET INSTRUCTOR'S PERMISSION before using any new or unfamiliar power tool, lift, or other shop equipment. Your instructor may need to give you a demonstration.

19. RESPECT THE HIGH FUEL PRESSURE in a diesel injection system. Diesel injection pressure can cause diesel fuel to pierce your skin or eyes. Blood poisoning or blindness could result.
20. REFER TO THE SERVICE MANUAL when needed. It will provide critical instructions and specs for safe fuel system service.

CLEANING EQUIPMENT

Various types of cleaning equipment are commonly used by the fuel system technician. Knowing how and when to use each type will make fuel system service easier and more profitable.

COLD SOAK TANK

A *cold soak tank* contains mild solvent for parts cleaning. One is shown in Fig. 2-6. A pump in the bottom of

Fig. 2-6. Cold soak tank. Parts are soaked in solvent and scrubbed with brush to remove grease and dirt. (Build-All)

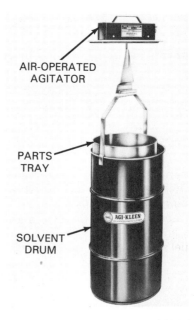

AIR-OPERATED
AGITATOR

PARTS
TRAY

SOLVENT
DRUM

Fig. 2-7. This powerful parts cleaner provides air agitation for cleaning carburetor or fuel injection parts. (BAC)

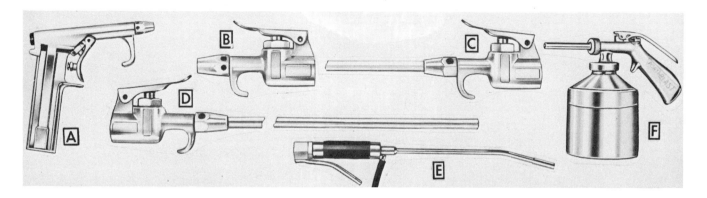

Fig. 2-8. An air gun and various nozzle types. A—Pistol grip air gun. B—Air gun. C and D—Air guns with long snouts. E—Gun draws up solvent and forces it over parts. F—Cleaning gun with solvent container.

the unit circulates solvent up through a flexible pipe. A spout directs a stream of solvent over parts.

CARBURETOR CLEANER

Carburetor cleaner is a powerful solvent for cleaning fuel system components. It will remove gum, hard carbon, oil, grease, and most other deposits.

As shown in Fig. 2-7, a tray is usually provided for dipping parts into the solvent. This keeps your hands out of the solution and avoids loss of small parts.

DANGER! Wear rubber gloves and eye protection when working with carburetor cleaner. It is a caustic substance that can cause severe skin and eye burns. Follow label directions if chemically burned by carburetor cleaner.

AIR GUNS

Air guns are frequently used by the fuel system technician. Various styles are available, Fig. 2-8.

An *air gun* is commonly used to air dry parts after washing them in solvent. By directing air into small passages, the inside of a component (carburetor or throttle body for example) can also be blown clean and dry.

OTHER CLEANING TOOLS AND EQUIPMENT

Other more specialized cleaning tools will be covered in later text chapters. Refer to the index to find information as needed.

FUEL SYSTEM SERVICE TOOLS AND EQUIPMENT

General service tools and equipment for the fuel system technician include the VOM (volt-ohm-milliammeter), compression gauge, vacuum gauge, stethoscope, and many tools. These tools are used to check fuel system and engine performance.

Modern fuel systems, unlike systems of the past, have many electrical sensors, solenoids, and other electrical components. They require special electrical test equipment.

The fuel system mechanic must also be able to determine engine condition. An engine with mechanical problems can produce symptoms similar to those produced

by the fuel system.

This section of the chapter will quickly review general tools and equipment relating to engine, fuel system, and electrical testing.

TESTING SAFETY RULES

When using test equipment, you should follow general safety rules given earlier. In addition, there are several special rules to remember.

1. Read the operating instructions for the test equipment. Failure to follow correct procedures could cause bodily injury, and damage to the part or instrument.
2. If the engine must run during your tests, set the parking brake. Block the wheels. Place an automatic transmission in park or manual transmission in neutral. Connect an exhaust vent hose to the tailpipe if you are working in an enclosed shop.
3. Keep test equipment leads (wires) away from engine belts, the fan, and hot engine parts.
4. Refer to the auto manufacturer's service manual or specific procedures.

SPARK TESTER

A *spark tester* is used to check the basic operation of the engine ignition system. One is shown in Fig. 2-9. The spark tester is connected between the end of a secondary (spark plug or coil) wire and ground. When the engine is cranked or started, a "hot" spark should jump across the tester gap. A fuel system technician might use a spark tester to help isolate whether a no-start problem is in either the ignition system or the fuel system.

COMPRESSION GAUGE

A *compression gauge* measures the amount of pressure during the engine compression stroke. See Fig. 2-10. It provides a means of testing the mechanical condition of the engine. If the compression gauge readings are NOT within specs, something is mechanically wrong in the engine.

A compression test should be performed when symptoms (engine miss, rough idle, puffing noise in induction

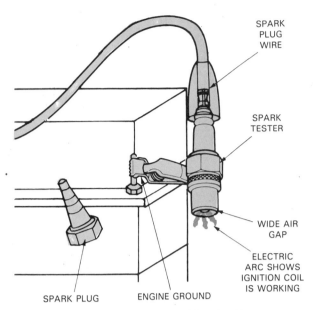

Fig. 2-9. Spark tester is inexpensive tool for checking presence and strength of electric arc produced by ignition coil.

Fig. 2-11. This vacuum analyzer also has meter for checking sensors and other components. (Peerless)

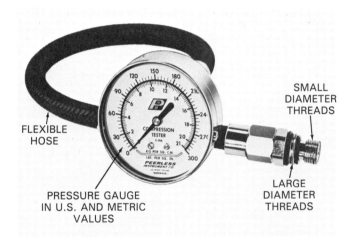

Fig. 2-10. Compression gauge checks condition of seal in cylinder to isolate engine problems. (Peerless)

system or exhaust) point to major engine problems.

Compression testing

To do a *compression test,* remove all of the spark plugs (gasoline engine) or glow plugs (diesel engine). Disable the ignition or injection system. Block open the throttle valve (gasoline engine).

Screw the compression tester into a spark plug or glow plug hole. Crank the engine at least four compression strokes while reading the gauge. Measure and record the pressure for each cylinder. Repeat on other cylinders.

A normal compression reading will make the gauge increase evenly to specs. The pressure in each cylinder should not vary more than about 10 percent.

VACUUM GAUGE

A *vacuum gauge,* Fig. 2-11, measures negative pressure (suction) produced by the engine, fuel pump,

vacuum pump, or other component. It can be used to determine engine condition and to check vacuum devices.

The vacuum gauge is connected to a vacuum fitting on the engine intake manifold or a component. The gauge readings can then be compared to normal readings.

PRESSURE GAUGE

A *pressure gauge,* Fig. 2-12, measures psi (pounds per square inch) and/or kPa (kilopascals). It is used to check various pressures in a car, such as fuel pump pressure, pressure regulator pressure, and turbocharger boost pressure.

A mechanic will usually have several different pressure gauges. Each will be designed for a specific job.

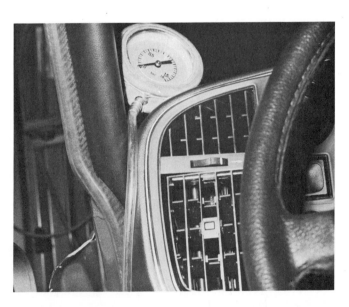

Fig. 2-12. Pressure gauge can be used to check condition of many pressurized systems on a car.

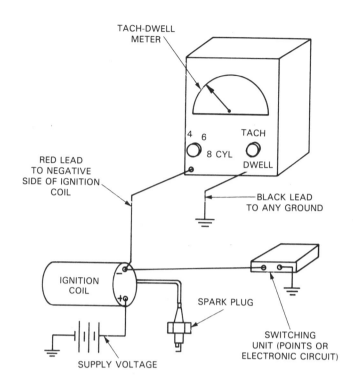

Fig. 2-13. A tach-dwell meter is commonly used during tune-ups.

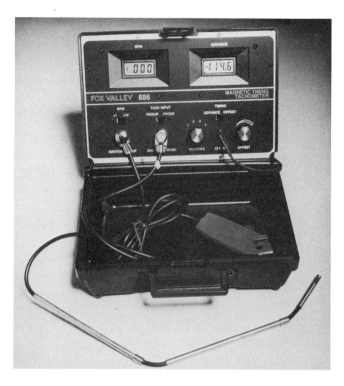

Fig. 2-14. This mag-tach tester displays engine speed and timing advance in digital (number) form. (Hennessy Industries)

TACH-DWELL METER

A *tach-dwell meter,* shown in Fig. 2-13, is a tachometer and a dwell meter combined. The tach is for measuring engine rpm (speed). The dwell meter measures in degrees for contact point adjustment or computer controlled carburetor calibration.

Follow operating instructions when connecting a tach-dwell meter. Procedures vary. To measure rpm for an idle speed adjustment, you must, typically, connect the red lead to the negative coil terminal or tach terminal. The black tester lead connects to ground.

A tach-dwell meter can also be used to check the operation of some computer controlled carburetor systems. It will measure the on-off cycles (dwell) going from the computer to the fuel mixture control solenoid in the carburetor. This is explained in later chapters.

MAG-TACH

A *mag-tach* is a magnetically triggered tachometer for measuring engine rpm. See Fig. 2-14. It is used on both diesel and gasoline engines. A diesel engine does not have an electrically operated ignition system to power a conventional tachometer.

The mag-tach is operated by a magnet that senses a notch in the engine flywheel or damper. In this way, engine speed can be measured without a connection to the ignition system.

VOM

A *VOM* (volt-ohm-milliammeter), also called a multimeter, is commonly used to check the condition of numerous electrically-related engine components. It will

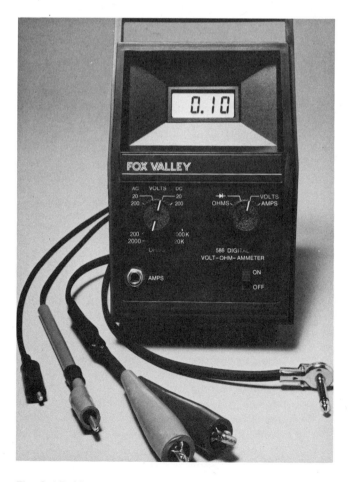

Fig. 2-15. Measuring electrical values at component or in circuit can tell you much about component condition. Pictured is a digital readout VOM. (Hennessy Industries)

measure voltage, current, and resistance. By comparing the meter readings to specs, you can determine the condition of electrical components or circuits. If an electrical value is too high or too low, there is a problem.

Look at Fig. 2-15. When using a VOM, make sure all selector settings are correct. Calibrate the meter if needed. Place the meter in a safe location, where it cannot fall or be damaged by moving or hot parts.

To repair modern fuel systems, you must know how to use a VOM. It will test engine sensors, wiring, actuators (control devices), and other components found in today's computer controlled fuel systems.

TEST LIGHT

A *test light* is commonly used to check for electrical continuity (that circuit is unbroken and can conduct current). It is a slightly faster method to check a circuit for power than a voltmeter.

A test light can be used for numerous circuit tests in a fuel system. It will easily check for power to a fuel injector, Fig. 2-16, or diesel engine glow plug, for example.

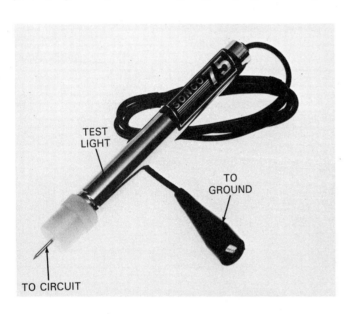

Fig. 2-16. Test light will tell you quickly whether circuit has electrical continuity. Connect clip to ground and touch the tip to circuit. Light will glow if circuit has power. (Sonco Mfg.)

NOTE: As will be explained later, there are many components on a late model car that can be damaged if improperly checked with a test light. A test light will draw more current than a high impedence (resistance) digital meter and can overload a circuit. This is extremely important when testing computer systems with delicate electronic components.

Only use a test light on a computer control system when told to do so by service manual instructions. Otherwise, use a digital VOM.

EXHAUST ANALYZER

An *exhaust analyzer* is a testing instrument that measures the chemical content of engine exhaust gases. See Fig. 2-17.

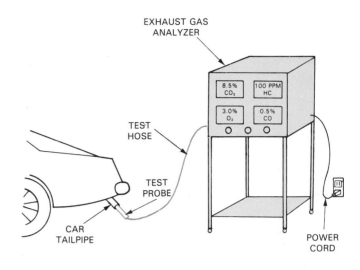

Fig. 2-17. Exhaust gas analyzer measures chemical content of engine's exhaust. This enables mechanic to measure combustion efficiency of engine and amount of pollutants it produces.

The analyzer probe (sensor) is placed in the car's tailpipe. With the engine running, the exhaust analyzer will indicate the amount of pollutants and other gases in the exhaust. The mechanic can use this information to determine the condition of the engine and other systems affecting emissions.

An exhaust gas analyzer is an excellent diagnostic tool that will indicate:

1. Carburetor or fuel injection problems.
2. Engine mechanical problems.
3. Vacuum leaks.
4. Ignition system problems.
5. PCV troubles.
6. Clogged air filter.
7. Faulty air injection system.
8. Evaporative control system problems.
9. Computer control system troubles.
10. Catalytic converter condition.

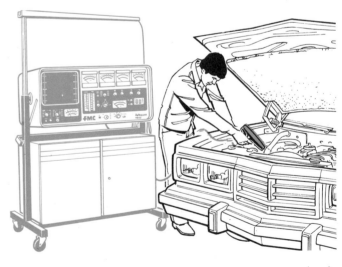

Fig. 2-18. Engine analyzer normally includes timing light, tachometer, dwell meter, vacuum-pressure gauge, VOM, exhaust gas analyzer, and oscilloscope. (FMC)

ENGINE ANALYZER

An *engine analyzer* is a group of different testing instruments mounted in one assembly. One is pictured in Fig. 2-18.

DYNAMOMETER

A *dynamometer,* also called a *dyno,* is used to measure the power output and performance of an engine. There are two basic types of dynamometers: the engine dynamometer (engine removed and mounted on dyno) and the chassis dynamometer (car's drive wheels turn dyno).

By loading the engine, the dynamometer can be used to check engine acceleration, maximum power output, fuel consumption, and on-the-road performance characteristics. Fig. 2-19 shows a chassis dynamometer.

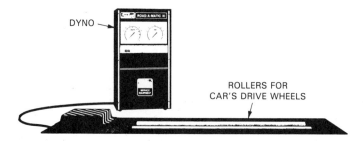

Fig. 2-19. Chassis dynamometer creates "open road" conditions so mechanic can check car's response under different speed and load conditions. (Sun Electric)

Fig. 2-20. Stethoscope intensifies sounds in operating parts to find trouble.

STETHOSCOPE

A *stethoscope* is a listening device for checking the operation of various components. As shown in Fig. 2-20, if it is touched on a component, any sound in the component will be amplified (increased in intensity).

A stethoscope is useful for troubleshooting sounds in fuel pumps, fuel injectors, vacuum pumps, emission control air pumps, etc. It helps pinpont mechanical or electrical problems.

VACUUM PUMP-GAUGE

A *vacuum pump-gauge* combines a conventional vacuum gauge with a hand-operated pump, Fig. 2-21. It is a handy tool for testing numerous vacuum devices found in fuel and engine systems.

The vacuum hose can be connected to a part suspected of being faulty. Then, the mechanic applies

Fig. 2-21. Testing vacuum diaphragm. Pump gauge shows whether component can hold vacuum.

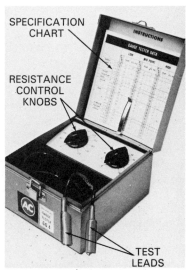

Fig. 2-22. Two types of fuel gauge testers. Units check out fuel gauge circuit, sending unit, and gauge itself, to locate problems. (AC-Delco)

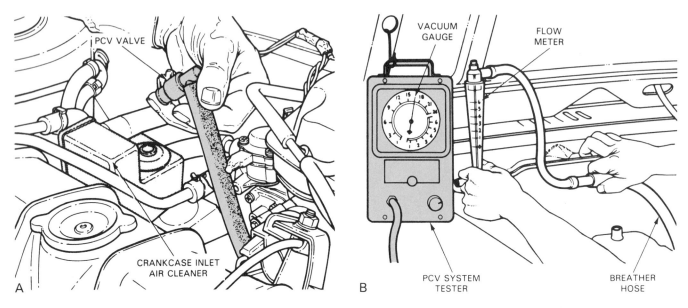

Fig. 2-23. Checking out a positive crankcase ventilation (PVC) system. A—First, with engine running, you should place a finger over PCV valve to determine if system is working. B—Then, if first test indicates trouble, you can test further with PCV valve tester. Compare specs in the tester instructions to determine condition. (Chrysler Corp. and AMC)

vacuum to check its operation. For example, if a vacuum diaphragm fails to hold a steady vacuum, the diaphragm is leaking and must be replaced. The gauge on the tester will let you compare test vacuum with specs. If a vacuum-operated device does NOT function within prescribed vacuum levels, it is bad.

FUEL GAUGE TESTER

A *fuel gauge tester* will check the operation of the dash gauge, gauge circuit, and the fuel gauge sending unit in the fuel tank. Two types are pictured in Fig. 2-22.

The tester is connected into the fuel gauge circuit. The tester can be set on full (low resistance), for example, and the fuel gauge in the car's dash should read full.

PCV SYSTEM TESTER

It is important to have a properly operating PCV system. If the system is not operating properly it can cause engine sludging, and wear, a rough idle, as well as other problems. If the system is leaking, it could cause a vacuum leak or might produce a lean air-fuel mixture when the engine is idling. If the airflow is restricted through the PCV system, it could result in an overly rich fuel mixture. Again, this would affect engine idle adversely.

While an instrument check is the most effective way to test the system, a quick test is recommended to see if the system is clogged or stuck open. Simply pull the PCV valve out of the fitting on the top of the engine. The engine should be idling during this procedure. Place your finger over the end of the valve, as shown in Fig. 2-23A. You should feel suction on your finger and the engine idle speed should drop about 40-80 rpm.

If you feel no vacuum, there may be a restriction in the hose to the valve. However, if the engine speed drops still farther than 40-80 rpm, the PCV valve may be stuck

open. It should be replaced.

A PCV system tester, however, provides a more complete check of a positive crankcase ventilation system. A testing hookup and the proper tool is shown in Fig. 2-23B. It will provide more conclusive evidence of how well the system is working and will indicate whether the system is pulling toxic fumes out of the engine crankcase as it should.

Before condemning a PCV valve as faulty, make certain the malfunction is not caused by other PCV system parts. For example, a cracked hose leading to the PCV valve may be leaking vacuum.

PCV valves cannot always be identified by their appearance. When replacing them, always refer to the part number listed in the shop manual for the specific make and model of car.

CARBURETOR SERVICE TOOLS AND EQUIPMENT

Carburetor service tools are more specialized than the tools just discussed. Becoming familiar with them now will prepare you to understand later text chapters.

Fig. 2-24 shows a service manual illustration of special carburetor service tools. The manual will refer to these tools by PART NUMBER.

CARBURETOR SERVICE HAND TOOLS

To make carburetor repairs easier, a fuel system technician will have a wide assortment of special wrenches and drivers. Fig. 2-25 shows a few of these.

A *flex driver* is commonly used for making carburetor adjustments, Fig. 2-25A and 2-25B. It is also used on gasoline injection and diesel injection systems. It will reach around a corner to turn an adjustment screw or nut, for example.

A *straightedge,* view C, may be used to check a carburetor or throttle body for warpage. If a feeler gauge

SPECIAL TOOLS CARBURETOR

DWELL METER

USED TO MONITOR THE FUEL CONTROL DELIVERY DETERMINED BY THE ECM COMMAND. (SET ON 6 CYL. SCALE)

CARBURETOR GAGE SET

USED TO PERFORM CARBURETOR GAGE AND ANGLE SETTING

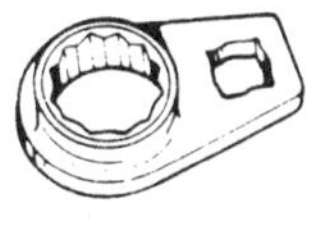

J9789-C/BT3005

OXYGEN SENSOR WRENCH

USED TO REMOVE OR INSTALL THE OXYGEN SENSOR

J29533A/BT8127

BENDING TOOL (PART OF J9789-C)

USED TO BEND CARBURETOR LINKAGE.

J9789-111/BT8231C

M/C SOLENOID GAGING TOOL

USED TO ADJUST THE MIXTURE CONTROL SOLENOID PLUNGER ON E2ME OR E4ME CARBURETOR

J33815-1/BT8253-A

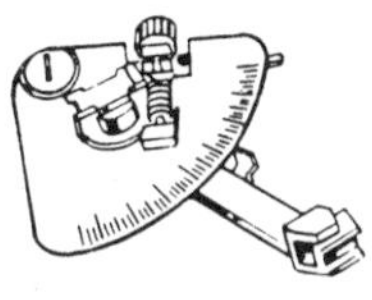

CHOKE ANGLE GAGE (PART OF J9789-C)

USED TO SET CHOKE ANGLE TO ADJUSTMENT SPECIFICATION

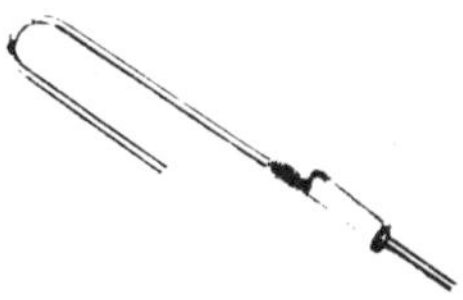

J26701-A/BT7704

AIR BLEED VALVE GAGING TOOL

USED TO ADJUST IDLE AIR BLEED VALVE ON E2ME OR E4ME CARBURETOR

J33815-2/BT8253-A

CARBURETOR FLOAT LEVEL GAGE (PART OF J9789-C)

USED TO CHECK FLOAT LEVEL ON M/C SOLENOID PLUNGER TRAVEL ON E2ME OR E4ME CARBURETOR

J9789-130/BT7720

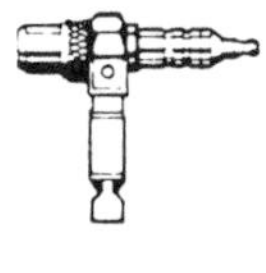

HEI SPARK TESTER

USE TO CHECK HEI SPARK VOLTAGE.

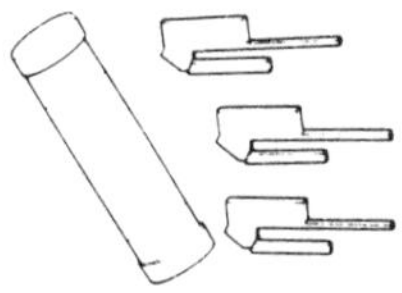

J26795/BT7220-1

FLOAT LEVEL GAGE SET

USED TO CHECK FLOAT LEVEL ON 2SE OR E2SE CARBURETOR

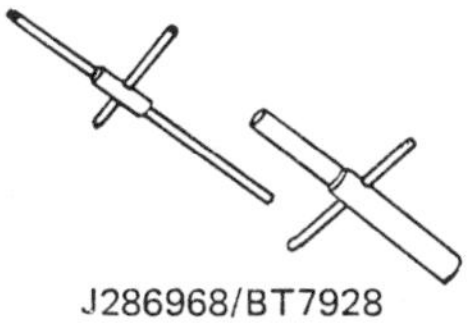

J9789-135/BT8104

MIXTURE ADJUSTMENT TOOL

USED TO ADJUST LEAN MIXTURE AND RICH MIXTURE STOP SCREWS ON E2SE, E2ME OR E4ME CARBURETOR

J286968/BT7928

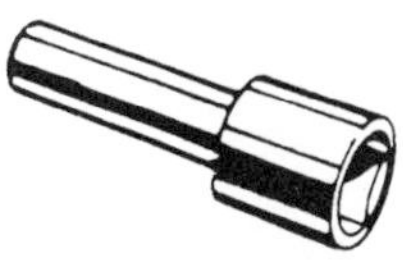

IDLE MIXTURE SOCKET

USED TO ADJUST IDLE MIXTURE NEEDLE ON A E2SE CARBURETOR

J29030-B/BT7610B

PUMP LEVER PIN PUNCH

USED TO DRIVE PUMP LEVER PIN INWARD TO ALLOW REMOVAL OF THE PUMP LEVER.

J25322/BT7523

ISC ADJUSTING WRENCH

USED TO ADJUST ISC PLUNGER TO OBTAIN MAXIMUM SPECIFICATION RPM SPEED.

J29607/BT8022

CARBURETOR ADJUSTMENT WRENCH

USED TO REMOTE ADJUST IDLE MIXTURE NEEDLE ON THE VEHICLE.

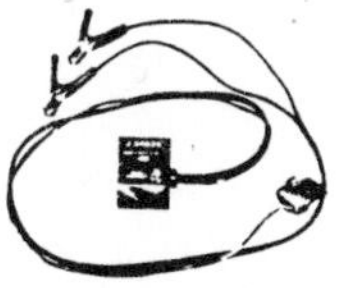

J22646-02

ISC MOTOR TESTER

USED TO TEST OPERATION OF ISC MOTOR IN EITHER DIRECTION AND CONDITION OF THE INTERNAL SWITCH

J34025/BT8256A

Fig. 2-24. Page from shop manual showing special carburetor tools and their uses. Part numbers can be looked up elsewhere in manual for more tool information. (Pontiac)

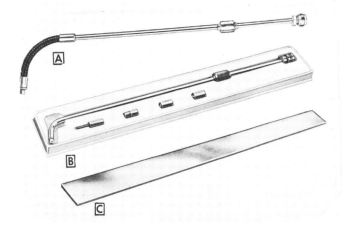

Fig. 2-25. Carburetor hand tools. A and B—Flex-drivers allow mechanic to work in tight places. C—Straightedge checks surfaces for flatness. (Snap-On Tools)

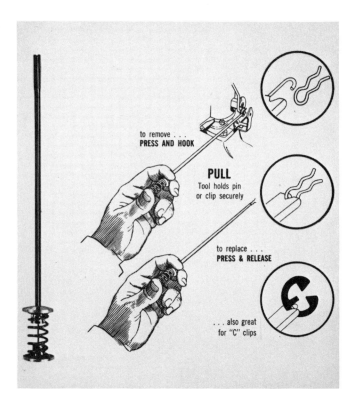

Fig. 2-28. Special hook is designed to grip clips in hard to reach places. (Lyle Tools)

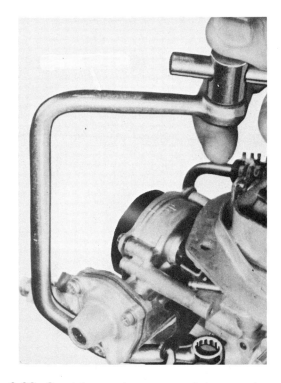

Fig. 2-26. Special wrenches are needed to reach around obstructions. (Ford)

will fit between the body and the straightedge, the body is warped.

Special wrenches, with odd bends and curves, are also needed for reaching hidden fasteners that secure a carburetor or throttle body. Fig. 2-26 shows a mechanic using one of these wrenches to remove a hold-down nut.

Fig. 2-27 illustrates a special screwdriver that will turn a screw from an angle. This is handy when making carburetor or injection system adjustments.

A special *hook tool* is shown in Fig. 2-28. It will grasp and remove tiny clips that are used on carburetor linkages or throttle body.

When working, you must always be thinking about whether another tool will do a better job. Even though a tool might work, maybe you could think of a tool that will be a little faster or more efficient. This kind of thinking will help you develop into a successful fuel system technician.

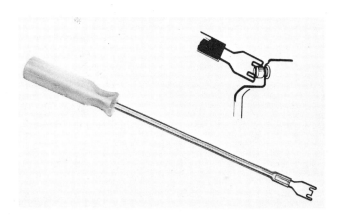

Fig. 2-27. Forked blade on special screwdriver grips slot at any angle. (Lyle Tools)

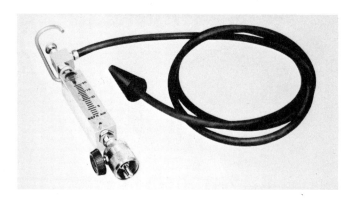

Fig. 2-29. Propane enrichment device injects propane into intake manifold during idle mixture adjustment. (Kent-Moore Tools)

Fig. 2-30. Carburetor gauge set includes float level scales, carburetor stand, bending tool, various float level gauges, and choke angle gauge. (Kent-Moore Tools)

PROPANE ENRICHMENT DEVICE

A *propane enrichment device* is used to inject propane gas into the engine during carburetor fuel mixture adjustment. One is shown in Fig. 2-29.

Some carburetors require propane enrichment during adjustment. When a service manual requires one, the device is attached to a bottle of propane gas and to the intake manifold or carburetor. Then, a regulated amount of gas is injected into the engine. This richens the fuel mixture and allows you to accurately adjust the carburetor for a lean, low emissions setting.

CARBURETOR GAUGE SET

A *carburetor gauge set* contains various sizes of rods or blades for making carburetor adjustments, Fig. 2-30. The gauges can be used to measure choke opening, throttle plate opening, and many other carburetor adjustments very accurately.

Fig. 2-31. Choke angle gauge is used on some carburetors to adjust choke rod, vacuum break, and unloader angle settings. (Kent-Moore Tools)

CHOKE ANGLE GAUGE

A *choke angle gauge* measures choke opening in degrees. See Fig. 2-31. It is needed when manufacturer specs give choke adjustment settings in degrees.

The gauge has a magnetic or clip-on mount that attaches to the choke. A leveling bubble and screw allow for quick setup of the gauge. A pointer is used to read degrees of choke opening.

T-SCALE

A *T-scale* can be used to adjust carburetor float drop and float level. As pictured in Fig. 2-32, the T-scale will accurately measure distances between the float and carburetor body. It is also handy for many other measurements.

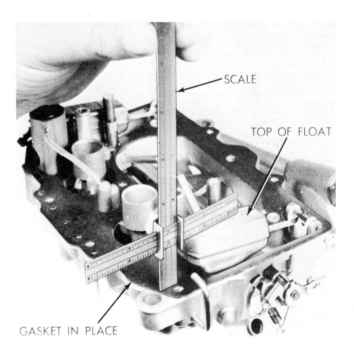

Fig. 2-32. This T-scale is being used to adjust carburetor float. (Carter)

CARBURETOR OR THROTTLE BODY STAND

When servicing a carburetor or throttle body assembly, a *carburetor stand,* also called *throttle body stand,* makes service easier.

The stand fits into the holes in the base plate, Fig. 2-33. This holds the bottom of the carburetor or throttle body up away from the work surface. It should always be used since the slightest dent or nick in a throttle plate can affect fuel system operation.

GASOLINE INJECTION SERVICE TOOLS

Since new cars are commonly equipped with fuel injection, it is extremely important for today's mechanic to be familiar with the most common service tools for gasoline injection.

Fig. 2-33. Always use carburetor stand when working on carburetor removed from vehicle. It protects throttle plates from damage. (Offenhauser Equipment Corp.)

FUEL INJECTION PRESSURE GAUGE

A *fuel injection pressure gauge* is similar to a conventional pressure gauge for a carburetor fuel system. However, it will read higher pressures and may have special adapters for connection to the fuel system. See Fig. 2-34.

A fuel injection pressure gauge is a commonly used tool. It can provide information about the condition of the fuel pump, pressure regulator, injectors, fuel lines, and other components.

GRADUATED CONTAINERS

A *graduated container(s)* is needed to measure the amount of fuel volume injected during various tests. Look at Fig. 2-35. A graduated container is marked in cc's (cm³ or cubic centimeters). A single large container may be marked in ounces.

A single graduated container is commonly used to measure fuel pump output. Several small graduated containers may be used to check gasoline injector output for locating problems with the injectors, injection lines, computer circuit, or fuel distributor.

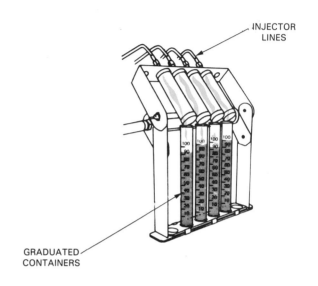

Fig. 2-35. Graduated containers measure injector output in cubic centimeters (cc or cm³). Great variance in volume (7 or more cm³) between injectors would indicate faulty injector, computer, wiring, or fuel distributor. (Volkswagen)

FUEL INJECTION ANALYZER

A *fuel injection tester* or *analyzer* is an electronic device for troubleshooting problems in a modern computer-controlled fuel injection system, Fig. 2-36. By following tester instructions, the analyzer can be used to pinpoint problems with engine sensors, actuators, in-

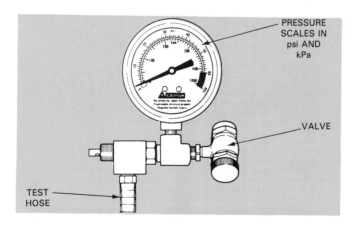

Fig. 2-34. Fuel injection pressure gauge reads higher pressure. (Ford)

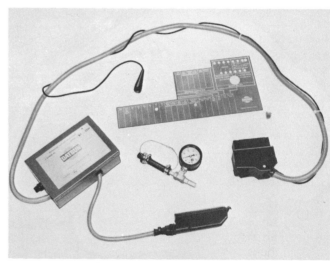

Fig. 2-36. Fuel injection analyzer normally plugs into test connection in fuel system wiring harness. (Kent-Moore Tools)

Fig. 2-37. State-of-art analyzer is being used to check out multi-point gasoline injection system. Note array of controls. (Microtron Products, Inc.)

SMOKE METER

A *smoke meter* is a testing device for measuring the amount of smoke (ash or soot) in diesel exhaust. See Fig. 2-38. The smoke meter measures the amount of light that can shine through an exhaust sample. If the exhaust smoke blocks too much light, engine or injection system repairs or adjustments are needed.

DIESEL INJECTION TESTER

A *diesel injection tester* is a set of pressure gauges and valves for measuring system pressure. Refer to Fig. 2-39. This tester will check feed pump pressure, fuel volume, injector operation, and other functions.

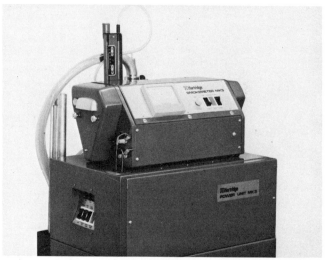

Fig. 2-38. Smoke meter measures amount of particulates in diesel exhaust to help determine combustion efficiency. (Hartridge)

jectors, or the control computer. Also see Fig. 2-37.

EFI test equipment varies. Make sure you follow the operation instructions carefully. Frequently, auto makers recommend one brand of tester for their vehicles. The service manual will explain the use of this tester.

DIESEL INJECTION SERVICE TOOLS

Since a diesel injection system is primarily mechanical, different methods are needed to check the condition of the injection system. There are also additional safety precautions.

1. Even though diesel fuel is not as flammable as gasoline, it still poses a serious FIRE HAZARD. Follow all safety precautions that apply to gasoline.
2. NEVER attempt to remove an injection system component with the engine running. With 6000 to 8000 psi (42 000 to 56 000 kPa) fuel pressure, fuel could squirt out and puncture your skin or eyes.
3. Double-check all torqued fittings before starting the engine.
4. Never attempt to stop a diesel engine by covering the air inlet opening. Since there is no throttle valve, there is enough suction to cause INJURY or to suck rags and other objects into the engine.

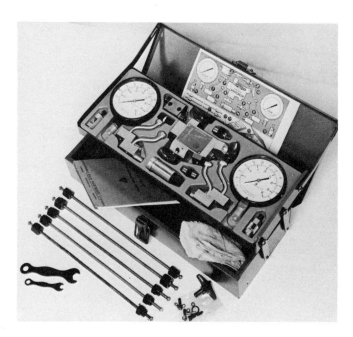

Fig. 2-39. Spotchecker will test injection pump and nozzles on-car. (Hartridge)

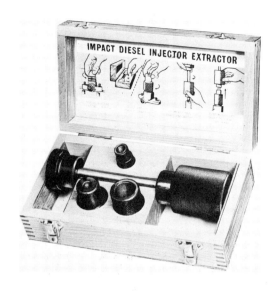

Fig. 2-40. Special extractor removes injectors which are pressed into cylinder head. It is commonly needed on large diesel engines. (Hartridge)

If diesel injection pressures are not within specs, repairs or adjustments are needed. Again, specific operating instructions vary.

INJECTOR EXTRACTOR

An *injector extractor* is a slide-hammer puller designed to remove press-fitted diesel injectors. Several truck type diesel injectors require this tool, Fig. 2-40. Most automotive injectors screw into the cylinder head.

DIESEL INJECTION SERVICE SET

A *diesel injection service set* includes special brass tools for cleaning and repairing injectors and other fuel system components. One is pictured in Fig. 2-41.

Since the clearances in a diesel injector or injection

pump are so "tight," care must be taken not to scratch or mar components. Never use a steel brush to clean diesel injection parts.

POP TESTER

A *pop tester* is a device for checking the condition of a diesel injector nozzle, Fig. 2-42. The injector is mounted on the tester. By pumping the tester handle and watching the pressure gauge, you can check opening pressure, spray pattern, and injector leakage.

DANGER! Extremely high pressures are developed when pop testing a diesel injector nozzle. Wear eye protection and keep your hands away from the fuel spraying out of the nozzle.

GLOW PLUG RESISTANCE TESTER

A *glow plug resistance tester* combines a special wiring harness with an ohmmeter. See Fig. 2-43. It is commonly used to troubleshoot a rough idle problem with a diesel engine.

Fig. 2-42. Pop tester is shown testing diesel injector nozzle. Paper or cloth beneath catches spray from nozzle, showing spray pattern. (OTC Tools)

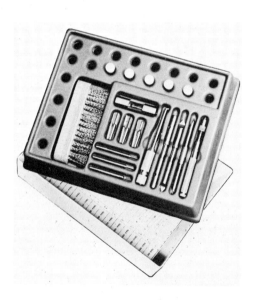

Fig. 2-41. Diesel injection service kit includes soft brass brushes and other special tools.

Fig. 2-43. Tester may pinpoint source of rough diesel engine idle using resistance test of glow plugs. (OTC Tools)

The tester harness has leads that connect to each of the glow plugs. Then, the ohmmeter will read the resistance of each glow plug.

After a period of engine operation, combustion will increase the temperature of the glow plugs, also affecting glow plug resistance. A comparison of glow plug resistance may indicate that one or more cylinders are not firing properly. The tester helps the mechanic find a "dead cylinder" (cylinder not burning fuel charge).

DIESEL INJECTION TIMER

A *diesel injection timer* is a device for adjusting diesel INJECTION TIMING (when fuel is injected in relation to piston position in cylinder). An illuminosity (light detecting) type of timing device is illustrated in Fig. 2-44.

The timer is threaded into a glow plug hole. It can then detect the light produced by the burning diesel fuel on the combustion stroke. This will let the mechanic rotate the injection pump to set injection timing properly.

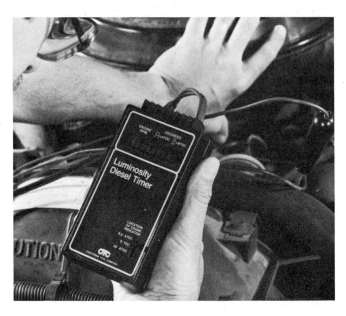

Fig. 2-44. Diesel injection timer reacts to firing or burning of diesel fuel in cylinder by sending signal instant fuel charge ignites. (OTC Tools)

INJECTION PUMP TESTER

An *injection pump tester* is used by specialty shops to check the operation of a diesel injection pump. As shown in Fig. 2-45, it is a very large and complex machine. Most auto shops do NOT have one.

An injection pump tester will spin the injection pump input shaft while measuring output pressures, injection advance, etc. This tester is used when an injection pump is returned to the factory or rebuilder for service.

OTHER FUEL SYSTEM SERVICE TOOLS AND EQUIPMENT

There are many other service tools and equipment used by fuel system technicians. Later textbook chapters will show their use.

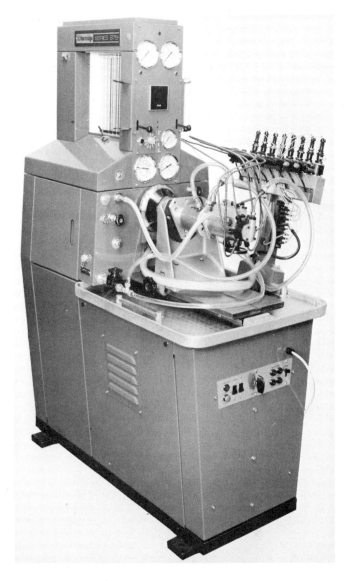

Fig. 2-45. Device for testing diesel injection pump. It is a tool primarily used by rebuilders of pumps. (Hartridge)

Also, later text chapters will give more details for using the tools and equipment introduced in this chapter.

Remember! When using any tool or piece of test equipment, follow instructions. Fuel system and equipment variations make it imperative that specific steps are followed when testing. The slightest mistake could ruin the equipment or cause part damage.

SUMMARY

An auto shop can be a safe place to work if safety rules are carefully followed. The mechanic must guard against fires, explosions, asphyxiation, chemical burns, shock, and physical injuries. A careful mechanic will wear eye protection. He or she will avoid clowning and those who clown, use proper tools, work carefully, and provide good lighting and sufficient ventilation.

The fuel system mechanic will need general service tools to properly test and repair fuel systems. A few of these include a VOM, compression gauge, vacuum gauge, stethoscope, tach-dwell meter, mag-tach, test light, exhaust analyzer, engine analyzer, fuel gauge tester, and PCV system tester.

In addition to special wrenches, certain specialty tools may be needed to perform fuel system service. Among these are: propane enrichment device, carburetor gauge set, choke angle gauge, T-scale, carburetor stand, gasoline injection service tools, fuel injection pressure gauge, graduated container, fuel injection analyzer, diesel injection tester, diesel injection timer, pop tester, and glow plug resistance tester.

Diesel injection service calls for special attention to safety. Extreme care must be used. Injection pressures are so high that spurting fuel striking the skin or eyes can cause serious injury or even death.

KNOW THESE TERMS

Safety can, Asphyxiation, Ground prong, Actuators, Cold soak tank, Tach-dwell meter, Exhaust analyzer, Flex driver, Propane enrichment device, Choke angle gauge, Graduated container, Fuel injection analyzer, Pop tester, Glow plug resistance tester, Diesel injection timer.

REVIEW QUESTIONS—CHAPTER 2

1. List the six different types of accidents that can occur in an auto shop.
2. _____ is the most dangerous material in the shop.
3. The car's battery need not be disconnected during fuel system service and repair. True or false?
4. What precautions should be taken if you must work on a car with the engine running?
5. List seven uses of an exhaust analyzer.
6. Since diesel fuel is not as flammable as gasoline, it is not a fire hazard. True or False?
7. Why should you never attempt to stop a diesel engine by covering the air inlet opening?

Match the statements and terms by placing the correct letter in each blank.

8. ____ Circulates solvent for parts cleaning.
9. ____ Powerful solvent for cleaning fuel system parts.
10. ____ Uses a stream of air to clean parts.
11. ____ Checks basic operation of engine ignition system.
12. ____ Measures engine rpm and correct adjustment for points.
13. ____ Measures engine rpm magnetically.
14. ____ Group of different testing instruments in one assembly.
15. ____ Measures output and performance of an engine.
16. ____ Measures chemical content of engine exhaust.
17. ____ Used to check carburetor or throttle body for warpage.
18. ____ Injects a gas into some types of carburetors during carburetor adjustment.
19. ____ Measures choke opening in degrees.
20. ____ Pressure gauge that can be used to check condition of fuel pump, pressure regulator, injectors, and other fuel system parts.

a. Air gun.
b. Spark tester.
c. Cold soak tank.
d. Carburetor cleaner.
e. Compression gauge.
f. Engine analyzer.
g. Mag-tach.
h. Tach-dwell meter.
i. Dynamometer.
j. Exhaust analyzer.
k. Fuel injection pressure gauge.
l. Choke angle gauge.
m. Straightedge.
n. Propane enrichment device.

3
PRINCIPLES, MEASUREMENTS, FORMULAS

After studying this chapter, you will be able to:
- *Explain basic principles that relate to fuel system operation.*
- *Describe how forces in nature affect fuel system operation.*
- *Use basic measurements relating to engines and fuel systems.*
- *Explain different engine horsepower measurements.*
- *Summarize engine thermal, mechanical, volumetric, and overall efficiency.*
- *Describe the three states of matter.*
- *Summarize heat transfer principles.*

The modern motor car has evolved into a highly complicated, interacting network of systems. It is no longer the simple, easy-to-fix, machine of yesterday. To quickly find and correct a fault in today's vehicles, you must be able to form a ''mental picture'' of what is going on inside a part or system. This chapter is designed to help you acquire this essential knowledge.

By knowing how an auto part or system functions, you will be much better prepared to troubleshoot and repair a car. Without this knowledge, you may be lost on more difficult problems. Trial and error parts replacement can be expensive and wasteful.

Automobile fuel systems are engineered to use some of the forces of nature. These forces include: gravity, atmospheric pressure, centrifugal force, and heat transfer.

GRAVITY

Gravity is the invisible force that holds everything to the surface of the earth. Actually, it is a mutual energy of attraction between all matter.

When you throw a ball high into the air, gravity pulls it back to the ground. Without earth's gravitational pull, the ball would continue to travel into space. Only the friction of the atmosphere (air) would slow it down.

WEIGHT

The force of gravity on an object is measured in *weight* (tons, pounds, ounces, kilograms, grams). Gravity's

strength or pull depends upon mass and distance. For instance, a bowling ball is heavier than a basketball, but only because of gravity. Gravity pulls harder on the bowling ball's greater mass and gives it more weight.

Examples of gravity

Gravity affects the operation of a car in many ways. By giving air weight, it forces air through the air cleaner and into the engine for combustion. It causes engine oil to drain down into the oil pan. It also presses the car onto the road.

Gravity feed fuel system

Some early automobiles did NOT use a fuel pump. A *gravity feed* fuel system forced the fuel from the fuel tank into the carburetor. Many motorcycles and lawnmowers still use this system. Illustrated in Fig. 3-1, the gas tank must be located higher than the carburetor to produce fuel flow.

ATMOSPHERIC PRESSURE

The natural pressure formed by the weight of air in the atmosphere is known as *atmospheric pressure.* Air is a

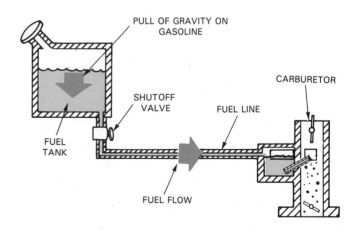

Fig. 3-1. Gravity pulls gasoline down through tank, lines, and into carburetor. Force of gravity actually causes pressure in system.

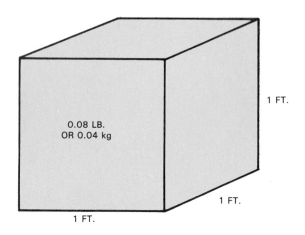

Fig. 3-2. Air has weight because of force of gravity. Cubic foot of air weighs 0.08 lb.

fluid. It has weight, mass, and will flow. It is attracted to the earth like all matter.

A cubic foot of air weighs about .08 lb. (0.04 kg), Fig. 3-2. The weight of air surrounding us forms pressure that pushes equally in all directions. In fact, at sea level, air exerts 14.7 psi (100 kPa) of force per square inch. This is a substantial amount of air pressure, Fig. 3-3. A square foot of area has well over a ton of pressure pushing against it. Our bodies would easily be crushed if an equal amount of pressure were not present on the inside of our bodies.

Earth's atmosphere extends approximately 120 miles (193 km) above the surface of the earth. See Fig. 3-4. Pressure is greatest on the surface of the earth. It decreases with increased altitude.

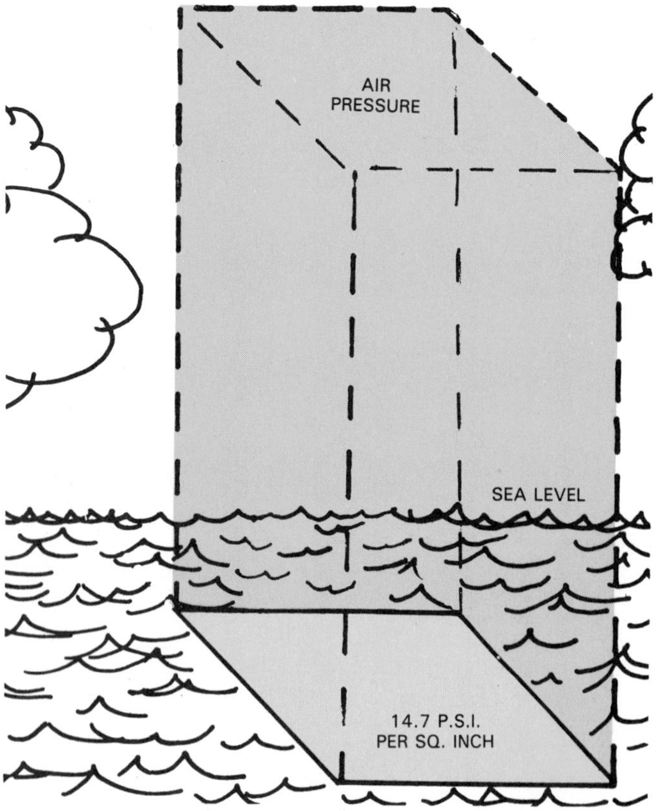

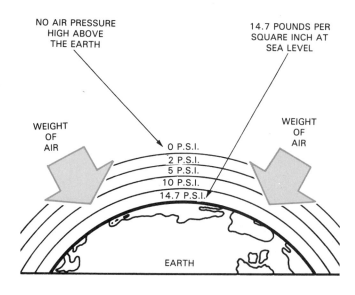

Fig. 3-4. Picture yourself being at bottom of an ocean of air. As you travel upward, toward surface (outer space), pressure decreases. As you go deeper or closer to bottom (earth's surface), air pressure increases.

Air pressure and the automobile

This atmospheric pressure change with altitude affects the operation of an automobile. For instance, at sea level, where air pressure is high, an engine can take in more air and develop more power. The higher pressure forces more air into the cylinders. Further, the air is more compact or dense and contains more oxygen per unit of volume. High in the mountains, air pressure and density will be less than at sea level. As altitude increases, the carburetor or throttle body has less air passing through it. This tends to make the fuel mixture richer.

VACUUM

A *vacuum* can be defined as any area having less than the surrounding atmospheric pressure. For instance, at sea level, any space with LESS THAN 14.7 psi (100 kPa) has a vacuum.

Fluid flow example

Flow will occur in a fluid whenever a PRESSURE DIFFERENCE is present. Pictured in Fig. 3-5, the direction of flow is always towards low pressure. As you suck soda through a straw, a vacuum or low pressure area is formed at the mouth of the straw. Outside air pressure will then push on the soda forcing it through the straw and into your mouth. On the other hand, if you blow into

Fig. 3-3. Atmospheric pressure exerts 14.7 pounds per square inch of pressure at sea level. It presses on everything and in all directions.

44

Fig. 3-5. When you blow into a straw, pressure in your mouth is above atmospheric pressure; flow will be away from your mouth. If you suck on a straw, higher outside air pressure pushes soda into your mouth. Difference in pressure causes flow. (Carter Carburetor)

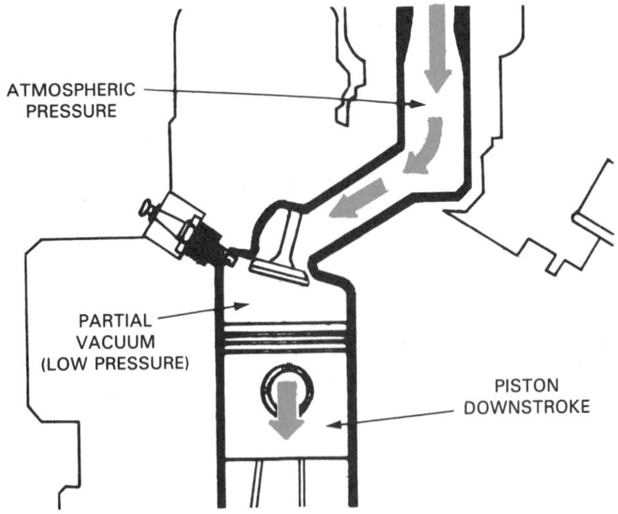

Fig. 3-7. On intake stroke, engine acts as vacuum pump. It produces low pressure area so that outside air pressure can push air and fuel into cylinder. (AFC Carter)

the straw and increase pressure above atmospheric pressure, flow will be reversed.

Simple vacuum pump

Vacuum can be produced mechanically, as shown in Fig. 3-6. In A, a simple vacuum pump (piston, cylinder, and valve) is surrounded by natural outside air pressure. Atmospheric pressure is pushing equally in all directions on the outside and inside of the pump. In B, the piston has traveled down in the cylinder with the valve closed. As a result, cylinder area has increased while the amount of air in the cylinder has remained the same. This reduces

pressure inside cylinder and forms a vacuum. In drawing C, valve has been opened and outside air pressure rushes in to fill low pressure area. Airflow into cylinder will continue until outside and inside pressure are equal.

An engine as a vacuum pump

An automobile engine acts as a *vacuum pump* during its intake stroke. This is illustrated in Fig. 3-7. When the engine's intake valve is opened and the piston is moving down, a partial vacuum is produced in the engine cylinder. Higher outside atmospheric air pressure forces air through the air cleaner, carburetor or throttle body,

FORMING A VACUUM (SIMPLE VACUUM PUMP)

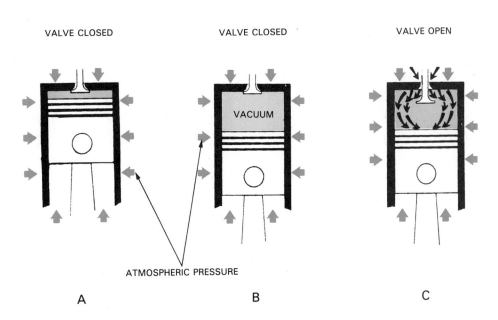

Fig. 3-6. A—Atmospheric pressure pushes equally on all surfaces of engine. B—Valve is closed and cylinder is sealed. When piston moves down, vacuum or low pressure area is produced in cylinder (area is increased, but air quantity stays same). C—Air pressure rushes in to fill cylinder vacuum as valve opens. (Echlin)

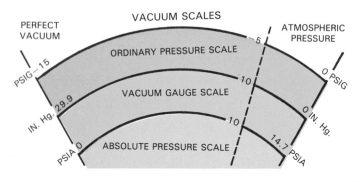

Fig. 3-8. Study difference in three scales. Absolute scale, not commonly used in auto repair, illustrates that we are constantly subjected to atmospheric pressure.

intake manifold, and into the combustion chamber. As you can see, the piston does not really "suck" the fuel and air into the engine; air pressure PUSHES it in.

Measuring vacuum

In auto mechanics, *vacuum* is commonly measured with a pressure or a vacuum gauge. Both operate the same way, but their scales or faces are different. Fig. 3-8 shows these two gauge scales as well as an absolute scale.

The absolute pressure scale is not frequently used on a car. It registers 15 psi at atmospheric pressure and zero at a perfect vacuum. This illustrates that the weight of air forms natural pressure and that a perfect vacuum (0 psi) exists when all of the air has been removed from an area.

Pressure and vacuum gauge scales are commonly used in auto repair. Both register zero at atmospheric pressure. The vacuum gauge reads in inches of mercury (or millimeters of mercury). Quite often, a pressure and

Fig. 3-9. A vacuum gauge is used to measure vacuum (negative air pressure). It is commonly used to check engine operation. If intake manifold vacuum is low or fluctuates, there may be an engine problem. This gauge is a compound gauge because it also has a pressure scale. By connecting it to fuel pump, pump pressure can be measured.

vacuum gauge will be combined into one gauge, as in Fig. 3-9. This is very handy because it makes the gauge dual purpose. On the left side of zero, the scale registers pressure in psi (pounds per square inch) or kPa (kilopascals). To the right of zero, the scale registers vacuum in in./Hg (inches of mercury) or mm/Hg (millimeters of mercury).

The pressure scale is commonly used to test fuel pump pressure, fuel pressure regulator settings, etc. The vacuum scale can be used to measure engine intake manifold vacuum, detect bad engine valves, poor carburetor adjustment, improper ignition timing, and faulty distributor vacuum advance, as well as take other critical measurements.

WORK

Technically, *work* occurs when any force moves an object. In other words, work is performed when anything is moved from one place to another. When you stand up, you have worked by moving your body a specific distance.

The purpose of any engine is to accomplish work. It uses the force (expansion) of the burning fuel to move the pistons, rods, crankshaft, and other components.

Work formula

Work is measured in foot-pounds (ft.-lb.) or Newton meters (N·m). The amount of work performed can be calculated by the simple formula:

$$\text{WORK} = \text{FORCE} \times \text{DISTANCE, or}$$
$$W = F \times D$$

If you lift a 5 lb. weight a distance of 2 ft., how many units of work have you done? Simply multiply force (5 lb.) by distance (2 ft.). The answer is 10 ft.-lb. Work is performed when shafts are rotated, springs compressed, or when objects are slid.

POWER

Power is the speed or rate at which work is completed. It is measured in foot pounds per second or horsepower. If a great amount of work is done in a short period of time, a large amount of power must be used. Conversely, when only little work is accomplished very slowly, little power is needed. The dimension of *time* has been added to the concept of work.

Power formula

The formula for power, the rate of doing work, is:

$$\text{POWER} = \frac{\text{FORCE} \times \text{DISTANCE}}{\text{TIME}}$$

To use this formula, simply "plug" in the values, then multiply and divide. For example, if a small gas engine, equipped with a pulley mechanism, lifts a 100 lb. weight 5 ft. in 60 seconds, how much power was used?

$$\text{POWER} = \frac{100 \text{ lb.} \times 5 \text{ ft.}}{60 \text{ sec.}} = \frac{500}{60} = 8.33 \text{ ft.-lb./sec.}$$

As you can see from the above calculation, the engine produced 8.33 ft.-lb./sec. of power.

46

ENERGY

Anything capable of doing work has *energy*. You have energy. An operating engine has energy. There are several classifications of energy. Important to the operation of a car are: mechanical energy, thermal or heat energy, chemical energy, and electrical energy.

Mechanical energy

Mechanical energy is the energy of motion. It is commonly used to do work in an automobile. The two classifications of mechanical energy are potential and kinetic energy.

Potential energy is mechanical energy held in check or at rest. It is not doing work, but has the ability to do work as a result of an object's position or form. For instance, if you use a jack to lift the front of a car, as in Fig. 3-10, you have created energy in the car's weight. The suspended car could perform work (cause movement over a distance) by falling back to the ground. The raised car has potential energy because of its position.

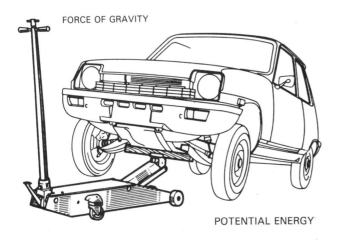

Fig. 3-10. Suspended car has a tremendous amount of potential energy. In fact, enough to cripple or even kill if it were to fall on someone. (Renault)

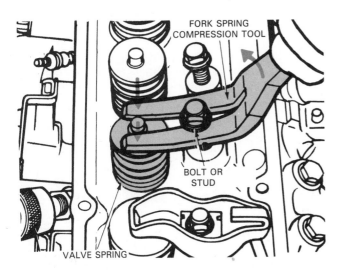

Fig. 3-11. A compressed valve spring has potential energy to close valve. This is obvious when compressing a valve spring for valve seal replacement. (Ford Motor Co.)

A compressed spring has potential energy of form as it tries to return to its normal shape. The tendency (to return to its original shape) makes it able to move or support a component on a car, or close an engine valve. Look at Fig. 3-11.

Air under pressure in an air compressor tank has the potential energy to fill and expand a tire, to spin and power an air impact gun, or to blow solvent off of a part.

Kinetic energy is the energy of a body or object in motion. It is the same as mechanical energy. A speeding car or spinning engine crankshaft are examples of kinetic energy. Rotating kinetic energy, for instance, can be used with gears, pulleys, belts, and chains, to move a car and operate its various systems.

Chemical energy

Chemical energy is normally released whenever molecules of different substances react with each other. Gasoline, coal, diesel fuel, LP gas, and alcohol all have chemical energy. The energy is released as heat during combustion.

A *battery* is a good example of chemical energy. The acid and metals inside a car battery can react to produce electricity, another form of energy. In turn, when the alternator charges the battery, electrical energy is changed back into chemical energy.

Electrical energy

Electrical energy, in simple terms, is the flow of electrons (parts of an atom) through a conductor (commonly a piece of wire or metal component). This flow of electricity is due to a difference in electrical potential. If one end of a wire is positively charged while the other is negatively charged, a pressure difference exists between them. Electricity will flow and electrical energy will be present in the wire.

Energy conversion

An alternator or generator is commonly connected to the engine crankshaft by a fan belt. As you learned earlier, the spinning crankshaft has kinetic energy. By turning the alternator, this kinetic energy is changed into electrical energy to power the lights, ignition system, radio, etc., and to recharge the battery, Fig. 3-12.

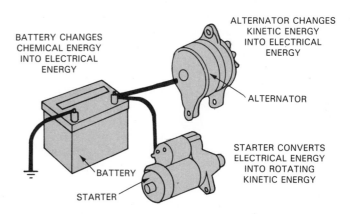

Fig. 3-12. An alternator converts engine's rotating kinetic energy into electric energy. Battery changes chemical energy into electrical energy for engine starting. Starting motor changes electrical energy into kinetic energy. (Mazda)

In turn, electrical energy can be used to produce kinetic energy. When starting an automobile engine, battery electricity is changed back into rotating kinetic energy by the starting motor. This kinetic energy "cranks" or turns the engine for starting.

INERTIA

Anything that moves has *inertia.* This means that an object in motion or at rest will resist any change in its motion, speed, or direction. Also, anything that moves, wants to move in a straight line. The amount of inertia depends upon the speed and weight of the moving object. An increase in either will increase inertia. For example, a baseball struck by a bat has some weight but great speed or velocity. As shown in Fig. 3-13, when the baseball strikes a stationary (standing still) object, the ball's inertia will oppose a sudden change in motion. As a result, the ball will pass through the window. The high speed gives the ball its high inertia.

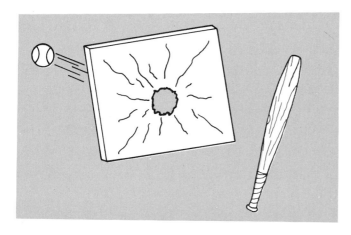

Fig. 3-13. Baseball gets high inertia from its high speed. A car gains most of its inertia from weight.

An automobile, in contrast, can move more slowly than a baseball but has many times the weight. It will gain its inertia more from weight than from speed. When a car hits a brick wall, it is mainly inertia from mass that crushes the car's body.

Flywheel inertia

An engine flywheel is an excellent example of how inertia is put to use. Remember, a four cycle, single cylinder engine only produces power 25 percent of the time, when the piston moves downward on the power stroke. During the other three strokes, power is consumed by the engine. This engine would run very roughly without a flywheel. It would speed up during the power stroke and then slow down during the other strokes. By adding a heavy flywheel to the crankshaft, inertia will help resist this slowdown. It will help keep the crankshaft spinning at the same speed during all four strokes.

CENTRIFUGAL FORCE

Centrifugal force is that force generated by an object moving in a curved path. The object tends to move out-

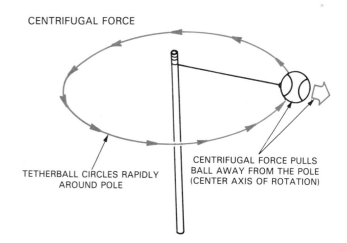

Fig. 3-14. When you hit a tetherball and make it spin, centrifugal force pulls ball away from pole.

ward from its center of rotation. The amount of pull or force depends upon the spinning object's speed, weight, and distance from center (rpm held constant). As these factors are increased (object spins faster, is heavier, or is moved farther from center of rotation), centrifugal force is increased. As they are reduced, centrifugal force decreases.

When you hit and spin a tetherball as in Fig. 3-14, centrifugal force pulls the ball away from the pole. When you round a corner in a speeding car, centrifugal force tends to push you over to one side of the car seat.

Centrifugal force operates in an automobile in many other ways. An out-of-balance car tire bounces up and down on the road because of centrifugal force. This can make the unbalanced tire shake the whole car. Some electric fuel pumps, many water pumps, and turbochargers utilize centrifugal force. A spinning impeller causes liquid or air to fly away from the center of rotation. This outward movement moves or pumps the fluids, air or water through each system. Another example, an

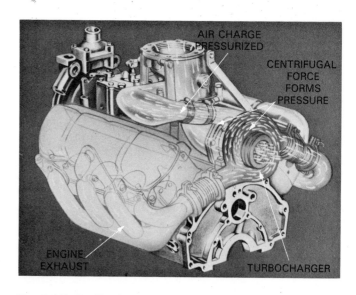

Fig. 3-15. A turbocharger uses centrifugal force. When incoming air is spun by compressor, it is thrown outward and forced into engine under pressure. (Detroit Diesel)

48

automotive distributor, uses a centrifugal advance mechanism to control the spark timing at the spark plugs. Pictured in Fig. 3-15 is a turbocharger which utilizes centrifugal force.

HORSEPOWER (hp)

As mentioned, the rate of doing work is called power. *Horsepower,* in automotive terms, is a measure of an engine's ability to perform work. The more horsepower, the more work performed.

At one time, one horsepower was the average strength of a horse. A 300 hp engine could, theoretically, perform the work of 300 horses.

Mathematically, one horsepower equals 33,000 foot-pounds of work per minute (or 550 lb. per sec.). For instance, a horse should be capable of lifting a 33,000 lb. weight 1 ft. in one minute, Fig. 3-16.

Formula for horsepower

To calculate engine horsepower, you simply use the following formula:

$$\text{HORSEPOWER} = \frac{\text{DISTANCE (ft.)} \times \text{WEIGHT (lb.)}}{33,000}$$

or

$$\frac{\text{WORK (ft.} \times \text{lb.)}}{33,000}$$

Suppose that a small engine lifted 500 lb. 700 ft. in one minute. About how much horsepower was used?

To calculate the answer, apply these values to the horsepower formula.

$$\frac{500 \text{ lb.} \times 700 \text{ ft.}}{33,000} = 10.6 \text{ hp}$$

Factory horsepower ratings

Automobile engines are rated from the manufacturer at a specified horsepower for a particular rpm (revolutions per minute). For instance, a high performance V-8 engine might be rated at 400 hp at 5000 rpm. This engine power rating is commonly given in a shop manual. Notice that it is stated in *brake horsepower,* Fig. 3-17. Some manuals include a metric value of engine power. Metric power is normally given in kilowatts (kW).

Usually, the *factory hp rating* will be maximum at the stated rpm. Engine power will be reduced at slow engine speeds because the engine will be burning less fuel. Also, engine horsepower will decrease when engine speed is increased beyond the maximum factory rpm rating. At excessive speeds, an engine will no longer be capable of taking in enough fuel and air. There will not be enough time for a full charge of fuel mixture to enter the cylinders. As a result, engine power will drop at very high rpm levels.

Brake horsepower

Brake horsepower, abbreviated bhp, is engine power output measured at the crankshaft. It is the actual or useful power output. Brake horsepower is commonly used to rate the power developed by automobile engines.

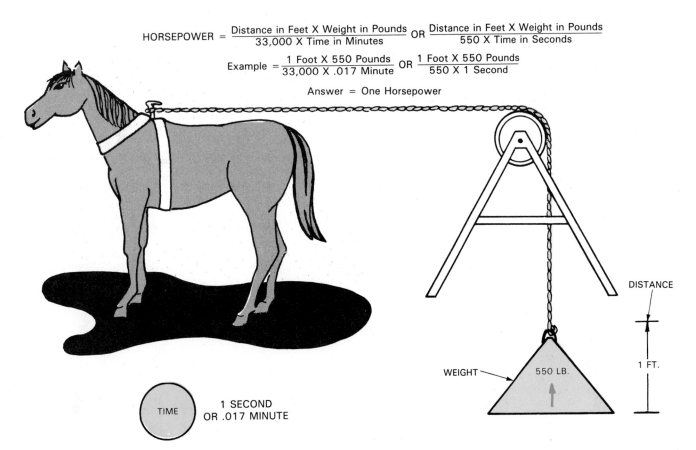

$$\text{HORSEPOWER} = \frac{\text{Distance in Feet X Weight in Pounds}}{33,000 \text{ X Time in Minutes}} \text{ OR } \frac{\text{Distance in Feet X Weight in Pounds}}{550 \text{ X Time in Seconds}}$$

$$\text{Example} = \frac{1 \text{ Foot X 550 Pounds}}{33,000 \text{ X .017 Minute}} \text{ OR } \frac{1 \text{ Foot X 550 Pounds}}{550 \text{ X 1 Second}}$$

Answer = One Horsepower

TIME 1 SECOND OR .017 MINUTE

DISTANCE

1 FT.

WEIGHT 550 LB.

Fig. 3-16. One horsepower equals 33,000 lb. lifted a foot in a minute or 550 pounds one foot in a second.

ENGINE BRAKE HORSEPOWER SPECIFICATIONS

Engine	Compression Ratio	Bore and Stroke	Taxable Horsepower	Brake Horsepower
429-2V	10.5:1	4.36 x 3.590	60.82	320 @ 4400
429-4V	11.0:1			360 @ 4600
460	10.5:1	4.36 x 3.850	60.83	365 @ 4600

Fig. 3-17. Brake horsepower is stated in a shop manual at a specific rpm or engine speed. It is usually determined at sea level where power will be maximum. (Ford)

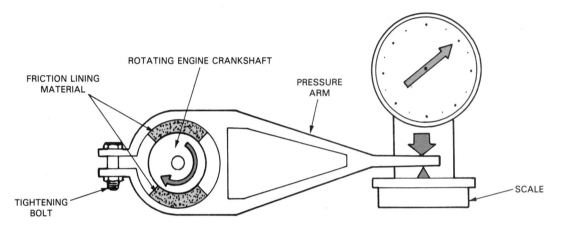

Fig. 3-18. By tightening bolt on prony brake, a drag is placed upon spinning engine crankshaft. This makes pressure arm rotate, pushing down on scale. Reading on scale can then be used to calculate engine brake horsepower.

The term was derived from a device called a *prony brake,* Fig. 3-18. A friction brake was tightened around an engine-powered shaft. This placed a drag on the engine. As a result, the pressure arm was forced down on the weight scale. The scale reading could then be used to calculate bhp.

Dynamometer

A *dynamometer* is used to measure engine brake horsepower. Usually referred to as a *dyno,* it will load an engine and measure its power output. It sometimes uses a large electric motor or generator. The motor connects to the crankshaft and produces a high electric current when the engine is operated. The maximum amount of current developed by the electric generator will indicate the maximum engine horsepower. Sometimes, a water brake is utilized.

Brake horsepower is the most commonly used engine horsepower rating method. It is referred to in service manuals and owner's manuals. Fig. 3-19 shows engine horsepower "specs" from a service manual.

Indicated horsepower

Indicated horsepower (ihp) refers to the amount of power or pressure formed inside the engine's combustion chambers. During this type of horsepower measurement, a special pressure sensing device is placed through the cylinder head to measure cylinder pressures. See Fig. 3-20. These pressure readings are used to calculate indicated horsepower.

To perform an indicated horsepower test, the pressure during all four engine strokes must be measured and averaged. Then the average pressures of the intake, compression, and exhaust strokes are subtracted from

Fig. 3-19. An actual engine dynamometer measuring engine horsepower. Elaborate computer system monitors engine's performance. Dynamometer simulates driving conditions and puts load on engine. (AC Spark Plug)

the pressure of the power stroke. The result is the mean effective pressure (mep). This pressure can be used in the following formula to calculate indicated horsepower:

$$\text{INDICATED HORSEPOWER} = \frac{PLANK}{33,000}$$

P = mean effective pressure in psi
L = length of piston stroke in feet
A = cylinder area in square inches
N = number of strokes per minute
K = number of cylinders

To use this formula, multiply the PLANK values together and divide by 33,000.

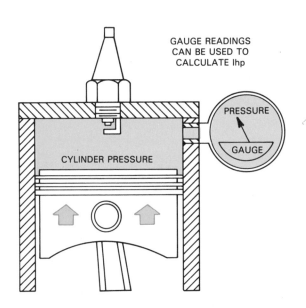

Fig. 3-20. Indicated horsepower averages cylinder pressures of all four strokes in order to find engine power. It does not, however, consider engine friction.

Fig. 3-21. A net horsepower rating measures power with all accessories installed on engine. It indicates real or usable power of an engine installed in a car. (Ford Motor Co.)

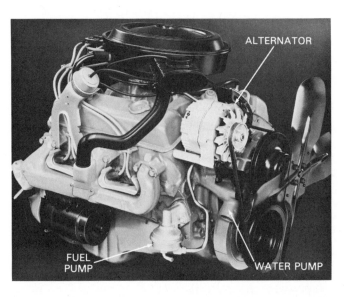

Fig. 3-22. Gross horsepower measures horsepower of a "basic engine." A basic engine includes only the necessities such as fuel pump, water pump, oil pump, and fan. An alternator can only be included when one of needed engine accessories is electrically powered. (General Motors Co.)

Frictional horsepower

Frictional horsepower (fhp) refers to the power needed to overcome engine friction. It is a measure of the friction between the moving parts in an engine. It is the power lost to friction.

Usually, the sliding friction between the spring-loaded piston rings and the cylinder walls causes the highest percentage of drag and horsepower loss. However, every moving component (timing chain, valves, crankshaft, camshaft, and lifters) contributes to frictional horsepower.

Measuring engine friction

Frictional losses in an engine can be measured by spinning an engine with an external power source. An electric motor (starting motor) can be used to rotate the engine. Then, the amount of electricity needed to turn the engine can be used to calculate *frictional horsepower.*

During this test, the engine must be fully warmed (160° to 195° F) and the carburetor butterflies should be wide open. Also, to prevent wash-down of the cylinders (oil removal), the carburetor should be empty of gasoline or the injection system disabled.

The following formula shows another way to find frictional horsepower losses.

FRICTIONAL HORSEPOWER = Indicated Horsepower − Brake Horsepower

To calculate fhp, simply subtract bhp from ihp. This will give you a number representing the amount of power it takes to overcome engine friction.

Net horsepower

Net horsepower is the maximum horsepower developed when an engine is loaded down by all engine powered accessories. These accessories include the water pump, fuel pump, fan, muffler, air injection pump for emission control, alternator, and, if so equipped, air conditioner (compressor), and power steering pump. Net

horsepower is the amount of useful horsepower with the engine installed in an automobile, Fig. 3-21. As a result, the net horsepower value is very informative to the automobile owner. It shows the engine power remaining after all of the horsepower-robbing accessories have been added. Net horsepower is lower than either brake or indicated horsepower.

NET HORSEPOWER = Brake Horsepower − Frictional Drag of Engine Accessories

Gross horsepower

Gross horsepower is similar to net horsepower. However, it measures engine power with only the basic or essential accessories installed. It does not include the power lost to the air conditioning compressor, power steering pump, air injection pump, or other accessories. Logically, gross horsepower will be higher, in most cases, than net horsepower. See Fig. 3-22.

51

TAXABLE HORSEPOWER

GENERAL SPECIFICATIONS

Engine	Compression Ratio	Bore and Stroke	Taxable Horsepower	Brake Horsepower	Gross Torque Ft-Lbs.
170	8.7:1	3.50 x 2.94	29.4	100 @ 4200	148 @ 2600
200	8.7:1	3.68 x 3.13	32.5	115 @ 4000	180 @ 2200
250	9.0:1	3.68 x 3.91	32.5	145 @ 4000	232 @ 1600

Fig. 3-23. Notice that taxable horsepowers of 200 and 250 cu. in. engines are same while their brake horsepower ratings are different. Evidently, cylinder bores of these two engines are same but their strokes are different. Taxable horsepower is not very accurate because it does not consider engine displacement (size) or actual power. (Ford)

Taxable horsepower

In many states, the amount of tax placed on a car is partially dependent upon a car's taxable horsepower. Abbreviated thp, taxable horsepower is actually a measurement of engine size rather than engine power. The formula for finding taxable horsepower is:

TAXABLE HORSEPOWER = Cylinder Bore ×
Cylinder Bore × No.
of Cylinders × .4

Service manual values for three different engines are given in Fig. 3-23.

TORQUE

Torque is a turning or twisting force that is either moving or stationary. When you open a door, for example, you must apply torque to the door knob. Even if the door was locked and the knob would not turn, torque was still applied. When you pull on one side of a steering wheel as shown in Fig. 3-24, torque rotates the steering shaft.

Engine torque is turning force at the crankshaft. When combustion pressure pushes down on the piston and connecting rod, a powerful turning force is placed upon the crankshaft. Torque is transmitted through the drive train (transmission, drive shaft, differential, or "rear end", and axles) to the rear wheels. The wheels apply torque to the road, and the car is propelled forward.

Torque may be measured in either pound-feet (lb.-ft.) or in foot-pound (ft.-lb.). Engine torque, however, is normally stated in POUND-FEET. One pound-foot is illustrated in Fig. 3-25.

To calculate torque, multiply turning force by the length of the lever arm. This is shown by the formula:

TORQUE = Force in Pounds ×
Lever Arm Length in Feet

Engine torque

Engine torque specifications, as given in service or shop manuals, are usually stated at a particular engine speed (rpm). For instance, 78 lb.-ft. @ 3000 rpm. This means that the engine is capable of producing maximum torque of 78 lb.-ft. when operating at 3000 revolutions per minute. In metric, torque is given in Newton meters (N·m). See Fig. 3-26. Engine torque normally increases with engine speed, up to a certain point.

VOLUMETRIC EFFICIENCY

Volumetric efficiency is the ratio between actual air drawn into the cylinder and the maximum volume of air with the cylinder completely filled. It refers to how well an engine can "breathe" or take in air during the intake stroke. As more air is drawn into the cylinder, volumetric efficiency and engine power will improve.

Volumetric efficiency is normally stated as a percentage. Ideal volumetric efficiency is 100 percent. Naturally, engines are far from perfect and 80 to 85 percent volumetric efficiency is good at high engine speeds.

MEASURING TORQUE

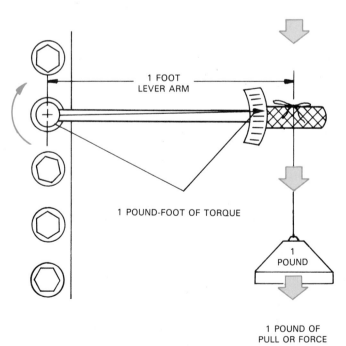

1 FOOT LEVER ARM

1 POUND-FOOT OF TORQUE

1 POUND

1 POUND OF PULL OR FORCE

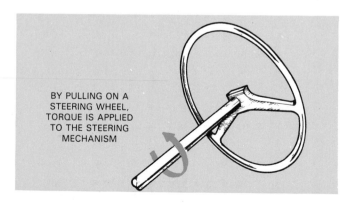

BY PULLING ON A STEERING WHEEL, TORQUE IS APPLIED TO THE STEERING MECHANISM

Fig. 3-24. Torque is simply a turning effort. It is present in thousands of places in an automobile. (Renault)

Fig. 3-25. One pound-foot equals one pound of pull on a 1 ft. long lever arm.

ENGINE TORQUE SPECIFICATIONS

	1400 CC		1600 CC		2000 CC		2600 CC	
	POWER (HP/kW)	TORQUE (lb.ft./N•m)	POWER (HP/kW)	TORQUE (lb.ft./N•m)	POWER (HP/kW)	TORQUE (lb.ft./N•m)	POWER (HP/kW)	TORQUE (lb.ft./N•m)
Federal (Under 4000') and California	70/52 @ 5200	78/106 @ 3000	80/60 @ 5200	87/118 @ 3000	93/69 @ 5200	108/146 @ 3000	105/78 @ 5000	139/188 @ 2500

Fig. 3-26. Torque, like horsepower, is normally stated at a specified rpm or speed. This engine has 78 lb.-ft. or 106 Newton-meters of torque at 3000 rpm. Newton-meter is a metric measurement of torque. (Chrysler Corp.)

The following formula can be used to find volumetric efficiency:

VOLUMETRIC EFFICIENCY =
$$\frac{\text{ACTUAL VOLUME OF AIR TAKEN INTO THE ENGINE}}{\text{MAX. POSSIBLE VOLUME IN ENGINE CYLINDERS}}$$

For instance, what would the volumetric efficiency of an engine be if it was capable of displacing 500 cubic inches of air per minute but only used 400 cu. in.? Values are given in cubic feet per minute, abbreviated cfm. To calculate the induction system efficiency or volumetric efficiency:

$$\text{VOLUMETRIC EFFICIENCY} = \frac{400 \text{ cfm}}{500 \text{ cfm}} = \frac{80\%}{\text{EFFICIENCY}}$$

Factors affecting volumetric efficiency

Volumetric efficiency is affected by engine speed, engine load, carburetor throttle position, engine temperature, and inlet air temperature. When accelerating from a "dead" stop at full throttle (throttle plates wide open), volumetric efficiency would be extremely high, possibly at maximum. There would be plenty of time for a full charge of fuel mixture to enter and fill the engine cylinders. Also, with the carburetor or TBI throttle plates wide open, there would be little restriction upon airflow. It could enter the engine easily.

When engine speed is increased, however, there is less time during each intake stroke for air to enter the cylinders. The pistons and valves are operating so fast that cylinder filling is incomplete. The engine "starves" for air and fuel.

Air temperature affects volumetric efficiency and power

Heat causes expansion of a gas. If you heat a balloon filled with air, it will swell up. The same amount of air is still in the balloon. It just takes up more space. Its molecules are farther apart. For this reason, the temperature of the inlet air (air entering the air cleaner) is very important to volumetric efficiency and engine power. If the air entering the engine is hot, it will have less weight and will contain less oxygen for combustion. Engine volumetric efficiency and horsepower will be reduced. Cool inlet air would noticeably increase volumetric efficiency and power.

Airflow restrictions reduce volumetric efficiency

All engines have some restriction upon the flow of fuel and air into their cylinders. Thus, 80 to 85 percent volumetric efficiency is normal. Many of the induction system parts—valves, valve stems, cylinder head port irregularities, or carburetor—can block or restrict fuel mixture entry. To improve efficiency, engines designed for high speed, power, and efficiency (racing engines) should have large valves, large, straight, and smooth cylinder head ports, a large-area air filter, cool air duct, and high lift, long duration camshaft. All of these factors help air to enter the engine at higher rpms.

As a consequence low speed engine efficiency is usually reduced. At low speeds, air velocity (speed) is very important. It helps to force air into the cylinders. It also helps to form combustion chamber turbulence and mixing, which is important to combustion. An engine with extremely high volumetric efficiency will often suffer from poor low-speed power and efficiency. Its fuel mixture velocity will be too slow to provide thorough mixing, turbulence, and combustion.

A "full bore" racing engine, for example, will not develop good power until it reaches very high rpm. At lower speeds, it may run rough, miss, and have less horsepower. It may even be possible for an unmodified, "stock" street driven engine with lower volumetric efficiency to have more low-speed horsepower (2000 rpm) than a racing engine.

ENGINE EFFICIENCY

Engine efficiency is the ratio of power obtained (brake horsepower) to power supplied (heat content of fuel). By comparing fuel consumption to engine power, you can find engine efficiency.

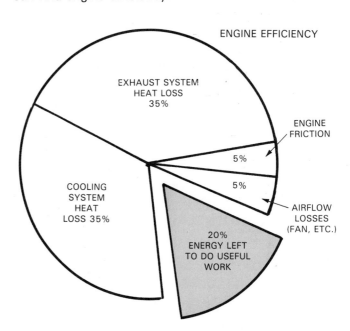

Fig. 3-27. Pie chart illustrates how only a small amount of fuel's heat energy is used to propel car.

If all of the heat energy in gasoline were converted into useful work (crankshaft rotation), an engine would be 100 percent efficient. This is not possible with a reciprocating engine. Today's internal combustion piston engines are only about 20-28 percent efficient.

As illustrated in Fig. 3-27, most of the energy used by an engine is wasted. In fact, around 70 percent of the heat energy in gasoline flows into the cooling system and the exhaust system as unused power. See Fig. 3-28. Only 20 to 28 percent of the fuel's energy is left over to propel the car.

Engine efficiency can be divided into three basic categories: mechanical efficiency, thermal efficiency, and practical efficiency.

MECHANICAL EFFICIENCY

Mechanical efficiency is a comparison between brake horsepower and indicated horsepower. It is a measurement of mechanical friction. Remember, indicated horsepower refers to the theoretical power formed by combustion. Brake horsepower is the actual power supplied at the engine crankshaft. The difference between the two horsepower readings is the loss due to friction.

Mechanical efficiency of around 85 percent is normal. This means that only about 15 percent of the engine's power is lost to mechanical inefficiency or friction.

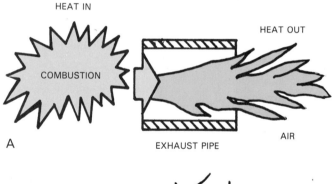

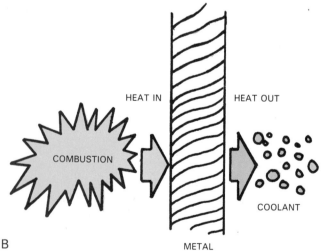

Fig. 3-28. A—About 35 percent of combustion heat blows out exhaust as waste. B—Another 35 percent is conducted into engine and cooling system as unused power.
(Gates Rubber Co.)

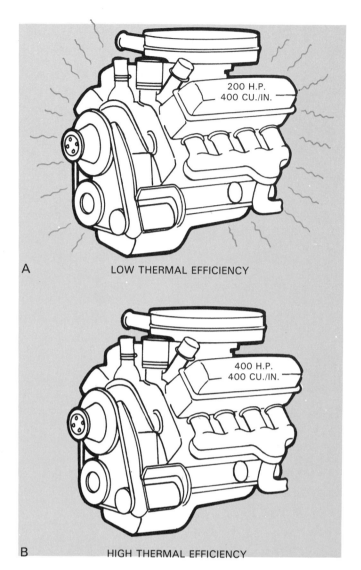

Fig. 3-29. Engine A has low thermal efficiency, a lot of its combustion heat is tranferring into cooling and exhaust systems. Engine B has better thermal efficiency and more power. (Gates Rubber Co.)

THERMAL EFFICIENCY

An engine's *thermal* or *heat efficiency* is determined by comparing the amount of fuel burned to horsepower output. It indicates how well an engine can utilize a fuel's heat energy. It measures the amount of heat energy converted into useful crankshaft rotation.

As an example of thermal efficiency, let us compare two similar engines shown in Fig. 3-29. The two engines are equal in mechanical efficiency and fuel consumption. However, engine A does not produce as much horsepower as B. Since everything else is equal it is easy to see that engine B has a lower thermal efficiency. More of its combustion heat is apparently leaving the engine unharnessed. Engine A must be using more of the fuel's heat to produce piston movement.

To find the thermal efficiency of an engine, you must know a few values. A gallon of gasoline has around 19,100 Btus of heat energy. Also, one horsepower equals about 42.4 Btus of heat energy per minute. With this information, and the following formula, you can find thermal efficiency:

THERMAL EFFICIENCY =
$$\frac{Brake\ Horsepower \times 42.4\ Btu/minute}{19,100\ Btu/gallon \times gallons\ per\ minute}$$

Generally, engine thermal efficiency is around 25 to 30 percent. The other 70 to 75 percent of the fuel's heat energy is lost.

COMPRESSION RATIO

As was discussed in the two previous chapters, an engine's *compression ratio* controls how tightly the fuel mixture is squeezed during the compression stroke. Up to a point, a high compression ratio tends to improve engine efficiency, fuel mileage, and horsepower.

As pictured in Fig. 3-30, the ratio between maximum cylinder volume and minimum cylinder volume regulates compression pressure and the compression ratio of an engine. It is a comparison of the largest possible cylinder volume (piston at the bottom of its stroke) with the smallest possible cylinder volume (piston at the very top of its stroke). Today's gasoline powered cars use a compression ratio of around 8 to 1 (also stated 8:1). See Fig. 30-30 again.

To calculate compression ratio, divide the maximum cylinder volume (piston at BDC) by the minimum cylinder volume (piston at TDC). Look at Fig. 3-31. If the cylinder volume is 50 cubic inches at BDC and only 5 cubic inches at TDC, the compression ratio would equal 50 divided by 5.

$$COMPRESSION\ RATIO = \frac{BDC\ CYLINDER\ VOLUME}{TDC\ CYLINDER\ VOLUME}$$
$$= \frac{50}{5}$$
$$= 10:1\ COMPRESSION\ RATIO$$

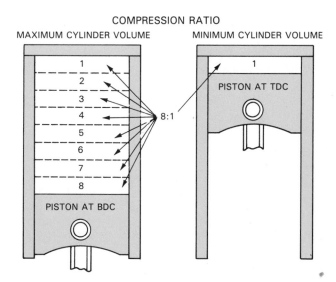

COMPRESSION RATIO

MAXIMUM CYLINDER VOLUME   MINIMUM CYLINDER VOLUME

Fig. 3-30. On left, piston is at BDC and cylinder volume is maximum at eight units. When piston reaches TDC, cylinder volume is minimum or one unit. Engine's compression ratio is a comparison of two volumes, maximum and minimum. Hence, 8 to 1 is engine's compression ratio.

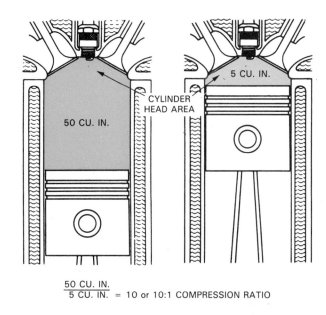

COMPRESSION RATIO 10 TO 1

$$\frac{50\ CU.\ IN.}{5\ CU.\ IN.} = 10\ or\ 10:1\ COMPRESSION\ RATIO$$

Fig. 3-31. By dividing maximum cylinder volume by minimum volume, compression ratio can be calculated. Most of TDC or minimum volume is in cylinder head portion of combustion chamber. (Ethyl Corp.)

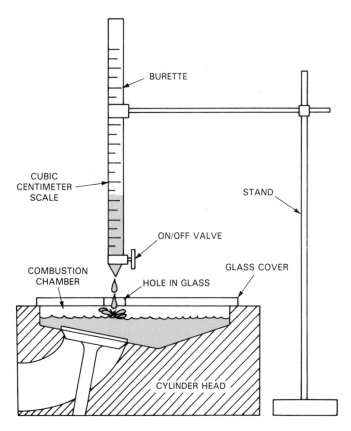

Fig. 3-32. Using "CC" method to find volume of combustion chamber. Cover and seal combustion chamber with special piece of glass with hole cut into it. Use burette (scaled glass tube for measuring liquid volumes) to fill chamber with liquid. By noting cubic centimeters (cm³) required to fill chamber, volume can be determined.

MEASURING COMPRESSION PRESSURE

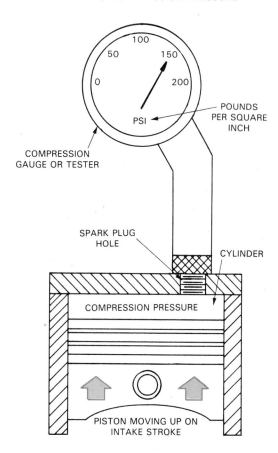

inches of piston displacement, divide the number of cylinders (8) into cylinder displacement (427).

Finding combustion chamber volume

The volume of the combustion chamber (piston at TDC) is a little harder to find. You would have to "CC" the cylinder head. Fig. 3-32 shows the equipment needed. The top of the combustion chamber must be covered and sealed with a special piece of drilled glass. Then, a chemist's burette can be used to fill the combustion chamber with liquid. The amount of liquid needed to fill the chamber is equal to the chamber volume.

Compression pressure

A high compression ratio will produce high pressure in the combustion chamber. This is known as *compression pressure.* It may vary from around 110 psi to over 200 psi in some high performance engines.

When you use a compression gauge to check engine compression, as shown in Fig. 3-33, you are measuring compression pressure. Service manuals give the recommended compression stroke pressure for various engines. See Fig. 3-34. It can be used to help determine the general condition of an engine.

For example, if an engine only produces 90 lb. of pressure in one cylinder and specifications call for 150 lb., something is mechanically wrong. Pressure is leaking out of the cylinder. The piston rings and cylinders could be worn. A valve could be "burned." A head gasket could be "blown."

Fig. 3-33. During engine compression test, pressure gauge is placed into spark plug hole. Pressure, produced by compression strokes of "cranking" engine, can be measured. Low reading would indicate engine problems (burned valves, worn rings or cylinder wall, or blown head gasket).

Modifying engine compression

An engine's compression ratio and compression pressure can be increased in several ways. The cylinder heads can be machined or milled to reduce combustion chamber volume. Special domed or "pop-up" pistons can be installed. A thinner-than-normal steel head gasket can be used. Cylinder reboring will also increase compression slightly. This is because cylinder volume increases (larger diameter) while combustion chamber (TDC) volume remains the same.

If the formula values just given are not "stock" or factory original and they cannot be found in a service manual, they will have to be found as follows.

Finding cylinder volume

The *volume* in a cylinder can be found mathematically by multiplying bore area times stroke. It may also be found by dividing the engine displacement by the number of cylinders. For example, if a V-8 engine has 427 cubic

Reducing engine compression

To reduce an engine's compression ratio, you could install a thicker head gasket, dished pistons, shorter connecting rods, or a different crankshaft with a shorter stroke.

ENGINE SPECIFICATIONS

Type...................................... 90° V-8 O.H. Valve	Production Engine No. Location .... Front Face
Bore and Stroke............................ 4.00" x 3.00"	of Right Cylinder Bank
Compression Ratio 8.2	Left Bank................................... 1-3-5-7
Compression Pressure at Cranking Speed (Wide Open	Right Bank................................ 2-4-6-8
Throttle)..................... 120-160 PSI @ 155-175 RPM	Firing Order 1-8-4-3-6-5-7-2
Car-Engine Serial No. Location .... Front Face	
of Right Cylinder Bank	

Fig. 3-34. Shop manual will give exact compression pressure specifications for each engine. Sometimes, additional instructions are given (wide open throttle for example). (Buick)

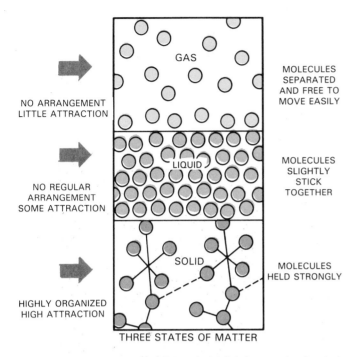

NO ARRANGEMENT
LITTLE ATTRACTION

GAS

MOLECULES
SEPARATED
AND FREE TO
MOVE EASILY

NO REGULAR
ARRANGEMENT
SOME ATTRACTION

LIQUID

MOLECULES
SLIGHTLY
STICK
TOGETHER

HIGHLY ORGANIZED
HIGH ATTRACTION

SOLID

MOLECULES
HELD STRONGLY

THREE STATES OF MATTER

Fig. 3-35. Molecules in a solid are held tightly together by their attraction for each other. Fluids only have slight attraction between molecules. Gases have little or no molecular attraction.

MOLECULAR THEORY

When several atoms of matter are bound together in a group, a molecule is formed. There are basically three arrangements of molecules. These arrangements form the three basic states of matter. These states are illustrated in Fig. 3-35 and are briefly described as follows:

1. *Gas:* molecules are scattered randomly. They are only lightly bound together because of the very great distance and weak attraction between them.
2. *Liquid:* molecules tend to stick together, but without much force. They are only somewhat organized.
3. *Solid:* molecules are held together with a strong force of attraction. They can either be randomly or symmetrically arranged. The symmetrical organization of molecules is the stronger.

HEAT

Heat is actually caused by the agitation and increased movement between molecules. Heat is due to increased molecular motion and friction between molecules. When you heat a piece of metal with a ''gas torch,'' Fig. 3-36, you have made the molecules in the metal move around violently. Their collisions will cause more heat and the vibrating molecules will move still farther apart. This makes the metal expand slightly and even melt as more heat is applied.

The heated molecules bump into each other causing molecular separation. When the metal molecules are first heated and knocked apart, the metal increases in size. After enough heating, the metal molecules can move far enough apart that they will not stay locked together. The metal becomes a liquid at this point.

Effects of heat

Heat can affect the operation of an automobile in thousands of ways. As discussed in Chapter 2, overheated gasoline can boil and cause vapor lock. An overheated engine can develop melted pistons, rings, and cylinders.

On the other hand, heat is needed to power an automobile. Without heat, a car would be useless. Heat is needed to vaporize and burn gasoline. Heat is also used in the headlights and taillights.

Heat expansion

Heat causes expansion. When a solid, liquid, or gas is heated, it will become larger in size. It is this type of expansion in the cylinder that powers an engine. As heat is removed from an object (made colder), it will shrink or become smaller.

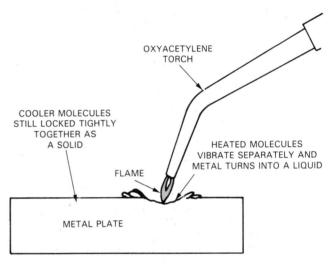

OXYACETYLENE
TORCH

COOLER MOLECULES
STILL LOCKED TIGHTLY
TOGETHER AS
A SOLID

FLAME

HEATED MOLECULES
VIBRATE SEPARATELY AND
METAL TURNS INTO A LIQUID

METAL PLATE

Fig. 3-36. Heat of a ''gas torch'' can change state of metal. It can increase movement between molecules until they separate. When this happens, metal will melt.

Fig. 3-37. With conduction, heat will travel through matter such as iron as heated molecules bump into cooler ones passing along some of their heat.

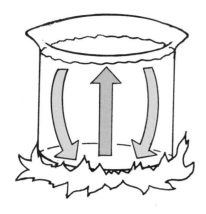

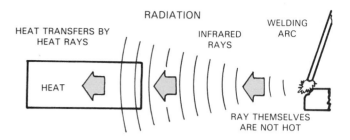

Fig. 3-39. Heat rays or infrared rays are not really hot to the touch. They cause heat transfer by increasing molecular movement in objects.

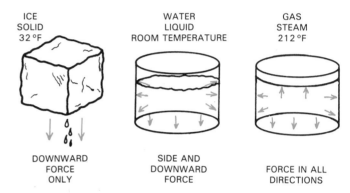

Fig. 3-40. Notice difference in forces between a solid, a liquid, and a gas. A gas pushes on its container equally in all directions. A liquid pushes more on bottom and less on sides. A solid only has downward force caused by gravity.

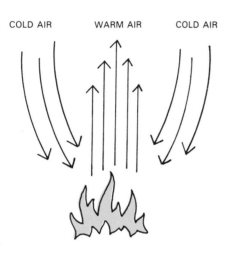

Fig. 3-38. During convection, air or liquid is needed for heat to transfer into other object.

Heat transfer

Heat can be transferred or sent from one object to another in three ways: conduction, convection, and radiation.

1. *Conduction* of heat, as in Fig. 3-37, requires actual contact between the two objects. They must touch. Molecules of matter bump into cooler molecules and heat them up. Thus the hotter object will send heat into the cooler object until their temperatures are equal. When you touch a hot frying pan with your hand, you have experienced the effect of conduction.

2. *Convection* of heat from one object to another, as shown in Fig. 3-38, uses air as a medium or method of transfer. The hot object transfers heat into the surrounding air. Then, the heated air increases the temperature of the second object. When you sit near a camp fire, you can feel the convection of heat.

3. *Radiation* of heat, Fig. 3-39, occurs when infrared rays penetrate an object. Also called heat rays, infrared radiation does not contain heat but it causes heat. The rays are magnetic waves that increase molecular motion in an object. As you learned, increased molecular movement causes heat.

CHANGE OF STATE

A changing of matter from solid, liquid, and gas is called a *change of state* or *phase change.* You are familiar with the three states of water: solid (ice), liquid (water), and gas or vapor (steam). These three states are illustrated in Fig. 3-40.

Both temperature and pressure can control the state of matter. At room temperature, water is normally a liquid. When extremely cold (below 32°F), water changes into a solid. Water turns into a gas when heated to 212°F (100°C). In most cases, heat tends to change matter from solid to a liquid and liquid to gas.

Pressure also affects the state of matter. It increases the boiling point of water. It keeps water in a liquid state rather than changing into a gas. A cooling system on a car is pressurized for this reason. Conversely, low pressure has just the opposite effect on the state of matter. It tends to help matter change from a solid to a liquid and liquid to gas.

Remember gasoline vapor lock? The gasoline boils into a vapor whenever the fuel pump is on the vacuum stroke. When gasoline is under low pressure, it will boil at a lower temperature.

EFFECT OF PRESSURE ON GASES

Pressure affects gases in several ways. See Fig. 3-41. Pressure temporarily increases the temperature of a gas. A diesel engine squeezes air until it is hot enough to ig-

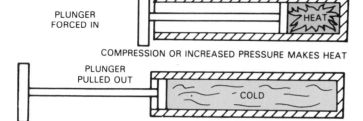

PLUNGER
FORCED IN

HEAT

COMPRESSION OR INCREASED PRESSURE MAKES HEAT

PLUNGER
PULLED OUT

COLD

VACUUM OR DECREASED PRESSURE PRODUCES LOWER TEMPERATURE

Fig. 3-41. Plunger in upper cylinder has been rammed into cylinder and has squeezed gas. Gas temperature is increased. Other plunger has been pulled back to lower gas pressure. Gas becomes cooler.

nite the fuel. Pressure is used to start combustion. When pressure is decreased, the temperature of a gas will drop.

Pressure also affects the vaporization (gasification) and condensation (liquifying) temperatures of a gas. High pressure tends to help condensation and reduce the chances of vaporization. The opposite is also true as pressure is decreased.

SUMMARY

Gravity gives objects weight by pulling them towards the earth.

Atmospheric pressure is caused by the weight of air. It pushes upon everything equally with 14.7 psi at sea level. Air pressure decreases with altitude. A vacuum exists in an area where pressure is below outside atmospheric pressure. A pressure difference between two points causes the flow of a fluid. Pressure and vacuum are commonly measured in auto mechanics with a compound gauge.

When anything moves, work has been performed. Work is measured in foot-pounds. Power is the rate at which work is completed (ft.-lb. x time). Power is measured in horsepower or watts. Anything capable of doing work has energy. There are different kinds, including: potential and kinetic or mechanical, chemical, and electrical. Energy conversion involves a change from one form of energy to another.

When an object is in motion or at rest, it has inertia because it will resist change in motion. When an object is spun around an axis, centrifugal force will pull the object away from the axis (center) of rotation.

Horsepower is a measure of an engine's rate of performing work. One horsepower equals 550 lb. lifted one ft. in one second.

Brake horsepower is a common measure of actual engine power. It measures the horsepower output of an engine at the crankshaft. Indicated horsepower measures the average pressure developed in the engine cylinder and then uses average pressure to calculate engine power. Frictional horsepower measures the power needed to overcome engine friction. Net horsepower is a rating which includes all of the power reducing accessories.

Gross horsepower only includes the basic engine accessories and usually is higher than net horsepower. Taxable horsepower measures the general size of an engine,

mainly the bore. It does not specify actual engine displacement or power.

Torque is a turning force. Engine torque is the rotating force at the engine crankshaft. It is usually stated in pound-feet.

Volumetric efficiency measures the amount of air actually drawn in by an engine, compared to a theoretically possible air consumption. Normally, at low engine speeds, volumetric efficiency is good because there is plenty of time for air to flow into the cylinders.

Engine efficiency is a comparison between power produced (bhp) and power supplied (heat content of fuel). Today's engines are only about 20 percent efficient.

Mechanical efficiency is a measurement of engine friction. Thermal efficiency is found by comparing the amount of fuel burned to the engine's power output.

An engine's compression ratio is the ratio between maximum and minimum cylinder volumes, piston at TDC versus BDC. Presently, engines use a compression ratio of around 8:1. Compression pressure is dependent upon compression ratio. A compression gauge is commonly used to measure the actual compression pressure in each cylinder.

Heat is caused by increased movement and friction between molecules. Heat can transfer by either conduction (objects touch), convection (heat transferred through air), and by radiation (infrared rays).

The three basic states of matter are: solid, liquid, and gas. Heat and pressure control the state of matter. Generally, heat changes matter from a solid to a liquid and liquid to a gas. Cold does just the opposite.

KNOW THESE TERMS

Gravity, Weight, Atmospheric pressure, Vacuum, Vacuum gauge, Work, Absolute pressure, Psi, Power, Energy, Mechanical energy, Potential energy, Kinetic energy, Chemical energy, Electrical energy, Inertia, Centrifugal force, Horsepower, Brake horsepower, Dynamometer, Indicated horsepower, Frictional horsepower, Net horsepower, Gross horsepower, Taxable horsepower, Torque, Volumetric efficiency, Mechanical efficiency, Thermal efficiency, Compression ratio, Molecular theory, Heat, Conduction, Convection, Radiation, Phase change.

REVIEW QUESTIONS—CHAPTER 3

1. _____ is an invisible force or attraction between bodies that tends to hold all such objects to the earth.
2. Natural air pressure surrounding everything on earth is called:
   a. Fluid.
   b. Inflation.
   c. Atmospheric pressure.
   d. Gauge pressure.
3. At sea level, air pressure is about _____ per square inch.
4. As altitude increases, atmospheric pressure also increases. True or false?
5. What is a vacuum?
6. Explain how an engine works as a vacuum pump.

7. Sketch and label a combination pressure and vacuum scale.
8. Which of the following are uses for a compound vacuum-pressure gauge?
   a. Test fuel pump pressure.
   b. Measure intake manifold vacuum.
   c. Detect bad engine valves.
   d. Detect poor carburetor adjustment.
   e. Find improper ignition timing.
   f. Detect faulty distributor vacuum advance.
   g. All of the above.
   h. None of the above.
9. When any force moves an object any distance, _____ is performed.
10. What is the formula for power?
11. Anything capable of doing work has energy. True or false?
12. Which of the following is *not* an example of potential energy?
    a. Compressed air.
    b. Air pressure.
    c. Automobile parked at top of hill.
    d. Compressed spring.
    e. Spinning flywheel.
13. Describe kinetic energy.
14. The acid and metal in an automobile _____ react to produce electricity.
15. Electrical energy is (select best answer):
    a. Energy flowing through a wire.
    b. Flow of electrons through a conductor.
    c. A type of pressure causing voltage that will move a current through a conductor.
16. An alternator converts potential energy into electrical energy. True or false?
17. Anything that moves or is at rest has _____, a force that resists any change in motion.
18. A flywheel (does, does not) have inertia.
19. Centrifugal force is (select correct answer):
    a. Force tending to pull rotating object away from center of its orbit.
    b. Force which tends to keep a spinning object rotating in the direction it is moving.
    c. Force generated by objects on a curved path which tends to move the objects outward from their center of rotation.
20. State the formula for engine horsepower.
21. An engine moves a 2000 lb. automobile 1000 ft. in 60 seconds. How much horsepower was required?
22. Brake horsepower is engine power output measured at the _____.
23. Which of the following phrases applies to torque (more than one may apply)?
    a. Another word for horsepower.
    b. Force applied in a circular path.
    c. A twisting or turning force.
    d. Another term for centrifugal force.
24. An engine which is capable of displacing 500 cu. in. of air per minute actually uses 400 cu. in. per minute. What is its volumetric efficiency? (Use the formula for volumetric efficiency.)
25. An engine's _____ _____ expresses how tightly the fuel mixture is squeezed during the compression stroke of one of its pistons.
26. Heat is caused by the rapid movement and friction between molecules. True or false?
27. Describe the three methods of heat transfer.
28. Explain how compression ratio of an engine might be modified.
29. List the three basic states of matter.
30. Going from a solid, to liquid, to gas (in any direction) is called _____ _____ _____.

4 AUTOMOTIVE FUELS

After studying this chapter, you will be able to:
- *Explain how crude oil is converted into automotive fuels.*
- *Describe the properties of gasoline.*
- *Describe the properties of diesel fuel.*
- *Summarize the properties of alternate fuels.*
- *Explain gasoline octane and diesel fuel cetane ratings.*

This chapter will help you understand not only automotive fuels, but fuel systems, carburetion, fuel injection, emission control, and internal combustion engines. It will give you an essential foundation for fully understanding later chapters.

PETROLEUM

Petroleum or *crude oil* is oil in its natural state. It is a mixture of semisolids, liquids, and gases. Normally, heavier elements (tars and asphalts) of crude oil must be heated and broken down into lighter liquids and gases

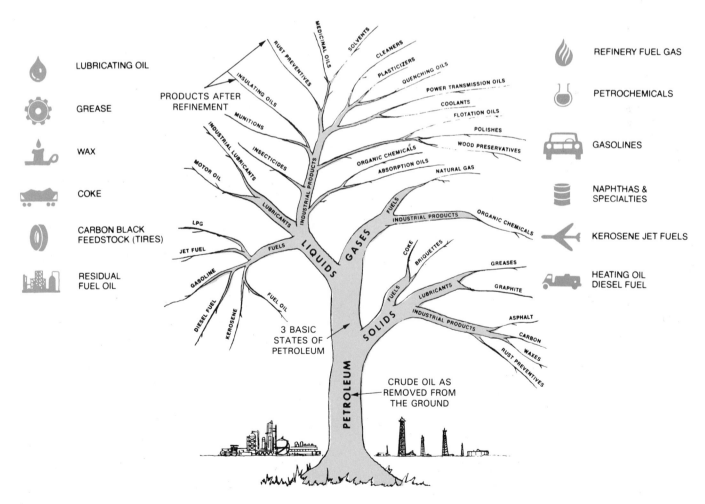

Fig. 4-1. Through refining, petroleum or crude oil is broken down into numerous other products besides gasoline. (Ethyl Corp. and Gulf Oil Corp.)

such as gasoline, diesel fuel, and LP gas. Petroleum can be changed into hundreds of other useful substances, including lubricating oils, plastics, and insecticides (bug killers), etc. See Fig. 4-1.

Crude oil is largely made up of flammable hydrocarbons. *Hydrocarbons* are chemical mixtures of around 12 percent HYDROGEN (light gas vapor) and 82 percent CARBON (heavy black solid). Hydrocarbons include thousands of different compounds.

Crude oil impurities

Crude oil normally contains low concentrations of sulfur (0 to 7%), oxygen (0 to 2%), nitrogen (0 to 1%), various metals (0 to .01%), including nickel, iron, chromium, vanadium, and sometimes, salt water. Sulfur, salt, and some metals can cause undesirable fuel characteristics and must be removed from crude oil.

ORIGIN OF OIL

Among several theories, the most scientific is that petroleum originated through the *decomposition* (long term decay) of animal and vegetable matter. At temperatures around 200 to 400 °F (93 to 204 °C), in the presence of tremendous pressures, natural chemicals, and salt water, organic reactions formed crude oil deposits deep underground. This is the BIOGENIC THEORY. See Fig. 4-2.

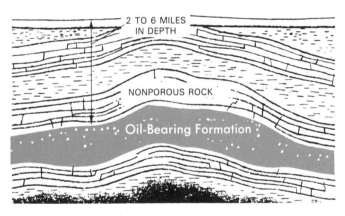

Fig. 4-2. Oil deposits will usually be found under a nonporous or solid rock formation. (Shell Oil Co.)

EXPLORATION

The search for oil deposits is known as *exploration.* The oil may lie two to eight miles below the earth's surface. Various tests are used to locate it.

Seismic studies

Usually, oil deposits are found under solid rock formations. One of the best methods of oil prospecting, therefore, is to look for these types of rock. Geologists conduct *seismic studies,* a process similar to those used to measure earthquakes.

During a seismic study, Fig. 4-3, an explosion is set off. The shock waves will travel into the earth until they encounter layers of hard rock. The waves bounce off the rock mass and return to the surface. Sensors pick up the

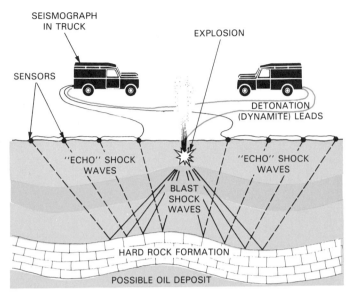

Fig. 4-3. A seismic study is used to help locate oil by determining characteristics of solid rock formations. Oil deposits are normally found near particular types of rock. (Shell Oil Co.)

"echo" of the shock wave and send electrical pulses into the SEISMOGRAPH. Experts study the strength and duration of the shock wave and calculate the likelihood of oil beneath the rock formation.

Magnetic inspections

Another method for finding oil is *magnetic inspections.* A magnetometer can be used to detect rock formations with iron deposits in them, an indicator of oil deposits.

Gravity measurement

A third method of exploring for oil involves the use of a *gravimeter.* This instrument measures the force of gravity over spots likely to have oil deposits. Gravity will increase as the mass of the underground rock increases. See Fig. 4-4.

SEISMOGRAPH SURFACE MAPPING GRAVIMETER

MAGNETOMETER CORES FROM WELL FOSSIL STUDY

Fig. 4-4. These are the six most common methods of locating oil hidden inside earth.

Test drilling

Geologists also do *test drilling* or core sampling as another indicator of oil deposits. A core sample is a small plug of rock taken by drilling deep within the earth. It is examined for fossil remains to estimate the chances of finding oil.

DRILLING FOR OIL

Special *drilling rigs,* capable of boring holes to a depth of five miles, are used to drill oil wells. A drill rig is shown in Fig. 4-5.

A steel framework, called a *derrick,* some 15 stories high, is erected over the drilling site. It supports block and tackle which are used to raise and lower the drill bit and pipe. It also provides temporary storage for the long sections of pipe attached to the drill bit.

RECOVERING OIL

When the drilling operation strikes an oil formation there may be enough pressure from natural gas to force the oil to the surface. When natural pressure is low, or when an "old" well loses pressure, a pump somewhat

Fig. 4-6. Pump like this one must be used after natural gas pressure in well is exhausted. (Shell Oil Co.)

like the one in Fig. 4-6 is used to draw the oil from the ground.

Injection pumping is another recovery method used when a producing well has exhausted its own natural pressure. In this method, a second hole is drilled so the well can be artificially pressurized. As pictured in Fig. 4-7, dry gas, water, steam, carbon dioxide, acid, or solvent type chemicals are pumped into the second hole by a huge compressor pump. The resulting pressure pushes the oil up through the well pipe.

REFINING PETROLEUM

Crude oil is moved to the refinery by pipeline, rail, truck, or ship. At the refinery, it is cleaned to remove

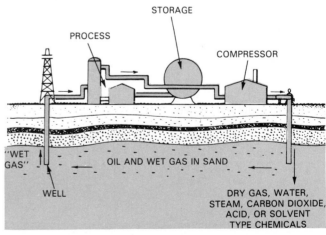

Fig. 4-7. During injection pumping, huge compressor forces oil through sand and rock to base of well. it can then be drawn to surface.

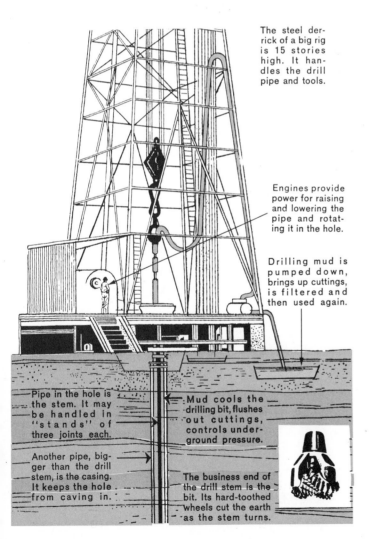

The steel derrick of a big rig is 15 stories high. It handles the drill pipe and tools.

Engines provide power for raising and lowering the pipe and rotating it in the hole.

Drilling mud is pumped down, brings up cuttings, is filtered and then used again.

Pipe in the hole is the stem. It may be handled in "stands" of three joints each.

Another pipe, bigger than the drill stem, is the casing. It keeps the hole from caving in.

Mud cools the drilling bit, flushes out cuttings, controls underground pressure.

The business end of the drill stem is the bit. Its hard-toothed wheels cut the earth as the stem turns.

Fig. 4-5. This illustration explains, in detail, how oil drill rig operates.

impurities such as salt and sulfur. Then, through the refining process, it is broken down into smaller parts, called fractions.

Distillation

Refining begins with distillation. This process uses heat, pressure, and a huge device called a "fractionating tower." Crude oil separates into natural gas, gasoline, motor oil, asphalt, and other products.

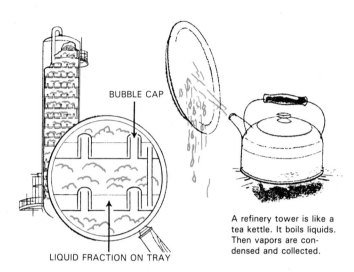

BUBBLE CAP

LIQUID FRACTION ON TRAY

A refinery tower is like a tea kettle. It boils liquids. Then vapors are condensed and collected.

Fig. 4-8. To separate oil into different parts, oil is first heated and vaporized and then allowed to cool and condense onto trays. (Shell Oil Co.)

Petroleum is made up of many different sizes of hydrocarbon compounds. Size determines at what temperatures boiling and condensing will take place. The larger, heavier hydrocarbons (tar, asphalt, lubricating oils) for example, have very high boiling temperatures. The lighter compounds (LP gas, gasoline, and kerosene) have lower or cooler boiling and condensing temperatures. This difference, Fig. 4-8, makes it possible for the fractionating tower to separate crude oil into more useful products.

Fractionating tower

Before distillation, crude oil is pumped out of storage tanks and into a desalter, Fig. 4-9. The desalter removes corrosive salt from the oil. Then the oil enters a huge furnace which heats it to around 800 °F (427 °C). This partially vaporizes the oil.

The heated petroleum and petroleum vapors enter the base of the fractionating tower where they separate by weight or thickness. The temperature of the *fractionating tower* or *pipe still* is hottest near the bottom and becomes progressively cooler at higher levels. When the hot oil enters the tower, it instantly "flashes" or vaporizes and begins to rise in the tower. As the oil vapors rise, they condense (return to a liquid) and settle out on the trays in the tower.

GASOLINE REFINEMENT

Distillation alone will not produce a large percentage of gasoline (about 1/5 barrel of gasoline per barrel of oil).

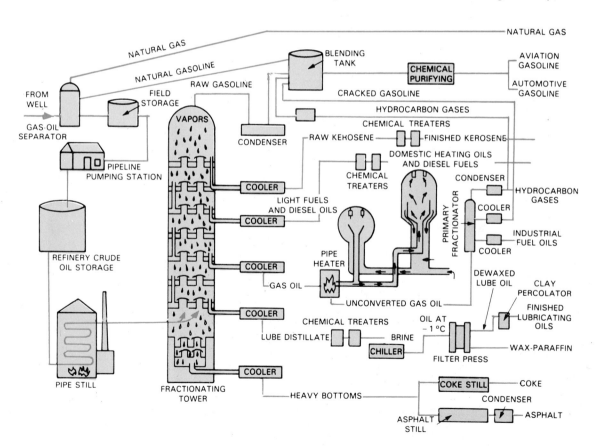

Fig. 4-9. Distilling process. Lighter fractions condense near top of tower, heavier fractions at bottom. (Ethyl Corp.)

Moreover, gasoline produced by distillation, called "straight-run gasoline," does not have all of the qualities needed for today's engines. To increase the amount of high quality gasoline produced from a given amount of crude oil, we need other refining processes.

Catalytic cracking

Catalytic cracking is a very common refinery process that uses moderate heat and a chemical catalyst (usually a special clay powder) to change heavy fractions of crude oil into lighter ones (gasoline). A catalyst is any substance that will assist in a chemical reaction without, itself, entering into the reaction.

Thermal cracking

Thermal cracking, which has been almost totally replaced by catalytic cracking, was one of the first methods of producing extra gasoline from the fractions of distillation. It used extreme heat—above 1000 °F (538 °C)—and extreme pressures of around 1000 psi (6 894 kPa) to break large hydrocarbon compounds into smaller gasoline type hydrocarbons.

Hydrocracking

Hydrocracking is a commonly used refinement process in which hydrocarbon cracking (separation) occurs in the presence of hydrogen gas and a catalyst. It produces large percentages of hydrocarbon components useful in unleaded premium gasoline.

Alkylation

Alkylation, the opposite of cracking, changes small hydrocarbon compounds into larger ones. In other words, it changes LIGHT-GASES (bottled types) into gasoline. Alkylation is one of the few refining processes that can produce high quality (high octane) gasoline in large quantities.

Polymerization

Like alkylation, *polymerization* produces gasoline out of very light gases. However, the resulting "polymerized" gasoline has inferior antiknock qualities when compared to fuel produced through alkylation.

Reforming

Referring to a category of refining processes (thermal reforming, catalytic reforming, etc.), *reforming* generally has the purpose of changing low octane gasolines into higher octane gasolines. It is commonly used to upgrade "straight run" gasoline (gasoline as it comes from the fractionating tower). Light gases (bottled types) are not suitable for reforming. Look at Fig. 4-10.

GASOLINE ADDITIVES

Numerous chemicals, called *additives,* are mixed into gasoline to improve performance and storage characteristics. The most common of these will be described. Fig. 4-11 shows characteristics desired in gasoline.

Antioxidants

Antioxidants are chemicals added to gasoline to make it more stable against evaporation. They help prevent the formation of gum and precipitates (small solid particles) in gasoline.

When gasoline sits for a long period of time (generally over six months), some of the chemicals evaporate into the air. The heavier substances that do not evaporate remain as gum and precipitates.

REQUIREMENT	REASON
Form vapors at low temperature.	For quick starting.
Vaporize increasingly as carburetor and manifold temperatures rise.	For fast warm-up, smooth acceleration and even fuel distribution among the cylinders.
Have proper vaporizing characteristics for the climate and altitude where it will be used.	To prevent vapor lock (Subject 37) and boiling of the fuel in carburetors, fuel pumps, and fuel lines.
Contains very little excessively high-boiling hydrocarbons.	To insure good fuel distribution and freedom from crankcase dilution and deposits. Reduce air pollution.
Have high antiknock quality (octane number) throughout the fuel's entire boiling range.	For freedom from fuel knock (detonation) at all engine speeds and loads.
Be low in gum content.	To prevent valve sticking, carburetor difficulties, and deposits in the engine and intake manifolds.
Be low in sulfur content.	To prevent corrosion and unpleasant odor.
Have good stability (against oxidation).	To prevent deterioration and gum formation in storage.
Have satisfactory odor.	For public acceptance.

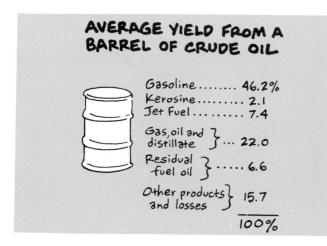

AVERAGE YIELD FROM A BARREL OF CRUDE OIL

Gasoline........ 46.2%
Kerosine......... 2.1
Jet Fuel......... 7.4
Gas, oil and distillate } ... 22.0
Residual fuel oil } 6.6
Other products and losses } 15.7
_____
100%

Fig. 4-10. With today's refining methods, almost 50 percent of each crude oil barrel can be made into automotive gasolines. A barrel of oil equals 42 gallons. (Phillips 66)

Fig. 4-11. Characteristics of good gasoline. After a period of time, stored gasoline can spoil (partially evaporate). It will oxidize, begin to smell like paint thinner, and produces a gum deposit. (Ethyl Corp.)

Gasoline gum is a thick, sticky, varnish-like substance. It looks like the varnish or shellac used as coating on wood furniture. This gum can collect inside a fuel system and cause a variety of problems (carburetor flooding, fuel filter stoppage, and sticking engine valves).

Metal deactivators

Metal deactivators are commonly used with antioxidants to help prevent formation of gum in gasoline. When copper (a metal used in some fuel systems) comes in contact with the chemicals in gasoline, the copper can act as a catalyst that actually speeds up oxidation.

Deposit modifier

A *deposit modifier* is a fuel additive that prevents carbon deposits from building up in an engine. This additive is usually an organic or metallic compound containing phosphorous.

Combustion chamber deposits are very undesirable. They can make an engine "knock" or "ping." Also, they can cause an engine to "diesel" (continue to run after shutoff). Engine deposits can cause cylinder "hot spots" (red hot bits of carbon) or increased compression pressures. These can ignite the fuel mixture prematurely. Additionally, deposit modifiers, by reducing deposits, will help prevent spark plug fouling. See Fig. 4-12. Fig. 4-13 shows how fuel without additives build up carbon deposits on engine valves.

Scavengers

Scavengers are included in antiknock additives to help remove lead compounds or carbon deposits from the combustion chambers of an engine. The scavengers and combustion heat easily change lead deposits into gaseous vapors. Then, the carbon can be drawn out of the engine during the exhaust stroke.

Antirust agents

Antirust agents in gasoline, as their name implies, reduce corrosion and rusting in the fuel system and fuel handling facilities. An additive of this nature will help prevent tiny rust and corrosion particles from plugging fuel filters, causing engine wear problems, and fuel pump damage.

Detergents

Detergents are mainly used in gasoline to prevent deposits around the throttle section of the carburetor or throttle body injection unit. See Fig. 4-14. This section of the carburetor subjects gasoline to a very unpleasant environment (heat, cold, moisture, crankcase vapors, and airborne particles). Without detergents, deposits can upset the operation of a carburetor or TBI unit, Fig. 4-15. These deposits can cause driveability and performance problems, decreased fuel economy, increased emissions, and a rough idle.

Anti-icing additives

Anti-icing agents or *additives* act like antifreeze. In cold weather, air can contain a considerable amount of water (.4 lb. per 1000 cu. ft.). When this cold air rushes into the carburetor, ice can form on the carburetor throttle

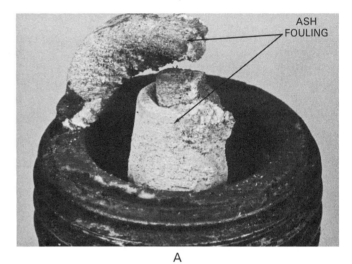

A

B

Fig. 4-12. A—Ash fouled spark plug, caused by ash in the fuel, is often confused with carbon fouled plug. B—Carbon fouled plug. (Champion Spark Plug Co.)

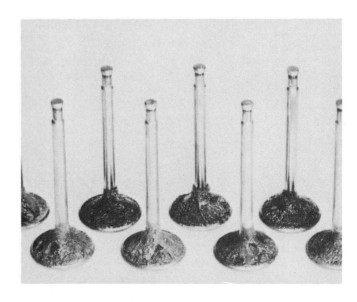

Fig. 4-13. Note heavy deposits. These valves were operated in an engine using fuel without deposit modifiers. (Texaco Inc.)

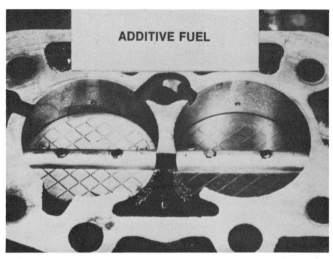

Fig. 4-14. Carburetor is likely to build up deposits as at left without detergent additive in gasoline.

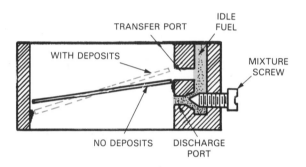

Fig. 4-15. Carburetor deposit of this nature would prevent proper carburetor adjustment and operation. Throttle plates would have to be opened too wide, which could upset engine idle. (Ethyl Corp.)

Fig. 4-16. Generally, around a teaspoon of lead is added to a gallon of leaded gasoline. This can increase octane and antiknock value of fuel as much as 7 or 8 octane numbers. (Ethyl Corp.)

plates. If this happens, the ice can interfere with the air and fuel entering the engine. As a result, engine stalling or ''dying'' may occur and continue for up to 15 minutes.

To help prevent carburetor icing, one of two types of anti-icing additives are mixed into gasoline. One is a solvent type which serves as a gas line antifreeze agent. The other is a surface coating type that prevents ice from sticking to the parts of the carburetor.

Lubricants

Gasoline *lubricants,* usually lead compounds, reduce engine valve face wear, valve seat wear, piston ring and upper cylinder wear. Lead, even under the extreme heat conditions of combustion, will coat and protect these engine components.

Unleaded gasoline contains little or no lead and, in some engines not designed for unleaded fuel, can increase valve-face-to-valve-seat wear. Engines designed for no-lead gasoline have special hardened valve seats that can withstand the added friction, heat, and wear caused by the removal of the lead lubricant.

Dyes

Dyes are blended into gasolines simply for identification purposes. They indicate grade, brand, or type of

gasoline. This color coding helps to prevent accidental mix-up of gasoline types.

Antiknock additives

Antiknock additives, usually tetraethyl lead (TEL) or tetramethyl lead (TML) are used, Fig. 4-16, as antiknock compounds. They increase the OCTANE RATING of gasoline. Engine ''knock,'' will be discussed in detail later. It refers to the sound produced when gasoline is ignited too soon (preignition, surface ignition, etc.) or when the burning of the gasoline is too violent (detonation). Lead additives help to prevent this problem by raising the temperature at which a gasoline begins to burn and by slowing down the speed of combustion. Look at Fig. 4-17.

Another additive, made from a metal *compound of manganese* (MMT), is the primary antiknock additive in lead-free gasoline. It can raise the octane rating of the ''base stock'' gasoline one or two numbers without any undesirable effects on emission control devices.

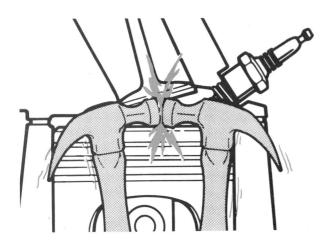

Fig. 4-17. The sound of detonation. This condition occurs when a portion of fuel-air mixture explodes during combustion. Certain additives will reduce detonation or increase gasoline's antiknock qualities. (Ford Motor Co.)

Aftermarket additives

Numerous gasoline additives are available from auto parts outlets, large department stores, etc. As with "factory additives," aftermarket additives are designed to prevent and cure various fuel system problems—carburetor icing, gas line freezing, gumming, and spark knock. Depending upon purpose, some of these additives are poured directly into the gas tank while others are poured into the carburetor or throttle body air horn.

Several automobile manufacturers caution against the use of some aftermarket additives because they can cause fuel system corrosion or deterioration of rubber and neoprene parts. Read the claims and directions on the additive.

GASOLINE TESTS AND RATINGS

To assure that gasolines produced by different companies in different locations meet uniform standards of quality and performance, numerous gasoline tests are used. Various ratings are assigned on the basis of testing.

Volatility test

Volatility is a measure of the temperature at which a gasoline will change from a liquid to a vapor. This is very important because gasoline must vaporize before it can ignite and burn in an engine. Further, gasoline with high volatility will generally exhibit good acceleration and cold starting characteristics. Low volatility is desirable for control of vapor lock and to increase fuel economy.

Volatility tests such as vapor-liquid ratio, vapor pressure, and distillation are commonly used to determine the vaporization as well as operating characteristics of different gasoline blends.

Impurities tests

Gasoline *impurities* can cause a wide range of problems. Water and sediment (dirt) are the most common gasoline impurities. Water is heavier than gasoline and will not mix with gasoline. However, water can still be drawn into the fuel system.

Sulfur test

Sulfur is one of the most undesirable compounds in petroleum and automotive fuels. Normally, it is removed for several reasons. Sulfur can cause an unpleasant "rotten egg smell." It can cause corrosion of metal parts, assist in formation of fuel system gum and combustion chamber deposits. Sulfur may react unfavorably with lead additives (tetraethyl lead) to lower the octane or antiknock characteristics of gasoline. A sulfur test determines the amount of sulfur in a fuel.

Hydrocarbon type analysis

As we mentioned earlier, the basic *hydrocarbons* found in gasoline are paraffins, napthenes, and aromatics. The percentage of each will affect the operating characteristics of a fuel.

In this test, dyed gasoline is passed through a special column of silica gel which separates the gasoline into the three major hydrocarbons. The amount (percentage) of the three hydrocarbon types can then be compared in order to determine the characteristics of the fuel.

OCTANE RATINGS

The *octane rating* of a gasoline indicates its ability to resist knocking or pinging. For example, if the fuel mixture ignites too soon (excessive spark advance), the burning and rapidly expanding gases may actually try to force the piston backwards (against normal direction of piston movement). This violent action of the piston slamming against the expanding gases can cause a loud knocking noise. In general, the higher the fuel's octane-rating or number, the greater the amount of heat and compression the gasoline can withstand before abnormal combustion or knocking occurs.

Fig. 4-18. Gasoline is often sold in three grades.

68

Gasoline grades

Presently, gasoline is sold in three general grades—PREMIUM, UNLEADED, and REGULAR. See Fig. 4-18. Naturally, premium fuel, often called high-test, super-test or ethyl, has the highest octane rating. It would have the higher antiknock number and should be used in a high compression engine. For example, if an engine knocks or pings in one fuel, a fuel with a higher octane rating may be needed.

Octane numbers

While *octane numbers* are fairly common terms, they are often misunderstood. This confusion occurs because there are basically four methods of determining the octane number of gasoline—Research Method (RON), Motor Method (MON), Road Method, and the Research Motor Average Method.

COMPARISON OF LABORATORY OCTANE NUMBERS		
TEST METHOD	ASTM MOTOR	ASTM RESEARCH
TYPICAL GASOLINE COMPONENTS	OCTANE NUMBER	OCTANE NUMBER
STRAIGHTRUN	62	64
NATURAL	71	72
CRACKED		
THERMAL	70	80
CATALYTIC	80	92
POLYMER	80	95
REFORMATE	80	90
ALKYLATE	89	92

Fig. 4-19. Notice that the Research Method of rating gasoline stocks is slightly higher than the Motor Method. (Ethyl Corp.)

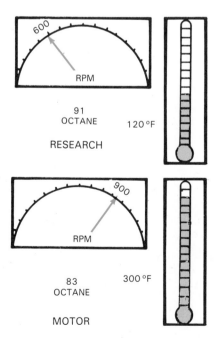

Fig. 4-20. The "Research" rating method is not as severe as the "Motor" method. (Marathon Oil Co.)

Gasoline classification (Research, Motor)

The *Research* and *Motor* methods of classifying gasoline octane ratings are both performed in the laboratory using a test engine. Quite often, it is a special adjustable single cylinder engine. Fuel engineers can precisely control the compression ratio, ignition timing, fuel mixture, speed, outside air temperature, engine temperature, surrounding humidity, and oil pressure of the engine. By comparing the knocking qualities of the unknown test fuel to that of the reference fuel, usually 90 octane, engineers can arrive at an octane number that indicates the knock value of various gasoline blends.

In most instances, the Motor and Research methods will have different octane numbers for the same gasoline. This is illustrated in Fig. 4-19. The Research rating test is usually performed at lower engine speeds and temperatures. As a result, the Research Method is less severe and will rate most gasolines with a higher octane number than the Motor Method, Fig. 4-20.

The difference between the Research and Motor ratings is called SENSITIVITY. For example, a gasoline rated at 95 octane Research and 80 octane Motor has a difference or sensitivity of 15.

Road octane number

To provide a more meaningful octane rating than that produced by laboratory octane ratings, oil and automotive companies have developed techniques for testing fuels under normal driving conditions. This octane rating method is known as *road octane number.* It may be performed in a vehicle operating on a road or on a test dynamometer (device that simulates road conditions of acceleration, load, horsepower output, etc.).

Basically during a road octane test, the car is driven under various conditions of acceleration, pull, speed, ignition timing, temperature, and humidity. Engineers, using special test equipment, record exactly when and under what conditions knock occurs. By comparing the knocking tendencies of the test fuel to that of a reference fuel (90 octane blend), the octane value of the test fuel can be determined. See Fig. 4-21.

The Road Method will usually rate a fuel somewhere between the Research and Motor Methods. For example, 100 octane Research gasoline might equal only 90 octane Motor but 98 octane Road Rating.

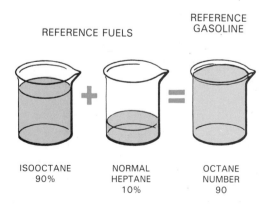

Fig. 4-21. Octane rating of an unknown gasoline can be determined by comparing it to a blend of two reference fuels. (Ethyl Corp.)

69

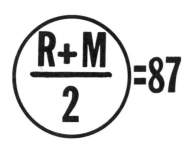

Fig. 4-22. This simple formula is now being used to identify octane ratings of different gasolines. (Marathon Oil Co.)

Average method

The fourth method used to rate the antiknock value of gasoline is the *Research Motor Average Method.* It is commonly seen on gas station pumps as (R + M)/2, usually on a yellow and black sticker. The purpose of the average octane method is to make fuel rating numbers more reliable and understandable to the average consumer.

This gasoline rating system simply averages the laboratory performed Research and Motor Methods that we discussed earlier. See Fig. 4-22.

Octane requirements

Many factors affect *octane requirements.* Remember, high octane fuel is basically designed to prevent engine knock. It allows the engine to be operated with more spark advance and higher compression ratios. This increases gas mileage and engine power. As listed in Fig.

4-23, several factors can raise or lower the octane requirements of an engine.

Octane selection

It is very important to use fuel with the manufacturer's specified antiknock rating. There are many engine design variables (compression, combustion chamber shape, ignition timing, operating temperature, etc.) that effect engine knock and octane ratings. For this reason, every automobile engine has been factory tested in order to determine its minimum and most efficient octane requirement.

A lower fuel octane number will hurt overall engine performance. It can reduce power, lower fuel economy, cause run-on or dieseling (engine continues to run with the ignition key off), and spark knock. Under extreme conditions, a low-octane fuel can actually cause physical damage to the engine. Detonation, the most severe and harmful type of engine knock, can burn or knock holes in the top of pistons, bend connecting rods, burn valves, and finally destroy the engine.

UNLEADED GASOLINE

As the name implies, *unleaded gasoline* does not contain lead antiknock additives. By legal definition, unleaded gasoline must not have more than 0.05 grams of lead per gallon and not more than 0.005 grams of phosphorous per gallon. Also, lead-free gasoline must meet Environmental Protection Agency (EPA) standards of minimum octane.

Generally, since only manganese antiknock additives and not lead ones are used, unleaded gasoline must be

---

OCTANE REQUIREMENT FACTORS
Some Engine Design and Operating Conditions Affecting Octane-Number Requirement

Octane-number requirement tends to go UP when:	Octane-number requirement tends to go DOWN when:
1. Ignition timing is advanced. 2. Air density rises due to supercharging or a larger throttle opening, or higher barometric pressures. 3. Humidity or moisture content of the air decreases. 4. Carburetor air temperature goes up. 5. Lean fuel-air ratios (near max knock) are employed. 6. Compression ratio is increased. 7. Coolant temperature is raised. 8. Antifreeze (glycol) engine coolant is used. 9. Combustion chamber design provides little or no quench area. 10. Vehicle weight is increased. 11. Engine loading is increased such as climbing a grade, pulling a trailer, or increasing wind resistance with a car-top carrier.	1. Car is operated at higher altitudes (lower barometric pressure). 2. Fuel-air ratio is richer or leaner than that producing maximum knock. 3. Spark-plug location in the combustion chamber provides shortest path of flame travel. 4. Combustion-chamber design gives maximum turbulence of the fuel-air charge. 5. Compression ratio is lowered. 6. Exhaust gas recycle system operates at part-throttle. 7. Ignition timing retard devices are used. 8. Humidity of the air increases. 9. Ignition timing is retarded. 10. Carburetor air temperature is decreased. 11. Reduced engine loads are employed.

Fig. 4-23. Many factors, as shown, affect engine's fuel octane requirements. (Ethyl Corp.)

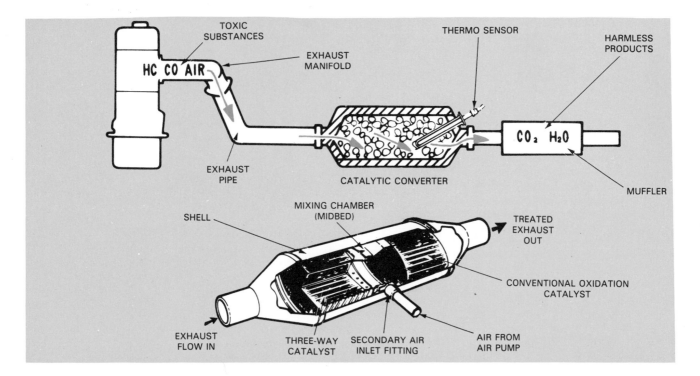

Fig. 4-24. Normally, a catalytic converter is placed forward in the exhaust system near engine. Top. Simplified drawing of single element converter. Bottom. Dual catalytic converter. (Ford Motor Co.)

a blend (mixture) of extremely high quality crude oil fractions. The fuels that are used in unleaded gasoline must be of premium grade quality in order to equal the antiknock level of regular leaded gasoline (91 Research octane). With lead additives, relatively low octane fuels (regular) can be boosted into high octane (premium) gasoline. Without lead additives, oil companies must use extra refining processes to produce naturally high octane fuel.

Catalytic converter

A *catalytic converter,* Fig. 4-24, is a device placed in the exhaust system to help reduce exhaust pollution. It chemically changes the poisonous parts of exhaust gases into less harmful substances.

Lead gasoline additives are extremely destructive to the catalyst substance in the catalytic converter.

One tankful of leaded gasoline used in a car designed for lead-free gasoline can cut down the pollution-reducing action of the catalytic converter by almost 50 percent. Fig. 4-25 illustrates this fact. The lead additives coat the inside of the converter and prevent the catalyst from functioning.

Unleaded fuel advantages

Besides helping to limit air pollution, lead-free gasoline increases the life of spark plugs and exhaust systems. Unleaded gasoline has also been proven to reduce combustion chamber *deposits.* By reducing deposits in an engine, lead-free gasoline will improve emission output levels.

Unleaded fuel disadvantages

In some cases, unleaded gasoline can increase combustion temperatures and engine knock tendencies. It sometimes decreases valve and valve seat life, reduces engine power, and lowers fuel economy.

Low-lead gasoline

Low-lead gasoline is not the same as no-lead, unleaded, lead-free gasoline. Low-lead gasoline can contain about half as much lead as leaded gasoline. For this

Fig. 4-25. Element in this catalytic converter has been clogged and rendered inoperable by lead additives and overheating damage. (Champion Spark Plug Co.)

71

reason, low-lead gasoline should not be used in cars designed for lead-free gasoline.

Fuel delivery problems

Various conditions in the fuel system can affect the *delivery of fuel* to a carburetor or injection system. There are basically three problems: vapor lock, percolation, and carburetor icing.

VAPOR LOCK

Vapor bubbles will form in hot gasoline just as bubbles form in boiling water. In a fuel system (fuel pump or lines), these bubbles can restrict or even stop gasoline flow into the carburetor and engine. This problem is called *vapor lock.*

Vapor lock effects

Partial vapor lock merely *reduces engine top speed* and *power* by leaning out the fuel mixture. The vapor bubbles take up space and leave less room for liquid fuel. See Fig. 4-26. If only a few bubbles form, some liquid fuel continues to flow out of the carburetor. The engine will continue to run, but poorly.

The lean mixture caused by mild vapor lock, can make the engine knock, hesitate upon acceleration, backfire, idle roughly, stall at idle, or surge (uneven speed) during steady driving.

Severe and complete vapor lock may stop an engine completely and make it very hard to restart. The engine may need to cool down before it will run.

Putting fuel under *pressure* will affect the temperature at which vapor bubbles form. For example, water normally boils at 212°F (100°C) at atmospheric pressure (14.7 psi or 101 kPa). However, if the water is pressurized (as in a cooling system or pressure cooker) its boiling point will rise above 212°F. More pressure, Fig. 4-27, produces a higher boiling temperature. By keeping fuel pressure relatively high and steady, the tendency towards vapor lock or bubbling will be reduced.

Vapor bubbles are more likely to form in a mechanical fuel pump than anywhere else in the fuel system. This is because the fuel pump alternates from pressure to vacuum (low pressure). Whenever the pump is on the

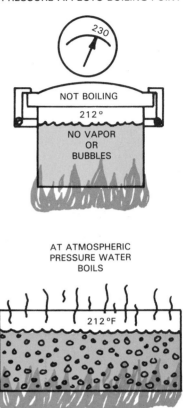

PRESSURE AFFECTS BOILING POINT

AT ATMOSPHERIC PRESSURE WATER BOILS

Fig. 4-27. Unpressurized liquid boils at 212°F while pressurized liquid does not.

intake or suction stroke, the fuel pressure in the pump drops and hot fuel "boils" easily.

A mechanical fuel pump is usually bolted to the side of the engine block near engine heat. Vapor bubbles displace some of the liquid fuel. When enough bubbles form, the pump does not develop the pressure needed to move sufficient fuel.

For instance, a car with air conditioning, power steering, a high-temperature thermostat for emissions, a large engine and a small engine compartment, could have an extremely high temperature around the engine. This is shown in Fig. 4-28. Also, the under-hood air circulation could be restricted.

Controlling vapor lock

Auto makers have devised several methods of cooling the gasoline in the engine compartment area of the fuel system. They are using fuel pump insulating gaskets and heat shields to prevent heat transfer from the hot engine into the gasoline.

Other measures are used to stop vapor lock. Fuel lines are routed away from engine heat. In addition, a fuel tank return line is often added to help prevent heat buildup in the fuel. This extra line allows cool fuel to constantly flow through the fuel pump and lines.

On modern cars, an electric fuel pump is commonly located in or near the fuel tank. These pumps also reduce pressure pulsations and vapor lock tendencies.

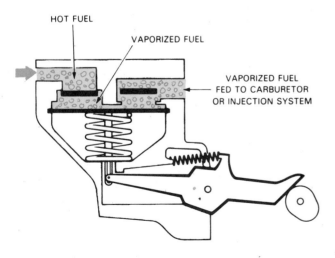

Fig. 4-26. Note effect of vapor lock. (Ford Motor Co.)

Fig. 4-28. Considering that fuel pump is buried deep inside this maze of accessories and is bolted against hot engine block, it is easy to understand how vapor lock can occur. (Champion Spark Plug Co.)

Troubleshooting vapor lock

When vapor lock does occur, look for:
1. Excessively high engine operating temperature (bad thermostat, faulty clutch fan, low coolant level, and extended periods of engine idling).
2. Fuel line touching or too close to hot engine part (exhaust manifold, cylinder head, etc.).
3. Air leak on the suction side of the fuel system allowing air to mix with the fuel (cracked fuel line or rubber hose, loose connection, etc.).
4. Partially clogged fuel filter, especially in gas tank or suction side of the fuel pump (lowers fuel pressure).
5. Defective fuel pump (low output pressure, bad bleed valve, etc.).
6. Low quality gasoline (low vapor pressure rating).
7. Abnormal weather for the season or hot weather when gasoline is a winter blend.

CARBURETOR PERCOLATION

As with vapor lock, *percolation* is a result of the vapor bubbles that can form in excessively hot gasoline. When

CARBURETOR PERCOLATION

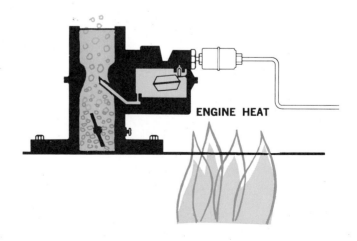

Fig. 4-29. Carburetor percolation (fuel boiling) can flood engine as bubbles open carburetor float. (Ford Motor Co.)

73

an engine is idled slowly for a period of time or when shut off after operation, the temperature of the engine can increase considerably. Then, the carburetor can absorb enough heat to make the fuel in the carburetor boil. The vapor bubbles can push gasoline out of the carburetor passages and into the engine. Fuel can pour or flood into the carburetor throat and into the engine's intake manifold, Fig. 4-29.

The fuel vapors usually exit through the vent and travel to a storage canister filled with charcoal. Later, the vapors are drawn back into the engine and burned.

Troubleshooting percolation problems

When an engine floods due to carburetor percolation, check for:
1. Overheated engine (bad thermostat, radiator, etc.).
2. Inoperative fuel bowl vent.
3. Missing carburetor insulator gasket.
4. Winter blend of fuel being used in the summer (stored in tank over the winter).

CARBURETOR ICING

When the outside air temperature entering the carburetor is cold and humidity is between 70 and 100 percent, ice can collect inside the carburetor. Called *carburetor icing,* this can upset (increase or decrease)

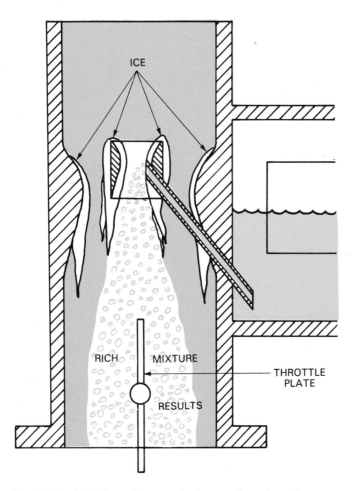

Fig. 4-30. At high engine speeds, ice can form in carburetor as illustrated. This draws extra fuel into engine and can cause engine flooding (excessively rich mixture).

the amount of fuel entering the intake manifold.

Note that the outside temperature does not have to be below freezing for carburetor icing to occur. When the cool, moist air flows through the carburetor and mixes with the vaporizing gasoline, the temperature of the mixture can drop as much as 25 °F (14 °C). The vaporizing fuel draws heat from surrounding air and metal parts. Vapor in the air rapidly drops below freezing and causes ice to form on the chilled edges of the throttle plates and throttle body, Fig. 4-30. The blockage can limit or even stop the flow of gasoline through the idling circuit of the carburetor. Then the engine will run poorly at idle.

With carburetor icing, the engine may restart easily after pumping the gas pedal. It may run fine at high engine speeds but "die" again when returned to an idle.

Anti-icing additives

To prevent carburetor icing, many oil companies add *anti-icing chemicals* or *additives* to their gasolines. Usually, these agents are only mixed into gasoline used in cool climates or during the winter months. One of these anti-icing additives is called a SURFACE ACTIVE AGENT. This additive coats the metal surfaces in the fuel system and keeps ice from sticking to them.

High speed carburetor icing

Another instance of carburetor icing, but less common, occurs after a car has been driven at highway speeds for a period of time. At high speeds, a huge volume of air is pulled through the carburetor. If the carburetor body is cool enough, ice can form and make the carburetor throat smaller, richening the mixture.

FUEL SYSTEM ICING

When a car is operated in extremely cold climates (outside air temperature below 32 °F [0 °]C) water in the gas tank, fuel filters, or fuel lines can freeze. The ice can then block the flow of gasoline through the fuel system. This is known as *fuel system icing* or *freeze-up.*

Solid pieces of ice can restrict or plug the fuel system and cause various symptoms. The engine may stall every few minutes, not start at all, miss at high speeds, or may be limited to a specific top speed. It will cause problems that are almost identical to those of a dirt-clogged fuel filter. The engine will starve for gasoline.

Whenever a car suffers from fuel system icing, fuel filters should be replaced. The vehicle may have to be taken into a warm garage or shop so that the ice in the system can thaw. Once melted, the fuel system (tank filters, and lines) should be drained and flushed with fresh fuel. As a safety measure, an aftermarket type gasoline antifreeze additive can be used. Then the fuel system will not freeze up again.

Oil companies add chemicals (usually alcohol and glycol) to gasoline used in cold climates. These additives serve as antifreeze agents.

FUEL BLENDING

As noted earlier, gasoline refinement processes (distillation, cracking, reformation, etc.), produce several

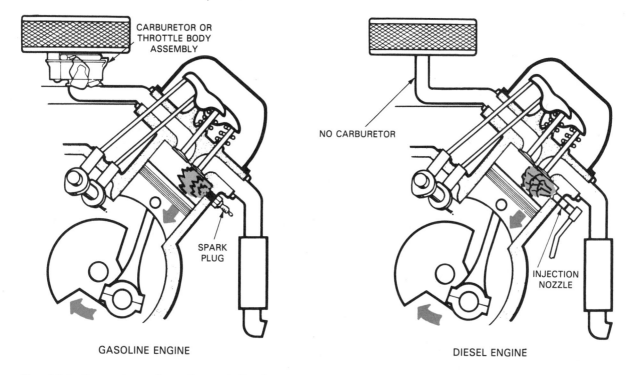

Fig. 4-31. Comparison of gasoline and diesel engines show difference in fuel delivery systems. (Oldsmobile)

different grades or qualities of gasoline. Some are high in octane value while others are very low. To prevent waste and to meet more closely the needs of the market, oil companies *blend* or mix gasoline with different characteristics of octane rating, vaporization, weight, etc. This produces a fuel that meets a particular specification.

DIESEL FUEL

Next to gasoline, *diesel fuel* is the most popular automotive fuel. Once used mainly in trucks and trains, diesel engines are growing in popularity in automobiles because of their high fuel economy.

A thicker fraction (part) of crude oil, diesel fuel, gallon for gallon, delivers more heat energy than gasoline. It can deliver more cylinder pressure and vehicle movement than an equal amount of gasoline. Diesel fuel is thicker and has different burning characteristics than gasoline. It, therefore, requires a different engine design and fuel system.

Diesel fuel will not vaporize as readily as gasoline. If it were to enter the intake manifold of an engine, it would collect as a liquid on the manifold walls. Therefore, diesel engines have an injection system that puts the diesel fuel directly into the combustion chamber as shown in Fig. 4-31.

DIESEL FUEL GRADE

Diesel fuel is classified into *grades.* These grades assure that the diesel fuel sold all over the country has uniform standards of service, including weight, cleanliness, ignition qualities, etc.

As described in Fig. 4-32, there are three diesel fuel grades: Number 1-D (D stands for diesel), Number 2-D,

Grade No. 1-D:

Grade No. 1-D comprises the class of volatile fuel oils from kerosene to the intermediate distillates. Fuels within this classification are applicable for use in high-speed engines in services involving frequent and relatively wide variations in loads and speeds, and also for use in cases where abnormally low fuel temperatures are encountered.

Grade No. 2-D:

Grade No. 2-D includes the class of distillate gas oils of lower volatility. These fuels are applicable for use in high-speed engines in services involving relatively high loads and uniform speeds, or in engines not requiring fuels having the higher volatility or other properties specified for Grade No. 1-D.

Grade No. 4-D:

Grade No. 4-D covers the class of more viscous distillates and blends of these distillates with residual fuel oils. These fuels are applicable for use in low-speed and medium-speed engines employed in services involving sustained loads at substantially constant speed.

Fig. 4-32. Diesel fuel grades. Note that each is designed for certain conditions. (Ethyl Corp.)

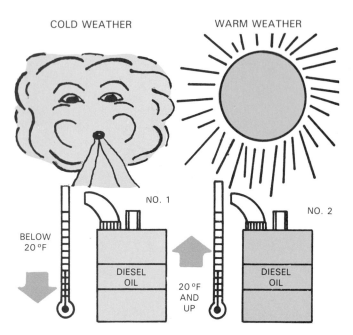

Fig. 4-33. Generally, Grade No. 1-D is for cold weather use while No. 2-D is for moderate to hot weather.

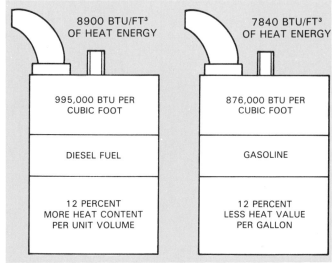

Fig. 4-34. Diesel fuel contains more heat energy by volume than gasoline.

and Number 4-D. Number 3-D has been discontinued.

Usually, No. 2-D fuel oil is recommended for automotive diesel engines. It is the only diesel fuel available in some areas. No. 1-D and No. 4-D can normally be purchased from special distributors.

Winter blends of diesel fuel

Sometimes, since No. 1-D fuel is thinner or less viscous than No. 2-D, it is recommended as a "winter fuel" or for "stop and go" driving. It can more frequently be bought in cold climates.

In extremely cold weather, No. 2-D may thicken and upset engine operation. This condition causes hard starting and poor fuel delivery. Being thinner, No. 1-D will function better in cold weather.

Some auto makers designate No. 2-D for temperatures above 20°F (−7°C) and No. 1-D for outside temperatures below 20°F. See Fig. 4-33.

Special "winterized blends" of No. 2-D are also offered during winter weather. Fuel additives are used to improve cold operating characteristics. Some engine manufacturers caution against use of winter blends for their engines. No. 1-D fuel oil does not have the lubricating qualities needed for many engines. Being thin, it may not provide the wear protection for the precision parts of the injection system. This could lead to early engine failure.

No. 1-D will not give as high fuel mileage or economy as No. 2-D. This is because No. 2-D contains more heat energy per gallon.

Caution! Do not try to use home heating oil or gasoline in a diesel engine. Major engine or injection system damage can occur.

Gasoline should never be used in a diesel engine. It could keep a diesel engine from running and would require a complete drain and flushing of the fuel system.

Always use the grade of diesel fuel oil recommended by the engine manufacturer.

HEAT VALUE

The *heat value* of a fuel indicates the amount of heat and power developed during combustion. It is an important property that indicates the heat content in a fuel.

Diesel fuel heat content is approximately 12 percent higher than gasoline, Fig. 4-34. In general, a heavy diesel fuel (higher grade number, high density, high viscosity, low cetane number, etc.), is better. The fuel will have more heat energy per gallon. Commonly, heat value is measured in BTUs (British Thermal Units).

FUEL VISCOSITY

The *viscosity* of a diesel fuel refers to the fuel's ability to flow under varying conditions. A high viscosity fuel oil is thick and resists flow. A low viscosity fuel is thin and runny, like water. See Fig. 4-35.

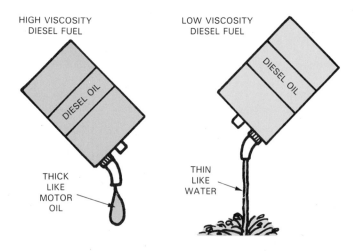

Fig. 4-35. Note how easily low viscosity fuel pours compared to high viscosity fuel.

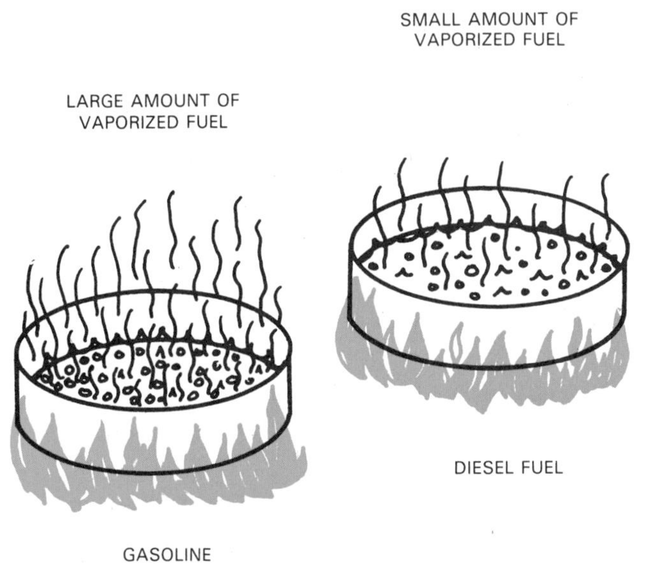

SMALL AMOUNT OF
VAPORIZED FUEL

LARGE AMOUNT OF
VAPORIZED FUEL

DIESEL FUEL

GASOLINE

Fig. 4-36. Vapors easily rise off highly volatile gasoline. Not so with low volatility diesel fuel. (Ford Motor Co.)

Viscosity has an important effect on how diesel fuel burns and how well it sprays into the combustion chamber. It also affects the fuel's ability to lubricate the injection system and engine components. It must be thick enough to coat and protect engine parts, but it must be thin enough to flow through the system properly.

VOLATILITY

As with gasoline, the *volatility* of diesel fuel is its ability to change from a liquid to a vapor. See Fig. 4-36. The volatility of diesel fuel is generally related to its weight, thickness, or viscosity. A light, thin diesel fuel, near the kerosene "end," would have a higher volatility rating. It would evaporate or vaporize easily. A thick, heavy fuel would have a lower volatility rating and would resist vaporization.

High volatility improves cold weather starting, speeds engine warm up, and reduces "white" exhaust smoke. A high fuel volatility could, however, hurt fuel economy and engine power slightly.

CETANE RATING

All diesel fuels have a *cetane* number or *rating* which indicates the starting quality of the diesel fuel. The higher the cetane number, the better the cold starting characteristics and the quieter the engine will operate, especially when cold. Most auto makers recommend a cetane rating of about 45 for diesel engines. This is the average cetane value of No. 2-D fuel. The cetane rating of a diesel fuel is similar, in some ways, to gasoline octane ratings. Both affect the knock tendencies of a fuel.

In general, high cetane fuels permit a diesel engine to start properly at low air temperatures and provide faster warm-up without misfiring. They limit or reduce the emission of white smoke from the exhaust during warm-up. They also reduce diesel knock and roughness and slow the rate of varnish (gum) or deposit formation in an engine.

CETANE VERSUS OCTANE RATING

The *cetane rating* of diesel fuel and the *octane rating* of gasoline should not be confused. A cetane number is, in a sense, the OPPOSITE of an octane number. This is shown in Fig. 4-37.

As you have learned, a high gasoline octane number indicates a lowered tendency for auto-ignition (knock) or spontaneous combustion (ignition without a spark from the spark plug or heat from an overheated surface). In contrast, a high cetane number means that the fuel will self-ignite easily. In a diesel engine, the fuel should ignite as soon as it comes in contact with the hot air in the cylinder in order to reduce the severity of engine noise (detonation).

In a mild way, combustion in a diesel engine is more like detonation than normal gasoline combustion. The idea is to limit the severity of detonation or auto-ignition by limiting the amount of fuel in the cylinder at one time. A diesel engine will knock and detonate when too much fuel ignites in the combustion chamber.

CETANE AND OCTANE COMPARISON

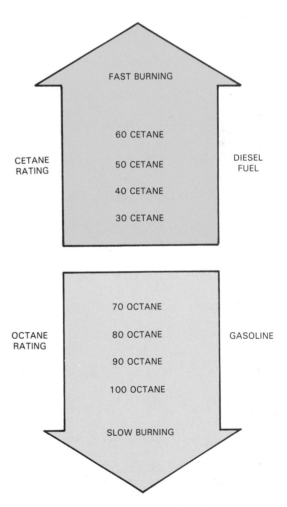

FAST BURNING

60 CETANE

CETANE
RATING

50 CETANE

DIESEL
FUEL

40 CETANE

30 CETANE

70 OCTANE

OCTANE
RATING

80 OCTANE

GASOLINE

90 OCTANE

100 OCTANE

SLOW BURNING

Fig. 4-37. Gasoline octane ratings and diesel fuel cetane ratings are, in a sense, the opposite. As cetane rating increases, diesel fuel burns faster. As octane rating increases, rate of burn slows in gasoline.

77

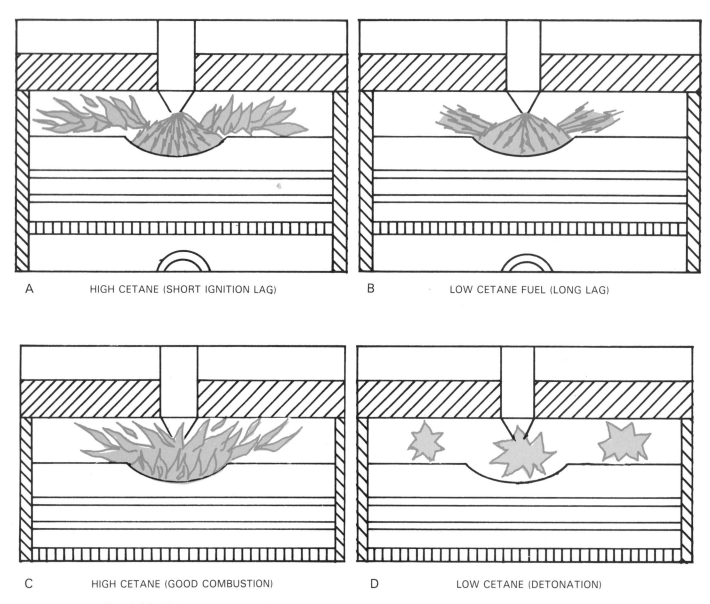

A HIGH CETANE (SHORT IGNITION LAG)	B LOW CETANE FUEL (LONG LAG)
C HIGH CETANE (GOOD COMBUSTION)	D LOW CETANE (DETONATION)

Fig. 4-38. High cetane diesel fuel begins to burn almost as soon as it is sprayed into cylinder.

Ignition lag

As illustrated in Fig. 4-38, *ignition lag* refers to the time between fuel injection and the start of combustion. Ignition lag or ignition delay is the time it takes the fuel to vaporize and "super-heat" before combustion. A short delay period is desirable while a long delay is undesirable. It will increase combustion knock and engine noise.

A diesel engine will tend to knock more when cold, than when hot. What happens is this: the cool engine surfaces in the cylinder prevent fuel from vaporizing and "super-heating" properly. As a consequence, unburned fuel collects in the combustion chamber. When all of this fuel does ignite, it burns too violently and causes a sharp pressure "spike" and knock. As this happens in all cylinders, it sounds like a low pitched engine rattle or clattering. Raising the compression ratio of a diesel engine will shorten the ignition lag period to reduce the severity of engine knock.

Cetane test fuel

Cetane itself is a fast-burning laboratory fuel used for comparison purposes. By comparing cetane and an unknown fuel in a test engine, like the one described in gasoline octane rating, a cetane number (ignition quality rating) can be assigned to the new fuel.

The rate of diesel fuel combustion (ignition lag period) is measured in terms of cetane numbers. A 100 cetane number has very rapid ignition.

Like diesel trucks, most diesel-powered cars require fuel with a cetane rating around 45. This is considered a happy middle range which still allows good starting and full-temperature diesel engine operation.

Flash point

The *flash point* of a diesel fuel does not relate to engine performance, but to the fuel as a *fire hazard* while handling and storing it. The flash point is the temperature at which the fuel vaporizes enough to ignite and burn when in contact with an open flame.

The flash temperature of diesel fuel must be high enough not to pose too much danger from flammable vapors during shipment and storage. See Fig. 4-39. An

FUEL GRADES	FLASH TEMPERATURE
No. 1-D	100 °F/38 °C
No. 2-D	125 °F/52 °C
No. 4-D	130 °F/54 °C

Fig. 4-39. Flash temperatures of diesel fuel.

unusually LOW temperature flash point may indicate that the diesel fuel has been contaminated with gasoline or kerosene.

Cloud point

Diesel fuel contains a wax called paraffin. At very cold temperatures, this wax separates from the other components in the fuel. As the wax separates, the fuel turns cloudy or milky. The temperature at which this occurs is the *cloud point* of the fuel.

When a fuel clouds, the wax can clog fuel filters. Engine operation becomes difficult or even impossible. The wax buildup can restrict or stop the flow of fuel. To avoid fuel flow or clouding problems, diesel fuel with a low cloud point is sold during winter in some areas.

In extremely cold weather, some auto manufacturers recommend adding either kerosene or regular gasoline to diesel fuel to lower cloud point, Fig. 4-40. This will lighten the fuel to prevent wax separation and fuel system stoppage.

CAUTION! Addition of gasoline or kerosene to diesel fuel *may void* some car warranties. Always check a service manual for specific information.

Pour point

The *pour point* of a fuel is the temperature at which it will no longer flow. When near its pour point, the fuel can become very thick. Pumping thickened fuel through the filters and injection system is hard. Since pour point temperature is lower than the cloud point, the pour point specification of diesel fuel has been replaced with the cloud point rating.

RATIO OF MIXTURE (PERCENT BY VOLUME) OF DIESEL FUEL TO KEROSENE OR REGULAR GASOLINE			
Ambient Temperature		Summer Diesel fuel #2**	Gasoline Kerosene Other*
°C	(°F)		
±0 to −10	(32 to 14)	70	30
−10 to −15	(14 to 5)	50	50***
−15 to −20	(5 to −4)	—	—
−20 to −22	(−4 to −8)	—	—

Fig. 4-40. Note, never use over 30 percent gasoline. Kerosene must be used for clouding protection when temperatures are below 14 °F (−10 °C). In either case, power and economy may drop according to proportion added to the diesel fuel. (Mercedes Benz)

Diesel fuel gravity

The *gravity* (specific gravity) of diesel fuel means the density or weight of the fuel per unit of volume (gallon or barrel). It is used in determining the quality and the characteristics of any blend of fuel oil.

Sulfur content

Sulfur in diesel fuel can cause a wide range of problems. The worst of these are engine corrosion and air pollution. Combined with water (in the fuel, or formed by combustion), it will produce sulfuric acid. This acid can corrode parts of an engine, especially those made of brass, bronze, or copper. Mixed with engine oil it will speed the formation of engine sludge. (This is a thick emulsion of oil and contaminants.) Its appearance is somewhat like chocolate pudding. Sulfuric acid also adds to air pollution as part of the engine exhaust.

Ash content

Diesel fuel will usually contain small amounts of *ash* (unburnable metal and abrasive solids in the form of microscopic particles). These impurities, if not limited, are another cause of rapid wear and deterioration of engine parts—injectors, injection pump, engine valves, cylinders, and even turbochargers.

Water and sediment contamination

Also common and troublesome is contamination of diesel fuel by water and sediment (rust, dirt, and other particles). Either can clog fuel filters and increase engine wear. In cold weather, water can freeze and block fuel flow into the engine. Water mixed with diesel fuel is also very corrosive.

CAUTION! Always make sure that the fuel cap, fuel cans, and fuel nozzles on gas station pumps are clean when fueling a diesel engine. A gasoline engine can tolerate and pass a certain amount of water but a diesel engine may not.

Carbon residue

Carbon deposits in a diesel engine are affected by the amount of carbon residue in the diesel fuel. The amount of residue can be measured by burning a given amount of fuel in a lab test. The quantity of solid carbon remaining is the fuel carbon content.

The chart in Fig. 4-41 shows all requirements for diesel fuel. Study it well.

Diesel exhaust smoke

The characteristics of diesel fuel have little effect on diesel smoke. Most often, dirty exhaust is caused by mechanical faults (bad injector, low engine compression, etc.), or a service or adjustment problem (rich fuel mixture, improper injection timing, etc.).

Basically, there are three types of diesel exhaust smoke: black (gray), blue, and white.

Black smoke is mainly made up of carbon particles chiefly due to incomplete fuel combustion. Usually, black smoke is caused by an extremely RICH FUEL MIXTURE (not enough air) and less often by a lean mixture.

Since there is not enough oxygen in a rich fuel mixture, some of the fuel droplets do not vaporize. They are "coked" (turned into a substance like charcoal powder).

ASTM LIMITING REQUIREMENTS FOR DIESEL FUEL OILS

Diesel Fuel Oil	Flash Point, °F	Cloud Point, °F	Water & Sediment, vol %	Carbon Residue on Residuum, %	Ash, wt %	90% Distillation Temperature, °F		Kinematic Viscosity at 100°F, centistokes		Sulfur, wt %	Copper Strip Corrosion	Cetane Number
	Min	Max	Max	Max	Max	Min	Max	Min	Max	Max	Max	Min
No. 1-D	100 or legal	*	0.05	0.15	0.01	—	550	1.4	2.5	0.50 or legal	No. 3	40**
No. 2-D	125 or legal	*	0.05	0.35	0.01	540†	640	2.0†	4.3	0.50 or legal	No. 3	40**
No. 4-D	130 or legal	*	0.50	—	0.10	—	—	5.8	26.4	2.0 or legal	—	30**

* For satisfactory cold-weather operation in most cases, the cloud point should be specified 10°F above the tenth percentile minimum ambient temperature at which the engine is to be operated.

† When cloud point less than 0°F is specified, the minimum viscosity shall be 1.8 cs and the minimum 90% point shall be waived.

** Low atmospheric temperatures or engine operation at high altitudes may require fuels with higher cetane ratings.

Fig. 4-41. These specifications are recommended for all diesel fuel. (Ethyl Corp.)

The "coked" fuel droplets can be seen as dark gray or black smoke blowing out the tail pipe.

Blue smoke indicates engine oil (crankcase oil) is getting into the cylinders where it burns. This condition may be due to worn piston rings, scored cylinders, loose valve guides, bad valve seals, or overfilled oil bath air cleaner.

White or *cold smoke*, is actually "raw" unburned diesel fuel blowing out the exhaust. It may have a light blue tint and commonly occurs before the engine has warmed to full operating temperature. A cold engine (cold cylinder walls and combustion chamber surfaces) can prevent some of the fuel from vaporizing and burning.

A "misfiring" engine (usually cold) can push unburned diesel fuel out of the tailpipe as white smoke. If the white smoke problem is severe or prolonged, the engine may have a thermostat stuck open, a bad injector, low compression, or the wrong fuel (high viscosity, for example).

Several methods are used to measure the amount of smoke leaving the exhaust pipe of a diesel engine. One of the newest is a light beam projected through the smoke as it leaves the engine. The light is reflected into a photocell, the more smoke, the less light and vice-versa. The photocell electrically measures the exact amount of light passing through the smoke. If too much smoke is detected, an adjustment or repair is needed.

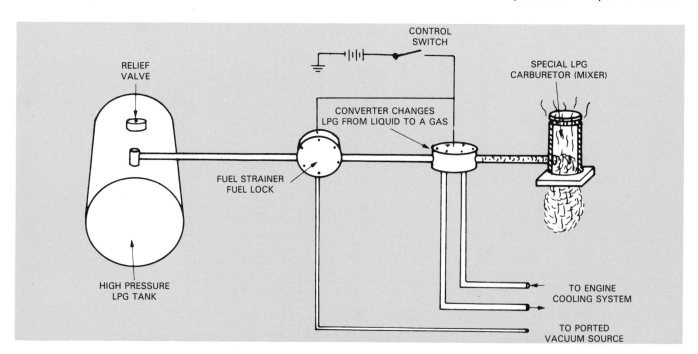

Fig. 4-42. Simple LPG fuel system.

LIQUIFIED PETROLEUM GAS (LPG)

As you learned earlier, *natural gas* is one of the lightest fractions (parts) of crude oil. Chemically, it is very similar to gasoline. Unlike gasoline, natural gas contains several light gases, heavy gases, and impurities. For natural gas to be used as an automotive fuel, it must be refined into *liquified petroleum gas* (LPG), a more pure fuel.

LPG is mainly propane and butane, along with small amounts of other gases. It has combustion qualities equal to or better than high-test gasoline.

The gases in LPG are made of carbon and hydrogen compounds, as is gasoline. LPG, however, contains a larger percentage of hydrogen and less carbon.

Hydrogen is an ideal fuel that burns well while producing little pollution. It is the main reason LPG is a very desirable fuel. It produces good power, economy, and low exhaust pollution levels.

LPG is a vapor or gas at normal room temperature and atmospheric pressure. This poses some design problems. In fact, since LPG is not a liquid, the entire fuel system must be redesigned to handle it. See Fig. 4-42.

LPG has operating characteristics almost identical to those of gasoline. However, because it is already a gas, the problem of breaking up liquid fuel is eliminated. Since the fuel enters the intake manifold and combustion chambers as a vapor, combustion is much more efficient. For this reason, LPG is generally an excellent automotive fuel.

BENZOL

Benzol (C_6H_6), also called benzene, is an extremely high octane liquid hydrocarbon extracted from coal tar. Benzol can be blended with gasoline to increase octane rating, fuel mileage, and flame temperature. Also, benzol can be mixed with alcohol to make high performance racing fuel.

Caution! Benzol is a colorless, flammable liquid that is very toxic. Prolonged breathing can lead to severe sickness or even death.

ALCOHOL

Alcohol is an excellent alternate fuel for automobiles. The two types used to power internal combustion engines are *ethyl alcohol* (ethanol) and *methyl alcohol* (methanol). Alcohols are especially desirable as an automotive fuel because they can be manufactured from sources other than crude oil. Alcohol intended as an automotive fuel must be almost pure. Quite often, several refining steps are needed to approach this purity.

Grain alcohol

Ethanol or "grain alcohol" is colorless, harsh tasting, and highly flammable. It can be made from numerous farm crops such as wheat, corn, sugar cane, potatoes, fruit, oats, soy beans, or any crop rich in carbohydrates. Crop wastes are also a source.

Methanol

Methanol or "wood alcohol," can be made from wood chips, coal, oil shale, tar sands, cornstalks, garbage, or even manure. Like ethanol, methyl alcohol is a colorless, smelly, flammable liquid.

Caution! Methanol is highly poisonous and can be fatal if taken internally.

Methanol is manufactured by first turning the materials shown in Fig. 4-43 into a gas. Then the gas is passed over a catalyst which converts the gas into methyl alcohol. Fig. 4-44 illustrates an experimental gasifier for converting the energy materials into a gas.

METHANOL PRODUCTION

Wood Chips Manure Coal Oil Shale Tar Sands Cornstalks Garbage Petroleum	Can Be Used To Make	(Methanol) Alcohol)

Fig. 4-43. Methanol (methyl alcohol) or "wood alcohol" can be made from a wide variety of substances.

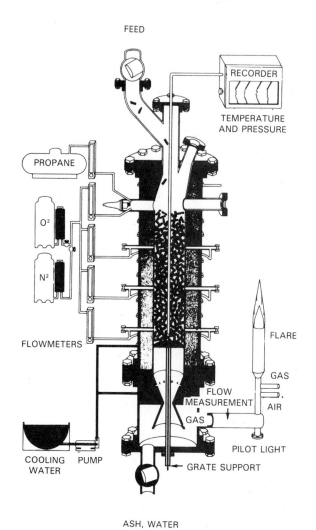

Fig. 4-44. Cutaway of pilot gasifier for producing methanol from coal, crop residue, and other biomass materials. (SERI)

Generally, both alcohols are similar in performance and operating characteristics. They have a very high octane or antiknock rating. In fact, they perform best in a high compression engine. Alcohol must be tightly compressed to produce maximum power.

Design considerations

With an alcohol fuel, engine compression ratios of between 11:1 and 13:1 are not unusual. Remember, today's engines use a compression ratio of around 7:1 or 9:1, much too low for *pure* alcohol. Power and economy would suffer. On the other hand, in a properly designed engine and fuel system, alcohol produces fewer harmful exhaust emissions.

Use of pure alcohol would require several engine modifications. Since alcohol contains about half the heat energy of gasoline per gallon, the alcohol-air mixture must be much richer than for gasoline. See Figs. 4-45 and 4-46.

To provide a proper fuel-air mix, a carburetor's fuel passages should be doubled in diameter to allow extra fuel flow.

Alcohol does not vaporize as easily as gasoline. This affects cold weather starting. The alcohol liquifies in the engine and will not burn properly. Thus, the engine may be difficult or even impossible to start in extremely cold weather. One solution is to introduce gasoline into the carburetor until the engine starts and warms up. Once the engine temperature is up, the alcohol will vaporize and burn normally.

Flame speed

Alcohol burns at about half the speed of gasoline. For this reason, ignition timing or the distributor advance curve must be changed so that more spark advance is provided. This will give the slow-burning alcohol more time to develop pressure and power in the cylinder.

GASOHOL

Gasohol, as the name indicates, is usually a mixture of 87 octane unleaded gasoline and grain alcohol. While the mixture can range from 2 to 20 percent alcohol, most is 10 percent. See Fig. 4-47. This mixture requires little or no engine modification or adjustment.

Gasohol has a lower heat content than gasoline but will usually have a higher octane rating. The alcohol acts like a lead antiknock additive. It can make low octane fuel perform like high octane premium. For example, 10 percent ethyl alcohol added to 87 octane unleaded gasoline can increase the fuel's octane value to around 92 to 93 octane. This is equal to the octane value of premium gasoline. An excellent fuel for older, high compression, high horsepower engines of the late 60s, gasohol can prevent engine knock, ping, and dieseling.

SYNTHETIC FUELS

Automotive fuels made from coal, shale oil, and tar sands, are known as *synthetic fuels* because they are synthesized (changed) from a solid hydrocarbon state to a liquid or gaseous state. Synthetic fuels are presently being developed to help replace petroleum as a major source of motor fuel.

RICHENED ALCOHOL FUEL MIXTURE

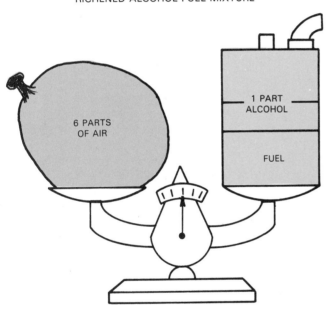

Fig. 4-45. Compared to gasoline, about two times as much alcohol must be mixed with a given amount of air.

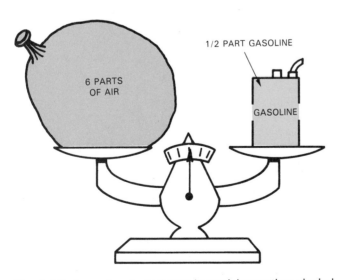

Fig. 4-46. A gasoline-fuel mixture is much leaner than alcohol mixture shown in previous illustration.

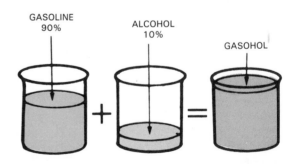

Fig. 4-47. Usually, gasohol is a mixture of 90 percent gasoline and 10 percent alcohol. This ratio requires little or no engine modification.

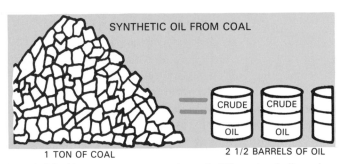

Fig. 4-48. About two to two-and-one-half barrels of oil can be produced from a ton of coal.

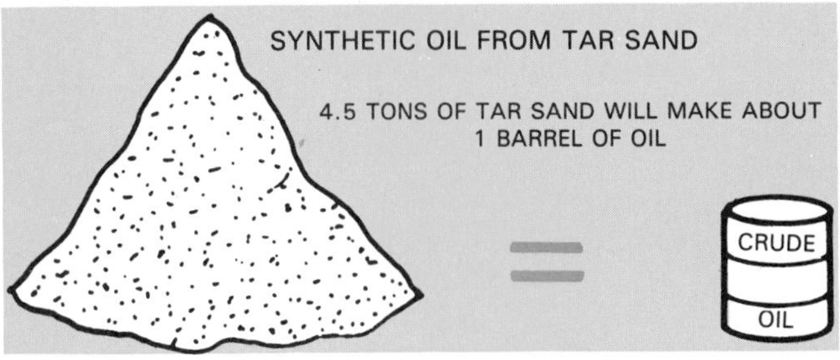

SYNTHETIC OIL FROM TAR SAND

4.5 TONS OF TAR SAND WILL MAKE ABOUT 1 BARREL OF OIL

Fig. 4-50. Between four and five tons of tar sand must be processed to make one barrel of oil.

Coal synthesized fuel

Through various processes, coal can be converted into either a hydrocarbon gas or a liquid, suitable for use in internal combustion engines. Changing coal into a gas is called *gasification.* Changing coal into a liquid is known as *liquification.*

Coal can be changed into an automotive fuel. One process is called PYROLYSIS. The coal is heated in an enclosed container with oxygen. This breaks the coal down into hydrocarbon-based oil, gas, and a charcoal. These products can then be refined and cleaned to make gasoline, diesel fuel, and gaseous petroleum.

A ton of coal yields around 2 1/2 barrels of synthetic oil suitable for refinement into gasoline, Fig. 4-48.

Oil shale

Oil shale is another natural resource that can be altered and made into oil and automotive fuels. Oil shale is a sedimentary type rock which contains tar-like substance called KEROGEN. The shale looks very much like chocolate marble ice cream, the "chocolate" being the thick kerogen hydrocarbon and the "vanilla" being the sedimentary rock.

When the shale is heated to around 1000 °F (538 °C), the kerogen melts into liquids and vapors that can be used to make gasoline, diesel oil, LP gas, etc.

As does coal, oil shale contains a relatively large amount of sulfur and other impurities that must be removed through refinement.

The amount of oil trapped in oil shale can vary from around 75 gallons (almost 2 barrels) per ton of rock to a low of around 20 gallons (1/2 barrel) per ton, Fig. 4-49.

One experimental method of producing oil from oil shale is called "in-place" or *in-situ* extraction. Natural gas is pumped into a pulverized (dynamited) oil shale

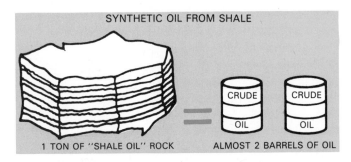

Fig. 4-49. A little less than two barrels of oil can be made from a ton of oil shale.

deposit and burned. The heat melts the kerogen so that it can be pumped out. Another promising method of extracting oil from shale uses high intensity microwave energy to heat and melt the kerogen.

Tar sand

As implied, *tar sands* are heavy hydrocarbons mixed with sand and dirt. There are several methods of removing these tars from the sand. One is to submerge the tar sand in hot water and steam. This forms a hot slurry that melts and liquifies the tar. Since oil is lighter than water, the petroleum based oils float to the top of the slurry where it can be easily removed.

Usually, tar sands contain about 12 1/2 percent petroleum compounds by weight. Around 4 to 5 tons of tar sand will produce a barrel of oil, Fig. 4-50.

HYDROGEN

Many fuel and energy experts believe that, in the long run, *hydrogen* is the most promising fuel. Most important, it is the most common element in the universe. When burned, it produces little or no pollutants. In this respect, it is superior to petroleum based fuels.

One of the best sources of hydrogen is water. Each molecule of water is made up of two atoms of hydrogen and one of oxygen. One simple method used to separate the hydrogen is by sending electric current through water, Fig. 4-51. The process is called *electrolysis.*

When burned with pure oxygen, hydrogen produces nothing but pure water. When burned in an engine using air instead of oxygen, its combustion releases water and a very small amount of pollution (NOx), but this is only a fraction of the pollutants formed by the combustion of most other fuels.

When technology overcomes the problems of hydrogen storage and the high cost of production, hydrogen may very well become an excellent automotive fuel.

SUMMARY

Petroleum is a mixture of numerous hydrocarbon compounds. The lighter ones are suitable as automotive fuels. A process of distillation and refinement is used to separate and purify crude oil into fractions. Fractions include gasoline, diesel oil, LP gas, and several petrochemical products.

Several additives are blended into gasoline to reduce and prevent fuel system corrosion, "gumming," rusting, icing, and carbon deposits. Lead additives help stop

HYDROGEN PRODUCTION BY ELECTROLYSIS

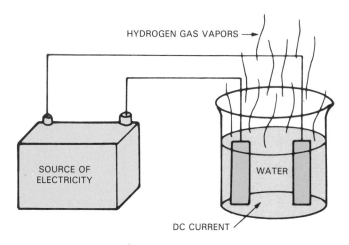

Fig. 4-51. Hydrogen gas is released from water when current is passed through the water.

engine knock, engine cylinder wear, and valve wear.

Gasoline tests are utilized to assure that certain standards of quality are maintained. Gasoline octane ratings indicate when, or under what conditions, gasoline will detonate, preignite, and knock. A high octane rating will indicate a high knock resistance. The Research-Motor Average Method is commonly used to rate gasoline octane values. It can be seen on most gas station pumps.

Unleaded gasoline is high test, high octane gasoline without lead additives. However, it is usually lower in octane than gasolines with lead additives. No-lead gasoline allows the use of a catalytic converter, an emission-reducing device in the exhaust system. Lead additives will quickly ruin a catalytic converter.

When gasoline is overheated in a fuel system, the fuel can boil. This boiling action forms tiny bubbles that can cause vapor lock. By reducing the flow of liquid fuel, vapor lock may upset the operation of the engine. Also caused by boiling gasoline, carburetor percolation occurs when vapor bubbles push fuel out of the carburetor and into the engine. This causes an overrich mixture and, possibly, engine flooding or stalling. Any method of lowering the temperature of the gasoline will help reduce vapor lock and carburetor percolation.

Fuel system icing or gas line freeze occurs when water in the fuel system, frequently in the fuel filters, freezes on a very cold day. System icing can completely stop fuel flow and engine operation. Alcohol, which absorbs water, is a common gasoline antifreeze agent.

Diesel fuel is similar to gasoline, but it is a heavier, thicker, and less volatile hydrocarbon. As a result, it has different operating characteristics than gasoline. Diesel fuel must be ignited by the heat of extreme air compression. A spark plug will not adequately ignite and burn diesel oil. Most diesel automobiles require grade No. 2-diesel fuel, a medium grade fuel, commonly sold at gas stations. Grade No. 1-D is sometimes recommended as a winter or cold weather fuel.

Diesel oil has several specifications: viscosity, volatility, cetane rating, flash point, pour point, cloud point, gravity, cleanliness, etc. All of these ratings or ''specs'' assure a high quality motor fuel.

Several fuels other than gasoline and diesel fuel can be used to power an automobile. Some of these alternate fuels include: LP gas, alcohol, gasohol, hydrogen, and synethetic fuels made from coal, tar sands, oil shale, garbage, animal manure, etc. Gasohol is presently one of the most promising alternate type automotive fuels. It is usually a mixture of 10 percent alcohol and 90 percent unleaded gasoline. Normally, gasohol does not require engine modifications for use in a passenger car.

KNOW THESE TERMS

Hydrocarbons, Hydrogen, Biogenic theory, Seismic studies, Seismograph, Drilling rigs, Derrick, Fractions, Distillation, Fractonating tower, Catalytic cracking, Alkylation, Polymerization, Antioxidants, Metal deactivators, Deposit modifier, Scavengers, Detergents, Anti-icing agents, Octane rating, Volatility, Research method, Motor method, Road octane number, Catalytic converter, Vapor lock, Percolation, Fuel viscosity, Cetane rating, Cloud point, Pour point, LPG, Benzol, Methanol, Ethanol, Gasification, Liquification, Pyrolysis, Oil shale, Kerogen, In-situ, Electrolysis.

REVIEW QUESTIONS—CHAPTER 4

1. Crude oil is largely flammable _____. They are a chemical mixture of _____ percent hydrogen and _____ percent carbon.
2. One theory states that petroleum came from the decomposition of animal and vegetable matter. True or false?
3. During the process of refining, crude oil is (check all that apply).
   a. Cleaned to remove salt and sulfur.
   b. Broken down into fractions.
   c. Heated to around 800 °F.
   d. Distilled.
   e. All of the above.
4. Explain the purpose of adding an antioxidant to gasoline.
5. What is volatility and what effect does it have on operation of an engine?
6. The _____ _____ of a fuel indicates its ability to resist knocking or pinging.
7. Lead additives act as _____ to prevent _____.
8. Unleaded gas may not have more than (0.5, 0.005, 0.05, 5) grams of lead per gallon.
9. A catalytic converter (select correct answer):
   a. Makes the engine exhaust resonate.
   b. Is part of the exhaust system which changes poisonous parts of the exhaust gases to help reduce air pollution or emissions.
10. Define vapor lock and explain its effect on operation of the engine.
11. List the seven things to look for in case of vapor lock. *
12. When there is enough heat in the carburetor to make the fuel boil, the resulting condition is called _____.
13. In the condition described above, list the four possible causes.

14. Describe the conditions that might lead to carburetor icing.
15. Fuel system icing is usually caused by _____ in the gasoline supply.
16. List the three grades of diesel fuel and indicate the conditions for which each is blended.
17. Why should gasoline NOT be used in a diesel engine?
18. _____ is the quality in a fuel which gives it the ability to flow under varying conditions.
19. The volatility of a fuel is generally related to its weight and _____.
20. In a sense, a cetane number is the same as an octane number. True or false?
21. Which of the following is the best definition of ignition lag in a diesel engine?
    a. The time between fuel injection and the start of combustion.
    b. The time it takes the fuel to vaporize and superheat before combustion.
    c. The time between spark and the fuel starting to burn.
    d. The amount of time it takes for combustion after the piston reaches top dead center.
22. The temperature at which a diesel fuel will no longer flow is known as its _____ _____.
23. What is the cloud point of diesel fuel and why is it important?
24. Cetane number or rating indicates the _____ _____ of a diesel fuel.
25. What is the makeup of LP-gas?
26. List two uses of Benzol.
27. To use pure alcohol as a fuel, an engine would need a compression ratio between _____ and _____.
28. What alterations would be necessary to a carburetor so it would deliver a suitable air-fuel mixture of alcohol to the intake manifold?
29. What effect does the addition of 2 to 20 percent alcohol have on gasoline?
30. List three synthetic fuels that could be used as motor fuel.
31. What advantage is there in burning hydrogen as an automotive fuel?
32. At present, what major problems prevent hydrogen from being used as a fuel?
33. A customer's gasoline engine starts well when cold but floods during a hot start.
    Mechanic A says that vapor lock is the cause.
    Mechanic B says percolation is likely the problem.
    Who is right?
    a. Mechanic A.
    b. Mechanic B.
    c. Both Mechanic A and B.
    d. Neither Mechanic A nor B.
34. A diesel engine will not start and run on a cold winter day. The glow plugs are functional.
    Mechanic A says that the injection pump might be bad since this is a common problem.
    Mechanic B says that fuel clouding could be clogging the fuel filters.
    Who is right?
    a. Mechanic A.
    b. Mechanic B.
    c. Both Mechanic A and B.
    d. Neither Mechanic A nor B.

5

GASOLINE AND
DIESEL COMBUSTION

After studying this chapter, you will be able to:
● *Explain the basic action of gasoline combustion.*
● *Describe the burning process for diesel fuel.*
● *Summarize the causes and resulting problems of abnormal combustion.*
● *Explain the factors affecting normal and abnormal combustion.*
● *Describe how air-fuel ratios affect combustion.*
● *Define the term "stoichiometric" fuel mixture.*
● *Summarize the major factors controlling combustion efficiency.*
● *Comprehend later text chapters more fully.*

This chapter describes the combustion process in an internal combustion engine. It also explains the many factors that affect combustion: ignition timing, air-fuel ratio, compression ratio, inlet air pressure, engine speed, and combustion chamber shape.

When you have developed a sound understanding of combustion, the power-producing force in both gasoline and diesel engines, you will be much better prepared to diagnose and repair automotive fuel systems and related components.

COMBUSTION

Combustion, as we said in Chapter 1, is the burning action above the engine piston that produces enough heat to cause expansion of the gases trapped there. Proper combustion of any automotive fuel requires that the fuel be mixed with air and compressed in the cylinder before it is ignited. See Fig. 5-1.

The fuel must be broken into tiny droplets and thoroughly *mixed* with air (oxygen). Liquid fuel will burn, but too slowly for adequate engine combustion.

AIR STREAM ACTION

On each engine intake stroke, the intake valve opens while the piston is sliding downward in the cylinder. This forms a vacuum in the cylinder, intake manifold, and carburetor or throttle body assembly. This is illustrated in Fig. 5-2. Outside air pressure rushes into the engine to fill the vacuum. An *air stream* is produced. The force of the air stream is used to mix fuel and air.

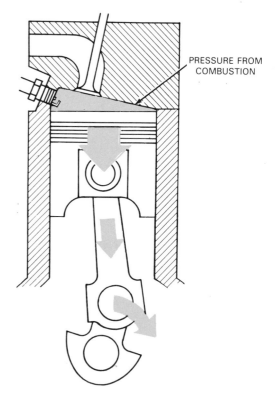

PRESSURE FROM COMBUSTION

Fig. 5-1. Combustion heat causes expansion. Expansion causes pressure. Then, pressure forces piston down and powers the engine. (Ford Motor Co.)

ATOMIZING THE FUEL

Atomization refers to how the carburetor or injector breaks up liquid gasoline into tiny droplets. For instance, an insect sprayer atomizes liquid insecticide. When the handle is pumped, Fig. 5-3, the sprayer sends out a fine mist of tiny drops.

Notice that combustion occurs at the end of the compression stroke and at the start of the power stroke. After the fuel mixture is ignited, the burning gases expand and increase the pressure in the combustion chamber two-and-one-half to three times.

For example, at the start of the compression stroke, the chamber temperature may only be around 575°F

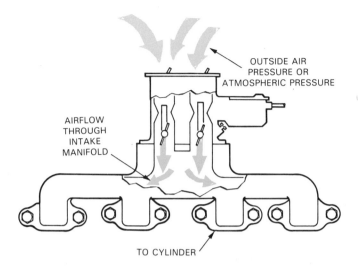

Fig. 5-2. Air stream action pulls air through carburetor throat intake manifold, and into cylinder. (Ford Motor Co.)

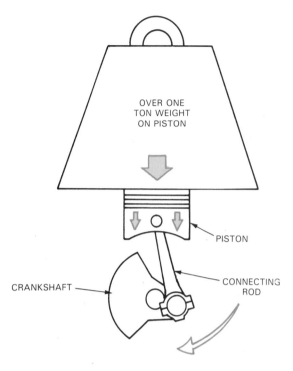

Fig. 5-4. Expanding combustion gases push down on piston with force equal to thousands of pounds.

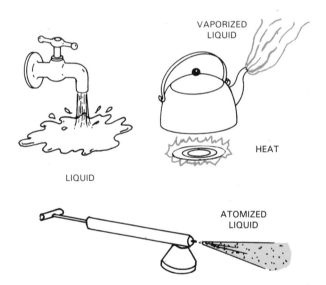

Fig. 5-3. Sprayer acts like carburetor to atomize and vaporize liquid.

(302 °C); after combustion the temperature may rise to as high as 4000° or 5000°F (2204° or 2760°C). This temperature increase will expand the gases in the cylinder to form *pressure.*

For instance, the cylinder pressure during the compression stroke may be around 150 psi (1 033 kPa); after combustion, the cylinder pressure can increase to almost 800 psi (5 512 kPa).

This is a tremendous amount of force pushing on the top of the piston. Every square inch has 800 pounds of pressure pushing on it. If a piston has an area of 12.6 sq. in. (around a 4 in. cylinder bore), the combustion pressure would be like having a 5 ton (4 525 kg) weight resting on the piston. This tremendous force, illustrated in Fig. 5-4, rams the piston down the cylinder and spins the crankshaft. This pressure, as you will learn in later chapters, can go much higher in supercharged or turbocharged engines.

COMBUSTION CHEMISTRY

Combustion is a chemical reaction (union of fuel and oxygen) that produces useful heat. It is a rapid oxidation (burning) of carbon and hydrogen. *Oxidation* occurs when any substance combines with oxygen. Rusting of metal, for instance, is slow combustion. Rapid oxidation or combustion of gasoline, in an ideal situation, occurs when the hydrogen in gasoline combines with oxygen to form steam, and the carbon in gasoline combines with oxygen to form carbon dioxide gas.

IDEAL COMBUSTION

Under *ideal* situations, *combustion* forms harmless water (H_2O) and carbon dioxide (CO_2). Carbon dioxide can be seen as bubbles in soda water.

Unfortunately, actual combustion in an engine is not this perfect or complete. Air is not pure oxygen, Fig. 5-5. As a consequence, real combustion produces other undesirable chemicals or pollutants, Fig. 5-6. As you can see, several harmful gases are created because of the other substances in air and because some of the gasoline does not burn.

Another interesting fact concerning combustion, Fig. 5-7, is that it takes around 9000 gallons of air to burn 1 gallon (6.2 lb.) of gasoline. Over a gallon of water is created. The water can be seen as a vapor cloud on a cold day or as a liquid dripping from the tailpipe.

NORMAL COMBUSTION

If all of the factors affecting combustion are favorable (air/fuel ratio, compression, spark timing, fuel

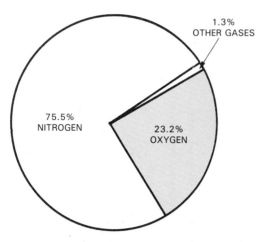

Fig. 5-5. Air is mostly nitrogen (75 percent) and 23 percent oxygen. About 1.3 percent is made up of other gases such as argon and carbon dioxide.

characteristics, etc.), the fuel burning process will be normal.

Normal combustion occurs in a gasoline engine when a single flame spreads evenly and uniformly away from the spark plug electrodes to form maximum pressure. This should occur when the piston is a few degrees before TDC. Fig. 5-8 illustrates normal combustion.

COMBUSTION PHASES

In Fig. 5-8A, a spark occurs at the spark plug and combustion starts. The fuel mixture slowly begins to burn and a small blue ball of flame forms around the plug electrodes. Notice that the piston has not yet reached TDC, but is rapidly moving up. In this phase of combustion, very little heat and expansion are present.

In Fig. 5-8B, the flame has spread about half the way through the combustion chamber. Generally, the flame is spreading in a uniform circle. At this point, the piston

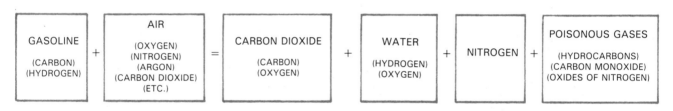

Fig. 5-6. Combustion, using air instead of pure oxygen, produces several poisonous gases.

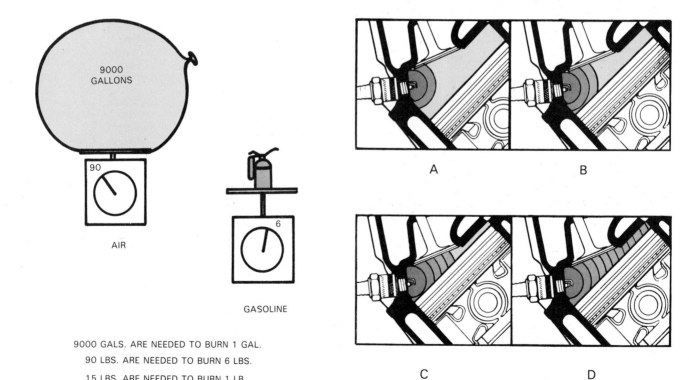

9000 GALS. ARE NEEDED TO BURN 1 GAL.

90 LBS. ARE NEEDED TO BURN 6 LBS.

15 LBS. ARE NEEDED TO BURN 1 LB.

Fig. 5-7. Combustion requires a very large quantity of air. (General Motors Trucks)

Fig. 5-8. Normal combustion has only one flame front which spreads away from spark plug electrodes. (Champion Spark Plug Co.)

is nearing the top of its stroke causing an increase in compression pressure and combustion speed. Combustion has moved far enough into the chamber so that it, too, is producing heat and pressure.

In Fig. 5-8C, the piston has reached TDC and the flame is rapidly picking up speed. However, it is still expanding in a uniform circle.

The flame in Fig. 5-8D has expanded rapidly to consume the rest of the fuel in the chamber. The flame has spread almost as far in this illustration as in the last three. As you can see, combustion has been completed while the piston moved only a short distance down the cylinder.

COMBUSTION SPEED AND POWER

This normal burning or combustion process takes around 3/1000 of a second, which is comparatively slow. Do not think of normal combustion as an explosion. An explosion, like that of dynamite, would take only around 1/50,000 of a second — a fraction of the time it takes for the combustion of gasoline. Note, however, the combustion of gasoline, pound for pound, produces more total energy than dynamite. Just a thimble full of gasoline has enough energy to produce tons of force.

The burning of gasoline in an engine is slower than the explosion of dynamite, but it is equally as powerful. Under some undesirable conditions, gasoline can be made to burn too fast, making part of combustion resemble an explosion.

ABNORMAL COMBUSTION (DETONATION)

Detonation is abnormal combustion during which the last unburned portion of the fuel mixture almost explodes in the combustion chamber. This action is pictured in Fig. 5-9. A part of detonation is, in fact, almost as fast as an explosion of dynamite.

The total process of engine detonation is about six times faster than normal combustion.

Detonation can cause a very rapid pressure rise in the engine cylinder. It can be so fast and extreme that parts of the engine (cylinders, cylinder heads, etc.) actually flex and vibrate. This vibration is extremely high pitched and can be heard as a knock or ping. A detonation knock will sound something like a ball peen hammer rapidly striking the pistons or engine block.

As you will learn, detonation is a very serious engine combustion problem. It reduces the efficiency and power of an engine. Extreme cases of detonation can cause major engine damage.

Detonation phases

Study the illustrations in Fig. 5-9 as they are mentioned.

At first, as in Fig. 5-9A, combustion is normal. The spark plug ignites the fuel mixture. A ball of flame forms. In Fig. 5-9B, you can see that the flame is spreading very slowly through the mixture. It is only about one-third of the way through the combustion chamber with the piston almost at TDC. Remember, during normal combustion, the flame had spread farther.

In this example, the lag in combustion could be caused

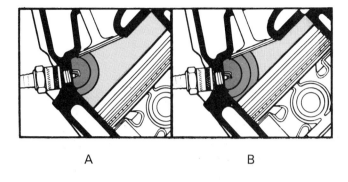

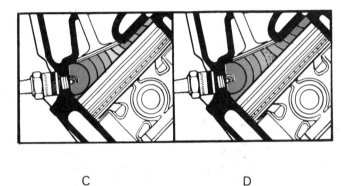

Fig. 5-9. During detonation, normal combustion is extremely slow. This causes unburned fuel in chamber to "self-ignite" and knock. (Champion Spark Plug Co.)

by an extremely lean fuel mixture, improper fuel distribution, or lack of turbulence. As a result, the unburned part of the fuel mixture, sometimes called the END GAS, is exposed to heat and pressure for a longer period of time.

This prolonged heating and squeezing of the unburned fuel mixture will make it very volatile. So explosive is the fuel, that in Fig. 5-9C, you can see where the end gas actually ignites spontaneously (on its own) and explodes. For an instance, there are TWO FLAME FRONTS shooting towards each other. When this happens, the rest of the combustion process is speeded up. Flame speed increases to supersonic levels (faster than the speed of sound) and a knocking noise is heard.

Detonation damage

During detonation, the pressure in the cylinder can climb 50 percent higher than normal. Engine power and performance will suffer and, under severe detonation, connecting rods may bend, cylinder heads crack, pistons are ruined, spark plugs shatter, Fig. 5-10, and bearings are battered. When detonation occurs over a long period, the extreme heat can partially melt and soften the top of a piston. Then the pressure can force a hole in the softened aluminum. See Fig. 5-11.

Detonation-caused knocking will usually be heard when you push the gas pedal all the way to the floorboard while the car is only traveling at moderate speeds (20 to 40 mph). Normally, at high speeds there will not be enough time for the end gas to heat up, ignite without a spark, and detonate.

Fig. 5-10. This spark plug insulator has been shattered by the extreme pressures of detonation. (Champion)

Conditions causing detonation are:

1. Slow-burning lean fuel mixture (faulty carburetor or injector, fuel pump, blocked fuel filter or line, vacuum leak at higher engine speeds caused by bad PCV valve or EGR valve).
2. Gasoline with low octane or antiknock rating. This is more common with unleaded gasoline.
3. Carbon deposits increasing compression ratio. This is the result of oil entering cylinders or poor detergent action of gasoline.
4. Engine operating at above-normal temperature due to low coolant level, water jacket blockage in head, etc.
5. Ignition timing too advanced (improper setting of initial ignition timing, inaccurate distributor advance curve, computer malfunction, etc.).
6. Bad rings and/or valve seals allowing oil (low octane hydrocarbon) to be burned in cylinders.
7. Air cleaner door stuck allowing too much hot exhaust manifold air to enter engine.
8. Excessive turbocharger boost pressure from a bad pressure-limiting valve.

PREIGNITION

Preignition is also known as "ping." It occurs when surface ignition causes the fuel mixture to burn before the spark plug "fires," Fig. 5-12.

Its effects can vary from slight loss of power and fuel economy to a very serious engine-damaging ping, knock, or "rattle." Severe or prolonged cases of preignition can even cause detonation, burned pistons, bent connecting

Fig. 5-11. Detonation heat and pressure has knocked a hole in head of this piston (Champion Spark Plug Co.)

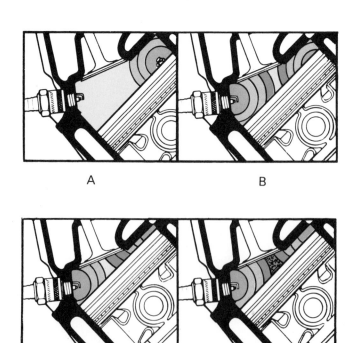

A   B

C   D

Fig. 5-12. During preignition, two flame fronts slam into each other. This causes knock or ping.
(Champion Spark Plug Co.)

Detonation factors

You can think of detonation as a "race"—a race between the normal flame front and the auto-ignition (burning without spark) of the end gas. If normal combustion is faster than the heating and squeezing on the unburned fuel, normal combustion will win and detonation will be prevented.

rods, cracked heads, scored cylinders, damaged bearings, blown head gaskets, and, in some cases, damaged spark plugs.

Surface ignition

To fully understand preignition, you must first understand its cause, surface ignition. Surface ignition is caused by:

1. Incandescent (glowing hot) piece of carbon in the combustion chamber.
2. Overheated engine from improper operation of cooling system.
3. Exhaust valve overheated by lean fuel mixture (lean carburetor setting, clogged injector strainer, vacuum leak, stuck EGR valve, etc.).
4. Overheated spark plug (heat range too high).
5. Exhaust valve overheated by leakage (insufficient tappet clearance, weak valve spring, sticking valve, etc.).
6. Sharp edges in the combustion chamber (overheated threads on spark plug, edge of head gasket, sharp, machined surface, etc.).
7. Excessively dry and hot atmospheric conditions or an air filter door stuck shut.

Postignition

Surface ignition can also occur after the spark plug fires. Since it occurs after normal spark ignition, it is termed *post-ignition* and its effects are minimal.

Preignition versus detonation

Preignition is similar to, yet quite different from, detonation; the processes of each are reversed. Remember, detonation occurs after the start of normal combustion and preignition begins before normal combustion. However, one can cause the other and both are capable of causing engine damage.

Preignition phases

The processes or phases of preignition are illustrated in Fig. 5-12. In view A, the fuel mixture is ignited by a "hot spot" (glowing bit of carbon, in this case) on top of the piston. A small ball of flame forms and preignition begins. Note, the piston is still far down in the cylinder and is traveling up.

In Fig. 5-12B, the abnormal preignition flame has spread a little farther into the fuel. The piston has reached a point in the cylinder where normal combustion should begin. The spark plug "fires," and the normal flame front also begins to burn.

Fig. 5-12C shows the two flame fronts rapidly traveling toward each other. The closer they get, the faster they move. This is because the two flames are building up heat and pressure in the chamber. The piston is near TDC which also increases the pressure, temperature, and speed of the two flames.

Finally, in Fig. 5-12D, the two combustion fronts or flames slam into each other. The collision of the flames causes the remaining fuel to ignite with a sudden shock wave. The shock wave can be so intense that parts of the engine vibrate and send out a ping or knocking sound. Engine power and fuel economy will be reduced. If repeated for too long, the engine can be damaged.

Wild knock

When carbon breaks loose from the combustion chamber wall it can cause *wild knock.* When this happens, the carbon particle bounces around in the cylinder. While suspended in the burning fuel it will heat up and ignite the fuel mixture each time fuel enters that cylinder. This can cause a pinging sound that can last for a few moments or come and go for no apparent reason.

Dieseling (after running, run-on)

Surface ignition (preignition or postignition) can also cause an engine to run after the ignition switch is turned off. A knocking, coughing, or fluttering noise can be heard. In some cases, an unpleasant "rotten egg" odor may also be evident.

The term *dieseling* is used to identify this condition because, like a diesel engine, the gasoline engine runs without spark. The ignition switch is off and no voltage is present in the ignition system or at the plugs. The terms "run-on" and "after running" are also used.

Usually, the engine continues to run because "hot spots" are igniting the fuel in the cylinders. The fluttering or coughing sound, common to after-run, usually occurs right before the engine stops. This coughing sound can be a result of the engine actually "kicking" backwards. Air can blow back through the intake ports.

After-running places undue strain on pistons, cylinders, valves, throttle body or carburetor gaskets, and the air cleaner.

Dieseling can be halted by leaving the automatic transmission in drive, or by dragging a manual clutch to lug the engine. However, for a permanent fix, the source of the problem must be found and corrected.

Reasons an engine may diesel include:

1. Engine idle set too high.
2. Carbon deposits increasing engine compression ratio.
3. Inadequate or low gasoline octane rating.
4. Engine overheating (low coolant level, bad thermostat, etc.).
5. Spark plug heat range too high.
6. Idle fuel mixture adjustment incorrect (usually too lean), which results in improper idle speed screw setting.
7. Sticking throttle or gas pedal linkage.
8. Engine needs a tune-up (plugs, timing, carburetor adjustments, etc.).
9. Driver not allowing engine enough time to return to curb idle speed before turning off the ignition switch.
10. Oil entry into cylinders from engine mechanical problem.

Spark knock

Spark knock is an engine ping caused by the electric arc occurring too early at the spark plugs. It is an ignition timing problem. The timing is advanced too far and is causing combustion pressures to slam into the upward moving piston (maximum cylinder pressure before TDC).

The spark knock sequence of events is shown in Fig. 5-13. In A, notice that the spark occurs at the plug too soon, while the piston is still far down in the cylinder. As seen in Fig. 5-13B and 5-13C, the single flame

SPARK PLUG "FIRES" TOO SOON

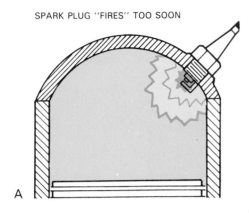

A

PISTON MOVES TOWARD
FLAME FRONT

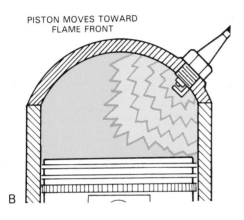

B

PRESSURE BUILDS AS PISTON
SLAMS INTO COMBUSTION FLAME

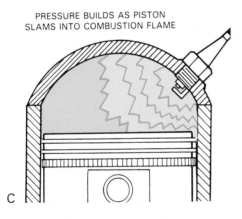

C

SPARK KNOCK OCCURS BECAUSE OF
EXCESSIVE PRESSURE IN CYLINDER

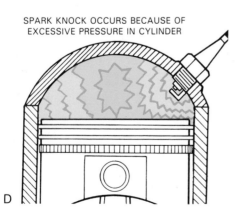

D

Fig. 5-13. Spark knock is initially caused by over-advanced
ignition spark timing.

spreads through the chamber and the piston moves farther up. When the piston gets closer to the top, early combustion spreads farther and combustion gets steadily faster and more violent. There is more heat and pressure to speed up the burning process.

Finally, in Fig. 5-13D, combustion is rapidly completed and maximum pressure is developed in the cylinder. Unfortunately, the piston is still traveling up in the cylinder. The premature combustion tries to push the piston down against normal rotation. A pressure wave of several tons striking the piston can rattle engine parts and produce a ping or knocking sound.

Sometimes, as mentioned earlier, early spark timing can cause detonation. It is hard to distinguish the various types of engine knocks by sound alone. If retarding the ignition timing cures or corrects a pinging problem in an engine, your problem was primarily spark knock.

FACTORS AFFECTING COMBUSTION

Numerous factors control combustion and determine whether it will take place normally or abnormally. In general, the objective of good combustion is to utilize as much of the fuel's heat energy as possible, to burn the fuel as fast as possible (short of detonation), and to produce as little exhaust emissions as possible. As you will learn, there are many interacting factors affecting combustion.

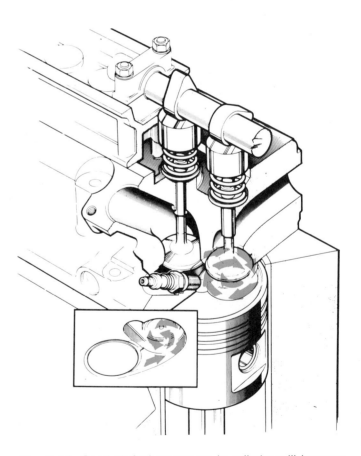

Fig. 5-14. Good air-fuel movement in cylinder will improve
combustion. (Jaguar)

Combustion chamber turbulence (swirl)

Air and fuel movement or *turbulence* in the cylinder will normally speed up and improve combustion efficiency, Fig. 5-14. Turbulence will mix or stir the fuel mixture and expose more of the unvaporized (unbroken) fuel droplets to the combustion flame. With more fuel area touching the hot flame, combustion will be faster and cleaner (less exhaust pollution).

For instance, if you stir up a pile of smoking leaves (slow combustion), a flame can reappear and speed up the burning process. You have exposed more of the unburned leaves to the hot materials and oxygen.

Spark plug location

Ideally, a spark plug should be centrally located in the combustion chamber. With the plug in the center of the chamber, the distance from the plug electrodes (start of combustion) to the farthest point in the chamber will be at a minimum. This speeds up combustion because the flame will have the shortest possible distance to travel.

For example, Fig. 5-15 shows that in an engine with a 4 in. cylinder bore, a centrally located spark plug would have a maximum flame travel distance of around 2 1/4 in. If the plug were located on one side of the combustion chamber, the distance would be increased up to 3 in. more. As a consequence, combustion would take longer and more heat energy would be absorbed into the metal walls of the cylinder, combustion chamber, and piston head.

Combustion chamber surface area

A *small surface area* in a combustion chamber, even with a constant compression ratio, will improve the efficiency of combustion. For example, look at Fig. 5-16. A sphere or dome would be an excellent combustion chamber shape because it would have a small surface area. By having less area to absorb heat, there will be more heat remaining in the burning fuel to cause expansion, pressure, and power. Notice that the irregular chamber has more surface area than the smoother domed chamber.

Combustion chamber temperature

Combustion is improved, in general, as the *temperature* of the *combustion chamber,* Fig. 5-17, is increased. This is true as long as the temperature is not

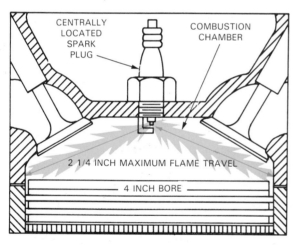

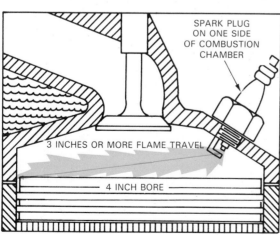

Fig. 5-15. Upper chamber has much shorter flame travel distance than lower one. A short distance improves and speeds up combustion.

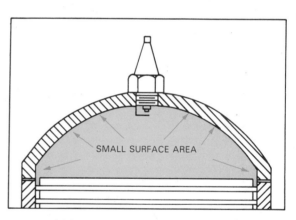

GOOD COMBUSTION CHAMBER SHAPE

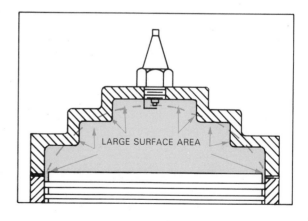

POOR COMBUSTION CHAMBER SHAPE

Fig. 5-16. Domed chamber (hemispherical) in upper illustration would naturally have less heat-absorbing surface area than lower, irregular combustion chamber.

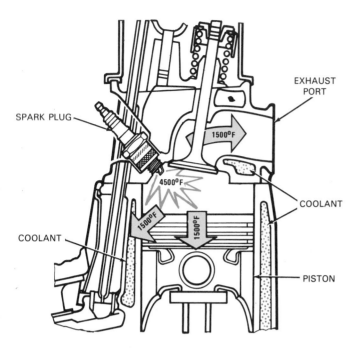

Fig. 5-17. Combustion chamber temperatures are kept high for efficient engine operation. Temperatures reach 4500 °F at combustion but average 2000 °F. (Chrysler Corp.)

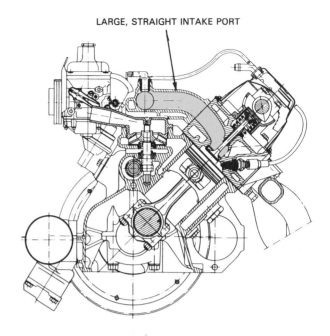

CROSS SECTION OF ENGINE AT WATER PUMP

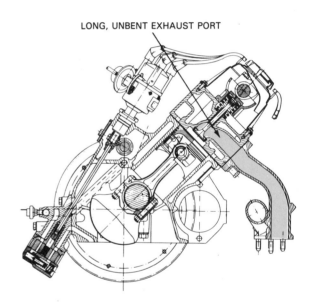

CROSS SECTION OF ENGINE AT DISTRIBUTOR

Fig. 5-18. This engine is designed for power. It has large, straight intake and exhaust ports that allow the engine to "breathe" at high speeds. (Saab)

high enough to cause surface ignition. If the chamber wall is cool, fuel near the cold metal surface will not vaporize and burn properly. This unburned fuel can leave the engine as wasted energy and pollution.

Combustion chamber temperatures are affected by compression ratio, cooling system design, air-fuel ratios, and ignition timing. Normally, a combustion chamber will average a little under 1500 °F (816 °C) at the inner wall of the cylinder during moderate engine operation.

Compression ratio

The compression pressure or *compression ratio* of an engine is very important to combustion. Generally, a high compression ratio will increase engine power and fuel economy, but will increase harmful exhaust emission (NOX) levels.

Valve and port design

To speed up combustion, improve fuel economy, and increase power at low speeds, engine *valves* and cylinder head *ports* should be designed to accelerate flow in and out of the combustion chamber. The increased velocity (speed) will improve the fuel and air mix, Fig. 5-18. It will also help pack more fuel mixture into the combustion chamber.

Piston head shape

Since the piston *head* serves as one surface of the combustion chamber, it is very important that it be *shaped* to work with the cylinder head. A critical period of combustion occurs while the piston is at TDC. If the piston head gets too close to the cylinder, it can restrict combustion by either blocking off areas or by quenching

(cooling) a portion of the fuel mixture.

Any area not easily reached by the flame can lower combustion efficiency. Also, these areas can get an uneven mixing of fuel. Thus, they may ignite and burn poorly.

The two combustion chamber illustrations in Fig. 5-19 show the effects of piston designs. Another example of a piston reshaped to a smooth "rolling surface" with no sharp edges or deep sharp grooves is pictured in Figs. 5-20 and 5-21.

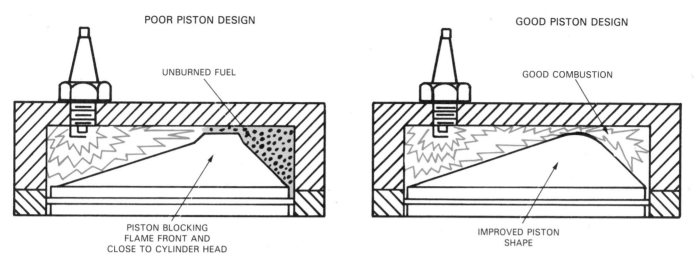

Fig. 5-19. The poor piston-to-chamber design temporarily prevents combustion from spreading through all of the fuel mixture. The chamber on the right has been improved by reshaping the piston.

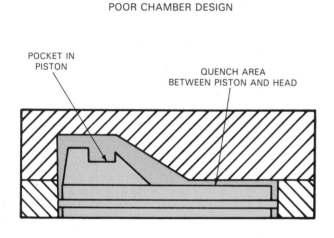

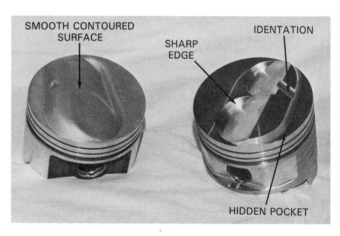

Fig. 5-21. Notice how piston on left has been machined to a smooth contour. This will normally improve combustion. (Zorian)

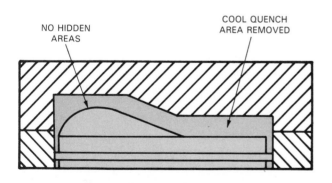

Fig. 5-20. Top. This combustion chamber has a quench area between piston and cylinder head where combustion can be incomplete. It also has a pocket in piston that can suffer from poor combustion. Bottom. Same combustion chamber is improved by reshaping piston. Cool quench area and hidden pockets have been removed.

Spark timing

Spark or *ignition timing* is another important condition controlling combustion efficiency. Generally, when ignition timing is advanced without harmful knock, engine power will be at maximum.

At the other extreme, too little spark advance, called *spark retard,* will waste energy letting combustion heat leave the engine unused, Fig. 5-22.

With *late spark timing,* much of the heat will be passed into the engine block and exhaust system instead of being used to power the engine. In fact, late ignition timing can cause an engine to overheat so much that the exhaust manifolds crack. Some spark retard, however, is helpful in reducing exhaust pollution.

Spark intensity and duration

The *size* and *duration* of the *spark* at the spark plug electrodes, Fig. 5-23, is very important to combustion,

Fig. 5-22. When spark timing is late (retarded), much of the combustion heat and flame can blow out exhaust port as wasted energy.

especially with today's lean fuel mixtures. Such mixtures are hard to ignite and keep burning. This is the main reason for wider spark plug gaps and high ignition voltages. A wider plug gap requires more secondary ignition voltage (40,000 V plus) to sustain an arc across the spark plug electrodes. The resulting spark has more energy or intensity to start and maintain the combustion of a lean fuel mixture.

Spark plugs gapped too narrow would cause the engine to have a lean "misfire" (commonly a miss at moderate to high engine speeds). The smaller plug gap could not produce a "hot" enough spark to ignite the lean mixture.

For complete fuel burning, it is important to "hold" or sustain a spark for a definite period of time (about 1/3 of time it takes for combustion). Normally, a spark should occur at a spark plug for at least one millisecond (1/1000 of a second); .9 millisecond may be too short.

Note! Some engine analyzers are capable of measuring the duration of the spark at a spark plug.

RICH FUEL MIXTURES

When gas molecules are closely packed together, as with richer fuel mixtures, the spark needs to start only a few burning. Each burning gas molecule will likely propagate to the next and so forth until most or all gas molecules have been ignited.

This self-propagation only requires that the spark occurs at the right time. The combustion process will take care of itself.

LEAN FUEL MIXTURES

With lean fuel mixtures the gas molecules are scattered farther apart in the combustion chamber. When the spark occurs a few gas molecules will ignite. However, if the spark immediately quits those gas molecules that have been ignited will likely burn out before they have had a chance to transmit the flame to others. The result is obvious . .
    A MISFIRE

If the spark is sustained, the turbulence designed into the combustion chamber will keep pushing gas molecules into the spark until enough are burning to complete the combustion.

FOR THIS REASON PLUG FIRING TIME NEEDS TO BE MEASURED.

Fig. 5-23. Plug firing time is important where fuel-air mixture is lean.

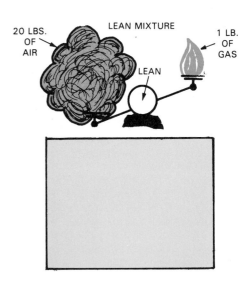

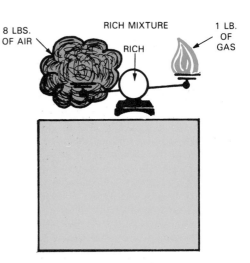

Fig. 5-24. Top. Lean fuel mixture has fewer molecules of fuel in suspension. Bottom. Rich fuel mixture has many more fuel molecules in suspension.

Air-fuel ratio

The amount of fuel compared to air entering an engine is called the *air-fuel ratio.* The proportions can be adjusted to meet the engine's changing needs.

A *lean fuel mixture,* containing a lot of air compared to fuel, will give better fuel economy and fewer exhaust emissions.

A *rich fuel mixture,* with a larger percentage of fuel in it, improves engine power and starting (cold engine operation); however, it will increase emissions, Fig. 5-24.

Air-fuel ratios affect the speed of flame travel during combustion. Flame travel is quite slow whenever the mixture is extremely lean or rich.

A chemically correct fuel ratio, is called a *stoichiometric mixture.* It is a theoretically ideal ratio of around 14.9:1 (14.9 parts air to 1 part fuel). Under steady-state engine conditions (unchanged engine load, temperature, fuel distribution, etc.) this ratio of fuel to air would assure that all of the fuel will blend with all of the air and be burned.

Multicylinder automobile engines, however, do not operate under ideal conditions. They suffer from variations in fuel delivery from cylinder to cylinder, as well as differences in compression, temperature, and timing. For this reason, a stoichiometric mixture is not always desirable as shown in Fig. 5-25.

Intake manifold pressure

The amount of *negative pressure* (vacuum) inside the intake manifold of an engine also affects combustion. It regulates how much fuel enters the engine. A low intake manifold pressure (high vacuum) pulls large amounts of fuel into the combustion chamber, Fig. 5-26. This increases power output. On the other hand, high intake manifold pressure (low vacuum) reduces the power developed by decreasing the quantity of fuel and air entering the cylinders.

A high negative intake manifold or inlet pressure occurs when the gas pedal of a car is pushed to the floor. This opens the throttle or air valves completely permitting full atmospheric pressure (14.7 psi at sea level) to

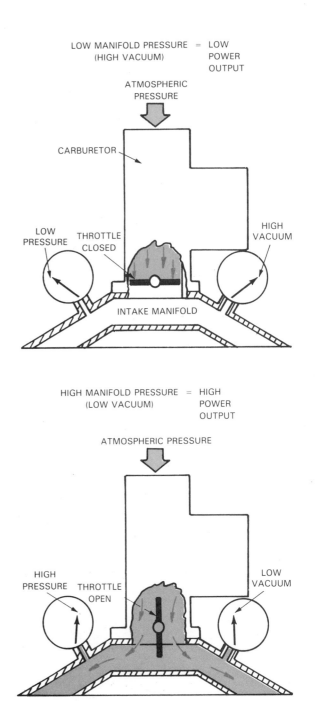

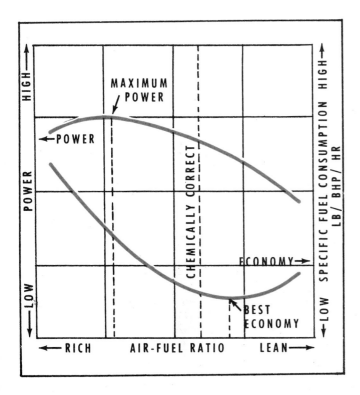

Fig. 5-25. Power is improved with a rich mixture. Economy is improved with a lean fuel mixture. A chemically correct mixture would be between these two. (Ethyl Corp.)

Fig. 5-26. As manifold pressure increases, more fuel is drawn into combustion chamber. Engine speed and power then increase.

push the fuel into the engine cylinders.

As will be discussed in Chapter 22, supercharging raises the maximum intake manifold pressure higher than at atmospheric pressure. This increases the fuel charge entering the engine and raises the compression pressure before combustion. As a result, supercharging can increase engine power by as much as 50 percent.

Cylinder bore and stroke

To a certain degree, a smaller cylinder bore and short stroke will improve combustion efficiency. With a smaller bore, the distance the combustion flame must travel is reduced. As a result, combustion speed increases, heat loss is cut, and efficiency is improved.

With a short stroke, less energy is lost to friction (rings rubbing against cylinder walls). More power is available to spin the crankshaft.

Fuel delivery system

The type of *fuel delivery system* will have a slight effect on combustion. Fuel injection improves combustion for two basic reasons. Injection systems allow more air to enter the cylinders, and have more control over the amount of fuel entering each cylinder. Gasoline injection is covered in later chapters.

Valve timing

Valve timing is another important factor controlling combustion. For instance, the intake valve is quite often held open a little past BDC, during the start of the compression stroke. The inertia (tends to keep flowing) of the fuel mixture in the intake manifold port will help to

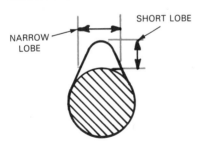

SMALL LIFT, SHORT DURATION "CAM"

SHORT LOBE

NARROW LOBE

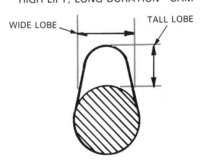

HIGH LIFT, LONG DURATION "CAM"

WIDE LOBE

TALL LOBE

Fig. 5-27. The "stock" cam in the upper drawing opens the valves less and for a shorter period of time than the lower "racing" cam.

force more fuel and air into the cylinder. This increases combustion power.

This principle is also used by the exhaust valve to draw out a little more of the burned gases. This is called *scavenging.* It allows more room for the fresh fuel mixture. Also, valve timing can be used to "clean up" combustion by retaining some of the exhaust gases in the engine cylinders where they can limit peak combustion temperature.

Valve lift and duration

Valve lift (how far the engine valves open), and *valve duration* (how long the valves stay open) can have a pronounced effect on combustion, engine power, fuel economy, and drivability. Generally, lift and duration determine at what engine speed best performance will be developed.

In Fig. 5-27, valve lift is determined by the height of the camshaft lobe (centerline of cam to lobe tip). A tall lobe will have a high lift. Valve duration is controlled by the width of the cam lobe; the wider the lobe, the longer it will keep the valve open.

Stock cam

In most *stock,* unmodified engines, lift and duration specs are designed to give maximum combustion efficiency (combustion power, fuel economy, and cleanliness) between idle speed (around 800 rpm) and highway speed (around 55 mph or 3000 rpm).

Race cam

In a race engine designed for maximum power and maximum engine speeds, valve lift and duration will be increased. This will open the valves wider and longer to help the engine "breathe" (take in more fuel mixture) at high speeds (6000 to 10,000 rpm). It will increase high speed horsepower, but at the cost of fuel economy and emissions.

Note, however, that a *race cam* will usually DECREASE engine power, fuel economy, and exhaust cleanliness at normal engine speeds. The engine will operate inefficiently from idle up to normal highway speed. In fact, if a race cam is installed without other engine modifications (exhaust headers, increased compression ratio, etc.), usable engine power will normally be reduced at normal operating rpm.

Intake air temperature

The temperature of the air entering the air cleaner has some effect upon combustion. Generally, cool air increases engine performance. Cool air is more dense or compact than hot air. Therefore, it carries more power-giving oxygen into the cylinders.

Most automobiles use a thermostatically controlled air cleaner which maintains a constant inlet air temperature of between 80 °F and 110 °F (27 °C and 43 °C). This intake air temperature will provide a happy medium between engine power and low exhaust emissions.

COMBUSTION CHAMBER DESIGN

The *combustion chamber* is the space above the piston where fuel is burned. Various combustion chamber designs are in present use. All try to burn as much of the fuel as possible. They try to reach maximum combustion efficiency and exert maximum combustion pressure on the top of the piston. The most frequently used designs are the wedge, the hemispherical, the jet, and the stratified charge combustion chambers.

Wedge combustion chamber

As its name implies, the *wedge combustion chamber* takes on the shape of a wedge or triangle, Fig. 5-28. It is a commonly used chamber which uses ''squish'' turbulence to assist in fuel mixing, knock reduction, and low speed emission control. When the piston is at TDC, part of the piston comes very close to the cylinder head. This forms a squish or squeeze area. The fuel mixture in the squish area is squirted or blown into the main chamber which swirls the gases in the chamber and improves burning.

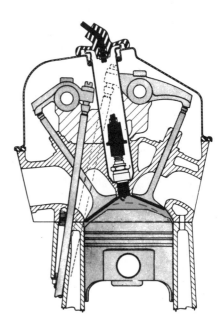

Fig. 5-29. A ''hemi'' combustion chamber takes shape of dome. (Chrysler Corp.)

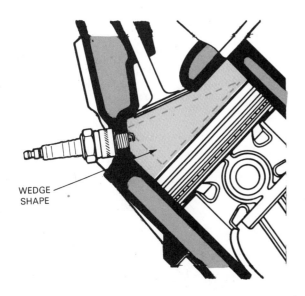

Fig. 5-28. Wedge type combustion chamber is actually shaped like a wedge. (Champion Spark Plug Co.)

Hemispherical combustion chamber

A *hemispherical combustion chamber,* or ''HEMI,'' is a dome shaped combustion chamber. See Fig. 5-29.

The smooth, symmetrical or uniform shape is extremely resistant to knock, even under high pressure conditions. A hemispherical chamber allows the use of a centrally located spark plug. It produces a relatively short, straight, and equal path for combustion. This speeds up combution to increase power and economy. It has no hidden pockets where fuel can hide from the flame front. This also improves efficiency, power, fuel mileage, and exhaust emissions.

The hemi is one of the most popular chamber configurations for all-out racing engines. Auto makers are beginning to take advantage of its operating characteristics in their small, high mileage, unleaded fuel-

burning, low octane engines. With the valves located on each side of the chamber at an angle, it is an ideal chamber for an overhead cam engine.

''Jet'' combustion chamber

A *jet combustion chamber* uses an extra valve to increase chamber turbulence, fuel mixing, combustion cleanliness, fuel economy, and to reduce knock tendencies. As pictured in Fig. 5-30, the engine's third valve injects a high velocity stream of air/fuel mixture into the cylinder. The high speed blast of fuel mixture blows into and stirs up the conventional mixture already there. The air jet action causes mixture swirling and a more even distribution of fuel in the chamber. This prevents lean or rich areas that can reduce combustion efficiency.

Stratified charge combustion chamber

A *stratified charge engine* utilizes two distinct combustion chambers and fuel mixtures to improve combustion. Fig. 5-31 illustrates a stratified charge engine. The combustion chamber has a small cup-shaped cavity known as a prechamber. It contains a very small intake valve and is connected to the main combustion chamber. This prechamber has a passage running up to the carburetor. It supplies a rich, easy-to-ignite fuel mixture of around six parts air to one part fuel.

In engines having a stratified charge chamber, the main combustion chamber operates on a fuel mixture of around 20 to 1. Normally, this extremely lean mixture would be very hard, if not impossible, to ignite and burn. However, when the spark plug ignites the rich mixture in the prechamber, flames shoot out into the main combustion chamber and easily ignite the lean mixture. The combustion efficiency of a stratified charge type engine is excellent. Exhaust emissions are very low. Power is good and fuel economy is outstanding.

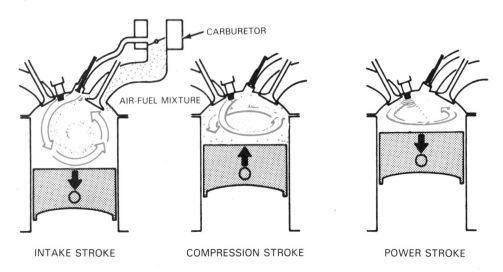

Fig. 5-30. Small ''jet'' valve in this chamber squirts a stream of fuel mixture into the cylinder. This helps mix the fuel and air. (Chrysler Corp.)

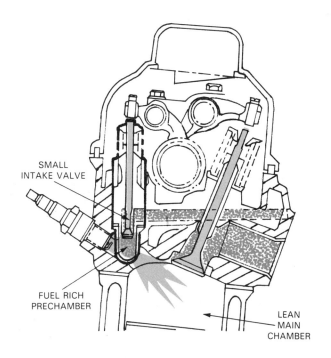

Fig. 5-31. This engine has two separate combustion chambers. They allow use of ''super-lean'' mixture. (Champion Spark Plug Co. and Honda)

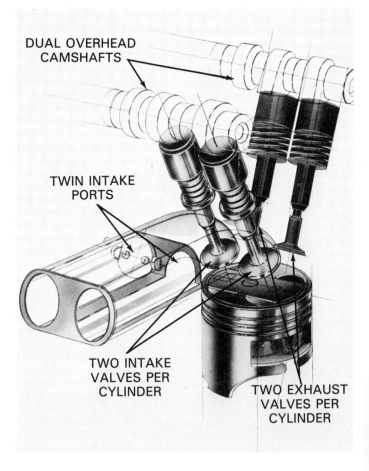

Fig. 5-32. This modern engine uses four valves per cylinder. (Toyota)

Four-valve combustion chamber

A *four-valve combustion chamber* uses two intake valves and two exhaust valves per engine cylinder. This setup is used in a few late model, high performance engines to increase combustion efficiency and power output. One is shown in Fig. 5-32.

Four valves and dual ports are used for several reasons. They increase engine breathing ability without sacrificing velocity. Very large single valves and ports would be needed to flow the same amount of air. This would decrease velocity and hurt mixture atomization.

Also, four, smaller valves allow a better combustion chamber shape. The valves do not have to open as wide and the pistons do not have to have deep flycuts to allow adequate valve-to-piston clearance. This produces a compact ''pocket'' (combustion chamber) directly over the center of the piston to utilize more combustion pressure and to produce more complete combustion.

DIESEL COMBUSTION

In many respects, a four-stroke cycle automotive diesel engine is very similar to a four-stroke cycle gasoline engine, Fig. 5-33. Since diesel engines burn a diesel oil, their fuel must be ignited by the heat of extreme air compression. A spark plug cannot adequately ignite diesel fuel.

A

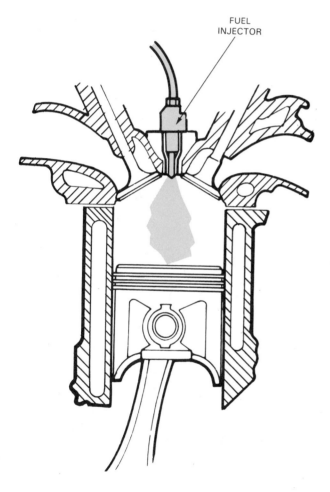

Fig. 5-33. Cutaway of diesel shows hemispherical combustion chamber and placement of fuel injector in center of chamber. (Federal Mogul)

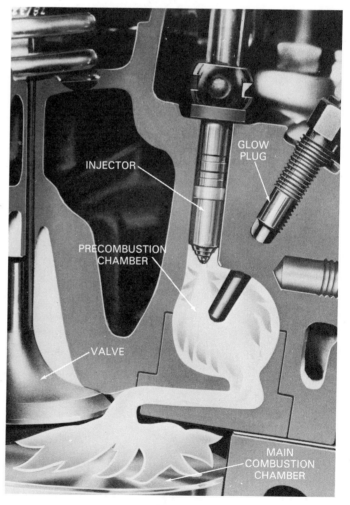

B

Fig. 5-34. A—Note precombustion chamber and flat, high compression combustion chamber. B—Combustion taking place in diesel combustion chamber. Heated air in precombustion chamber ignites fuel as it is injected. Combustion travels down into main combustion chamber. (Fel-Pro and GM Trucks)

Air alone is compressed in the cylinder until it is hotter than the ignition point of the fuel. Then, the fuel is injected or sprayed into the hot air which causes ignition. Termed compression ignition, this assures thorough combustion of the diesel fuel, Fig. 5-34. Gasoline cannot be ignited by compression ignition because it would ignite, detonate, and knock.

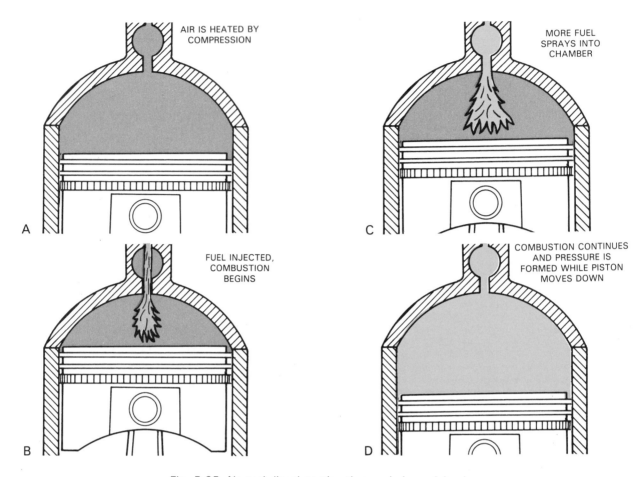

Fig. 5-35. Normal diesel combustion cycle is explained.

NORMAL DIESEL COMBUSTION PHASES

The phases of diesel combustion are illustrated in Fig. 5-35. In Fig. 5-35A, air with no fuel has been compressed in both the precombustion chamber and main cylinder by the upward movement of the piston. The pressure has heated the air to around 1000 °F (538 °C). (The small chamber in the cylinder head is used in most automotive diesel engines to quiet and smooth out engine operation.)

In Fig. 5-35B, fuel injection begins. As soon as the fuel is heated enough by the hot air, it vaporizes and combustion begins. As the fuel begins to burn and expand in the small precombustion chamber, extreme pressure is formed.

Then, in view C, prechamber pressure has "blown" flames and raw fuel into the main combustion chamber. The explosion of flames and expanding gas causes swirling or turbulence in the main chamber. Simultaneously, more fuel is being injected into the chamber. As a result, combustion continues until all of the fuel has been consumed, Fig. 5-35D.

Note that the piston is a considerable distance down in the cylinder while combustion is still taking place. This continued combustion tends to hold a steady or constant pressure on the top of the piston. It smooths out and quiets combustion noise.

ABNORMAL COMBUSTION (DIESEL KNOCK)

The combustion process in a diesel must be carefully controlled. If all of the fuel were ignited in the cylinder at one time, an explosion and knock would result.

IGNITION LAG (the time it takes the fuel to heat and vaporize before combustion) is one factor controlling knocking and detonation in a diesel engine. If the lag is too long and the fuel vaporizes too slowly, a large amount of fuel can be present in the cylinder at one time. As a consequence, when the fuel does ignite, detonation and a violent explosion can occur. A high cetane fuel with a short lag period helps prevent diesel knock.

DETONATION DAMAGE

Detonation in a diesel is not quite as damaging as in a gasoline engine because of the heavy design of a diesel engine. However, detonation will lower engine power, fuel economy, and shorten engine life.

DIESEL DETONATION PHASES

Fig. 5-36 illustrates how knock or *detonation* occurs in a diesel engine. It is quite different from gasoline engine knock or ping.

In Fig. 5-36A, air is being squeezed and heated. There

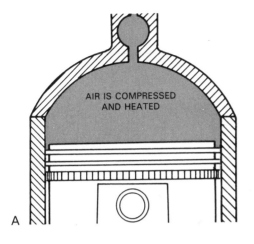

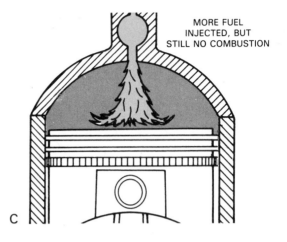

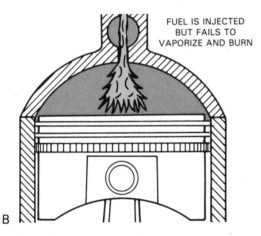

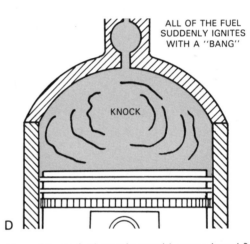

Fig. 5-36. Phases of diesel detonation. Do you see why ignition of large fuel supply would cause knock?

is no fuel in the combustion chamber at this time.

Fig. 5-36B shows the fuel being injected into the cylinder. Notice that the fuel is neither vaporizing nor burning, but is having IGNITION LAG or ignition delay.

In Fig. 5-36C, more fuel is sprayed into the engine. Yet, it still does not begin to burn.

Finally, in Fig. 5-36D, all of this fuel ignites with a loud bang or knock. Detonation was caused by the ignition and burning of *too much fuel.* This situation could have been created by a low cetane fuel, an improper fuel spray pattern (clogged injector nozzle), improper injection timing, etc.

DIESEL COMBUSTION FACTORS

Many factors affect combustion in a diesel engine. Some are very similar to gasoline combustion while others are not. In the following paragraphs, we will discuss these factors and how they change diesel engine power, economy, and exhaust emissions.

Combustion power

The power and speed of a diesel engine is controlled by the amount of fuel injected into the cylinders. More fuel produces more power and speed. The amount of air entering the engine stays about the same at all engine speeds. This allows an excess or extra amount of air to enter the cylinders at all times, except top speed. Since oxygen is one of the limiting factors on the power and completeness of combustion, the extra supply of air during diesel combustion is one of the factors increasing diesel engine efficiency over a gasoline engine.

Combustion efficiency

The other major factors that increase the fuel economy of a diesel engine include: a "super lean" air-fuel ratio (little fuel compared to air), extremely high compression ratio, and high heat value of diesel fuel.

Air-fuel ratios in a diesel engine vary from around 100:1 (100 parts air to 1 part fuel) at idle speed to 20:1 at full load. By comparison, a gasoline engine has an air/fuel ratio of between 18:1 and 12:1 which is much richer and less fuel efficient.

This large quantity of air in a diesel assures that all of the fuel is consumed. It also aids combustion by absorbing some of the heat of the burning fuel mixture. The extra air, when heated, expands greatly to help form more pressure, engine power, and fuel economy.

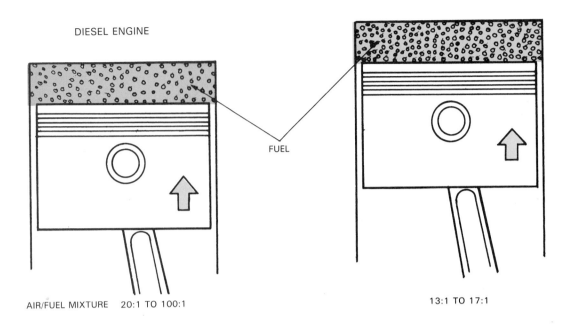

DIESEL ENGINE

GASOLINE ENGINE

FUEL

AIR/FUEL MIXTURE   20:1 TO 100:1

13:1 TO 17:1

Fig. 5-37. Lean diesel fuel mixture accounts for low hydrocarbon and carbon monoxide emissions.

The exhaust emissions of a diesel engine are also quite different from that of a gasoline engine. In basic terms, a diesel engine produces few hydrocarbon and carbon monoxide emissions but a large quantity of smoke and oxides of nitrogen. See Fig. 5-37. Many of the emission control systems used on gasoline engines are not needed on a diesel.

SUMMARY

Fuel must be mixed with air for good combustion. The piston sliding downward on the intake stroke sets up an airstream into the combustion chamber which is designed to mix the air and fuel. Then, when the mixture is compressed and ignited, power is produced. The piston, under force from expanding gases, spins the crankshaft and propels the auto.

Chemically, combustion is fast oxidation of fuel and oxygen. The process is far from perfect in the engine and causes air pollution. Combustion must be carefully controlled to reduce these emissions.

Combustion will be normal if all of the factors are correct. A single flame will spread evenly and uniformly away from the spark plug electrode to form maximum pressure when the piston is slightly past TDC.

Abnormal combustion (ping or detonation) is the result of uneven burning of the fuel-air mixture. This condition can cause engine damage if severe enough and if the condition continues over a period of time.

Factors important for good combustion are spark timing, spark intensity and duration, air-fuel ratio, intake manifold pressure, cylinder bore and stroke, fuel delivery system, valve timing, valve lift and duration, and intake air temperature.

Combustion chamber design can also improve fuel combustion in an engine. Important designs include the wedge, hemispherical, jet, stratified charge combustion chamber, and four-valve chamber. Piston head shape also influences combustion.

Diesel engines do not use spark plugs or a carburetor. Rather, they compress the air in the cylinder until it is hot enough to ignite the diesel oil. Diesel combustion begins when the fuel is sprayed into this hot air.

Diesel knock is caused by too much fuel igniting in the cylinder at one time. When ignition lag is too long, or fuel has a low cetane value, the large quantity of fuel can ignite violently and vibrate engine parts. A pronounced knock or rattle can be heard. Since a diesel engine is constructed much heavier than a gasoline engine, detonation damage is not usually very severe.

The power and speed of a diesel engine is controlled by the amount of fuel sprayed into the combustion chamber. More fuel produces more power. The amount of air entering a diesel cylinder is almost constant from high to low speeds. Generally, combustion efficiency in a diesel is high because of the lean fuel mixture, high compression ratio, and because of the high heat content of diesel oil.

KNOW THESE WORDS

Air stream, Atomization, Ideal combustion, Detonation, End gas, preignition, Postignition, Wild knock, Dieseling, Spark knock, Stoichiometric mixture, Scavenging, Wedge combustion chamber, Hemispherical combustion chamber, Jet combustion chamber, Stratified charge combustion chamber, Four-valve chamber.

REVIEW QUESTIONS—CHAPTER 5

1. Proper combustion of any automotive fuel requires that the fuel be mixed with _____ and _____ in the cylinder before it is _____.

2. It is the job of the carburetor or injector to break the fuel into small droplets that will mix with _____. This process is called _____.

3. The temperature increase and expansion during combustion will form (vacuum, pressure).

4. A piston with about a 4 in. diameter will have tremendous pressure exerted on it by the expansion of the ignited air-fuel mixture. The pressure could be equal to:
   a. 500 lb.
   b. 5000 lb.
   c. 5 ton.
   d. 50 ton.

5. Pollution from engine exhaust is the result of:
   a. Unburned fuel.
   b. Nitrogen and other gases (besides oxygen) in the air that mixes with the fuel.
   c. Both of the above.
   d. None of the above.

6. _____ combustion occurs in a gasoline engine when a single flame spreads evenly and uniformly away from the spark plug electrodes to form maximum pressure.

7. Normal combustion takes about _____ of a second and is much (faster, slower) than exploding dynamite.

8. Define detonation.

9. A _____ _____ sounds somewhat like a small ball peen hammer rapidly striking the pistons or engine block.

10. Detonation is not at all harmful to an engine. True or false?

11. What is end gas and what role does it play in detonation?

12. During detonation, cylinder pressure can increase by _____ percent.

13. Severe engine knock is most evident when:
   a. Engine is idling.
   b. During maximum acceleration.
   c. During deceleration.
   d. At high cruising speed.

14. List eight causes of detonation.

15. What is preignition?

16. Which of the following cannot be caused by severe preignition:
   a. Bent connecting rod.
   b. Bent push rod.
   c. Burned piston.
   d. Cracked cylinder head.

17. Spark knock is (check all correct answers):
   a. The same as preignition.
   b. A ping in the engine caused by the electric arc occuring too early at the spark plugs.
   c. Caused by timing being retarded.
   d. Caused by timing being advanced too far.
   e. All of the above.
   f. None of the above.

18. What is the difference between spark knock and preignition?

19. List eight causes of preignition.

20. Wild knock is commonly caused by a hot _____ of _____ bouncing around in the cylinder.

21. What is dieseling and what causes it?

22. The most common cause of engine dieseling is a _____ _____ setting.

23. Check which of the following are objectives of getting good combustion.
   a. Get all the heat energy possible out of the fuel.
   b. Burn the fuel as fast as possible.
   c. Produce as few exhaust pollutants as possible.
   d. None of the above.

24. Air and fuel turbulence in the cylinder will (speed up and improve, interfere with) combustion efficiency.

25. Ideally, a spark plug should be located in the _____ of the combustion chamber.

26. A large combustion chamber area cuts down on the efficiency of combustion. True or false?

27. A piston head which gets too close to the cylinder head at TDC will _____ _____ by either blocking off areas of the chamber or by quenching (_____) a portion of the fuel mixture.

28. Which of the following conditions will result from late timing of the spark?
   a. Engine power will be reduced.
   b. Engine will overheat and may even cause damage to the exhaust manifold.
   c. Exhaust pollution is reduced.
   d. Gas mileage will be poor.
   e. Preignition will cause engine damage.
   f. All of the above.
   g. None of the above.

29. A _____ air-fuel mixture is one theoretically ideal; that is, it assures that all the fuel will blend with the air and will be burned during combustion.

30. _____ _____ _____ refers to the amount of vacuum inside the intake manifold.

31. A racing cam in an engine will usually (decrease, increase) engine power, fuel economy, and exhaust cleanliness at normal engine speeds.

32. Explain how a wedge type combustion chamber operates.

33. A combustion chamber utilizing two separate chambers and mixtures of fuel is called a _____ _____ combustion chamber.

34. Ignition _____ is one factor controlling knocking and detonation in a diesel engine.

35. Describe the detonation sequence in a diesel engine.

36. The power and speed of a diesel is controlled by the amount of _____ _____ into the cylinders.

37. List three major factors that increase the fuel economy of a diesel engine over that of a gasoline engine.

NORMAL COMBUSTION

A combustion process which is initiated solely by a timed spark and in which the flame front moves completely across the combustion chamber in a uniform manner at a normal velocity. In such a process there is no sudden release of energy from the fuel-air mixture, nor are there any auxiliary sources of ignition from combustion-chamber deposits, hot spark plugs, overheated valves, or other hot surfaces within the combustion chamber. Engine roughness associated with high gas loads and mechanical deflections of engine components can accompany normal combustion.

ABNORMAL COMBUSTION

A combustion process in which a flame front may be started by hot combustion-chamber surfaces either prior to or after spark ignition, or a process in which some part, or all, of the charge may be consumed at extremely high rates. This term, therefore, includes any surface ignitions of the charge, and it includes ordinary knock or knock which is induced by surface ignition phenomena, either prior to or after spark.

SPARK KNOCK*

A knock which is recurrent and repeatable in terms of audibility. It is controllable by the spark advance; advancing the spark increases the knock intensity and retarding the spark reduces the intensity. This definition does not include surface-ignition induced knock.

SURFACE IGNITION
(HOT SPOTS—COMBUSTION-CHAMBER DEPOSITS)

Surface ignition is the initiation of a flame front by any hot surface other than the spark discharge prior to the arrival of the normal flame front. The flame front or fronts so established propagate at normal velocities. This phenomenon can be further subdivided into preignition and postignition.

PREIGNITION

Surface ignition _before_ the occurrence of normal spark.

POSTIGNITION

Surface ignition which occurs _after_ the passage of the normal spark.

KNOCKING*
SURFACE IGNITION

Knock which has been preceded by surface ignition. It is not controllable by spark advance. It may or may not be recurrent and repeatable.

RUN-ON

Continuation of engine firing after the electrical ignition is cut.

NON-KNOCKING
SURFACE IGNITION

Surface ignition which does not result in knock.

RUNAWAY SURFACE IGNITION

Surface ignition which occurs earlier and earlier in the cycle. This phenomenon is generally caused by overheated spark plugs, valves, or other combustion-chamber surfaces. Generally it is not caused by floating deposit particles or deposits adhering loosely to the combustion-chamber walls. This is the most destructive type of surface ignition. It can lead to serious overheating and structural damage to the engine.

WILD PING

Knocking surface ignition characterized by one or more erratic sharp cracks. It probably is the result of comparatively early surface ignition from deposit particles.

RUMBLE

A low-pitched thudding noise different from knock and accompanied by engine roughness. One of the causes probably is the high rates of pressure rise associated with very early ignition or multiple surface ignition.

---

*KNOCK—The noise associated with autoignition[1] of a portion of the fuel-air mixture ahead of the advancing flame front. The flame front is presupposed to be moving at normal velocity. With this definition the source of the normal flame front is immaterial—it may be the result of surface ignition or spark ignition.

[1]AUTOIGNITION—The spontaneous ignition and the resulting very rapid reaction of a portion or all of the fuel-air mixture. The flame speed is many, many times greater than that which follows normal spark ignition. There is no time reference for autoignition.

Chart of abnormal diesel combustion symptoms and causes. (Coordination Research Council)

6

FUEL TANKS AND GAUGES—
OPERATION, SERVICE, REPAIR

After studying this chapter, you will be able to:
- *Describe the construction of modern fuel tanks.*
- *Explain fuel tank design variations.*
- *Describe fuel tank filler neck construction.*
- *Explain fuel cap design variations.*
- *Summarize service procedures for fuel tanks and related components.*
- *List safety rules for fuel tank service.*
- *Describe fuel tank sending unit operation.*
- *Explain how to test and replace a fuel tank sending unit.*
- *Describe the operation of a fuel return system.*
- *Summarize the operating principles and service procedures for fuel gauges.*

This chapter will introduce fuel tanks, gauges and related parts, sealed fuel caps, vapor separators, rollover check valves, fuel return lines, vapor storage canisters, tank expansion chambers, vent systems, fuel gauge sending units, and low-fuel-level indicators. Since these components frequently require service, it is important for you to study this chapter carefully.

FUEL TANKS

As you can see in Fig. 6-1, fuel tanks are no longer the simple vented containers of a few years ago. They have evolved into a rather complicated system.

A *fuel tank* holds and stores the auto's fuel supply. The tank must be large enough to provide fuel for prolonged engine operation. Its size limits *''driving range''* (maximum miles a car can be driven between refills).

Normal fuel tank capacity averages around 12 to 25 U.S. gallons (45 to 95 liters) depending on vehicle size.

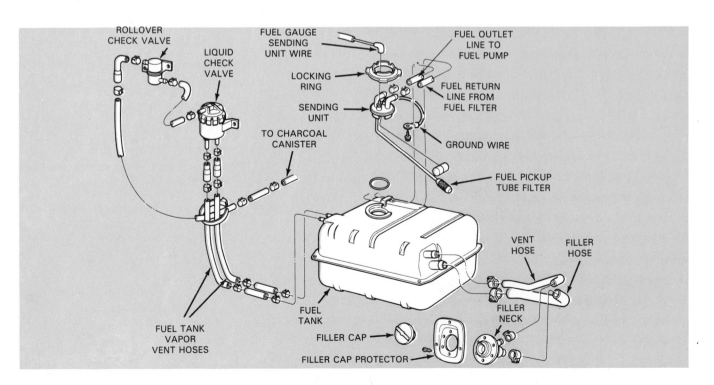

Fig. 6-1. Notice ''hardware'' attached to modern fuel tank. Most of these components help prevent gasoline fumes from entering atmosphere. (Chrysler Corp.)

107

FUEL TANK DESIGN

A well designed fuel tank should have the following features:

1. It should be light in weight, yet strong.
2. The tank should be shaped and located so that it uses a minimum amount of passenger and storage space.
3. It must be made of material that resists rust and corrosion.
4. The inside of the tank should contain metal separators or baffles to prevent splashing or sloshing.
5. It should be capable of relieving excessive pressure and vacuum.
6. There should be a system of trapping and storing fuel vapors.
7. The tank should let the fuel expand (heat expansion) without spilling.
8. It should have a filler opening and a connection for fuel pickup (removal).
9. A gas tank should have enough capacity (size or volume) for extended engine operation.
10. It must be safe from rupture during an auto accident.

FUEL TANK CONSTRUCTION

Gasoline tanks are usually made of thin sheet metal. Normally, two pieces of sheet metal form the body of the tank. These halves and other parts (filler neck, baffles, vent tubes, and expansion chamber) are welded or soldered together to form the complete fuel tank assembly. See Fig. 6-2. Sheet metal gasoline tanks are often coated with lead-tin alloy to prevent rust and corrosion, especially inside the tank.

A few fuel tanks are made out of plastic or aluminum. These materials are almost totally resistant to rust and corrosion. Also, they are very light and strong. Some plastic tanks are soft and pliable and can be severely

dented without leakage or permanent damage. A flexible plastic tank is pictured in Fig. 6-3.

Baffles

Fuel tank *baffles* or *surge plates,* pictured in Fig. 6-2, are commonly used in metal fuel tanks. They keep the fuel from splashing around inside the tank. Sheet metal baffles are normally welded inside the tank body. Holes and slots in them allow some fuel to flow from one end of the tank to the other.

Besides slowing down fuel movement, baffles help keep the fuel pickup tube submersed in fuel during hard braking and acceleration.

Air expansion space

Most new car gas tanks are shaped so that they cannot be completely filled. Around 10 percent *expansion space* is reserved to prevent fuel spillage from expansion of heated fuel. If a tank is full, it would overflow whenever the temperature of the fuel increased.

Fig. 6-3. An aftermarket plastic fuel tank. It is often used as an extra or "auxiliary" tank. (Econo-Tank)

Air expansion chambers

There are several methods of designing an air expansion chamber into a fuel tank. One of the more popular methods is shown in Fig. 6-4. The tank filler neck is simply located partway down from the top of the tank. It is positioned so that the tank cannot be completely filled. This leaves an air pocket in the top of the tank. This pocket is the fuel expansion chamber.

Sometimes a *FILL CONTROL TUBE* is added to this fuel level control. The fill control tube, shown in Fig. 6-5, will spill gasoline into the top of the filler neck. This action shuts off the gas pump.

External and internal air chambers

Some fuel tanks use a completely separate air chamber to allow for fuel expansion. An *external expansion chamber* is pictured in Fig. 6-6.

Illustrated in Fig. 6-7, an *internal expansion tank* is sometimes used. The smaller tank fits inside the main fuel tank. Its operation is basically the same as other types.

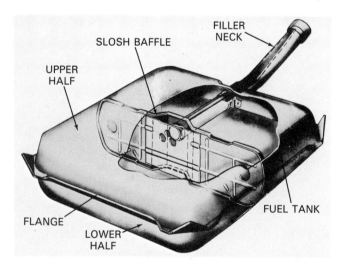

Fig. 6-2. Normally, the two halves, baffles, and filler neck are either welded or soldered together to form fuel tank assembly. (Cadillac)

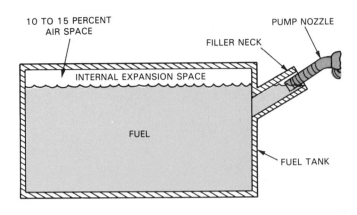

Fig. 6-4. Location of this fuel filler tube prevents upper portion of tank from filling. Air pocket in top of tank allows fuel expansion.

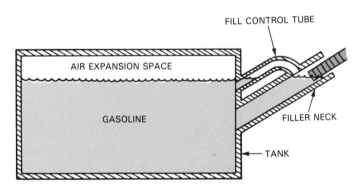

Fig. 6-5. Fill control tube is sometimes added to filler neck. It spills fuel onto pump nozzle and helps assure that air space is formed.

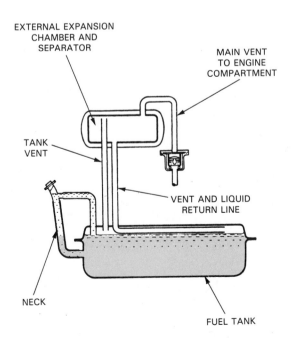

Fig. 6-6. This fuel tank design has separate or external air expansion chamber. Extra tank serves, in this case, as both fuel vapor separator as well as expansion chamber. (Chrysler)

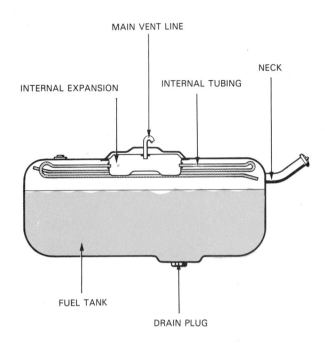

Fig. 6-7. Internal expansion tank also allows for heat and fuel enlargement, without fuel spillage. (Merdeces Benz)

Vent tubes may run to the expansion tank. Small orifices (holes) normally keep the expansion tank from filling. Tiny holes let fuel slowly expand and enter the expansion tank. However, they will not let fuel enter during tank filling. The gas station pump will shut off before the expansion chamber can fill.

FUEL FILLER PIPE

A gas tank normally has a pipe welded or soldered to it. Called a *fuel filler pipe,* this pipe or neck extends toward the outside of the car and permits the tank to be filled with fuel. It may extend all the way to the out-

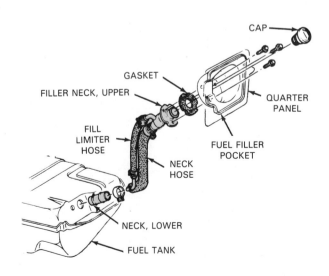

Fig. 6-8. This fuel filler neck uses neoprene connecting hose between upper and lower neck sections. To remove tank, simply remove lower hose clamp. (Chevrolet)

side of the body as one piece, or it may be only a few inches long.

A neoprene hose is sometimes used to connect the tank neck to the upper filler neck. This type is pictured in Fig. 6-8. A flexible hose allows slight movement between the tank and car body without stress on the tank neck. Usually, the neoprene hose is fastened to the tank and upper filler necks by hose clamps. Some filler pipes have a small fill control tube paralleling them.

FILLER NECK RESTRICTOR

All cars using a catalytic converter (emission reducing device in the exhaust system) and unleaded fuel, have a *filler neck restrictor.* This restrictor, Fig. 6-9, prevents leaded gasoline from being pumped into a car designed for no-lead fuel.

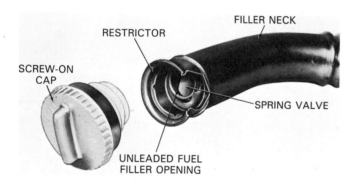

Fig. 6-9. Automobiles using unleaded fuel will have fuel neck restrictor plate and usually a vapor flap or valve.   (Cadillac)

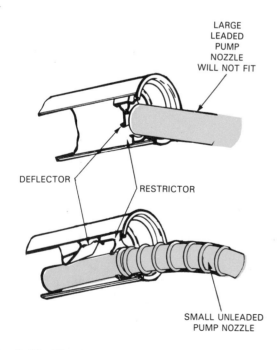

Fig. 6-10. The inside of unleaded type fuel filler neck has restrictor plate that will not accept larger diameter ''leaded fuel'' pump nozzle (top).   (Oldsmobile)

Gas station pump nozzles are of two different sizes. Nozzles on leaded fuel pumps are larger than unleaded pump nozzles. See Fig. 6-10.

Besides the restrictor, a spring-loaded valve or flap is usually located under the restriction hole, Fig. 6-11. The flap is easily opened by the small unleaded fuel nozzle and closes as soon as the nozzle is removed. This *filler tube flap* helps seal the fuel tank whenever the gas cap is off. It prevents gasoline vapors from escaping into the atmosphere.

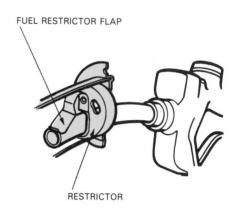

Fig. 6-11. Restrictor valve or flap also helps stop escape of gasoline fumes whenever fuel cap is removed.   (Ford)

FUEL TANK FILLER CAPS

The main function of a *fuel filler cap* is to keep gasoline from sloshing out of the tank filler pipe and to prevent dirt and water from entering. There are two basic types of fuel caps: the pre-emission type vented fuel cap and the modern sealed or pressure type cap. The sealed fuel tank cap has been in use since around 1970.

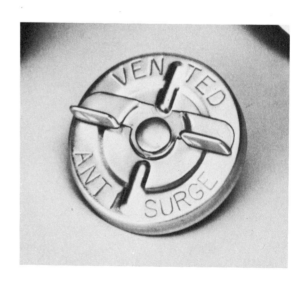

Fig. 6-12. Vented fuel cap, used on ''pre-emission control cars,'' allows fuel vapors to enter atmosphere through cap.   (Gates Rubber Co.)

Vented gasoline cap

A *vented cap* was used before autos had evaporative control emission systems. One is shown in Fig. 6-12. It simply allowed pressure or vacuum to flow through small holes in the cap. For instance, as the fuel pump removed fuel from the tank, air entered through the cap. If outside air could not flow into the tank, a strong vacuum could form. The vacuum could collapse the fuel tank, or stop fuel flow and engine operation.

Sealed fuel cap (pressure-vacuum type)

To contain gasoline fumes, a *sealed gas cap* is now used on most cars. See Fig. 6-13. It uses safety pressure and vacuum valves that normally stop the escape of fumes to the air.

CAP VACUUM VALVE ACTION

ANTI-SURGE VENTED CAP

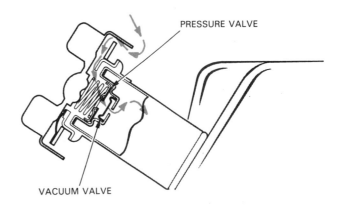

Fig. 6-14. Vacuum of around 1/4 to 1/2 in. of mercury pulls vacuum valve away from its seat and outside air pressure enters tank. This can prevent fuel tank collapse and slow fuel flow to engine. (Ford)

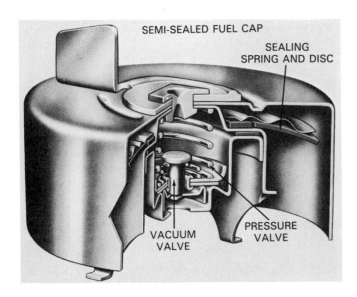

Fig. 6-13. Pressure fuel cap normally prevents gasoline vapors from polluting air. It uses safety pressure and vacuum valves that open only under extreme conditions.

CAP PRESSURE VALVE ACTION

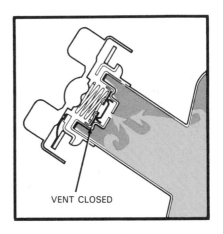

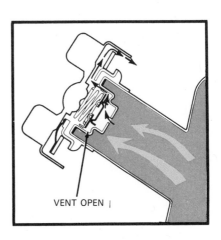

When a sealed or pressure cap is used, the fuel tank will often have another method of releasing abnormal pressure and trapping gasoline fumes. The two valves in the sealed fuel cap are actually only safety valves. They release excessive pressure and vacuum whenever the normal tank venting system is damaged, clogged, or inoperative. See Fig. 6-14 and Fig. 6-15.

The cap shown in Fig. 6-16 uses a rollover check valve in addition to pressure and vacuum valves. The *rollover check valve* assures that fuel cannot leak out through the pressure cap should the auto overturn.

A fuel cap pressure valve will usually open between 1/4 and 1 psi (2.8 and 6.9 kPa). The vacuum valve will release at around 1/4 to 1/2 in. of mercury (about 0.85 kPa).

Many pressure caps have two pairs of tangs which lock onto the filler neck. The tangs function much like the tangs on a radiator cap. They prevent system pressure from pushing liquid out of the system. By turning a sealed fuel cap a half turn, tank pressure is released

Fig. 6-15. Normally, fuel cap pressure valve is closed (top). However, excessive pressure in the tank (around 1/4 to 1 psi) will open valve. This may prevent fuel from flooding into engine. (Ford)

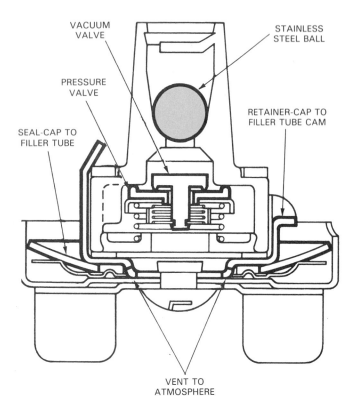

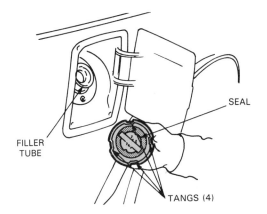

Fig. 6-17. Double locking fuel cap has two sets of tangs. They keep fuel from spraying out of pressurized fuel tank.   (Volvo)

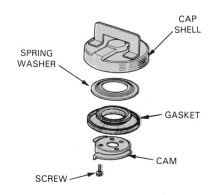

Fig. 6-16. Cap with rollover check valve. It assures that fuel cannot leak out of cap's pressure valve after an auto accident. Gravity causes steel ball to close off vent in cap.   (AMC)

Fig. 6-18. This type fuel tank cap can be disassembled. Gasket can then be replaced.   (Toyota)

slowly. It keeps pressure from blasting out of the tank all at once. Then, with another quarter turn, the cap can be removed.

Note! With some systems, there may be an occasional pressure release when the cap is loosened. Pressure or vacuum may make a hissing noise. Unless excessive, this is normal with some vehicles. If in doubt, check a service manual.

Fuel cap inspection

Always inspect the seal or gasket on a pressure-vacuum fuel cap. It must be in good condition or excessive emissions will occur. The seal should not be torn, split, hardened, or cracked, Fig. 6-17. Look at the seal or gasket closely to make sure that it has a full compression print.

A *COMPRESSION PRINT* will show up as a shiny, indented imprint or line all the way around the gasket. It is a mark made by the contact between the gasket and the filler neck lip. If the imprint is not visible all the way around the gasket, check the neck and cap for physical damage or misalignment.

Fuel cap replacement

Most fuel caps must be replaced when the gasket or seal becomes damaged. However, a few fuel caps have replaceable gaskets. One is shown in Fig. 6-18. The fuel cap can be disassembled so that a new seal or gasket can be installed. Some auto manufacturers recommend that their fuel caps or cap gaskets be replaced periodically.

When purchasing a new fuel cap, make sure that you get the same type. For instance, if a vented cap is used instead of a sealed cap, excessive air pollution will result because of the escaping gasoline fumes. See Fig. 6-19.

On the other hand, if a vented cap is replaced with a sealed cap, excessive pressure and vacuum can build in the fuel system.

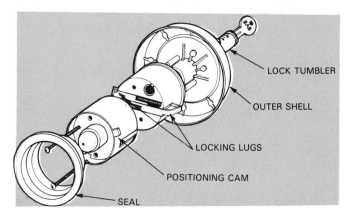

Fig. 6-19. When purchasing aftermarket, locking gas cap, make sure it is same type (vented or sealed) as old one.   (Gates Rubber Co.)

FUEL TANK PICKUP ASSEMBLY

Most modern *fuel tank pickup* assemblies have at least three basic parts:
1. A metal tube reaching nearly to the bottom of the gasoline tank for fuel removal by the fuel pump.
2. A sending unit or control mechanism for the dashboard fuel gauge.
3. A filter to prevent dirt and water from entering the fuel system.

A common type of tank pickup assembly is shown in Fig. 6-20. In some cases, a vent tube and fuel return line are included as part of the assembly. An electric fuel pump may be mounted on the pickup pipe.

The fuel pickup unit is installed through a hole in the top or one of the sides of the gas tank, Fig. 6-21. It is secured by a lock ring and sealed with an O-ring. This is pictured in Fig. 6-22.

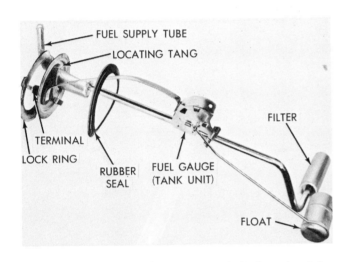

Fig. 6-20. These are basic parts of fuel tank pickup assembly. (Chrysler)

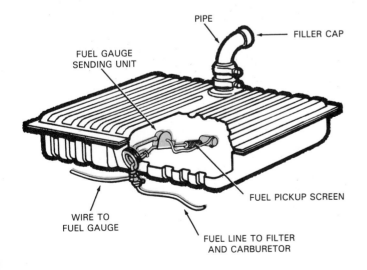

Fig. 6-21. Notice how fuel tank sending unit fits into tank. This one has connections for fuel gauge circuit and main fuel line. (Ford)

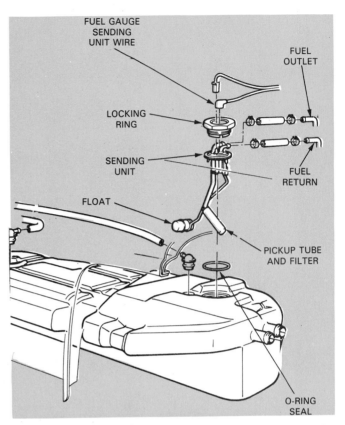

Fig. 6-22. Exploded view of tank sending unit. Neoprene O-ring seal must fit perfectly between tank and fuel pickup assembly. If not, a serious fuel leak and possibly fire could result. (Oldsmobile)

The pickup pipe extends nearly, but not all the way, to the bottom of the tank. Dirt, rust, water, and sediment can collect in the bottom of the fuel tank without being drawn into the fuel tank filter. This prevents the filter from becoming clogged. A ground wire is often attached to the tank unit.

Tank pickup assembly removal

Often, a tank sending unit can be serviced without removing the tank. However, some units are located between the car body and tank, requiring removal of the tank. Later in this chapter, we will discuss this procedure.

CAUTION: Do NOT attempt to remove a fuel tank sending unit before draining the fuel from the tank. If fuel is not drained, it may pour out of the pickup opening. A TREMENDOUS FIRE COULD RESULT!

As pictured in Fig. 6-23, a special tool is made for pickup assembly removal. It can be used with a ratchet or a small breaker bar. The tool fits over the metal tabs on the lock ring. About a quarter turn will loosen the lock ring, and the pickup unit will come out.

If you do not have one of these special tools, a large full shank screwdriver or drift punch, Fig. 6-24, will work. Place the screwdriver or drift punch against one of the tangs and tap lightly in a counterclockwise direction. The lock ring will loosen.

While pulling the tank sending unit from the tank, be careful not to bend the float arm or float, Fig. 6-25. Damage may cause a fuel gauge malfunction.

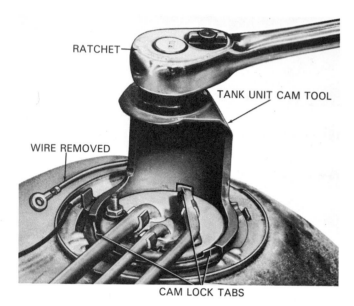

Fig. 6-23. With tank unit tool fitted on ratchet, lock ring can easily be turned and removed. Check that tank is drained of fuel and battery is disconnected. (Cadillac)

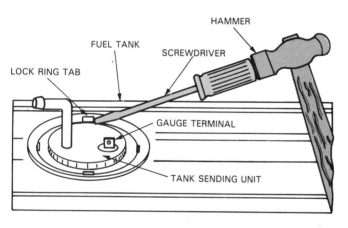

Fig. 6-24. In an emergency or when lock ring tool is not available, screwdriver or drift punch and ballpeen hammer can be used for tank unit removal. A light tap on lock ring will free it from tank.

Pickup unit inspection

Whenever the fuel tank pickup pipe assembly is removed, it should be closely inspected. If the filter is clogged, replace it. The hollow float should not be dented or leaking. Shake the float next to your ear. If you can hear liquid or fuel inside, it is leaking and must be replaced. Make sure that the float arm is not bent. Also, check that the O-ring seal is in good shape. Usually, it is wise simply to replace the O-ring.

If the filter is reusable, you may direct a very light air blast down through the pickup pipe in reverse direction. This will dislodge any dirt that might be caught in the fuel filter.

Note! If an excessive amount of dirt is found in the

fuel tank filter or tank, the tank should be cleaned. If this is not done, the filter may soon become clogged.

Pickup pipe-sending unit installation

Take extreme care when fitting the pickup assembly into the gas tank. Avoid bending the float arm or denting the float. Guide it slowly and carefully into the tank as shown in Fig. 6-25.

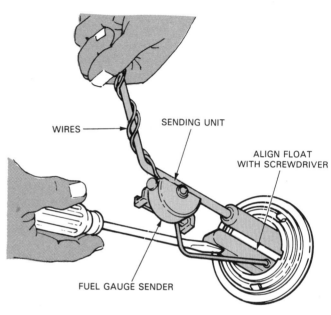

Fig. 6-25. Be extremely careful not to damage float, float arm, or screen when installing tank sending unit. See how screwdriver is used to align float so unit will slide through opening. Also, make sure that O-ring seal is in position. Some manufacturers recommend adhesive to hold seal. (Ford)

Be certain that your O-ring seal is in proper position. It must fit completely into the indentation or small groove in the tank opening. If this seal is even slightly out of place, the tank may leak.

In some cases, it may be helpful to place a small amount of non-hardening adhesive or gasket cement on the seal. This will hold the seal in place. With the O-ring and unit held into position with your hand, fit the lock ring into place. Then, give the ring a clockwise turn until it is locked. Use the special tool or screwdriver and hammer.

After filling the fuel tank with gasoline, check the pickup pipe assembly for leaks. NEVER allow a car to leave the shop after a pickup unit repair without looking for fuel leaks. A leak could cause a very serious fire. In any event, an unhappy customer will have to return the car to have you fix the leak.

FUEL TANK LOCATIONS

A car's gasoline tank is usually located in the rear for improved safety. However, in a rear engine vehicle, it can be placed in the front of the car.

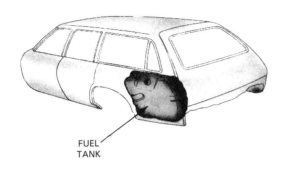

FUEL TANK

FUEL TANK IN REAR QUARTER PANEL

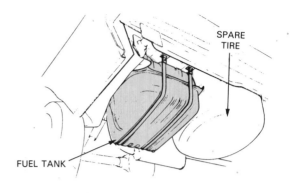

SPARE TIRE

FUEL TANK

FUEL TANK UNDER TRUNK ON ONE SIDE

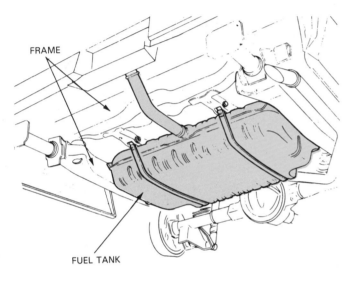

FRAME

FUEL TANK

FUEL TANK INSIDE FRAME

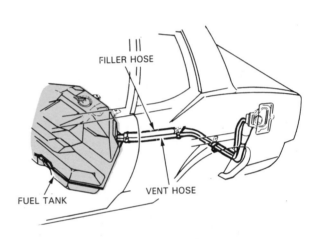

FILLER HOSE

FUEL TANK

VENT HOSE

FUEL TANK UNDER SEAT

Fig. 6-26. Gasoline tanks can be located almost anywhere that is safe from damage. (Oldsmobile)

Fig. 6-26 shows rear tanks in various locations. In any case, the gas tank should be located where it will receive minimal damage in a collision. A ruptured gas tank can cause a serious fire.

FUEL TANK STRAPS

A gasoline tank is usually held in place with large metal *straps,* like those pictured in Fig. 6-27. The straps can be bolted either to the frame or sheet metal subframe. Some tanks have heavy flanges that bolt directly to the car without straps, Fig. 6-28.

GAS TANK INSULATORS

Felt or rubber spacers are often fitted between the fuel tank and the tank straps. The insulator straps serve two basic functions. They quiet tank noise and also help keep the tank from shifting and vibrating inside the straps.

Never replace a fuel tank without installing all of the tank insulator straps. Comebacks, or even worse, tank leakage, could result.

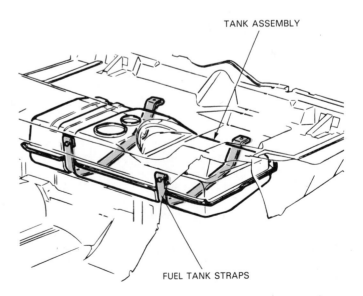

TANK ASSEMBLY

FUEL TANK STRAPS

Fig. 6-27. Tank filled with fuel is very heavy and must be held securely. Metal straps are often used to fasten fuel tank to car. (Chevrolet)

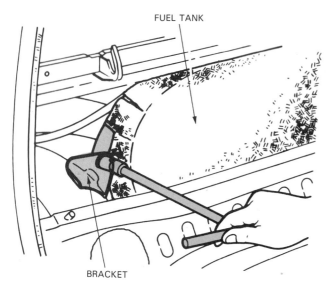

FUEL TANK

BRACKET

Fig. 6-28. Some fuel tanks use a heavy metal flange which permits the tank to be bolted directly to the automobile. This is especially common in compact cars using "subframes." (Subaru)

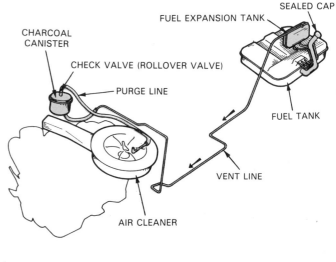

CHARCOAL CANISTER

CHECK VALVE (ROLLOVER VALVE)

PURGE LINE

FUEL EXPANSION TANK

SEALED CAP

FUEL TANK

VENT LINE

AIR CLEANER

Fig. 6-29. Study parts of complete fuel tank venting system. Note that even carburetor fuel bowl is vented to charcoal canister. (Chrysler)

FUEL TANK VENTING

As mentioned, gasoline tanks must have some method of *venting* or releasing pressure and vacuum without allowing gasoline fumes to enter the atmosphere.

A temperature increase of the fuel in the tank can cause the fuel to expand and form pressure. If this pressure is not relieved, fuel can be forced through the fuel system. Excessive tank pressure could flood an engine with fuel and prevent its operation. On the other hand, if vacuum builds up, the gas tank could collapse.

CLOSED TYPE FUEL TANK VENT

To stop fuel tank vapors from polluting the atmosphere, *closed vent systems* were developed. Part of the fuel evaporative control system is shown in Fig. 6-29. The sealed tank vent no longer allows gasoline fumes to leave the vent system.

Basically, a sealed fuel tank vent system, part of the evaporative control system, consists of six components which will be described next.

Sealed vent tubes

Normally, a sealed gas tank has several *vent tubes* designed into it. This assures that at least one vent is always ABOVE the fuel in the tank. For instance, if a car is parked on a steep hill, the fuel can submerge some of the vents. However, as shown in Fig. 6-30, the other vents on the opposite side of the tank will still be above the fuel level and capable of venting off vapors. The vent tubes connect to a liquid-vapor separator.

Vapor separator

Fig. 6-31 shows how the tank vents connect to the bottom of the vapor separator. The *vapor separator* is a metal container where the tank vent lines join the main

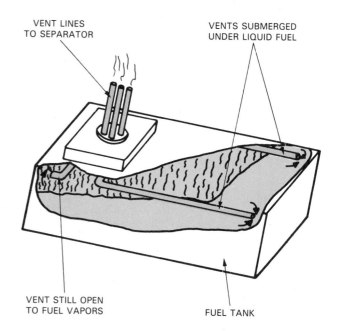

VENT LINES TO SEPARATOR

VENTS SUBMERGED UNDER LIQUID FUEL

VENT STILL OPEN TO FUEL VAPORS

FUEL TANK

Fig. 6-30. With vent lines in several corners of tank, at least one of vent tube openings is always above fuel level. This is true even with tank tilting at an extreme angle.

vent line. As you can see, the separator is designed so that liquid fuel CANNOT enter the main vent running to the front of the car. Liquid fuel entering the separator will settle to the bottom and flow back into the gas tank. Only fuel vapors can reach and flow into the vapor recovery line (main vent line).

Vapor recovering line

The *vapor recovery line* is a metal tube that carries fuel vapors from the separator to a charcoal canister in the engine compartment. It should NOT be confused with the main fuel line and the fuel return line.

VAPOR SEPARATOR

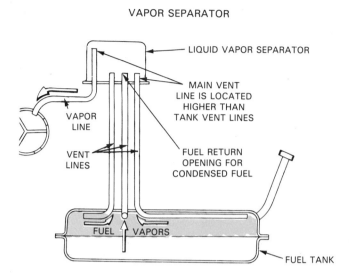

Fig. 6-31. Fuel tank vapor separator allows some of fuel vapors to condense back into liquid and return to tank. Only vapors can normally enter higher main vent tube opening. (Fiat)

CHARCOAL CANISTER

Pictured in Fig. 6-32, the charcoal canister is a plastic or metal container filled with activated charcoal. Charcoal has the ability to absorb and store a large amount of fuel vapors. As the vapors leave the tank, separator, and vapor line, they enter this container where they are kept until the engine starts. Another type of charcoal canister is shown in Fig. 6-33.

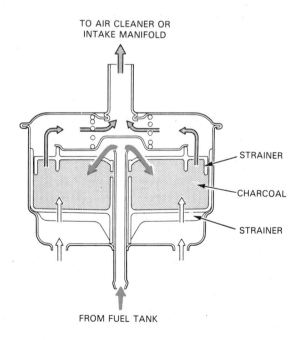

Fig. 6-32. Gasoline vapors are absorbed and stored by activated charcoal in canister. When engine is running, these vapors are drawn out of the charcoal and into engine where they are burned. (Chrysler)

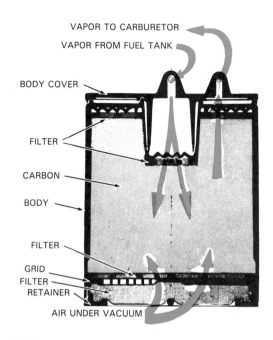

Fig. 6-33. Purge line is nothing more than vacuum line connected between charcoal canister and intake manifold. Engine vacuum draws vapors out of canister and into engine cylinders. (General Motors Corp.)

Purge line

A purge line or hose, Fig. 6-34, is connected between the charcoal canister and the engine intake manifold. This *purge line* allows engine vacuum to draw vapors out of the charcoal canister.

When the engine is started, the gasoline vapors in the canister are pulled from the charcoal into the engine for burning. The bottom of the charcoal canister is open. Air flows up through the bottom of the canister, through the bed of charcoal, into the engine.

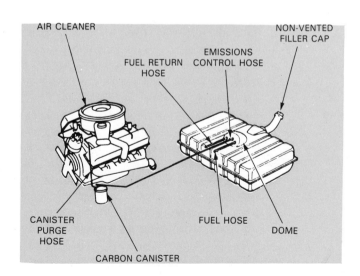

Fig. 6-34. During purge action, carburetor or intake manifold vacuum draws flow of air through bottom of canister. This air flow carries gasoline fumes into engine. (General Motors Corp.)

Rollover valve or vapor valve

Usually located between the fuel tank and the charcoal canister, the *rollover valve* is a safety device. It prevents gasoline from pouring out the vent system when a car rolls or turns upside down in an accident. See Fig. 6-35.

The rollover valve may have a float and needle mechanism that closes as soon as liquid fuel enters the valve, Fig. 6-36. The valve will pass only vapor, not liquid. Sometimes, the rollover valve is located in the main vent line or in the top of the fuel tank. In some applications, the rollover valve is combined with the liquid-vapor separator forming one unit.

Usually, the carburetor bowl is also vented into the charcoal canister. Like the tank vapors, carburetor bowl

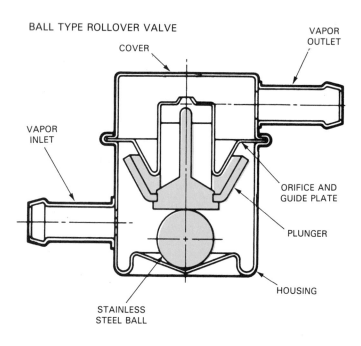

Fig. 6-35. If car flips upside down during accident, rollover valve stops liquid fuel from pouring out of system. Weight of steel ball closes plunger and seals off vent line. (American Motors Corp.)

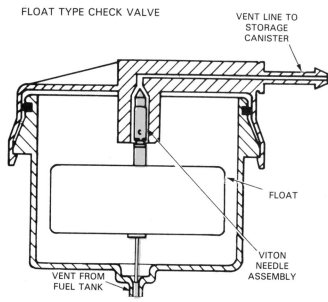

Fig. 6-36. Float type check valve stops liquid from flowing through vent system with car in any position. Hollow float rides on top of any liquid gasoline and closes needle valve. (AMC)

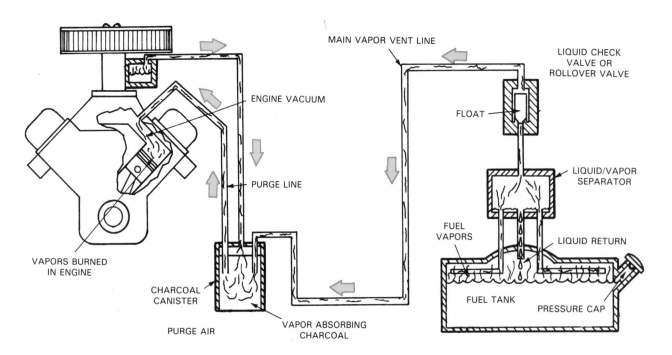

Fig. 6-37. Basic components and flow pattern of vapors through fuel tank vent system. Try to develop mental image of this evaporative control system. This will help you in troubleshooting and repair.

vapors are routed to the engine and burned. Carburetor bowl venting will be covered more completely in the chapter on carburetors. A complete tank and carburetor vent system is illustrated in Fig. 6-37.

FUEL RETURN SYSTEM

As explained in earlier chapters, overheated fuel can boil and form bubbles. The bubbles can upset engine operation by changing an engine's fuel mixture. The *fuel return system* (sometimes called *vapor return system*) lowers the temperature of the fuel in the fuel pump and reduces chances of vapor lock.

Fuel return system components

Basically, a return system consists of a special fuel pump equipped with an extra outlet fitting, a steel fuel line running from the pump back to the gas tank, sections of flexible hose, and a fitting in the tank or pickup unit. A fuel return system is pictured in Fig. 6-38.

The fuel return line usually runs alongside the conven-

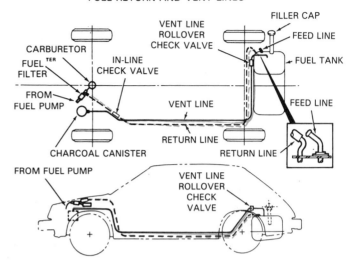

FUEL RETURN AND VENT LINES

Fig. 6-39. Notice parallel routing of fuel return line and vapor line. (American Motors Corp.)

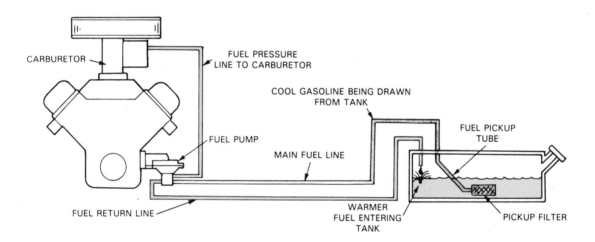

Fig. 6-38. Note flow of liquid gasoline from tank through main fuel system, and fuel return system. Gasoline is drawn from tank into fuel pump. From pump, gasoline flows to carburetor as well as back to fuel tank. This tends to cool gasoline in pump and reduces chances of vapor lock.

tional fuel line, Fig. 6-39. It may even look like the regular fuel line or the vent line. In the fuel return line, however, flow would be in the opposite direction, toward the gas tank, NOT toward the engine. The fuel return system constantly allows a metered amount of cool gasoline to circulate through the tank and fuel pump.

A vapor separator is sometimes connected to the output side of the fuel pump, Fig. 6-40. It has fittings that go to the engine and to the fuel tank. The separator helps prevent gasoline vapor bubbles from reaching the engine and possibly upsetting the fuel mixture.

In some cases, the separator has a dual purpose. It filters fuel and returns fuel vapor bubbles to the gas tank. This is illustrated in Fig. 6-41. The bubbles gather in the top of the separator and then flow through the small orifice in the return tube. These bubbles are carried back to the tank.

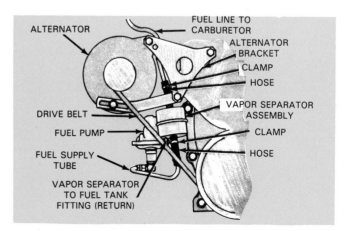

Fig. 6-40. A vapor separator is sometimes confused with a fuel filter.   (Chrysler)

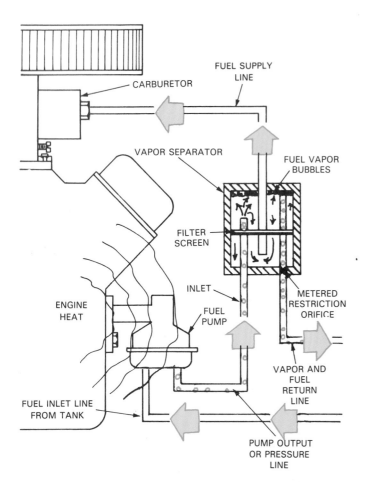

Fig. 6-41. Several fuel return systems use an in-line vapor separator. It allows vapor bubbles to collect in top of separator before returning to tank with liquid fuel. The small orifice keeps fuel pressure from dropping. Only metered amount of fuel can flow back to fuel tank.

Fuel return system problems

If a problem ever arises that appears to be vapor lock related (engine stalling in hot weather, hard to start after the hot engine is shut off, etc.), you should check the operation of the fuel return system. The small orifice in the vapor separator of the fuel return line may be clogged. Air should pass through the return line with little resistance.

A plugged separator should be replaced. A plugged return line should be flushed and cleaned. Air pressure may force the dirt from the line. Always disconnect both ends of a fuel line before blowing air into it. This will avoid blowing air and dirt into the fuel tank or pump.

FUEL TANK SERVICE

When a fuel tank is leaking, dirty, or has water in it, the tank should be removed for repair or cleaning.

DANGER! Be careful when servicing fuel tanks. Fire is always possible.

DANGER! Remember that even an empty gas tank may have enough gasoline vapors to cause a tremendous explosion. Treat an empty fuel tank with respect.

DRAIN TANK FIRST

Before starting, disconnect the battery. This will prevent an electrical spark (tank sending unit wire, etc.) from igniting vapors. Drain the tank. A full tank is extremely heavy and dangerous. If the tank is equipped with a drain plug, remove it. If not, use one of the following drain techniques.

SIPHONING FUEL

Caution! NEVER use your mouth to start a siphoning action. Gasoline is poisonous!

A siphoning tool can be made very easily out of a short length of metal tubing or neoprene hose, Fig. 6-42. Cut an angled slot in the side of the tube or hose with a hacksaw. Blowing air into the slot with an air gun will form a vacuum in the tube.

Push a section of neoprene hose over the siphon device and stick the hose into the fuel tank. Blow air into the slot. Once the fuel has begun to flow, air pressure is no longer needed. As long as the slot is held so it is lower than the fuel level in the tank, gasoline will siphon out of the tank.

Drain the fuel into an approved safety type gasoline can, never an open container.

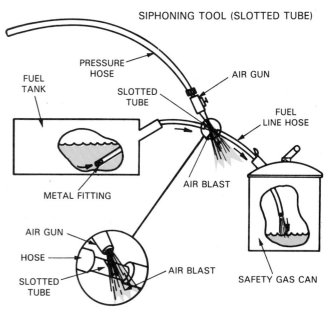

Fig. 6-42. A home-made siphon can be made to drain tank.

Draining fuel tank with electric pump

As shown in Fig. 6-43, an automotive electric fuel pump can be used to drain a gasoline tank quickly, easily, and with few fuel vapors. Connect sections of hose to the output and intake fittings on the electric fuel pump. Hang the fuel pump in a safe place where gasoline cannot spill onto it. Securely connect long jumper leads or wires to the fuel pump terminals.

Connect the intake hose of the pump to the tank

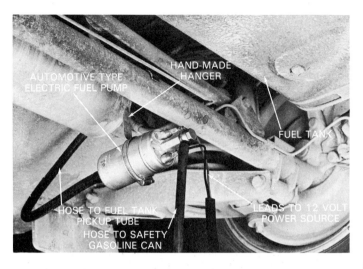

Fig. 6-43. An automotive electric fuel pump can be used to transfer gasoline out of fuel tank and into gasoline can. (Saab)

pickup tube in place of the main fuel line hose.

CAUTION! Remove the tank fuel cap before removing any fuel lines. Any pressure in the tank will force gasoline out of the tank.

Stick the output hose of the pump into a gasoline can. Connect the fuel pump wires to a 12V power supply (battery, battery charger, or any "hot" wire with sufficient current rating). Depending upon the volume rating of the pump, the tank will soon be drained.

Fig. 6-44. This commercial fuel pump is operated by hand. It has an extra large storage container for fuel. By turning a valve on pump, fuel can be pumped back into car's fuel tank. (Chrysler)

Commercial fuel tank pump

A *commercial fuel tank pump* equipped with a large storage tank is shown in Fig. 6-44. It will pump fuel out of a gas tank much like the electric fuel pump previously mentioned. The suction hose on the pump may be connected to the pickup pipe on the tank or the hose can be stuck into the filler neck. By operating the handle on the pump, gasoline will be drawn out of the fuel tank. After the repair, reverse the valve on the pump and the fuel (if good) can be hand-pumped back into the tank.

TANK REMOVAL

When the tank is drained, it can be taken off the car. Double-check that the battery is disconnected and that the tank is completely empty. Filler cap must be off. Unplug the wires going to the tank pickup assembly.

Note! You should be wearing eye protection, as in Fig. 6-45.

Remove the fuel line, vent, and vapor return lines. Carefully look around the tank to make sure that all wires

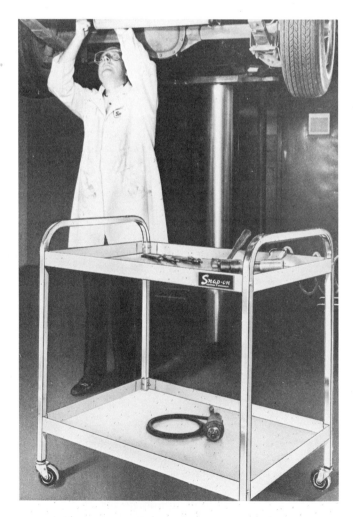

Fig. 6-45. When removing a fuel tank and working under car, wear eye protection. Gasoline, dirt, rust, and other debris can drop into your eyes. A roll-around cart, like one shown, is a great time-saver when working away from your main tool box. (Snap-on Tools)

and hoses have been disconnected.

Next, unfasten the filler neck from the tank. If it is one piece, you will need to remove the screws around the outside of the filler neck near the filler cap. If a neoprene hose and a two-piece neck are used, you will have to loosen the clamps and remove the hose. You may need to grasp the hose twisting and pulling firmly to remove it from the neck. Sometimes, the neck protrudes into the fuel tank and is sealed with a neoprene O-ring. If so, you may have to pull the neck out of the tank while lowering the tank.

When you are sure that everything is either off or loosened, unscrew the bolts or nuts securing the tank straps. Shake the tank to make sure it is empty and does not weigh too much. A small amount of fuel will probably still be in the tank. Remove the straps and carefully lower the tank to the shop floor. While lowering, watch that all wires and tubes are unhooked. There may be a ground wire or vent line hidden on top of the tank.

FUEL TANK REPAIRS

If the proper safety precautions are taken, a gasoline tank can be repaired by soldering or welding (brazing).

NOTE! Do not attempt to weld or solder a fuel tank without experienced supervision or training.

Before putting any heat on the tank, you must take the following safety measures:

1. Clean the fuel tank. If the tank is metal, thoroughly steam clean the tank, inside and out. Another cleaning method involves using a special emulsifying agent. The agent should be mixed with water and sloshed around in the tank for about 10 minutes. Follow specific emulsifying agent directions.

   Do not steam clean a plastic tank. Use detergent and warm water. Check your service manual before using an emulsifying type cleaner. A plastic fuel tank would be damaged by the steam or by strong cleaning agents.
2. Fill tank with nonflammable substance. Fill the sheet metal tank with water or inert gas (nonflammable carbon dioxide, nitrogen, etc.). This will keep the tank from exploding during welding.
3. Work in a ventilated area. A fan is recommended because it will help blow any fuel vapors out of the area. If available, an explosion meter should be used to check in or near the tank for fuel vapors.
4. Have a fire extinguisher ready. Place a fire extinguisher nearby and warn the other workers in your area of the danger. Be careful!

Repairing a fuel tank leak

A leak in a sheet metal gas tank can either be soldered or brazed. Acid core solder should be used when soldering. A bronze type brazing rod and borax type flux are good when brazing.

If you forgot to fill the gas tank with water or inert gas, you are taking a great risk. An empty fuel tank will very likely explode when heated. Mechanics

have been injured or even killed when trying to weld a fuel tank improperly.

Fiberglass can be used to repair both metal and plastic fuel tanks. To seal a leak properly with fiberglass, a large area around the leak must be cleaned. Wire brush the area. Sand it with sandpaper and then clean it with solvent.

If the area around the leak is not properly prepared, the fiberglass will not adhere and the tank may leak after only a short period of operation. Always follow the detailed directions supplied with the fiberglass patch kit.

Test your tank repair for leaks. Coat the repair with soapy lather and apply very low air pressure to the tank. If the soapy water does not bubble, the tank should be all right.

Repairing fuel tank dent

A thin sheet metal gas tank can be dented very easily. If damage is not excessive, a dent can be taken out of a fuel tank using WATER and AIR PRESSURE.

After removing the tank from the car, completely fill the tank with water, Fig. 6-46. Plug all of the vent tubes and fuel line fittings and install a nonvented fuel cap. The vents can be plugged by clamping short pieces of hose over each vent line. The hoses can be plugged by clamping bolts into the ends of the hoses.

Using an air nozzle, blow air into the tank pickup pipe or tube. As you force air into the tank, the dent should pop out. Be careful not to apply too much pressure, just enough to remove most of the dent. Excessive pressure is dangerous. The seams of the tank could give way or the vent plugs could blow out.

CAUTION! Do not apply air pressure to a gas tank unless it is filled with water. Air alone can cause the tank to explode violently!

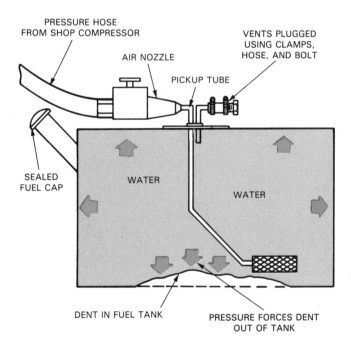

PRESSURE HOSE FROM SHOP COMPRESSOR
AIR NOZZLE
VENTS PLUGGED USING CLAMPS, HOSE, AND BOLT
PICKUP TUBE
SEALED FUEL CAP
WATER
WATER
DENT IN FUEL TANK
PRESSURE FORCES DENT OUT OF TANK

Fig. 6-46. After sealing filler neck (nonvented cap), fuel line, and vent lines, fill tank with water. Air pressure can be used to force out dent.

Fuel tank replacement

Before installing a fuel tank, always check that the inside of the tank is clean. Pour in a little gasoline and slosh it around. Then, pour the gas into a clean container or through a paint filter (paper and cloth filter used to remove hardened particles from paint). Inspect for dirt or impurities. If present, clean the fuel tank thoroughly.

Make sure that all of the felt or rubber insulators are in place. You may need to glue them to the tank or to the holding straps with weatherstrip cement.

With the tank straps in place, slide the tank into position. As necessary, place the filler neck into the tank hole. Loosely fit the tank straps around the tank, but do not tighten them. Check that the tank is properly positioned. Look all around the tank to make sure that none of the vent tubes, hoses, or wires are pinched. You may have to connect some of the vent lines before clamping the tank into place. Also, examine the filler neck for alignment, and for full insertion into the tank. If a hose coupling is used on the filler neck, it, also, should not be kinked or in a bind.

Tighten the strap bolts and secure the tank to the car. Install all of the tank accessories (vent line, sending unit wires, ground wire, fill control tube, etc.) Fill the tank with fuel and check it for leaks, especially around the filler neck and the pickup assembly. Reconnect the battery and check the fuel gauge for proper operation.

FUEL CELL OR BLADDER TANK

Fuel cell tanks are normally constructed of an impregnated fabric called ballistic nylon. This material is extremely rugged and is sometimes self-sealing in case of a puncture. Fuel cells or bladder tanks are available in capacities ranging from around 8 gallons up to 45 gallons. Usually, a fuel cell will be larger than its maximum capacity. This allows for expansion due to heat.

Because a fuel cell is soft and flexible, it is less likely to be torn open during an auto accident. It can bend and flex to allow for high impact, Fig. 6-47.

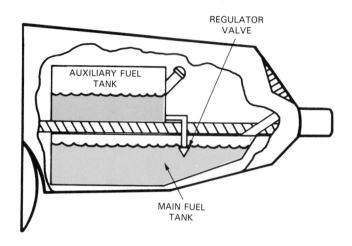

Fig. 6-48. An auxiliary fuel tank increases vehicle's driving range.

trunk above the conventional tank. In trucks, an auxiliary tank is often called a saddle tank.

An auxiliary tank increases the DRIVING RANGE of a vehicle. Great care must be used in choosing a safe location and proper tank installation and venting. Observe all applicable emission regulations.

FUEL GAUGES

A fuel gauge shows the driver how much gasoline or fuel is in the fuel tank. Fig. 6-49 illustrates how they work. Fuel gauges are normally dashboard or console mounted and are electrically powered.

A fuel gauge circuit includes a power source (battery or charging system), the ignition switch, dash gauge assembly, fuel tank sending unit (float and variable resistor mechanism on the pickup tube), connecting

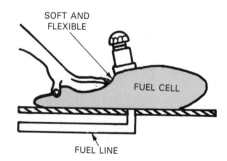

Fig. 6-47. Fuel cell was first used in racing cars. It was safer during a crash.

AUXILIARY FUEL TANK

An *auxiliary fuel tank* is an extra gas tank. One is shown in Fig. 6-48. In cars, it is often mounted in the

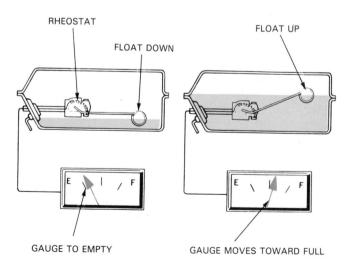

Fig. 6-49. Sending unit of fuel gauge is a rheostat operated by float. As float moves, rheostat sends electrical signal to fuel gauge needle. (Ford)

wires, and sometimes a voltage regulator. A basic fuel gauge circuit is pictured in Fig. 6-50.

The amount of current flowing through the tank resistor unit controls the dash gauge needle movement. Any change in the tank's fuel level changes the sending unit resistance, circuit current, and fuel gauge reading.

BALANCING COIL (MAGNETIC) FUEL GAUGE

A balancing coil fuel gauge is illustrated in Fig. 6-53. The dashboard gauge consists of two electric coils, an integral (combined) pointer and armature, connecting wires, and terminal connections. The coils are set at 90

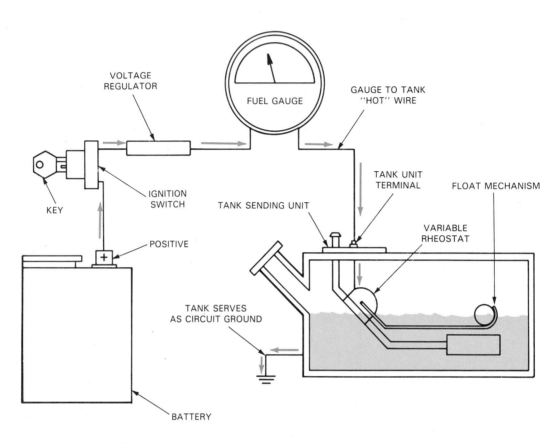

Fig. 6-50. Study this basic fuel gauge circuit and trace flow of current.

BASIC FUEL TANK SENDING UNIT

Most cars use a *tank unit* consisting of a hollow float, a float arm, variable resistor (wiping contact and resistance element), metal housing around the resistor, "hot" wire terminal or lead wire, and, sometimes, a ground wire. See Fig. 6-51. The tank control unit is normally mounted on the pickup pipe. The complete tank assembly is secured in the gasoline tank with a lock ring.

Sending unit operation

When the tank is empty, the float and float arm swing down to the bottom of the tank, Fig. 6-52. This will change the resistance in the fuel gauge circuit so that the dash gauge reads empty. When the tank is half full, the float and arm will move the sliding contact on the resistance element to about the middle of the element. This will allow a moderate amount of current to flow through the fuel gauge. The needle deflects to show a half tank of fuel. With the tank full, resistance will change so that the fuel gauge reads "full."

degrees from each other to make them pull on the gauge needle armature. To prevent erratic movement of the gauge pointer, a dampening device is also used.

When current flows through one of the coils, a strong magnetic field forms around that coil. The coil becomes an electromagnet. Magnetism in the metal core or the coil pulls on the fuel gauge needle. It will swing toward the coil with the stronger magnetic field.

The balancing coil gauge is wired so that the tank unit controls the amount of current in each coil. Look at Fig. 6-54. When the fuel tank is empty, the tank sending unit has low resistance. It draws or shunts much of the current through the "empty coil" of the fuel gauge. Thus, the empty coil will have a strong magnetic attraction while the "full" coil has a weak magnetic field. This will pull the gauge armature to the left or empty side of the gauge.

When the tank is filled, the tank float rises and the sliding contact moves to the high resistance position on the resistance wire. Current can no longer shunt through the tank unit. It must flow through the "empty" coil

TANK FUEL METER UNIT

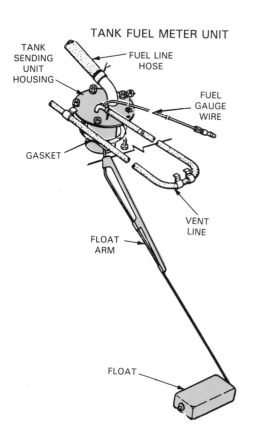

Fig. 6-51. Tank pickup unit or sending unit consists of float mounted on float arm, unit housing, variable resistor, hose fitting, etc. (Subaru)

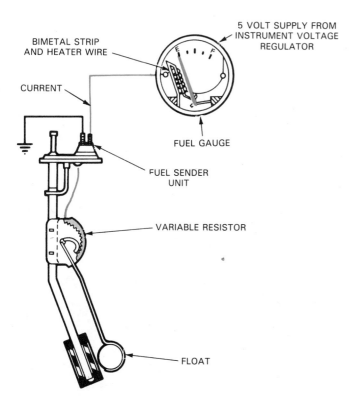

Fig. 6-52. As hollow float rides up and down on fuel in tank, the tank resistor's resistance value changes and so does current. This change in current flow causes an equal deflection in gauge pointer. (Ford)

BALANCING COIL TYPE FUEL GAUGE

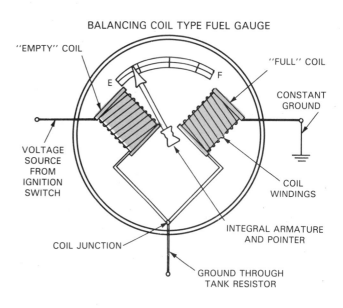

Fig. 6-53. Gauge coils are capable of producing a magnetic field. The field will attract and pull upon armature. Coil with strongest magnetic field will attract armature the most. When ignition switch is off, pointer may not return to empty.

BALANCED COIL GAUGE (EMPTY TANK)

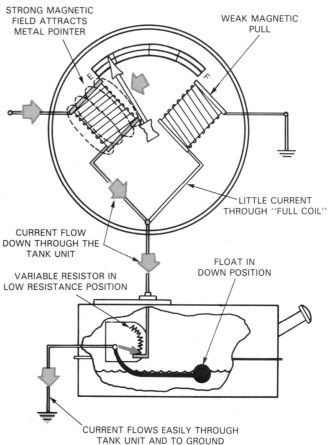

Fig. 6-54. When fuel level is low, tank unit has low resistance. As a result, current shunts down through empty coil and tank resistor; little current flows through full coil. The magnetic field around empty coil, working on armature, moves pointer to left.

and then through the "full" coil. This is illustrated in Fig. 6-55. Current will flow through both of the coils and not through the tank unit. The full coil on the right normally has the stronger magnetic attraction. It will overpower the empty coil on the left. The fuel gauge pointer will swing to full.

When there is only a half tank of gas, some current shunts into the tank unit. An equal field is developed at both of the coils. The needle points to the middle of the scale. A balanced coil fuel gauge is not affected by either a change in system voltage or temperature.

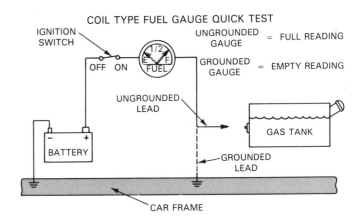

COIL TYPE FUEL GAUGE QUICK TEST

Fig. 6-56. Quick test for coil type fuel gauge. When hot wire to fuel gauge sending unit is grounded, gauge pointer should swing to empty. When wire is not grounded, pointer should move to full.

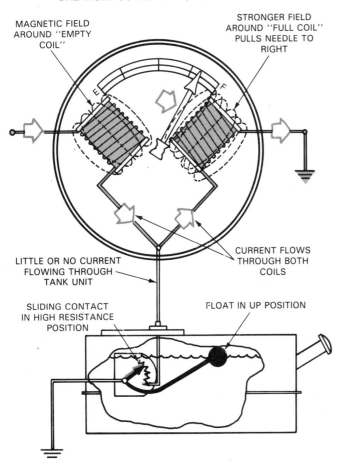

BALANCED COIL GAUGE (FULL FUEL TANK)

Fig. 6-55. When tank is full, sending unit resistance is high. This causes current to flow through both coils. Since full coil is designed to have strongest magnetic field, it will overpower empty coil and move gauge pointer to right.

Balanced coil fuel gauge troubleshooting and repair

To isolate a problem in a magnetic fuel gauge, disconnect and GROUND the "hot wire" going into the gas tank. Then turn on the ignition switch. This procedure is illustrated in Fig. 6-56.

With the tank wire GROUNDED, the magnetic fuel gauge pointer should read EMPTY. With the tank unit wire DISCONNECTED, the fuel gauge pointer should swing all the way to FULL.

Bad tank unit or ground

If the fuel gauge operates as described during the test, the fuel gauge in the dash and the circuit must be good. The problem is in the tank unit or the tank ground. First, try externally grounding the tank with a jumper wire, Fig. 6-57. If the fuel gauge now operates, the tank has lost its ground (rusted tank, broken ground lead, etc.) If a bad ground is not the problem, the tank float unit or sending unit may be defective. It should be removed, tested, and replaced as needed.

Bad circuit or gauge

If the gauge did NOT swing to empty and full as the tank lead wire was grounded and ungrounded, the circuit wiring or dash gauge may be at fault. To test further, unhook the dashboard lead wire at the dash. This is the wire running from the gas tank to the fuel gauge. With the ignition ON, ground and unground the wire as you did at the tank. Again, the gauge should swing back and forth. If not, the fuel gauge may be defective.

One other possible cause of fuel gauge problems can

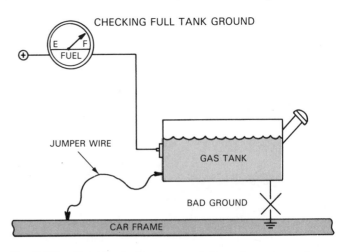

CHECKING FULL TANK GROUND

Fig. 6-57. Checking fuel tank ground. Sometimes gas tank can lose its ground. If so, adding jumper to ground should make fuel gauge operate.

be in the supply voltage. Any of the following: bad ignition switch, loose connection, or broken circuit board, can keep a fuel gauge from functioning.

Fuel gauge continuity test

An ohmmeter or a special fuel gauge tester can be used to check the fuel gauge. A gauge tester is shown in Fig. 6-58.

The tester is designed to test a coil or magnetic type fuel gauge. Instead of grounding the tank unit lead wire directly to ground, you can ground through the tester. The tester can be adjusted to specific resistance values in order to pinpoint faults or inaccuracy in the gauge, tank unit, or circuit. Operating instructions are included with this type of fuel gauge tester.

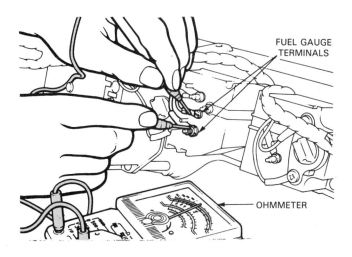

Fig. 6-59. Making fuel gauge continuity test. On some cars, a resistance value will be given for gas gauge. An ohmmeter check should read within these "specs." If not, gauge is defective. Whenever gauge has either zero or infinite resistance, it is defective.

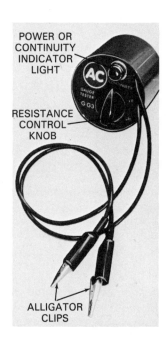

Fig. 6-58. This fuel gauge tester can be connected into fuel gauge circuit. Indicator light will show power to gauge circuit. Also, internal resistance of tester can be changed to check gauge action.

Fig. 6-59 shows how to use an ohmmeter to test a fuel gauge. Simply measure the resistance across the terminals of the gauge and compare them to specifications.

THERMOSTATIC FUEL GAUGE

A *thermostatic fuel gauge* uses an electric heating unit which operates a special thermostatic or bimetallic strip. A *bimetallic strip* is a metal assembly that becomes warped or bent by heat.

The fuel gauge needle is controlled by the resistance of the fuel tank sending unit. When the tank is empty, Fig. 6-60, the float mechanism swings down and moves the sliding contact to the high resistance position. This high resistance lets very little current flow through the gauge circuit. The gauge heating element stays cool and

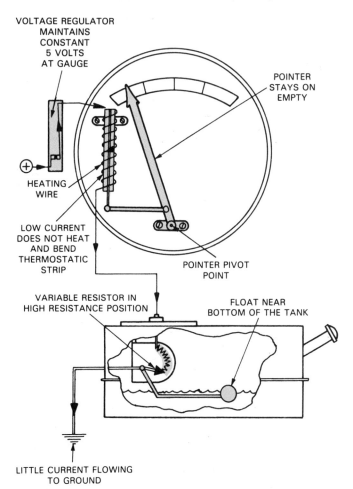

Fig. 6-60. With fuel tank empty, tank resistor has high resistance. Very little current can flow through circuit. Heating element in gauge stays cool. Thermostatic arm and pointer will not move off empty.

the bimetallic strip is not affected. As a result, the fuel gauge pointer does not move. It stays on empty.

As the tank fills, Fig. 6-61, the float will rise and move the sliding contact to the low resistance position of the resistor. In turn, high current flows through the circuit and heating element. This heats and bends the bimetallic strip. As the thermostatic arm bends, it moves the fuel gauge pointer to the right or full side of the scale.

A thermostatic type fuel gauge requires the use of a *dash voltage regulator* because it does not have the self-correcting action of a magnetic type gauge. Without a voltage regulator, the fuel gauge would be affected or thrown off by any change in supply voltage (alternator output, battery condition, etc.). A dash gauge voltage regulator should hold a steady input voltage.

THERMOSTATIC FUEL GAUGE (FULL)

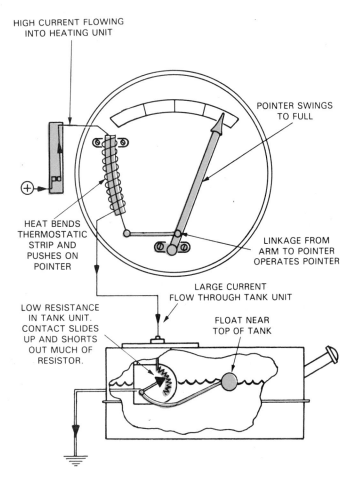

Fig. 6-61. Full tank moves tank sliding contact to low resistance position. As a result, a large amount of current flows through gauge. Current heats element. Thermostatic arm bends and pushes gauge pointer toward full.

Thermostatic fuel gauge service and repair

A thermostatic fuel gauge test (Ford, Chrysler, and some foreign type gauges) is made in the same way as described for a coil type fuel gauge circuit test. The fuel gauge needle action, however, is REVERSED.

Remove and ground the wire going into the gas tank unit. See Fig. 6-62. This should make the gauge swing to FULL. When the wire is unhooked, the fuel gauge should swing to empty. If the gauge moves as described, then the circuit and gauge should be all right. The problem is in the tank sending unit or the tank ground. Test them as needed.

CAUTION! Use care in making this test. If maintained for too long, ground contact could damage the gauge. Hold the ground only momentarily. Also, certain types of gauges, particularly digital, could be damaged by grounding the hot wire. Check a service manual for the vehicle for proper instructions.

If the gauge needle does NOT move as the tank wire is grounded, the problem is in either the circuit or the fuel gauge. Try grounding the gauge wire at the dash. If the gauge begins to operate, an open or break exists in the wiring between the tank and dash. If the gauge still fails to move, test the gauge output from the dash voltage regulator. If the gauge still does not work, you may need to replace it.

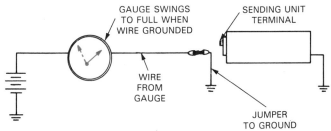

Fig. 6-62. With fuel gauge hot wire grounded, gauge should read full. When not grounded, it should read empty.

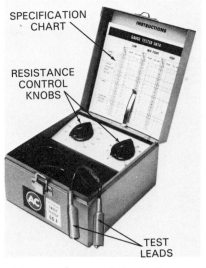

Fig. 6-63. This gauge tester will check all types of fuel gauges. By setting internal resistance of tester to given value, accuracy of gauge can be tested.

Fuel gauge tester

A fuel gauge tester is pictured in Fig. 6-63. Though designed to test most types of fuel gauges, it will also test the voltage regulator that feeds power to the dash gauges. Use a gauge tester if one is available!

For instance, if a fuel gauge operates but is inaccurate, this type of tester can help determine the problem. It is connected into the fuel gauge circuit and adjusted to factory resistance settings. If the gauge reads low or high, the tank unit arm can be bent to correct the problem, the gauge can be replaced, bad connection repaired, etc.

Fig. 6-64 shows how a gauge tester is connected and used. When the tester is set to the correct values, the fuel gauge needle should reflect the amount specified.

Note! Different fuel gauges will have different resistance specifications. These values can be found in a shop manual.

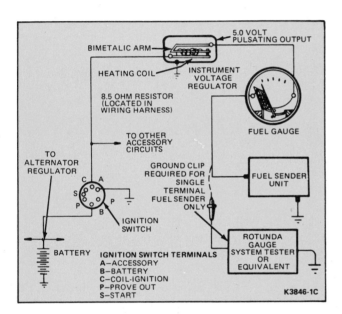

A

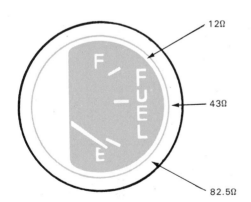

B

Fig. 6-64. Shop manuals should be checked for proper procedure and resistance values when testing fuel indicator systems. A—Shop manual diagram shows how to hook up tester. (Ford Motor Co.) B—Tester resistance values, when set on tester, should cause gauge indicator to point to proper level. (Mazda)

Testing the tank unit

Some auto makers give specifications for the fuel tank sending unit. As shown in Fig. 6-65, an ohmmeter can be used to measure the resistance of the tank unit. Resistance, measured across the terminals, is compared to specifications. There will be a resistance value (in ohms) for both empty and full positions. The float arm must be swung up for full and down for empty.

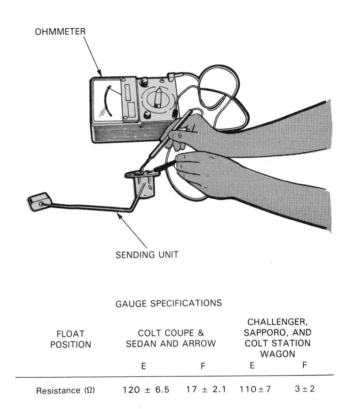

GAUGE SPECIFICATIONS

FLOAT POSITION	COLT COUPE & SEDAN AND ARROW		CHALLENGER, SAPPORO, AND COLT STATION WAGON	
	E	F	E	F
Resistance (Ω)	120 ± 6.5	17 ± 2.1	110 ± 7	3 ± 2

Fig. 6-65. Tank sending unit can sometimes be tested with ohmmeter. It should meet specifications. Resistance is measured between terminal and ground. Note the "specs" for this unit. Readings are taken at both "full" and "empty" position of the float. (Chrysler Corp.)

Testing the voltage limiter

The *voltage limiter* or *dash voltage regulator* adjusts the gauge supply voltage to allow for any change in battery or alternator output voltage. It keeps the gauge readings accurate.

Fig. 6-66 illustrates how a voltmeter can be used to test a gauge voltage regulator. The voltmeter is used to measure the voltage coming out of the dash regulator. If too high or low, an adjustment or replacement must be made.

Fuel gauge removal and installation

Fig. 6-67 is typical of steps taken to remove and install a fuel gauge. Procedures vary. Follow manufacturers instructions.

To remove a fuel gauge from the dashboard, first disconnect the battery negative cable, Fig. 6-67A. This prevents the possibility of an electrical fire.

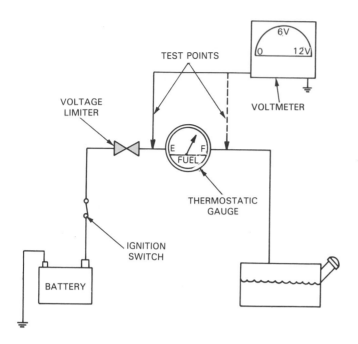

Reach behind the dash and unhook the speedometer cable. It will often have a plastic fitting that must be pinched, Fig. 6-67B. When squeezed, the fitting will release and the cable can be pulled free. On some cars, it may be easier to disconnect the speedometer cable at the transmission or engine compartment or to release the cable after the instrument panel is loose.

Next, remove the screws from the instrument cluster as in Fig. 6-67C. Carefully pull the cluster away from the dash a few inches. Unplug the wiring harness. There may also be several other wires that must be taken off. Lift the instrument cluster completely out of the dash, as in Fig. 6-67D.

As shown in Fig. 6-67E, remove the clear plastic mask (lens) from the cluster. This will expose the center gauge mounting plate. Remove the necessary screws and nuts so that the fuel gauge can be removed, Fig. 6-67F.

Install the new gauge and reassemble the dashboard in reverse order. Reconnect the battery. Check all gauges and speedometer for proper operation.

LOW FUEL WARNING SYSTEMS

Some cars have a *low fuel indicator light* mounted on the face of the fuel gauge or in the dash instrument

Fig. 6-66. Checking voltage limiter output. Voltage limiter or regulator is tested with voltmeter. Bad unit will usually affect all dash gauges, not just fuel gauge.

FUEL GAUGE REMOVAL AND INSTALLATION

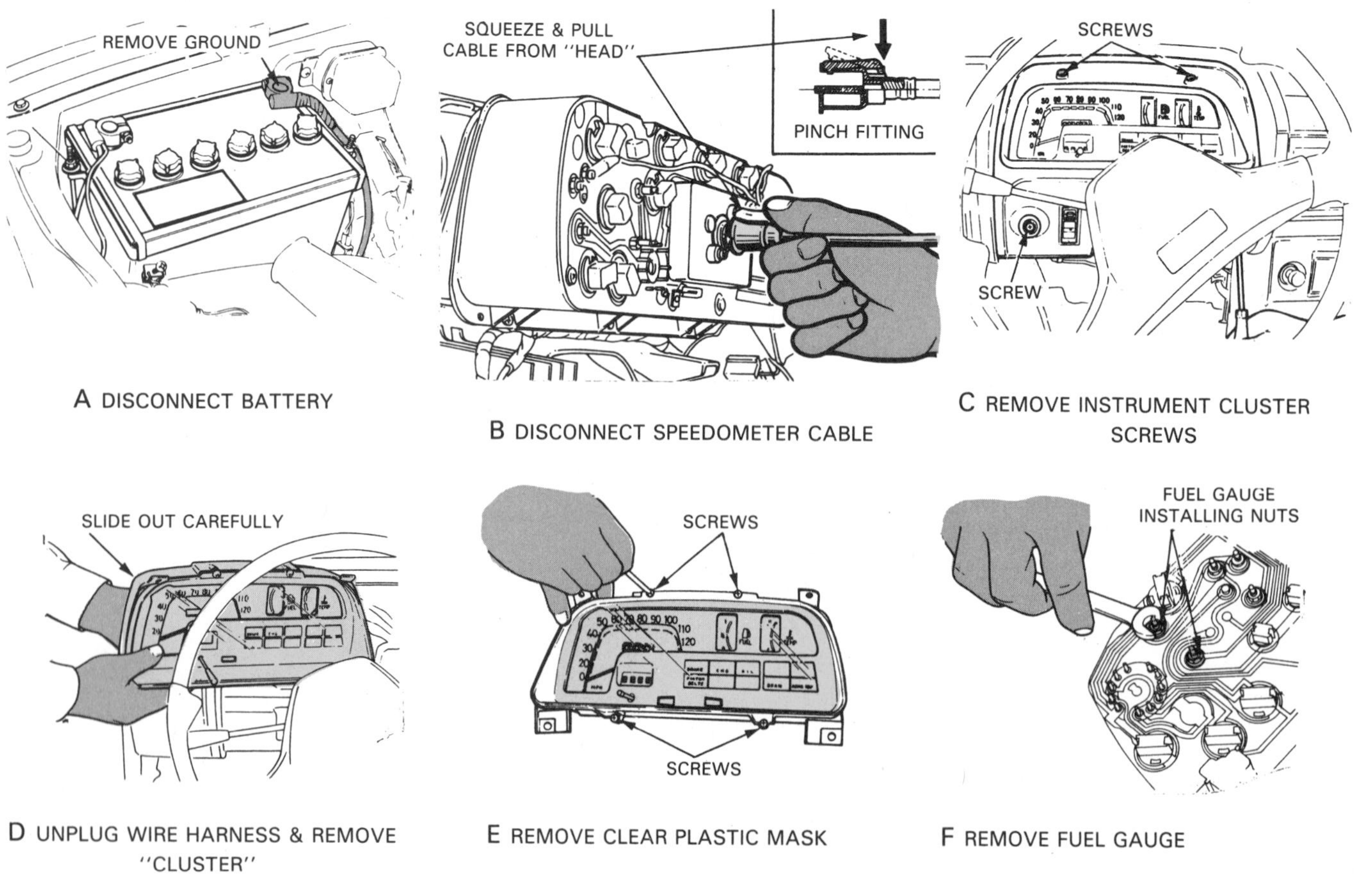

A DISCONNECT BATTERY

B DISCONNECT SPEEDOMETER CABLE

C REMOVE INSTRUMENT CLUSTER SCREWS

D UNPLUG WIRE HARNESS & REMOVE "CLUSTER"

E REMOVE CLEAR PLASTIC MASK

F REMOVE FUEL GAUGE

Fig. 6-67. Follow these six basic steps to remove a fuel gauge assembly. You will have to modify these directions slightly from car to car.

cluster. One type of low fuel gauge is shown in Fig. 6-68. It turns ON when around 3 to 5 gallons of fuel remain in the tank. There are three basic types of low fuel warning systems: a light emitting diode type, a switch type, and a thermistor type.

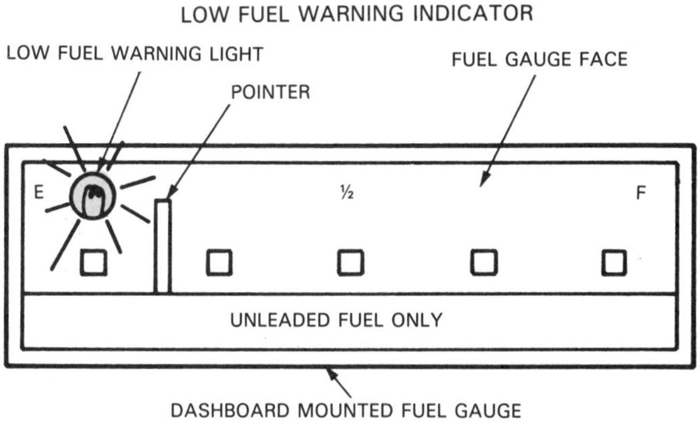

LOW FUEL WARNING INDICATOR

Fig. 6-68. Light on fuel gauge comes on to warn driver of low fuel level in tank.

LED fuel warning system

A LED (light emitting diode) *low fuel level indicating system* is illustrated in Fig. 6-69. It uses a balanced coil type fuel gauge, a light emitting diode circuit, and the conventional fuel tank sending unit. The tank unit controls both the fuel gauge pointer and the indicator light. Notice that the light emitting diode is connected across the "empty coil" in the gauge. When the current flow through the empty coil and tank sending unit becomes high enough, the voltage across the diode becomes high enough to illuminate the diode lamp.

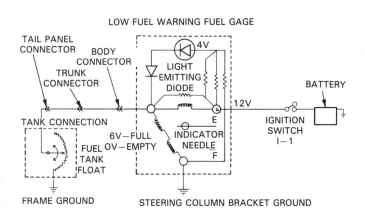

Fig. 6-69. When current flows through tank unit and empty coil is high, current flows through light emitting diode. With tank filled, current flows through full coil and voltage potential across the diode is reduced. This turns off light emitting diode. (Cadillac)

Diode circuit test

The low fuel warning diode can be tested using a common test light. The test light should be a 12 volt type requiring an external power source. It should NOT use batteries.

To test, first turn the ignition switch key to the ON position. This will supply electricity to the fuel gauge assembly.

Disconnect the "hot" wire from the fuel tank sending unit. There should be a connector at the tank. As shown in Fig. 6-70, ground the test light prong and connect the test light lead to the gauge wire. The test light should glow and the fuel gauge pointer should swing to empty. The low fuel warning light should operate.

If the warning light does glow, then the diode circuit and wires going to the tank are all right. The problem may be in the fuel tank sending unit. The tank unit is probably defective and should be replaced. Perform the tank sending unit test previously described.

If the test light does not glow when you ground the fuel gauge lead, there is an open (disconnected plug, broken wire, etc.) in the gauge circuit running to the tank. Trace back through the circuit until the open is found.

If the test LIGHT FUNCTIONS but the fuel gauge POINTER FAILS TO MOVE over to empty, the fuel gauge is defective. When the gauge pointer swings to empty but the warning light does not turn on, then the LED, itself, is burned out or the diode circuit is defective. Replace the diode, diode circuit, or gauge as needed.

TESTING DIODE TYPE LOW FUEL WARNING SYSTEM

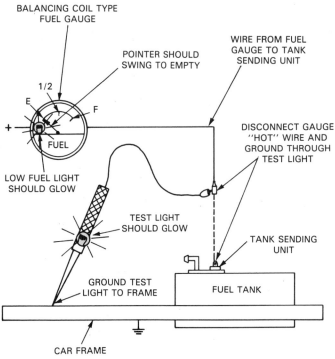

Fig. 6-70. To check operation of low fuel light, ground sending unit wire through a 12 volt test light. If low fuel light glows, then tank unit or tank ground may be bad.

Switch type low fuel warning system

The *electronic switch fuel warning system* consists of a transistorized low fuel level switch, an indicator light mounted inside the fuel gauge, console, or instrument panel, and connecting wires.

Fig. 6-71 shows the electronic switch connected across the terminals of the fuel gauge. The switch is controlled by the difference in voltage between the two gauge terminals.

For instance, when the fuel tank is nearly empty, the circuit inside the electronic switch is closed because very little current is flowing through the gauge. The voltage across the terminals is small. The switch can sense this and the warning light is illuminated.

When the tank is filled, a large amount of current will flow through the gauge and to ground in the tank sending unit. This causes a stronger voltage difference across the two gauge terminals. As a result, the electronic switch turns off the indicator light.

This type of low fuel warning system commonly has a "proveout" feature. When the ignition switch is first turned on, the low fuel bulb should glow. This tells the driver that the bulb and circuit are probably in working order.

Note: Some older cars use a relay type switch instead of an electronic switch. Basically, however, their operation is the same.

Switch type low fuel indicator testing

If the low fuel warning light does not glow with the key switch first turned on, check the indicator bulb. It may be burned out.

If the bulb is good, perform a low fuel switch test. The switch may also be at fault whenever the light stays on with more than 1/4 of a tank of fuel. It may also cause the indicator bulb to stay off all the time. Using a shop manual for the particular switch circuit, go through each step of a detailed switch test.

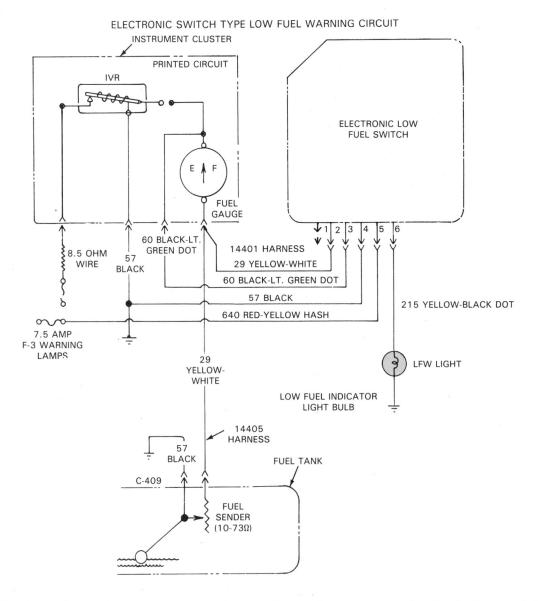

Fig. 6-71. The electronic switch can sense amount of current flowing through gauge and tank unit. When tank fuel level and current flow becomes low, switch turns on low fuel warning light.   (Ford)

If the switch is good, there may be a problem in the gauge circuit wiring. Trace each wire and check for an open (disconnected wire) or a short (wire touching ground).

Thermistor low fuel warning system

A *thermistor* is a temperature-sensitive resistor or semiconductor. Its internal resistance changes with temperature.

A thermistor type low fuel warning indicator circuit includes a dash light, a fuel tank thermistor, and connecting wires.

Illustrated in Fig. 6-72, the thermistor is fastened to the tank pickup tube. It is positioned an inch or so from the bottom of the tank. Whenever the ignition key is turned on, current flows through the thermistor. This heats the thermistor.

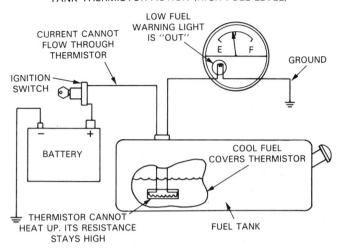

Fig. 6-73. When fuel covers thermistor, the thermistor cannot heat up. This keeps its resistance high and the low fuel light off.

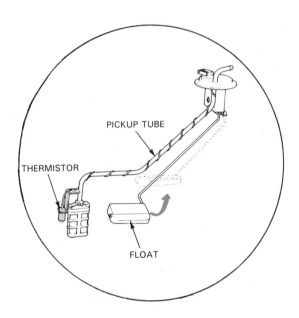

Fig. 6-72. Thermistor is connected to tank pickup tube. It operates low fuel warning light. (Honda)

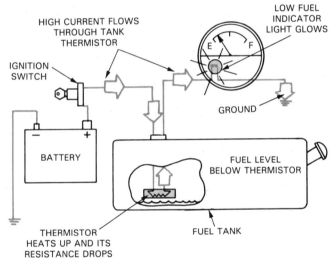

Fig. 6-74. Thermistor heats up when not covered by gasoline. As a result, thermistor's internal resistance drops and current flow increases. This turns on low fuel light in dash.

When hot, the thermistor's internal reistance is low. When it is cool, resistance is high. This principle is used to operate the low fuel warning light.

When the tank is full, the thermistor is covered with cool gasoline. See Fig. 6-73. It cannot be heated by the current flow. The cool thermistor will have a very high resistance which permits little current flow in the light circuit. The low fuel light will not glow. As the tank nears empty, Fig. 6-74, the thermistor is no longer covered by the cool gasoline. It can then heat up. The resistance of the thermistor is lowered and current can flow to the low fuel indicator bulb.

Troubleshooting thermistor low fuel warning system

If the indicator light fails to glow with the tank empty, isolate parts of the system to see whether the problem is in the bulb circuit or in the tank thermistor. Remove the wire from the thermistor at the tank. Ground the wire

through a 12 volt test light. (Do not ground the gauge wire too long or the fuel gauge could be damaged.) The dash warning light should glow. If not, check the relay, wire connectors, and other components.

If the indicator light functions with the thermistor lead grounded, the gauge circuit is functional and the thermistor may be defective. To test a thermistor for operation, remove the tank sending unit from the tank. Connect the thermistor gauge unit to the battery with your test light, Fig. 6-75. Then immerse the thermistor in a container of water. The test light should not glow with the thermistor in the cool water. When the thermistor is lifted from the water, it should glow. If the thermistor cannot turn the test light on and off, replace it.

TESTING FUEL TANK THERMISTOR

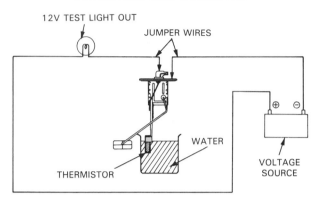

Fig. 6-75. To test thermistor, connect 12 volt test light across thermistor terminal and unit housing. Test light should glow with thermistor out of water and turn off as thermistor is immersed in cool water. If not, replace thermistor. (Chrysler)

MILES-TO-EMPTY INDICATOR

A *miles-to-empty* gauge predicts the maximum distance a car can be driven on the fuel remaining in the tank. This distance is computed by comparing the average fuel consumption of the engine with the amount of fuel in the tank.

As shown in Fig. 6-76, the gauge is operated by pressing the ON button. Mileage readout will flash on the face of the gauge and remain for a few seconds. Also, when mileage-to-empty is less then 50 miles, the readout may come up automatically as a reminder or warning.

MILES-TO-EMPTY GAUGE

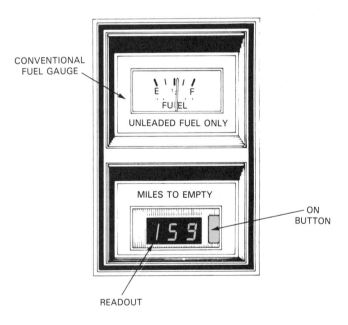

Fig. 6-76. Pressing button on miles-to-empty gauge, will generate readout for miles that can be driven on fuel remaining in tank. (Ford Motor Co.)

Miles-to-empty indicator operation

The number display is usually operated by a completely separate variable resistor in the fuel tank. Also in the circuit are a vehicle speed sensor, and an electronic control unit. See Fig. 6-77. The tank sending unit has two individual rheostats (variable resistors), one for the fuel gauge and one for the miles-to-empty indicator.

The electronic module uses the current flow to the fuel tank and the current flow through the speed sensor to calculate the proper miles-to-empty number. It then displays this computer-calculated number on the face of the gauge. To test and troubleshoot this system, check a shop manual for specific directions.

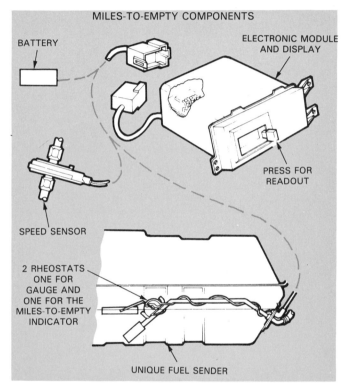

Fig. 6-77. The electrical signals from speed sensor and tank unit are fed into electronic module. Module processes these signals and operates miles-to-empty number readout. (Ford)

FUEL CONSUMPTION INDICATOR

A *fuel consumption* or *fuel economy indicator* tells how much fuel the engine is using. A dash light or lights, Fig. 6-78, tell the driver about basic fuel consumption. By measuring engine intake manifold vacuum, the fuel economy indicator can tell the driver when to ease up on the gas pedal or even when to shift transmission gears. It helps the driver use ideal driving habits for best gas mileage.

A green light may glow when gas mileage (for that specific vehicle) is good and a red one comes on when gas mileage is poor.

Engine intake manifold vacuum is an excellent indicator of fuel consumption. See Fig. 6-79. For instance, when you push the gas pedal all the way to the floor for

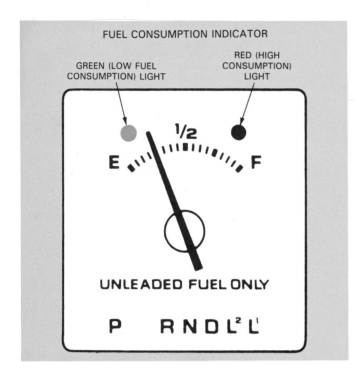

FUEL CONSUMPTION INDICATOR

GREEN (LOW FUEL CONSUMPTION) LIGHT

RED (HIGH CONSUMPTION) LIGHT

1/2

E   F

UNLEADED FUEL ONLY

P R N D L² L'

Fig. 6-78. When driving techniques are efficient, green light will glow. If gas pedal is depressed too rapidly, red light will come on to indicate high fuel consumption. (Buick Div., General Motors Corp.)

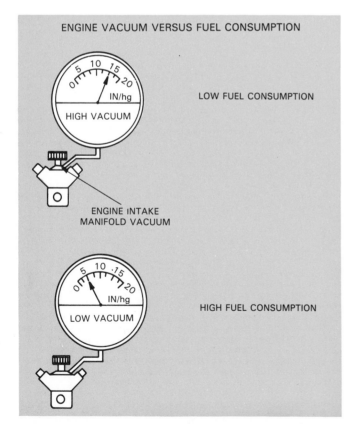

ENGINE VACUUM VERSUS FUEL CONSUMPTION

5 10 15
0   20
IN/hg
HIGH VACUUM

LOW FUEL CONSUMPTION

ENGINE INTAKE MANIFOLD VACUUM

5 10 15
0   20
IN/hg
LOW VACUUM

HIGH FUEL CONSUMPTION

Fig. 6-79. Engine vacuum is excellent indicator of fuel use. Generally, high vacuum indicates good gas mileage while low engine vacuum indicates poor gas mileage. This principle is used by fuel consumption indicator.

fast acceleration, engine vacuum drops very low. The throttle is wide open and outside air is rushing into the engine trying to fill the vacuum. This rush of air draws in a large amount of fuel. As a result, fuel economy is reduced.

Low engine vacuum indicates that the throttle or air valves are open too far for ideal economy and efficiency. On the other hand, when the gas pedal is depressed slowly and easily, intake manifold vacuum stays high, and so does gas mileage. A fuel economy system circuit is pictured in Fig. 6-80. Trace current flow.

FUEL ECONOMY (CONSUMPTION) INDICATOR CIRCUIT

ECONOMY LIGHT WIRING

GREEN AMBER

TELL-TALE ASSEMBLY

TELL TALE CONNECTOR

NORMALLY OPEN CONTACT

NORMALLY CLOSED CONTACT

IGNITION SWITCH I-1

ECONOMY LIGHT SWITCH

BATTERY

ENGINE VACUUM

GROUNDS THROUGH MOUNTING

Fig. 6-80. Amber or red light will glow with engine off. This feature shows that high consumption indicator is operational. Study components of this circuit. (Cadillac)

Troubleshooting fuel consumption indicator

If the fuel economy lights fail to function properly, turn the ignition switch to ON and ground each of the wires going to the vacuum or economy switch. As each wire is grounded, the corresponding dash light should glow. If not, check for loose connections, burned out bulbs, or broken wires.

If the dash lights glow when the lead wires are grounded, the vacuum switch may be bad. To check the switch, connect a 12-volt test light. This is pictured in Fig. 6-81. The engine should be stopped. With the test light connected to power (battery positive post, fuse box, etc.), touch the test light to each terminal of the vacuum switch. The light should glow when touching the normally closed contact terminal and should stay off when touching the other.

Next, connect a hand vacuum pump to the switch, Fig. 6-82. Pump about 4 to 6 in. of vacuum into the switch. This should operate the switch and close the set of contacts. Your test light should now glow when touching the normally open terminal. If not, replace the switch. Note: if a vacuum pump is not available you can start the engine in order to apply a vacuum to the switch.

Reconnect the switch to the engine and observe its operation. Start and idle the engine. The green light should glow at idle. Push the gas pedal quickly down to the floor and release it. This should make the red light

TESTING FUEL CONSUMPTION VACUUM SWITCH

NO VACUUM APPLIED (HIGH CONSUMPTION SWITCH POSITION)

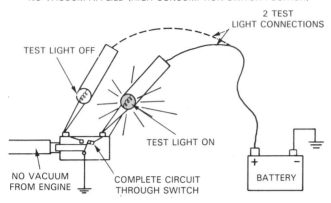

Fig. 6-81. When engine is off or when no vacuum is applied, test light should glow when touching normally closed (high consumption) side of switch. It should stay off when touching other terminal.

VACUUM APPLIED (HIGH ECONOMY SWITCH POSITION)

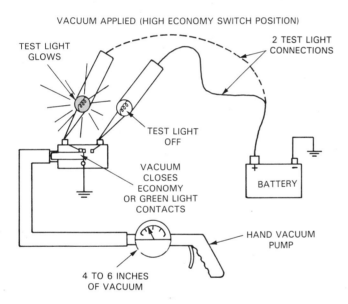

Fig. 6-82. As vacuum is applied to switch, test light should glow when touching the normally open contact terminal (low fuel consumption terminal). It should not glow when touching other terminal. If switch fails any part of this test, it should be replaced.

flash on and off.

If the lights still fail to function, check the vacuum line running to the engine. Feel the end of the hose. Is vacuum reaching the switch? The vacuum line may be cracked, kinked, or disconnected.

TRIP COMPUTER

A *trip computer* can be used to perform several calculations and find:
1. Miles to empty tank.

2. Estimated time of arrival at destination.
3. Distance to destination.
4. Time, day, month, and date.
5. Present and average fuel consumption.
6. Fuel consumed.
7. Average speed.
8. Miles traveled.
9. Elapsed driving time.

The number of calculations can vary with the particular trip computer system. Some cannot do as many things.

Like the miles to empty indicator, a trip computer also uses sensors and a computer to display the information in the dash. Fig. 6-83 shows a block diagram of one type of trip computer. Note how it uses a speed sensor, injector pulse width sensor, fuel tank sensor, and other sensors to feed data to the computer.

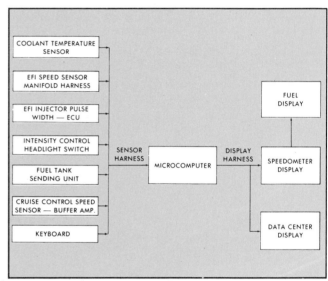

Fig. 6-83. Microcomputer is heart of trip computer system. Sensors located at various critical points feed information to computer.

Fig. 6-84 gives a wiring diagram for another trip computer system. A keyboard is provided for requesting data and making some of the calculations. See Fig. 6-85.

Trip computer operation

At cruising speeds with the engine at full operating temperature, vehicle speed (usually sensor in speed control system), engine speed (sensor in ignition system), engine temperature (coolant sensor), fuel used (injector pulse width sensor), and fuel in tank (in-tank sensor) information is fed to computer. The driver can then use the computer to find mpg, miles to empty, average speed, and other information. The computer is preprogrammed (preset) to analyze this data and provides the requested output at the display panel.

Trip computer service

Most modern trip computer systems have self-diagnostic capabilities. When the computer is activated,

136

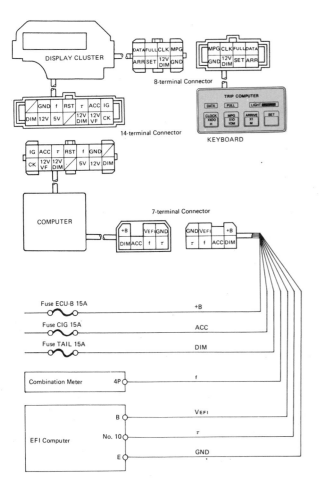

Fig. 6-84. Diagram of another trip computer. Note keyboard for making calculations and calling up data.

codes will show in the display. These number codes will be "broken down" (answer given) in the service manual. Usually, a chart and detailed procedures for checking system operation will also be given.

Listed below are a few service rules to remember when working on a trip computer system:

1. Before troubleshooting, check operation of system to detect trouble symptoms.
2. Check wiring harness connections. Some may be loose, causing a poor electrical connection.
3. When measuring voltage, current, resistance, use a high quality, high impedance (resistance) digital meter. A low quality, needle type meter could draw too much current and ruin electronic components in the system.
4. Do NOT attempt to disassemble and repair individual components. Normally, just replace them.
5. Do NOT make tests by applying direct battery voltage to components unless instructed to do so by the service manual.
6. Do not let your fingers or tools touch electronic components. Static electricity or oil on your fingers may upset their operation.
7. Never reverse battery connections. Component damage may result in the trip computer system.
8. Do NOT disconnect the battery when the engine is running. This could cause instant reverse charge of up to 100 volts which could burn out components.
9. Always disconnect the battery before disconnecting parts in the trip computer system.
10. Refer to the service manual for detailed directions for testing the system. System designs vary and so do testing methods.

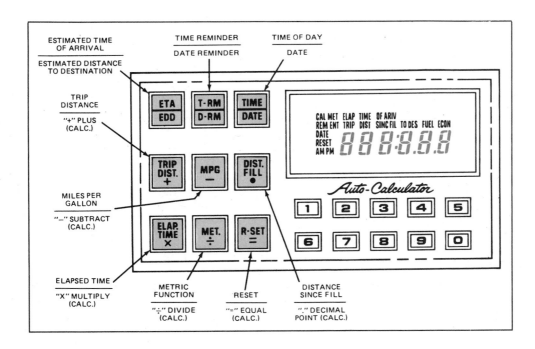

Fig. 6-85. Close-up of keyboard for a trip computer. Note buttons for functions.   (Oldsmobile)

SUMMARY

A gasoline tank should be light and strong. It should use a minimum amount of passenger and storage space. A metal tank should resist rust and corrosion and should have internal baffles to limit fuel surging. Also, a method of trapping fuel vapors should be provided.

Cars designed for unleaded fuel use a filler neck restrictor. It prevents accidental pumping of leaded fuel into a vehicle designed for unleaded gasoline.

Most fuel tank caps have pressure and vacuum valves built into them. These valves are normally closed, and seal gasoline vapors into the tank.

The fuel tank pickup unit or sending unit has a tube extending nearly to the bottom of the tank for fuel removal. It has a variable resistor unit and float mechanism for controlling the fuel gauge pointer. It also has a micro-mesh filter to prevent dirt and water from entering the fuel system.

A fuel tank is held in place by large metal straps or it may be bolted directly to the car by its flanges. Felt or rubber insulators are fitted between the tank, straps, and car body.

Today's automobiles use a closed fuel tank vent system, part of the evaporative control system. Vent lines from the top of the tank connect to a liquid vapor separator. A main vent line leaves the separator and runs to the charcoal canister in the engine compartment. The canister is filled with activated charcoal which absorbs and stores fuel vapors.

A fuel return system helps prevent excessive heating of gasoline in the fuel system and resulting vapor lock. It constantly circulates cool fuel from the gas tank through the fuel pump and back to the gas tank.

A fuel tank must be drained before servicing. The fuel can be siphoned or it can be pumped out with an electric or commercial pump.

Never attempt to repair a fuel tank without first receiving detailed instruction. A fuel tank must be cleaned, filled with water or inert gas before welding or soldering. Fiberglass kits are also available for repairing fuel tank leaks.

The two basic types of fuel gauges are the balancing coil (magnetic) type and the thermostatic strip (thermal) type. The magnetic type reads full when the tank unit has a high resistance and empty with low resistance. The thermal type is just the opposite.

A low fuel warning system uses a small dash mounted light which alerts the driver of the low fuel situation. The three basic types of low fuel system include: the diode type, electronic switch type, and the thermistor type.

A miles-to-empty gauge is an optional replacement for the low fuel warning system. It displays a number read out which indicates how many miles can be driven on the fuel remaining in the tank.

A fuel consumption indicator or fuel economy indicator uses engine vacuum, a vacuum switch, and light circuit to inform the driver of general fuel use. A vacuum switch, operated by engine vacuum, controls indicator lights in the dashboard.

KNOW THESE TERMS

Driving range, Filler neck, Baffles, Vent tubes, Expansion chamber, Fill control tube, External expansion chamber, Internal expansion chamber, Filler neck restrictor, Rollover check valve, Fuel tank pickup assembly. Pickup pipe sending unit, Vapor separator, Vapor recovery line, Charcoal canister, Purge line, Fuel cell tank, Float, Float arm, Magnetic fuel gauge. Thermostatic fuel gauge, Voltage limiter, LED fuel warning system, Low fuel warning system, Thermistor.

REVIEW QUESTIONS—CHAPTER 6

1. Which of these terms does NOT relate to fuel tanks:
   a. Capacity.
   b. Baffles.
   c. Vents.
   d. Area.
2. List five desirable features of a well designed fuel tank.
3. Fuel tanks are usually made of thin aluminum. True or false?
4. To prevent fuel surging or splashing, fuel tank _____ are used.
5. Most new car fuel tanks CANNOT be completely filled with fuel; they have a _____ percent _____ _____.
6. List the four basic methods of producing fuel expansion chambers.
7. Some filler pipes have a smaller filler control tube attached and running parallel to them. True or false?
8. A _____ _____ _____ prevents accidentally pumping leaded fuel into cars with catalytic converters.
9. A vented fuel tank cap (should/should not) be used on modern cars with emission controls.
10. Why are pressure-vacuum fuel caps used?
11. The pressure valve in a fuel cap will open around _____ and the vacuum valve opens at _____.
12. How do you inspect a fuel cap?
13. Which of the following are functions of the fuel tank pickup unit?
    a. Prevent overfilling the fuel tank.
    b. Provide a means for drawing fuel from the tank.
    c. Provide a sensing unit to control dashboard fuel gauge.
    d. Filter fuel going to engine.
    e. All of the above.
14. A tank pickup assembly (sometimes/never) includes the fuel pump.
15. A sending tank unit is held in place by a _____ and sealed with neoprene _____.
16. Fuel tanks (must/need not) be drained before removal of the fuel tank pickup assembly.
17. Describe how you should visually inspect a tank sending unit.
18. A gasoline tank may be held in place by large _____ _____ or by bolting the tank's

heavy _____ directly to the car.

19. What may happen if the fuel tank insulators are not replaced?

20. Which of the following problems may result from clogged vents in a fuel tank?
    a. Expansion of fuel in tank could force fuel through fuel system.
    b. Vacuum buildup could collapse tank.
    c. Vacuum in tank could prevent fuel flow.
    d. Fumes could escape to the air.
    e. Gasoline could leak out onto ground.

21. Closed type tank vents are used to stop fuel _____ from entering and _____ the _____.

22. How does a fuel tank vapor-liquid separator operate?

23. This unit is used to store excess gasoline fumes.
    a. Vents.
    b. Purge line.
    c. Charcoal canister.
    d. Fuel bowl.

24. A _____ _____ is a hose which allows engine vacuum to draw fuel vapor out of charcoal canister.

25. To prevent gasoline from leaking out of a car after an accident, a _____ _____ is often used somewhere in the fuel transfer system.

26. A fuel return system (check only statements that are NOT true):
    a. Reduces chance of vapor lock.
    b. Consists of a special outlet on the fuel pump to attach a return line between the pump and tank.
    c. Prevents gas line freeze.
    d. Recirculates certain amounts of fuel between engine and the fuel pump to prevent vapor lock.

27. An in-line vapor separator connected to the output side of a fuel pump helps keep _____ _____ from reaching the engine.

28. Give three methods of draining a fuel tank, besides removing a drain plug.

29. One method of cleaning a tank is to thoroughly _____ clean it inside and out; another method is to fill the tank with a mixture of water and a special _____ _____ sloshing the mixture around in the tank for 10 minutes.

30. The two basic types of fuel gauges are the thermal and the magnetic. True or false?

31. How does the balancing coil type fuel gauge cause pointer movement?

32. The magnetic or balancing coil type fuel gauge reads empty when the tank unit resistance is _____.

33. A balancing coil type gauge requires a voltage regulator. True or false?

34. First step in troubleshooting a balanced coil fuel gauge is to _____ and _____ the hot wire

to the tank. Then turn (on/off) the ignition switch.

35. Explain how a thermostatic type fuel gauge moves its pointer toward full.

36. A thermostatic fuel gauge (does/does not) require the use of a voltage regulator.

37. How do you quickly disgnose a faulty thermal type fuel gauge?

38. The three basic types of low fuel warning systems include: _____ _____, _____ and _____.

39. Which of these tools would most commonly be used to quickly troubleshoot a low fuel warning system?
    a. Diode tester.
    b. Voltmeter.
    c. Test light.
    d. Ammeter.

40. How does a thermistor function?

41. When vacuum drops below _____ or _____ in. of mercury in a fuel economy indicator system, the _____ fuel consumption light is turned on by the _____ _____.

42. Why should fingers or tools be kept from touching electronic components of a trip computer?

43. Explain why a quality high impedance digital meter should be used when checking out a trip computer.

44. One should (always/never) disconnect the battery when working on a trip computer.

45. A metal fuel tank is going to be soldered to fix a small leak. Mechanic A says that after a thorough steam cleaning, it is safe to braze or solder the tank. Mechanic B thinks that, in addition to the steam cleaning, the tank should be filled with water or some other nonflammable fluid (inert gas) before attempting repairs.
    Who is right?
    a. Mechanic A.
    b. Mechanic B.
    c. Neither Mechanic A nor Mechanic B.
    d. Both Mechanic A and Mechanic B.

46. Acting on a customer's complaint that a thermostatic fuel gauge is not working, the mechanic disconnected and grounded the hot wire going into the tank. When the ignition was turned on, the fuel gauge indicator moved to full. When the tank unit was disconnected, the fuel gauge indicator moved to empty.
    Mechanic A says that the trouble is not in the fuel gauge but must be in either the sending unit or the tank ground. Mechanic B says that both the dash fuel gauge and the sending unit are functioning properly and the tank has lost its ground.
    Which mechanic is correct?
    a. Mechanic A.
    b. Mechanic B.
    c. Neither Mechanic A nor Mechanic B.
    d. Both Mechanic A and Mechanic B.

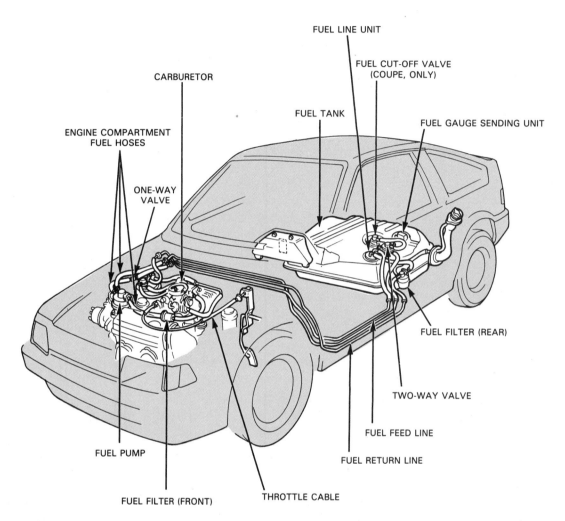

FUEL LINE UNIT

FUEL CUT-OFF VALVE
(COUPE, ONLY)

CARBURETOR

FUEL TANK

FUEL GAUGE SENDING UNIT

ENGINE COMPARTMENT
FUEL HOSES

ONE-WAY
VALVE

FUEL FILTER (REAR)

TWO-WAY VALVE

FUEL FEED LINE

FUEL RETURN LINE

FUEL PUMP

FUEL FILTER (FRONT)

THROTTLE CABLE

Check shop manual for service information on fuel tanks, gauges and fuel pumps.   (Honda)

FUEL LINES, HOSES, FILTERS—OPERATION, CONSTRUCTION, SERVICE

7

After studying this chapter, you will be able to:
* *Explain how to select and install double-wrapped steel tubing in a fuel system.*
* *Describe how to form a double-lap flare.*
* *Identify common types of fittings used in fuel systems.*
* *Explain the types of hoses found in a fuel system.*
* *Properly install fuel systems hoses.*
* *Describe fuel filters and strainers.*
* *List the typical locations for fuel filters.*
* *Summarize how to replace or clean a fuel filter.*
* *Explain a diesel two-stage fuel filter system.*
* *Describe a diesel water-in-fuel detection system.*
* *Test, drain, and repair a water-in-fuel detection system.*
* *Clean and flush a fuel system.*

It is almost impossible to service any part of a modern fuel system without handling a fuel line or a fuel filter. A fuel system mechanic must have a thorough understanding of fuel filters and lines and possess basic tube-working skills (cutting, reaming, bending, and flaring). You should be able to select, inspect, diagnose, remove, install, and repair these components. This chapter explains the most essential information related to fuel line and filter operation, service, and repair.

FUEL SYSTEM TUBING

Basically, there are three general types of tubing: RIGID (thick-wall steel), SEMIRIGID (copper, brass, aluminum, and thin-wall steel), and FLEXIBLE (hose, plastic tubing, etc.). Fuel systems commonly use only rigid tubing and flexible hoses. See Fig. 7-1.

DOUBLE-WRAPPED, BRAZED STEEL TUBING

Fuel lines, Fig. 7-2, are normally made of *double-wrapped, brazed steel tubing* for safety reasons. This type of tubing is made by wrapping two layers of thin metal together. Layers are then brazed (welded) together.

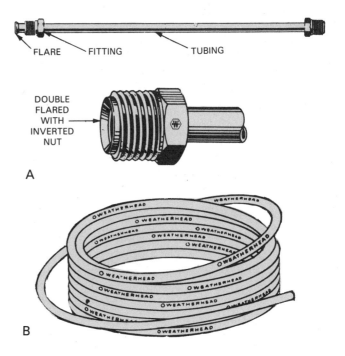

Fig. 7-1. Types of tubing. A—Rigid steel tubing is made in four sizes (3/16, 1/4, 5/16, 3/8 in. O.D.). It comes assembled with inverted nuts on each end. Ends are already flared. B—Plastic tubing is used in many under-the-hood applications for low pressure air or fluid lines. (Weatherhead Co.)

A coating of copper alloy is usually added to the outer surfaces of the tube to protect it against rust and corrosion. The result is a fuel line that is very strong and resistant to vibration, heat, pressure, and rusting.

Never use copper, brass, aluminum, or thin-wall steel tubing when repairing or replacing a fuel line. Vibration and movement can easily fatigue (weaken) these types of tubing and cause failure or breakage. A ruptured fuel line can cause fire. ONLY USE DOUBLE-WRAPPED STEEL TUBING!

141

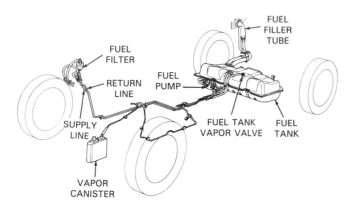

Fig. 7-2. For safety, lines carrying fuel must be double-wrapped steel. (Ford)

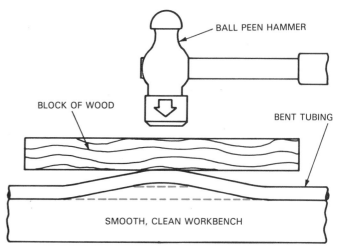

Fig. 7-4. A block and hammer can be used to take small bend out of metal tubing.

Handling roll tubing

Tubing is sometimes coiled into a roll and must be straightened before use. Only short pieces (2 to about 6 ft. long) can be purchased in lengths. Never try to unroll coiled tubing by pulling the tubing straight out sideways from the coil. In its spiral shape, this will kink the tubing.

To unroll coiled tubing, first determine how much tubing is needed. A flexible tape measure can be used to accurately measure the old fuel line. Place the coil of tubing on a clean bench as in Fig. 7-3. While holding the end of the tube down against the bench, unroll the coil with your other hand. The tubing will unroll with a minimum of curves, twists, and bends.

Try to avoid unnecessary bending of the tubing as metal fatigue and weakening will result.

Store tubing where it will not be damaged by heavy tools or equipment. Keep it out of busy work areas. Dents or bends will have to be straightened before the tubing can be used. A piece of masking tape wrapped over the end of the tubing will keep out dirt.

Straightening tubing

Bends, kinks, and dents can be straightened as illustrated in Fig. 7-4. Place the bent section of tubing on a smooth, clean bench top with the convex part of the bend up. Lay a block of wood lengthwise over the bent section. A few light taps with a hammer will straighten it. Avoid hitting the block too hard; you can flatten the tubing.

When out-of-round or flattened, the tubing can be clamped in either a special rounding and crimping tool or a common flaring bar, Fig. 7-5. Clamp the bent part of the tubing in the appropriate sized opening. Tighten the two bars to reshape the tubing.

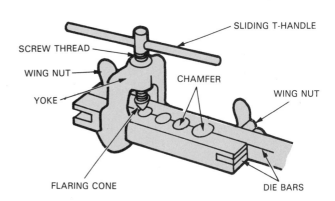

Fig. 7-5. Flattened, tubing can be rounded using this flaring tool with flaring cone assembly removed. (Florida Dept. Voc. Ed.)

Cutting tubing

A *tubing cutter* should be used to cut off tubing square. Squareness is essential when the tubing must be flared.

A tubing cutter is, by far, the fastest, safest, and most precise tool for this purpose. A fine-toothed hacksaw can be used, but the cut will not be as square nor as smooth. A saw cut will usually require hand filing and excessive reaming to prepare the tube end for flaring. A tube cutter will leave very little burr on the inside of the tube and usually no burr at all on the outside.

Using a tubing cutter

To use a tubing cutter, place the section of tube inside the cutter as shown in Fig. 7-6. Tighten the thumb screw

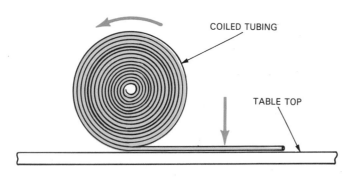

Fig. 7-3. When uncoiling tubing, be very careful not to bend, kink, or twist the line. Hold coil vertically against work surface.

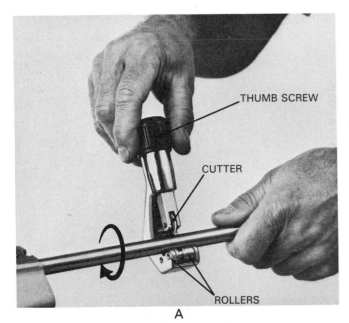

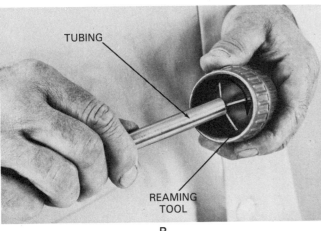

Fig. 7-6. Cutting tubing. A—Tubing cutter is, by far, the fastest and most accurate way of cutting metal tubing. B—Reaming tool is used to remove burrs from inside tubing. (Imperial)

or knob until the tube is lightly clamped into position between the cutter and the rollers. Rotate the cutter as shown. Steadily tighten the cutter down on the tube.

Be careful not to overtighten or the tube will be dented and an unwanted lip will be formed on the inside of the tubing. Tighten the cutter only enough to produce a slight drag on rotation. Continue turning and tightening the cutter until the cut is completed.

Reaming off burrs

The end of the tube should be reamed to remove burrs or rough pieces of metal that result from cutting. Small pieces of metal left hanging on the inside of the tubing could break loose while the tubing is in use and cause a fuel system malfunction. Also, a burred inner edge will be smaller in diameter than the rest of the tubing. Such reductions can restrict fuel flow through the line.

Quite often, a tube cutter will be equipped with a *reaming knife.* It can be swung from the safety position into

the reaming position.

Insert the cutter knife or reamer into the tubing. Hold the tube opening down so that the metal shavings will NOT fall into the tubing. Push the reamer into the tube squarely and rotate, Fig. 7-6B. Watch the cut closely and be careful not to over-ream. Over-reaming will reduce the thickness of the tubing wall and can cause poor flares (weak, cracked, etc.).

Cleaning tubing

Tubing should be cleaned after cutting and reaming. Injecting a blast of compressed air into each end of the line will usually remove foreign matter. In extreme cases, (dirt, oil, metal shaving, etc.) clean solvent or gasoline should be run through the line. Then compressed air should be used. NEVER install any fuel line without cleaning it.

Bending tubing

Automotive fuel lines will usually have numerous compound bends and curves. The new tubing must be bent to match the old fuel line.

Bending spring

For most jobs, a *bending spring* will do the job. It is acceptable when accuracy is not critical. To use a bending spring, insert the tube into the spring. Then, as pictured in Fig. 7-7A, carefully bend the tubing and spring together. The spring will spread the bending force over a longer section of tubing and will help prevent kinking or over-bending. (You may have to bend the spring

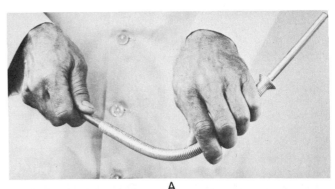

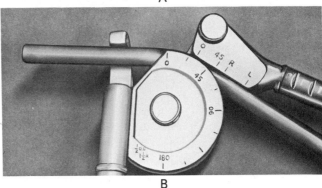

Fig. 7-7. Two kinds of tubing bender. A—Bending spring will help prevent kinks when bending tubing. Slide tubing into spring as shown. B—Bending bar is especially useful when several identical bends are required. Scale on tool indicates angle of bend. (Imperial)

a little farther than the desired bend.) The tubing and spring may flex back slightly. Remove the spring and compare the new bend with the old one.

Bending bar

To make a bend, slip the bender over the tubing. See Fig. 7-7B. Then, pull the two arms slowly together. Stop when the tubing is properly bent. Check that the tubing does not spring back. Once bend is completed, simply release the handles and remove the tubing. The degree scale is useful when making a number of identical bends.

Note! Never make a bend TOO CLOSE TO THE END of the tubing. There should be enough room for a fitting. If the tubing is to be flared, the flare fitting will have to slide onto the tubing. Also, you will need room to clamp the flaring tool bar around the tubing.

Flaring tubing

Tubing is flared by forcing or spreading the end of the tubing outward. This funnel shaped end is used with a special tube fitting to form a leakproof connection. The two flare angles are 37 and 45 degrees. The 45 degree flare is more commonly used on automotive fuel lines.

Flaring tools

A *flaring tool* is used to form or expand the outer edge of the tubing into a flare. Two basic types of flaring tools are available. The older design is a COMPRESSION TYPE and the new one is a GENERATING TYPE. See Fig. 7-8.

The older tool has a flaring cone which spreads the tube end until its outer diameter is pinched against the profile die. With the newer generating type, the flaring bar holds the tube end slightly above the bar surface. Then, the flare is spread and formed.

Types of flares

The two basic types of flares are the *single-lap* and *double-lap* flares. They are illustrated in Fig. 7-9. Since automotive fuel lines use double-wrapped steel tubing, a double-lapped flare MUST be used. If NOT, the flared end may split or leak and cause a fire.

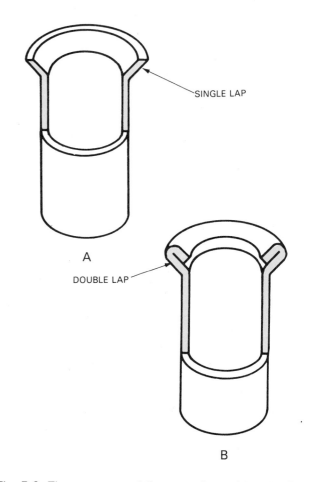

Fig. 7-9. The two types of flares used on tubing. A—Single lap flare is used on aluminum tubing. B—Double lap flare is used on steel tubing.

Forming a double-lapped flare

The most widely used double-lap flaring tool is one that consists of a bar, yoke, and set of adapters. This type is shown in Fig. 7-10.

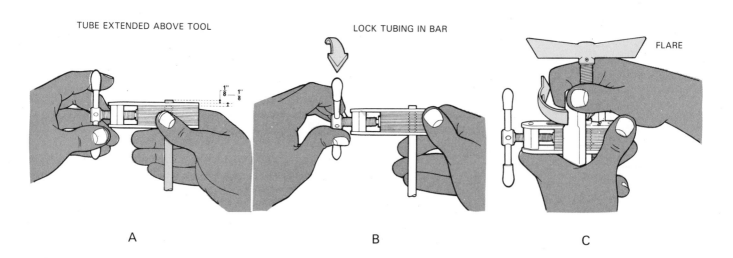

Fig. 7-8. Generating type flare tool. A—Position and lock tube about 1/8 in. above the flare bar. B—Lock bar on tube. C—Position flaring cone over tube. Tighten yoke handle. This will form flare. (Imperial-Eastman)

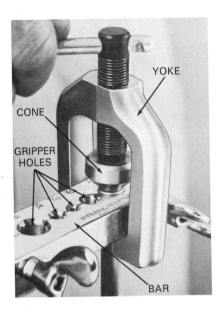

Fig. 7-10. Flaring tool is used to expand end of tubing outward. Then it can be used with flare type fitting to form leakproof connection. (Imperial-Eastman)

To make a double flare, clamp the tube tightly in the *flaring bar.* Select the right size gripper hole and extend the end of the tubing about 1/8 in. (3.2 mm) above the bar. Check with your specific tool instructions if in doubt. After clamping, fit the right adapter into the end of the tubing. See Fig. 7-11.

Next, slip the screw and yoke mechanism into place. The cone should be directly over the tubing and the tool arms around the bar. Screw the cone down on the tubing and adapter until it bottoms against the flaring bar as in view A. Do not force it, but bottom it snugly.

Now, back off the cone and remove the adapter. The tubing should be folded inward and belled outward. If not, re-cut and repeat the operation. Finally, screw the cone, without the adapter, down onto the tubing, view B. This will fold down the inner edge of the tubing to produce a very strong and accurate double-lap flare.

Note! Do not forget to slide your fittings onto the tubing before flaring both ends of the tubing. The fittings cannot be installed after flaring.

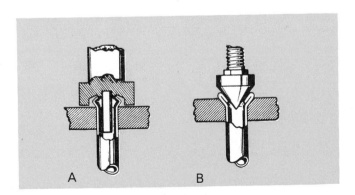

Fig. 7-11. Making double-lap flare. A—First force adaptor and cone down onto tubing. B—Remove the adapter and screw just the cone into the tube. (Bendix)

FLARE FITTINGS

Flare fittings are used with flared tubing ends to connect a fuel line to another component such as a carburetor, injection system pressure regulator, fuel pump, or other section of tubing. The correct selection of fittings is important, not only to the quality of the connection but to the amount of time taken to complete the repair.

The two basic types of flare fittings are the SAE type and the inverted flare type.

The *SAE fitting,* shown in Fig. 7-12A, uses a long female nut that fits over the fitting body. In contrast, the *inverted flare fitting,* Fig. 7-12B, has a male nut fitted into a female body.

The inverted flare may be somewhat more common than the SAE type in automotive fuel systems.

PIPE FITTINGS

Pipe fittings are parts designed to make connections on pipes. They have tapered threads as illustrated in Fig. 7-12C. As a result of this design, when two pipe fittings (male and female) are threaded together, a leakproof seal is formed. The threads jam together to prevent leakage.

Pipe threads are sometimes used in automotive fuel systems. They can commonly be found in soft aluminum or pot metal components (carburetors, fuel pumps, etc.), where a metal fuel line must be attached.

Pipe fitting assembly

When connecting a pipe fitting, make sure the two threads are perfectly aligned. Run the fitting in with your fingers. You may have to hand bend the steel tubing slightly because perfect alignment is critical. If the pipe fitting is started with a wrench, the fitting can crossthread and strip very easily. After proper starting, use a tubing wrench to snug up the fitting. Usually, about two to three turns will seal the fitting.

CAUTION: Pipe fittings require very little torque. They will seal very easily, with very little pull on the wrench. Overtightening can strip threads. This is especially true with aluminum parts.

It may be desirable to use sealer on fuel system pipe fittings. Two basic types of sealer are available. One is a soft paste (pipe dope) sealer. It can be purchased in a can or tube. The second is a special teflon tape pipe thread sealer.

When using a *paste type sealer,* wipe a little sealer onto the pipe threads. Avoid using too much as the sealer could get into the fuel system.

The *tape type sealer* is simply wrapped around the male pipe threads. When the fittings are screwed together, the tape is smashed between the threads. This helps seal the threads against the fuel system pressure.

COMPRESSION FITTINGS

As implied, a *compression fitting* seals a connection by compressing or squeezing the tubing against the fitting. A compression fitting may use either a separate or

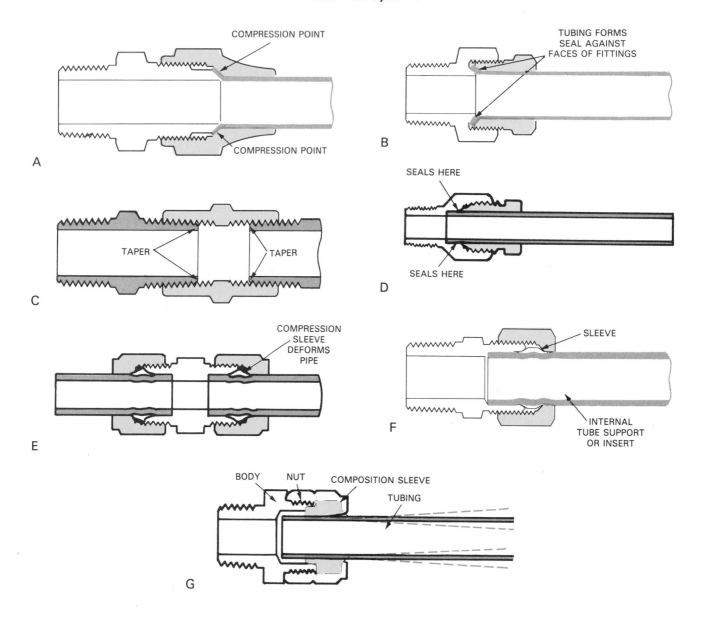

Fig. 7-12. Cutaway views of flare, pipe, and compression fittings. A—SAE fitting. (The Weatherhead Co.) B—Inverted flare fitting. (Note how flared tubing seats against fitting.) C—Pipe fitting has tapered threads. D—Double compression fitting. E—Compression fitting with separate sleeve. (Gould) F—Flareless compression fitting. G—Flexible compression fitting.

an integral (built-in) sleeve. By tightening the fittings together, the tubing is pinched, held, and sealed by the collapsing action of the sleeve.

Double-compression fittings

With a *double-compression fitting,* Fig. 7-12D, the nose of the male fitting is pressed inward against the tubing. This seals the connection.

Separate sleeve compression fitting

A compression fitting with a separate sleeve is illustrated in Fig. 7-12E. The compression sleeve is not part of the fitting nut. The sleeve is slipped over the tubing after the nut. When tightened, the nut presses the sleeve against the fitting body. This smashes the sleeve inward against the tubing to produce a leakproof seal.

NOTE! Sleeve type fittings are acceptable for use in a fuel system only when vibration is not excessive. They should not be used on or near the engine.

Since flaring is not needed, sleeve fittings are very handy and fast. They do not require the more time-consuming flaring process of other fitting types.

CAUTION! Never use compression fittings on high pressure applications (fuel injection lines, diesel injection lines, etc.). They are only acceptable for use on low pressure systems.

Assembling compression fittings

First, check that the tubing is in perfect alignment with the fitting. Then, slide on the nut and a sleeve (if used). Insert the tubing all the way into the fitting body and hold it in. Be careful not to release the tubing and let it slide out of the fitting or you will ruin the job. Keep this in mind at all times.

While holding the tubing, tighten the nut until it just grasps the tubing. Then, tighten it about 1 1/2 turn. This should lock the tubing into the fitting and form a leakproof joint.

When reassembling an old compression fitting, do not try to turn the fitting nut 1 1/2 turn beyond initial snugging. A used compression fitting should be tightened until snug, no specific number of turns can be recommended.

Compression fittings for plastic tubing

Rigid or hard plastic tubing can be treated much like soft metal tubing with respect to fitting selection. Rigid tubing can be connected with either separate or double-compression fittings. Soft, flexible tubing, however, requires the use of an inner support sleeve or insert, Fig. 7-12F. The insert fits inside the soft tubing giving it strength and support. Flexible tubing, for example, can be used to route oil to a turbocharger.

Flexible compression fitting

A *flexible compression fitting* can withstand severe vibration. One is illustrated in Fig. 7-12G. The special composition (multi-material) sleeve allows the tubing to move sideways in the fitting. The composition sleeve is flexible, not rigid as with a conventional metal sleeve. The tubing can vibrate and move without bending the tubing or loosening the connection. This type of fitting should be used only on low pressure applications. It is not commonly installed as a factory item.

Ermeto fitting

An *Ermeto compression fitting* is designed to withstand extremely high pressures. The sleeve actually cuts into the side of the metal tubing. This produces a powerful clamping action and seal. It can be used in place of a flare fitting with comparable dependability.

Nut length

Always select a nut length that matches the job. Nuts are available in standard (short) and long lengths. The longer nut should be used where vibration or movement will be pronounced. It will provide better support of the tubing and prevent stress.

FITTING DESIGN

Numerous fittings are designed to meet almost any need, Fig. 7-13. As a fuel system mechanic, you should be familiar with various types and how they are used.

T-fittings

T-fittings are used when one line must branch or run off of another. They come in both male and female types and in various sizes.

Unions

As its name implies, a *union* can be used to connect or join two sections of tubing. Straight unions are commonly used to splice together two pieces of tubing, as with a fuel line repair. It saves the cost and labor of replacing the complete section of fuel line.

Connectors

Connectors are commonly utilized to fasten tubing to fuel system components (carburetors, fuel pumps, etc.). They often have male pipe threads on one end and a tube fitting on the other.

Elbows

Elbows are designed to allow a fuel line to leave the fitting at an angle. Elbows are available in 45 and 90 degree angles with either male or female threads.

Pipe fittings

Pipe fittings use tapered threads. The thread diameter varies so that a force fit is produced.

NOTE! Never thread a fitting with a tapered pipe thread onto a conventional thread. The two will fasten together, BUT SEVERE THREAD DAMAGE AND A POOR SEAL WILL RESULT.

O-Ring fittings

With an *O-ring fitting,* a neoprene (rubber) O-ring is pressed between the fitting nut and the fitting body. The compressed O-ring forms a leakproof seal.

When installing this type of fitting, always push the O-ring all the way up against the body of the fitting. If the O-ring is left out on the threads of the fitting, it can be cut. A poor seal and leak could occur.

Whenever an O-ring type fitting is disassembled, check the rubber seal for cuts, cracks, splits, or hardening. If the seal is bad, replace it.

Swivel fittings

Though not commonly used on ''stock,'' unmodified cars, a *swivel fitting* is useful in situations where there is extreme movement. It actually allows the attached fuel line to rotate around the fitting. Some off-road, or racing vehicles, for example, may use swivel fittings to allow for the severe twisting and flexing between different parts of the vehicle (engine and body or frame, fuel tank and frame, etc.).

Distribution blocks

Where several lines must be joined at a common location, a *distribution block,* also called a *junction block,* is used. It allows one main feed line to supply fuel to several lines. For example, special distribution blocks are commonly used to supply fuel for multicarburetor setups.

Drain cocks

A *drain cock* is simply a valve used to remove or drain the contents of a system. For instance, they are sometimes used in the bottom of diesel fuel filters. When the cock is opened, any water settling to the bottom of the filter can be drained out.

Shutoff cocks

Not to be confused with a drain cock, a *shutoff cock,* shown in Fig. 7-13, is used to stop flow through a line or system. It is connected in-line so that it can seal off pressure and flow.

Push connect fittings

Push connect fittings use a special clip and O-rings to

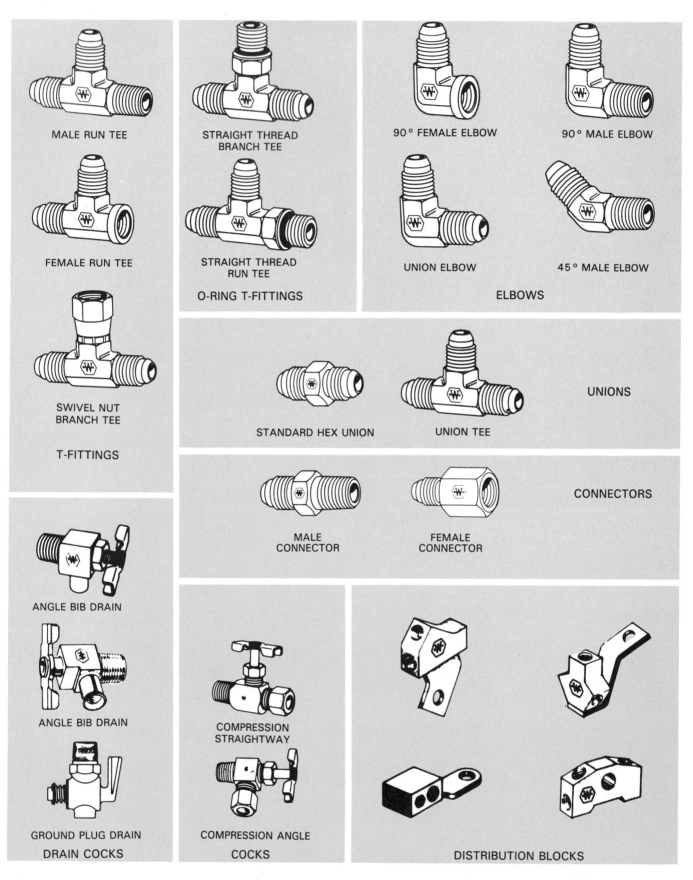

MALE RUN TEE

STRAIGHT THREAD
BRANCH TEE

90° FEMALE ELBOW

90° MALE ELBOW

FEMALE RUN TEE

STRAIGHT THREAD
RUN TEE

O-RING T-FITTINGS

UNION ELBOW

45° MALE ELBOW

ELBOWS

SWIVEL NUT
BRANCH TEE

T-FITTINGS

STANDARD HEX UNION

UNION TEE

UNIONS

MALE
CONNECTOR

FEMALE
CONNECTOR

CONNECTORS

ANGLE BIB DRAIN

ANGLE BIB DRAIN

GROUND PLUG DRAIN

DRAIN COCKS

COMPRESSION
STRAIGHTWAY

COMPRESSION ANGLE

COCKS

DISTRIBUTION BLOCKS

Fig. 7-13. Fittings are designed for different purposes. T-fittings are needed where line divides and goes to two different locations. Connectors join tubing with system components. Elbows are used when line must exit accessory at angle. Unions are designed to connect sections of tubing. They are handy when damaged sections of tubing must be replaced. O-ring fittings form leakproof seals. Distribution blocks allow one line to feed several lines. Drain cocks provide for drainage of liquids. Shutoff cocks control flow through system. (The Weatherhead Co., Gould, and Imperial-Eastman)

form a leakproof fuel line connection, Fig. 7-14. Two O-rings fit between the steel fuel line and the fitting body. The fitting slides together and then the clip is snapped into place, locking the line and fitting together.

Push connect fittings are sometimes used in modern electronic fuel (gasoline) injection systems. They make fuel line removal and installation easier. Tools are not normally needed for a disconnect. You can usually use your hands to remove the clips and pull the fitting apart.

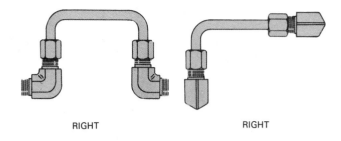

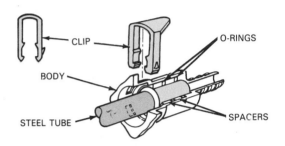

Fig. 7-14. Push connect fitting depends on O-rings to produce leakproof connection. (Ford Motor Co.)

TUBING INSTALLATION

After straightening, cutting, reaming, flaring, and bending operations are complete, it is equally important to properly *install* the line. ALIGNMENT IS CRITICAL. Tubing fittings can be stripped and damaged very easily if threaded improperly.

When installing a new fuel line, fit one end of the line securely. Start it BY HAND and snug it down lightly. Then, fit the other end of the line into position. Visually sight down the top and sides of the tubing to check for correct alignment. Usually, additional hand bending will be needed for perfect alignment. Bend the tubing until the line runs straight into the fitting.

You should be able to START the fitting nut with your FINGERS. Never use a wrench. After the nut is partially turned into the fitting, tighten it with a tubing wrench.

Normally, during fuel line replacement, the new line should be routed and bent about the same as the old one. If you change the routing of the line, problems can result. Auto manufacturers have vast experience and normally use the best path for the fuel lines. The next few paragraphs discuss a few common tube-working rules.

Avoid straight runs

Straight runs should NOT be used because they do not allow for movement (vibration, shifting, contraction, and expansion). If a line were to run straight from one component to another, the fittings or connections could loosen, leak, or even break.

To prevent these problems, form bends into fuel lines that avoid stress on joints. As illustrated in Fig. 7-15, the bends in the tubing will flex and relieve tension.

Avoid hot spots

Route fuel lines away from engine and exhaust system

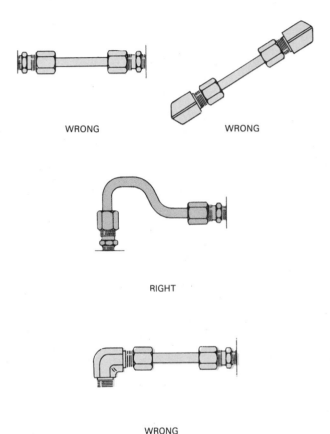

Fig. 7-15. Right and wrong way to connect two components. Expansion, contraction, and vibration could loosen or even break fittings on which there is too much strain. Tubing at top can bend and flex without failure. (The Weatherhead Co.)

heat. This will help eliminate the problem of:
1. Vapor lock.
2. Excessive expansion and contraction which could loosen or damage fuel lines.

See Fig. 7-16. If a line cannot easily be moved away from a hot spot, a sheet metal baffle is sometimes used as a shield.

Supporting long sections of tubing

Always support a long section of fuel line with small body clips. These clips, also called mounting clamps, hold the line tightly against the car body and frame. Long lengths of unsupported line can VIBRATE and quickly fail. A few examples of how support clamps are used are given in Fig. 7-17. In general, support clips should be used on any line over 2 or 3 ft. long.

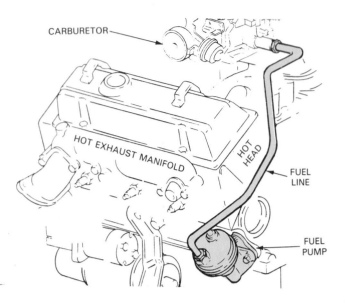

Fig. 7-16. Route fuel lines away from HOT parts of engine. (GM Corp.)

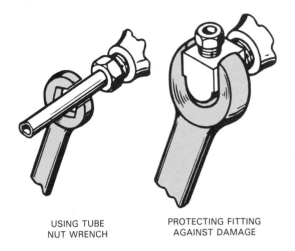

USING TUBE NUT WRENCH

PROTECTING FITTING AGAINST DAMAGE

Fig. 7-18. Left. Tubing wrench should be used on tube or line fittings. Right. Open-end wrench is used on some elbow fittings. (Bendix)

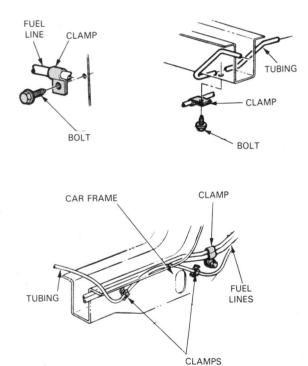

Fig. 7-17. Fuel line must be clamped securely to car. If not, it will vibrate and quickly fail. (Ford Motor Co.)

Tubing wrench

When tightening or loosening a tube fitting, use a tubing wrench. A *tubing wrench,* Fig. 7-18, is a 6-point box end wrench with a clearance slot cut into it. The slot allows the wrench to slide over the tubing and then onto the fitting nut. Logically, a 6-point wrench produces maximum holding power and will help keep the soft fitting nut from rounding off. An open end wrench can strip or round off the flats of a tubing nut very easily.

Support for fitting body and nut

It is important, particularly when loosening a tubing fitting, to use two wrenches. One wrench keeps the fitting body from turning while the other wrench loosens the fitting nut.

If only one wrench is used to loosen the nut, both the fitting body and nut can turn together. The tubing will remain locked between the two and will be twisted.

When installing fittings, always check that the body FITTING IS TIGHT in the component (carburetor, pressure regulator, fuel pump, etc.) before tightening the outer nut. If not, the body fitting may leak.

When the component is made of soft aluminum or pot metal, two wrenches should be used. The wrench on the body fitting will prevent overtightening and possible thread damage.

Checking clearance

Moving parts such as fan belts, throttle linkage, and steering components can rub holes in improperly routed fuel lines. A serious leak and fire could result. Operate these moving components while checking for adequate clearance.

Check for leaks

After the fuel line is installed, inspect it for leaks. Start the engine to pressurize the line and examine all of the connections. Rub a finger under the connections to check for wetness and a fuel odor.

If a leak is found, tighten the fitting slightly. If moderate tightening does not correct the leak, disassemble the fitting and check for flare or fitting problems. Repair the line or fitting as required.

Replacing diesel injection lines

Normally, it is best to replace a bad diesel injection line with a new, factory bent unit. See Fig. 7-19. Line diameter and length is very critical to assure correct injection timing and fuel quantity. If you make up your own line, it may not be the exact length or diameter. This could reduce diesel engine efficiency and power.

150

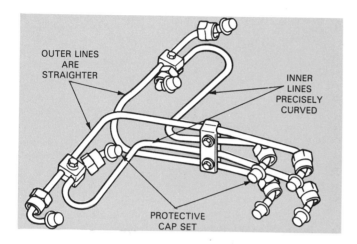

Fig. 7-19. Normally, new, factory-shaped diesel injection line should be installed if old line is damaged. Note how two inner lines are curved to keep all four lines same length. This assures correct timing and fuel injection quantity. (Ford)

FUEL SYSTEM HOSE

Metal lines are unsuitable where they will be subjected to great movement or vibration. In such cases, flexible hose is used.

Engines, being attached to the frame by rubber mounts, rock and vibrate and can move several inches. Metal line connected between the frame and the fuel pump would soon break. A section of flexible hose, as shown in Fig. 7-20, can absorb such motion.

Flexible hose is also used between the fuel tank and the main fuel line, between vent lines and the tank, and between in-line filters and main lines.

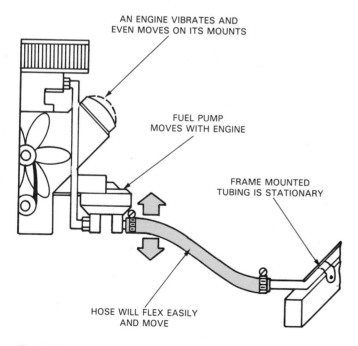

Fig. 7-20. Section of flexible hose will allow engine to move up and down on its mounts without damage to fuel lines.

Hose material

Various types of hoses are used on a car. Selecting the right type for use in a fuel system is very important. Buna N, neoprene, polyethylene, and some plastic hoses are a few of the types acceptable for use with gasoline, diesel fuel, gasohol, etc.

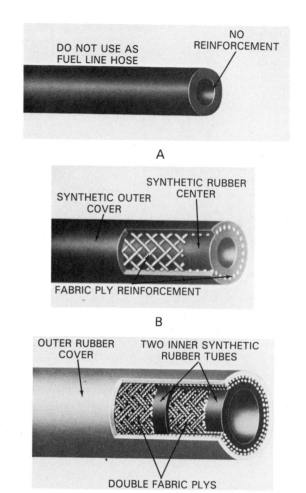

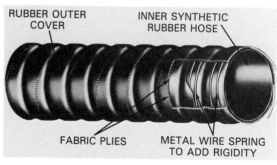

Fig. 7-21. A—Vacuum type hose has no reinforcement to give it strength. It must not be used to carry gasoline. Chemical action of gas will ruin vacuum hose and cause it to fail. B—Fuel line hose is strengthened with at least one fabric ply. It is made of synthetic rubber that is not affected by gasoline. (Gates Rubber Co.) C—This two-ply reinforced fuel hose is ideal for heavy duty diesel applications. D—This type hose is commonly used on filler neck of fuel tanks. It is heavily reinforced. (Gates Rubber Co.)

It is very important that the hose material does not chemically react to the fuel. Many plain or synthetic rubber hoses are satisfactory as vacuum lines, but will be quickly destroyed by the chemical action of fuel. See Fig. 7-21A. They can harden, crack, soften, and rupture.

> Only use hose that is SPECIFICALLY DESIGNED for use in a fuel system.

Hose construction

Most cars are equipped with reinforced type fuel line hoses, Fig. 7-21B. The inside of the hose has a layer or layers of cloth braid which greatly adds to its strength and durability. It can withstand higher pressure and vacuum than unreinforced hose. Just as important, it will have a much longer service life.

A two-ply braided reinforced hose is shown in Fig. 7-21C. Its extra strength (200 psi rating) and durability are excellent for heavy duty systems.

Remember our discussion of fuel tank filler necks? Two-piece necks use a hose connection. This type of hose, Fig. 7-21D, must be resistant to chemical action, vibration, pressure, and vacuum. It is usually reinforced with either cloth or metal strands.

> Danger! A ruptured fuel line hose on the pressure side of the fuel pump could lose gasoline rapidly (about a quart a minute). The engine will CONTINUE TO RUN on the fuel remaining in the system. An electric pump could EMPTY the fuel tank! If ignited by a spark from the ignition, the spilled gasoline could engulf an automobile in flames.

Hose inspection

Old, used fuel line hoses should always be checked for deterioration during fuel system service or repair. After a few years, neoprene hose can become hard and brittle. The hardened hose can crack and rupture.

To test fuel line hose, bend the hose as illustrated in Fig. 7-22. The hose should be replaced if cracks show. Even though the inner layer of the hose may seem alright, the hose is bad.

Also, fuel line hose should be checked for EXCESSIVE

Fig. 7-22. If hose shows cracks when kinked, it should be replaced. (Airtex)

SOFTNESS. Under some conditions (oil covered, transmission fluid leakage on hose, etc.) a hose can break down and become soft. A softened hose is also dangerous because it can rupture at any time. Always squeeze fuel line hoses as a second check of their condition.

Cutting hose

Any of several tools will cut hose: sharp knife, diagonal cutting pliers, or side cutters. The best method, however, may be the special hose-cutting pliers shown in Fig. 7-23. They are similar to shrubbery shears. Place the hose between the jaws of the shears and squeeze the handles. The cutter will force the hose down against the dull jaw. The cutting jaw will slice the hose quickly and squarely. Diagonal cutting pliers and side cutters can be used in much the same way.

> When cutting hose with a knife, be extremely careful NOT to cut towards any part of your body.

With any method of cutting fuel line hose, it is very important that the cut is square. The connection will look and fit better.

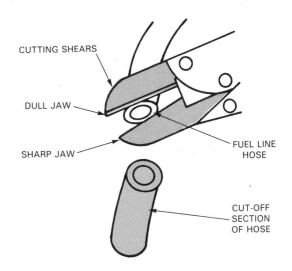

Fig. 7-23. Large hose-cutting pliers will cut hose off cleanly and squarely.

Hose routing

Proper routing of fuel system hose is even more important than routing of metal fuel lines. Flexible hoses rupture more easily than metal lines. A neoprene hose touching a hot or moving engine part will develop a hole very quickly. Hose too near heat will harden and break.

Hose length

Cut hose to a proper length. A hose cut off too short may be pinched shut at a sharp bend. This can restrict or even stop fuel flow. Also, it can cause the tubing to fail prematurely (split, crack, etc.).

A fuel line hose that is cut too long will look bad and will be more likely to touch hot or moving parts. When replacing a fuel system hose, try to cut it to the same length as the original hose.

Hose end fittings

Several types of hose end fittings are in use on automobiles. Three basic types are shown in Fig. 7-24. Try to replace a hose end fitting with one of a similar or identical design. The same type will usually give the best service.

When possible, barbed, beaded, or ribbed type fitting should be used with a fuel hose. The raised ribs or barbs and hose clamps hold the hose securely in place. Fig. 7-25 shows a fuel pump assembly with barbed fittings. Clamps are used to secure the hose connections between the fuel line and fuel pump.

Hose clamps

Several types of small fuel system hose clamps are also available. They come as screw, spring wire, spring strap, and crimp or clamp types.

Crimp hose clamp

As illustrated in Fig. 7-26A, a crimp type hose clamp should be installed with special crimping tool if available. After positioning the hose and clamp over the barb on the line, the special pliers are used to compress the clamp around the hose. When this tool is not available, diagonal cutting pliers will work, though not as well.

Note! Crimp clamps cannot be reused. Always replace them if reused, they could form an imperfect seal and hose could leak.

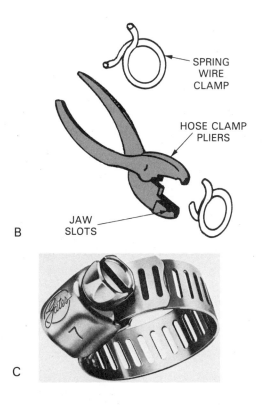

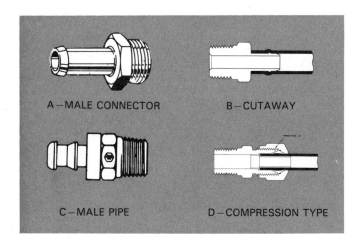

Fig. 7-24. Barbed hose end fittings are very handy. A hose clamp will quickly and easily secure the hose to the fitting. A—Single barb fitting. B—Cutaway of barb fitting showing barb holding action. C—Double barb fitting. D—This type uses compression sleeve to clamp hose to body. (The Weatherhead Co.)

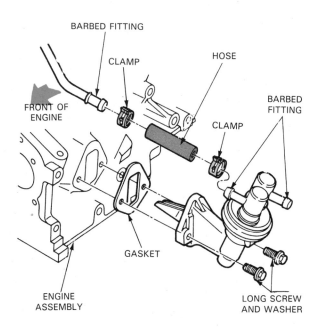

Fig. 7-25. Note how small section of hose is used to connect this fuel pump to a metal fuel line. (Ford)

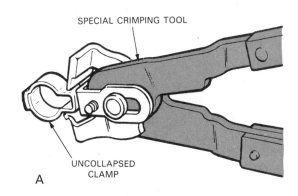

Fig. 7-26. A—Special pliers are available for squeezing a crimp type hose clamp. Diagonal cutting pliers will also work. (Ford) B—Spring wire clamps are best installed and removed with special hose clamp pliers. Small notches are cut into the jaws to hold clamp tangs securely. (Gates Rubber Co.) C—Screw or minex clamp is often used as a replacement item. It will adjust to various sizes. (Gates Rubber Co.)

Spring type hose clamps

Spring type hose clamps are made of either spring wire or spring strap metal. They are removed and installed with pliers. Common slip-joint pliers work well on strap type spring clamps.

Special hose clamp pliers, pictured in Fig. 7-26B, should be used on spring wire type clamps. Slots in the jaws will lock around the clamp and keep it from popping out of the pliers.

Screw type hose clamps

Screw type hose clamps are very popular as replacement items. See Fig. 7-26C. Sometimes called aviation type clamps, screw clamps assure uniform pressure around the outside of the hose. They will generally produce a better seal than other types of clamps. A screw clamp is also reusable.

Clamp and hose installation

To install fuel line hose, first slide the opened clamp over the hose. Then slide the hose over the fitting as shown in Fig. 7-27. Make sure that the clamp is on the CORRECT SIDE of the fitting barb or bead. If a double-bead line is used, the clamp should be installed between the two beads.

Hose compression fittings

A compression type fitting can be used on high pressure lines. One was shown in Fig. 7-12F. A compression fitting should only be used with high pressure hose. There is no reason to use a high pressure compression fitting with a low pressure hose. Its ability to withstand the pressure would be greater than the hose.

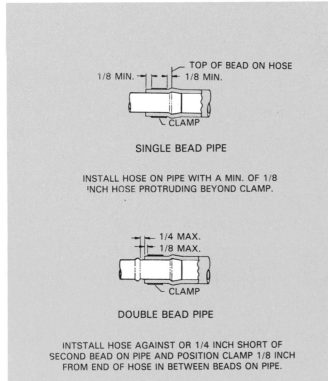

SINGLE BEAD PIPE

INSTALL HOSE ON PIPE WITH A MIN. OF 1/8 INCH HOSE PROTRUDING BEYOND CLAMP.

DOUBLE BEAD PIPE

INTSTALL HOSE AGAINST OR 1/4 INCH SHORT OF SECOND BEAD ON PIPE AND POSITION CLAMP 1/8 INCH FROM END OF HOSE IN BETWEEN BEADS ON PIPE.

Fig. 7-27. Always make sure clamp is located on inside of a fitting barb. If not, hose may be forced off, possibly starting a fire. (General Motors)

Hose end installation

When installing a fuel line hose, as with any hose, avoid twists, sharp bends, too much slack, etc. Before tightening down the hose clamps, double-check that the hose is not twisted between connections. Make sure the hose is pushed completely onto the fitting or at least 2 inches over an unbeaded fuel line.

CLEANING A FUEL SYSTEM

When fuel filters continue to clog with foreign matter, it may be necessary to clean the fuel lines and the tank. Repeated and excessive deposits in the fuel filters or strainers are indications of contamination.

To clean a fuel system, first disconnect the battery ground cable and the ignition system feed wires. This will help prevent sparks that could start a fire. Drain and clean the fuel tank as described in the previous chapter. Replace or clean the tank filter and reinstall the fuel tank. Do not, however, connect the fuel lines to the tank.

NOTE! If the inside of the fuel tank is rusted, the tank must normally be replaced. Once the protective coating over the sheet metal has been penetrated, the fuel tank can no longer be used.

Next, remove the main fuel line at the inlet of the fuel pump or at the engine (electric pump type system). Remove the fuel return line at the pump or liquid vapor separator. With both ends of the fuel lines disconnected, force compressed air through the lines in both directions. This will blow out dirt, water, or other contaminants.

If a fuel strainer-separator is used, it should be replaced or cleaned. Reconnect all of the fuel lines and put about 5 or 6 gallons of clean fuel into the fuel tank. Connect a hose to the supply line at the engine and run it into an approved gasoline can. Reconnect the battery, but not the ignition system voltage supply wire.

Crank the engine with the starting motor until around 2 quarts of fuel are pumped into the can. This will purge any remaining contaminants. Clean or replace line filters. Reinstall the fuel line and the ignition system lead. Start the engine and check for leaks.

REPLACING A DAMAGED SECTION OF FUEL LINE

It may be too costly or time-consuming to replace a complete section of fuel line when a small part of it is damaged. Either a section of hose or steel tubing, or both, can be used to replace the damaged section.

When replacing a section of line shorter than 6 in., you can use a piece of HOSE to make the repair. Fuel line hose should never be longer than 10 in. (6 in. from tubing end to tubing end). This is illustrated in Fig. 7-28.

Cut out the bad section of fuel line and measure its length. If it is shorter than 6 in., cut a piece of fuel hose that is 4 in. longer than the removed length of tubing. Slide the fuel line hose, with clamps, over the tubing ends. The hose should extend at least 2 in. over the ends of the metal tubing. Place the hose clamps near the ends of the hose and tighten.

If the damaged section is longer than 6 in., a new piece of metal line must be used. Either hose connections or fittings can be used to connect the new metal line to the existing line.

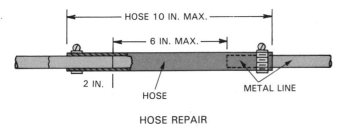

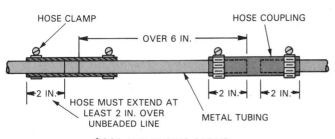

Fig. 7-28. Both metal tubing and hose must be used when replacing sections of fuel line over 6 in. long. (General Motors)

CAUTION: Hose should not be used within 4 inches of any hot engine or exhaust system component. A metal line must be used.

FUEL LINE AND HOSE SERVICE RULES

When working with fuel lines and hoses:

1. Place a shop rag around the fuel line fitting during removal. This will keep fuel from spraying over you or onto the hot engine. Use a flare nut or tubing wrench on fuel system fittings.
2. Only use approved double-wall steel tubing for fuel lines. Never use copper or plastic tubing.
3. Make smooth bends when forming a new fuel line. Use a bending spring or bending tool.
4. Form double-lap flares on the ends of the fuel line. A single lap flare is not approved for fuel lines. Refer to Fig. 7-11 again.
5. Reinstall fuel line hold-down clamps and brackets. If not properly supported, the fuel line can vibrate and fail again.
6. Route all fuel lines and hoses away from hot or moving parts. Double-check clearance after installation.
7. Use only approved synthetic rubber hoses in a fuel system. If vacuum type rubber hose is accidentally used, the fuel can chemically deteriorate the hose. A leak could result.
8. Make sure a fuel hose fully covers its fitting or line before installing the clamps. Pressure in the fuel system could force the hose off if not installed properly.
9. Double-check all fittings for leaks. Start the engine and inspect the connections closely.

DANGER! Most fuel injection systems have very high fuel pressure. Follow recommended procedures for bleeding or releasing pressure before disconnect-

ing a fuel line or fitting. This will prevent fuel spray from possibly causing injury or a fire!

FUEL STRAINERS AND FILTERS

Fuel strainers and *filters* prevent contaminants such as road dirt, dust, rust, scale, water, solder, and oxidation gum from passing through the fuel system.

It is very important that only clean fuel enter the system. Many fuel system parts can be rendered inoperable or even ruined by foreign matter. Carburetors, for example, have many small passages, jets, and air bleeds that can be clogged by the smallest speck of dirt.

Diesel and gasoline injection systems are even more sensitive to fuel contaminants. Their parts are machined to extremely close tolerances. Solid particles in the fuel act as abrasives (grinding compound) which can cause rapid wear of the injector pump and injectors.

Dirty fuel can also speed up wear on engine cylinders, rings, and pistons. Fuel filters must block and trap fuel impurities before they can cause fuel system or engine problems.

FUEL CONTAMINANTS

Contaminants or "dirt" can enter the fuel at almost any time. Fuel is normally clean immediately after refinement. However, it can pick up dirt, rust, water, leaves, etc. during storage, shipping, transfer, and even while in a car's fuel system.

Condensation (water) can collect inside an automotive fuel tank much like it does on a glass of cold water. When the tank is partially empty, a difference in temperature between the outside and inside of the tank can cause this condensation. These water droplets can then enter the fuel.

Water alone can clog fuel filters, corrode some metals, and cause poor engine performance (missing, stalling, no-starts, etc.). Water in fuel can produce solid contaminants by causing fuel system corrosion and rusting. Small chips of rust, for example, can flake off the inside of the fuel tank and enter the fuel.

LOCATION AND USES OF STRAINERS

A strainer should not be confused with a filter. A *strainer* is normally much COARSER. It will stop larger particles but pass smaller ones.

A strainer is usually located on the suction, not pressure, side of the fuel pump. See Fig. 7-29. Strainers are designed to remove the bigger contaminants, protect the fuel pump, and relieve some work load from the main filter.

NOTE: Some fuel injection systems have fuel strainers in each injector, Fig. 7-30. This keeps debris from clogging the injector spray nozzle opening.

Generally, a strainer will stop particles as small as 0.002 to 0.003 in. (0.05 to 0.076 mm) in diameter. A filter, on the other hand, can stop particles as small as 0.001 in. (0.025 mm) in gasoline systems and even 0.00004 in. (0.001 mm) in some diesel fuel systems.

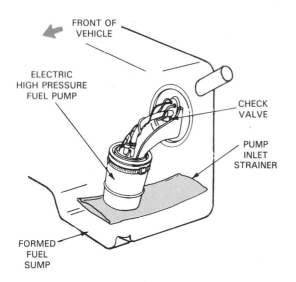

Fig. 7-29. A fuel strainer looks very much like extra-fine window screen. (Ford)

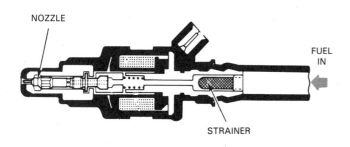

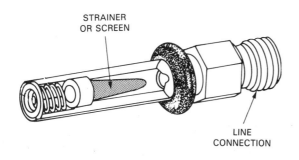

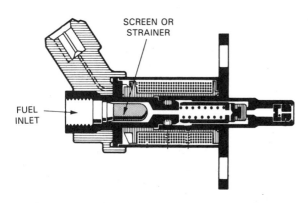

Fig. 7-30. Some injection systems use strainers in each injector.

FILTERING MATERIALS

Several types of material are used in the construction of fuel filters. These include metal screen, sintered bronze, ceramic, treated paper, and woven saran.

Screen filters

A *metal* (copper, brass, coated steel, etc.) *screen filter,* often called a strainer, is sometimes used in automobile fuel systems. If coated with resin, a screen filter can filter out water.

Screen filter material resembles door and window screening, except the mesh is much finer. Screen filters are commonly used in the fuel tank, fuel pump, injectors, and fuel return vapor separator.

Sintered bronze filters

A *sintered bronze filter* is shown in Fig. 7-31A. It is made by placing powdered bronze, copper, cast iron, or other types of metal under extreme pressure. This fuses the powdered metal into a very porous substance. Tiny holes and passages exist between each powdered metal particle. Fuel can pass through the tiny pores, but foreign matter cannot.

Generally, a sintered bronze type filter will trap and stop matter around 0.001 in. (0.025 mm) in diameter.

Sintered bronze type filters are commonly used inside the inlet fitting of carburetors. They serve as a final filter that prevents dirt from entering tiny carburetor passages. For their small size, they are capable of stopping a large quantity of contaminants without clogging.

Ceramic filter material

A *ceramic filter* operates in much the same way as the sintered bronze filter.

Treated paper filter materials

Treated paper fuel filters, shown in Fig. 7-31B, are common in automotive fuel systems. They are usually made of pleated (folded) paper coated with a special resin.

The resin coating helps stop water and any smaller impurities. Pleating increases the filtration surface area. A larger filter area reduces flow restriction and allows the filter to trap more contaminants before clogging.

Paper filter elements can be found in the carburetor, fuel pump, and in-line filters, but not inside the fuel tank. They are popular because of their high efficiency, dependability, and low cost.

Woven saran filter elements

A *woven saran filter* looks very much like a screen mesh strainer made out of fiberglass. Normally much finer than screen type filters, it is soft and flexible to the touch. A woven saran type filter is commonly used on the end of the fuel tank pickup tube. It can usually filter out some water and fairly small solid particles. Refer, again, to Fig. 7-29.

FILTER ELEMENT SERVICE

Usually, ceramic and sintered bronze type fuel filters are located in the main fuel line between the engine and

Fig. 7-31. A—Sintered bronze filter can be cleaned if not too dirty, clogged with rust, or coated with gum. B—Paper filters cannot be cleaned. They must be replaced when dirty. (AC Delco)

Treated paper, saran, and other fuel filter types are often disposable. They cannot be cleaned. They may be found almost anywhere in the fuel system—the tank, fuel line, fuel pump, or carburetor. They should be replaced with filters of equal specifications (general size, surface area, flow capacity, and filtration ability).

Clogged tank filter

A clogged tank filter may be hard to diagnose. Its symptoms can resemble a severe fuel starvation or an ignition short. It may appear only at certain speeds. Partial clogging builds up a suction that may collapse the soft material, cutting off the fuel supply. The engine will stop abruptly, restart in a few minutes, and run normally at low speeds.

The clogging condition may not show up on a fuel pump volume test, as the filter material is not then collapsed. When the symptoms described appear, the filter should be removed and inspected. If found to be partially clogged, replace it.

IN-LINE FILTER

In-line filters are usually located between the fuel pump and the engine. They are often attached to the metal fuel line with short sections of hose. See Fig. 7-32.

In most cases, the outside of the filter will be marked inlet and outlet or it may have a directional arrow. The arrow, shown in Fig. 7-33, must point in the direction of the fuel flow.

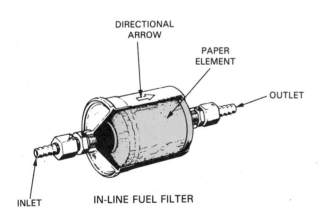

Fig. 7-32. An in-line fuel filter is generally positioned between carburetor and fuel pump. Unit shown is disposable canister filter. It is connected to metal fuel line with two short sections of fuel hose which is fastened with hose clamps. Such filters are replaced at recommended intervals. (Saab)

fuel pump. To clean them, force compressed air through the filter element in REVERSE direction. This will dislodge and remove dirt, rust, and water. If the filter is coated with gasoline gum it should be replaced.

Filter test

To test a filter, blow through it. A clean filter should offer little or no air restriction.

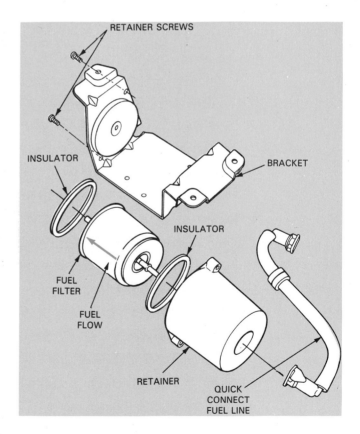

Fig. 7-33. Cartridge filter fits into retainer housing. Arrow points out open end of retainer. (Ford)

Sediment-bowl type fuel filter

A *sediment-bowl filter,* as its name implies, has a bowl-shaped housing around the filter element. This type may use a ceramic, sintered bronze, screen, or treated paper type element. The bowl may be either metal or glass.

The *glass bowl filter* is handy because it allows you to see when the element is dirty. A sediment bowl type filter is very efficient and is often cleanable, Fig. 7-34. It is often mounted in the engine compartment as an in-line filter or on the fuel pump.

SEDIMENT-BOWL FILTER CLEANING

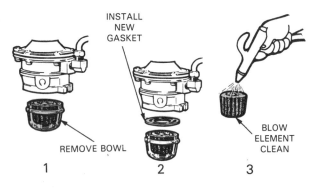

Fig. 7-34. Some types of bowl filters can be cleaned and reused. This one is mounted on fuel pump. (Ford)

FUEL PUMP FILTER

Sometimes, the fuel pump has a fuel strainer or filter built in. The filter is connected to the output side of the pump and may replace the in-line mounted filter. Also, when a fuel return line is used, the filter may double as a vapor separator.

A large screw-on type fuel pump filter can be removed and installed with a small diameter oil filter type wrench.

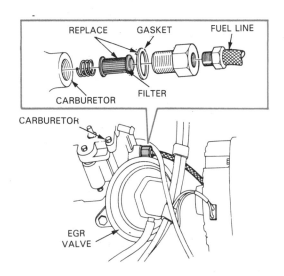

Fig. 7-35. Many cars have a fuel filter mounted inside carburetor inlet fitting. (AMC)

When replacing this filter, always inspect the seal closely and replace when needed. The filter should be hand tightened only. Never use the filter wrench.

Carburetor filter

Fig. 7-35 shows a fuel filter located on or inside the fuel inlet fitting. The internal filter assembly consists of a special fuel inlet fitting, gasket or O-ring, filter element (usually paper or sintered bronze), and a spring.

Usually, the filter element must be installed in only one direction. One end of the filter may be sealed and the other open. The open end should face the incoming fuel. The fuel flows into the center of the element, through the filter and then into the carburetor. If reversed, the filter could restrict or even stop the fuel flow.

Note: TWO WRENCHES are usually needed to replace this filter. See Fig. 7-36.

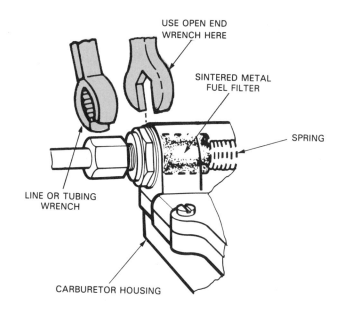

Fig. 7-36. Small diameter tubing wrenches are available for fuel filter removal.

Filter check valve

Some fuel filters are equipped with a *check valve,* as pictured in Fig. 7-37. The valve prevents fuel from leaking out of the system in the event of a roll-over. During normal operation, fuel pump pressure forces the valve open. To prevent fire, the valve closes and remains closed when the car is upside down.

In some fuel systems using a fuel return line, this check valve can also stop fuel flow back to the fuel pump or fuel tank. This will prevent fuel from draining out of the fuel line when the car is not in operation. Were drain-back to occur, an air pocket could form in the line. This would result in temporary fuel starvation and an engine stalling problem during start up.

FILTER CLEAN AND CHANGE SCHEDULE

Auto makers recommend periodic fuel filter change. Generally, carburetor mounted type fuel filters have smaller surface areas. They should be replaced every 12 months or 12,000 miles (about 19 000 km), whichever comes first. With larger in-line filters, the filter should be cleaned or replaced around every two years or 24,000 miles (38 600 km). When severe or repeated filter contamination is found, the fuel filter or filters should be replaced more often, or as needed.

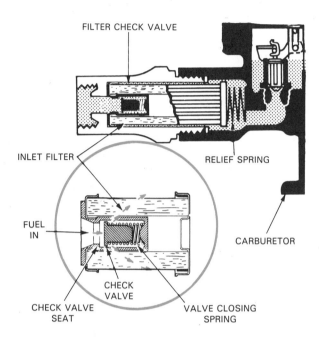

Fig. 7-37. This fuel filter contains a safety check valve which stops fuel leakage during an auto accident. (General Motors)

FUEL FILTER PROBLEMS

A *clogged fuel filter* can cause a variety of performance problems. In many cases, the flow restriction can reduce engine power, cause engine power pulsations (power surging), high speed missing, or it can stop the engine completely. Since the restricted filter may allow some fuel to reach the engine, the engine may run normally at low speeds. However, at higher engine speeds, when fuel demand is high, the engine starves for fuel. It could miss, stumble, cut out or "die."

FUEL FILTER TESTING

Once a fuel filter is removed, it can be tested by simply blowing air through it, Fig. 7-38. In-line and pump mounted filters can be flow tested. A fuel filter flow test is nothing more than a FUEL PUMP FLOW TEST, which will be discussed in the next chapter.

The basic procedure is to unhook the fuel line at the engine and attach a section of hose running into an approved container. Crank the engine for 30 seconds. The fuel pump should put out about a pint of fuel. This value will vary depending upon engine size. If in doubt, check the engine specifications. When an appropriate amount of fuel is pumped out of the system, the filters should be clean and in good condition.

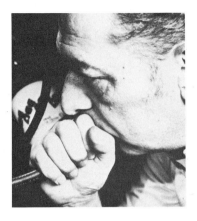

Fig. 7-38. If you cannot blow easily through fuel filter, it is clogged. (Airtex Co.)

Note! Do not forget about the collapsing action of some in-tank filters. This was discussed earlier. They may work normally during a flow test even though they are actually clogged.

DIESEL FUEL SYSTEM FILTERS

Diesel engine fuel filters must be even more efficient than those used on gasoline type fuel systems. A typical fuel injector nozzle may be assembled to tolerances measured in microns (millionths of a meter). A micron equals 0.000039 in. Considering that these injectors are lubricated by the diesel oil flowing through the fuel system, it is easy to understand why diesel fuel must be kept clean.

Any dirt, dust, or water in the fuel can ruin expensive injector components. Any abrasive solids in the fuel can cause part wear and surface scoring. Water, when mixed with diesel oil, becomes very CORROSIVE. The water-oil mixture can corrode and pit parts.

Diesel fuel filters must be capable of trapping even the smallest fuel impurities. In some systems, filtration is as small as 30 millionths of a meter.

Single-stage filtration

Sometimes, a *single-stage filter system*, Fig. 7-39, is used with an automotive diesel engine. A tank strainer may be used with a main or final filter. The low-restriction strainer is on the suction side of the fuel pump and it stops larger particles. The main filter, on the pressure side of the feed pump, must be a compromise between low restriction and high efficiency. It must handle a large flow of fuel, yet still protect all components.

Two-stage filtration

Some diesel fuel systems use what is called a *two-stage filtering system*, Fig. 7-40. It has a coarse primary filter between the tank and the fuel pump and a fine secondary filter between the pump and the injector pump. The *primary filter* stops the larger particles. Then, the *secondary filter* traps the smallest impurities. As a result, clean fuel enters the injector pump and injectors.

Almost all large diesel truck engines use two-stage filtering systems. Some automotive diesel engines also

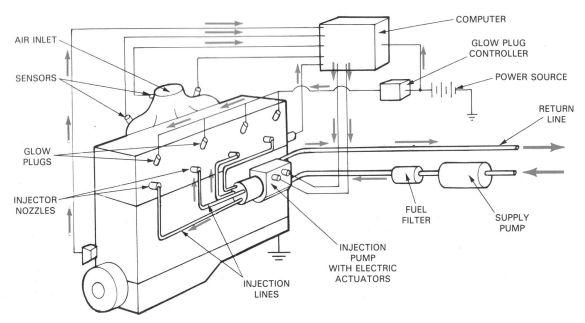

Fig. 7-39. This is a simple drawing of modern computerized diesel system. Supply pump feeds fuel to injection pump. Injection pump then forces fuel out nozzles on power strokes. Computer uses sensors to check engine conditions. Computer then operates small DC motors on injection pump.

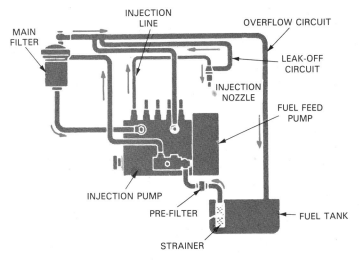

Fig. 7-40. Two-stage filtering system. Diesel fuel is filtered as it leaves tank and again before it goes to injectors. (Toyota)

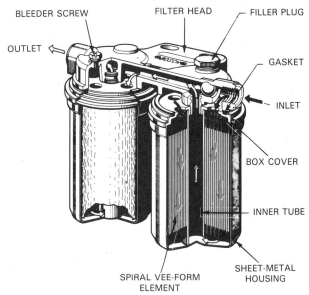

Fig. 7-41. Two-stage filter. Primary filter at right is coarser and will stop larger contaminants. Secondary filter is finer because it must purify fuel before it reaches the precision injection pump. Arrows show path through first element. (Bosch)

have them. Truck filters are usually much larger than automotive fuel filters. Fig. 7-41 is a sectioned view of a two-stage filter.

Water and sediment removal

Some diesel fuel filters are equipped with a *drain cock* for removing moisture that collects in the filter. One is shown in Fig. 7-42. Since water is heavier than diesel oil, it settles to the bottom. Collected water can be drained from the system. Also, any sediment will be purged along with the water.

Fuel circulation systems

Most modern automotive diesel fuel systems are circulating types. The fuel constantly flows, Fig. 7-43, from the tank, through the filter, injector pump, injector, and

then back to the fuel tank. The steady flow of fuel cools and lubricates the injector pump and injectors. Fuel circulation also warms the fuel in the tank slightly to help prevent clouding (waxing) in cold weather.

Changing diesel fuel filters

For normal diesel engine life and service, fuel filters MUST be changed or cleaned at regular intervals. Auto makers typically recommend service intervals of 24,000

160

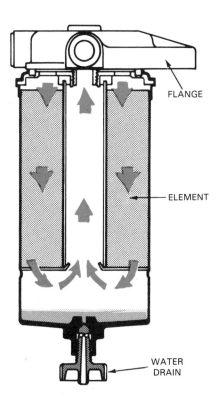

Fig. 7-42. Water settling to bottom of this filter can be drawn off using drain cock. (VW)

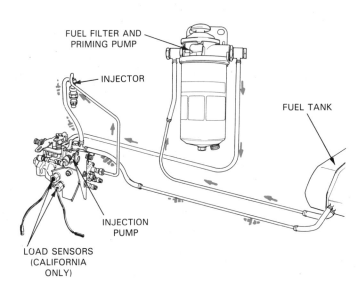

Fig. 7-43. Fuel circulation system. Fuel flow is constant, from the tank, through the filter and injection pump. Some of it flows back to the fuel tank. This action helps to cool and lubricate injection pump. (Peugeot)

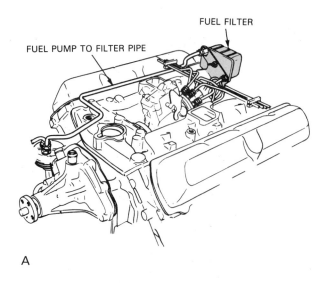

A

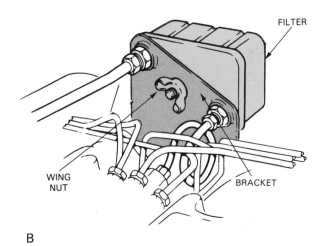

B

Fig. 7-44. On this diesel, fuel filter is mounted on engine intake manifold. A—Location of primary filter. B—Wing nut fastens filter to bracket. (Oldsmobile)

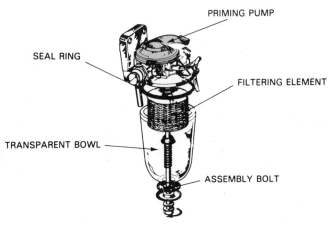

BOWL TYPE DIESEL FUEL FILTER

Fig. 7-45. Study parts of bowl type diesel fuel filter. Note priming pump lever. (Peugeot)

miles (about 36 600 km).

As shown in Fig. 7-44, the main or secondary filter is sometimes mounted on the engine. However, it can also be found elsewhere in the engine compartment.

Normally, with a single stage filtering system only, the main filter must be serviced during normal system care. The tank strainer is usually not serviced. With a two-stage filtering system, both the primary and the secondary filter may have to be serviced. If in doubt, check the car's shop manual for filter change recommendations.

Sediment-bowl type diesel fuel filter

A diesel bowl filter is similar to a gasoline sediment-bowl filter. The one in Fig. 7-45 has a priming pump mounted on top. Pumping the lever forces fuel through the lines. Since the bowl is transparent, water settling in it can be detected and removed.

To replace the filter, loosen and remove the bolt, or thumb screw which secures the bowl. Replace or clean the element and wipe out the inside of the bowl. Check the seal ring for cracks, splits, or hardening. Replace if needed. Finally, reassemble and reinstall the filter.

Cartridge type filter replacement

To replace a cartridge filter, Fig. 7-46, stop the engine. Clean the outside of the filter and surrounding area with clean fuel. This will keep dirt from falling into the system. Drain the filter. Open the filter housing and remove the used cartridge.

Install the new filter. Inspect the condition of the cover gasket and replace if needed. Close the drain and vent or bleed the filter according to manufacturer's recommendations. Finally, run the engine and check the filter for leaks.

clean a gasoline type fuel system. Refer to Chapter 20 for proper procedure.

Bleeding

Some diesel systems must be bled after servicing the fuel filter or other parts of the system. *Bleeding* the filter will remove POCKETS OF AIR in the fuel and prevent possible part damage.

A drain valve is opened, often on the filter, and fuel is forced through the system. When all of the air bubbles are out and there is a solid stream of fuel, close the drain. The diesel is now ready for operation.

Special fuel delivery system cautions

CAUTION! There is a very real and ever-present danger of fire and/or explosion when working on a fuel system. As you have learned, a small amount of gasoline or diesel fuel contains as much heat energy as a stick of dynamite. Observe all safety rules when servicing or repairing a fuel system. Ask yourself these questions:
1. Are all fittings properly mated and tightened?
2. Have you properly supported all lines?
3. Are fuel lines protected from excessive heat, flying objects, or interference with moving parts?
4. Is the fuel tank properly secured, filler neck attached?
5. Have you used the appropriate hose, tubing, and fitting?

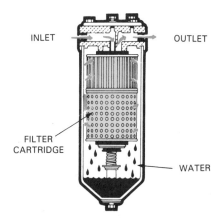

Fig. 7-46. Cartridge style filter is very efficient at removing water from fuel oil. (Fram)

Spin-on type filter replacement

When replacing a spin-on type fuel filter, clean the area above and around the filter. Then, using a filter wrench, unscrew the old filter and remove the spud gasket or O-ring seal. Fill the new filter with diesel oil and install. Follow the directions on the filter box or in a service manual. Finally, run the engine for about five minutes and watch for leaks.

Cleaning a diesel fuel system

When water, gasoline, or too many impurities are found in a diesel fuel system, the system should be purged (cleaned). The method is similar to that used to

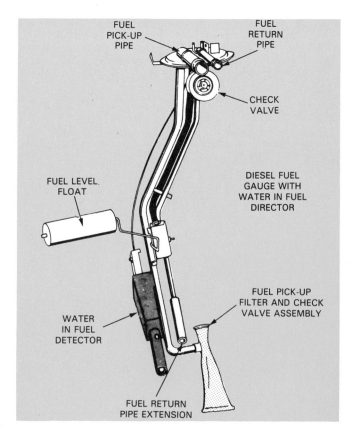

Fig. 7-47. Diesel fuel gauge with water detector. Sensor, being near bottom of tank, completes an electric circuit when water is present and turns on light at instrument panel. (Buick)

6. Are you draining fuel into an approved safety container?
7. Did you check your components for leaks after the repair?

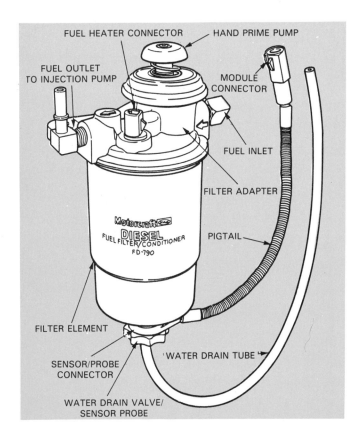

FUEL HEATER CONNECTOR   HAND PRIME PUMP

FUEL OUTLET
TO INJECTION PUMP   MODULE
CONNECTOR

FUEL INLET

FILTER ADAPTER

Motorcraft
DIESEL
FUEL FILTER/CONDITIONER
FD-790   PIGTAIL

FILTER ELEMENT

SENSOR/PROBE
CONNECTOR   WATER DRAIN TUBE

WATER DRAIN VALVE/
SENSOR PROBE

Fig. 7-48. Diesel filter/conditioner with water sensor unit. When water reaches predetermined level, sensor causes lamp to flash and warn operator. (Ford)

WATER-IN-FUEL DETECTOR

A *water-in-fuel detector* turns on a warning light when water is present in the diesel fuel tank. A sensor on the end of the pickup tube, Fig. 7-47, activates the light.

Water contamination can occur from being pumped into the tank at the service station. Condensation could also allow water to form in the tank.

Another water-in-fuel detector system places the sensor in the bottom of an in-line fuel filter. See Fig. 7-48. It works like the system with an in-tank sensor.

Testing water-in-fuel sensor

Fig. 7-49 shows how to test one type of sensor for a water-in-fuel detector system. When the sensor is submersed in a container of water, the test light should glow. It should not glow when the sensor is removed from the water.

The water sensor will only conduct electricity when in the water. If the sensor does not work properly, replace it.

Fig. 7-50 shows a wiring diagram for a diesel fuel gauge and water-in-fuel detector. Note how a fuse feeds current to the indicator light and to the solid-state, in-tank sensor.

Fuel line heater

A *fuel line heater* is sometimes used to prevent diesel fuel clouding (wax formation) in cold weather. It is simply an electric heating element formed around the fuel line. See Fig. 7-51.

In cold weather, high current flow through the heater warms the fuel. This prevents paraffin wax from separating out of the fuel and clogging the fuel filter.

DUAL FUEL TANK SYSTEM

A *dual fuel tank system* uses two separate fuel tanks to increase a car's driving range. This system is com-

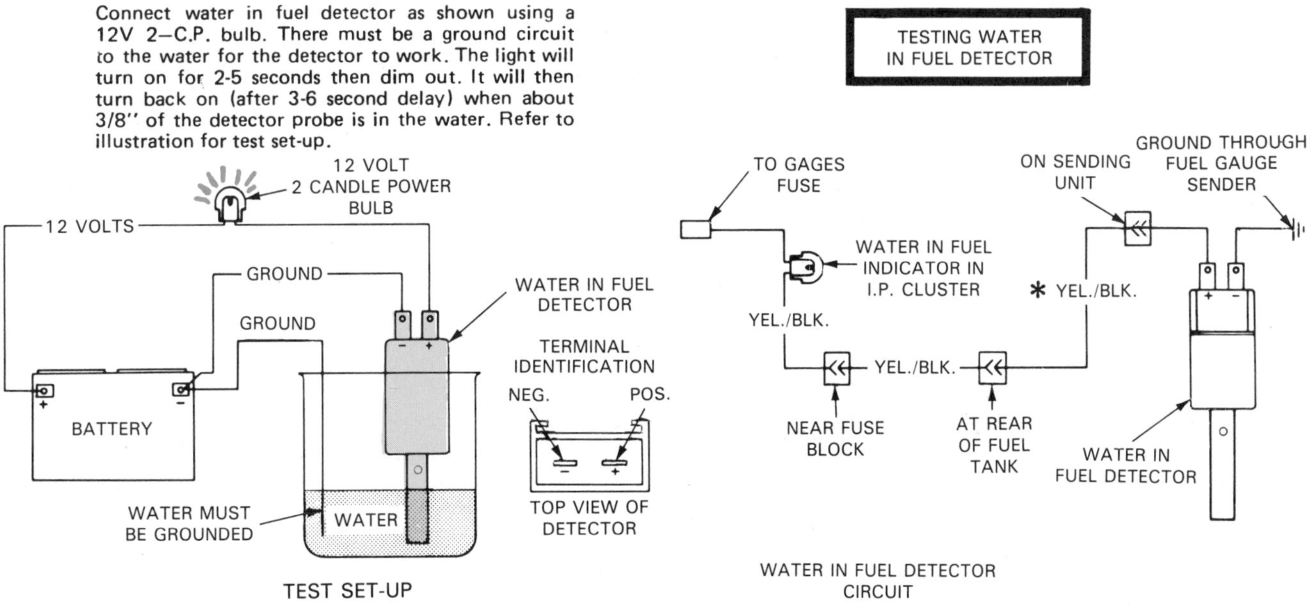

Connect water in fuel detector as shown using a 12V 2-C.P. bulb. There must be a ground circuit to the water for the detector to work. The light will turn on for 2-5 seconds then dim out. It will then turn back on (after 3-6 second delay) when about 3/8" of the detector probe is in the water. Refer to illustration for test set-up.

12 VOLT
2 CANDLE POWER
BULB

12 VOLTS

GROUND

GROUND

WATER IN FUEL
DETECTOR

TERMINAL
IDENTIFICATION

NEG.   POS.

BATTERY

TOP VIEW OF
DETECTOR

WATER MUST
BE GROUNDED   WATER

TEST SET-UP

TESTING WATER
IN FUEL DETECTOR

TO GAGES
FUSE   ON SENDING
UNIT   GROUND THROUGH
FUEL GAUGE
SENDER

WATER IN FUEL
INDICATOR IN
I.P. CLUSTER

YEL./BLK.   * YEL./BLK.

YEL./BLK.

NEAR FUSE
BLOCK   AT REAR
OF FUEL
TANK   WATER IN
FUEL DETECTOR

WATER IN FUEL DETECTOR
CIRCUIT

Fig. 7-49. Water-in-fuel detector and method for testing water-in-fuel sensor. (Buick)

163

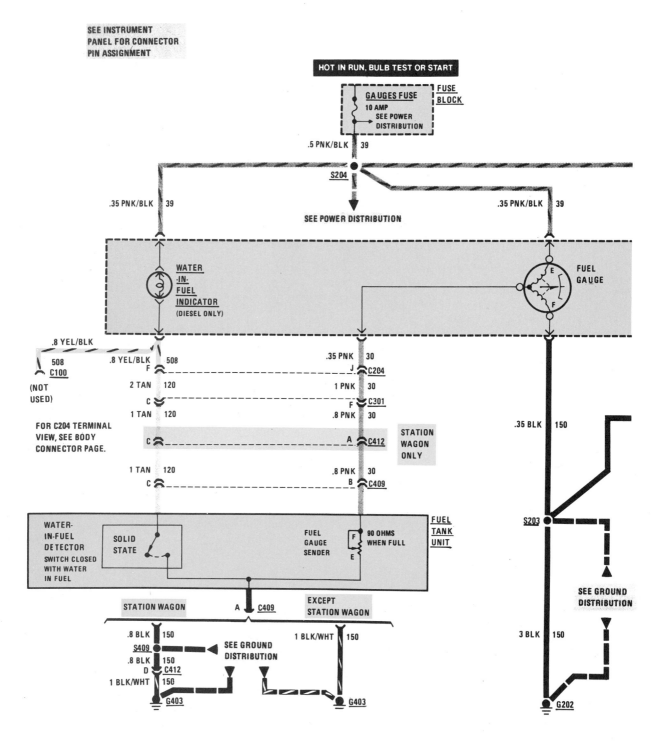

Fig. 7-50. Electrical wiring diagram for diesel fuel gauge and water-in-fuel detector.

monly found on pickup trucks and some vans. Refer to the illustration in Fig. 7-52.

A selector valve controls which fuel tank feeds fuel to the engine. A passenger compartment switch can be used to activate valves in the selector mechanism. As shown in Fig. 7-52, the fuel selector valve is normally mounted under the car and is in-line with the fuel lines from each tank.

When problems are encountered, check the electrical connections and fuse for the selector valve. Make sure

power is reaching the valve. Use service manual directions if needed.

SUMMARY

Generally, only rigid, double-wrapped steel tubing and flexible hose are used to carry fuel from the tank to the engine. For safety reasons, semirigid tubing of copper, brass, aluminum, etc., should not be used as fuel line.

A tubing cutter is the best method for cutting tubing.

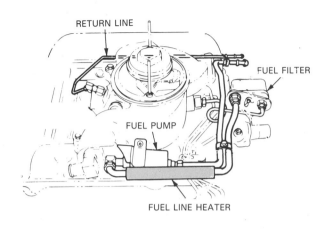

Fig. 7-51. Fuel line heater warms diesel fuel in cold weather. (Oldsmobile)

It will cut squarely with little or no burr.

Various types of bending tools are available for making fuel lines, including bending springs and lever arm benders. Avoid bending tubing too close to the end. Allow enough room to slide a fitting onto the tube.

A flaring tool can be used to shape the end of tubing into a flare. A flared end can be used with a flare fitting to form a leakproof seal and connection. Since automotive fuel lines use double-wrapped steel tubing, flares must be double-lapped. A single lap flare could split and leak.

Before installing a new fuel line, clean it by blowing out the line with compressed air.

Numerous types of fittings are available to meet almost every need.

When installing a fuel line, alignment of threaded mating parts is critical. If the fitting is started at an angle, threads can be damaged. Start the fitting with your

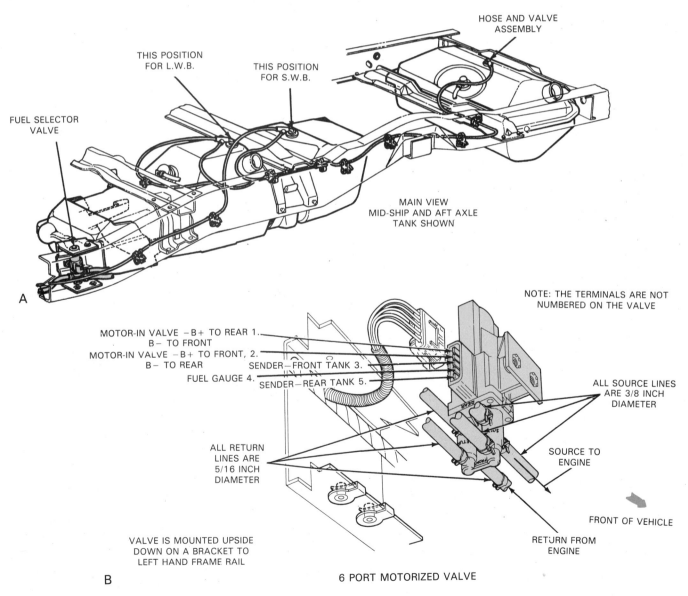

Fig. 7-52. A—Dual tank system increases range of certain pickup trucks. Note location of fuel selector valve. B—Diagram of fuel tank selector valve. (Mercury)

fingers before tightening it with a tubing wrench. Remember the basic tubeworking rules: no straight runs; avoid hot spots; support long sections of line; double check fitting alignment; use two wrenches when needed; check for clearance with moving parts; and check the line for fuel leaks.

Hoses are used at points of high vibration or great movement. Install only recommended fuel line hose. Do NOT use nonreinforced vacuum type hose. Always inspect used fuel line hose closely for cracks, splits, hardness, softness, swelling. Replace as necessary. Hose can be cut using various tools (knife, shears, diagonal pliers, etc.). In any case, the cut should be made squarely. In general, follow the same rules with hose installations as with tubing installation.

Screw, spring, and crimp type clamps are used to secure hose to their fittings. Make sure that the clamp is installed on the inside of the bead or barb of the fitting. If not, the hose could be forced off, causing a fire. When a line is being repaired and there is no bead on the metal connection, the hose must overlap at least two inches and be clamped.

A contaminated fuel system should be cleaned. Drain and clean the fuel tank. Blow out all of the fuel lines with compressed air. Run a couple of quarts of clean fuel through the system and into a safety can. Reinstall the line and check for leaks.

Hose can be used to replace a damaged section of gas line, so long as it is shorter than six in. When the section of line is over six inches long, steel tubing, along with hose, couplers, or fittings, must be used.

Fuel filters keep road dirt, dust, rust, scale, water, solder, and gum out of the fuel system components. Fuel contaminants (dirt) can enter the fuel during transportation, storage, transfer, and even while in a car's tank.

A strainer removes coarse debris from fuel. It is commonly used on the suction side of the fuel pump.

A filter is much finer than a strainer. It is usually located on the pressure side of the fuel pump where flow resistance is not as harmful. The main filter protects the carburetor, fuel injectors, injector pump, and other parts from clogging or damage. Some filter elements (sintered bronze, ceramic, etc.) can be cleaned and reused while others (treated paper), must be replaced.

Fuel filters should be changed at regular intervals. A small fuel filter, (in-carburetor, for instance), should be replaced every 12 months or 12,000 miles (19 312 km). A larger in-line filter can go as long as two years or 24,000 miles (38 600 km) without service.

A clogged filter will often cause engine performance problems at a SPECIFIC ENGINE SPEED. The engine may miss, surge, or lose power at a particular engine speed. Check filter condition by blowing through them. You should be able to blow easily through a clean fuel filter.

Diesel fuel filters must be even more efficient than gasoline type fuel filters. Diesel injection systems can be damaged by the smallest amount of dirt or water. Water is extremely harmful to diesel fuel systems because of its corrosive action. Both single-stage and two-stage filtering systems are utilized in automobiles. Some diesel fuel filters have a drain which permits water and sediment removal.

Most diesel fuel systems are circulating types. They are designed so that fuel constantly flows from the tank, through the injection mechanism, and back to the tank. The diesel oil lubricates the injection pump and injectors. Also, this action helps warm the fuel in the tank and prevents cold weather fuel clouding.

Always be aware of the constant danger when working on a fuel line or filter. The slightest fuel leak or fuel spill can cause a serious fire.

KNOW THESE TERMS

Rigid tubing, Semirigid tubing, Flexible tubing, Double-walled steel tubing, Reaming knife, Bending spring, Single-lapped flare, Double-lapped flare, SAE flare fitting, Inverted flare fitting, Compression fittings, Double-compression fitting, Separate sleeve fitting, Ermeto fittings, T-fittings, Unions, Elbows, O-ring fittings, Swivel fittings, Distribution blocks, Push connection fittings, Barb type fittings, Crimp hose clamp, Spring type hose clamp, Screw type hose clamp, Sintered bronze filter, Filter check valve.

REVIEW QUESTIONS—CHAPTER 7

1. _____ _____ _____ tubing must be used to make a fuel line. It is strong enough to withstand the _____ and _____ of an auto fuel system.
2. The best tool for cutting a piece of tubing is a hacksaw. True or false?
3. In your own words, explain how to use a tubing cutter.
4. Which of these problems CANNOT be caused by tubing burrs?
   a. Flooded engine.
   b. Restricted fuel flow.
   c. Ruined water pump.
   d. Scored cylinder wall.
   e. Harm fuel system.
5. List two types of tubing benders and explain their characteristics.
6. Why must tubing never be bent too close to the end?
7. What is a tubing flare?
8. Explain how to form a tubing flare using a generating type flaring tool.
9. On a fuel line (double-wrapped tubing), a _____ type flare must be used to prevent a possible _____, _____ and _____.
10. The three basic parts of a double-lap flaring tool are _____, _____, and _____.
11. Flare fittings are used with _____ ends to connect a fuel line to another component such as a _____.
12. Describe the two basic types of flare fittings.
13. This type of fitting uses tapered threads.
   a. SAE.
   b. Inverted flare.
   c. Compression.
   d. Pipe.
14. By tightening a compression fitting, the tubing is _____.
15. A T-fitting is commonly used to splice together two

sections of tubing. True or false?

16. This fitting is needed to route tubing around an obstacle.
    a. Pipe.
    b. Elbow.
    c. Compression.
    d. Hose.
17. An _____ is one that uses a soft neoprene seal to form a leakproof connection.
18. What is the difference between a drain cock and a shutoff cock.
19. List the eight major "tubeworking" rules.
20. Why are fuel system hoses needed?
21. If a vacuum type hose is used as a fuel line hose, what may happen?
22. Describe the construction of a fuel line hose.
23. If a pressurized fuel line or hose ruptures with the engine running, what could happen?
24. Describe how to inspect a fuel line hose.
25. List the four types of hose clamps.
26. Describe, in your own words, how to clean a fuel system.
27. Hose can be used to repair a fuel line when the removed section of metal line is shorter than _____. The hose itself should not be longer than _____.
28. Hose must extend at least _____ over the end of an unbeaded metal line.
29. What is the general purpose of a fuel filter?
30. Explain the difference between a fuel strainer and a filter.

31. This type fuel filter CAN be cleaned.
    a. Paper.
    b. Treated paper.
    c. Sintered bronze.
    d. Disposable.
32. A treated paper fuel filter is _____ to provide more filtering. It is treated with special _____ to stop water and other impurities.
33. List four fuel filter locations.
34. Smaller fuel filters should be changed about every _____ or _____; large filters should be changed every _____ or _____.
35. What performance problems can be caused by partially clogged fuel filter?
36. Explain two simple methods of testing for a plugged fuel filter.
37. Why must diesel fuel filters be so efficient?
38. With a single-stage fuel filter, the filter must handle a _____ of fuel, yet still _____ all of the system components.
39. In your own words, explain a two-stage diesel fuel filtering system.
40. Why is water so harmful to a diesel fuel system?
41. Indicate which of the following is NOT a purpose of a circulating type diesel fuel system.
    a. Cool the injection pump.
    b. Lubricate the pump.
    c. Carry dirt back to tank filter.
    d. Prevent fuel clouding.
42. List the seven basic safety rules for servicing and repairing a fuel system.

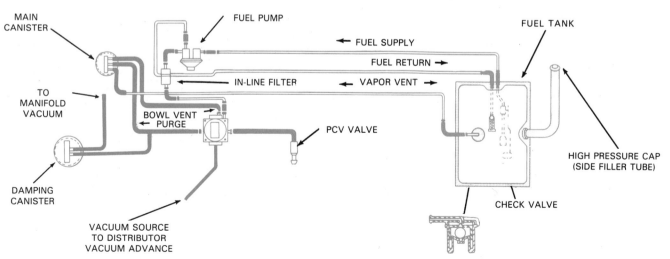

Schematic of modern fuel system with 360 degree rollover fuel leakage protection. (Chrysler Corp.)

Auto Fuel Systems

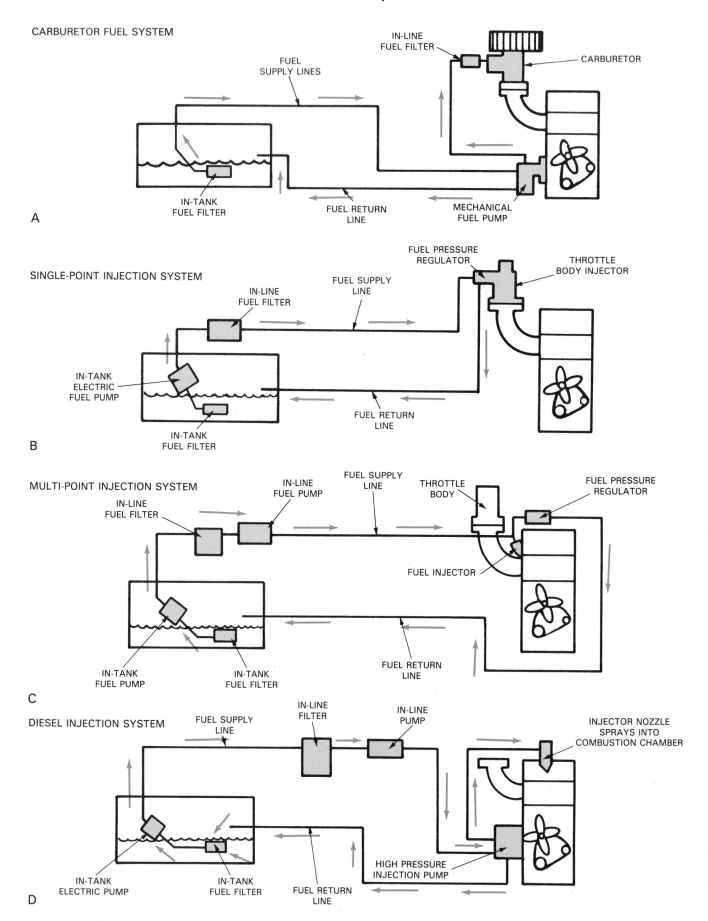

CARBURETOR FUEL SYSTEM

IN-LINE
FUEL FILTER

CARBURETOR

FUEL
SUPPLY LINES

IN-TANK
FUEL FILTER

A

FUEL RETURN
LINE

MECHANICAL
FUEL PUMP

SINGLE-POINT INJECTION SYSTEM

FUEL PRESSURE
REGULATOR

THROTTLE
BODY INJECTOR

IN-LINE
FUEL FILTER

FUEL SUPPLY
LINE

IN-TANK
ELECTRIC
FUEL PUMP

IN-TANK
FUEL FILTER

FUEL RETURN
LINE

B

MULTI-POINT INJECTION SYSTEM

IN-LINE
FUEL PUMP

FUEL SUPPLY
LINE

THROTTLE
BODY

FUEL PRESSURE
REGULATOR

IN-LINE
FUEL FILTER

FUEL INJECTOR

IN-TANK
FUEL PUMP

IN-TANK
FUEL FILTER

FUEL RETURN
LINE

C

DIESEL INJECTION SYSTEM

FUEL SUPPLY
LINE

IN-LINE
FILTER

IN-LINE
PUMP

INJECTOR NOZZLE
SPRAYS INTO
COMBUSTION CHAMBER

IN-TANK
ELECTRIC PUMP

IN-TANK
FUEL FILTER

FUEL RETURN
LINE

HIGH PRESSURE
INJECTION PUMP

D

Fig. 8-1. Schematics show relationship of basic parts of four different fuel delivery systems. A—Carburetor fuel system. B—Throttle body injection system. C—Port fuel injection system. D—Diesel fuel system.

8 FUEL PUMPS—DESIGN, CONSTRUCTION, OPERATION

After studying this chapter, you will be able to:
- *List the functions of a fuel pump.*
- *Name the different types of fuel pumps.*
- *Summarize mechanical fuel pump operation.*
- *List the major parts of a mechanical fuel pump.*
- *Compare mechanical fuel pump design differences.*
- *Name the different types of electric fuel pumps.*
- *Explain electric fuel pump operation.*
- *List some advantages of an electric fuel pump.*
- *Summarize the operation of a low oil pressure safety circuit.*
- *Describe how a computer can be used to control electric fuel pump operation.*

This chapter will introduce the most common fuel pump types. It will explain basic design, construction, and operation from a theoretical viewpoint.

The next chapter will cover fuel pump diagnosis, testing, removal, repair, and installation. By studying both theory and "hands-on" repair, you will be well prepared for actual service and repair procedures.

FUEL PUMP FUNCTIONS

The main purpose of the fuel pump is to supply adequate fuel flow to the engine, Fig. 8-1. The pump must force gasoline or diesel fuel out of the tank, through the lines, filters, and into the carburetor or injection mechanism.

GRAVITY FEED

As illustrated in Fig. 8-2, early cars did not use a fuel pump. Instead, the gas tank was above the carburetor. Known as a *gravity feed fuel system,* this arrangement is still used on most lawnmowers and motorcycles. The force of gravity moves fuel through the system. No fuel pump is needed.

Gravity systems are fine so long as the carburetor is kept below the fuel tank. However, when driving up a steep hill, Fig. 8-3, fuel flow may stop.

MODERN PUMP SYSTEM

Fuel tanks in today's automobiles are normally lower than the engine. This requires a pump that can lift fuel

GRAVITY FEED FUEL SYSTEM

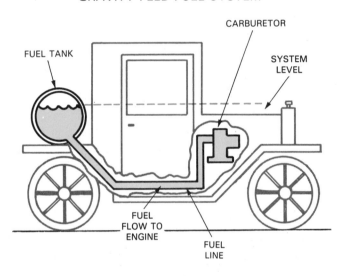

Fig. 8-2. Early gravity fuel system. With car level, fuel is gravity fed from fuel tank to carburetor.

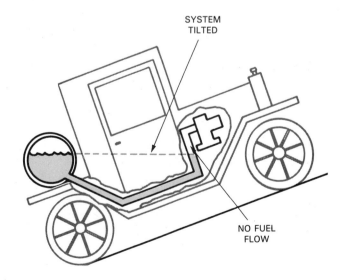

Fig. 8-3. On a hill, with tank lower than carburetor, car could starve for fuel and stall.

169

from the tank and push it into the carburetor or injectors. To overcome the weight or pressure of the "head" of fuel, the pump must be able to produce more pressure than the downward force produced by the weight of fuel in the lines.

A fuel pump can have several basic functions:
1. Provide fuel flow to the carburetor or injection mechanism.
2. Prevent fuel pressure fluctuations or surging (changes) during operation.
3. Separate vapor bubbles from liquid fuel to prevent vapor lock.
4. Supply vacuum to various accessories.
5. Pump fuel through fuel return line, again, to prevent vapor lock.
6. Remove or filter impurities from fuel.

In some cases, the fuel pump provides vacuum to augment (assist or replace) engine intake manifold vacuum. Extra vacuum may be needed to operate various accessories. A second vacuum pump assembly can be combined with the conventional fuel pump. Diesel engines have no engine vacuum and an auxiliary vacuum pump is needed to operate vacuum powered devices.

TYPES OF FUEL PUMPS

Fuel pumps are either mechanically or electrically operated. An engine-powered eccentric or cam lobe drives a mechanical pump. An electric fuel pump needs battery or alternator electricity to power either an electric motor or an electromagnet. Within these two fundamental fuel pump types are numerous variations.

Reciprocating fuel pumps

A *reciprocating* fuel pump works with a back and forth motion. See Fig. 8-4. Movement of parts inside the pump is linear (straight line).

All mechanical fuel pumps and a few electric pumps are reciprocating. A drawback of all reciprocating pumps are the pressure pulsations caused by their action.

Rotary fuel pumps

As the name implies, a *rotary fuel pump* uses a spinning motion. See Fig. 8-5.

Generally, a rotary pump produces a smoother fuel flow than a reciprocating pump. Line pressure pulsations are almost completely eliminated. Only motor-driven electric fuel pumps are rotary.

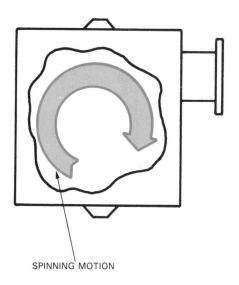

SPINNING MOTION

Fig. 8-5. Rotary fuel pump uses spinning motion to produce fuel flow.

Positive displacement fuel pumps

Positive displacement fuel pumps (sliding vane, roller vane, diaphragm and plunger types) trap fuel in a sealed pump enclosure to compress or pressurize it. This forces a definite amount of fuel out of the pump on each revolution or stroke.

Centrifugal fuel pumps

A *centrifugal fuel pump* (stationary impeller) works on the same principle as a water pump or a blower fan. It has spinning blades which use centrifugal force to throw the fuel outward. This outward thrust is directed at an outlet opening. As a result, fuel flow and pressure are developed.

A centrifugal pump is different from a positive displacement type. There is clearance or space between the impeller blades and the housing of the pump.

Mechanical fuel pumps

A *mechanical fuel pump*, Fig. 8-6, normally bolts to the side of the engine block. Usually, it is powered by an ECCENTRIC (egg shaped lobe) on the engine camshaft. The eccentric's out-of-round shape produces a pumping action when the camshaft turns. Sometimes, the fuel pump eccentric is mounted on the front of the

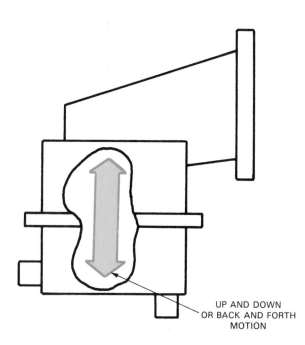

UP AND DOWN OR BACK AND FORTH MOTION

Fig. 8-4. Reciprocating type pump uses back and forth pumping movement.

cam sprocket, Fig. 8-7, or on the distributor shaft.

As the engine runs, the rotating lobe repeatedly pushes on the pump arm powering the fuel pump, Fig. 8-8. For this reason, an eccentric-powered pump is known as a mechanical fuel pump.

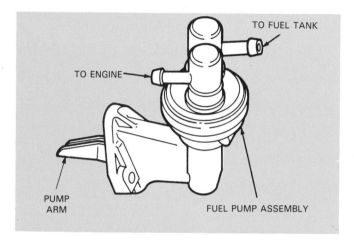

Fig. 8-6. A mechanical pump is mounted to engine block. It is powered by the action of a camshaft eccentric or lobe which pushes on pump arm. (AC-Delco)

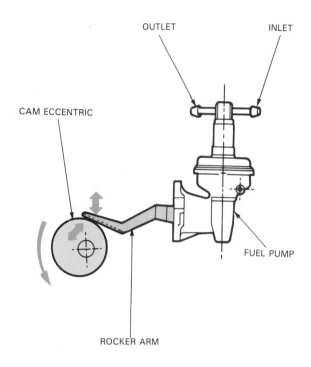

Fig. 8-8. Rotating eccentric repeatedly pushes up on rocker arm. Spring on other end of arm returns arm for other half of pumping action. (Ford Motor Co.)

Fig. 8-7. On V-8 type engines, fuel pump eccentric is normally mounted on front of cam gear. (General Motors)

MECHANICAL FUEL PUMP CONSTRUCTION

Even though mechanical fuel pumps are available in hundreds of different designs, shapes, and sizes, they still use the same fundamental parts.

Pump diaphragm

The "heart" of the mechanical fuel pump is a *rubber diaphragm.* Its movement pushes fuel through the fuel system. Refer to Fig. 8-9. The flexible diaphragm is clamped between the two halves of the pump body.

For strength and durability, the rubber in the diaphragm is normally reinforced with several layers of cloth. During the life of a fuel pump, the diaphragm may be stretched back and forth millions of times. It is usually one of the first fuel pump components to fail.

Pull rod and rocker arm

A small *pull rod,* Fig. 8-10, is attached to the center of the diaphragm and to the pump rocker arm. When the pump rocker arm moves up and down, the pull rod will cause the diaphragm to flex or stretch up and down.

A *seal* is normally fitted around the pull rod to keep hot engine oil off the diaphragm.

Diaphragm spring

A large spring is attached to and presses against the back of the rubber pump diaphragm. This spring is shown in Fig. 8-10. When compressed by the action of the rocker arm and pull rod, the spring forces the diaphragm back down into the pumping chamber. During pump operation, the *diaphragm spring* forces fuel out of the pump under pressure.

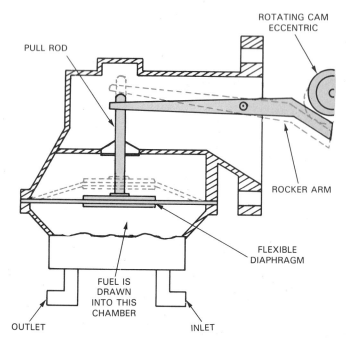

Fig. 8-9. In mechanical pump, a soft, flexible diaphragm is used to push fuel through system. Notice how movement of eccentric, rocker arm, and pull rod produce diaphragm movement.

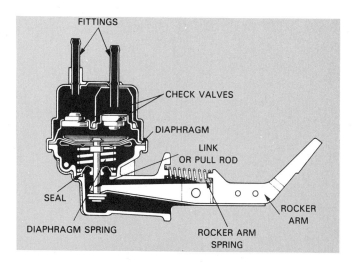

Fig. 8-10. Cutaway shows interior parts and assembly of mechanical fuel pump. Diaphragm spring tension causes fuel pressure and fuel flow. Return spring keeps rocker arm snug against eccentric. (Toyota)

Rocker arm

The *rocker arm* is a lever which transfers the motion of the camshaft and cam lobe to the pull rod. Also called the pump lever, it pulls on the pull rod to compress the diaphragm spring.

Rocker arm return spring

A smaller spring, also pictured in Fig. 8-10, fits between the rocker arm and the pump body. It keeps the rocker arm riding on the cam lobe.

Sometimes, the *return spring* is positioned around the pull rod, on the end of the rocker arm. Fig. 8-11 shows this spring in a fuel pump cutaway section.

Without a functional rocker arm return spring, a loud CLATTERING OR KNOCKING SOUND can be produced. The sound is much like that made by a defective hydraulic lifter.

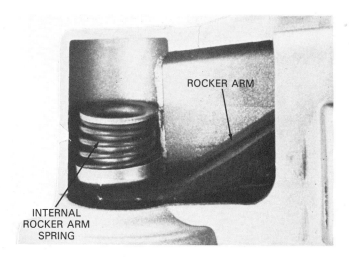

Fig. 8-11. If internal rocker arm spring is bad, a loud clattering can be heard. It will sound very much like a noisy hydraulic lifter. (Airtex)

Pumping chamber and vent

The area in the pump opposite the pull rod is called the *pumping chamber.* The pull rod chamber on the backside of the diaphragm contains only air and an air *vent.* The vent prevents pressure or vacuum buildup when the diaphragm flexes.

Check valves

Two check valves are normally mounted in the pumping chamber. Illustrated in Fig. 8-12, a *check valve* permits fuel flow in only one direction.

The fuel pump *inlet check valve* faces in a direction that allows fuel to flow only into the pump from the line. The *outlet check valve* faces the other way. It allows fuel to leave, but not enter, the pump.

These valves have a flat washer which is normally held closed by a fine, weak spring. As fuel pushes against the face of the valve, the spring easily compresses to allow fuel flow. In the opposite direction however, fuel forces the washer against its seat and the valve seals the opening. As more pressure is applied, the check valve closes even tighter.

Pulsation diaphragm

As a mechanical fuel pump operates, it reciprocates during intake and output strokes. This action produces spurts of pressure and flow. These rapid pressure pulsations can cause a mild but very rapid vibration in the column of fuel going to the engine. If uncontrolled, fuel pump pressure pulsations can cause undue wear and

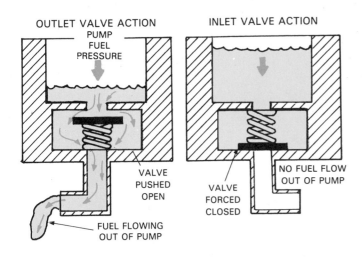

OUTLET VALVE ACTION

PUMP FUEL PRESSURE

VALVE PUSHED OPEN

FUEL FLOWING OUT OF PUMP

INLET VALVE ACTION

VALVE FORCED CLOSED

NO FUEL FLOW OUT OF PUMP

Fig. 8-12. Observe action of two fuel pump check valves. They allow fuel flow in one direction only.

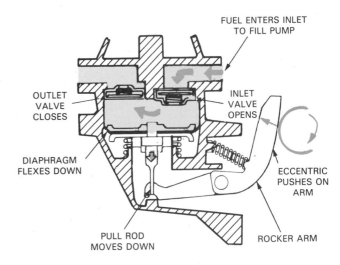

FUEL ENTERS INLET TO FILL PUMP

OUTLET VALVE CLOSES

INLET VALVE OPENS

DIAPHRAGM FLEXES DOWN

ECCENTRIC PUSHES ON ARM

PULL ROD MOVES DOWN

ROCKER ARM

Fig. 8-14. Mechanical fuel pump intake stroke. Cam lobe pushes rocker arm which moves pull rod and diaphragm, drawing in fuel.

fatigue inside the carburetor (inlet needle, needle seat, float hinge, etc.).

To reduce and control this undesirable condition, a *pulsation diaphragm* is sometimes installed in the pump or fuel line. In Fig. 8-13, the dampener is nothing more than a flexible diaphragm stretched across an enclosed air chamber.

PULSATION DIAPHRAGM

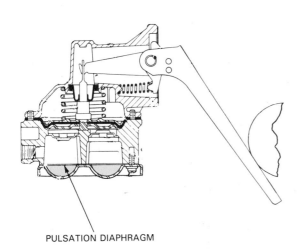

PULSATION DIAPHRAGM

Fig. 8-13. Flexible pulsation diaphragm helps smooth flow of fuel into carburetor. (GMC)

MECHANICAL FUEL PUMP OPERATION (INTAKE STROKE)

During the *pump intake stroke,* Fig. 8-14, the camshaft lobe pushes on the fuel pump rocker arm. In turn, the pull rod movement stretches the diaphragm and compresses the diaphragm spring. This increases the volume

in the fuel pump chamber and produces a low pressure area or vacuum. The inlet check valve opens and the output valve closes.

As a result, fuel flows in, pushed by the outside atmospheric pressure. The fuel pump fills with fuel and is ready for the output stroke.

Fuel pump output stroke

When the engine camshaft or distributor eccentric rotates away from the fuel pump rocker arm, the pump arm is released. This starts the fuel *pump output stroke,* Fig. 8-15. The compressed diaphragm spring then pushes on the diaphragm and the fuel in the chamber. The inlet check valve closes, the outlet valve opens and fuel is forced into the carburetor fuel line.

The amount of pressure depends upon the stiffness or strength of the diaphragm spring. The volume or quantity of fuel depends upon the stroke (distance of travel) and diameter of the diaphragm.

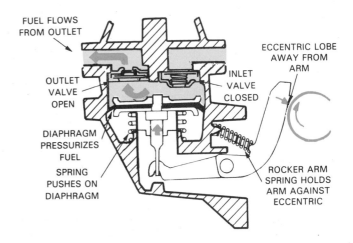

FUEL FLOWS FROM OUTLET

OUTLET VALVE OPEN

ECCENTRIC LOBE AWAY FROM ARM

INLET VALVE CLOSED

DIAPHRAGM PRESSURIZES FUEL

SPRING PUSHES ON DIAPHRAGM

ROCKER ARM SPRING HOLDS ARM AGAINST ECCENTRIC

Fig. 8-15. Mechanical fuel pump in output stroke. Diaphragm pushes fuel out outlet valve while intake valve closes.

Fuel pump idling

A fuel pump normally can provide more fuel than the engine needs. Most of the time, fuel pump output does not have to be at maximum. To reduce fuel volume, the rocker arm is often designed so that it can slide up and down on the pull rod. This action, shown in Fig. 8-16, is called *pump idling.*

When the carburetor is full of fuel and pump output must be reduced, the pump rocker can move up and down in a full stroke while the pull rod and diaphragm make only a partial stroke. This lets the camshaft eccentric and rocker arm function normally without unecessary diaphragm action. Fuel line pressure, however, is maintained by the compressed diaphragm spring.

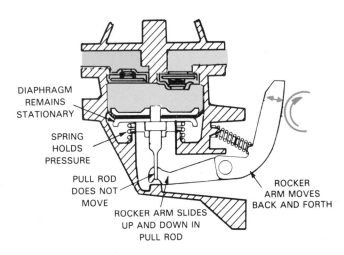

Fig. 8-16. With fuel pressure too high for spring to overcome, pump idles. Rocker arm pumps but diaphragm does not move.

Fuel return systems

A *fuel return system* consists of a valve in the fuel pump and a second fuel line from the fuel pump back to the fuel tank. It serves a dual purpose. First, it allows excess fuel to return to the fuel tank when the engine does not need it. Secondly, it allows cooled fuel to recirculate through the fuel pump to prevent vapor lock. Fig. 8-17 shows a simplified fuel return system.

Vapor return valve

With some fuel return systems, a vapor and hot fuel return valve or a bypass hole is fitted inside the fuel pump body. See Fig. 8-18.

As you know, heated fuel can form vapor bubbles that upset carburetor operation. To help end this problem, the pump's *fuel return valve* will open whenever fuel pressure is high or fuel flow is low. It permits fuel to flow through the return line back into the fuel tank. The circulation of COOL FUEL from the tank reduces the chance of vapor lock.

Should fuel pressure drop, (rapid acceleration, high engine speed, empty carburetor bowl, etc.), the spring in the fuel return valve will close. This keeps fuel from

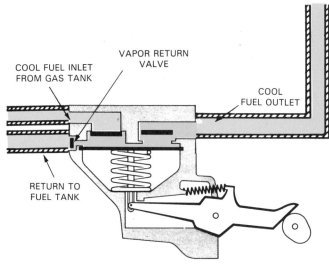

Fig. 8-18. Vapor return valve lets extra fuel flow out of system back into fuel tank. (Ford Motor Co.)

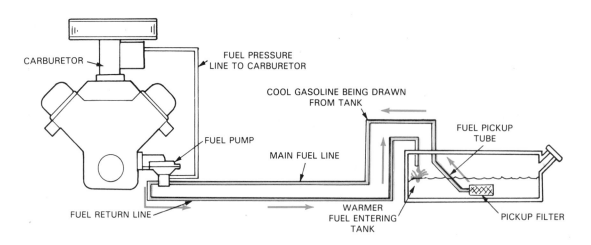

Fig. 8-17. Simplified example of fuel return system. Note extra fuel line between fuel pump and fuel tank.

flowing into the return line and assures adequate fuel pressure and volume. The valve opens only when line pressure is near maximum.

Fuel pump bypass hole

Sometimes, a *fuel pump bypass hole,* illustrated in Fig. 8-19, allows fuel pressure to be released into the inlet side of the pump. This action prevents *percolation* (fuel bubbling into intake manifold because of heat). It relieves fuel line pressure so that the carburetor float needle will not be forced open by line pressure.

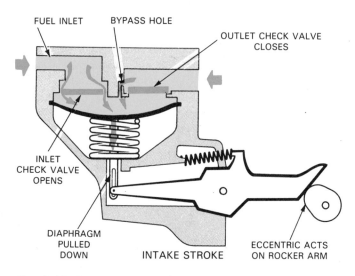

Fig. 8-19. Bypass hole relieves excessive pressure by allowing return flow into inlet side of pump. (Ford)

Pump pressure and volume

All fuel pumps have factory specifications for both output pressure and output volume. Pressure and volume values generally increase as engine size increases.

Volume is measured in ounces, pints, quarts, liters, etc. A volume rating is the amount of fuel delivered in a given amount of time.

As a rule-of-thumb, the fuel pump of a small four cylinder engine should pump about 1/2 pt. (236 mL) of fuel every 30 seconds. On a slightly larger 6-cylinder engine, the fuel pump should pump around 3/4 pt. (355 mL) in 30 seconds. A large V-8 engine's fuel pump should pump close to a full pint (473 mL) or more every 30 seconds.

In general, a fuel pump for a carburetor should produce around 4 to 7 psi of PRESSURE (28 to 48 kPa). A fuel pump for a gasoline injection system usually produces higher pressures—10 to 40 psi (69 to 276 kPa).

As we discussed under fuel pump testing, always check factory specifications when a fuel pump's pressure or volume is near its lower "spec" limit.

SERVICEABLE FUEL PUMP

A *serviceable fuel pump* is one that can be taken apart and repaired. It can be disassembled and rebuilt. Screws are commonly used to hold the two halves of the pump body together, Fig. 8-20. The exploded view in Fig. 8-21 shows the parts of a serviceable fuel pump.

Fig. 8-20. The halves of a serviceable fuel pump are fastened together with screws. Pump can be disassembled and rebuilt. (AC-Delco)

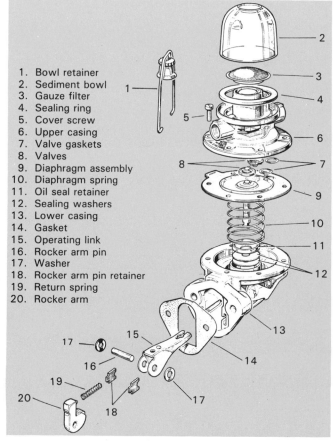

1. Bowl retainer
2. Sediment bowl
3. Gauze filter
4. Sealing ring
5. Cover screw
6. Upper casing
7. Valve gaskets
8. Valves
9. Diaphragm assembly
10. Diaphragm spring
11. Oil seal retainer
12. Sealing washers
13. Lower casing
14. Gasket
15. Operating link
16. Rocker arm pin
17. Washer
18. Rocker arm pin retainer
19. Return spring
20. Rocker arm

Fig. 8-21. Study parts of fuel pump in this exploded view.

NONSERVICEABLE FUEL PUMP

A *nonserviceable fuel pump* cannot be disassembled and rebuilt. The pump body or chamber halves, Fig. 8-22, are permanently assembled at the factory. A folded metal lip holds the halves together.

Because of the high cost of labor and parts, it is becoming more economical to remove and replace a fuel pump than to repair or rebuild it. A new nonserviceable pump can be installed very quickly, with little added cost.

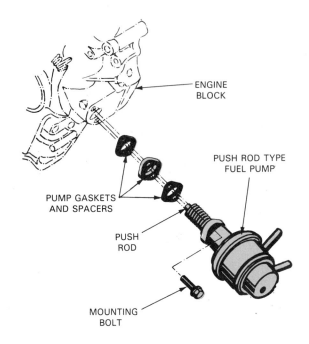

Fig. 8-23. In this pump, push rod is used instead of rocker arm. When pump is bolted to engine, push rod rides on eccentric. (Buick)

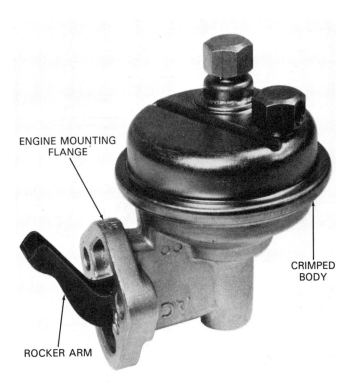

Fig. 8-22. Nonserviceable pump cannot be taken apart. It is permanently crimped together at factory.

PUSH ROD TYPE MECHANICAL FUEL PUMP

Some engines use a *mechanical push rod pump.* Instead of a rocker arm, a straight metal rod rides on the cam eccentric to provide in and out pump motion.

This type bolts to the side of the engine block. One type of push rod fuel pump is pictured in Fig. 8-23. It is often used on small displacement engines because of the short distance between the camshaft and outer block surface.

The internal parts and operation of a push rod fuel pump are basically the same as a rocker arm type fuel pump. See Fig. 8-24.

FUEL PUMP LUBRICATION

Moving parts of a mechanical fuel pump are lubricated by engine oil. Normally, the fuel pump housing around the rocker arm is open to the engine crankcase. This allows oil droplets and vapors to enter. The oil reduces wear on the rocker arm, rocker arm pivot, and pull rod

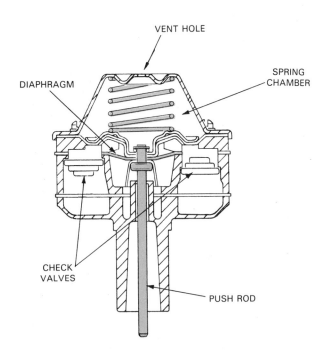

Fig. 8-24. Push rod acts directly upon pump diaphragm. (General Motors)

connection. As mentioned earlier, the pull rod seal keeps oil away from the rubber diaphragm.

COMBINATION FUEL AND VACUUM BOOSTER PUMP

In older cars and some trucks, a vacuum pump is combined with the conventional mechanical fuel pump. It is called a *vacuum booster pump* because it sup-

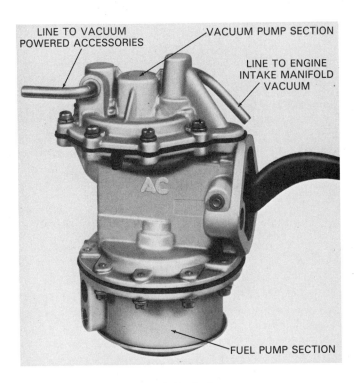

LINE TO VACUUM
POWERED ACCESSORIES

VACUUM PUMP SECTION

LINE TO ENGINE
INTAKE MANIFOLD
VACUUM

AC

FUEL PUMP SECTION

Fig. 8-25. Upper section of this fuel pump serves as vacuum booster pump. (AC-Delco)

plements engine vacuum. See Fig. 8-25.

Basically, a second diaphragm and pumping chamber is attached to the opposite end of the fuel pump body. A long pull rod joins the vacuum diaphragm to the regular fuel pumping diaphragm. When the pump rocker arm moves up and down, both diaphragms flex together. To control the direction of airflow, check valves are installed in the inlet and outlet openings in the pumping chamber.

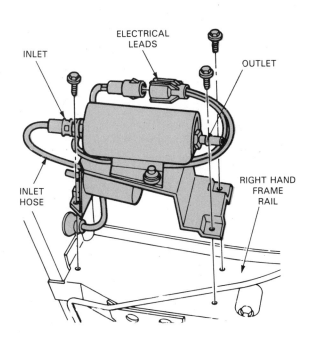

ELECTRICAL
LEADS

INLET

OUTLET

INLET
HOSE

RIGHT HAND
FRAME
RAIL

Fig. 8-26. This electric fuel pump is located on car frame between engine and fuel tank.

Usually, the intake of the vacuum booster pump is connected to the vacuum accessories and the outlet is connected to the engine intake manifold.

Vacuum booster pumps are known to cause excessive engine oil consumption. If the vacuum pump diaphragm ruptures or develops a leak, it is possible for large amounts of engine oil to be sucked into the pump and intake manifold.

ELECTRIC FUEL PUMPS

There are two basic types of electric fuel pumps: reciprocating and rotary. Electric pumps may be located somewhere between the engine and the fuel tank or in the fuel tank itself, Figs. 8-26 and 8-27.

Within these two types, are numerous variations. Bellows, diaphragm, and plunger types can be classified as reciprocating pumps. The most common rotary pumps are the sliding vane, impeller, and roller vane.

The advantages of an electric fuel pump will usually greatly outweigh any disadvantages. For this reason, electric fuel pumps are becoming more and more popular.

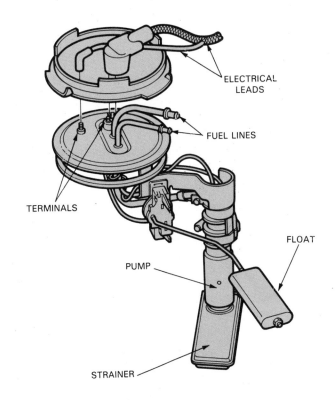

ELECTRICAL
LEADS

FUEL LINES

TERMINALS

FLOAT

PUMP

STRAINER

Fig. 8-27. This electric fuel pump is located inside fuel tank. Pump motor is sealed so fuel will not short out windings. (Ford)

Elimination of vapor lock

Unlike a mechanical pump which is attached to a hot engine block, an electric fuel pump is normally mounted near the fuel tank in the rear of the car. See Fig. 8-28. The pump remains cool and the fuel in the hot engine compartment is always pressurized. This helps prevent the formation of vapor bubbles.

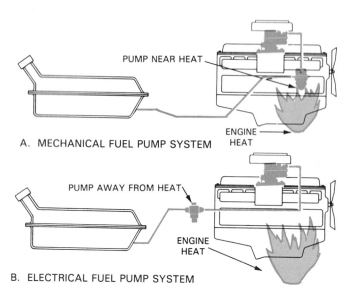

A. MECHANICAL FUEL PUMP SYSTEM

B. ELECTRICAL FUEL PUMP SYSTEM

Fig. 8-28. Relative locations of mechanical and electric fuel pumps. Electric pumps, unlike mechanical ones, can be far away from heat. (Ford Motor Co.)

Instant fuel pump pressure

With a mechanical fuel pump, fuel pressure may not develop until the engine has been "turning over" for a few moments. It may take a few seconds before the engine will start and run. This means additional wear on the starting motor and the battery.

With an electric pump, fuel system pressure is produced as soon as the ignition key is turned ON. Starting motor operation is not needed to fill the carburetor or injectors with fuel. With the key turned to ON, the electric pump will pressurize the fuel system.

Reduced fuel pressure pulsations

As mentioned, a reciprocating mechanical fuel pump develops tiny surges (spurts) of pressure and flow. Several kinds of electric fuel pumps, especially rotary types, can reduce or almost eliminate fuel pulsations. A rotary pump (sliding vane, roller vane, stationary impeller, etc.) does not use a back and forth reciprocating motion.

High volume output

Some electric fuel pumps, especially in-line types, are capable of moving large amounts of fuel in a short period of time. High volume electric fuel pumps are well suited to many high performance applications. When fuel consumption is extremely rapid (multiple carburetors or throttle body injectors, or turbocharged engines), a high volume electric pump may be essential. It can prevent fuel starvation at high engine speeds.

Saves energy

The rocker arm or push rod of a mechanical fuel pump operates constantly when the engine is running. The camshaft lobe moves the rocker arm and compresses the return spring hundreds of times each minute, even when fuel is not needed. This wastes energy.

An electric fuel pump shuts off completely (no parts moving, no friction, and no energy consumption) when

fuel flow is not needed to the engine.

An electric fuel pump, however, does have a few undesirable characteristics. These disadvantages include: noise, high cost, and engine flooding.

Electric fuel pump noise

A diaphragm or bellows type electric pump can have a noticeable clicking sound. A positive displacement, rotary pump can whir or whine. To some people, this noise is objectionable.

High cost

The electric motor or solenoid add to the cost of an electric pump. The added price can be undesirable when replacing a defective fuel pump.

Engine flooding

Under certain circumstances (ignition ON and injector needle valve leaking), an electric fuel pump can flood an engine with gasoline. It can empty the fuel tank into the engine. The fuel can then leak out of the bad injector and into the engine. This is a rare problem, however. Since the operation of a mechanical pump depends upon engine operation, a mechanical fuel pump cannot cause this dangerous situation.

BELLOWS ELECTRIC FUEL PUMP

A *bellows fuel pump*, Fig. 8-29, consists of a flexible bellows, an electromagnet (solenoid), return spring, set of points, inlet check valve, and an outlet check valve.

The *bellows* is like the folded section of an accordion. It is nothing more than a chamber with pleated or folded walls. The folds make the walls expandable.

Intake stroke

When the ignition is ON, voltage is supplied to the pump terminal. Current causes the electromagnet to attract the metal armature with great force. Being fastened to the bellows, the armature expands the bellows as it is pulled toward the windings of the electromagnet. Low

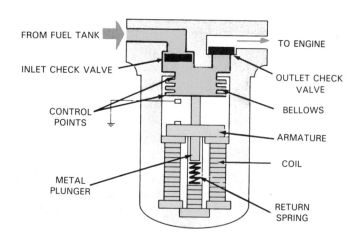

Fig. 8-29. As electromagnet and spring repeatedly force bellows up and down, check valve action causes fuel flow through pump.

pressure in the bellows causes the inlet valve to open. Fuel flows into the bellows from the fuel tank.

Output stroke

As soon as the plunger is pulled far enough into the windings, it strikes and opens a set of contact points. The opened points break the current flow into the pump electromagnet and the plunger is no longer attracted. Then, the compressed spring pushes on the fuel-filled bellows. Fuel forces the outlet valve open and fuel flows to the engine under pressure. This action is repeated over and over during pump operation.

Automatic pump shutoff

With little demand for more fuel (engine idling, etc.), the electric fuel pump may shut off. If enough pressure exists in the lines, the bellows may be held in the enlarged position. The spring may not be able to collapse the bellows. In turn, the contact points can be held open to stop the fuel pump. As soon as more fuel is needed, pressure will drop and the pump will begin to run again.

DIAPHRAGM ELECTRIC FUEL PUMP

Except for its source of power, an *electric diaphragm fuel pump* is like a reciprocating type. It generally is the same as a mechanical diaphragm fuel pump. An electric

solenoid moves the diaphragm up and down in a pumping motion. Fig. 8-30 shows the parts of a diaphragm type electric fuel pump in a cutaway view.

It is not very common because of its generally low output volume. Since it is a reciprocating type, some fuel pressure pulsations and noise may be produced.

PLUNGER ELECTRIC FUEL PUMP

A *plunger electric fuel pump* operates much in the same way as the bellows and diaphragm type electric fuel pumps. A switching transistor is used to magnetize and then shut off the solenoid field. Instead of a bellows or diaphragm, this pump uses a special stainless steel plunger mounted on a brass cylinder. See Fig. 8-31.

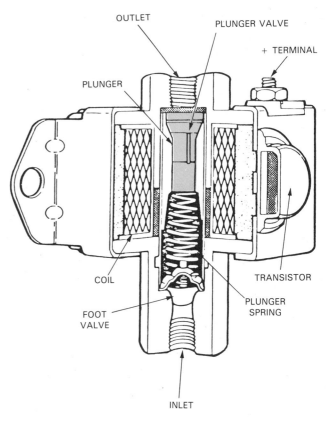

Fig. 8-31. Plunger type solid state pump. Metal plunger moves up and down to cause pressure and flow. (KEM)

SLIDING VANE ELECTRIC FUEL PUMP

A *sliding vane fuel pump* uses movable or sliding blades that fit into a central rotor. The rotor is intentionally offset to one side of the pumping chamber. A 12 volt electric motor is commonly used to spin the rotor and sliding vanes.

During pump operation, the spinning motor shaft and rotor causes the sliding vanes to be thrown and held

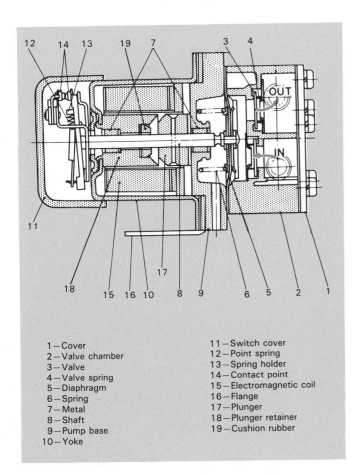

1—Cover
2—Valve chamber
3—Valve
4—Valve spring
5—Diaphragm
6—Spring
7—Metal
8—Shaft
9—Pump base
10—Yoke
11—Switch cover
12—Point spring
13—Spring holder
14—Contact point
15—Electromagnetic coil
16—Flange
17—Plunger
18—Plunger retainer
19—Cushion rubber

Fig. 8-30. Cutaway view of diaphragm type electric fuel pump. Note names of various parts. (Datsun)

against the wall of the pumping chamber by centrifugal force. Since the rotor is offset to one side, fuel is drawn into the larger inlet side of the chamber. Then, fuel is trapped between the sliding vanes and is carried toward the smaller outlet section of the chamber. This action causes the fuel to be compressed and spurt out.

Pressure relief valve

A *pressure relief valve* is normally fitted inside the pump body to limit pressure developed by the pump. The valve is simply a spring-loaded plunger. When the desired pressure is reached, the spring is compressed and the plunger is pushed back in its cylinder. This uncovers a passage running back to the inlet side of the pump. As a result, some of the fuel recirculates inside the pump. Both fuel pump pressure and volume are reduced when the relief valve opens.

The valve is sometimes adjustable. Loosening the relief valve screw reduces spring tension and causes lower fuel pump pressure. Tightening increases pump pressure.

ROLLER VANE ELECTRIC FUEL PUMP

The *roller vane electric pump,* Figs. 8-32 and 8-33, is driven by a 12 volt electric motor. Very similar to the sliding vane pump, its steel rollers are loosely held in slots cut in the rotor. As the motor spins, the roller vanes are thrown outward by centrifugal force. This holds and SEALS them against the walls of the pressure chamber. With the rotor set off to one side, the fuel is squeezed as it is carried toward the smaller outlet side of the chamber. This action forces fuel through the outlet.

Normally, a roller vane pump is used as an in-line type fuel pump for a fuel injection system. It may be teamed up with an in-tank, pusher type pump. The in-tank pump fills or primes the roller vane pump. Then, the roller vane pump produces sufficient pressure and flow to feed the fuel injection system.

Relief and back-flow valves are normally fitted into the roller vane fuel pump. See Fig. 8-34.

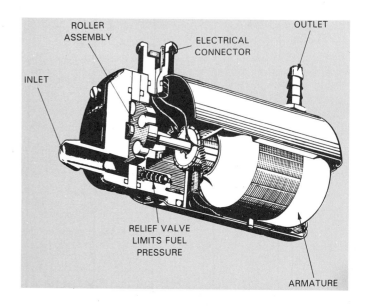

Fig. 8-33. Note parts of roller vane fuel pump. (Volvo)

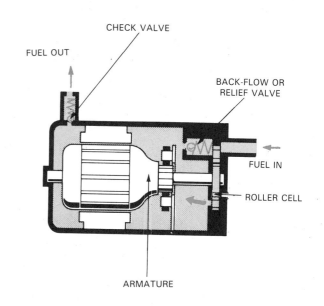

Fig. 8-34. This electric fuel pump contains a relief valve. Valve lets excessive pressure flow back to fuel tank. (Volkswagen)

IN-TANK (VANE) ELECTRIC FUEL PUMP

Some cars have the fuel pump mounted inside the gasoline tank. Refer again to Fig. 8-27B. An *in-tank pump* is often called a "pusher" type fuel pump. It pushes fuel toward the engine rather than pulling or sucking it, like an engine mounted mechanical pump.

An in-tank pump can be removed for service by releasing the tank sending unit lock ring. The pump is normally attached to the tank's pickup tube. The pump, sending unit, pickup tube, and strainer are removed as an assembly.

An in-tank pump is often a stationary impeller type pump. One is shown in Fig. 8-35. It uses a 12V electric motor to spin a winged impeller. The impeller is a fan

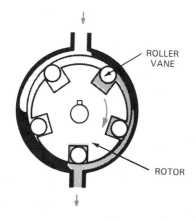

Fig. 8-32. Simplified drawing of roller vane action. Centrifugal force throws roller vanes outward. This action seals rollers against pressure chamber wall. (Volkswagen)

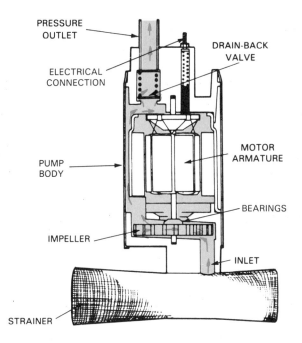

PRESSURE OUTLET

ELECTRICAL CONNECTION

DRAIN-BACK VALVE

PUMP BODY

MOTOR ARMATURE

BEARINGS

IMPELLER

INLET

STRAINER

Fig. 8-35. Fuel flows through strainer, impeller chamber, motor and drain-back valve. Hence, in-tank pumps are often known as submersed fuel pumps. (Volvo)

Back-flow valve

A *back-flow check valve* is often used in the pump outlet. The check valve prevents fuel from draining back into the tank when the pump is not in operation. If the fuel system lines were allowed to empty whenever the car sat idle, engine starting problems could result. A fuel strainer is normally fitted over the inlet of the pump.

In-tank fuel pump operation

Quite often, an in-tank fuel pump operates whenever the ignition switch is turned to ON or is set to the START position. With the electric motor functioning, fuel is drawn into the pump area between the spinning impeller blades and stepped wall of the pumping chamber. The fuel is then spun around in the chamber and centrifugal force throws the fuel outward. The shape of the pumping chamber then directs the fuel out of the pump under pressure, as in Fig. 8-36.

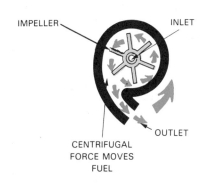

IMPELLER

INLET

OUTLET

CENTRIFUGAL FORCE MOVES FUEL

Fig. 8-36. Impeller type electric fuel pump is really rotary centrifugal pump. Motor driven impeller moves fuel from inlet to outlet on outer edge. (Ford Motor Co.)

blade type device which, in a sense, blows the fuel out the pump like a window fan blows air.

Sealed pump

Since the electric pump motor is submersed in highly flammable gasoline, it must be a *sealed* unit. Sometimes, fuel actually circulates through the motor itself. By sealing air out of the motor, the fuel cannot be ignited.

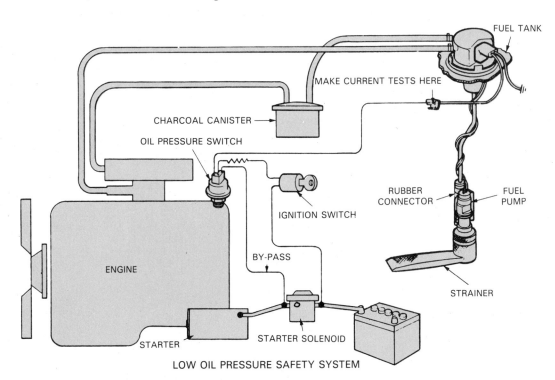

FUEL TANK

MAKE CURRENT TESTS HERE

CHARCOAL CANISTER

OIL PRESSURE SWITCH

IGNITION SWITCH

RUBBER CONNECTOR

FUEL PUMP

ENGINE

BY-PASS

STRAINER

STARTER

STARTER SOLENOID

LOW OIL PRESSURE SAFETY SYSTEM

Fig. 8-37. Oil pressure switch is "wired" so that it controls power to fuel pump. Low oil pressure shuts OFF pump. (Ford Motor Co.)

With a rotary, centrifugal fuel pump, inlet and outlet directional check valves are NOT needed, nor is a pulsation diaphragm. Not having a reciprocating (up and down) motion of a diaphragm, bellows or plunger, fuel pressure fluctuations are almost eliminated. Fuel flow to the engine is extremely smooth.

LOW OIL PRESSURE SAFETY CIRCUIT

A *low oil pressure safety circuit* automatically shuts OFF the fuel pump and engine whenever engine oil pressure drops to dangerously low levels. It is sometimes used with an electric fuel pump. Shown in the schematic of Fig. 8-37, the engine oil pressure switch is wired to a fuel pump relay.

Upon starting, current flows directly to the pump. This turns the fuel pump ON while cranking the engine with the starting motor. With the ignition key switch in the start position, the fuel pump will operate even without engine oil pressure.

With the switch in the ON or RUN position, Fig. 8-38, fuel pump operation DEPENDS upon engine oil pressure. When oil pressure is normal, current does not flow through the oil pressure switch to ground. Instead it flows through the fuel pump relay and to the fuel pump. The relay closes and the fuel pump operates. However, if engine oil pressure DROPS too low, the oil pressure switch will close. Current will flow through the switch and the pump relay will open. This turns the fuel pump OFF and protects the engine from extensive damage.

DUAL ELECTRIC FUEL PUMP CIRCUITS

Many injection systems use two electric fuel pumps: an in-tank pump and an in-line pump, Fig. 8-39.

An in-tank fuel pump for an EFI (electronic fuel injection) system is usually a low pressure lift or supply pump. It pulls fuel out of the tank and feeds it to the in-line pump. The in-tank pump usually will produce only 3 to 7 psi (21 to 48 kPa). Normally, the in-tank pump is an

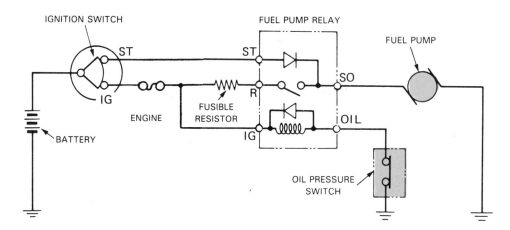

Fig. 8-38. This is an actual service manual schematic of a fuel pump control circuit. Notice how oil pressure switch is connected. (Toyota)

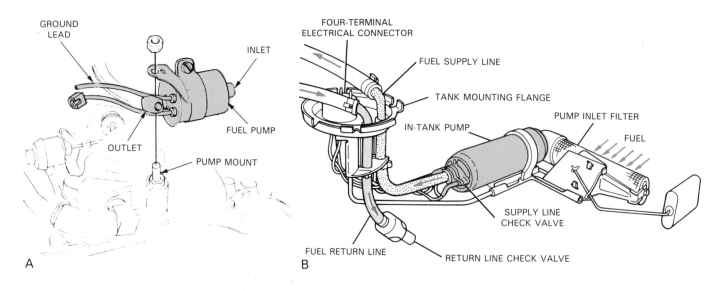

Fig. 8-39. Some injection systems use two electric fuel pumps. A—In-line pump. B—In-tank pump. (Oldsmobile)

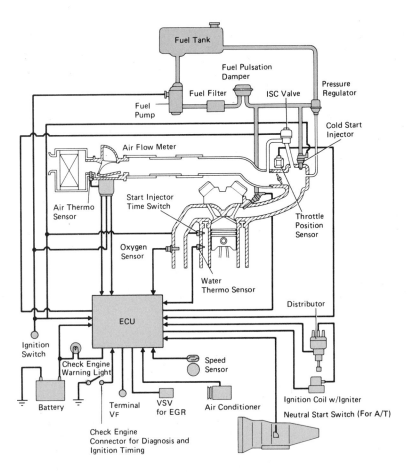

Fig. 8-40. Schematic of electronically controlled fuel system. Pressure regulator at upper right controls preset fuel pressure. ECU receives data from sensors about speed, throttle position, air temperature, and coolant temperature. These data are used to control pressure regulator, fuel pump, and other systems. (Toyota)

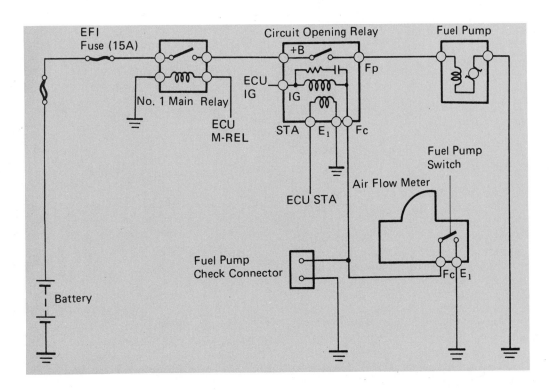

Fig. 8-41. Fuel pump circuit diagram. Relay controls fuel pump switch. (Toyota)

impeller type.

An in-line fuel pump for an EFI system is usually a high pressure, positive displacement pump. The in-line pump for an injection system is generally capable of producing adequate pressure for fuel injection into the intake manifold. Pressure outputs range from 15 to 40 psi (103 to 275 kPa).

A *pressure regulator* is used in a fuel injection system to maintain a preset fuel pressure. It bleeds excess fuel pump output back to the tank. See Fig. 8-40.

Fuel pressure regulators and related subjects are covered in later chapters on fuel injection. Use the index to find these topics, if needed.

COMPUTER CONTROL OF ELECTRIC FUEL PUMPS

Most modern cars use an on-board computer to control the operation of electric fuel pumps. Basically, the computer allows the fuel pump to run only when the engine is running and during starting. Refer to Fig. 8-40.

When the ignition key is turned to start, the computer activates the electric fuel pump to build system pressure. If the engine does not start in a brief period, the computer deactivates the fuel pump. When the engine starts, the computer again supplies current to the fuel pump.

Fig. 8-41 shows a wiring diagram for another electric fuel pump circuit. Note how this circuit uses a main relay to supply current to the fuel pump relay. The computer or ECU (electronic control unit) can then send current to the fuel pump relay, closing the relay to turn on the fuel pump.

A fuel pump switch is mounted on the airflow meter. It only closes and allows pump operation when the engine is running. If the engine stalls, the switch opens and shuts off the fuel pump.

This circuit also has a fuel pump check connector. It can be used to activate the fuel pump, bypassing the control of the computer.

SUMMARY

Fuel pumps are used to draw fuel from the fuel tank and force it through lines and filters to carburetor or fuel injection mechanism.

Fuel pumps are powered either by the mechanical action of the engine or by electricity. Mechanical pumps are normally bolted to the side of the engine. Electric pumps can be located well away from the engine. Beyond this fundamental difference, there are many variations in fuel pump design. However, they can be classified under four types:
1. Reciprocating pumps which have a back and forth, straight-line pumping motion.
2. Rotary pump which depends on spinning action of a fan blade to produce fuel flow.
3. Positive displacement pump which traps fuel in a sealed enclosure and then pressurizes it. Types include sliding vane, roller vane, diaphragm, and plunger.
4. Centrifugal pump which works on the same principle as a fan. Spinning blades throw the fuel to the outside of the pump housing.

The mechanical fuel pump consists of a rubber diaphragm whose back and forth movement provides the pumping action to draw fuel into the pump cavity. A pull rod and a rocker arm transfer motion from the cam to the diaphragm. A diaphragm spring pushes the diaphragm back into the pump chamber after the action of the rocker arm has moved it out of the chamber. Check valves permit flow of fuel in only one direction.

A fuel return system consists of a special valve in the fuel pump and a second fuel line from the pump back to the fuel tank. It allows excess fuel to return to the tank and helps prevent vapor lock through circulation of cooled fuel between the tank and the fuel pump.

Pressure and volume capabilities of a fuel pump depend on the size of the engine. As a rule, the fuel pump of a four cylinder engine will deliver a half pint of fuel every 30 seconds. For a small six cylinder engine, the pump should probably be able to produce 3/4 pint of fuel in 30 seconds. Pressure requirements for a carburetor fuel system range from 4 to 7 psi. Gasoline injection systems require from 10 to 40 psi.

A serviceable fuel pump is one that can be disassembled and rebuilt. A nonserviceable pump is permanently assembled and must be discarded and replaced when it becomes defective.

There are many design variations in electric fuel pumps. The reciprocating designs include the bellows, diaphragm, and plunger types. The most common rotary designs include the sliding vane, impeller, and roller vane types.

A sliding vane pump has sliding blades set into a central rotor. As the rotor turns, the blades are thrown against the wall of the pumping chamber by centrifugal force. The rotor is offset so that large amounts of fuel are trapped and then compressed by the blades.

A roller vane pump is similar in operation to the sliding vane. However, it uses rollers in place of blades. This type is normally used as an in-line pump for fuel injection systems.

An impeller type pump works like a fan. Blades use centrifugal force to sling fuel toward an outlet on the outside of the housing.

A low oil pressure safety circuit is designed to stop the engine in case of low oil pressure. An oil pressure switch is placed in the fuel pump relay circuit. Low oil pressure opens the circuit and the fuel pump will stop operating.

Often, gasoline injection fuel systems will have two electric fuel pumps. One is in the fuel tank; the other is in-line.

An on-board computer controls operation of electric fuel pumps in most modern cars. The computer will activate the fuel pump when the ignition switch is turned to ON. If the engine does not start within a short time, the computer will automatically turn the fuel pump off.

KNOW THESE TERMS

Gravity feed fuel system, Reciprocating fuel pump, Rotary fuel pump, Positive displacement fuel pump, Pulsation diaphragm, Pump idling, Serviceable fuel pump, Vacuum booster pump, Sliding vane fuel pump, Roller vane fuel pump, Impeller type fuel pump, Pressure regulator.

REVIEW QUESTIONS—CHAPTER 8

1. The main purpose of the fuel pump is to _____ _____.
2. Instead of using a fuel pump, a _____ system places the fuel tank above the level of the carburetor to produce pressure and flow.
3. List six possible functions of a fuel pump.
4. Which of these pumps uses a reciprocating motion?
   a. Rotary.
   b. Centrifugal.
   c. Mechanical.
   d. None of these.
5. The "heart" of a mechanical fuel pump is a _____ _____; its movement _____ _____ through the system.
6. When and how are pressure and flow produced inside a mechanical fuel pump?
7. What will usually happen if a fuel pump has a broken rocker arm return spring?
   a. Low fuel pressure.
   b. Improper fuel flow.
   c. Damage the camshaft.
   d. Clattering sound will be produced.
8. What is a check valve?
9. To reduce fuel pressure pulsations, a _____ is sometimes used inside a fuel pump.
10. What does the cam eccentric do on the fuel pump intake stroke?
11. Describe mechanical fuel pump idling.
12. A vapor return valve permits fuel flow into the _____ _____ and back to the _____.
13. Sometimes, to prevent pressure buildup, fuel pressure is released to the inlet side of the pump by means of a _____ _____.
14. As a rule of thumb, how much volume output should 4, 6, and 8 cylinder engine fuel pumps produce?
15. In general, how much pressure should a fuel pump develop?
16. A fuel pump which can be repaired is called a _____ type fuel pump.
17. A booster type vacuum pump is sometimes used to _____.
18. Which of the following fuel pumps has a folded, flexible pumping chamber?
   a. Diaphragm.
   b. Bellows.
   c. Impeller.
   d. Plunger.
19. Diaphragm type electric pumps are not very common because of their _____ and since they suffer from some pressure _____.
20. Why is an in-tank fuel pump often called a "pusher pump"?
21. Unlike a stationary impeller pump, a sliding vane pump is positive _____.
22. Normally, a roller vane fuel pump is used as an in-line pump for a _____ _____ _____.
23. How does a low oil pressure safety circuit operate?
24. List the advantages and disadvantages of an electric fuel pump.

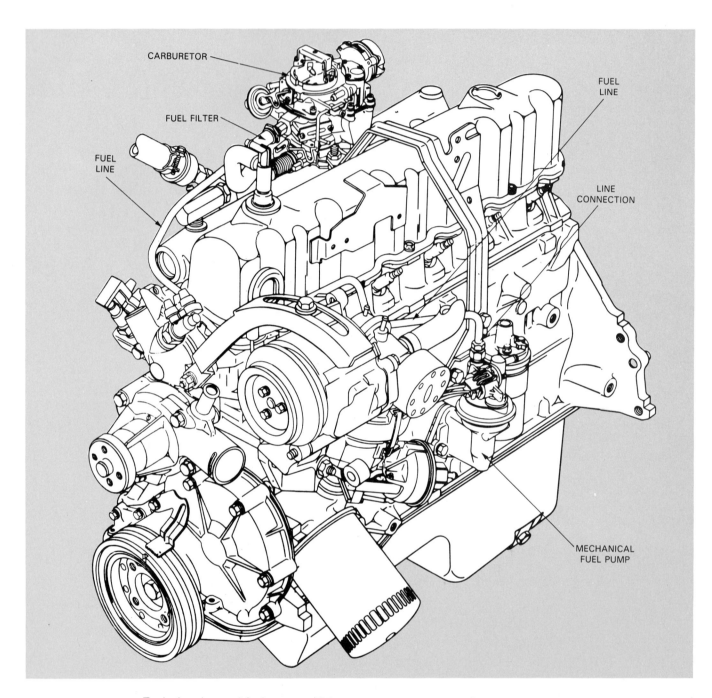

Typical carbureted fuel system. This system uses a mechanical fuel pump. (Ford)

9

FUEL PUMPS—PROBLEM DIAGNOSIS, TESTING, REPAIR

After studying this chapter, you will be able to:
● *List safety rules for fuel pump service.*
● *Explain the relationship between fuel pump vacuum, volume, and pressure.*
● *List and describe common fuel pump problems.*
● *Inspect for fuel pump problems.*
● *Test mechanical and electric fuel pumps.*
● *Use fuel pump test results to pinpoint troubles.*
● *Remove and replace a fuel pump.*

Faulty fuel pumps can be difficult to diagnose because other parts in the fuel system (clogged fuel filter, corroded electrical connections, kinked fuel line, defective ignition coil, faulty on-board computer) can produce similar symptoms.

By studying this chapter, you will learn to diagnose, test, and repair fuel pump related problems. Figs. 9-1A and 9-1B show some common fuel pump problems.

SAFETY PRECAUTIONS

The dangers of FIRE and EXPLOSION are always present when testing or replacing fuel pumps. Follow these rules:

1. Never smoke, weld, sand, grind, or drill near a disconnected fuel line. The slightest spark or flame could start a fire or explosion.
2. Always disconnect the car's BATTERY whenever any of the fuel lines to the pump are removed. Again, any electrical spark could be fatal.
3. Drain fuel into approved containers. Avoid using glass jars, open cans, or other unsafe containers. They can break, rupture, or spill.
4. Be certain that a suitable and operable FIRE EXTINGUISHER is close by. A few seconds time, with a fuel-fed fire, can mean the difference between LIFE and DEATH.
5. Some fuel injection systems have extremely high operating pressures. Loosening a fuel line WITHOUT FIRST BLEEDING PRESSURE, can allow jets of fuel to squirt dangerously across the shop. Highly pressurized fuel can cause an eye injury or a fire.
6. When loosening a fuel line, WRAP A SHOP CLOTH around the fitting. This will keep fuel from spraying.
7. After working on a fuel pump, or any fuel system

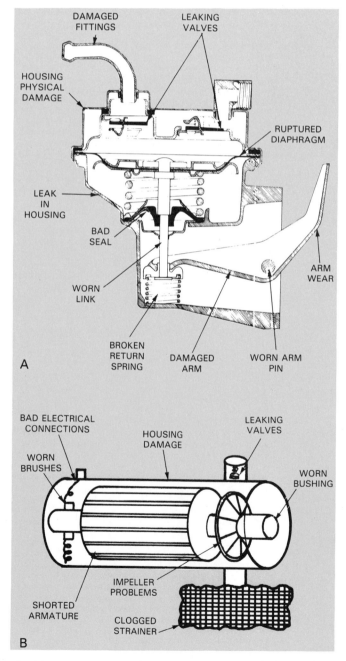

Fig. 9-1. Note typical problems with mechanical and electric fuel pumps. (KEM)

187

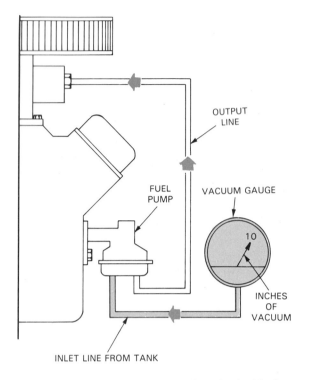

Fig. 9-2. This hookup checks vacuum of mechanical fuel pump at inlet.

component, CHECK CONNECTIONS FOR LEAKS. Fuel leaks are not only wasteful and expensive, but they are dangerous. Small fuel leaks have caused many fires.

FUEL PUMP PROBLEMS

Fuel pump problems usually show up as low fuel pressure, inadequate fuel flow, abnormal pump noise, or fuel leaks at the pump. Both mechanical and electric fuel pumps can fail after prolonged operation.

Low fuel pump pressure

A weak diaphragm spring, ruptured diaphragm, leaking check valves, or physical wear of moving parts can produce *low pump pressure.* The engine is starved for fuel at higher engine speeds.

High fuel pump pressure

High pump pressure, more frequently a problem with electric pumps, indicates an inoperative pressure relief valve. Both pressure and volume can be above normal, producing a rich fuel mixture or even engine flooding.

Fuel pump noise

Clacking sounds from inside a mechanical fuel pump are a sign of trouble. This noise can be easily mistaken as valve or tappet clatter.

NOTE! Most electric fuel pumps make a whirring sound when running. Only when the pump noise is abnormally loud should an electric fuel pump be considered faulty.

A clogged tank strainer can cause excess electric pump noise. Pump speed may increase because fuel is not entering the pump properly.

Fuel pump leaks

Fuel pump leaks are caused by physical damage to the pump body or deterioration of the diaphragm or gaskets. Most mechanical fuel pumps have a small vent hole in the pump body. When the diaphragm is ruptured, fuel will leak from this hole.

Fuel leaking from a ruptured mechanical fuel pump diaphragm can contaminate the engine oil. The fuel enters the oil sump through the side of the block.

As you learned earlier, several other problems can produce symptoms similar to those caused by a bad fuel pump. Before testing a fuel pump, check for:
1. Restricted fuel filters.
2. Smashed, kinked, or collapsed fuel lines or hoses.
3. Air leak into vacuum side of pump or line.
4. Carburetor or injection system troubles.
5. Ignition system problems.
6. Low engine compression.

FUEL PUMP VACUUM, VOLUME, AND PRESSURE RELATIONSHIP

The three output specifications of a fuel pump: vacuum, volume, and pressure, are all related. Each has an effect on the others. Look at Figs. 9-2 and 9-3.

If, for example, fuel pump pressure is low, fuel flow or volume will also be low. With low pressure, the pump may not be able to push fuel from the tank to the engine. Fuel flow may even stop. As you will learn, it is important to understand these relationships.

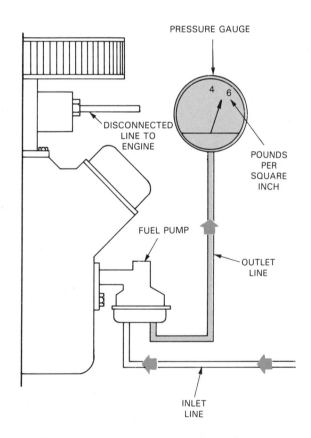

Fig. 9-3. Test static pressure of mechanical fuel pump at pump outlet using this hookup.

Fuel pump vacuum

Fuel pump vacuum refers to the amount of suction (negative pressure) produced at the inlet fitting of the pump, Fig. 9-2. A fuel pump must suck fuel out of the gas tank and through the fuel lines. With low vacuum, a fuel pump cannot draw fuel.

Generally, an in-line fuel pump should "pull" about 10 in. hg (33.77 kPa), commonly termed 10 inches of vacuum. Specs will vary, however. See Fig. 9-4.

Fuel pump volume

Fuel pump volume is a measurement of the amount of fuel flow from the pump during a specific time span. Pump volume is important, especially when fuel use is high. Fuel flow into the engine must be sufficient under hard acceleration or high engine speeds.

If fuel pump volume is low, an engine may starve for fuel. It may miss or stall and never reach higher speeds. This may be due to a bad pump or line restriction.

FUEL PUMP SPECIFICATIONS				
	Volume (30 seconds)	Pressure (PSI) (kPa)	Vacuum (Hg)	
			Direct	Indirect
Four Cylinder	0.6 pint (0.28 liters)	4 to 6 (28.58 to 41.36)	7 23.64 kPa	3 10.13 kPa
Six Cylinder	1 pint (0.47 liters)	4 to 5 28.58 to 34.47	10 33.77 kPa	3 10.13 kPa
Eight Cylinder	1 pint (0.47 liters)	5 to 6.5 34.47 to 44.42	10 33.77 kPa	3 10.13 kPa

Fig. 9-4. These fuel pump specifications are typical. (AMC)

The output specs for pump volume will vary slightly with engine size and pump type. A larger engine will normally require a higher pump rating than a smaller engine.

As a very general rule, a fuel pump should move about "a quart a minute." This will be explained shortly.

Fuel pump pressure

Fuel pump pressure indicates the amount of pushing power produced by the action of the pump, Fig. 9-3. Fuel pressure is needed to force fuel out of the pump, through the fuel line, and to the fuel-metering section.

Fuel pump pressure must be neither too high nor too low. If low, fuel supply to the engine will not be adequate. Engine performance will suffer. If fuel pressure is too high, the engine may become flooded.

Fuel pressure for a CARBURETOR fuel system should be about 4 to 6 psi (28 to 41 kPa) pressure, Fig. 9-5.

A GASOLINE INJECTION system will usually have a higher pressure output. Fuel pressure can run from 15 to 40 psi (103 to 276 kPa) and higher.

A DIESEL SUPPLY PUMP should produce around 6-10 psi (41-69 kPa). It feeds fuel to the high-pressure, mechanical injection pump.

INITIAL FUEL PUMP INSPECTION

Always visually inspect the fuel pump and fuel system before performing any detailed fuel pump tests. Several fuel-pump-related problems can be found by simply "looking over" the fuel system.

Fuel leaks

First, check the fuel pump, fittings, hoses, and pump body for leaks, Fig. 9-5. Rub your finger under these points and check for a fuel smell.

Gasoline leaks are hard to find because a fuel leak may not leave a puddle of liquid. The gasoline may vaporize almost as soon as it leaks out of the system.

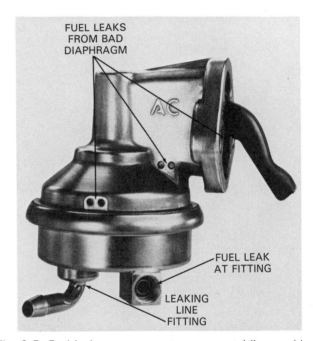

Fig. 9-5. Fuel leaks can occur at pump, metal lines, rubber hoses, and tank.

Air leaks

A rupture in the fuel line running to the fuel pump can reduce fuel pump output. Air, drawn into the line, can displace some of the liquid fuel and upset fuel pump operation. This is shown in Fig. 9-6.

Inspect fuel lines between the tank and pump with the engine stopped. Fuel may be leaking out of the rupture. Check closely where lines are clamped to the frame. They tend to rust more rapidly around clamps or brackets.

Oil leaks

Look for oil leaks on or around a mechanical fuel pump's mounting gasket or body. This indicates loose bolts or mechanical damage.

Damaged fuel lines

Check for kinked or deteriorated fuel hoses or lines. Look over the hoses at the tank, pump, and carburetor or injectors. Also feel the hoses. They should be neither too hard nor too soft.

AIR LEAK IN FUEL SYSTEM

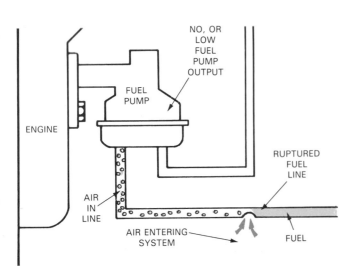

Fig. 9-6. An air leak is caused by a hole in fuel line on suction side of pump. Vacuum draws air into line, instead of fuel.

Fig. 9-7. Place stethoscope or screwdriver on various engine parts. Component making loudest noise is the source of trouble. It may be defective. (Lisle Tools)

Excessive heat

Make certain that HEAT SHIELDS are in place around the pump, engine, and fuel lines. Fuel lines MUST NOT TOUCH THE HOT ENGINE. This is a common cause of vapor lock and reduced fuel flow to the engine.

Clogged filter

If the car has a transparent fuel filter, check whether the filter element is dirty. Some in-line and bowl type filters have "see through" housings.

Remember! A clogged fuel filter has symptoms almost identical to that of a defective fuel pump.

Mechanical fuel pump noise

The metallic clattering or tapping sound mentioned earlier may be due to a worn pump rocker arm, rocker arm pivot pin, or bad arm return spring. A loose or badly worn cam eccentric may be another cause.

The knocking will occur at the same rate or speed as the rotation of the camshaft. In fact, the sound produced by a defective fuel pump is almost identical to the sound of a faulty hydraulic lifter.

A stethoscope, Fig. 9-7, or a long handled screwdriver can be used to listen to the fuel pump. The sound inside the pump will be amplified. If the rapping noise is loudest in the pump, the pump, rather than a tappet, may be the problem.

Gasoline in engine oil

Inspect the engine oil for signs of gasoline. Pull out the dipstick and smell the oil. If a strong gasoline odor is present, the mechanical fuel pump diaphragm may be leaking. Fuel can leak through the diaphragm and enter the upper section of the pump body. From there, it can enter the crankcase.

DANGER! Engines have literally exploded when fuel has filled the oil pan. Drain and replace the oil immediately.

Empty gas tank

With an inoperable fuel gauge, it is possible for the dash gauge to read full when the tank is actually empty. To check the tank, dip a length of clean mechanic's wire into the tank. Inspect the wire for fuel. If the tank is empty or nearly so, the fuel gauge or its circuit is bad and the fuel pump may be all right.

Electric fuel pump inspection

Electric fuel pumps develop many of the same problems as mechanical pumps. Look for fuel leaks, air leaks, damaged lines, cross threaded fittings, kinked, cracked, or softened hoses and lines touching hot engine components, and abnormal pump noise. Any of these situations could affect operation.

Bad fuel pump ground

The mounting bracket for an electric fuel pump often serves as a ground connection. If this bracket is loose or rusted, the electric fuel pump will not function properly. Note Fig. 9-8.

Poor electrical connections

Inspect all electrical plug type connections going to the fuel pump. They should be tight and clean. Pull the connections apart and check for corrosion or burning.

With the pump turned ON, wiggle the wires and connections. If movement affects pump operation, the wires or connections are faulty. A voltage drop will also detect poor electrical connections.

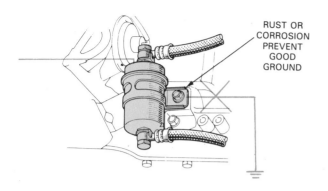

Fig. 9-8. A loose or corroded fuel pump bracket can sometimes cause erratic pump operation. It may serve as a ground for pump motor. (Toyota)

Electric fuel pump noise

Turn ON the electric pump and observe its operation. Listen for odd noises. Carefully feel the pump. It should not be too hot. A rotary type pump should have little or no vibration.

Bad pump bearings can cause an electric pump to drag, overheat, vibrate, or make a whining or grinding sound.

High fuel pressure (line restricted, clogged fuel filter, stuck pressure regulator) can also cause excess electric pump noise and heat.

If an in-tank filter is clogged, the electric pump can starve for fuel. The pump motor can race (speed increases), making a loud buzzing or whirring sound.

Note! Some noise from an electric fuel pump is normal. A rotary type may have a very quiet buzz. If it is a reciprocating type, the pump may click.

Electric fuel pump relay

Some electric fuel pumps use a cutoff relay that controls current and voltage to the pump. See Fig. 9-9.

If fuel pump problems are indicated, first check that the wires to the relay are secure. Also, check the voltage entering and leaving the relay. Fig. 9-10 shows a voltmeter being used to check voltage at fuel pump.

Bad fuse

If an electric fuel pump will not turn ON, always check its fuse. It may be "blown." If its condition cannot easily be seen, touch each side of the fuse with a test light. The light should glow both times. If the light only glows on one side, the fuse is burned out and should be replaced.

After fuse replacement, check pump operation. If the fuse blows a second time, there is a SHORT somewhere in the fuel pump circuit. Find and correct the short.

Oil pressure switch

Some electric fuel pump circuits use an *oil pressure control switch.* When engine oil pressure drops dangerously low, Fig. 9-11, the switch shuts off the fuel pump so engine will not be ruined.

When a fuel pump will not operate, check the oil pressure switch. It may be unplugged. To test the switch, connect a jumper wire across the switch terminal. This will bypass the switch and send current

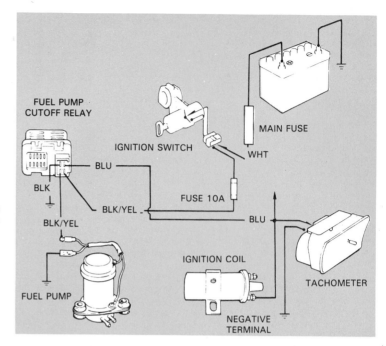

Fig. 9-9. Underdash relay often controls current to pump. If it is disconnected or faulty, electric fuel pump will not function properly. (Honda)

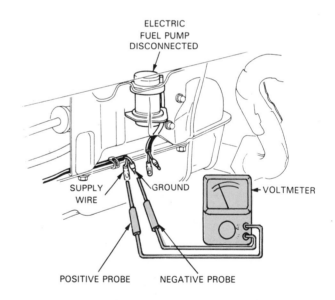

Fig. 9-10. Checking for proper voltage to electric fuel pump. (Honda)

directly to the electric fuel pump. If the jumper turns the pump ON, then the oil pressure switch is BAD. If not, then another problem is preventing pump operation.

Inertia switch

An *inertia switch* is a safety device that prevents electric fuel pump operation after an automobile collision. After an impact, inertia (tendency to keep moving) is used to actuate the switch and break the circuit to the fuel pump.

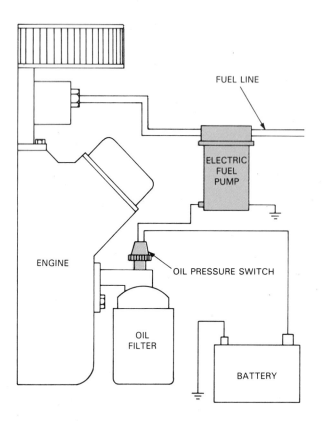

Fig. 9-11. An oil pressure switch is sometimes used to protect engine from damage. It controls electrical current to fuel pump. If engine oil pressure drops, pump is shut off.

Static pressure test procedures (mechanical pump)

During a static pressure test, the fuel line leading to the engine is disconnected. The rubber hose on the pressure gauge is fitted over the fuel line. See Fig. 9-12. Then, the engine is started and runs on the fuel remaining in the carburetor fuel bowl. The engine will run for about a minute.

The pressure gauge reading is compared to "specs." If either low or high, further tests are needed to isolate the problem.

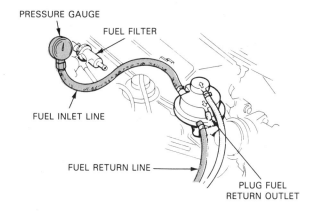

Fig. 9-12. Disconnect carburetor inlet line and fasten it to the pressure gauge. Service manual may also tell you to disconnect and plug, or pinch off, fuel return line as shown. (Toyota)

A reset button is normally provided on the inertia switch. Pushing this button will reset the switch and allow fuel pump operation.

The inertia switch can be located in the trunk or engine compartment. Check a manual to find its location.

Electric fuel pump computer control

Although not common, a *computer malfunction* can prevent electric fuel pump operation. If the feed wire between the computer and fuel pump is "dead" (no voltage output), the circuitry in the computer may be damaged. Always check electrical connections between the pump and computer before condemning the computer, however.

Computer testing and self-diagnosis are covered in later chapters.

FUEL PUMP TESTING

The following are general procedures for testing various types of fuel pumps. In most cases, they will be sufficient. However, equipment hookups and procedures may vary. When in doubt, always follow specific instructions provided with test equipment or service manual.

Static pressure test (mechanical pump)

This test measures the output pressure of a fuel pump at the pump outlet. A pressure gauge is connected DIRECTLY to the outlet line of the pump. Since fuel is NOT flowing through the system, this is termed a *static test.* Fuel flow is static or at rest.

Dynamic pressure test (mechanical pump)

A dynamic pressure test overcomes many of the weaknesses common to a static pressure test. A *dynamic fuel pump test* is performed with all of the fuel lines and fittings intact. The fuel line is still connected to the carburetor or injection system. This allows normal fuel flow to the engine while measuring pump pressure, Fig. 9-13.

The pressure gauge registers fuel pressure in the functioning system. This accounts for any effect of fuel flow upon pressure. A dynamic test is better than a static test.

Note! The hose going to the gauge should be as short as possible. This will prevent any pressure drop before the gauge.

Dynamic pressure test procedures (mechanical pump)

To perform a dynamic pressure test, connect the pressure gauge as in Fig. 9-13. Start the engine and let it idle. Observe the pressure reading on the gauge. It should be within specifications.

With this hookup, the engine can run for extended periods of time. It can be raced to determine whether higher fuel consumption has any effect on fuel pressure.

Low pump pressure may indicate a defective pump, clogged fuel filter, or restricted fuel line. ALWAYS check fuel filters and lines before condemning a fuel pump.

Pressure test (electric fuel pump)

Basically, electric fuel pump pressure output is measured in the same way as a mechanical pump. A

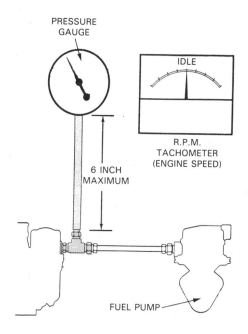

Fig. 9-13. For a dynamic pressure test, use T-fitting to connect gauge. This lets fuel flow to engine normally.

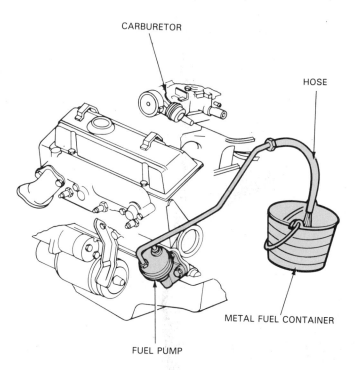

Fig. 9-14. To measure pump flow or volume, run fuel pump output line into approved container. (Chrysler)

pressure gauge can be connected directly to the output line of the pump. This will measure the static pressure of the pump. Or, the pressure gauge may be fitted to the fuel system with a T-fitting. This will check the dynamic or actual pressure in the system.

With an electric pump, the engine does NOT always have to be started to determine fuel pump output. Turning the ignition key ON should activate the electric fuel

pump. If not, use a jumper wire and/or special test plug to energize the pump.

Electric pump pressure test procedures

After connecting the pressure gauge to the outlet of the electric pump, activate the pump. Note the pressure gauge reading and compare its specifications.

Low fuel pump pressure may indicate a dragging fuel pump, low supply voltage, poor pump ground, clogged fuel strainer, or a similar problem. High fuel pressure can be due to a stuck relief valve or pressure regulator valve (covered in chapters on fuel injection).

DANGER! When testing an electric fuel pump, do not let fuel contact electrical wiring. The smallest spark (electric arc) could ignite the fuel, causing a DEADLY FIRE!

Volume test (mechanical fuel pump)

A fuel pump *volume* or *flow test* measures how much fuel can be pumped through the system in a given amount of time. A flow test is also called a pump CAPACITY TEST.

Basically, during a volume test, the output line of the pump is disconnected and inserted into a graduated container. Look at Fig. 9-14. A graduated container is marked in ounces (cm³), much like a kitchen measuring cup. The engine is started and allowed to run for a specific time. Then, the quantity of fuel forced into the container is observed.

The primary purpose of a volume test is to determine the condition of the pump diaphragm, check valves, pump linkage mechanism, cam eccentric, fuel filters, and tank strainer. It finds out whether a pump can move enough fuel to meet high fuel demands. Fuel flow requirements increase during hard acceleration or high-speed driving.

Volume test procedures (mechanical fuel pump)

When performing a fuel pump volume or capacity test, BE VERY CAREFUL. A fuel pump can spray fuel across the shop. Check all of your connections and that the output hose is safely in the container. Place the container where it CANNOT be spilled.

After performing all of the normal pretest inspections, remove the fuel line leading to the engine. Slide a length of rubber hose over the metal fuel line and run it into a container. Pinch the hose shut with a pair of vise-grip pliers or special clip tool. Start and idle the engine. While timing the test with a wristwatch, open the pliers or clip and let fuel flow into the container. After 30-45 seconds, clamp the hose shut again and stop flow. Shut off the engine and check the amount of fuel entering the container. Compare this amount to specifications.

Test accuracy

Note! For precise flow measurements, the output hose should not be too long nor too small in diameter. Its inside diameter should be the same as the metal fuel lines.

Interpreting volume test results (mechanical pump)

Generally, during the 30-45 second test, a fuel pump should be capable of forcing at least a pint of fuel into the container. This quantity can vary with engine size. A fuel pump for a large-displacement engine may have a higher output. Check the manual for precise values.

Air bubbles

Note! BUBBLES in the fuel coming out of the pump indicate an air leak. Pump suction is pulling air into the system through a hole in the pump, hose, or fuel line. Check the system thoroughly.

A low flow output may indicate a badly ruptured pump diaphragm, worn pump arm or pivot, clogged filter, clogged tank strainer, or a smashed fuel line.

Low flow problem isolation (mechanical pump)

When a fuel pump fails a capacity or flow test, further tests are needed to isolate the exact problem. Do NOT condemn and replace the fuel pump yet. Other problems may be affecting pump output.

Auxiliary fuel supply test

An *auxiliary fuel supply flow test* will determine whether there are any faults in the car's fuel supply system (clogged filters, clogged tank strainer, softened hoses, crushed lines, etc.).

With the flow test hose still in the container, disconnect the inlet hose going INTO the pump. Run the pump inlet hose into a second container full of fuel, Fig. 9-15.

To bleed the setup of air, start and run the engine for a few moments. Empty the output container. You are now ready for the test.

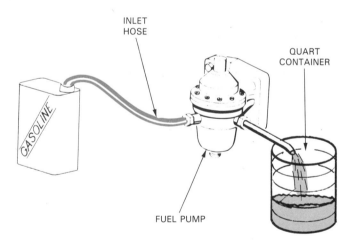

Fig. 9-15. To help pinpoint problems in fuel system (clogged filter, kinked line, etc.), perform an auxiliary supply flow test.

Restart the engine. Time the fuel flow into the container. Stop flow after 30-45 seconds. Measure the amount of fuel.

If the auxiliary supply flow test IMPROVES PUMP OUTPUT, something in the system has reduced flow. Check this out.

If flow REMAINED THE SAME, lower than specifications, then the fuel supply system is good and the pump is probably defective.

Volume test (electric fuel pump)

An electric fuel pump's volume is tested in the same way as a mechanical pump. With an electric fuel pump, the engine DOES NOT have to be started to test the pump. Turn the ignition key to ''ON'' or use a jumper wire, Fig. 9-16, to provide voltage to the pump motor.

Volume test results (electric pump)

Electric fuel pump output should usually be about a quart a minute or 1/2 pint every 30 seconds. This may vary depending on engine size and fuel system requirements.

Check fuel supply system

If pump delivery is low, check everything ''before the pump.'' Inspect fuel lines, hoses, and in-line filters. Restrictions or leaks could reduce pump capacity. With an in-tank pump, Fig. 9-17, the only thing before the pump is the strainer.

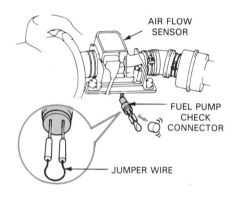

Fig. 9-16. Some electric fuel pumps have special plug for activating electric fuel pump. When jumper wire is inserted across plug terminals, voltage is applied to pump motor. (Toyota)

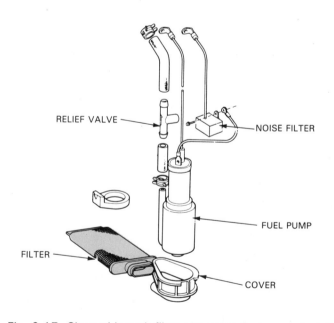

Fig. 9-17. Clogged in-tank filter can cause fuel pump to fail both pressure and flow tests.

Check for clogged tank strainer

Low electric fuel pump volume may be caused by a dirty fuel tank strainer. To check the strainer, you will have to remove the pump from the tank.

Note! Being able to blow through the strainer does not mean it is good. A partially clogged strainer MAY COLLAPSE when flow is towards the pump, but not when flow is reversed.

Measure electric fuel pump supply voltage

Defective electrical supply can lower output of electrical fuel pumps simply because they do not operate at full speed. Reduced voltage, poor ground, and excessive current draw will affect pump operation.

To check supply voltage to the electric fuel pump, disconnect the wires to the fuel pump. With a voltmeter, measure the voltage at the wires, Fig. 9-18.

Normally, fuel pump supply voltage should NOT be below 11 volts. If lower, check all the wires and plug-in terminals. A high resistance in something is causing a voltage drop between the battery and the pump.

TESTING PUMP SUPPLY VOLTAGE

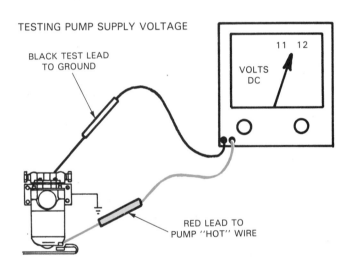

Fig. 9-18. Always check supply voltage when electric pump output is low. Low voltage will reduce pump pressure and volume.

Test fuel pump ground

To check pump ground, connect a jumper wire from the pump body or ground wire to the frame of the car. This will bypass a faulty fuel pump ground. Fig. 9-19 shows how to hook up the jumper wire.

If the jumper wire increases fuel pump speed, then the ground is bad. Measure fuel pump delivery with the wire connected or after cleaning off or repairing the poor pump ground connection.

Measure pump current draw

Worn pump bearings or a bad motor can cause a pump to drag, run slowly, and draw excessive current. Pump output will be reduced. An ammeter can be used to measure electric fuel pump current draw, an indicator of pump condition.

TESTING ELECTRIC PUMP GROUND

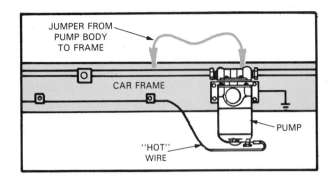

Fig. 9-19. Using a jumper wire, ground electric pump. If this action improves pump performance, pump ground is bad.

Disconnect the "hot" lead going to the electric pump, Fig. 9-20. Connect the ammeter BETWEEN the supply wire and the pump terminal. Turn ON the pump and read the meter. Current should not be too low or too high.

Again, high current draw would indicate low internal resistance (worn and dragging pump bearings, shorted motor windings, etc.). Low current draw would indicate high internal resistance (dirty armature commutator, worn brushes, bad connection, or broken wires).

Generally, electric pump current draw should be around 3 to 9 A. However, this will vary with pump size and design. Always check a shop manual for specs.

Non-return valve

A non-return valve is located in some electric in-line fuel pumps. Some pressure problems can be traced to this valve. Cleaning the valve may solve the problem. See Fig. 9-21.

Check for faulty pressure regulating valve

Many electric fuel pumps on gasoline injection systems have a pressure regulator valve. The valve controls the

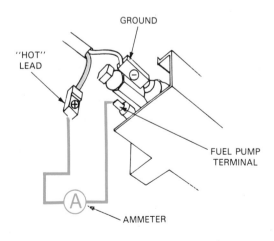

Fig. 9-20. To check condition of electric pump motor, perform a current draw test. Connect ammeter in series with pump. If current is excessively higher or lower than specs, pump is defective. Check voltage supply first.

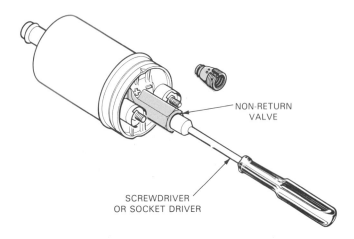

Fig. 9-21. Electric fuel pump check valve or non-return valve may be source of problem. Cleaning or replacing valve may be necessary.

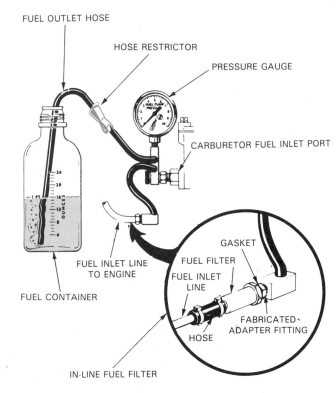

Fig. 9-22. To perform delivery flow test, open clip and allow fuel to enter container. (Sun)

output pressure and volume of the pump. If pump output volume is too HIGH, the regulator valve may be stuck shut. If pump output is too LOW, the pressure valve may be stuck open. Refer to Chapters 15, 16, and 17 for more information on troubleshooting fuel pressure regulators.

Detail of pressure regulators are given in the later chapters on fuel injection.

Bench testing electric fuel pump volume

An electric fuel pump can be bench flow tested. Prime the system to remove air. Then, time pump output.

WARNING! NEVER run an electric fuel pump DRY. Without the cooling action of gasoline flowing through the pump, an electric pump can quickly overheat and fail.

Fuel delivery system test

In most cases, a fuel delivery system test is the only test needed to determine the operating condition of a fuel pump. It is nothing more than a pressure and flow test performed at the same time.

A special fitting and gauge setup are available for a delivery test. The fuel pressure gauge is mounted on a special fitting. The fitting is attached between the incoming fuel line and the feed line to the carburetor or injectors. An outlet hose is connected to the fitting for flow testing.

With the output clip (restrictor) closed, the engine is started and fuel pressure measured. By opening the clip, fuel flow can be timed and determined. This is illustrated in Fig. 9-22.

NOTE: for accurate flow readings, air should first be pumped from the gauge and lines. Let the first few ounces of fuel pour into the container. Then, start the volume test.

Vacuum test (mechanical fuel pump)

When fuel pump volume or pressure is NOT up to specifications, a fuel pump vacuum test is needed. The fuel pump may not be at fault. An air leak or restriction between the fuel tank and the pump will produce the same symptoms.

Vacuum test procedures

Disconnect the hose between the main fuel line and the gas tank. Fuel may pour out of the tank so be careful. Plug the tank line. Then, connect a vacuum gauge to the main fuel line, Fig. 9-23.

Note! The output line of the fuel pump should still be routed into the flow test container. This will allow free fuel flow and maximum pump vacuum. Start the engine and let it idle while recording the highest vacuum reading.

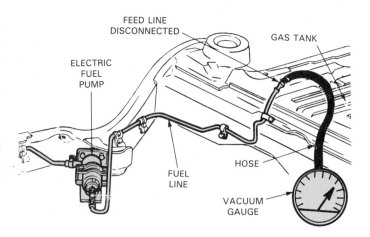

Fig. 9-23. Preparing for vacuum test of electric pump. First, connect vacuum gauge to main fuel line at gas tank. This will check pump as well as condition of lines and hoses ahead of pump. (Cadillac)

Interpreting vacuum test results

Normally a mechanical pump should have around 10 in./Hg. (about 34 kPa) of vacuum. If not, a problem exists in either the fuel line or the fuel pump.

If vacuum is low, measure the vacuum directly at the pump inlet. This will determine whether the problem is caused by the pump or lines.

If fuel pump vacuum reads low BOTH at the tank and directly on the pump, the fuel pump is the cause. The pump diaphragm may be ruptured or the check valves defective. Replace or repair the pump.

Vacuum test (electric fuel pump)

In-line electric fuel pumps can be vacuum tested in the same way as a mechanical pump. Leave the output line in the flow test container. Disconnect the inlet fuel line at the gas tank. Plug the line leaving the tank to prevent spillage. Then, connect the vacuum gauge to the main fuel line. Operate the pump. Record the highest vacuum gauge reading.

Interpreting electric pump vacuum test results

Normally, an electric fuel pump should also produce around 10 in./Hg. (34 kPa) of vacuum. Check a service manual for more exact figures.

FUEL PUMP SERVICE, REPAIR, AND REPLACEMENT

Most modern fuel pumps CANNOT be repaired. They are crimped and sealed at the factory and cannot be disassembled. Therefore, fast, yet efficient, fuel pump removal and replacement is of primary importance to an auto mechanic. Since a few fuel pumps, however, can still be disassembled and serviced, fuel pump repair will be discussed briefly.

Mechanical fuel pump removal

Before removing the pump, take precautions against fire. Disconnect the battery ground cable and ready a fire extinguisher. Place a shop rag around any line fittings to be loosened. The rag will catch and absorb fuel that may spray or leak from the system.

Using a tubing wrench, crack open the output line on the pump. Turn in a counterclockwise direction. Remember that, with some fittings, TWO WRENCHES ARE REQUIRED. One wrench must be used to hold the inner body fitting and one must be used to turn the outer line fitting.

On the inlet, remove the hose clamps and carefully slide off the hose. This hose, attached to the incoming, main fuel line should be plugged so fuel will not leak from the tank or line. Use a bolt or an old spark plug.

Using a socket, extension, and ratchet, loosen the bolts holding the fuel pump to the engine. Other tools (box wrench, swivel socket, etc.) may be needed on some pumps. Remove the fuel pump. If it sticks to the block, pull with a brisk up and down movement, or tap with a plastic hammer.

Rebuilding mechanical fuel pump

If the body halves of the fuel pump are held together with screws, the pump may be rebuilt, Fig. 9-24. First, make sure that a fuel pump REBUILD KIT is available from

Fig. 9-24. The pump on the left is not serviceable. The one on the right is serviceable. (AC-Delco)

a "parts house." If so, new gaskets, diaphragm, and other parts in the kit can be used to restore the pump to good operating condition.

Disassembly techniques (mechanical fuel pump)

Before disassembling the pump, wash off grease and road dirt from the outside of the pump. To assist in aligning the halves at reassembly, scribe lines on the side of the pump across the seam joining the two halves.

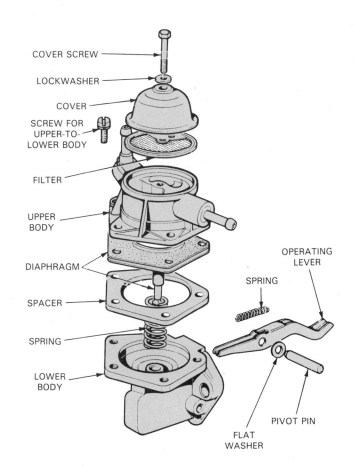

Fig. 9-25. During pump disassembly, note how all parts fit together. This will aid reassembly. Also study names of parts. (Fiat)

Read any instructions included with the rebuild kit. If available, also read a shop manual covering repair of your particular fuel pump. Rebuild procedures will vary with pump design and construction.

Remove the screws holding the pump body together. Pull the body apart carefully. Watch how ALL of the internal parts fit, Fig. 9-25. Mentally or on paper, note how the pump diaphragm, springs, pins, linkages, and valves are arranged in the pump body. Remove the diaphragm, pull rod, check valves, etc. Keep track of all parts.

Parts cleaning and inspection (mechanical pump)

Scrape old gasket material from the engine block and pump flange. To prevent foreign matter from entering the engine, stuff a small clean rag into the block opening.

Thoroughly clean all of the fuel pump parts in solvent, preferably carburetor cleaner. Double-check that all of the rubber or plastic parts are removed. They will be ruined by carburetor cleaner. Blow all of the components clean with compressed air.

Inspect all of the pump parts to be reused. Check the pump body for cracks or warpage. Check the rocker arm pin, rocker arm, and link for excessive wear. Arm and pivot wear can reduce pump stroke and pump output.

Fuel pump reassembly (mechanical pump)

Open the fuel pump rebuild kit and organize all of the new parts on your work bench. Throw away old parts being replaced. Generally, the fuel pump should be reassembled in reverse order of disassembly.

Be careful not to damage the soft housing. Hammer and clamp it lightly during the rebuild.

Make sure that the check valves are positioned in the right direction. Drive the valves into their seats with a small driving tool or socket. Check valves may either be held in place with screws or a staking (denting) process. To stake the valves in the pump, burr the edge of the pump around the valves. Using a punch and hammer, dent the pump body so that the valves cannot come out of their seats.

Lightly lubricate the diaphragm pull rod, rocker arm, and pivot with approved lubricant. Then fit together the internal parts of the pump. Double-check the diaphragm

spring, pull rod seal, link and rocker arm mechanism for proper assembly.

Refer again to the exploded view in Fig. 9-25 on how one type of mechanical pump should fit together. If in doubt, refer to a manufacturer's manual for an assembly drawing of your specific pump.

Fit the pump body halves together so that the diaphragm and scribe marks are correctly aligned. Start all of the fasteners. The screws should pass easily through the soft pump diaphragm. Run the screws in until just snug. Then, tighten in a crisscross pattern.

Never tighten the screws on one side of the pump all at once. This would cause an uneven clamping and sealing action. The pump body could be broken or warped.

Pump installation (mechanical fuel pump)

Using an approved sealant, glue the new pump gasket to the pump flange. As shown in Fig. 9-26, insert the bolts to assure good gasket alignment. Rub some grease on the end of the pump rocker arm and put a little oil into the pump rocker arm cavity.

To make installation easier, the cam eccentric should be rotated to the RELEASE POSITION (lobe away from pump arm). Fig. 9-27 illustrates this point.

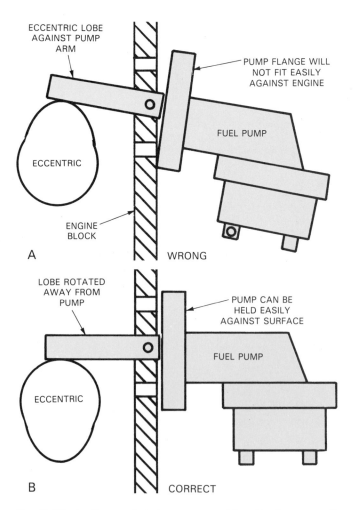

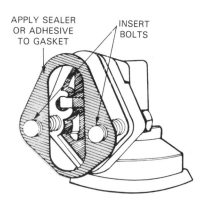

Fig. 9-26. Apply coating of nonhardening gasket cement to pump mounting gasket. This will help hold gasket in place during installation. Insertion of mounting bolts also helps hold the gasket. (KEM)

Fig. 9-27. A—Eccentric is in wrong position. Rocker arm will have to be depressed to fit the pump against the engine. B—Eccentric has been rotated away from pump arm. Pump will fit against engine block without forcing it.

To put the eccentric in the right position, turn the engine over while feeling the action of the eccentric. Insert a screwdriver through the opening or hold the pump in place. When the eccentric or lobe rotates away from the pump rocker arm, stop cranking the engine.

After the fuel pump bolts have been started by hand, alternately snug them down a little at a time. Overtightening one before the other can break the "ears" off the pump or cause an oil leak.

Reinstall the fuel lines and hoses. Start the engine and carefully feel under the connections. Any wetness indicates a leak. Test drive the car to assure proper engine performance.

Push rod type fuel pump installation

Some mechanical fuel pumps use a short, metal push rod that fits inside the engine block. The rod rides on the cam eccentric and transfers motion to the pump rocker arm, Fig. 9-28. This rod must be SUPPORTED during fuel pump installation. If not, pump damage may occur.

SERVICE, REPAIR, REPLACEMENT (ELECTRIC FUEL PUMPS)

Like mechanical fuel pumps, most electric pumps are NOT serviceable. They must be replaced when defective. However, a few can be disassembled and repaired.

Electric fuel pump removal (in-tank type)

Remember to disconnect the battery. This is especially important because of the wires running up to the pump. Have a fire extinguisher handy and use a rag to prevent fuel spillage.

When the electric fuel pump is mounted inside the tank, fuel must sometimes be drained BEFORE removing the pump. This is true whenever the pump is located on the side of the tank, NOT the top. In some cases, the fuel tank must be removed from the car to get at the electric fuel pump.

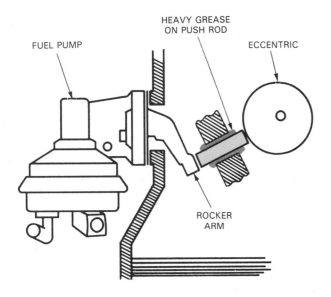

Fig. 9-28. Pump push rod must be supported when installing fuel pump. If not, rocker arm can jam into rod. Heavy grease or a hacksaw blade can be used to hold rod in place during pump installation on engine. (KEM)

To prevent dirt from entering the tank, clean the area around the fuel pump lock ring with a wire brush. Wash the cleaned area with solvent and blow dry.

Disconnect the wires and lines to the pump. Then, using a lock ring tool or full shank screwdriver and hammer, turn the lock ring counterclockwise. Rotate the ring about a quarter turn and pull the pump from the fuel tank as shown in Fig. 9-29.

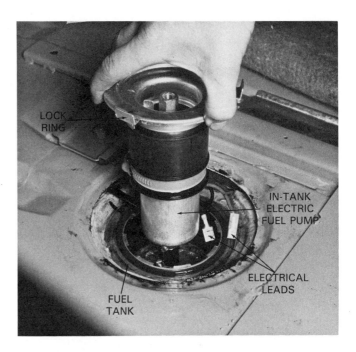

Fig. 9-29. Removing an in-tank fuel pump. Lock ring has been removed and electrical wire disconnected. (Saab)

Electric fuel pump installation (in-tank pump)

Compare the old pump with the new one. Make sure they are the same before installation.

Carefully fit the new pump into the fuel tank. Position the new O-ring seal. Then, fit and tighten the lock ring.

Install the wires and fuel lines on the pump. BEFORE reconnecting the car battery, fill the tank with fuel and check for leaks. If no leaks appear, the battery can safely be connected. Start the engine and observe fuel pump operation. Test drive the car to assure proper performance.

Electric fuel pump removal (chassis mounted type)

Again, remove a battery cable. Hold a "catch rag" under or around the line fittings as they are loosened. As shown in Fig. 9-30, a pan will also work. Remove the lines leading into and out of the electric pump. Then, plug the lines to keep fuel from leaking out.

Next, remove the bolts securing the fuel pump to the car. You may need to remove the pump bracket or at least loosen it to slide out the pump.

Electric fuel pump disassembly (chassis mounted type)

A few electric fuel pumps permit varying degrees of disassembly and repair. Before wasting time disassem-

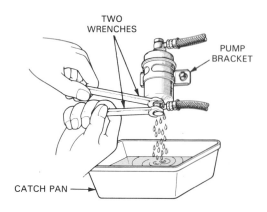

Fig. 9-30. Use container to catch fuel when removing an electric fuel pump. (Toyota)

bling an electric pump, first make sure you can get parts.

Many electric pumps have filters. It may be mounted on the pickup or inside the pump. Always check this filter when a faulty pump is removed. If it is clogged, cleaning may return the pump to normal operating condition.

Disassemble the fuel pump while referring to a detailed shop manual explaining repair of the particular electric pump. Remove the end cover bolts. Remove the covers while carefully watching the inner components of the pump. Note the arrangement of the parts.

Clean and closely inspect all of the pump components. See Fig. 9-31. Look for wear, scratches, scoring, or broken parts.

Check the fit of all of the critical parts. For example, fit the motor armature into the bearings. Feel for excessive wear. If the armature moves too freely on the shaft, either the bearings or armature shaft is worn.

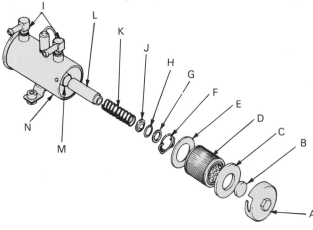

A. Cover	H. O-ring
B. Magnet	I. Fittings
C. Cover gasket	J. Inlet valve
D. Filter	K. Return spring
E. Gasket	L. Plunger
F. Spring retainer	M. Plunger cylinder
G. Washer	N. Body

Fig. 9-31. Look over all of the electric fuel pump components very carefully and note their names. Worn or damaged parts must be replaced. (Nissan)

Inspect the armature and field windings. They should not show signs of overheating or burning. Smell the windings. If they SMELL BURNED, then the pump should probably be replaced. Normally, it is not economical to replace the entire pump armature or field.

If the pump is a sliding or roller vane type, especially inspect the vanes and pumping chamber for wear or scoring. If the vanes are worn, replace them if available. A worn or scored pump chamber cylinder would normally warrant pump replacement.

On a diaphragm pump, check the soft rubber diaphragm for tears. It is commonly the first part to fail.

On a rotary type electric fuel pump, check the armature bearing closely. It frequently fails and causes the motor to drag. Also check the action of check valves. If used, make sure the pressure regulating valve is operating properly.

Electric fuel pump assembly (chassis mounted type)

After replacing defective components, reassemble the pump. It should go together in reverse order of disassembly. Lubricate the armature bearings, plunger, vanes, and other parts with white grease. This will protect these parts during initial pump starting.

Always consult a manufacturer's service manual when in doubt. The SMALLEST MISTAKE can prevent proper pump operation.

Install the pump on the car. Hook up fuel lines and wires. Check for fuel leaks. Then, test drive the car.

DUAL ELECTRIC FUEL PUMP SETUPS

A few fuel systems use *two electric pumps* (dual-pump system). Commonly, one pump is located in the fuel tank. It is known as an in-tank, "pusher" type pump. The other pump is usually an in-line, chassis mounted type fuel pump.

The two pumps work together to supply a high quantity of fuel to the engine. Generally, the in-tank pump primes the in-line pump. It keeps the chassis pump full of fuel. The in-line pump forces fuel to the engine under high pressure.

Basically, each pump of a dual system should be tested separately. Replace or repair each as needed.

Dual pump systems will be discussed further in the chapters on FUEL INJECTION and DIESELS. These systems normally use more than one pump.

SUMMARY

Diagnosis of fuel pump problems involves testing not only fuel pumps but many other parts of the fuel system. These parts, when malfunctioning, produce symptoms similar to faulty fuel pump operation.

Common fuel pump problems include low fuel pressure, high fuel pressure, inadequate flow, abnormal noise, or fuel leaks at and in the pump. Both mechanical and electrical pumps may fail after long service.

Before testing a fuel pump, make certain that the symptom or symptoms are not caused by clogged fuel filters, flattened, kinked, or collapsed fuel lines and hoses, air leaks through vacuum lines, carburetor or injection system problems, faulty ignition, or low engine compression.

A fuel pump has three output specifications: vacuum, pressure, and volume. Each must be checked. A problem with one has an effect on the others. Usually, a fuel pump should pull about 10 in. of vacuum. Fuel pressure for a carburetor system should be from 4 to 6 psi. An injection system may require pressures ranging from 15 to 40 psi at the injectors. A supply pump in a diesel system should produce about 6-10 psi of pressure. Volume of fuel pumps will vary according to engine size. As a general rule, a pump should deliver a quart of fuel a minute.

Fuel pumps and systems should be visually inspected before testing. This will often reveal leaks, clogged filters, and restrictions that affect pump operation.

A voltmeter is used to check the voltage to the electric fuel pump. Wires to the pump are disconnected and hooked up to the voltmeter. A jumper wire connected between the pump body and the car frame will test whether the pump is properly grounded. Worn pump bearings or a bad electric motor will create excessive current draw and can be detected with an ammeter.

A vacuum test, made on the inlet side of the fuel pump, determines whether there is a leak in the suction side of the fuel delivery system. A mechanical fuel pump should have about 10 in. of vacuum.

Some pumps can be serviced. Others must be replaced. Kits are available for repairing the serviceable types.

To install mechanical fuel pumps, make sure that the cam lobe is not in a position where it will put pressure on the lever. This could make installation very difficult.

A few electric fuel pumps can be disassembled and repaired. Be sure that you can get parts before disassembly. Inspect disassembled parts, looking for wear, scratches, scoring or breakage. Reassemble parts in the opposite order of disassembly.

KNOW THESE TERMS

Fuel evaporation, Inertia switch, Static pressure test, Dynamic pressure test, Capacity test, Auxiliary fuel supply flow test, Vacuum test, Dual-pump system.

REVIEW QUESTIONS—CHAPTER 9

1. List six safety rules for fuel pump service and repair.
2. Which of the following is NOT an output specification for a fuel pump?
   a. Pressure.
   b. Force.
   c. Vacuum.
   d. Volume.
3. A good fuel pump should produce at least _____ inches of vacuum.
4. Generally, fuel pump volume should be around a _____ a _____.
5. In general, fuel pressure for a carburetor fuel system should vary between:
   a. 10 to 20 psi.
   b. 4 to 6 psi.
   c. 1 to 4 psi.
   d. 8 to 10 psi.
6. List nine things to look for during an initial mechanical fuel pump inspection.
7. What 12 checks should be completed during the inspection of an electric fuel pump?
8. This type of fuel pump pressure test does not allow fuel to enter the carburetor.
   a. Dynamic test.
   b. Delivery test.
   c. Volume test.
   d. Static test.
9. Why is a dynamic pressure test better than a static test?
10. A _____ is used to connect a pressure gauge during a dynamic pressure test.
11. The engine must be started to measure the output of an electric fuel pump. True or false?
12. Give two other names for a fuel pump flow test.
13. How do you perform a fuel pump flow test?
14. Bubbles in the fuel leaving a pump indicate an _____ or a _____ in a fuel line.
15. Explain the purpose of an auxiliary fuel supply flow test.
16. Generally, 4, 6, and 8 cylinder engines have different fuel pump flow specifications. True or false, and why?
17. How do you check electric pump supply voltage?
18. Explain how to test for a bad electric pump ground.
19. When a special attachment is used to simultaneously measure pump pressure and volume in one operation, it is called a _____.
20. Normally, pump vacuum is only measured when _____.
21. During a vacuum test, the gauge should first be connected to the:
   a. Fuel pump.
   b. Carburetor.
   c. Fuel line.
   d. Strainer.
22. When pump vacuum is low at the tank, test _____ at the _____ inlet. This will help isolate the problem.
23. A majority of today's fuel pumps cannot be repaired or rebuilt. True or false, and why?
24. What kind of wrench should be used to loosen a fuel line fitting?
25. What is a fuel pump rebuild kit?
26. How should the eccentric be positioned when installing a mechanical fuel pump?
27. To hold up the push rod during pump installation, you can use either:
   a. Grease or a hacksaw blade.
   b. Grease or a hammer.
   c. Hacksaw blade or pliers.
   d. Saw blade or hammer.
28. When removing an electric fuel pump, it is especially important to disconnect the _____ because of the _____ running to the fuel pump.
29. A customer complaint indicates fuel starvation at higher engine speeds. Tests indicate that pressure and flow of fuel are below specs.
   Mechanic A says that the electric fuel pump is at fault.
   Mechanic B says that the in-tank fuel filter and fuel lines from tank to pump must be checked before

condemning the fuel pump.
Who is right?
a. Mechanic A.
b. Mechanic B.
c. Both Mechanic A and B.
d. Neither Mechanic A nor B.

30. A test of the feed wire between the on-board computer and the electric fuel pump indicates no voltage.

Mechanic A says that the computer is at fault because it supplies voltage to pump.

Mechanic B says that the possibility of a poor connection between the computer and the fuel pump could be the cause of the malfunction and should be checked out first.

Who is right?
a. Mechanic A.
b. Mechanic B.
c. Both Mechanic A and B.
d. Neither Mechanic A nor B.

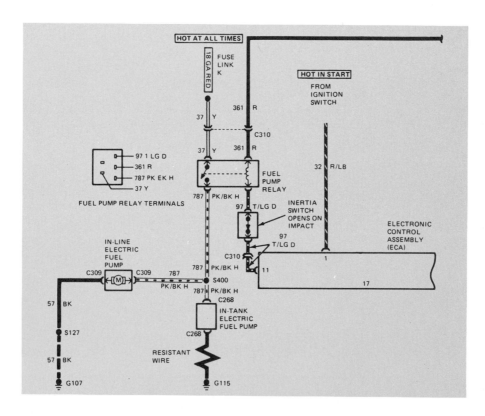

Shop manuals have many diagrams that will help the mechanic troubleshoot fuel pump systems. (Ford)

10

AIR CLEANER, INTAKE MANIFOLD, VACUUM PUMP CONSTRUCTION & OPERATION

After studying this chapter, you will be able to:
● *Define the term "induction system."*
● *Explain the importance of an air cleaner.*
● *List different types of air cleaners.*
● *Describe the purpose and operation of a thermostatically controlled air cleaner.*
● *List and explain the parts of a throttle cable assembly.*
● *Describe the construction of modern intake manifolds.*
● *Describe similarities and differences in intake manifolds for carburetor, gasoline injection, and diesel injection fuel systems.*
● *Describe the construction and operation of vacuum pumps.*

Air cleaners, intake manifolds, and vacuum pumps have changed considerably over the past few years. Air cleaners and related components now do more than simply filter air entering the engine. Intake manifolds have been redesigned for greater efficiency and power. Vacuum pumps are now commonly used on cars with diesel engines. This is an important chapter that will help you understand later chapters on service and repair. Study carefully!

INDUCTION SYSTEM

An *induction system* includes all of the parts of a fuel system that carry air or air-fuel vapor into the engine combustion chambers. See Fig. 10-1.

Induction systems can have several functions:
1. Filter the air entering the engine.
2. Control the temperature of the air entering the engine.
3. Monitor the temperature of the air entering the engine.

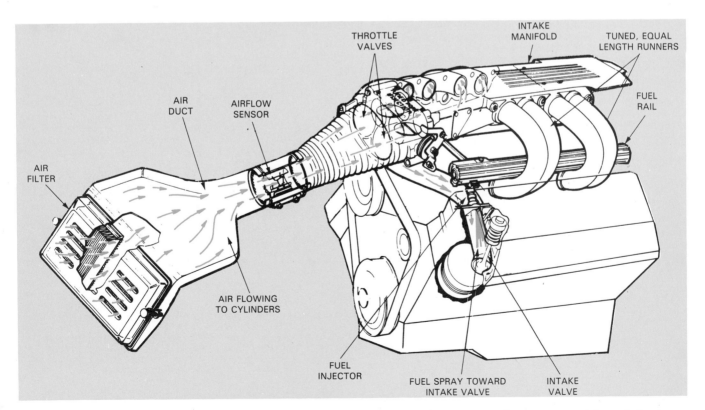

Fig. 10-1. Induction system for gasoline engine with fuel injection. Note names of components. (Chevrolet)

4. Monitor the amount of air flowing into the engine.
5. Work with the PCV (positive crankcase ventilation) system to burn crankcase fumes in the engine.
6. Control the amount of air-fuel mixture flowing into gasoline engine combustion chambers.

As you will learn, specific components are used to handle these functions.

Induction system for gasoline injection

A basic induction system for a gasoline injected engine includes an air cleaner, air ducts, airflow sensor, air temperature sensor, throttle valves, and intake manifold assembly.

Fuel is commonly injected (sprayed) close to the intake valves. With port or multi-point injection, only air passes through the throttle valves and main section of the intake manifold. With single-point injection, both air and fuel pass over the throttle valves and through the intake manifold.

Induction system (carburetion)

An induction system for a carbureted engine includes the basic parts illustrated in Fig. 10-2. Compare the differences with the system in Fig. 10-1. This is an older system that uses intake manifold vacuum (suction) to draw air and fuel into the intake manifold.

Fuel enters the system above the throttle valves. The atomized mixture is then pulled through the intake manifold, cylinder head ports, past the valves, and into the combustion chambers.

Induction system (diesel engine)

Diesel engine induction systems are normally simpler than those for gasoline engines. Only air passes through the induction system. Look at Fig. 10-3.

A diesel engine does NOT use throttle valves to control the amount of air and fuel entering the engine. A full charge of air is allowed to flow into the combustion chambers at all engine speeds.

A basic diesel engine induction system consists of an air cleaner and intake manifold assembly. An EGR (exhaust gas recirculation) valve and other components may be added to reduce harmful emissions.

Induction system components

The parts of an induction system, depending upon design, may include an air cleaner, air ducts, intake manifold, throttle valves, sensors, and a computer. Other components can be added to increase engine efficiency.

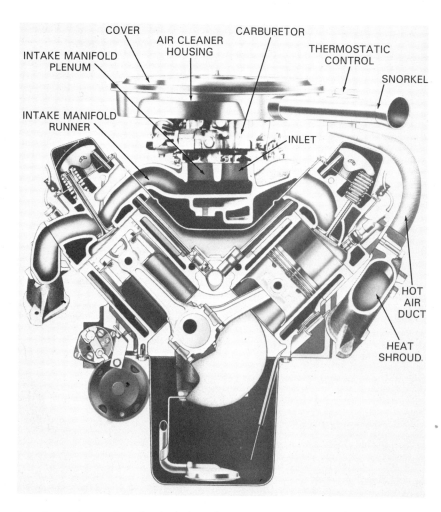

Fig. 10-2. Induction system for carbureted engine includes all components that provide pathway for air-fuel mixture to enter combustion chambers of engine. (Cadillac)

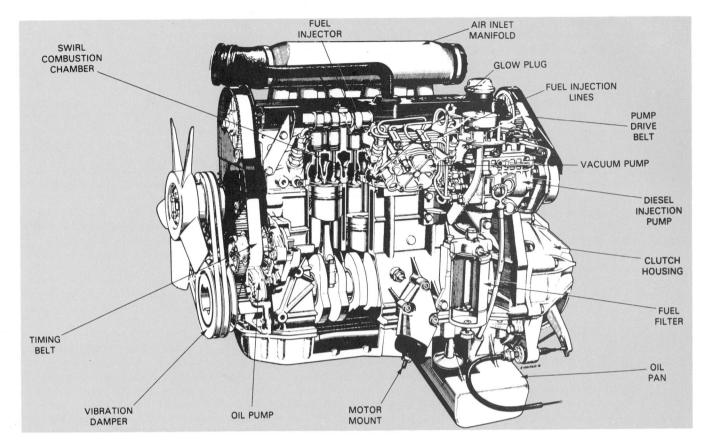

Fig. 10-3. Diesel fuel injection system. Air only is delivered through air inlet manifold. Fuel is delivered directly to each combustion chamber. (Volvo)

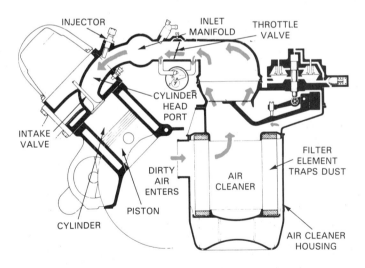

Fig. 10-4. Air cleaner action is shown for engine. Note movement of air through filter to cylinder. (Saab)

AIR CLEANERS

Air cleaners, also called *air filters,* prevent airborne debris (dust, dirt, etc.) from entering the intake manifold and combustion chambers. Some will also control temperature of air entering the engine.

A running engine draws huge amounts of air into the cylinders. Outside air contains many contaminants. Thousands of gallons of air are needed to burn just one gallon of fuel.

It is critical that this air is clean. Without a properly functioning air filter, debris can enter the combustion chambers and cause ABRASIVE WEAR on piston rings, pistons, cylinder walls, valves, and other parts. Fig. 10-4 shows the basic action of an air cleaner.

Air cleaner locations

The location of an air cleaner varies with engine and fuel system design. It may be at:
1. The carburetor or throttle body (gasoline engine).
2. The intake manifold (diesel engine).
3. Remote location away from engine (gas or diesel engine).

Locations will be discussed and illustrated later in the chapter.

Parts of an air cleaner

Depending upon design and location, the parts of an air cleaner assembly will vary. Typically, it will consist of:
1. AIR CLEANER HOUSING (metal or plastic container that holds element and supports other parts).
2. ELEMENT (filtering device).
3. AIR CLEANER COVER (removable lid that allows service of element).
4. AIR CLEANER GASKETS (rubber or fiber seals that prevent air leakage between components of air cleaner).
5. SNORKEL (rigid air inlet extending out from air cleaner housing).

6. AIR DUCTS (flexible plastic, metal, or metal-paper tubes that carry air into housing).

7. OTHER PARTS (other components, discussed shortly, for controlling and monitoring air temperature).
    Refer to Fig. 10-5. It shows most of these basic components.

Element types

There are several types of air cleaner elements: paper, oil soaked, and oil bath. The *paper element* is shown in Fig. 10-6. Also called a *dry type element,* it is made of special pleated paper held together by rubber end seals.

Usually it has a wire screen for support. All of the air drawn into the engine must be pulled through the paper element. The pores in the paper are small enough to trap debris, but will pass clean air.

The *oil soaked air cleaner* normally is made of polyurethane plastic and a metal ribbon mesh. The element is coated with oil, Fig. 10-7. The oil attracts and holds dust and dirt as it passes through the element. This is not a common type of air cleaner because it requires frequent cleaning.

The *oil bath air cleaner* has a reservoir of oil in the bottom of the housing to trap and hold airborne contaminants. As illustrated in Fig. 10-8, the shape of the air cleaner housing causes the air to quickly change directions as it moves over the oil. Centrifugal force throws debris into the oil. This type is not commonly used because idle or low speed efficiency is very poor.

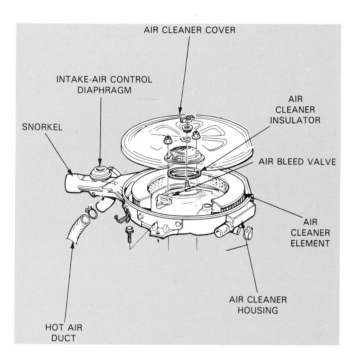

Fig. 10-5. Exploded view of air cleaner.

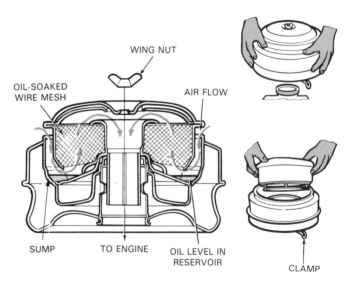

Fig. 10-7. Polyurethane elements can be cleaned, oiled, and reused.

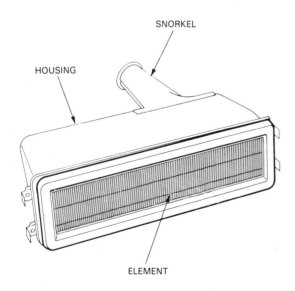

Fig. 10-6. Paper element or cartridge is most common one in use today.

Fig. 10-8. Oil bath cleaner. Oil traps dirt and holds it until filter is serviced.

Air cleaner ducts

Air cleaner ducts at the inlet and outlet of an air cleaner carry air to the element and engine. Fig. 10-9 shows a remote, mounted air cleaner using ducts.

Special plastic or metal clamps normally hold the ducts together. Plastic ducts are used where engine heat is not a problem. Special paper-metal ducts or all-metal ones are used when they will be exposed to excess engine heat. Look at Figs. 10-10 and 10-11.

Other air cleaner components

Fig. 10-12 shows an air cleaner with a computer. Primarily, it controls the ignition system. Note how the air duct extends to one side of the radiator. This allows cool air, not air warmed by radiator action, to flow into the engine.

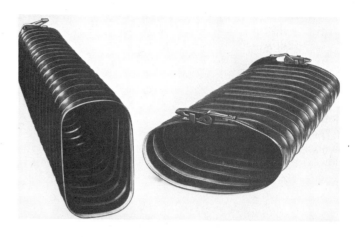

Fig. 10-10. Duct hose must be designed to withstand heat conditions under hood. This air intake hose is plastic supported by spiraled metal framework.

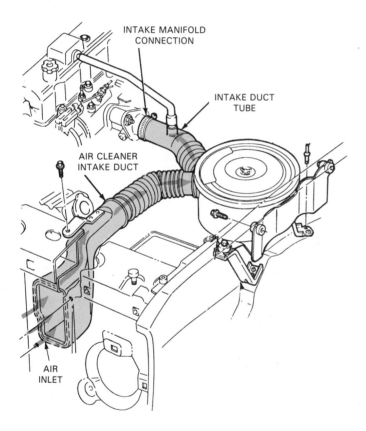

Fig. 10-9. Ducts are used for remote (away from engine) air cleaner on this diesel engine.

Fig. 10-11. Emission hose is made of flexible metal to withstand heat. (Gates)

Fig. 10-13 pictures a positive crankcase ventilation filter located in some air cleaners. It must be serviced after oil vapors collect in the element and reduce filtering action.

THERMOSTATIC AIR CLEANER

A *thermostatic air cleaner* controls the temperature of the air entering the intake manifold. Its basic parts include:

1. THERMAL VACUUM SWITCH (temperature sensing switch that controls vacuum to operate vacuum

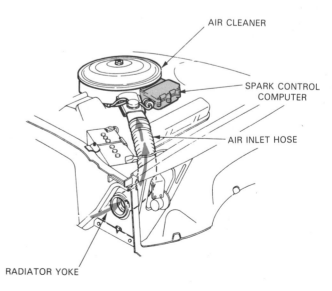

Fig. 10-12. Computer mounted on air cleaner housing controls ignition spark. (Echlin)

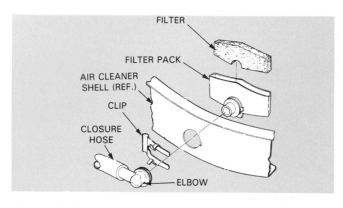

Fig. 10-13. Exploded view of crankcase ventilation filter pack and hose. Filter separates oil vapors from air being pulled from crankcase. (Ford)

motor), Fig. 10-14.
2. VACUUM MOTOR (vacuum diaphragm that opens and closes air door in air cleaner snorkel).
3. AIR DOOR (hinged metal flop inside air cleaner snorkel).
4. HEAT SHROUD (sheet metal enclosure around exhaust manifold to collect warm air).
5. VACUUM HOSE (connects intake manifold vacuum to system).
6. DUCTS (connect the heat shroud to the bottom of the snorkel. Another duct carries cool outside air into the main snorkel inlet).
   Fig. 10-15 shows typical location of several of these parts. Fig. 10-16 shows how the air door operates.

Thermostatic air cleaner operation

When vacuum is applied to the motor, the diaphragm

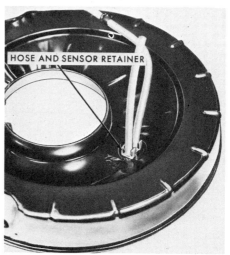

Fig. 10-14. Views of sensor assembly from top and bottom of air cleaner housing. Hoses connect to intake manifold. (Cadillac Div., GMC)

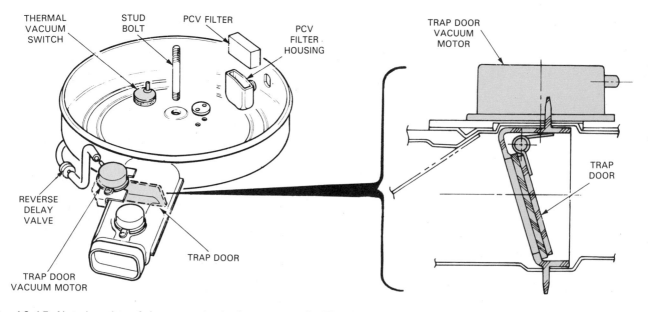

Fig. 10-15. Note location of thermostatic air cleaner controls. Thermal vacuum switch turns trap door vacuum motor on and off to close and open trap door. (AMC)

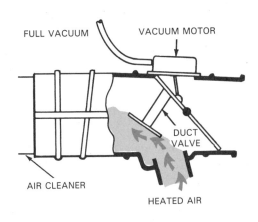

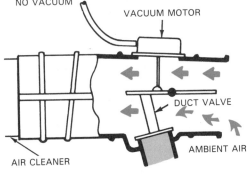

Fig. 10-16. Operation of the air door. A—During cold start full vacuum is applied to motor. Air door closes off snorkel. Only warmed air from exhaust manifold enters. B—As engine warms up, no vacuum is delivered to vacuum motor. Door opens and allows cool air into air cleaner. (Ford)

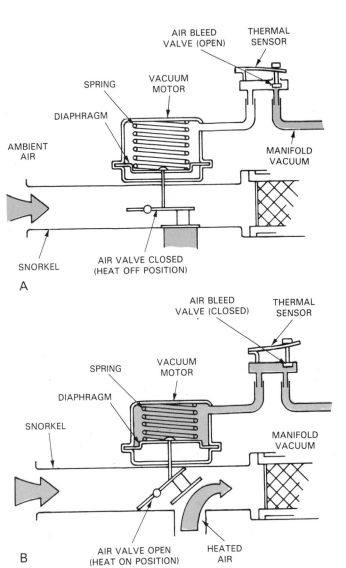

Fig. 10-17. Temperature vacuum switch operation. A—When under-hood air temperature is higher than 100°F, thermal sensor's bimetal strip will flex to hold air bleed valve open, shutting off vacuum to vacuum motor. B—At colder under-hood temperatures, bimetal strip will flex in opposite direction allowing vacuum to act on diaphragm of vacuum motor. (AMC)

and door are pulled up. This blocks the cool air and allows warm air from the exhaust heat shroud to enter the engine. With no vacuum applied, the diaphragm moves down. This blocks warm air and opens the cool air main snorkel passage.

The *temperature vacuum switch* is added to the circuit to control when and how much the vacuum door opens or closes. In this way, the system can maintain a preset air inlet temperature. This allows for higher engine efficiency by allowing a more precise setting of air-fuel mixture, ignition timing, and other engine functions.

Fig. 10-17 illustrates thermal vacuum switch operation. Note how it can bleed air (vacuum) to control the amount of suction applied to the vacuum motor diaphragm. A thermostatic air cleaner assembly for a late model, fuel injected engine is shown in Fig. 10-18.

Functions of thermostatic air cleaner

A thermostatic air cleaner has three basic functions. It:
1. Warms inlet air when engine is cold. This aids fuel vaporization and helps keep the engine from stalling when cold.
2. Allows cool air to enter engine in hot weather. This helps prevent vapor lock (fuel boils from excess heat) and percolation (fuel heats and expands, upsetting normal metering).
3. Maintains more constant temperature of inlet air. This reduces fuel consumption, emissions, and improves driveability by allowing the use of a more precise air-fuel mixture.

THROTTLE VALVE

The *throttle valve* is a butterfly type valve that allows the driver to control how much air flows into a gasoline engine. By regulating airflow, fuel metering and engine power are also controlled.

209

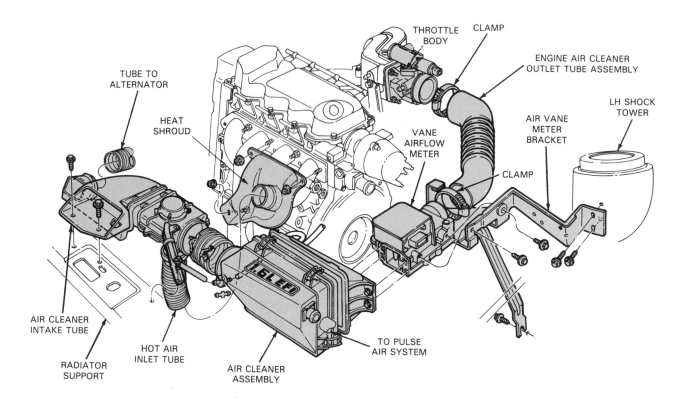

Fig. 10-18. Late model thermostatic air cleaner assembly. Note airflow meter between filter and throttle body. (Ford)

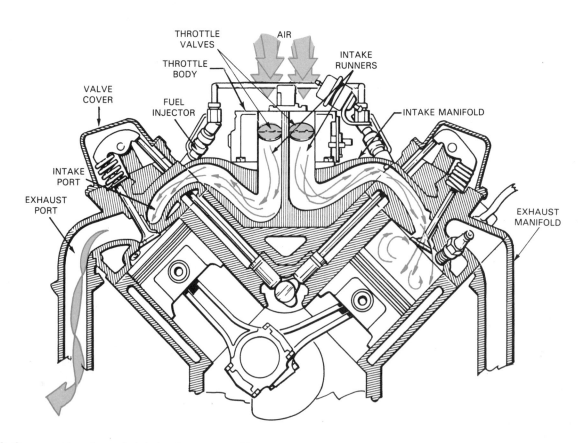

Fig. 10-19. Cross section view of air induction system. Throttle valves are like "doors" in throttle body. They regulate amount of air entering intake manifold. (Cadillac)

Refer to Fig. 10-19. The throttle valves are located in the induction system right before the intake manifold on most engines. They normally mount in the throttle body (gasoline injection) or in the carburetor. Sometimes, they are located in the intake manifold.

One or more throttle valves may be used. When the throttle is closed, airflow is almost stopped and the engine idles slowly. When the throttle is opened, more air can rush in to fill the vacuum in the intake manifold and cylinders. As a result, more fuel is also metered into the engine. This increases combustion pressure and power.

Fig. 10-20 shows a variation from the normal throttle valve location. The two throttle valves are centrally located in the intake manifold. They perform the same function however. Air flows down through the airflow sensor, through the throttle valves, into the intake manifold, and then to the cylinder heads.

THROTTLE CABLE AND LINKAGE

The *throttle cable* mechanism (or rod linkage mechanism on older cars) connects the gas pedal to the throttle valve assembly. See Fig. 10-21.

The throttle cable is a steel braided wire enclosed inside a metal and plastic housing. The steel cable is free to slide back and forth inside its housing.

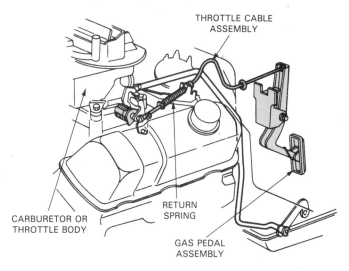

Fig. 10-21. Light metal braided cable links gas pedal to throttle valve. (Ford)

The older *throttle linkage* system consists of metal rods and swivel joints that transfer motion from the gas pedal to the throttle valve. This mechanism has been replaced by the lighter, more compact cable system.

On diesel engines, the throttle cable connects to the

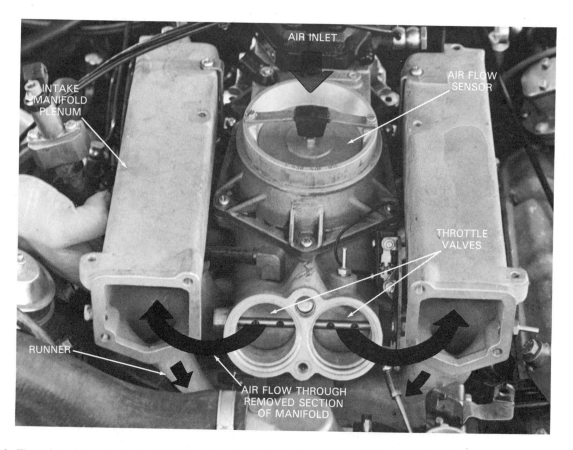

Fig. 10-20. Throttle valves can be located elsewhere in air induction system. With a section of intake manifold removed, they can be seen between air sensor and plenums. (Volvo)

diesel injection pump. The cable, by acting upon a throttle lever on the pump, increases or decreases fuel injection volume to control engine power.

Throttle cable and valve operation

Depressing the gas pedal pulls on the throttle cable. This causes the cable to slide in its housing. Since the other end of the cable is connected to the throttle lever, the cable pulls the throttle open and more air can rush into the gasoline engine. Engine power and speed increase.

When the driver releases the gas pedal, a *throttle return spring* pulls the throttle lever closed. The spring also pulls the cable and gas pedal into the idle or released position.

INTAKE MANIFOLDS

An engine *intake manifold* is a metal casting that bolts over and covers the intake ports in the cylinder head(s). See Fig. 10-22.

An intake manifold can have several functions:
1. Carry air-fuel mixture (gasoline engine) or air (diesel engine) into the cylinder head ports.
2. Evenly distribute air or air-fuel mixture to each engine cylinder.
3. Prevent the fuel from separating from the air.
4. Assist fuel vaporization and atomization.
5. Help warm the fuel charge when the engine is cold.
6. Assist in filling the engine combustion chambers with a fuel charge.

These functions vary with engine and intake manifold design. To be able to troubleshoot and repair today's fuel systems, you should have a sound understanding of intake manifold design and construction.

Intake manifold construction

Engine intake manifolds are usually cast from iron or aluminum. Aluminum is now more common because its lighter weight lowers fuel consumption. Surfaces are machined where they attach to the engine, throttle body

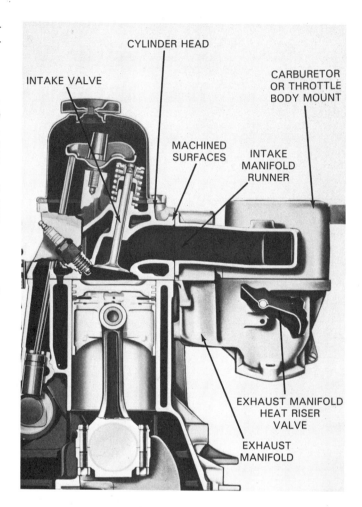

Fig. 10-23. Cutaway view of engine and intake manifold. Note how machined surface of intake manifold fits up to engine. Other machined surface shown is where carburetor or throttle body assembly attaches. (AMC)

or carburetor, and other components.

Fig. 10-23 shows a cutaway view of an engine intake manifold. Study how the intake fits on the engine and how a mounting surface is provided for a carburetor or throttle body assembly. The exhaust manifold is bolted to the bottom of the intake manifold with this design.

Fig. 10-24 shows intake manifolds for in-line and V-type engines. Note the differences. The in-line manifold positions the inlet to one side and passages run out to meet the ports in the cylinder head. Inlet of the V-type is in the middle with passages feeding out to each side.

Parts of an intake manifold

The exact parts of an intake manifold vary with design. However, most intake manifolds have:
1. RUNNERS (passages for air-fuel mixture or air), as shown in Fig. 10-25.
2. INLET or RISER BORE (opening for incoming air). Refer to Fig. 10-26.
3. PLENUM (large chamber below inlet or bore for equalizing pressure at openings to runners).
4. EXHAUST PASSAGE (separate passage for allowing engine exhaust gases to warm base of plenum).

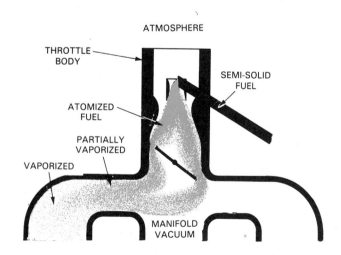

Fig. 10-22. Simplified sketch of intake manifold. It is designed to carry air or air-fuel mixture from throttle body to cylinder head ports. Note air-fuel flow. (GM Trucks)

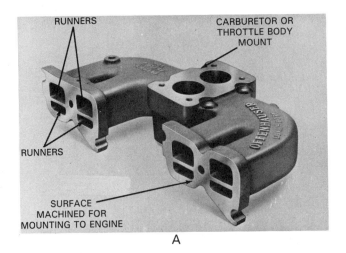

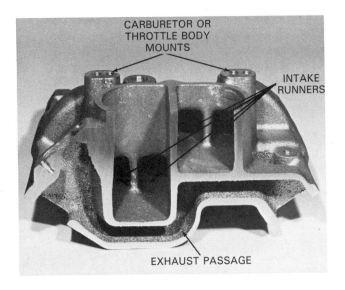

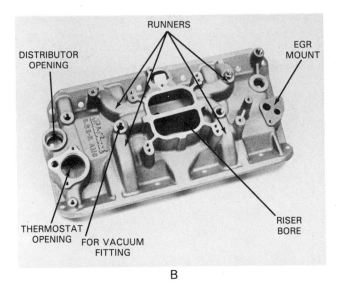

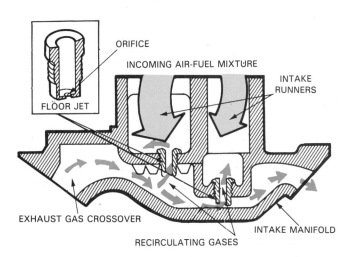

Fig. 10-24. Intake manifolds take various shapes, depending on engine type. A—For 4-cylinder, in-line engine. (Offenhauser) B—For 8-cylinder, V-type engine. (Edelbrock)

Fig. 10-25. Top. Cutaway shows intake runners which are passages that carry air-fuel mixture to combustion chambers. (Edelbrock) Bottom. Small passageways at bottom of runners are jets which allow exhaust gases to enter intake manifold. One or more surfaces are machined flat to mate with head or heads. (Chrysler Corp.)

5. EGR PASSAGE (opening that allows small amount of exhaust gases to enter plenum to reduce NO_x emissions).

6. WATER JACKETS (water passages for removing excess heat from intake manifold).

7. HEAD MATING SURFACE (machined flange that seals against cylinder head).

8. OTHER PROVISIONS (openings for vacuum fittings, EGR valve, fuel injectors, etc.).

Fuel distribution

Fuel distribution refers to how well an intake manifold routes an equal amount of air and fuel to each combustion chamber. Poor fuel distribution occurs when cylinders receive an UNEQUAL amount of fuel or air.

Fig. 10-27 illustrates fuel distribution. With the air-fuel mixture flowing through a common runner past each intake port, the last cylinder will receive less fuel charge than the others. The cylinders in front of it draw in most of the fuel charge on their intake strokes. An intake manifold must be designed to prevent this problem.

Intake manifold plenums must also be designed to counteract differences in temperature, flow, etc., to assure equal fuel distribution. The plenum wall can be heated by coolant. The air-fuel mixture can be redirected to collect heat from the manifold wall. This improves vaporization and distribution.

Intake manifold types

Intake manifold types differ in plenum design, runner length, port or runner configuration, general shape, and method of construction. There are other considerations, but these are the most important.

A *dual plane* intake manifold has a two-compartment plenum. This type is common on V-type passenger car engines. See Fig. 10-28. It provides good fuel vaporization and distribution at low engine speeds. This makes it very efficient for everyday driving.

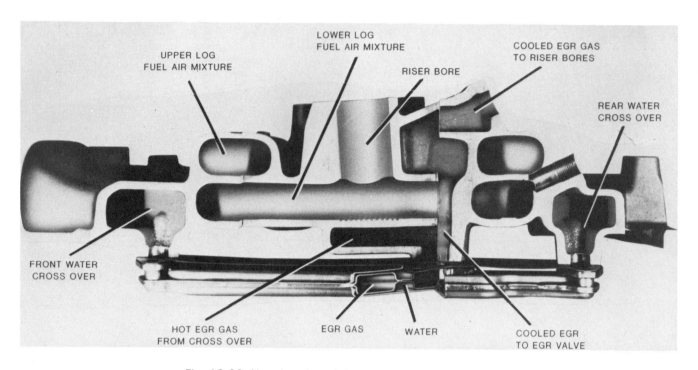

Fig. 10-26. Note location of riser bore at top center. (Ford)

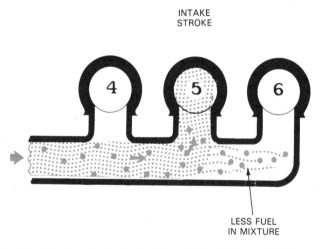

Fig. 10-27. How fuel gets distributed in typical manifold. Last cylinder (farthest away from supply) gets less air-fuel mixture than other cylinders. Not all the fuel is vaporized. Note larger droplets. Cool manifold walls can cause some vapor to condense. (Chevrolet)

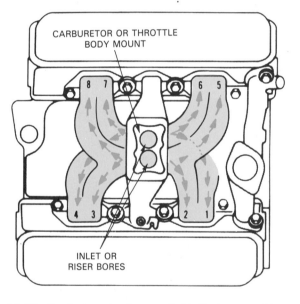

Fig. 10-28. Dual plane intake manifold. Each venturi in carburetor feeds separate passages. For example, one venturi feeds cylinders 1, 4, 6, and 7 while other feeds cylinders 2, 3, 5, and 8. (Ford)

In a *single plane* intake manifold, one common plenum feeds all runners. This type is used primarily on high performance engines. It is not as efficient as the dual plane type at lower engine rpm. A single plane, however, is capable of allowing greater flow at higher engine speeds, increasing engine horsepower capabilities.

Fig. 10-29 shows a dual plane intake manifold. Note that it also has dual runners. Each runner is divided in the center. This aids fuel distribution, atomization, and combustion efficiency.

A sectional intake manifold is made of two or more pieces that bolt together. See Fig. 10-30.

The trend is toward sectional intake manifold construction. This allows a more exotic shape for both runners and plenum.

Fig. 10-31 illustrates an engine using a two-piece intake manifold. The upper part of the intake manifold holds the throttle body. The lower section houses the four fuel injectors. A gasket fits between the two sections to prevent vacuum leakage.

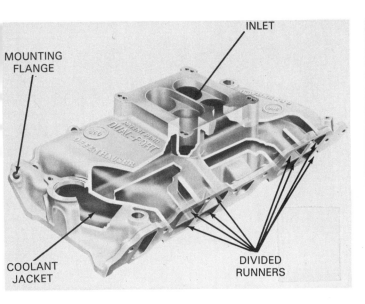

Fig. 10-29. Cutaway shows arrangement of dual plane, dual port intake manifold. Note how runners are shaped and divided to improve fuel atomization and distribution. (Offenhauser)

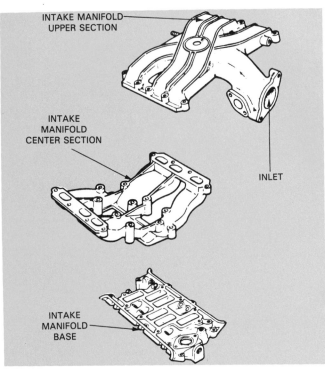

Fig. 10-30. Sectionalized intake manifolds allow greater variety in plenums and runner shapes. (Buick)

A *tuned runner intake manifold* uses long, equal-length runners to increase engine performance, Fig. 10-32.

Long runners tend to produce a "supercharging" effect by ramming more air into the combustion chambers at high engine speeds. The inertia of the air or air-fuel mixture flowing through the long runners tends to force extra air and fuel into the cylinders.

Fig. 10-33 illustrates the action of a tuned runner in-

take manifold on a multi-point fuel injected engine. Air entering the center of the plenum from the air cleaner is drawn through each runner and into the combustion

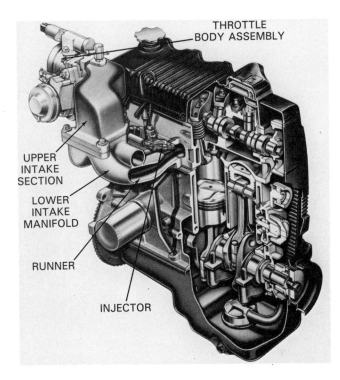

Fig. 10-31. This engine uses two-piece intake manifold. Upper half is throttle body. Note cutaway of injector located in lower half. (Ford)

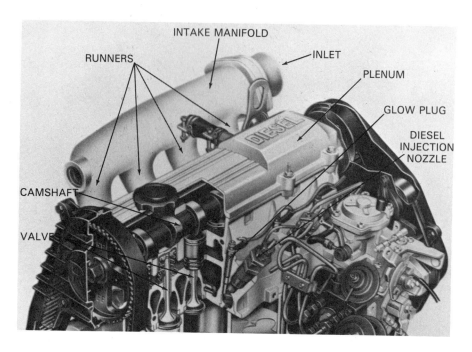

Fig. 10-32. Tuned runner intake manifold. Runners are long and of equal length. (Volvo)

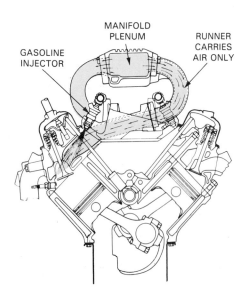

Fig. 10-33. Cutaway of air and fuel being drawn into fuel injected engine. Note length of runner. (Chevrolet)

Fig. 10-34. This intake manifold has dual plenums each connected to three long runners. Inertia in runners helps force larger charge into cylinder. (Volvo)

chambers. Fuel is sprayed into the airstream just before the intake valves.

Fig. 10-34 shows another intake manifold design using long runners. This intake, made of weight-reducing aluminum, is also built in sections that bolt together.

Fig. 10-35 shows the air inlet or intake for a diesel engine. Note how the EGR valve mounts onto the intake manifold. This part of the manifold bolts to a larger, lower section.

Jet air intake manifolds

A few engines use an *air jet valve* to admit a stream

of fresh air into the combustion chambers at idle. This aids fuel mixing and combustion. With this design, an extra air passage is normally formed in the intake manifold, Fig. 10-36.

As shown in Fig. 10-37, air is drawn through the extra passage on each intake stroke. Air-fuel mixture is pulled through the conventional runners.

Intake manifold accessory units

Various accessory units can be installed on an intake

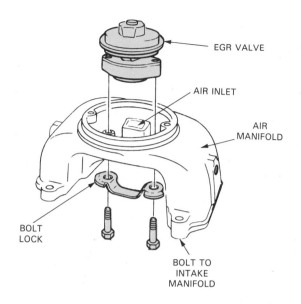

Fig. 10-35. Relationship of EGR valve and air inlet of intake manifold. Note how EGR valve is fastened to manifold. (GMC)

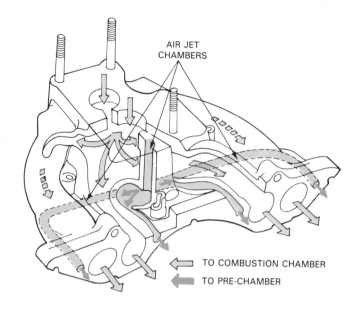

Fig. 10-37. Air flows into precombustion chamber of stratified charge engine through air jet chambers. (Champion Spark Plug)

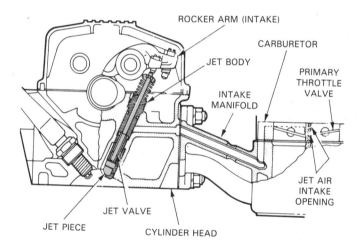

Fig. 10-36. In some engines, air jet valve admits additional air into combustion chamber at idle to lean air-fuel mixture. (Chrysler)

manifold, Fig. 10-38. Always refer to a service manual for information about any component. The manual will describe and illustrate the exact engine and intake on which you are working. Fig. 10-39 illustrates some of the sensor switches on an intake manifold.

FLOW BENCH

A *flow bench* is a test device used to measure the amount of airflow through various induction system components. Its basic action is illustrated in Fig. 10-40.

A large fan pulls air through the bench. A pressure measuring device can then be used to check the pressure difference across the component being tested. The pressure difference can then be used to calculate flow. For example, during a flow test, one runner or an intake

manifold might show more pressure drop than another runner. This would tell the operator that one runner (one with more pressure drop) has a restriction. By balancing the flow in each runner, engine power and efficiency can be increased.

A flow bench can also be used to improve airflow through carburetors, cylinder heads, etc. Ideally, the flow through all components carrying air to each cylinder should be balanced and unrestricted.

WATER INJECTION

Introducing water into the combustion chamber to help prevent knock and ping under high load, high cylinder pressure conditions, is known as *water injection.* One system is illustrated in Fig. 10-41.

Water injection is an aftermarket system commonly recommended for turbocharged engines or older high compression engines that do not run properly on today's low octane fuels.

During operation, a small amount of water is metered into the intake manifold. The water tends to remove carbon deposits and reduce peak combustion flame temperature, primary causes of engine knock.

Note! Before installing a water injection system, make sure the vehicle warranty will not be voided.

VACUUM PUMPS

A *vacuum pump* provides or supplements vacuum (suction) for various accessory units. It is commonly used on a diesel engine, Fig. 10-42. Without a throttle valve to restrict air inlet flow, a diesel does not produce strong vacuum in the intake manifold. Thus, there is no vacuum source for vacuum-assisted power brakes, vacuum diaphragms, and other components.

Vacuum pumps are either mechanical or electrical. Both are common, Fig. 10-43.

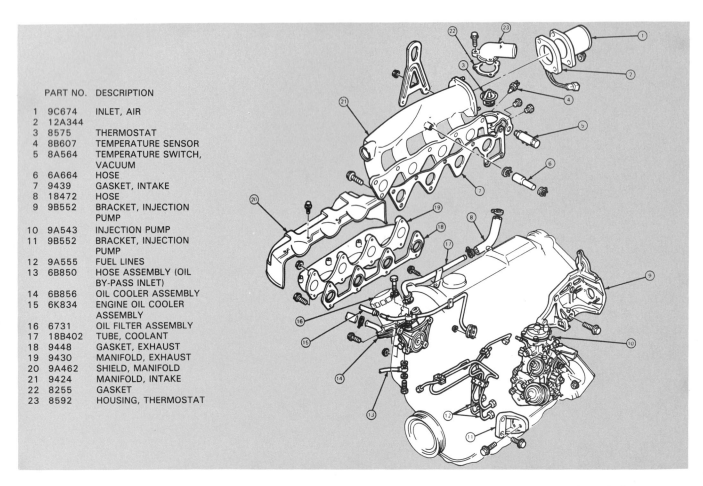

PART NO.	DESCRIPTION
1	9C674	INLET, AIR
2	12A344	
3	8575	THERMOSTAT
4	8B607	TEMPERATURE SENSOR
5	8A564	TEMPERATURE SWITCH, VACUUM
6	6A664	HOSE
7	9439	GASKET, INTAKE
8	18472	HOSE
9	9B552	BRACKET, INJECTION PUMP
10	9A543	INJECTION PUMP
11	9B552	BRACKET, INJECTION PUMP
12	9A555	FUEL LINES
13	6B850	HOSE ASSEMBLY (OIL BY-PASS INLET)
14	6B856	OIL COOLER ASSEMBLY
15	6K834	ENGINE OIL COOLER ASSEMBLY
16	6731	OIL FILTER ASSEMBLY
17	18B402	TUBE, COOLANT
18	9448	GASKET, EXHAUST
19	9430	MANIFOLD, EXHAUST
20	9A462	SHIELD, MANIFOLD
21	9424	MANIFOLD, INTAKE
22	8255	GASKET
23	8592	HOUSING, THERMOSTAT

Fig. 10-38. Other service manual illustrations identify accessories and provide part numbers for ordering.

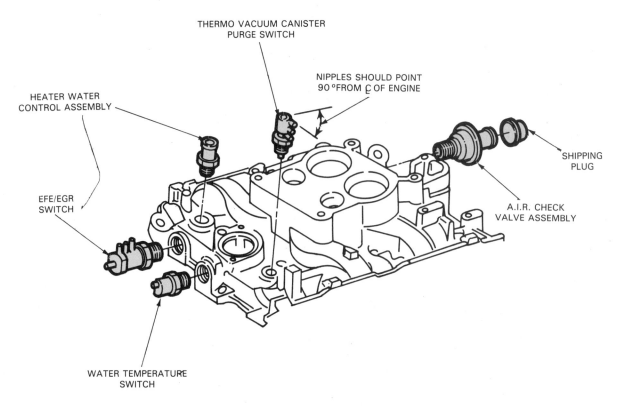

Fig. 10-39. Several types of sensor switches are found on intake manifold. (Oldsmobile)

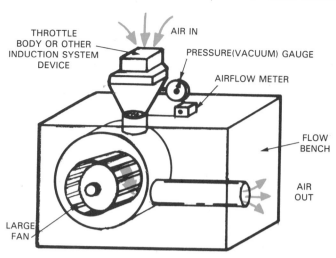

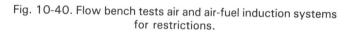

Fig. 10-40. Flow bench tests air and air-fuel induction systems for restrictions.

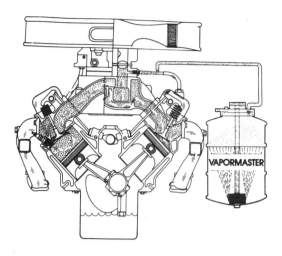

Fig. 10-41. Water injection systems are used to reduce knock and ping in high compression engines. (Edelbrock)

Fig. 10-42. Vacuum pump is common accessory on diesel engines. It provides vacuum for power brakes, and other vacuum powered accessories. (Mazda)

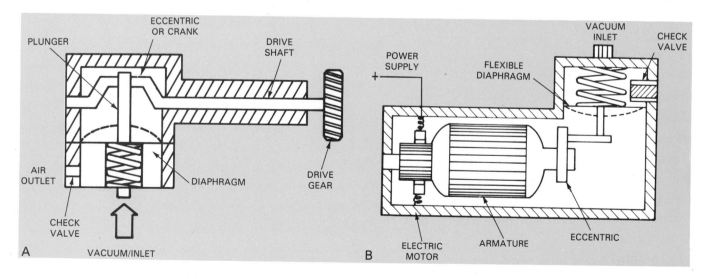

Fig. 10-43. Two types of vacuum pumps for automotive use. A—Mechanical pump is driven by belt or shaft. B—Pump driven by its own electric motor.

Mechanical vacuum pumps

A *mechanical vacuum pump* uses engine power to move a flexible diaphragm that produces negative pressure. It works something like a mechanical fuel pump. In fact, early mechanical vacuum pumps were part of the fuel pump. Modern belt driven mechanical fuel pumps have an eccentric inside the pump housing. This is pictured in Fig. 10-44.

A small rubber vacuum hose is connected to the output or suction outlet of the vacuum pump. This hose routes vacuum to all vacuum-operated devices.

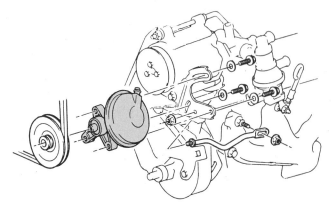

A

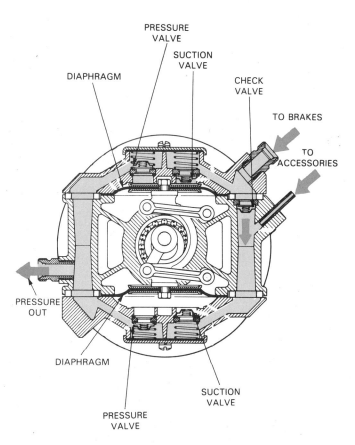

Fig. 10-44. Twin diaphragm vacuum pump has two vacuum circuits. One supplies braking system while other supplies vacuum for central lock system, key starting, automatic climate control, and modulator pressure for transmission and cruise control. (Mercedes Benz)

Mechanical vacuum pumps can be located on the side, front, or top-rear of the engine, Fig. 10-45. On some diesel engines, the vacuum pump is placed where the ignition distributor mounting hole is located on gasoline engines.

Fig. 10-46 shows a vacuum pump located in the rear of the charging system alternator. This particular pump is a rotary type. It works like a sliding vane electric fuel pump or power steering pump.

Electric vacuum pumps

Electric vacuum pumps, Fig. 10-47, use a small dc electric motor and, frequently, an eccentric and dia-

SECTION THROUGH VACUUM PUMP
AND OIL PUMP DRIVE

B

Fig. 10-45. Mechanical vacuum pump may be located in one of several different positions on the engine. A—Front mounted belt driven. B—Rear mounted mechanical vacuum pump. (Oldsmobile)

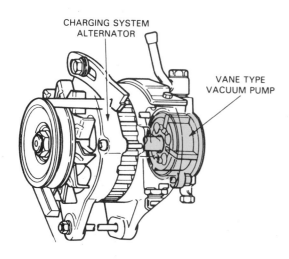

Fig. 10-46. This vacuum pump is located on same shaft as alternator. (Chrysler)

phragm action. They are becoming more common because they can provide vacuum even with the engine off. This provides a safety measure to operate the power brakes even when the engine ''dies.'' An electric vacuum pump can be located any place in the engine compartment. See Fig. 10-48.

Fig. 10-49 illustrates a vacuum circuit for a diesel engine. Note how the vacuum pump feeds to various components in the system.

SUMMARY

All of the components which carry air and air-fuel mist into the engine combustion chamber are included in the term, induction system.

Air cleaners or filters keep airborne debris out of the engine. They may be located in different spots: over the carburetor or throttle body, at the intake manifold, or at

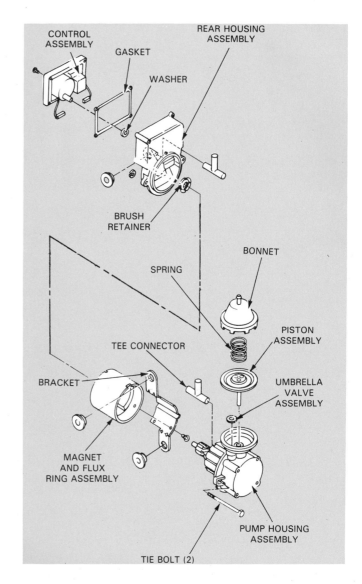

Fig. 10-47. Exploded view of an electric vacuum pump. Motor is located in pump housing assembly.

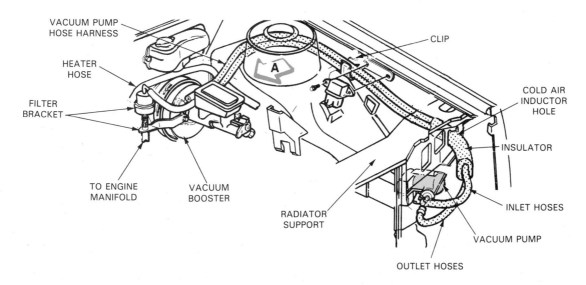

Fig. 10-48. Electric vacuum pumps have advantage of being located off the engine. Unit shown is on inner fender panel. Other possible locations are fire wall or under battery tray. (Oldsmobile)

221

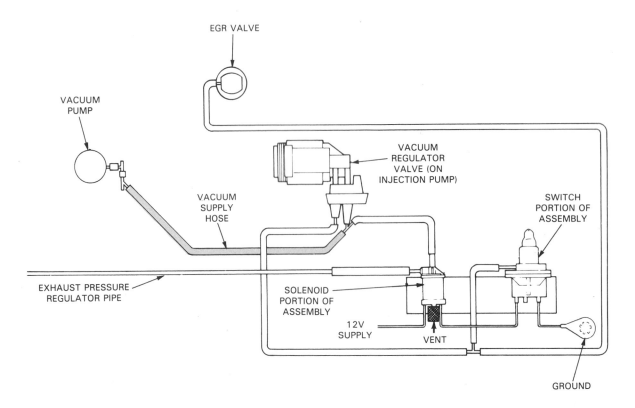

Fig. 10-49. Vacuum circuit is made up of hoses routed to various parts of the system including vacuum regulator and EGR valve. (Oldsmobile)

a distance from the engine. The cleaner consists of a housing, an element, a cover, gaskets, snorkel, air ducts, and other parts for controlling and monitoring air temperature.

A thermostatically controlled air cleaner not only cleans the incoming air but controls its temperature as well. A thermal vacuum switch controls the operation of a vacuum motor. This motor opens and closes an air door (flap) in the snorkel. The thermostatic air cleaner warms inlet air when the engine is cold, and allows cool air to enter the engine in hot weather.

Gasoline engines have a throttle valve which controls air volume into the intake manifold. This action affects fuel metering and engine power. One or more of these valves are normally located in the carburetor or throttle body at the entrance to the intake manifold. They are connected to and controlled by the gas pedal. Cables or rods link the gas pedal to the throttle valves. On diesels, the throttle cable connects to the injection pump.

Intake manifolds are enclosures or ducts that carry air-fuel mixture into the engine cylinder head ports. They are usually made of cast iron or aluminum. Surfaces mating against the engine and air-fuel delivery system are carefully machined for a good airtight fit. The intake manifold has several parts that may vary with design. Runners or passages are for air or air-fuel mixture.

Manifolds must be carefully designed to distribute fuel-air mixture equally to all cylinders. This has lead to development of several different types. They are classified by plenum design, runner length, port or runner configuration, general shape, and method of construction. Types include the dual plane, single plane,

sectional, tuned runner, and jet air intake manifold.

A flow bench is a test device that measures the airflow through various ports of the induction system. Air blown through the system is measured to detect restrictions, which can then be corrected to balance the airflow to each engine cylinder.

To help prevent knock and ping, water may be injected into certain engines. These systems are an aftermarket product. They tend to remove carbon deposits from combustion chambers and reduce peak combustion flame temperature.

Vacuum pumps, common to diesel engines, but also found on some gasoline engines, provide vacuum for various engine accessories. There are two types. Mechanical pumps are powered by the engine. They may be the reciprocating type or the rotary type. Electric pumps have small direct current motors.

KNOW THESE TERMS

Induction system, Air cleaner, Air ducts, Air flow sensor, Throttle valves, Air temperature sensor, Intake manifold assembly, Snorkel, Air cleaner element, Positive crankcase ventilation filter, Thermostatic air cleaner, Thermal vacuum switch, Heat shroud, Throttle cable, Throttle linkage, Throttle return spring, Intake runners, Inlet, Riser bore, Plenum, Exhaust passage, EGR passage, Water jackets, Head mating surface, Dual plane intake manifold, Single plane intake manifold, Sectional intake manifold, Tuned runner intake manifold, Jet air intake manifold, Water injection system, Flow bench, Vacuum pump.

REVIEW QUESTIONS—CHAPTER 10

1. Define an induction system and list its six functions.
2. Indicate which of the following are NOT a part of the induction system:
   a. Air cleaner.
   b. Air ducts.
   c. Air flow sensors.
   d. Air temperature sensor.
   e. Carburetor.
   f. Intake manifold assembly.
   g. PCV valve.
   h. Throttle valves.
3. An induction system is NOT used on a carbureted engine. True or false?
4. Compare the three induction systems—gasoline injection, carburetion, and diesel injection. Indicate their differences.
5. Without a (an) _____ _____, airborne debris can enter the combustion chamber and cause damage to piston rings, pistons, cylinder walls, valves, and other engine parts.
6. An air cleaner can be located at:
   a. Carburetor or throttle body (gasoline engines).
   b. Intake manifold (diesel engines).
   c. Any remote location away from the engine.
   d. All of the above.
   e. None of the above.

MATCHING TEST: On a separate sheet of paper, match the terms with the appropriate statements.

7. ____ Metal or plastic container to hold air cleaner element and support other parts.
8. ____ Device for separating debris from incoming engine air.
9. ____ Snout which runs outward from air cleaner housing and lets in air.
10. ____ Removable lid which allows servicing of element.
11. ____ Flexible tubes that carry air into housing.
12. ____ Seal which prevents air leakage between air cleaner components.

    a. Cover.
    b. Gasket.
    c. Air ducts.
    d. Housing.
    e. Snorkel.
    f. Element.

13. Name and describe the most popular type of air cleaner element used in automobiles today.
14. Explain the operation of a thermostatic air cleaner.
15. The throttle valve (indicate correct answers):
   a. Meters correct amount of fuel into intake manifold.
   b. Controls engine power.
   c. Allows driver to control airflow into a gasoline engine.
16. A throttle valve is found only on a gasoline engine. True or false?
17. List the six functions of the intake manifold.
18. Intake manifolds are usually cast from _____ or _____.
19. The _____ of an intake manifold is a large chamber located below the inlet or bore. Its purpose is to _____ _____ at openings to runners.
20. An intake manifold which distributes air-fuel mixture to cylinders unequally is said to have _____ _____ _____.
21. List four types of intake manifolds.
22. A _____ _____ is a test device used to measure the air flow through various induction system components.
23. The two types of vacuum pumps are _____ and _____.
24. Give the purpose of the vacuum pump on an automobile.

Routine service of air cleaners and other induction system parts is important for proper functioning and efficiency of engines. Factory representative demonstrates service features. (Ford Motor Co.)

11 AIR CLEANER, INTAKE MANIFOLD, VACUUM PUMP— DIAGNOSIS, SERVICE, REPAIR

After studying this chapter, you will be able to:
- *List and diagnose common air cleaner problems.*
- *Test a thermostatic air cleaner.*
- *Properly service an air filter.*
- *Replace faulty air cleaner components.*
- *Troubleshoot intake manifold-related problems.*
- *Service an intake manifold.*
- *Diagnose vacuum pump problems.*
- *Test and service a vacuum pump.*

This chapter discusses the most important topics relating to the maintenance, testing, and service of air cleaners and intake manifolds. If these components are not functioning properly, engine power and efficiency are reduced, while exhaust emissions increase. It is an important chapter that prepares you for on-the-job responsibilities.

AIR CLEANER PROBLEM DIAGNOSIS

Air cleaner problems do not usually cause obvious symptoms. Problems may go unnoticed until major damage has occurred. Typical problems, Fig. 11-1, include:
1. Dirty, restricted air cleaner element.
2. Inoperative door in thermostatic air cleaner.
3. Damaged hot air tube from exhaust manifold.
4. Missing or damaged base gasket between carburetor or throttle body and air cleaner.
5. Leaking or disconnected vacuum hoses on thermostatic air cleaner.

Air cleaner inspection
Begin diagnosis by close inspection of the air cleaner and related components. Remove the air cleaner lid. Look

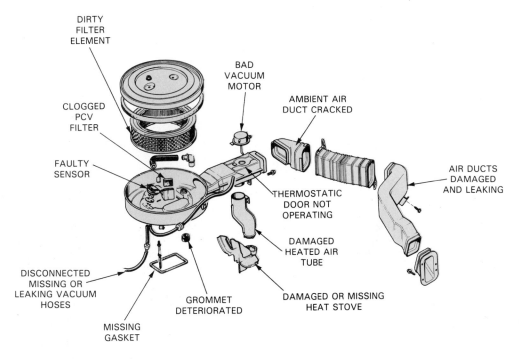

Fig. 11-1. Any of these air cleaner problems can affect engine performance. (AMC)

for any of the troubles just listed.

Check all vacuum hose ends for hardening and cracking. If old, the vacuum hoses may be deteriorated and leaking. The hot and cool air duct hoses should also be checked closely and replaced if damaged.

AIR FILTER ELEMENT SERVICE

Auto manufacturers recommend periodic replacement of the air filter element and the small PCV filter element.

Check service records for the car. You may find a service tag giving the date and mileage of the last air filter change. This could help you decide if the element needs replacement. Always check a service manual for exact service intervals. If the car is driven on dirt roads, the element should be changed more often than recommended intervals. See Fig. 11-2.

With practice, you may be able to use a drop light behind the element to determine if it is dirty. A dirty, clogged element will block more light from passing through than a fairly clean, usable element. This method is only acceptable when service records are not available.

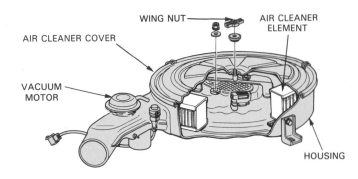

Fig. 11-3. Typical air cleaner design for carbureted engine. Remove wing nut and unhook rim clips to remove cover. New filter element must fit snugly in housing and form a seal against air leakage. (Honda)

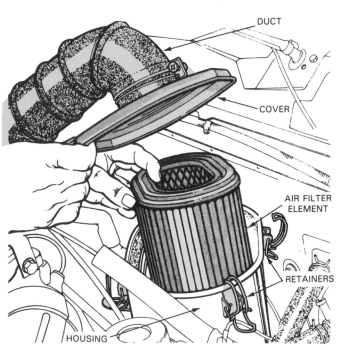

Fig. 11-4. This type of filter housing uses clips to secure cover. Note cool air duct is attached to cover. (Chrysler)

Fig. 11-2. Air filter element should usually be replaced every 30,000 miles (about 48 000 kilometers).

Element removal

To remove the air filter element, remove the cover from the cleaner housing. A wing nut, hex nuts, spring clips, or screws may hold the cover in place. Refer to Figs. 11-3 and 11-4. Lift the element out of the housing.

Use caution not to allow any parts or debris (rocks, leaves, etc.) to drop into the top of the carburetor, throttle body, or air inlet (diesel engine). Engine damage could result. Place a clean shop rag over the air inlet opening when servicing the element.

Element replacement

Before installing the new element, clean the bottom of the air cleaner housing, Fig. 11-5. Remove the housing from the engine. If needed, wash it out in cold-soak cleaner. Blow it dry and then wipe it out with a clean shop rag.

Compare the old and new filter elements. Make sure they are the same size and shape.

If the old filter is in good condition, you can use compressed air to remove light dust, as in Fig. 11-6. Be careful not to hold the air nozzle too close and thus damage the element.

Wash polyurethane foam elements in cold-soak cleaner and squeeze them out. Then carefully work some light motor oil into the foam so that it will trap dust more easily.

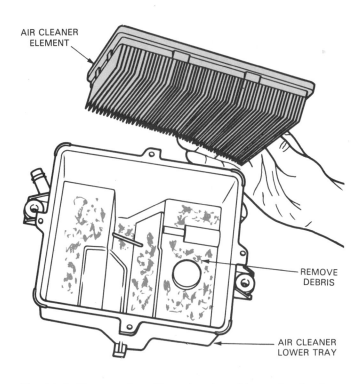

Fig. 11-5. Never replace filter element without cleaning bottom of housing. Debris could be drawn into engine causing damage. (Ford)

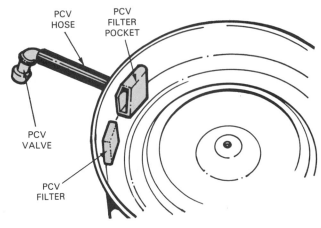

Fig. 11-7. On some vehicles, PCV filter is replaced by sliding new element into pocket inside air cleaner. (AMC)

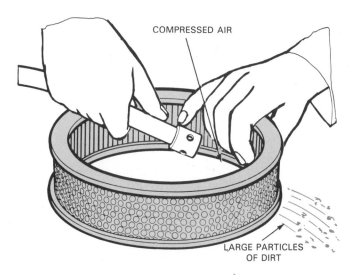

Fig. 11-6. Between service intervals, low pressure compressed air will dislodge dirt from paper filter element. Direct air blasts from inside or in opposite direction from normal intake airflow. (Chrysler)

components are installed properly. This is especially important on late model cars using computer controlled fuel injection. Faulty reconnection of hoses or wires could result in loss of computer control. Also, even a small air leak after the airflow sensor could affect the air-fuel mixture and engine operation. See Figs. 11-8 and 11-9.

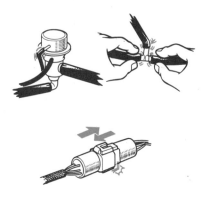

Fig. 11-8. Frequent causes of problems in air intake system include bad connections in wiring or vacuum hoses. Connections must be secure and correct.

When servicing the element, also service the PCV inlet filter, Fig. 11-7. If it is dirty and saturated with oil, replace it.

Check the condition of the air cleaner base gasket. Then, reinstall the housing, element, and cover, making sure the housing is properly aligned. Sometimes there is a notch in the housing opening that must align with a tube or other part on the carburetor or throttle body.

Double-check that all hoses, ducts, wires, and other

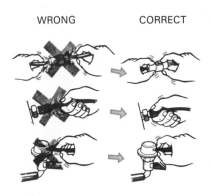

Fig. 11-9. Handle parts carefully. On wiring, pull on connectors only, never on wire. To disconnect vacuum hoses, pull on hose ends, never in the middle. (Toyota)

Look at Fig. 11-10. It shows the air cleaner components for a modern fuel injected, four-cylinder engine. Note how the airflow meter is mounted away from the throttle body.

The ducts between the sensor and throttle body must be airtight. Any leaks could upset engine and fuel injection system operation.

operating temperature, the door should swing open (cool air position), Fig. 11-11B. If not, something is wrong with the system. Further tests are needed.

Checking vacuum hoses

Check all vacuum hoses and hose connections. The hoses deteriorate and harden. Fig. 11-12 shows the

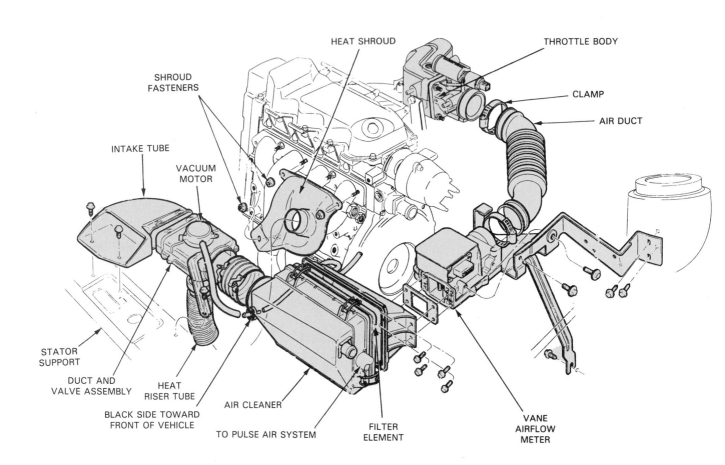

Fig. 11-10. Typical shop manual illustration for servicing air intake system for fuel injected engine. Check proper manual for vehicle being serviced and follow service instructions.

THERMOSTATIC AIR CLEANER SERVICE

The thermostatic air cleaner is important to engine performance. If it is not working properly, gas mileage, engine start-up, and engine power can be affected.

For example, an air cleaner door STUCK in the HOT AIR POSITION, could lead to high speed missing from an over-lean fuel mixture.

A door STUCK in the COOL AIR POSITION could cause stalling and poor power while the engine is warming up. A VACUUM LEAK in any of the air cleaner hoses could be the cause.

Checking thermostatic air cleaner operation

To check thermostatic air cleaner operation, observe the action of the air inlet door. When the engine is cool and first started, the door should swing to the closed (warm air) position, Fig. 11-11A. As the engine reaches

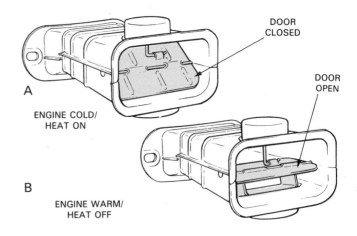

Fig. 11-11. Air cleaner thermostatic door positions. A—When engine is cold, door should be closed. B—Door should be open when engine is at operating temperature. (Ford)

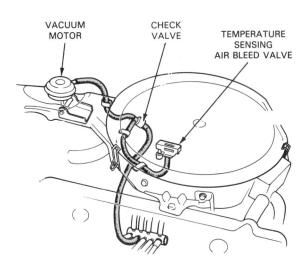

Fig. 11-12. These vacuum hoses should be checked for leaks during inspection of thermostatic air cleaners. (Honda)

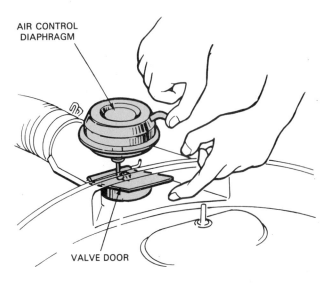

Fig. 11-14. To check vacuum motor, open air cleaner door and close vacuum port to motor with finger. Door should remain open.

typical hoses that should be checked.

Make certain that all hoses are connected to the right fittings. Refer to a shop manual for proper hookup. If a vacuum check valve is used, make sure it has not been installed backwards, Fig. 11-13.

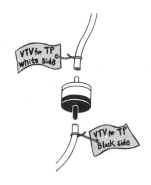

Fig. 11-13. Installation of vacuum valve, if used, must be carefully checked. Make sure it is not installed backward. When removing a valve, tag hoses for proper installation. (Toyota)

Checking vacuum motor

To quickly check the vacuum motor or diaphragm, remove the vacuum hose connected to it. Push up on the air door and then cover the opening in the motor's hose fitting with your finger. Look at Fig. 11-14. Release the door. If it stays up, the diaphragm is airtight and should be functional. If the door swings back down while your finger is making an airtight seal over the opening, the diaphragm is ruptured. Replace it.

A hand-operated vacuum pump can also be used to check the motor. Apply about 12 in. of vacuum (40 kPa) to the motor, as shown in Fig. 11-15. Watch the gauge to make sure vacuum (negative pressure) does not drop and that the door swings closed. If pressure drops or the door does not work, replace the diaphragm assembly.

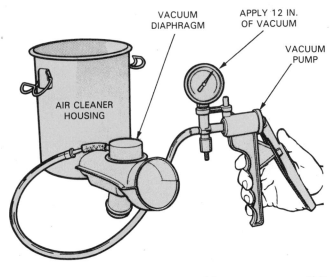

Fig. 11-15. Testing vacuum motor with vacuum pump. If it cannot hold 12 in. of vacuum, replace motor. (Chrysler)

Replacing air cleaner vacuum motor

Most air cleaner vacuum motors, Fig. 11-16, are riveted or spot welded to the air cleaner snorkel. To remove them, drill out the rivets or spot welds, Fig. 11-17. Then disconnect the rod going to the door. Lift off the old unit.

Compare the old and new diaphragm assemblies. Connect the rod. Then pop rivet the new unit onto the snorkel. Check for proper operation.

Air cleaner temperature sensor service

A *faulty bimetal sensor* can keep the air cleaner door from functioning, Fig. 11-18. It can cause the door to stay in either the cool or hot air positions. If your vacuum motor is good but the door does not function, check the temperature sensor.

To test the sensor, cool it down using an ice cube. A

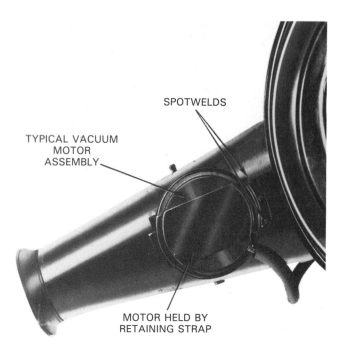

Fig. 11-16. Typically, vacuum motors are mounted on snorkel. (Chevrolet)

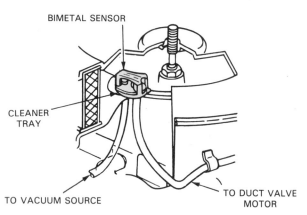

Fig. 11-18. Sensor is located where it is subjected to temperature changes inside air cleaner. At given increase in temperature, sensor bleeds off vacuum allowing vacuum motor to open duct door. (Ford)

open, allowing airflow, above this temperature. However, before condemning the sensor, check the service manual specs for an exact temperature.

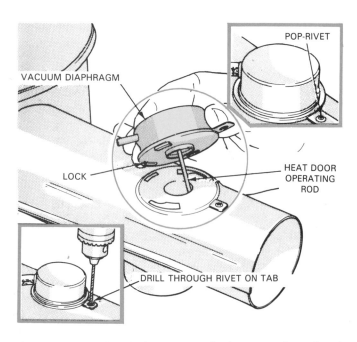

Fig. 11-17. Nonfunctioning vacuum diaphragm can be replaced as shown. Make sure new diaphragm is connected to operating rod. (Chrysler)

cold sensor should pass vacuum. To check the sensor, use a vacuum pump or your mouth.

Next, use a heat gun to warm the sensor. This should make the sensor valve close, blocking vacuum. Again, use the vacuum pump or your mouth to check the sensor valve.

Typically, the valve should block airflow when the temperature is below about 75°F (24°C). It should

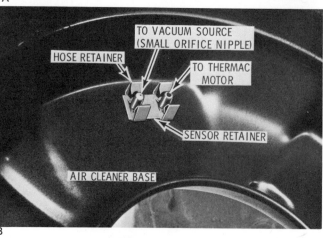

Fig. 11-19. Replacing faulty bimetal (temperature) sensor on air cleaner housing. A—Remove clips holding sensor to housing. B—Install new sensor and make proper hookup of vacuum hoses. (General Motors)

To remove the sensor, you must normally pry off a small clip on the bottom of the air cleaner housing. This is shown in Fig. 11-19A. After installing the new sensor, make sure the vacuum hoses are connected properly. Fig. 11-19B shows a service manual illustration for one type of car.

When reinstalling the air cleaner, double-check all hoses, wires, ducts, gaskets, and electrical wires.

INTAKE MANIFOLD SERVICE

Typical *intake manifold problems* include leaking intake-to-cylinder head gaskets, loose bolts, cracks, warpage, burned out heat crossover passage, etc. Fig. 11-20 illustrates some common troubles.

Inspecting intake manifold

Inspection of the intake manifold and related components will, sometimes, uncover developing troubles. Look for leaking vacuum hoses, loose or missing fasteners, or a stuck EGR valve.

As was explained in earlier chapters, a piece of vacuum hose held next to your ear can be used as a listening device. Move the free end around the intake.

When the hissing noise is loudest, you have found the vacuum leak. See Fig. 11-21.

Fig. 11-21. Using piece of vacuum hose to find vacuum leaks. Place one end next to ear; move free end around intake system. (Texaco)

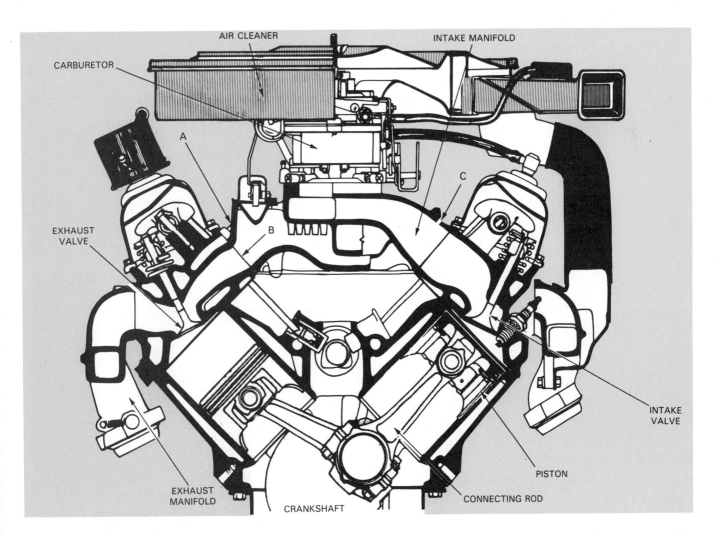

Fig. 11-20. Common intake manifold problems. A—Loose bolts connecting intake manifold to cylinder heads. B—Leaking gaskets or warpage between intake manifolds and cylinder heads. C—Cracks in intake manifold.

Vacuum gauge test for intake leaks

A vacuum gauge will help you find major intake manifold vacuum leaks. Connect the gauge to a vacuum fitting on the manifold, Fig. 11-22. Start and idle the engine.

At idle, a gasoline engine should have about 17 to 21 in./hg. (7 kPa) of vacuum. This can vary slightly, so check specs. If engine vacuum is steady and low and the engine idles roughly, it may indicate an intake manifold gasket leak.

Oil test for vacuum leaks

You can also use an oil can to check for intake manifold

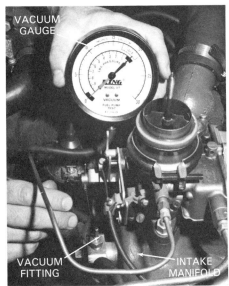

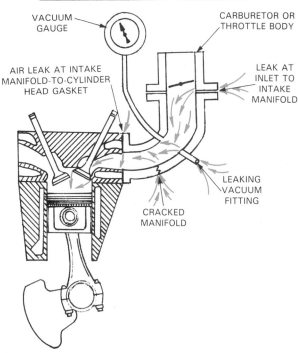

Fig. 11-22. Vacuum leaks rob engine of power and source must be found. Top.Leak can be detected by attaching vacuum gauge to one of vacuum hose fittings or manifold. (Champion) Bottom.Usual sources of manifold leakage.

gasket leaks. Squirt oil along the edge of the intake gasket with the engine idling. If the idle smooths out temporarily, the intake gasket is leaking.

Multiple-piece intake manifold leaks

Many modern intake manifolds for fuel injected gasoline engines are of two and three-piece construction. This increases the number of points that can develop vacuum leaks.

Fig. 11-23 shows an intake manifold for an engine with multi-point gasoline injection. A vacuum leak could upset the injection system by confusing the intake manifold pressure sensor, airflow sensor, oxygen sensor, and other system sensors. Keep vacuum leaks in mind when troubleshooting electronic fuel injection system troubles.

INTAKE MANIFOLD REMOVAL

Intake manifold REMOVAL is simple if you follow a few basic rules. These include:

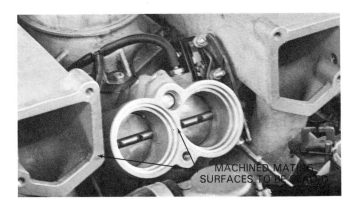

Fig. 11-23. Partially disassembled intake manifold for an engine with multi-point gasoline injection. Many surfaces must be sealed to prevent vacuum leakage. (Volvo)

1. Label all wires and vacuum hoses. Use masking tape or tags. Write down how the wires or hoses connect to each component. This can save time and trouble when trying to reconnect them. If you cross two wires or vacuum hoses, it can upset engine operation and cause problems that are very difficult to diagnose. See Fig. 11-24.
2. Carefully disconnect wiring. Some have special locking connectors. Do not try to force them off. Use service manual methods to release the connectors and prevent wiring damage.
3. Bleed fuel pressure before disconnecting fuel lines. This is especially important on gasoline injection systems that retain pressure with the engine shut off.
4. Cap off fuel lines as they are disconnected to keep out debris and prevent system contamination. Tape over the intake manifold air intake, Fig. 11-25.
5. Keep all fasteners organized by length. Also, if stud bolts are used in a few holes, remember where they

Fig. 11-24. Carefully label all wires and hoses before removal of intake manifold system.

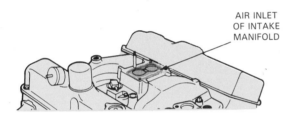

Fig. 11-25. During disassembly of intake manifold system, tape over air inlet.

install. This will save time upon reassembly.

6. Do not pry on machined sealing surface of an intake manifold. If you gouge the surface, a vacuum leak may result.
7. Lift the intake off the engine carefully without striking it against other parts or dropping it. See Fig. 11-26.
8. Cover the intake ports in the cylinder heads with rags, Fig. 11-27. This will help prevent anything from falling into the port where it could be drawn past the valve and into the combustion chamber. If anything falls into the combustion chamber, you will probably have to remove the cylinder head to retrieve it.
9. Immediately clean off all gasket materials. They will usually scrape off more easily when the parts are still a little warm.

INTAKE MANIFOLD INSTALLATION

To install an intake manifold, follow these rules and refer to a service manual.

1. Make sure all mating surfaces are clean and true. Use a straightedge and feeler gauge to check for surface warpage. Warped mating surfaces need milling.
2. Obtain the correct intake gaskets. Place the gaskets over the intake. Make sure all ports, oil holes, coolant passages, and bolt holes align properly.

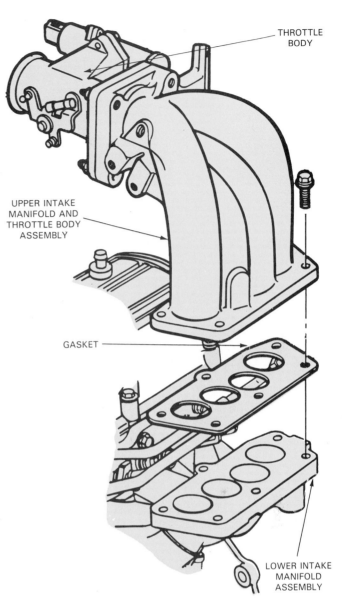

Fig. 11-26. One type of two-part intake manifold. Lower portion attaches to cylinder head. Handle carefully and make sure all joints are leakproof. (Ford)

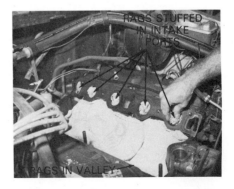

Fig. 11-27. Rags stuffed into intake ports will prevent dirt or small parts from falling into intake and damaging engine. (Edelbrock)

3. If studs are used in the cylinder head, Fig. 11-28, place the gasket over the studs. You may have to use gasket adhesive or RTV sealer to hold the gasket in place on some applications.

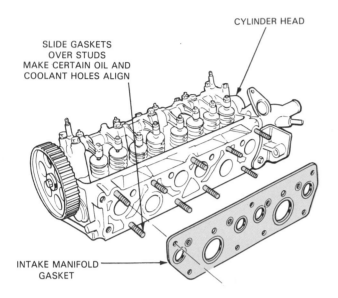

Fig. 11-28. Some engines have studs which make gasket installation easier. (Honda)

4. RTV or silicone sealer is normally recommended where two gaskets come together. As shown in Fig. 11-29, run a bead of sealant at the joint between the cylinder head and the valley gaskets on V-type engines.
5. Carefully position the intake manifold. Be careful not to bump and damage the intake gasket(s), Fig. 11-30.
6. Start all of the fasteners by hand before tightening down any of them. This will allow you to shift the intake slightly while starting each bolt.

Fig. 11-29. Bead of RTV sealant is run on valley gaskets of V-type engines to prevent oil seepage. This type of sealant is recommended where two different gaskets join. (Fel-Pro Gaskets)

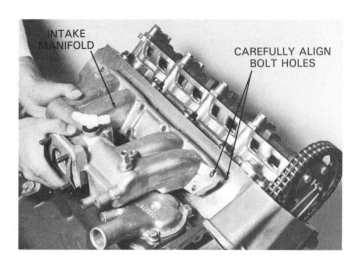

Fig. 11-30. Install intake manifold with care so gasket is not damaged. (Saab)

7. Make sure all bolts are located correctly according to their length. Also, if studs are used in a few holes, position them properly. They may be used to hold other parts. Refer to Fig. 11-31.
8. Torque the intake bolts to specs in the prescribed sequence. Generally, start in the center. Use a crisscross pattern, working your way outward, as in Fig. 11-32.
9. Install any remaining components. Start and warm engine. Retorque intake fasteners.

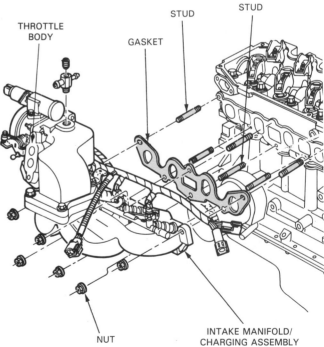

Fig. 11-31. Locate all studs in proper holes. They may be of different lengths, especially if some will hold additional parts. (Ford)

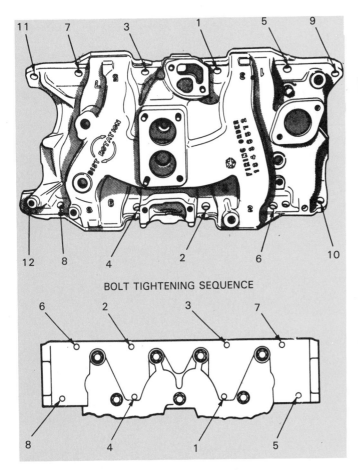

BOLT TIGHTENING SEQUENCE

Fig. 11-32. Follow recommended tightening patterns. Top. V-type intake manifold. Bottom. In-line engine intake manifold. (Chrysler and Ford)

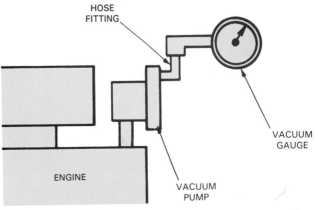

Fig. 11-33. Vacuum pump output is checked with a gauge.

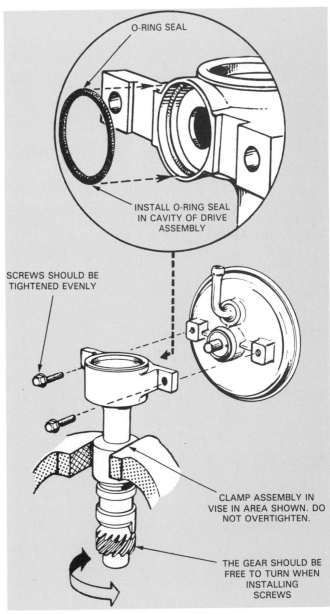

Fig. 11-34. Follow service manual instructions for proper vacuum pump service. (General Motors)

WARNING! When installing any engine component, do not use too much silicone sealer or other sealer. The sealer could drip into the engine. Chunks of rubber-like sealer could then circulate through the engine, possibly blocking oil passages or locking the oil pump.

Silicone sealer can also contaminate the oxygen sensor (exhaust sensor). Fumes from the sealer can be pulled into the engine cylinder, burned, and blown out the exhaust. This can coat the oxygen sensor and ruin it.

VACUUM PUMP SERVICE

Vacuum pump problems usually show up when vacuum-operated accessories fail to function. The diaphragm in the vacuum pump can rupture or check valves can stick. This will reduce or stop pumping action in the vacuum pump. Power brakes may not work properly or A/C and heater controls may fail to operate normally.

Vacuum pump testing

To test a vacuum pump, simply connect a vacuum gauge to the outlet fitting on the pump. Start the engine and read the gauge. Look at Fig. 11-33.

If vacuum output is low, use a hand-operated vacuum pump to apply vacuum to the unit. If vacuum drops, the diaphragm or a check valve is leaking. If the pump holds a vacuum, the drive gear pin on the pump shaft may be sheared. Remove the pump and inspect the gear and pin.

Vacuum pump replacement

Sometimes, the diaphragm chamber on the vacuum pump can be removed and replaced separately. As in Fig. 11-34, remove the bolts holding the chamber to the pump body. Install the new diaphragm chamber and check pump operation.

With other vacuum pump designs, the complete pump must be replaced. Fig. 11-35 shows how a front-mounted vacuum pump is mounted on the engine.

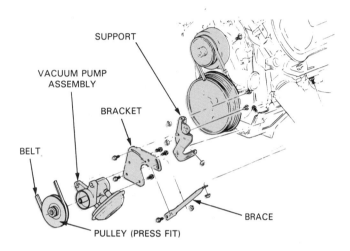

Fig. 11-35. Exploded view shows mounting arrangement for front-mounted vacuum pump. (Oldsmobile)

SUMMARY

Proper operation of all parts of the air intake system is important to the efficient operation and life of an automobile engine.

Air cleaner problems may include a dirty element, clogged PCV valve, hardened, cracked, missing or disconnected vacuum hoses, and missing gaskets.

Follow manufacturer's recommended intervals for changing the air filter element and the PCV filter element. When removing parts, use caution. Parts dropped into the carburetor throat, throttle body, or air inlet and into the intake manifold may cause severe damage if drawn into the engine. Removing them can be difficult and time-consuming.

Air filter element service includes cleaning dry elements with air pressure or replacing them. Some reusable elements can be washed, re-oiled, and replaced. PCV filters become oil clogged and should be replaced.

Inspect vacuum hoses for proper routing. Use a vacuum gauge or a section of vacuum hose to check for vacuum leaks.

Check the thermostatic air controls for proper operations. When a cool engine is started, the air inlet door

should close to allow warm air into the intake manifold. When the engine reaches operating temperature, the door should swing open to allow entry of cool air.

Faulty bimetal sensors may cause air cleaner doors to malfunction. They can be tested with ice cubes and a heat gun.

Typical intake manifold problems include leaking gaskets, loose bolts, warpage, cracks, and burned out heat crossover passages. A vacuum gauge will help in locating major leaks but the mechanic can also use a section of vacuum hose as a listening device to locate vacuum leaks.

Leaking gaskets where manifold joins the cylinder head can be located by squirting engine oil along the edge of the gasket. If engine roughness temporarily disappears, the gasket is leaking.

Some modern automobiles have two and three-piece intake manifolds which increase the points at which leakage may occur. On electronic fuel injected engines, vacuum leakage may upset the system by confusing the sensors.

When removing an intake manifold, carefully label all wires. Disconnect wires to avoid damaging the connectors. Bleed off fuel pressure. Cap disconnected fuel lines. Keep fasteners organized by length. Avoid prying on machined surfaces. Cover all exposed ports. Clean off gasket materials immediately.

To install an intake manifold, make sure that all mating surfaces are clean and true. Make sure you have the right gaskets. If studs are used, slide the gaskets over them. Where two gaskets join, use RTV or silicone sealer. Start all fasteners by hand. Make sure bolts are located correctly according to length. Torque all fasteners to specs using a crisscross pattern. When all components are installed, start and warm the engine; then, retorque intake fasteners.

When vacuum-operated accessories fail to work, the vacuum pump may be at fault. To test, attach a vacuum gauge to the pump outlet. If output is low, test further. Failure to hold vacuum indicates a bad diaphragm or check valve.

KNOW THESE TERMS

PCV filter element, Bimetal sensor, Heat gun, Heat crossover passage, Warpage, RTV sealer, Silicone sealer, Vacuum gauge.

REVIEW QUESTIONS—CHAPTER 11

1. Why is it important to keep air cleaners and intake manifolds in good operating condition?
2. List five common problems with air cleaners.
3. Service records of the vehicle are of no concern when servicing the air cleaner. True or false?
4. To pull apart electrical connectors:
   a. Pull on the wires just behind the connector.
   b. Pull on the ends of the connector itself.
   c. Either of the above.
5. Describe how to check a thermostatic air cleaner for proper operation.
6. The air inlet door of a thermostatic air cleaner is found to be inoperative.

Mechanic A says the problem could be a leaking vacuum hose between the vacuum motor and the vacuum source.

Mechanic B says that a damaged heat riser tube is the problem.

Who is right?
a. Mechanic A.
b. Mechanic B.
c. Both Mechanic A and B.
d. Neither Mechanic A nor B.

7. Regarding Question 6, suggest what procedure should be followed to discover the source of the trouble.

8. Vacuum motors are riveted or spot welded to the air cleaner housing. To replace them, you must:
a. Replace the entire air cleaner housing.
b. Cut them off with a welding torch and weld on new vacuum motor.
c. Drill out the rivets or welds and pop rivet on new vacuum motor.

9. List five common problems with intake manifolds.

10. At idle, a gasoline engine should have about:
a. 17 to 21 in./hg. of vacuum.
b. 12 to 16 in./hg. of vacuum.
c. 10 to 14 in./hg. of vacuum.
d. None of the above.

11. Describe a quick method of checking for a leak in an intake manifold gasket.

12. What problem is more likely to occur with a two- and three-piece intake manifold?

13. Why should all wires and vacuum hose be labeled during removal of an intake manifold?

14. Give the general rules for installing an intake manifold.

15. During installation of an intake manifold, silicone sealer or other types of sealer should be used sparingly because:
a. If too much is used, a good seal is not possible.
b. It can drip into the engine and possibly cause problems with oil circulation.
c. Silicone sealers can contaminate the oxygen sensor if they get into the combustion chamber and are burned.
d. All of the above.
e. None of the above.

16. A vacuum test on a vacuum pump which has a low vacuum reading indicates that the pump cannot hold vacuum applied with a hand-operated vacuum pump.

Mechanic A says that the problem could be a ruptured diaphragm.

Mechanic B says that the problem could be a leaking check valve.

Who is right?
a. Mechanic A.
b. Mechanic B.
c. Both Mechanic A and B.
d. Neither Mechanic A nor B.

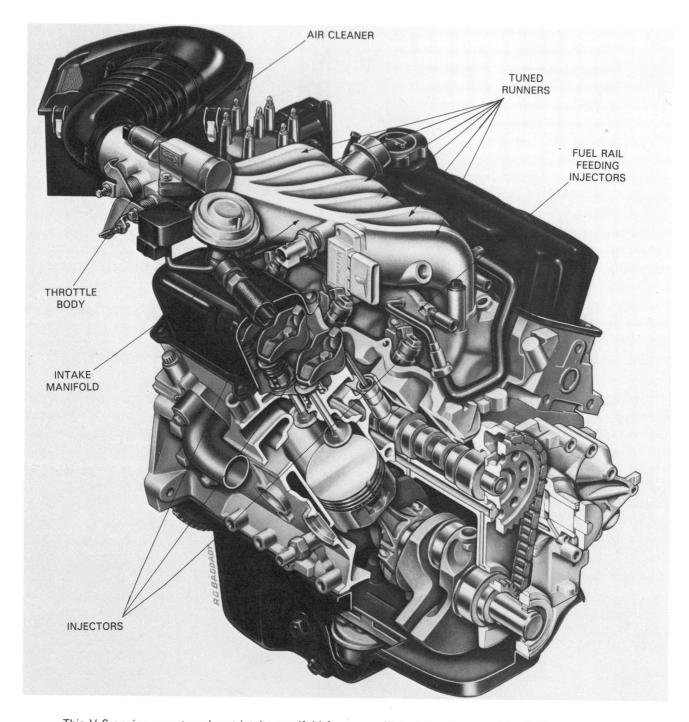

AIR CLEANER

TUNED
RUNNERS

FUEL RAIL
FEEDING
INJECTORS

THROTTLE
BODY

INTAKE
MANIFOLD

INJECTORS

This V-6 engine uses tuned-port intake manifold for more efficient "engine breathing." (Ford Motor Co.)

FUNDAMENTALS OF CARBURETION

After studying this chapter, you will be able to:
- *Explain the relationship between air-fuel ratios and engine performance.*
- *Describe and identify the basic parts of a carburetor.*
- *Compare carburetor design differences.*
- *List and explain the fundamental carburetor systems.*
- *Explain special carburetor devices.*
- *Describe the operation of computer controlled carburetors.*

In this chapter, you will learn about basic carburetor parts and their operation. You will discover how each of these parts make up the seven carburetor circuits. Then, you will study the purpose and action of these circuits. Finally, you will learn how on-board computers are linked to carburetors to control them.

NOTE! Chapters 1 through 5 on fuels, combustion, and physical principles should have been read BEFORE studying this chapter. The material covered in these earlier chapters is essential for complete understanding of carburetion.

PROPER FUEL MIXTURE NEEDED

To burn properly in an engine, gasoline must be broken into tiny droplets and mixed with air (oxygen). Liquid fuel will NOT burn with the explosive force needed for proper combustion in an engine.

AIRSTREAM ACTION

On each intake stroke of an engine, the intake valve opens while the piston is sliding down in the cylinder. As a result, a VACUUM (suction) is formed in the cylinder, intake manifold, and carburetor, Fig. 12-1. Outside air pressure rushes into the engine to fill this vacuum. The force of this airstream is used by the carburetor to mix fuel and air.

ATOMIZATION OF FUEL

Atomization means breaking up liquid fuel into tiny droplets. For instance, a pump type perfume bottle is sometimes called an atomizer. When the bulb is squeezed, Fig. 12-2, a fine mist of tiny drops sprays out.

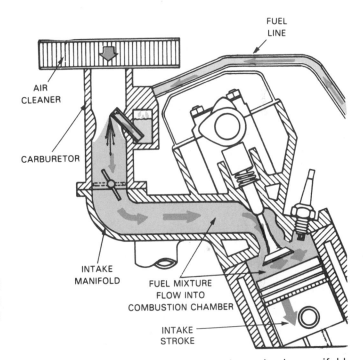

Fig. 12-1. Airstream is caused by engine or intake manifold vacuum (suction). Piston sliding down on intake stroke produces low pressure area in cylinder. Air above carburetor (atmospheric pressure) then flows through carburetor and manifold to fill vacuum. During engine operation, vacuum and air flow are continuous.

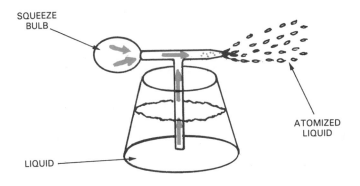

Fig. 12-2. Principle of simple atomizer is same as carburetor. Squeezing bulb creates airstream over tube, producing vacuum at top of tube which pulls liquid up tube and into airstream. Liquid, broken into tiny particles, comes out as fine mist.

VAPORIZATION OF FUEL

Vaporization is the process of changing liquid or atomized fuel into a VAPOR. For example, steam rising off boiling water is vaporized water, Fig. 12-3.

Automotive fuel must first be atomized and then vaporized. Several factors affect fuel vaporization:

1. Outside or inlet air temperature. Warm air improves vaporization.
2. Engine temperature. Vapors can condense on cold engine parts and not ignite.
3. Engine and airstream speed (more air moving faster improves vaporization).
4. Air turbulence and swirl improves vaporization.
5. Fuel characteristics (some fuels are more volatile).
6. Fuel discharge technique.

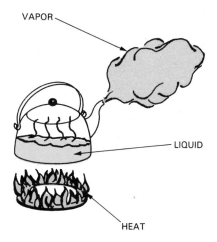

Fig. 12-3. In process of vaporization, liquid is changed to a gas or vapor. Steam is a vapor consisting of fine particles of moisture capable of floating and mixing with air.

AIR-FUEL MIXTURE RATIO

Air-fuel mixture ratio refers to how much fuel, and how much air are mixed together. This is also referred to as air-fuel proportion.

For proper combustion and engine performance, certain proportions of air and fuel are necessary. At highway speeds, for instance, an air-fuel ratio of 15:1 (15 lb. of air mixed with every 1 lb. of fuel) is ideal.

A *lean fuel mixture* is one containing a high ratio of air. For example, 20:1 is a very lean mixture. Oxygen is one of the major limiting factors of engine power. See Fig. 12-4. Normally, a *rich, fuel mixture* of around 12:1 produces more engine power than a lean mixture.

Why air-fuel ratios must vary

Engine speed, temperature, load, outside air temperature, altitude, engine design, all change the ideal fuel mixture ratio. The carburetor must adjust air-fuel ratios to meet engine demands as conditions change. Fig. 12-5 shows air-fuel ratios for maximum efficiency under different operating conditions.

EFFECTS OF VARYING FUEL MIXTURE	
CONDITION	**RESULTS**
TOO LEAN	POOR ENGINE POWER MISSING, ESPECIALLY AT CRUISING SPEEDS BURNED VALVES BURNED PISTONS SCORED CYLINDERS SPARK KNOCK OR PING
SLIGHTLY LEAN	HIGH GAS MILEAGE LOW EXHAUST EMISSIONS REDUCED ENGINE POWER SLIGHT TENDENCY TO PING OR KNOCK
STOICHIOMETRIC	BEST ALL-AROUND PERFORMANCE
SLIGHTLY RICH	MAXIMUM ENGINE POWER HIGHER EMISSIONS HIGHER FUEL CONSUMPTION LOWER TENDENCY TO KNOCK OR PING
TOO RICH	POOR GAS MILEAGE MISSING INCREASED AIR POLLUTION FOULED SPARK PLUGS OIL CONTAMINATION BLACK EXHAUST

Fig. 12-4. Proper fuel mixture is very important to good engine performance.

ENGINE OPERATING CONDITION	NORMAL AIR-FUEL RATIO
Idle or low speeds	Full power
Midrange speeds	Acceleration
Cruising speeds	Cold starting
12:1	13:1
16:1	12:1
17:1	8:1

Fig. 12-5. Air-fuel mixtures for best engine operation vary for different operating conditions.

Stoichiometric fuel mixture

A chemically perfect fuel mixture is called a *stoichiometric fuel mixture.* It is a ratio calculated to be around 14.9:1. Under perfect conditions of engine speed, load, temperature, fuel distribution to each cylinder, etc., a stoichiometric mixture would provide the best compromise for greatest combustion power and efficiency.

BASIC CARBURETOR

A *carburetor* is a device which uses hydraulic, mechanical, and vacuum devices for metering and mixing gasoline with air. It must make a good combustible mixture for the engine. Normally, it is bolted to the top of the engine intake manifold. It is covered by an air cleaner. See Fig. 12-6.

The carburetor has five basic functions:

1. Atomize the fuel into droplets.
2. Mix fuel and air.
3. Meter the correct amount of fuel into the airstream.
4. Control airflow into the engine and regulate engine speed and power.
5. Control vacuum operated devices such as the distributor vacuum advance, charcoal canister valve, and more.

Parts of carburetor

If we first look at a simple carburetor, understanding more complicated modern ones will be much easier. Refer to Fig. 12-6 as you study the list of parts.

1. AIR HORN or CARBURETOR THROAT. This is simply a large opening that passes through the carburetor body. All air entering the engine must pass through it.
2. FLOAT BOWL. Also called a *fuel bowl,* this is a storage area for fuel before being drawn into the air horn. Normally, it is part of the carburetor body.
3. FLOAT ASSEMBLY. Designed to control fuel entering the fuel bowl, this mechanism consists of a hollow float, a hinge, needle valve, and needle seat. As it rides on top of the fuel, the float opens and closes the needle valve.
4. BOWL VENT. This is an opening in the bowl to prevent pressure or vacuum from building up in the bowl. Often, the vent opens into the air horn.
5. MAIN DISCHARGE PASSAGE. This tube or opening connects the bottom of the fuel bowl with the center of the air horn. It allows fuel to move from the bowl to the air horn during high speed operation where the fuel mixes with air rushing through to the engine.
6. MAIN DISCHARGE JET. This small brass fitting, with a carefully sized hole, fits over the lower end of the main discharge tube. It controls how much fuel moves into the main discharge tube.
7. VENTURI. This is a narrowed portion of the air horn intended to create a vacuum as incoming air is speeded up, stretched, and swirled by the restriction.
8. SECONDARY or BOOSTER VENTURI. This is a smaller restriction in the middle of the primary venturi. It increases air speed, vacuum, and fuel flow.
9. THROTTLE VALVE. This round metal disc is mounted on a shaft running through the bottom of

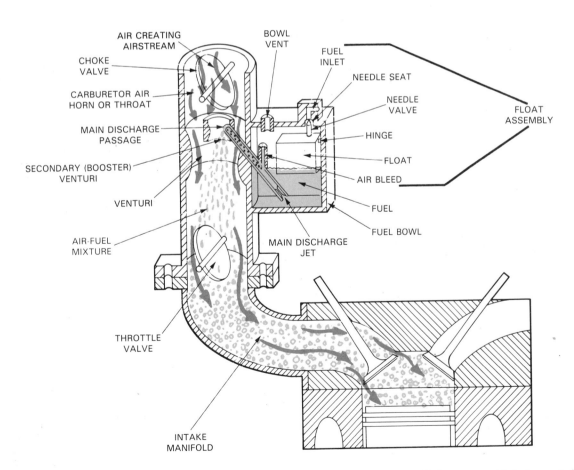

Fig. 12-6. All carburetors operate according to basic principles displayed by this simplified cutaway. Note how air-fuel vapors travel through intake manifold and into combustion chamber.

the carburetor air horn, Fig. 12-6. It rotates over 90 degrees to control airflow through the carburetor.

10. THROTTLE LINKAGE. Rods or cable connecting the throttle valve to the gas pedal make up this control mechanism. See Fig. 12-7.

11. CHOKE VALVE. This simple butterfly valve is attached to a shaft at the top of the air horn. It richens the fuel mixture for cold engine starting and operation by blocking airflow. Choke valves may be operated manually or automatically.

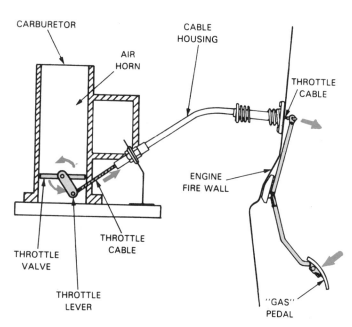

Fig. 12-7. Throttle linkage connects "gas pedal" to carburetor's throttle plate to control engine speed and power.

MODERN CARBURETOR SYSTEMS

The simple carburetor just described may function quite well on a single-cylinder engine, but not on today's automobile engine. Into and onto the carburetor body several carburetor SYSTEMS or CIRCUITS have been added. These make up a network of fuel passages, air passages, and related parts that provide the best possible fuel mixture during a specific engine operating condition.

There are seven basic carburetor systems:

1. FLOAT SYSTEM (maintains supply of fuel in carburetor bowl).
2. IDLE SYSTEM (provides a small amount of fuel for low speed engine operation).
3. OFF-IDLE SYSTEM (provides correct air-fuel mixture slightly above idle speeds).
4. ACCELERATION SYSTEM (squirts fuel into air horn when throttle valve opens and engine speed increases).
5. HIGH SPEED SYSTEM (supplies lean air-fuel mixture at cruising speeds).
6. FULL POWER SYSTEM (richens fuel mixture slightly when engine power demands are high).

7. CHOKE SYSTEM (provides extremely rich air-fuel mixture for cold engine starting).

Each of the carburetor circuits is easier to understand when studied separately. It is essential that you make a "mental picture" of the systems. Why are they needed? How do they work? What parts are used? This knowledge will make carburetor troubleshooting and repair easier.

FLOAT CIRCUIT

The *float circuit* is contained entirely within the fuel bowl. This system includes a hollow float, needle valve, needle seat, and hinge assembly. See Fig. 12-8.

Fuel bowls are generally cast as part of the carburetor body, often being placed in front of the air horn. This assures that fuel will not splash away from the main discharge jet during hard acceleration, starving the engine for fuel. Fuel bowl shapes, locations, and sizes vary considerably, as you will learn in the next chapter.

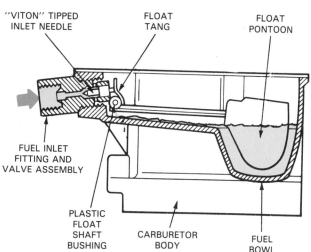

Fig. 12-8. Float circuit controls fuel level in carburetor bowl. Fuel passes through fitting, inlet needle, and drains into bowl. When bowl is full, float pushes needle closed and stops fuel entry. (Ford)

Float

A float is generally made of thin metal or plastic, Fig. 12-9. Since it is hollow, it is light and rests on top of the fuel. HANDLE WITH CARE; It is very fragile. If ruptured, dented, or damaged, it will leak, sink, and flood the engine.

Needle and seat

Inlet needle valves, usually made of brass, are attached to the float. Brass is easily machined and resists corrosion. The needle valve regulates the amount of fuel passing through the fuel inlet and needle seat.

Sometimes, the needle valve will have a soft viton rubber tip which seats better than a metal tip, especially if dirt lodges in the seat. The needle seat is a brass fitting that threads into the carburetor body.

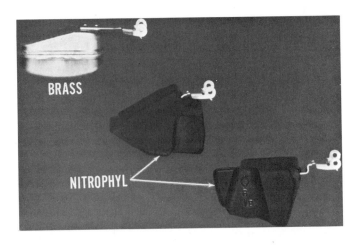

Fig. 12-9. Carburetor floats are constructed to be extremely lightweight. They are usually made of hollow, thin brass or nitrophyl, a plastic.

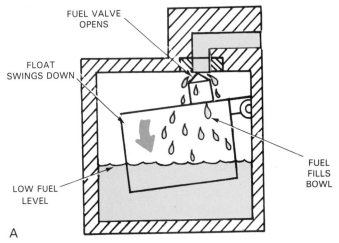

A

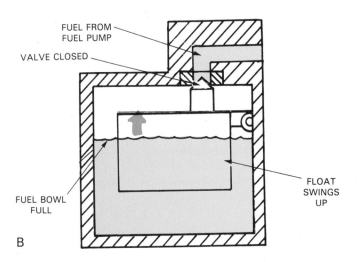

B

Fig. 12-10. Float operation. A—When fuel level is low in bowl, float drops and fuel valve moves off its seat allowing fuel to enter. B—When bowl is filled, needle moves back into inlet shutting off fuel flow.

Float system operation

When engine speed or load increases, fuel is rapidly drawn out of the fuel bowl. Refer to Fig. 12-10. As the fuel level and float drop, the needle valve moves off its seat. Fuel, under pressure from the fuel pump, sprays into and refills the bowl. As the fuel level rises, the float moves up and pushes the needle back into the seat. When enough gasoline has entered, float action presses the needle tightly into the seat, stopping flow completely.

Float level settings are very critical to carburetor operation. If either too high or too low, ALL of the other carburetor circuits can be adversely affected.

A high float setting allows gasoline to overfill the bowl and run out of the top of the carburetor onto the hot engine. Sometimes, the fuel will pour into the air horn and flood the engine. Engine missing, stalling, no-start, fluttering, and black exhaust smoke conditions result.

A too-low float setting can cause a lean or no-fuel situation. There may be too little fuel in the bowl to supply the needs of the carburetor circuits. The engine performs poorly, especially at higher speeds. It may miss, stumble, or even stop.

Bowl venting

Bowl venting, mentioned earlier, prevents a vacuum or pressure buildup as fuel leaves or enters the bowl. Refer to Fig. 12-11. Modern automobiles equipped with evaporation control type emission systems vent the fuel bowl into a charcoal canister.

IDLE SYSTEM

At very low engine speeds (below around 750 rpm or 20 mph), the *idle circuit,* Fig. 12-12, provides the correct air-fuel ratio. With the throttle valve closed, air CANNOT rush through the carburetor throat. Thus, venturi vacuum is not strong enough to pull fuel out of the main discharge passage. The high vacuum below the closed throttle plate must be used.

The idle circuit is a fuel passage that branches off the

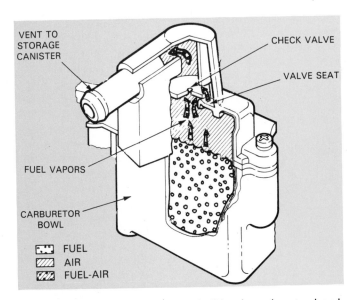

Fig. 12-11. Pressure or vacuum buildup in carburetor bowl could cause fuel starvation or flooding of engine. Venting vapors to charcoal canister prevents these conditions. (Ford)

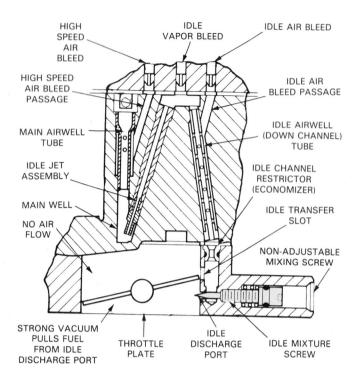

Fig. 12-12. Elements of idle circuit. Fuel is routed to enter carburetor throat below throttle valve at idle discharge port. (Ford)

main discharge passage, runs through a low speed jet, an economizer (restriction in passage), and down to the idle port. Beyond the idle screw restriction, it opens into the air horn just below the throttle valve. See Fig. 12-13.

Idle system operation

During idle system operation, the throttle plate is closed. Intake manifold vacuum, below the butterfly, sucks gasoline from the idle screw port. Fuel flows out the float bowl and through the low speed jet. This jet limits maximum fuel flow.

At the *bypass* (an idle air bleed port), outside air is pulled into the system. This partially atomizes the fuel. The amount of air entering is controlled by the bypass diameter. The bypass helps control flow and mixes tiny air bubbles with the liquid fuel. More air bubbles enter at the off-idle ports. This helps cut fuel consumption at idle speeds. Finally, the air bubbles and fuel pass the idle mixture screw and empty into the air horn.

Idle mixture screw

The *idle mixture screw,* Fig. 12-14, is fitted into the idle port. It can be adjusted to control the EXACT AMOUNT OF FUEL entering the engine at idle. Normally, turning the idle mixture screw IN leans the mixture and OUT richens the mixture.

Most new carburetors do not have ADJUSTABLE idle mixture screws. This helps prevent misadjustment and air pollution. See Fig. 12-15.

OFF-IDLE SYSTEM

The *off-idle circuit* consists of either holes or a slot in the air horn just above the throttle plate, Figs. 12-14 and 12-16. It is an extension of the idle circuit intended to provide extra fuel during transition from idle operation to higher speeds. Also called the IDLE TRANSFER PASSAGES, they open into the idle system fuel passage.

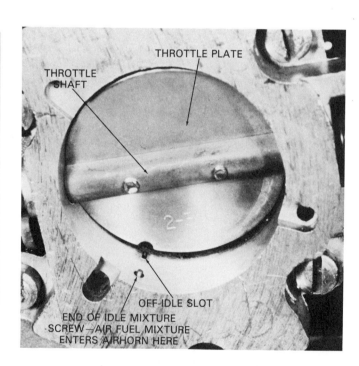

Fig. 12-13. Study parts and fuel flow in this basic idle circuit. Understanding idle circuit operation is very important to study of carburetion. Air mixes with fuel in idle tube. This assists fuel atomization and vaporization and engine efficiency at idle. Idle anti-syphon bleed prevents fuel syphoning (fuel continues to flow out of circuit when engine is shut off). (Ford)

Fig. 12-14. Bottom view of carburetor bore shows off-idle fuel openings. Below off-idle slot is end of mixture screw. Notch in throttle plate helps control both idle and off-idle circuit transition. (Carter Carburetor Div.)

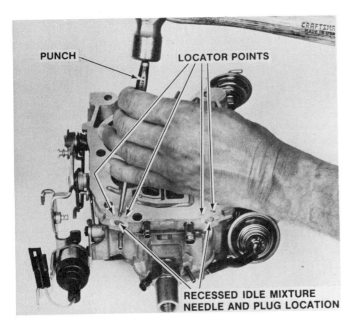

Fig. 12-15. On modern carburetors, idle mixture screws are covered with metal plugs to prevent tampering. Changing factory setting could upset mixture and increase exhaust emissions. (General Motors)

tional openings are EXPOSED to intake manifold vacuum. Then, they too begin to supply gasoline. As more air rushes past the opened butterfly, the added fuel is necessary to maintain correct air-fuel ratio.

The main discharge tube cannot provide fuel until a larger quantity of air is passing through the venturi. The off-idle system helps keep the engine from running lean, hesitating, or stumbling during initial acceleration from an idle.

FUEL CUTOFF SOLENOID

Carburetors can use a fuel cutoff solenoid to prevent engine dieseling or after-running. Look at Fig. 12-17.

An electric solenoid is wired to the ignition switch and battery. The plunger of the solenoid operates a needle valve inside the idle fuel passage.

Whenever the ignition is ON, the fuel cutoff solenoid

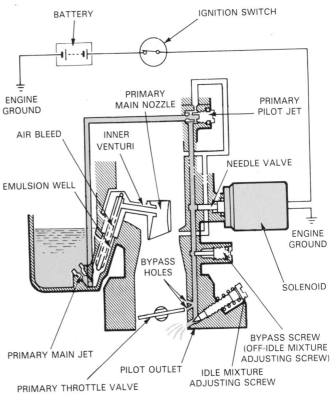

Fig. 12-17. Fuel shutoff solenoid prevents dieseling by closing idle circuit fuel passage. It blocks idle circuit whenever ignition switch is off. Fuel cannot drip from idle system to support combustion and dieseling. (Chrysler Corp.)

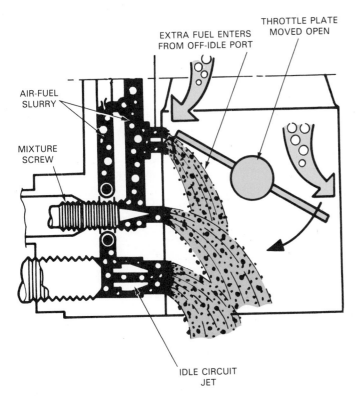

Fig. 12-16. Make a "mental picture" of an off-idle circuit. Note position of throttle plate and how fuel enters carburetor throat. (Renault)

When the throttle plate is closed, the off-idle opening (or openings) act as additional air bleeds for the idle circuit. When the throttle plate swings open, the addi-

is energized and retracted. This lets the idle circuit function normally. When the engine is turned off, the fuel shutoff solenoid is deenergized. The plunger and needle valve slide out to block the idle fuel passage.

Note! Some fuel cutoff solenoids are wired to the throttle position sensor through the computer. This allows the fuel cutoff solenoid to stop fuel flow when the engine is decelerating. This improves fuel economy.

HOT IDLE COMPENSATOR

A *hot idle compensator* is a temperature-sensitive air valve which prevents rough idle and stalling under high engine temperature conditions. It leans the fuel mixture and raises engine idle speed. With excessive heat, extra gasoline vapors can rise off the fuel in the carburetor bowl. These vapors can make the fuel mixture too rich.

A basic idle compensator is pictured in Fig. 12-18. It is made up of a bimetallic strip, air valve, and air passage running from the top of the carburetor to the bottom of the air horn.

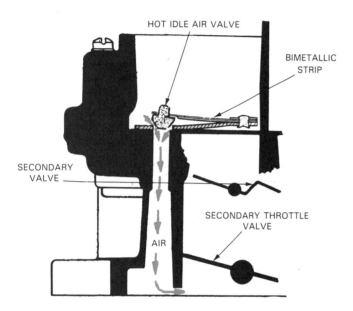

Fig. 12-18. Hot idle compensator offsets fuel vapors from excessively heated fuel bowl. When hot, valve opens to let outside air leak into carburetor bore. The small vacuum leak leans mixture and smooths engine idling. (Pontiac Motor Div.)

Idle compensator operation

With normal engine temperatures, the compensating valve remains closed. No extra air is needed to counteract extra fuel vapors. The idle circuit functions satisfactorily.

When engine temperatures are high, fuel in the carburetor bowl can vaporize too easily. These fuel vapors enter the airstream in the carburetor and richen the mixture. A rough, high-emission idle results.

To compensate for this over-rich mixture, the bimetallic spring bends upward when exposed to high temperatures, letting outside air flow through the passage and into the carburetor throat. The extra air leans the fuel mixture.

ACCELERATOR PUMP SYSTEM

When the throttle plate swings open upon acceleration, airflow increases suddenly in the air horn. Being heavier, fuel cannot respond as rapidly. Without some method to increase the fuel flow rapidly, the mixture entering the cylinders would be almost all air.

To remedy this temporary imbalance, an *accelerator pump system* is added to the carburetor. Like the off-idle system, it provides extra fuel as the fuel system moves from idle to high speed. See Fig. 12-19.

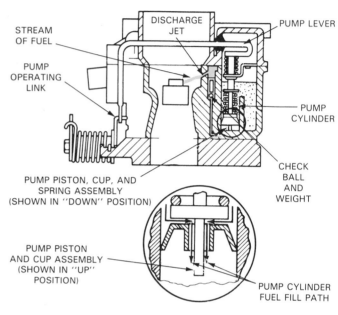

Fig. 12-19. Accelerator pump forces stream of fuel into carburetor throat every time gas pedal is pressed. Gasoline squirts out just like liquid from a toy squirt gun. Pump nozzle is normally aimed at center of booster venturi to aid atomization. (Ford)

Parts of the acceleration system include the pump jet, or nozzle, discharge check valve or weight, pump cylinder, pump piston, spring, rod, and actuating linkage, Fig. 12-20. Sometimes, instead of a piston or cup, a diaphragm is used.

Piston and diaphragm

A *piston type accelerator pump* has a round, cup-shaped pumping element. When the gas pedal is depressed, the linkage compresses pump spring, forcing pump piston down in cylinder. Fuel pressure closes the inlet check valve and opens the discharge weight. Fuel comes out of the pump nozzle in a steady stream. The gasoline is broken up and mixed with the air rushing through the carburetor throat. See Fig. 12-21.

As the gas pedal is released, the pump piston is forced back into its cylinder. This closes the discharge check valve and opens the inlet. Fuel from the bowl refills the pump cylinder.

A *diaphragm accelerator pump* has a square, flat pumping element, as shown in Fig. 12-22. Its operation is almost identical to a piston type, Fig. 12-23.

Vacuum-operated accelerator pump

Some carburetors utilize engine vacuum to help control and power the accelerator pump. This type uses a

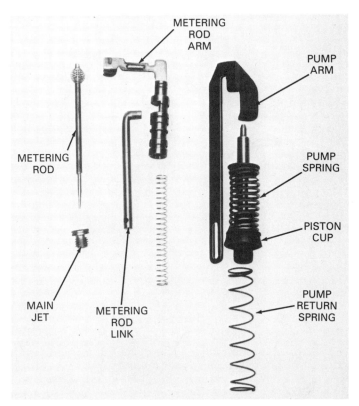

Fig. 12-20. These are parts of a piston type accelerator pump. (General Motors)

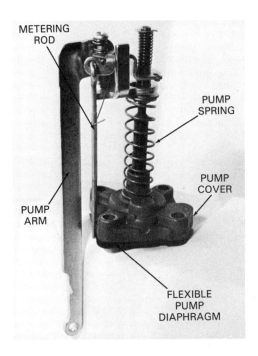

Fig. 12-22. Note parts of a diaphragm type accelerator pump. (AMC)

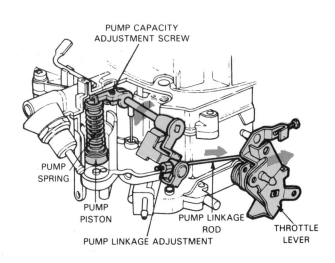

Fig. 12-21. Mechanical link to throttle lever activates most accelerator pump systems. (Ford)

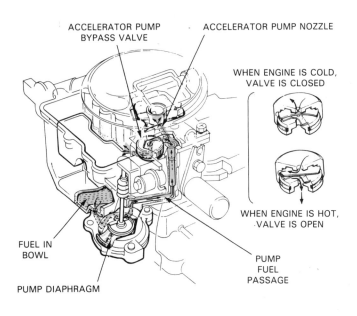

Fig. 12-23. Memorize fuel flow pattern and action of this temperature-sensitive accelerator pump. (Honda)

vacuum passage which runs from the intake manifold to the pump diaphragm. One side of the diaphragm is a vacuum chamber and the other a fuel pumping chamber. The center of the diaphragm is fastened to a rod and spring mechanism. Outlet and inlet check valves are positioned to control the direction of fuel flow.

When the engine is idling, strong engine vacuum pulls

on the diaphragm. As a result, the diaphragm pulls on and compresses the pump spring. This fills the pump chamber with fuel. Fuel flows through the inlet valve and into the chamber.

As the driver presses the gas pedal, the opening of the throttle plates causes intake manifold vacuum to drop. Without vacuum, the collapsed spring is free to push on the diaphragm. The spring and diaphragm action force fuel out of the pumping chamber. The inlet valve closes and the outlet opens. Fuel flows to the discharge nozzle

247

in the air horn. From there, it sprays into the airstream entering the engine.

When engine vacuum increases (steady speed driving, idling, deceleration, etc.), high vacuum again pulls on the diaphragm. The pump spring is compressed. Fuel flows out of the bowl and refills the pumping chamber.

Temperature compensated accelerator pump

A *temperature-sensitive bypass valve* is used on some carburetors to help regulate fuel flow from the accelerator pump. When the engine is cold, the bypass in the valve is closed. Maximum flow of fuel will be delivered to the pump nozzle. When the engine is at normal operating temperature, the bypass valve opens. This allows some of the fuel to bleed back into the carburetor float chamber. Refer again to Fig. 12-23.

HIGH SPEED (MAIN METERING) SYSTEM

The *main* or *high speed carburetor system* delivers the right air/fuel mix during part-throttle operation or cruising speeds (around 2000 to 3000 rpm). This circuit begins to function as airflow through the venturi causes enough vacuum to draw fuel out of the main discharge passage. As engine speed increases, more fuel flows out of the main system to match the airflow, Fig. 12-24.

The high speed system is in use during choke, acceleration, and full power. It operates during steady highway speeds (between 20 and 55 mph), providing the LEANEST, most efficient fuel mixture. For all practical purposes, the main system does not function during deceleration, at idle, etc. Fig. 12-25 shows location of parts and basic construction.

High speed system operation

During main system operation, the throttle plate must be opened wide enough to produce considerable airflow. Then a strong suction or vacuum is formed inside the venturi. This draws gasoline out of the fuel bowl, through

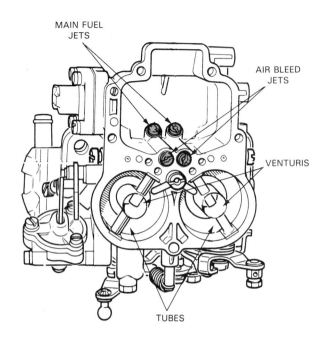

Fig. 12-25. Top view of high speed carburetor system. Construction is simple. Main discharge tube leads from bottom of fuel bowl to air horn venturi. Main jet (restriction in discharge tube) is normally threaded into bowl end of passage. Venturi is located at upper end. (Renault)

the high speed jet, main discharge tube, and into the center of the venturi. From there it is picked up in the airstream, atomized, and carried into the engine.

Because the throttle butterfly is open, there is more vacuum in the venturi than down near the idle ports. Thus, fuel no longer flows through the idle circuit.

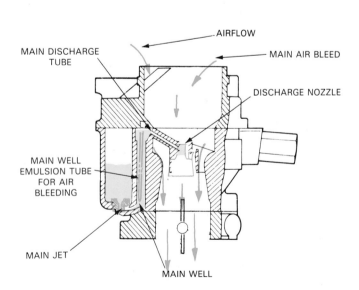

Fig. 12-24. Main jet, screwed into bottom of fuel bowl, regulates or meters fuel flow into main well and bore. (Holley Carburetors)

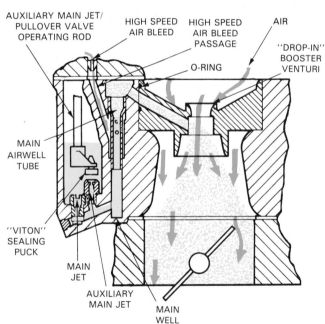

Fig. 12-26. Air bleeds into main passage to mix with fuel before it enters venturi. Vaporization of gasoline is, thus, made easier. (Ford)

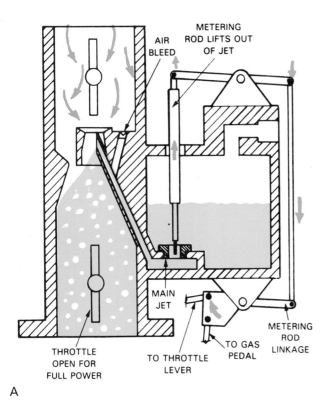

METERING
ROD LIFTS OUT
OF JET

AIR
BLEED

METERING
ROD
LINKAGE

TO GAS
PEDAL

TO THROTTLE
LEVER

MAIN
JET

THROTTLE
OPEN FOR
FULL POWER

A

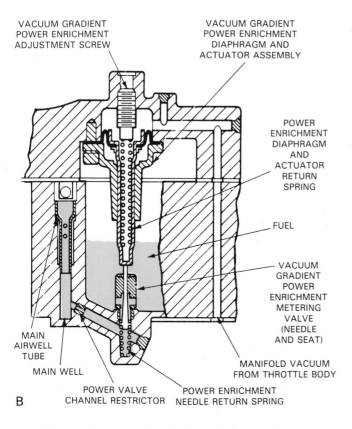

VACUUM GRADIENT
POWER ENRICHMENT
ADJUSTMENT SCREW

VACUUM GRADIENT
POWER ENRICHMENT
DIAPHRAGM AND
ACTUATOR ASSEMBLY

POWER
ENRICHMENT
DIAPHRAGM
AND
ACTUATOR
RETURN
SPRING

FUEL

VACUUM
GRADIENT
POWER
ENRICHMENT
METERING
VALVE
(NEEDLE
AND SEAT)

MANIFOLD VACUUM
FROM THROTTLE BODY

MAIN
AIRWELL
TUBE

MAIN WELL

POWER VALVE
CHANNEL RESTRICTOR

POWER ENRICHMENT
NEEDLE RETURN SPRING

B

Fig. 12-27. A full power circuit is just a high speed circuit with an additional means of richening fuel mixture for more engine power. Metering rod can be used to enlarge main jet area. A—Simplified drawing shows full power circuit activated by mechanical linkage. When metering rod is raised, more fuel enters through main jet into discharge tube. B—Cutaway of vacuum activated metering valve.

Small *air bleed holes* and *jets* are often designed into the main discharge passage, Fig. 12-26. They pre-mix air with the fuel.

FULL POWER (ENRICHMENT, ECONOMIZER) SYSTEM

The *full power system* is also called the *power enrichment* or *economizer circuit.* It supplies extra fuel to the main system for maximum power (highway passing, climbing steep hills, pulling heavy loads, etc.), Fig. 12-27.

The full power circuit produces a variable-ratio fuel mixture. It allows the mixture to change from lean to rich to meet engine demands.

Two types are common—the POWER VALVE and the METERING ROD. The rod type can use either mechanical, vacuum, or electrical powered actuation. The power valve is usually operated by vacuum.

METERING ROD

The main jet is enlarged to accept the end of the metering rod. When the rod is down inside the jet, the normal main jetting is in operation (main metering system). For full power and a richer mixture, the rod is pulled partway out of the main jet. This lets more fuel flow through the jet, up through the main passage, and into the air stream. The full power system begins to function when the metering rod is lifted out of the jet. The metering rod can be controlled by mechanical linkage, engine vacuum, or an electric solenoid and a computer.

Mechanical metering rod

An easy way to control the action of the metering rod is to link it to the throttle plate. When the throttle butterfly is NOT fully open, the linkage holds the rod IN the jet. Then, as the throttle is swung wide open for maximum engine power, the linkage lifts the rod OUT of the main jet.

Vacuum-operated metering rod

The same metering rod action is possible using engine vacuum. Intake manifold vacuum depends upon engine load and power demands. During steady-speed driving, power needs are low but engine vacuum is high. The opposite is true at wide open throttle. Engine vacuum is very low. This vacuum/load relationship is ideal for controlling a metering rod.

The metering rod is connected to a vacuum piston. The piston chamber has a small vacuum passage located below the throttle plate. Under normal speeds and loads, engine vacuum is high and a lean mixture is efficient. Intake vacuum pulls the piston down and compresses the spring. This holds the metering rod down inside the jet.

Electric metering rod

Some modern carburetors use an electric solenoid to control metering rod action. An exhaust gas sensor is located in the engine's exhaust system. An "on-board" computer evaluates signals from the exhaust sensor and then energizes the metering rod solenoid, Fig. 12-28. The computer control keeps down fuel waste and exhaust emissions.

Note! The last section of the chapter describes a computer controlled electric solenoid metering rod system.

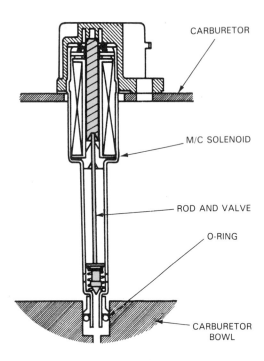

Fig. 12-28. Electrical impulses to solenoid control this electric metering rod. Solenoid pulls rod in and out of its jet to match fuel mixture to engine's needs. (Buick)

POWER VALVE (POWER JET)

In many carburetors, a *power valve* or *jet* serves the same purpose as a metering rod. In a sense, it also provides for better gas mileage at low power output. Hence, the power valve is sometimes called an ECONOMIZER VALVE, Fig. 12-29.

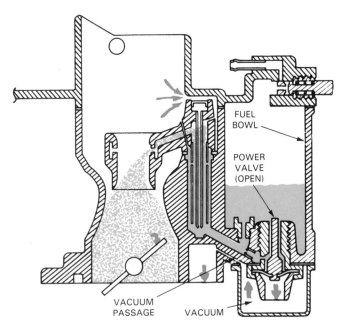

Fig. 12-29. Power valve or economizer valve richens the high speed fuel mixture. It is controlled by engine vacuum, an indicator of engine power demands. (AMC)

The power valve consists of a flexible diaphragm attached to a spring-loaded valve, Fig. 12-30. A vacuum passage goes to the bottom of the carburetor air horn. On one side of the power valve diaphragm is a supply of gasoline, on the other side, engine vacuum.

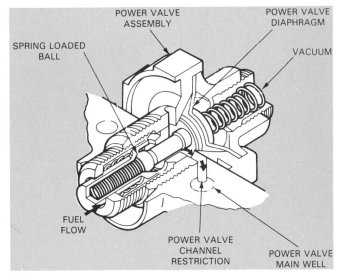

Fig. 12-30. Cutaway shows parts and construction of two-stage power valve. (Ford)

Two-stage power valve

A *two-stage power valve* reacts to different vacuum levels to meter fuel more precisely into the engine. Refer again to Fig. 12-30. With moderate vacuum, the valve is pulled midway so fuel is drawn through a first stage orifice. With full acceleration, the valve is pulled fully open to allow full fuel enrichment.

CHOKE SYSTEM

The *choke circuit,* Fig. 12-31, supplies an extremely rich fuel mixture for cold engine starting. A choke valve closes off flow of outside air into the engine. As a result, a very strong vacuum is formed inside the air horn by the downward movement of the piston on the intake stroke. The strong suction pulls a huge amount of gasoline out of the main discharge passage. The engine's intake manifold is, literally, flooded with raw fuel. Vapors rise off the liquid fuel in the intake and improve cold engine operation.

Parts of a choke system

Depending upon its design, a choke system will include a choke plate (or butterfly valve), thermostatic spring, and some linkage connecting its parts. Additional parts are found on specific choke types.

The *choke plate* is a disc-type valve with a rod through its middle supported by the top of the air horn in a way that allows the plate to swing through an arc of 90 degrees to block off the top of the air horn.

The *thermostatic spring* is an automatic device which

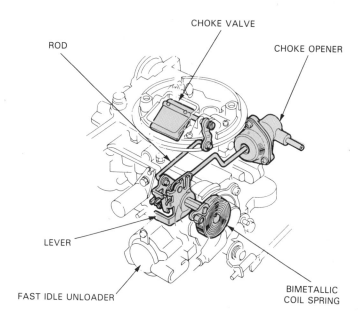

Fig. 12-31. Basic parts of choke circuit. Choke valve is also known as a butterfly valve. It is hinged to top of air horn. (Honda)

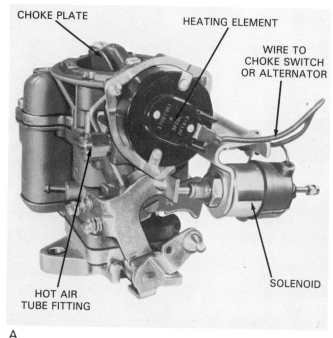

A

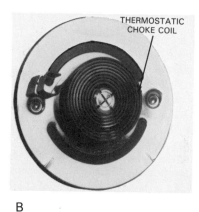

B

Fig. 12-32. A—Note parts of electric assist choke. Small heat element placed next to thermostatic spring speeds choke opening. B—View of thermostatic choke coil. (Ford and Chevrolet Motor Div.)

is linked to the choke plate. Its purpose is to open and close the choke plate as the spring reacts to the temperature of the engine and the air around it.

There are two basic types of carburetor chokes, MANUAL and AUTOMATIC.

Manual choke

As its name suggests, a *manual choke* is hand operated. It has a steel control cable which runs from the driver's compartment to the carburetor. When the choke knob on the dashboard is PULLED, the cable slides inside the stationary cable housing. The movement of the inner cable pulls on the choke plate arm and closes the choke.

Automatic choke

Many cars have an *automatic choke* that operates without driver assistance, Fig. 12-32A. Being more sensitive to engine temperature, an automatic choke is more efficient.

There are several types of automatic carburetor chokes:
1. An integral hot air type mounted on the carburetor.
2. A separate hot air choke mounted in the intake manifold and linked to the carburetor.
3. A combination hot air-electric choke.
4. An all electric choke.
5. A hot water (coolant) choke.

Thermostatic (bimetallic) spring

The primary control element of hot air type automatic chokes is a small *thermostatic spring,* Fig. 12-32B. The bimetallic spring is heat sensitive. When COLD, the spring coils up tightly. When HOT, the spring expands and unwinds into a larger spiral. This coiling and uncoiling action is used to open and close the thermostatic choke plate.

Integral hot air choke

Basically, the *integral hot air choke* consists of the butterfly valve, choke shaft, lever, bimetallic spring, spring housing, choke piston, vacuum passage, choke stove, and hot air tube.

With a cold engine, the coil spring winds up and closes the choke plate. When the engine starts, engine vacuum pulls on the choke piston and cracks open the choke to prevent engine flooding. The cold thermostatic spring keeps the choke from fully opening.

As the engine warms, hot air flows up from the exhaust manifold into the spring housing. The hot air causes the thermostatic spring to unwind, opening the choke butterfly. At engine operating temperature, the vacuum piston and the bimetallic spring keeps the choke fully open.

Integral hot water choke

This system simply routes engine water, instead of hot air, through the choke spring housing. When the engine water is cold, the choke spring holds the choke valve closed. As the engine warms, the heated coolant causes the thermostatic spring to open the choke butterfly.

Auxiliary mounted thermostatic spring

On some hot air type chokes, the bimetallic spring is mounted in the intake manifold. This type is similar to those mounted on the carburetor. However, a hot air tube is not needed and a vacuum diaphragm often replaces the vacuum piston.

A long linkage rod connects the spring to the choke shaft lever. When the engine starts, intake manifold vacuum causes the vacuum diaphragm to stretch away from the choke lever. Much like a vacuum piston, this pulls on and cracks open the choke.

As the engine warms, hot exhaust gases heat the choke spring chamber causing the thermostatic spring to unwind, steadily opening the choke plate.

Electric assisted hot air choke

Some cars use a conventional thermostatic spring choke assisted by a small electric heating element. The element reduces fuel waste and exhaust pollution by speeding up choke opening, Fig. 12-32.

A heat-sensing control switch is "wired" to the heater element. When the engine is cold, around 60 °F (16 °C) or below, the sensor switch opens and stops current to the heater. The choke operates normally and remains partially closed.

When engine temperature rises above specs, the temperature-sensing switch closes. Current flows to the heating element. The extra heat helps uncoil the thermostatic spring more quickly causing the choke to open.

Electric choke

A fully *electric choke* depends entirely upon a two-stage heating element to help control thermostatic spring action. See Fig. 12-33. The element is controlled by a temperature sensing device.

When temperature is below around 60 °F, only the first stage of the heating element receives current. The thermostatic spring uncoils very slowly. After the engine temperature increases (somewhere above 60 °F), the temperature-sensing switch turns ON the second stage of the heater element. This makes the choke open more quickly and leans the fuel mixture. As you will learn later, most choke systems use additional methods of controlling choke operation.

FAST IDLE CONTROL

To prevent cold engine stalling or "dying," it is necessary to increase idle speed during engine warmup. This is usually done by adding a *fast idle cam* mechanism to the choke shaft. Refer to Fig. 12-34.

When the choke plate fully closes, the fast idle cam swings into position. The idle speed screw then rests on top of the largest step of the cam. This holds the throttle plates open and increases engine speed several hundred rpm.

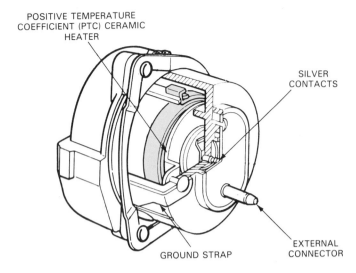

POSITIVE TEMPERATURE COEFFICIENT (PTC) CERAMIC HEATER

SILVER CONTACTS

GROUND STRAP

EXTERNAL CONNECTOR

Fig. 12-33. Voltage is applied to heating element in choke cap of electric choke system. This heat causes bimetal thermostatic coil to operate more quickly. (Mercury)

As the engine begins to warm and the choke opens, the fast idle cam swings to smaller steps. The throttle plates can close a little more and engine speed drops. Finally, when the choke opens all the way, engine speed returns to curb idle. A leaner mixture and slower idle speed can now provide smooth and dependable engine operation.

MECHANICAL CHOKE UNLOADER

The *mechanical choke unloader* is a mechanical device that partially opens the choke any time the "gas pedal" is pushed all the way to the floor. If the engine floods, air flowing past the opened choke will help evaporate and remove excess gasoline from the intake and cylinders, Fig. 12-35.

VACUUM CHOKE BREAK

As soon as a cold engine starts, the choke must be cracked open. As shown in Fig. 12-36, the *choke break* is linked to the choke shaft lever. A vacuum line runs from the dashpot to a vacuum supply nipple under the carburetor.

After the engine starts, vacuum is strong enough to suck on the diaphragm in the dashpot. The diaphragm stretches and pulls on the choke linkage causing the choke valve to open slightly.

THROTTLE RETURN DASHPOT

When engines equipped with automatic transmissions are quickly dropped to an idle, drag from the transmission can stall the engine. The addition of a *throttle return dashpot* to the carburetor linkage prevents this problem. This device, Fig. 12-37, also called an *anti-stall dashpot,* is a spring-loaded diaphragm inside a housing which is sealed except for a tiny air bleed hole in one end. A plunger, attached to the diaphragm, rests against the throttle linkage.

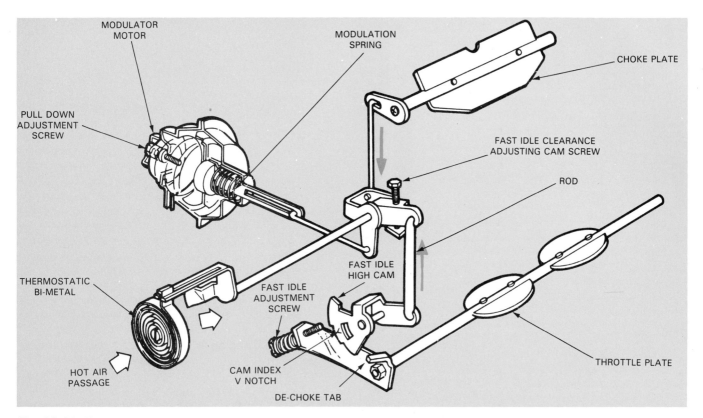

Fig. 12-34. Fast idle cam operation. When choke closes, its linkage pulls on cam, swinging it so fast-idle adjustment screw rests on its high point, keeping throttle plate slightly open. (Ford)

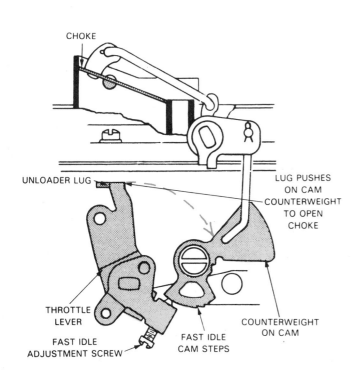

Fig. 12-35. Choke unloader opens choke any time gas pedal is "floored." This causes unloader lug to swing and strike fast idle cam weight pushing choke linkage and opening choke. (Carter)

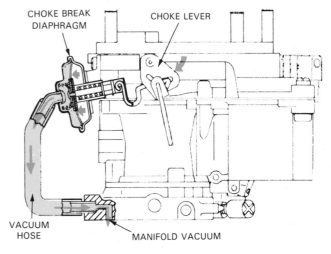

Fig. 12-36. Choke break automatically cracks open choke upon engine starting. Manifold vacuum sucks on diaphragm and linkage pulls choke valve partially open to prevent engine flooding. (Chrysler Corp.)

When the throttle is open, the spring on the plunger pushes it outward. When the engine returns to idle, the throttle lever strikes the extended plunger which can retract only as fast as air bleeds out of the small hole in the diaphragm. This gives the automatic transmission time to disengage (torque converter releases) before it can stall the engine.

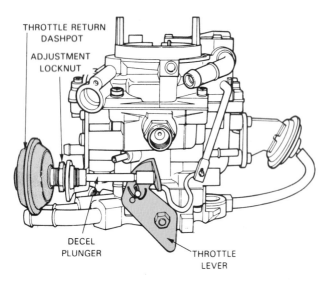

Fig. 12-37. Throttle dashpot keeps engine from stalling during rapid deceleration. Throttle lever strikes dashpot plunger which slowly returns throttle to curb idle. (Chrysler Corp.)

ANTI-DIESELING SOLENOID

Engines which must power automatic transmissions and air conditioning systems require higher idle speeds. The *anti-dieseling solenoid* is an electrically activated device designed to increase idle speed while the engine is running, but drop it to curb idle speed when the ignition is turned off. Refer to Fig. 12-38.

The solenoid is mounted like a throttle return dashpot, its plunger rod resting against the throttle linkage. Turning on the ignition energizes the solenoid which pushes

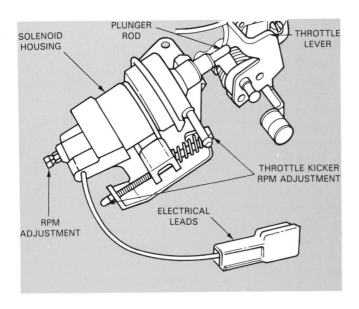

Fig. 12-38. Anti-dieseling solenoid increases idle speed when ignition is ON. It holds throttle linkage off curb idle screw to improve idle under added load of air conditioning and automatic transmission. Then solenoid prevents dieseling by fully closing throttle plates to curb idle when ignition is turned off. (Ford)

the rod outward, opening the throttle valve to a faster idle. Turning off the ignition de-energizes the ignition and the rod retracts. Too little fuel enters the carburetor throat to support combustion. Dieseling is avoided.

Most fast-idle solenoids are adjustable. The plunger rod or the mounting bracket can be turned IN or OUT to adjust engine speed.

ALTITUDE COMPENSATOR

An *altitude compensator* is a device that changes the carburetor's air-fuel mixture with changes in height above or below sea level. The compensator normally has an *aneroid* (bellows device that expands and contracts with changes in atmospheric pressure). See Fig. 12-39.

When a car is traveling up a mountain, for example, the density of the air decreases with height. This tends to make the air-fuel mixture richer. The reduced air pressure causes the aneroid to expand, opening an air valve. Extra air flows into the air horn producing a leaner air-fuel mixture. This action reverses when the car's height above sea level decreases.

CARBURETOR VACUUM CONNECTIONS

A modern carburetor has numerous vacuum connections, Fig. 12-40. When the vacuum connection or port is BELOW the carburetor throttle plate, the port always receives full intake manifold vacuum. However, when the vacuum port is above the throttle plate, vacuum is present only when the throttle is opened.

Typical components that operate off carburetor vacuum connections are:
1. EGR VALVE (exhaust emission control device).
2. DISTRIBUTOR VACUUM ADVANCE (diaphragm for advancing ignition timing).
3. CHARCOAL CANISTER (emission control container for storing fuel vapors).
4. CHOKE BREAK (diaphragm for partially opening choke when engine is running).

COMPUTER CONTROLLED CARBURETORS

A *computer controlled carburetor* normally requires a solenoid-operated valve to respond to commands from a microcomputer (electronic control unit). The system uses various sensors to send information to the computer. The computer then calculates how rich or lean to set the carburetor's air-fuel mixture. A typical system is shown in Fig. 12-41.

Computer system sensors
A computer controlled carburetor system, sometimes called a computer controlled emission system, consists of the components shown in Fig. 12-41. Included are:
An *oxygen sensor* or *exhaust gas sensor,* Fig. 12-42, signals any change in oxygen content (an indicator of air-fuel ratio) in the exhaust gases. An O_2 sensor produces a voltage output of about 0.2 volts when the mixture is lean and up to 1.0 volts when the mixture is rich. The oxygen sensor only functions when heated by the exhaust to about 600 °F (316 °C).

The *temperature sensor's* resistance changes with

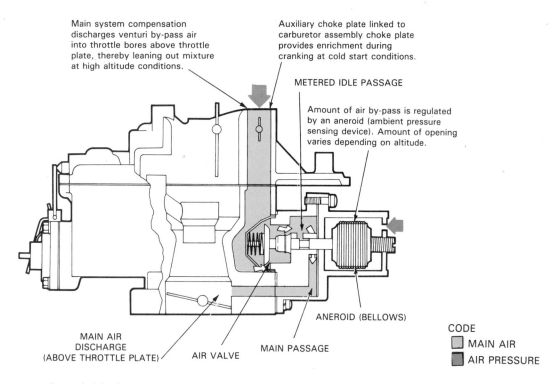

Main system compensation discharges venturi by-pass air into throttle bores above throttle plate, thereby leaning out mixture at high altitude conditions.

Auxiliary choke plate linked to carburetor assembly choke plate provides enrichment during cranking at cold start conditions.

METERED IDLE PASSAGE

Amount of air by-pass is regulated by an aneroid (ambient pressure sensing device). Amount of opening varies depending on altitude.

ANEROID (BELLOWS)

MAIN AIR DISCHARGE (ABOVE THROTTLE PLATE)

AIR VALVE

MAIN PASSAGE

CODE
☐ MAIN AIR
☐ AIR PRESSURE

Fig. 12-39. Cutaway shows aneroid device which expands and contracts like an accordian as it responds to atmospheric pressure. This action moves a rod which opens and closes air valve. It controls amount of air entering air horn just above throttle plate.

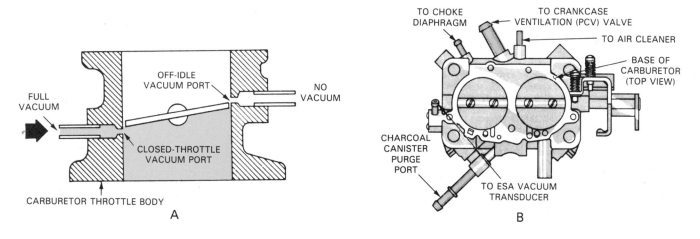

OFF-IDLE VACUUM PORT

NO VACUUM

FULL VACUUM

CLOSED-THROTTLE VACUUM PORT

CARBURETOR THROTTLE BODY

A

TO CHOKE DIAPHRAGM

TO CRANKCASE VENTILATION (PCV) VALVE

TO AIR CLEANER

BASE OF CARBURETOR (TOP VIEW)

CHARCOAL CANISTER PURGE PORT

TO ESA VACUUM TRANSDUCER

B

Fig. 12-40. Typical carburetor has several vacuum connections. A—Ports located below throttle valve have full vacuum at all times. Those located above throttle plate have vacuum only when throttle is open. B—Carburetor provides vacuum for many different accessories. (Ethyl Corp. and Chrysler Corp.)

the temperature of the engine coolant. This allows the computer to enrich the fuel mixture during cold engine operation and lean the mixture as the engine warms.

A *pressure sensor* measures intake manifold vacuum and engine load. High engine load or power output causes intake manifold vacuum to drop. The pressure sensor, in this way, can signal the computer with a change in resistance and current flow.

When the manifold pressure drops, the computer increases the air-fuel mixture for added power. If intake

manifold vacuum increases, the computer makes the carburetor setting leaner for improved economy.

A *throttle position sensor* is used on some carburetors to monitor the movement of the throttle plates. One is shown in Fig. 12-43.

As the throttle opens for acceleration, for example, the sensor changes resistance. Change in current flow signals the computer. The computer can then richen the mixture for more power. When the throttle is closed, the position sensor lets the computer know that a leaner mix-

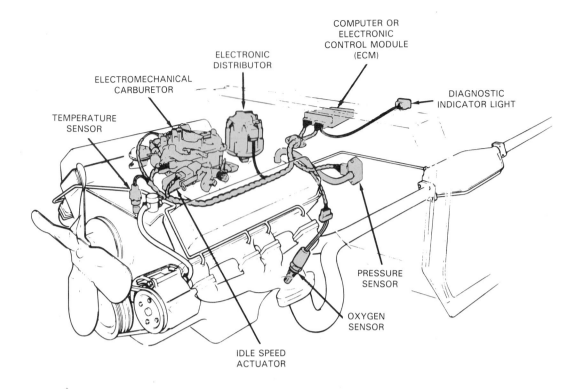

Fig. 12-41. Typical setup for computer controlled carburetor. Sensors feed information to computer (ECM). (General Motors Corp.)

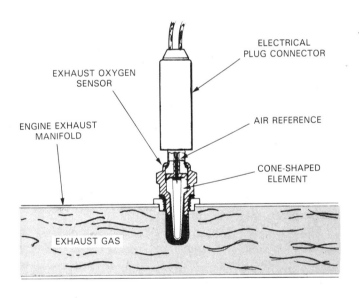

Fig. 12-42. Oxygen sensor detects amount of oxygen in exhaust by comparing sample to outside air. High oxygen means lean mixture—low oxygen, rich mixture. Computer corrects air-fuel mixture as needed. (AC)

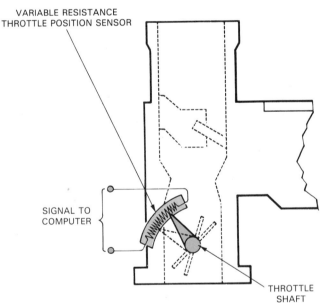

Fig. 12-43. Throttle position sensor is actually a potentiometer. Throttle shaft movement will change current flow in circuit to computer. Changes tell computer whether throttle plate is open or closed or somewhere in between. (Ford)

ture is needed.

A *WOT (wide open throttle) cutoff switch* is designed to signal the computer when the carburetor throttle plates are swung open for rapid acceleration. The computer may then shut off the air conditioning compressor and other power-robbing accessories to help the car increase its speed more quickly.

Note! Other sensors can also be used in a computer controlled carburetor system. These sensors are explained in Chapter 15, Gasoline Injection System Operation.

Electromechanical carburetor

An *electromechanical carburetor* is controlled by both electrical and mechanical devices. This type is used with a computer control system. It has electrical components (sensors and actuators) for interfacing with the computer and conventional mechanical parts as well.

Mixture control solenoid

A *mixture control solenoid* in the computer controlled carburetor responds to electrical signals (or dwell signals) from the computer to alter the air-fuel ratio. The signals trigger the solenoid to open and close air and fuel passages in the carburetor. See Fig. 12-44.

Fig. 12-45 shows two other types of mixture control solenoids. Fig. 12-46 illustrates mixture control solenoid action.

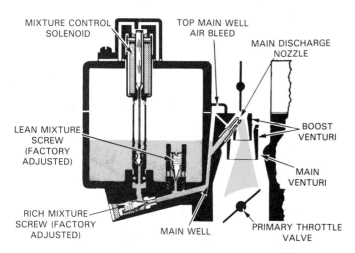

Fig. 12-44. Computer can activate mixture control solenoid in electromechanical carburetor to quickly open or close valve that changes carburetor's air-fuel ratio. Electric pulses from computer move plunger in solenoid. It operates mixture valve. Study how solenoid can regulate fuel flow. (Pontiac)

Carburetor control computer

The system computer, usually called the *electronic control unit,* is a small solid-state electronic device. One is shown in Fig. 12-47. This device has both logic circuits (artificial intelligence) and control circuits (power transistors) that are contained in one or more integrated circuit chips.

The electronic control unit (ECU) is capable of receiving information sent to it by sensors located in various places on the automobile. It can process this information and then electronically control the carburetor's mixture control solenoid.

For example, the ECU will receive reports from the oxygen sensor in the exhaust system about the amount of oxygen in the exhaust gas. The ECU, which can do very complicated math in a fraction of a second, compares this signal with the information stored in its memory. If the oxygen makeup of the exhaust gas is not correct, the memory sends a signal to other parts of the ECU. A control signal will then be sent out to the carburetor's mixture solenoid which is energized to adjust the air-fuel mixture for the correct oxygen content in the exhaust gas.

Computer controlled carburetor operation

In a computer controlled carburetor, the air-fuel ratio is maintained by CYCLING the mixture solenoid ON and OFF several times a second. Fig. 12-48 shows how the control signals from the computer can be used to meter different amounts of fuel out of the carburetor.

When the computer sends a rich command to the solenoid, the signal voltage to the mixture solenoid is usually OFF more than it is ON. This causes the solenoid to stay open more. Typically, the idle air bleed valve and main metering rod move up to increase fuel flow.

During a lean signal from the computer, the signal usually has more ON time. This causes less fuel to pass through the solenoid valve. The mixture becomes leaner. The idle air bleed valve and main metering rod move down to restrict fuel flow.

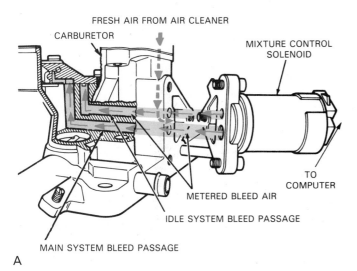
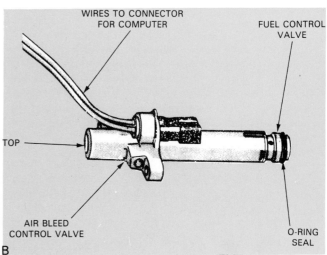

Fig. 12-45. Mixture control solenoids. A—This one opens and closes air passages to control mixture ratio. B—This one controls both air and fuel flow. (Ford and Chrysler)

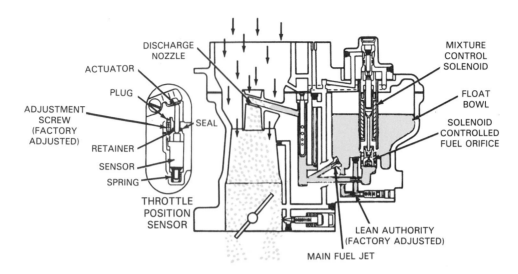

Fig. 12-46. Cutaway shows mixture control solenoid. Its up and down motion controls amount of fuel entering main discharge nozzle. (Pontiac)

Fig. 12-47. Electronic control unit has both logic circuits (artificial intelligence) and control circuits (power transistors) that are contained in one or more integrated-circuit chips. (Oldsmobile)

Open and closed loop operation

The term, *open loop,* means that the system is operating on preset values in the computer. For example, right after cold engine starting, the computer will operate open loop. The sensors, especially the oxygen sensor, CANNOT provide accurate information when cold.

Closed loop means that the computer is using information from the oxygen sensor and other sensors. The information forms an imaginary loop (circle) from the computer, through the engine fuel system, into the exhaust, and back to the computer through the oxygen sensor. Look at Fig. 12-49.

A computer controlled carburetor system normally operates in the closed loop mode after engine and oxygen sensor warm up. Using data from the engine sensors, a very precise air-fuel ratio can be maintained. The system monitors its own efficiency and the mixture can be changed with the slightest change in engine operating conditions.

Idle speed actuator (control motor)

An *idle speed actuator* or *control motor* may also be used to allow the computer to change engine idle speed. It is usually a tiny electric motor and gear mechanism. Responding to computer control current, it holds the carburetor throttle lever in the best position for idling. Fig. 12-50 illustrates an idle speed control motor mounted on a carburetor's linkage.

More information

Note! Computerized carburetor systems vary. For exact details, refer to a factory service manual. The manual will explain how the specific system functions.

As you will learn in later chapters, a computer controlled carburetor system is very similar to ELECTRONIC FUEL INJECTION. Both systems use engine sensors, a computer, and a solenoid action to control the amount of fuel entering the engine. Electronic fuel injection is covered in Chapters 15, 16, and 17.

Fig. 12-51 gives a wiring diagram for a typical computer controlled carburetor system.

SUMMARY

For good combustion, fuel must be mixed with air in exact proportions. Engine vacuum and the resulting airstream in the carburetor help atomize and vaporize the fuel. At highway speeds, an air-fuel ratio of 15 to 1 (15 parts air to 1 part of fuel) would be ideal. A mixture of 20:1 is a lean mixture and 8:1 rich. Either a too-rich or too-lean mixture results in poor engine performance.

A carburetor is a hydraulic, mechanical, vacuum

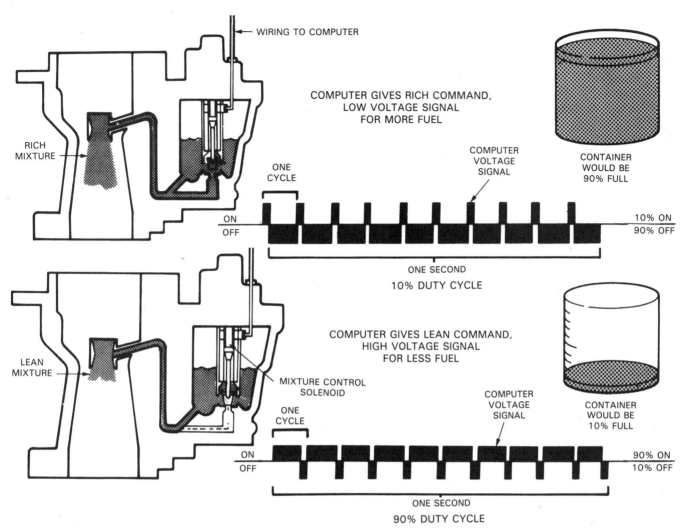

Fig. 12-48. Computer rapidly pulses mixture control solenoid OFF and ON to control air-fuel ratio. Note that, with this carburetor, longer OFF time enriches mixture. Shorter OFF time makes mixture leaner. (Chrysler Corp.)

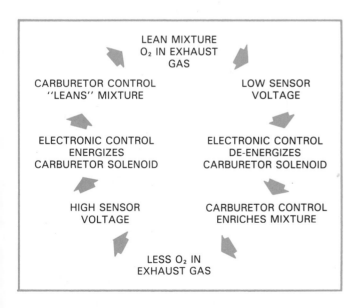

Fig. 12-49. Closed loop operation. Computer is continually receiving information from sensors and sending ''commands'' to solenoid as operating conditions change. (Buick)

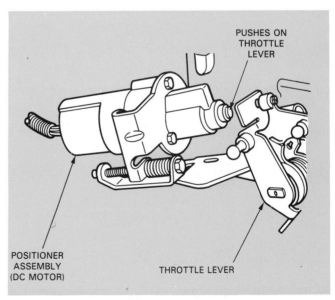

Fig. 12-50. DC motor, controlled by computer, can adjust idle speed of engine precisely. It moves throttle lever. (Ford)

259

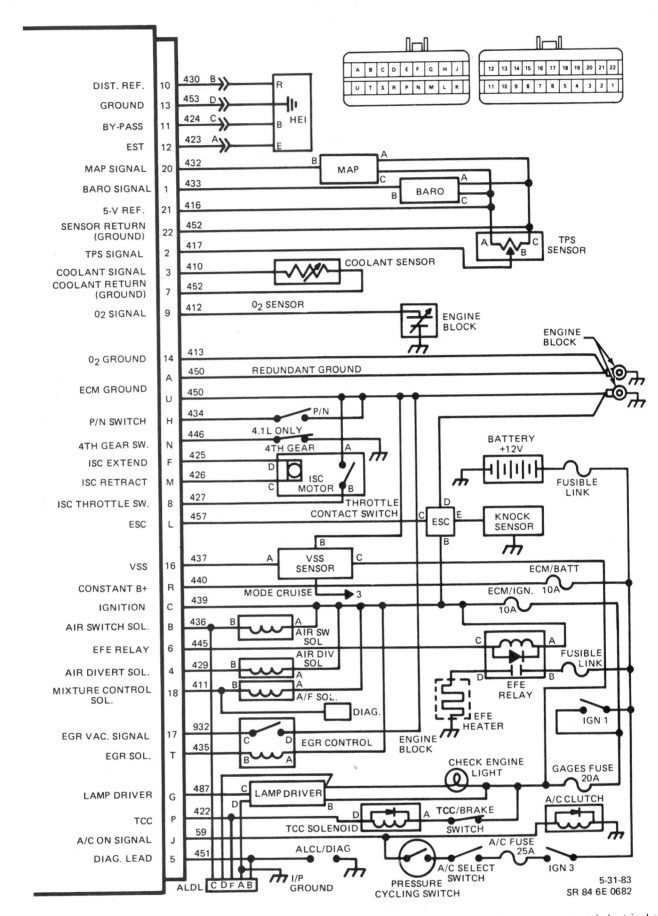

Fig. 12-51. This is a typical wiring diagram for a computer controlled carburetor system. Study components and electrical connections going to computer. (General Motors)

device for metering and mixing fuel and air in exact amounts. It has five basic functions: atomize fuel into tiny droplets, meter fuel, regulate airflow into engine, and control external vacuum powered components.

An air horn or throat is the metal passage through the carburetor body. The fuel bowl stores extra gasoline. To keep fuel in the bowl at a constant level, a float mechanism is used. The main discharge tube routes fuel from the bowl to the air horn.

Venturi action is used to increase air horn vacuum at low engine speeds. Air bleeds mix air bubbles with fuel to speed atomization. Jets control the amount of fuel or air flowing in a carburetor circuit. To control airflow in the carburetor, a throttle valve is positioned at the bottom of the throat. At the top of the air horn, a choke butterfly is utilized to richen the mixture for cold engine starting.

A carburetor system (or circuit) is a network of fuel passages, air passages, and related parts which supply the correct fuel mixture during a specific engine demand. There are seven basic carburetor systems: float, idle, off-idle, acceleration, high speed, full power, and choke.

An idle mixture screw can be adjusted to control the fuel mixture at idle. A bypass (air bleed) is located in the idle circuit to improve atomization. The accelerator pump squirts a stream of gasoline into the air horn whenever the throttle plate is opened. It prevents stalling during initial acceleration. A metering rod allows the high speed system to increase mixture richness for increased engine power output. A power or economizer valve also varies the high speed mixture ratio.

Chokes can be either manual or automatic. Automatic chokes can be integral hot air, separate hot air, combination hot air-electric, hot water or all electric. A fast idle cam is usually linked to the choke plate. During cold engine choke operation, the fast idle mechanism increases engine idle rpm to prevent stalling. A choke break or mechanical unloader assures choke opening upon engine starting. This prevents engine flooding and stalling.

A throttle return dashpot keeps an automatic-transmission-equipped car from stalling upon rapid deceleration. A hot idle compensator leans the idle fuel mixture during periods of excessive heat and fuel evaporation.

A vacuum choke break or air valve dashpot pulls the choke valve off full choke as soon as the cold engine starts. On engines with automatic transmissions, a throttle return dashpot brings the engine to idle speed slowly so that the drag of the transmission does not cause stalling.

Engines with today's higher idle speeds are very prone to "dieseling" (running on after ignition is turned off). An anti-dieseling solenoid increases idle speed when the engine is running but drops it to curb idle when the ignition is turned off.

An altitude compensator automatically changes the air-fuel mixture with changes in altitude. Normally an aneroid device is used. It expands and contracts with changes in atmospheric pressure.

Various vacuum ports located on the carburetor are intended to operate various devices: EGR valve, distributor vacuum advance, charcoal canister vapor storage, and choke break.

On computer controlled carburetors, various actuators respond to commands from the on-board computer to change settings that lean or richen the air-fuel mixture.

Sensors, located at various points, feed information to the computer. They measure oxygen content of the exhaust gas, temperature of the engine, manifold vacuum, engine load, and throttle position.

The on-board computer or electronic control unit (ECU) is a solid-state device which has both logic circuits and control circuits in one or more integrated-circuit chips. It can receive information, process it, and send out signals that control the operation of the fuel system.

KNOW THESE TERMS

Atmospheric pressure, Vacuum, Air horn, Throttle valve, Venturi, Main discharge tube, Carburetor system, Float, Needle Valve, Bowl vent, Idle mixture screw, Accelerator pump, Main jet, Metering rod, Power valve, Thermostatic spring, Mechanical choke unloader, Vacuum choke break, Fast idle cam, Fast idle solenoid, Throttle return dashpot, Hot idle compensator, Altitude compensator, Primary, Secondary, CFM, Variable venturi carburetor, Computer controlled carburetor, Mixture control solenoid, Engine sensors, Oxygen sensor, Electronic control unit, Open loop, Closed loop.

REVIEW QUESTIONS—CHAPTER 12

1. The force of the _____ entering the carburetor is used to mix _____ and _____.
2. During the process of atomization, liquid fuel (select best answer):
   a. Is turned into a vapor.
   b. Turns into tiny droplets.
   c. Becomes a gaseous mixture.
3. What six factors affect fuel vaporization?
4. A _____ fuel mixture contains a high ratio of air compared to fuel.
5. List and describe the function of each of the 11 essential parts of a basic carburetor.
6. A modern automobile would NOT run well on the basic carburetor described in the first section of this chapter. True or false?

MATCHING TEST: On a separate sheet of paper, match the terms with the appropriate statements.

7. ____ Provides small amount of fuel for low-speed engine operation.
8. ____ Supplies lean air-fuel mixture at cruising speed.
9. ____ Supplies correct air-fuel mixture slightly above idle speed.
10. ____ Supplies slightly richer air-fuel mixture when engine power demands are high.
11. ____ Gives an extremely rich air-fuel mixture for cold engine starts.
12. ____ Squirts fuel into air horn

a. Off-idle system.
b. Full power system.
c. Idle system.
d. Choke system.
e. Acceleration system.
f. High speed system.
g. Float circuit.

when throttle valve opens and engine speed increases.

13. ____ All other carburetor systems will be affected if float _____ _____ are either too high or too low.

14. At low engine speeds with the idle circuit working (indicate which of the following statements is NOT true):
    a. Venturi vacuum is NOT strong enough to pull fuel out of the main discharge tube.
    b. There is high vacuum below the throttle plate.
    c. The throttle plate is closed.
    d. Fuel is fed into the air horn from the low speed jet in the air horn just below the throttle plate.

15. Describe the off-idle circuit and explain its function.

16. A _____ _____ _____ is a temperature-sensitive air valve which prevents rough idle and stalling under high engine-temperature conditions.

17. On some carburetors, a temperature-sensitive _____ _____ is used to help regulate fuel flow from the accelerator pump.

18. The high speed system operates only when the engine is at cruising speeds. True or false?

19. Select all the TRUE statements about the operation of the full power system:
    a. It allows the fuel mixture to change from lean to rich as needed by the engine.
    b. It uses either a metering rod or a power valve to control fuel flow to the main discharge passage.
    c. The metering rod can be controlled by mechanical linkage, engine vacuum, or by a computer and an electric solenoid.
    d. The power valve, like the metering rod, controls flow of fuel through the main jet.
    e. Power valves send extra fuel through a second fuel passage that empties into the main, high speed discharge passage.

20. List the five types of chokes and explain how each operates.

21. To prevent cold engine stalling, a _____ _____ cam is used to increase idle speed during engine warmup. The stepped cam is linked to the choke _____.

22. A choke break:
    a. Mechanically opens the choke plate when the engine is flooded.
    b. Pulls the choke wide open as soon as the cold engine starts.
    c. Cracks the choke plate open to avoid flooding as soon as the engine starts.

23. The _____ _____ increases idle speed while the engine is running but drops it to curb idle speed when the ignition is turned off.

24. Describe how an on-board computer controls a carburetor.

This is one of the most fuel efficient vehicles in the world. It is powered by tiny, direct injection diesel engine that attained 3508 miles per gallon. How is that for ''gas mileage?'' (Volkswagen of America, Inc.)

OUTPUT DEVICES

THROTTLE KICKER SOLENOID

Receives signal from computer to adjust idle speed when:
–idle improvement is required during cold start.
–cooling is required during prolonged idle.
–idle improvement is required during air conditioning system operation.

FEEDBACK CARBURETOR ACTUATOR

Varies air/fuel ratio according to computer commands, based on oxygen measurement taken by EGO sensor during normal operation (closed loop).
Varies air/fuel ratio according to drive demands (open loop).

THERMACTOR AIR BYPASS SOLENOID VALVE

Computer signals solenoid valve to bypass (dump) Thermactor air to atmosphere under certain conditions, depending on engine calibration.
Receives signal from computer to direct Thermactor air to catalytic converter (downstream) during normal operation.
Signal from MCU to direct Thermactor air to exhaust manifold (upstream) during engine warm-up.

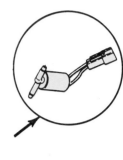

TAD

TAB

CANISTER PURGE SOLENOID VALVE

Opens on command from the computer to purge fuel vapor from charcoal canister during normal engine operation after delayed period following engine start.

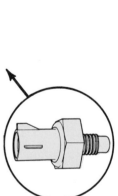

MICROPROCESSOR CONTROL UNIT

Receives electrical signals from the input devices monitoring specific engine functions. Processes information and determines what command signals are sent to output devices.

Sends command signals to the output devices to adjust engine operation, as required, for a specific operating condition and calibration.

INPUT DEVICES

EXHAUST GAS OXYGEN (EGO) SENSOR

Monitors exhaust emissions by measuring amount of oxygen in exhaust gases.
Provides electrical signal to computer to indicate lean or rich air-fuel mixture.

TACH INPUT FROM COIL CONNECTOR

Provides engine rpm signal to computer as a ground pulse from ignition coil primary circuit.

COOLANT TEMPERATURE SWITCHES

Switches monitor engine's coolant temperature.
Signals computer to adjust air/fuel ratio, ignition timing, idle speed, Thermactor air, and canister purge to compensate for coolant temperature.

VACUUM SWITCHES

Signals driver demands to computer during cruise, closed throttle, and wide-open throttle conditions.
Accordingly, electrical signals tell computer to adjust Thermactor air, air/fuel ratio, and ignition timing.

KNOCK SENSOR

Responds to spark knock caused by over-advanced ignition.
Signals computer to retard timing.

13 CARBURETOR TYPES

After studying this chapter, you will be able to:
● *List and describe carburetors by six different classifications.*
● *Give the basic formula for calculating carburetor size.*
● *Given the size of the engine and rpm requirements, select the correct size carburetor for the engine.*

This chapter will explain the most common automotive carburetors. Special attention will be given to new and important features. Finally, several typical carburetor types will be studied.

CLASSIFYING CARBURETORS

It is very important for an auto mechanic to be able to identify basic carburetor types on sight. If you can look at a carburetor and determine most of its operating characteristics, troubleshooting and repair will be relatively simple.

Carburetors can be classified in several ways:
1. Airflow direction.
2. Number of barrels (air horns).
3. Stage (design and method of opening throttle plates).
4. Venturi type.
5. Size (airflow capacity).
6. Fuel bowl configuration.

AIRFLOW DIRECTION

Three carburetor types can be identified by how air flows through the air horn. They are the downdraft, updraft, and sidedraft carburetors.

Downdraft carburetor

A *downdraft carburetor*, Fig. 13-1A, is normally bolted to the top of the engine's intake manifold. Fuel, drawn into the airstream, falls as it atomizes. Thus, a large-diameter air horn will still provide good low-speed effi-

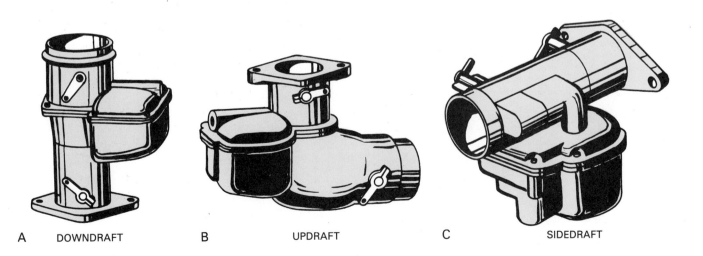

A   DOWNDRAFT   B   UPDRAFT   C   SIDEDRAFT

Fig. 13-1. Carburetor classifications for airflow direction. A—Downdraft carburetor, most common type, mounts on top of intake manifold. Air flows straight down horn. B—In updraft carburetor (no longer in use) air flows upward through horn. C—With sidedraft carburetor, air flows horizontally. (Keystone)

ciency. Fuel droplets tend to break up easily.

Downdraft carburetors are, by far, the most common. They provide both high fuel mileage and high power capabilities.

Updraft carburetor

An *updraft carburetor,* Fig. 13-1B, draws air up through the carburetor body. Since the fuel has to be lifted as it enters the air horn, air velocity must be very high. To form such a rapid airstream, the diameter of the carburetor throat is small. This severely limits engine power. For this reason, updraft carburetors are no longer used.

Sidedraft carburetor

A *sidedraft* or *crossdraft carburetor* directs airflow horizontally through the air horn. Fig. 13-1C shows the basic airstream action in a sidedraft carburetor.

This type is used when there is little clearance between the top of the engine and the car's hood. The carburetor and air cleaner sit to one side of the engine.

CARBURETOR BARRELS

A *barrel* is another name for carburetor throat or air horn. Carburetors may have one, two, three, or four barrels.

Single barrel

A *single* or *one-barrel carburetor,* Fig. 13-2A, has one throat. Commonly used on small four and six-cylinder engines, this type provides excellent fuel economy and dependability. However, with its small airflow capacity, engine power is usually low.

Two-barrel carburetor

A *two-barrel carburetor* has two air horns passing through the body, Fig. 13-2B. It is actually two single-barrel carburetors formed into one unit. Usually, each barrel is equipped with its own idle circuit, choke plate, and other parts.

Two-barrel carburetors represent an outstanding balance between engine power and economy. Their air horn diameters are small enough to give high fuel mileage. Yet, they are not so small that they cut engine horsepower greatly. Two-barrel carburetors are very common on all engines.

Three-barrel carburetor

A *three-barrel carburetor* has two small and one very large air horn. See Fig. 13-2C. The two forward barrels are small and round. The rear barrel is much larger and somewhat rectangular in shape. Three-barrel carburetors are NOT frequently used on production automobiles.

Four-barrel carburetor

A *four-barrel carburetor* has four air horns in a single body, Fig. 13-2D. The two front barrels operate like a two-barrel. They contain idle circuits and operate at low to medium engine speeds and loads. Normally, the rear barrels only open up when high engine power output is required or for top speed, ''hard'' acceleration, full throttle, etc.

Four-barrel carburetors are often used on large displacement V-8 or some V-6 engines. They have the potential of providing good fuel economy (only front two barrels open) as well as high horsepower output (all four barrels open).

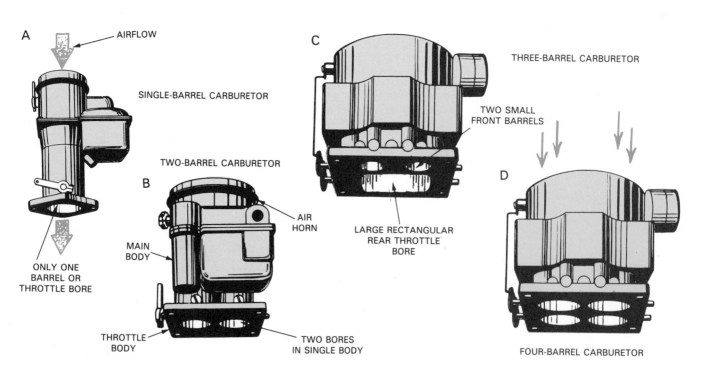

Fig. 13-2. Carburetors can use from one to four barrels in a single body. A—Single-barrel carburetor only has one throat or barrel. B—Two-barrel carburetor has two barrels on one carburetor body. It is used on both small and large engines. C—Rare three-barrel carburetor has two round, front bores and a larger rectangular rear bore. Rear throttle bore operates only during high power demands. D—Four-barrel carburetor is simply two two-barrel carburetors combined into one body. (Keystone)

Over the past few years, four-barrel carburetors have been partially phased out for more fuel efficient one-barrels, two-barrels, and fuel injection systems.

PRIMARY THROTTLES

The two front air horns in a four-barrel carburetor are called the *primaries.* They operate during idle and part throttle, Fig. 13-3D. The rear two barrels in a four-barrel are the *secondaries.* They open when the primaries go beyond part throttle.

The primaries will sometimes be smaller in diameter than the secondaries. The primary throttle bores act like a very small, fuel efficient carburetor under normal driving conditions. Then, under full throttle, the large secondaries "kick in" to supply a large amount of fuel mixture to the engine.

When the secondary throttle plates are made larger than the front primary plates, the carburetor is classified as a *spread bore.*

CARBURETOR STAGE

Carburetor stage refers to how the throttle butterflies are opened. They may open together, or one after the other. Throttle opening can be either mechanically or vacuum actuated.

Single stage

A *single stage carburetor's* throttle plates open at exactly the same time. As shown in Fig. 13-4, the two throttle plates are fastened side-by-side to a single throttle shaft. As the throttle lever swings open, both throttles open the same amount.

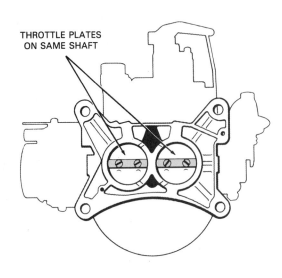

TO LEFT MANIFOLD   TO RIGHT MANIFOLD

Fig. 13-4. When both throttle plates open at same time, carburetor is called a single stage. (Ford Motor Co.)

1. PRIMARY THROTTLE OPENED ALMOST HALFWAY
2. MECHANICAL LINKAGE BETWEEN PRIMARIES AND SECONDARIES
3. SPREAD BORE HAS LARGE SECONDARIES.

Fig. 13-3. A—Primary throttles operate at all times and are usually the front barrels. Normally, rear or secondary barrels only function under high loads. B—Front barrels in spread bore carburetor are smaller than rear barrels. This improves gas mileage, yet allows for high power output. (Ford and Carter Carburetor Div.)

Two-stage carburetor

When the throttle butterflies do NOT open exactly together, the carburetor is *two-stage* or *double-stage.* Most passenger car four-barrel and many modern two-barrel carburetors are two-stage. See Fig. 13-5.

Normally, the front primary throttle opening is controlled by mechanical linkage and the rear secondary plates are opened by engine vacuum, Fig. 13-6.

A few two-barrel carburetors use this principle to open the rear butterfly. Only the front barrel functions during normal driving conditions. Then with hard acceleration, diaphragm action flips open the rear barrel.

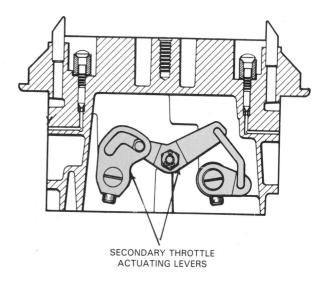

SECONDARY THROTTLE ACTUATING LEVERS

Fig. 13-7. Mechanical linkage is designed to delay opening of secondary throttles until engine vacuum builds and engine is ready for more fuel. (Holley)

Vacuum secondaries

Since engine vacuum is an indicator of engine load, the throttle plates open ONLY when there is a need for more power. The gas pedal can be floored, yet, the rear butterflies do NOT open until engine vacuum increases. This prevents the engine from *"bogging"* (severe, temporary loss of power). If all throttle plates were

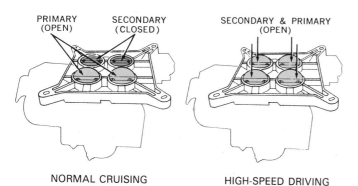

PRIMARY (OPEN) SECONDARY (CLOSED)   SECONDARY & PRIMARY (OPEN)

NORMAL CRUISING   HIGH-SPEED DRIVING

Fig. 13-5. When primary throttles open independently of secondary throttles, carburetor is classified as two-stage. Secondary opens when more power is needed. (Ford)

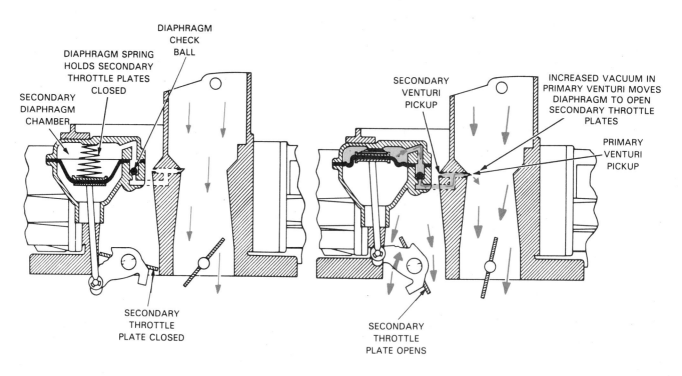

LOW SPEED/LIGHT LOAD   HIGH SPEED/HEAVY LOAD

Fig. 13-6. Primary throttle is partially opened to supply moderate power needs. Venturi vacuum is very low and rear butterfly stays closed. Primary throttles swing open for more power. Airstream forms a vacuum which pulls on secondary diaphragm. Then, diaphragm stretches up and pulls rear throttle plates open for more power. (Holley Carburetors)

swung open suddenly, the engine could not properly handle the large rush of air.

Mechanical secondaries

Usually, the secondary mechanical linkage is *progressive* (secondary throttle opening is delayed until primaries open about 45°). See Fig. 13-7. Progressive throttle linkage, like vacuum control, helps to prevent engine bogging and fuel waste.

Auxiliary air valve

Some four-barrel and a few two-barrel carburetors use an extra butterfly valve over the secondary throttle valves. It is called an *auxiliary air valve* or *auxiliary throttle valve.* One type air valve is shown in Fig. 13-8. It should not be confused with a choke valve.

This valve is designed to keep secondary barrel air velocity high enough to assure complete fuel mixing and atomization. It stops secondary throat operation until primary air speed is high enough to allow efficient operation. The auxiliary air valve is normally held closed by a light spring or counterweight and covers the air horn entrance.

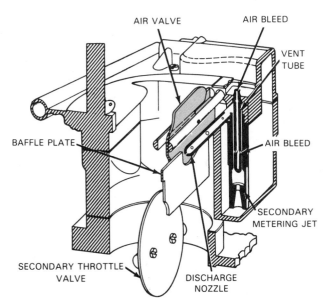

Fig. 13-8. A secondary air valve is sometimes placed over top of secondary throttle plates. It prevents secondary barrel operation until sufficient airflow is present in primary. (Chrysler)

VENTURI CLASSIFICATION

Various venturi designs are used in automotive carburetors. The size and shape of a venturi will depend upon carburetor size (barrel diameter), application (used for economy or power), and design.

The five major venturi types are:
1. Single venturi.
2. Double-venturi.
3. Triple-venturi.
4. Vaned venturi.
5. Variable venturi.

Single venturi

A *single venturi* is a restriction formed in the wall of the carburetor throat, Fig. 13-9. It is no longer used in the front barrels of a carburetor without additional secondary and booster venturis.

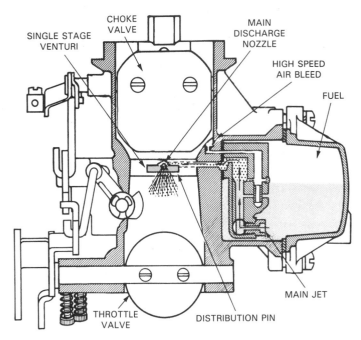

Fig. 13-9. Single stage venturi only has one restriction formed into wall of carburetor bore. (Holley Carburetors)

Double-venturi

A *double-venturi* carburetor has a primary venturi built into the wall of the throat and a secondary venturi suspended in the middle of the air horn, Fig. 13-10.

This type is generally used in the front or primary barrels of a carburetor but is sometimes found also in the secondary barrels.

A double-venturi provides good low and high speed efficiency. The extra venturi is capable of drawing in and breaking up fuel at very low engine speeds.

Triple-venturi

A *triple-venturi carburetor,* like the one shown in Fig. 13-11, has three venturis in each throttle bore. The third venturi further improves vacuum and atomization at low engine speeds.

Vaned venturi

A *vaned venturi* is designed to improve fuel economy over both dual and triple venturis. One is pictured in Fig. 13-12. It is a secondary venturi with fins or vanes running radially from the center to the outer edge of the venturi.

The *vanes* give turbulence to the airstream passing through the venturi. The turbulence helps break up the fuel droplets as they leave the main discharge tube. This increases atomization, vaporization, and fuel economy.

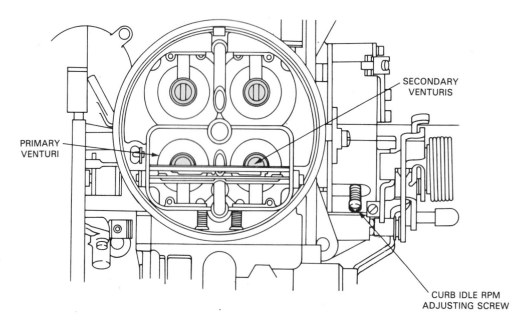

Fig. 13-10. A two-stage venturi uses both a primary venturi in bore wall and a secondary venturi suspended in center of air horn. (Ford)

Fig. 13-11. Three-stage venturi has primary venturi as well as two more secondary venturis mounted in middle of carburetor bore. (Rochester Carburetors)

Fig. 13-12. Vaned venturi is designed to increase fuel economy by improving fuel mixing (atomization) at low engine speeds. (Holley Carburetors)

However, the extra restriction of the vanes cuts engine power slightly.

Variable venturi

A *variable venturi* changes the carburetor throat area as engine speed and load change. The variable venturi increases throat air velocity matching air velocity and fuel entry with the exact needs of the engine. See Fig. 13-13.

Usually, a floatlike device (venturi valve) will be hinged to the side of the air horn. Metering rods are attached to the venturi valves. A large vacuum diaphragm is connected to the variable venturi valves. The diaphragm controls venturi opening and closing.

Whenever engine vacuum drops (high engine load conditions), the venturi diaphragm pulls the venturi valve open and the metering rods are pulled out of their jets. Air and fuel flow increase simultaneously, Fig. 13-14.

A variable venturi carburetor is so efficient in controlling fuel mixture ratios that idle and full power circuits are NOT required. The main metering system and

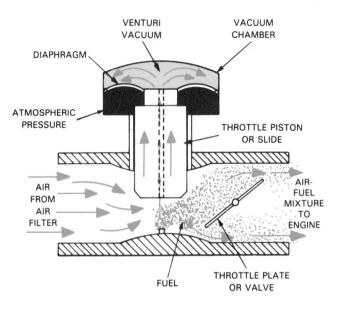

VENTURI VACUUM

VACUUM CHAMBER

DIAPHRAGM

ATMOSPHERIC PRESSURE

THROTTLE PISTON OR SLIDE

AIR FROM AIR FILTER

AIR-FUEL MIXTURE TO ENGINE

FUEL

THROTTLE PLATE OR VALVE

Fig. 13-13. Variable venturi tends to maintain constant air velocity independent of engine speed. It can vary its area to match engine speed, load, and vacuum.

the variable venturis and variable jets can supply the fuel mixture needs for all engine speed conditions. This provides the potential for better fuel economy.

FUEL BOWL CLASSIFICATIONS

Carburetor fuel bowls come in various shapes, sizes, locations, and numbers. They differ greatly from carburetor to carburetor. Normally, a fuel bowl will only have one inlet needle valve assembly. However, one or two floats can be used to operate the inlet valve.

The two basic types of fuel bowls are: integral fuel bowl and removable fuel bowl.

Integral fuel bowl

A bowl cast as part of the carburetor body is known as an *integral fuel bowl.* One is pictured in Fig. 13-15.

An integral fuel bowl can be located in front, to one side, or even in the center of the carburetor throat. See Fig. 13-16. The float may be hinged directly to the side of the bowl or to the air horn. Integral fuel bowls are used in one, two, and four-barrel carburetors.

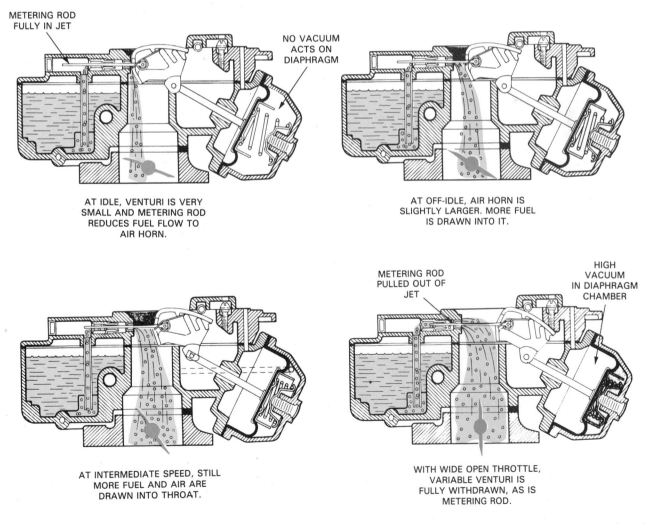

METERING ROD FULLY IN JET

NO VACUUM ACTS ON DIAPHRAGM

AT IDLE, VENTURI IS VERY SMALL AND METERING ROD REDUCES FUEL FLOW TO AIR HORN.

AT OFF-IDLE, AIR HORN IS SLIGHTLY LARGER. MORE FUEL IS DRAWN INTO IT.

METERING ROD PULLED OUT OF JET

HIGH VACUUM IN DIAPHRAGM CHAMBER

AT INTERMEDIATE SPEED, STILL MORE FUEL AND AIR ARE DRAWN INTO THROAT.

WITH WIDE OPEN THROTTLE, VARIABLE VENTURI IS FULLY WITHDRAWN, AS IS METERING ROD.

Fig. 13-14. Diaphragm controls variable venturis in this carburetor design. As engine power output increases, venturis swing out of air horn. (Ford)

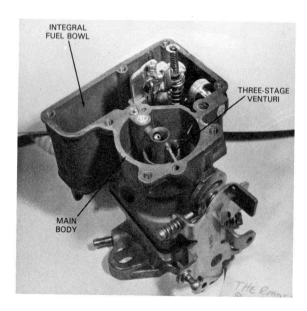

Fig. 13-15. An integral fuel bowl is cast as part of carburetor main body. Various shapes and sizes are utilized.

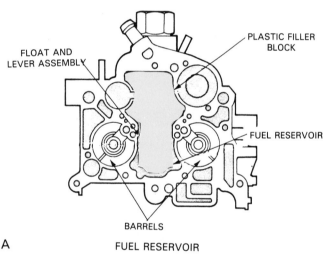

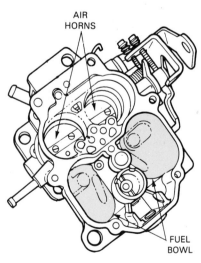

Fig. 13-16. Fuel bowl locations. A—In center of carburetor, between throttle bores. B—In front of bores. (Delco and Chrysler)

Removable fuel bowl

A *removable fuel bowl* is fastened to the main carburetor body. As shown in Fig. 13-17, a GASKET keeps fuel from leaking between the bowl and body.

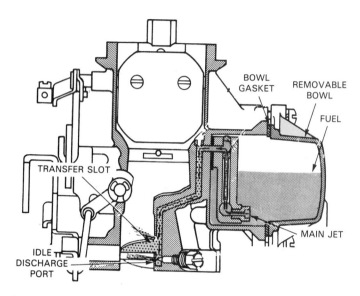

Fig. 13-17. Cutaway shows removable fuel bowl. Both two and four-barrel carburetors use them.

CARBURETOR SIZE

Generally speaking, carburetor size is given in *cfm* (cubic feet of air per minute), which is the amount of air that can flow through a carburetor at wide open throttle. A cfm rating indicates maximum airflow capacity.

A FLOW BENCH, Fig. 13-18, is used to measure carburetor capacity. It forces air through the carburetor barrels while measuring airflow.

Manufacturers utilize various testing procedures and pressures to arrive at carburetor cfm ratings. Thus, carburetor sizes are not always consistent but are usually excellent indicators.

Carburetor sizes vary from around 250 cfm on small, economy, one-barrel carburetors to well over 1000 cfm on high performance four-barrels.

Small cfm carburetors are more fuel efficient than larger carburetors. Air velocity, fuel mixing, and atomization are better with small throttle bores. A large cfm rating is desirable for high engine power output or racing at high engine rpm.

CARBURETOR SELECTION

Choosing an exact carburetor size is very complicated. Many technical factors (vehicle weight, differential gear ratio, engine rpm range, engine volumetric efficiency, power vs economy needs, etc.) play a role in determining the right carburetor cfm rating. In most cases, the carburetor installed by the auto manufacturer will be the ideal size. It will provide an excellent compromise between both economy and performance.

272

Fig. 13-18. Flow bench forces huge amounts of air through carburetor. It can be used to determine carburetor size in cfm. (Zorian Engineering)

A very basic formula allows carburetor size calculation. It is:

$$\frac{CID}{2} \times \frac{rpm}{1728} \times .80 = cfm$$

CID = engine cubic inch displacement
rpm = maximum engine speed in revolutions per minute

A low rpm value (3000) is needed if high fuel economy is the primary concern. A higher engine speed provides more power.

As an example, if you have a 350 CID engine and will be operating at only highway speeds (55 mph or around 3000 rpm):

$$\frac{350}{2} \times \frac{3000}{1728} \times .80 = 243 \text{ cfm}$$

You need a carburetor rated at about 250 cfm. It would provide excellent cruising speed economy and satisfactory low-speed power.

If your engine has been highly modified for performance and power output (racing cam, tube type exhaust headers, stiff valve springs, internally balanced engine assembly, increased compression ratio, etc.), then a higher rpm value could be used (8000 rpm).

$$\frac{350}{2} \times \frac{8000}{1728} \times .80 = 648 \text{ cfm}$$

A highly modified street engine would need a larger carburetor (around 650 cfm). The larger carburetor increases the engine's breathing ability and horsepower output. Low-speed power, however, would usually be lower because of the drop in throat velocity.

By far the most common problem with carburetor selection is OVER-CARBURETION. This is a carburetor with an excessively high cfm rating. Over-carburetion will not only cut gas mileage but will also REDUCE

ENGINE	ENGINE R.P.M.																
C.I.D.	1000	1500	2000	2500	3000	3500	4000	4500	5000	5500	6000	6500	7000	7500	8000	8500	9000
100	29	44	58	72	87	101	116	130	145	159	174	188	203	217	231	246	260
125	36	54	72	90	109	127	145	163	181	199	217	235	253	271	289	307	326
150	43	65	87	109	130	152	174	195	217	239	260	282	304	326	347	369	391
175	51	76	101	127	152	177	203	228	253	279	304	329	354	379	405	430	456
200	58	87	116	145	174	203	231	260	289	318	347	376	405	434	463	492	521
225	65	98	130	163	195	228	260	293	326	358	391	423	456	488	521	553	586
250	72	109	145	181	217	253	289	326	362	398	434	470	506	543	579	615	651
275	80	119	159	199	239	279	318	358	398	438	477	517	557	597	637	676	716
300	87	130	174	217	260	304	347	391	434	477	521	564	608	651	694	738	781
325	94	141	188	235	282	329	376	423	470	517	564	611	658	705	752	799	846
350	101	152	203	253	304	354	405	456	506	557	608	658	709	760	810	861	911
375	109	163	217	271	326	380	434	488	543	597	651	705	760	814	868	922	977
400	116	174	231	289	347	405	463	521	579	637	694	752	810	868	926	984	1042
425	123	184	246	307	369	430	492	553	615	676	738	799	861	922	984	1045	1107
450	130	195	260	326	391	456	521	586	651	716	781	846	911	977	1042	1107	1172
475	137	206	275	344	412	481	550	618	687	756	825	893	962	1031	1100	1168	1237
500	145	217	289	362	434	506	579	651	723	796	868	940	1013	1085	1157	1230	1302
525	152	228	304	380	456	532	608	684	760	836	911	987	1063	1139	1215	1291	1367
550	159	239	318	398	477	557	637	716	796	875	955	1034	1114	1194	1273	1353	1432
575	166	250	333	416	499	582	666	749	832	915	998	1081	1165	1248	1331	1414	1497
600	174	260	347	434	521	608	694	781	868	955	1042	1128	1215	1302	1389	1476	1563
625	181	271	362	452	543	633	723	814	904	995	1085	1175	1266	1356	1447	1537	1628
650	188	282	376	470	564	658	752	846	940	1034	1128	1223	1317	1411	1505	1599	1693
675	195	293	391	488	586	684	781	879	977	1074	1172	1270	1367	1465	1563	1660	1758
700	203	304	405	506	608	709	810	911	1013	1114	1215	1317	1418	1519	1620	1722	1823

CARBURETION GUIDE

Fig. 13-19. This chart, based upon single carburetor applications, is a basic guide for carburetor size. To use it, read down engine displacement (CID) column for displacement nearest size of your engine. Then read across to maximum usable rpm capability of engine. Size of carburetor will be given in cubic feet per minute. (Holley Carburetors)

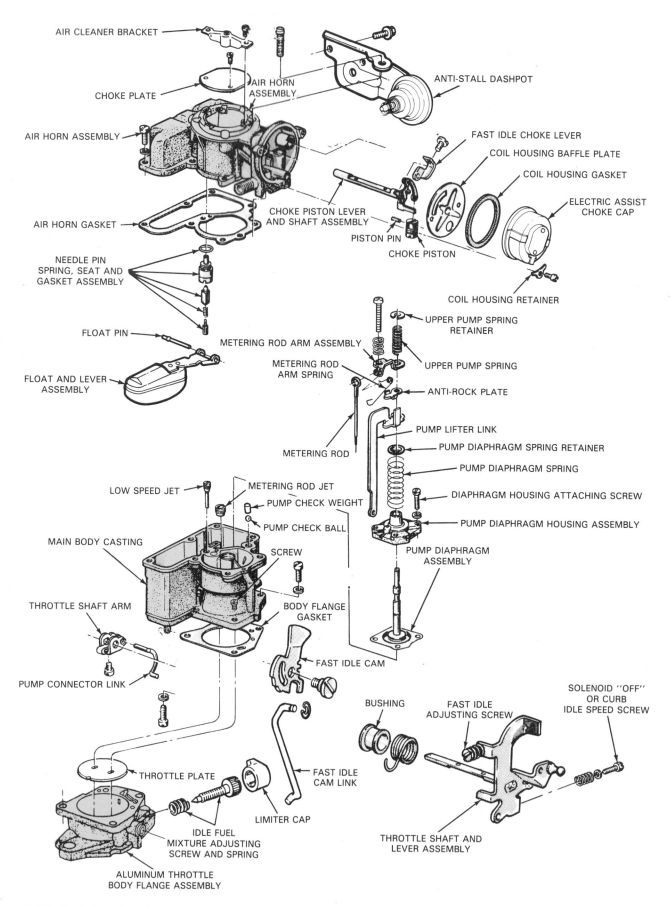

AIR CLEANER BRACKET

CHOKE PLATE

AIR HORN ASSEMBLY

AIR HORN GASKET

NEEDLE PIN
SPRING, SEAT AND
GASKET ASSEMBLY

FLOAT PIN

FLOAT AND LEVER
ASSEMBLY

LOW SPEED JET

MAIN BODY CASTING

THROTTLE SHAFT ARM

PUMP CONNECTOR LINK

THROTTLE PLATE

IDLE FUEL
MIXTURE ADJUSTING
SCREW AND SPRING

ALUMINUM THROTTLE
BODY FLANGE ASSEMBLY

AIR HORN
ASSEMBLY

ANTI-STALL DASHPOT

FAST IDLE CHOKE LEVER

COIL HOUSING BAFFLE PLATE

COIL HOUSING GASKET

ELECTRIC ASSIST
CHOKE CAP

CHOKE PISTON LEVER
AND SHAFT ASSEMBLY

PISTON PIN

CHOKE PISTON

COIL HOUSING RETAINER

UPPER PUMP SPRING
RETAINER

METERING ROD ARM ASSEMBLY

METERING ROD
ARM SPRING

UPPER PUMP SPRING

ANTI-ROCK PLATE

PUMP LIFTER LINK

METERING ROD

PUMP DIAPHRAGM SPRING RETAINER

PUMP DIAPHRAGM SPRING

DIAPHRAGM HOUSING ATTACHING SCREW

PUMP DIAPHRAGM HOUSING ASSEMBLY

PUMP DIAPHRAGM
ASSEMBLY

METERING ROD JET

PUMP CHECK WEIGHT

PUMP CHECK BALL

SCREW

BODY FLANGE
GASKET

FAST IDLE CAM

BUSHING

FAST IDLE
ADJUSTING SCREW

SOLENOID "OFF"
OR CURB
IDLE SPEED SCREW

FAST IDLE
CAM LINK

LIMITER CAP

THROTTLE SHAFT AND
LEVER ASSEMBLY

Fig. 13-20. Exploded view shows clearly three sections or bodies of carburetor. Note locations of gaskets and other typical parts. (GMC)

ENGINE POWER. Air velocity will be so low that fuel mixing, fuel entry, atomization, combustion power, and efficiency will all suffer drastically.

For high "street performance," a 300 to 400 CID engine requires a 650 to 750 cfm carburetor. This is the largest carburetor the engine could handle.

For fuel economy, select the smallest possible carburetor that will provide adequate engine speed and performance, the smaller the better. Fig. 13-19 shows typical carburetor sizes for different engine sizes.

CARBURETOR CONSTRUCTION MATERIALS

Various materials are used in the construction of carburetors. These materials include:
1. Aluminum (throttle, main body, and venturi).
2. Steel (throttle plates, choke, shafts, and fast idle cam).
3. Plastic (floats and baffles).
4. Cast iron (throttle body).
5. Synthetic rubber (pump pistons, seals, diaphrams, and needle tips).
6. Brass (floats, needle valves, valve seats, and jets).
7. Leather (older pump pistons).

You should become familiar with these materials and with the type of materials used in each carburetor part. This information will be especially useful when you begin to service and repair carburetors. Each substance must be handled differently during a carburetor rebuild.

MULTIPLE CARBURETORS

Before the energy crisis and concerns over pollution, *multiple carburetors* (more than one carburetor mounted on a single intake manifold and engine) were used on very high performance engines. They increased volumetric efficiency (air intake ability) and horsepower. The most common multiple carburetor arrangements are the three two-barrel and two four-barrel induction systems.

CARBURETOR BODY SECTIONS

There are three very important castings or sections to a carburetor. These sections, Fig. 13-20, include:
1. Air horn body.
2. Main body.
3. Throttle body.

As you read about these three body sections, refer to Fig. 13-20. It is very important to basic carburetion.

Air horn body
The *air horn body* serves as a lid for the main body and fuel bowl. It is held to the main body by screws. A gasket between the two stops air and fuel leakage.

Other parts are often fastened to the air horn body: choke, hot idle compensator, fast idle linkage rod, choke vacuum break, and, sometimes, the float and pump mechanisms.

Main body
The *main body* of the carburetor is the largest section. Usually, it will house the fuel bowl, main jets, air bleeds, power valve, pump checks, diaphragm type accelerator

pump, venturis, and circuit passages. Frequently, it also contains the float mechanism.

Throttle body
The carburetor's *throttle body* fastens to the bottom of the main body with screws. Again, a gasket makes the two air and fuel tight.

Found in or on a throttle body are the throttle butterflies, throttle shaft, throttle lever, idle mixture screws, idle speed screw, fast idle cam, dashpot, and fast idle solenoid.

Linkage rods for the choke, fast idle cam, metering rod, accelerator pump, temperature compensator, etc. run from the throttle shaft lever to the various components.

Note! The three body sections will vary somewhat from carburetor to carburetor. The next section will show some of these differences.

TYPICAL CARBURETORS

Figs. 13-21 through 13-28 show a number of carburetor designs. As you read the captions, study the special features of each carburetor. Try to visualize the operation and purpose of each component. This will help you later when trying to repair carburetors.

IDLE SPEED SETTING PROCEDURES

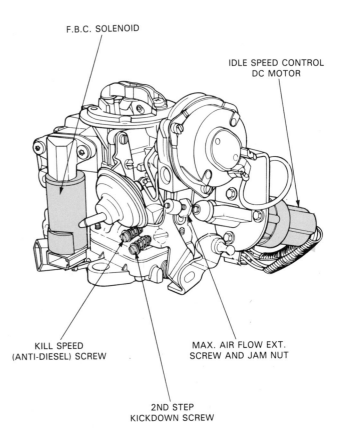

Fig. 13-21. This one-barrel carburetor uses an idle speed control motor. Also note feedback solenoid and adjustment screw locations. (Ford Motor Co.)

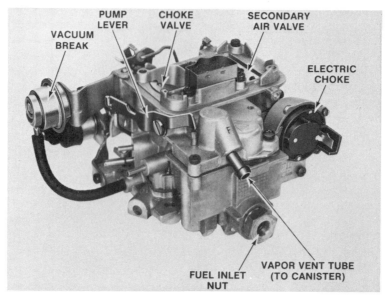

Fig. 13-22. Study two views of two-barrel carburetor. Locate secondary air valve, choke, vacuum break, and fast idle cam. (General Motors Corp.)

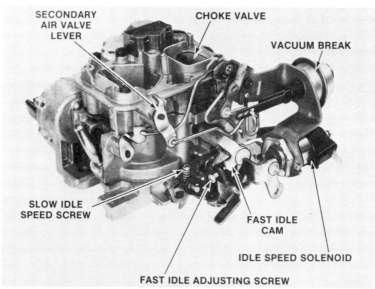

SUMMARY

Automotive carburetors can be classified in several ways—by airflow direction, number of barrels, stage, size rating, fuel bowl configuration, etc. It is very important for a mechanic to be able to identify these classifications. This will help during troubleshooting.

A downdraft carburetor is, by far, the most common type. Updraft and sidedraft are NOT commonly used. One, two, and four-barrel carburetors are frequently utilized. A one-barrel is very suitable on small displacement engines (4 and 6-cylinder). Two and four-barrel carburetors work well on slightly larger engines.

The primary or front air horn(s) of a 2 or 4-barrel carburetor supply air and fuel at low, medium, and high engine speeds and loads. They operate at all times. The secondary barrels function beyond half throttle.

The throttle plates in a single stage carburetor open

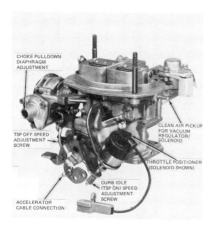

Fig. 13-23. This is another two-barrel carburetor. Compare it to the one shown in previous illustration. This one has barrels side-by-side, instead of one in front of other. (Ford)

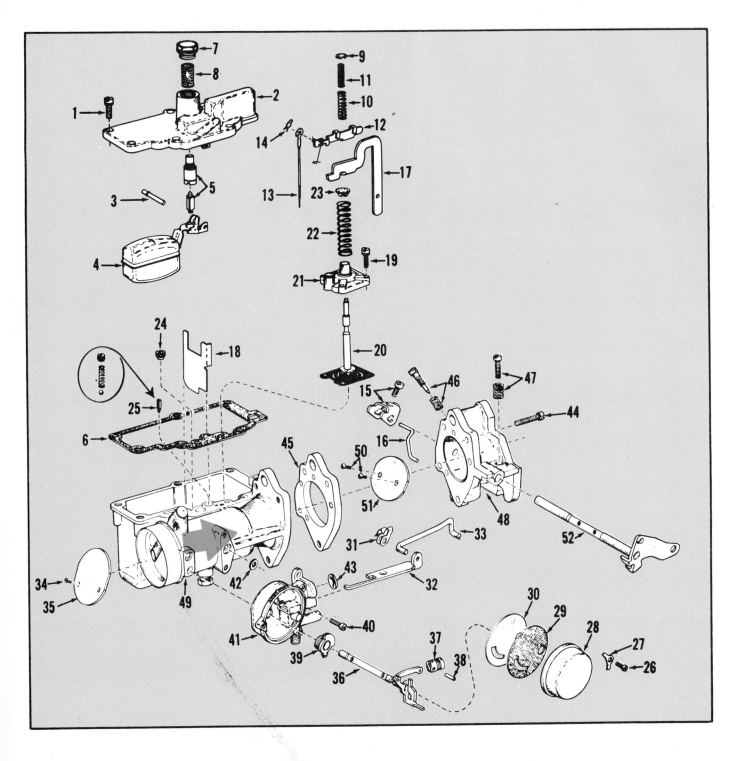

1. Bowl cover attaching screw and washer (6)
2. Bowl cover and strainer assembly
3. Float lever pin
4. Float and lever assembly
5. Needle and seat assembly
6. Bowl cover gasket
7. Strainer nut assembly
8. Bowl strainer
9. Upper pump spring retainer
10. Upper pump spring (outer)
11. Upper pump spring (inner)
12. Metering rod arm assembly
13. Metering rod
14. Pin spring
15. Throttle shaft arm assembly
16. Throttle shaft arm connector link
17. Pump lifter link
18. Fuel bowl baffle plate

19. Diaphragm housing attaching screw and washer (4)
20. Pump diaphragm assembly
21. Pump diaphragm housing assembly
22. Pump diaphragm spring
23. Pump diaphragm spring retainer
24. Metering jet
25. Pump check needle
26. Coil housing attaching screw (3)
27. Coil housing retainer (3)
28. Thermostatic coil housing
29. Thermostatic coil housing gasket
30. Baffle plate
31. Choke connector rod retainer
32. Fast idle link
33. Choke connector rod
34. Choke valve attaching screw (2)
35. Choke valve
36. Choke piston lever, link and shaft assembly

37. Choke piston
38. Choke piston pin
39. Fast idle cam and spring assembly
40. Piston housing attaching screw (3)
41. Piston housing and plug assembly
42. Piston housing gasket
43. Welsh plug
44. Body flange attaching screw and washer
45. Body flange gasket
46. Idle adjustment screw and spring
47. Throttle lever adjusting screw and spring
48. Body flange
49. Main casting
50. Throttle valve attaching screw (2)
51. Throttle valve
52. Throttle shaft and lever assembly

Fig. 13-24. This is an older side draft carburetor. Airflow (arrow) is horizontal instead of down. (Chevrolet)

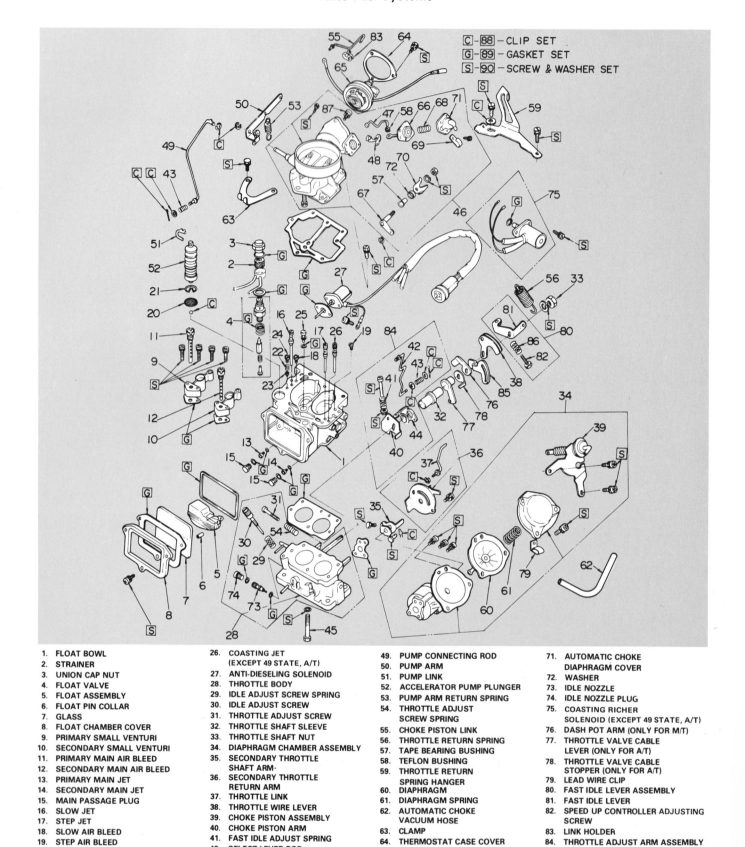

1. FLOAT BOWL	26. COASTING JET (EXCEPT 49 STATE, A/T)
2. STRAINER	27. ANTI-DIESELING SOLENOID
3. UNION CAP NUT	28. THROTTLE BODY
4. FLOAT VALVE	29. IDLE ADJUST SCREW SPRING
5. FLOAT ASSEMBLY	30. IDLE ADJUST SCREW
6. FLOAT PIN COLLAR	31. THROTTLE ADJUST SCREW
7. GLASS	32. THROTTLE SHAFT SLEEVE
8. FLOAT CHAMBER COVER	33. THROTTLE SHAFT NUT
9. PRIMARY SMALL VENTURI	34. DIAPHRAGM CHAMBER ASSEMBLY
10. SECONDARY SMALL VENTURI	35. SECONDARY THROTTLE SHAFT ARM-
11. PRIMARY MAIN AIR BLEED	36. SECONDARY THROTTLE RETURN ARM
12. SECONDARY MAIN AIR BLEED	37. THROTTLE LINK
13. PRIMARY MAIN JET	38. THROTTLE WIRE LEVER
14. SECONDARY MAIN JET	39. CHOKE PISTON ASSEMBLY
15. MAIN PASSAGE PLUG	40. CHOKE PISTON ARM
16. SLOW JET	41. FAST IDLE ADJUST SPRING
17. STEP JET	42. SELECT LEVER ROD
18. SLOW AIR BLEED	43. SELECT LEVER ROD SPRING
19. STEP AIR BLEED	44. SELECT LEVER
20. ACCELERATOR PUMP STRAINER	45. BOLT
21. STRAINER CLIP	46. AIR HORN ASSEMBLY
22. DISCHARGE CHECK VALVE SPRING	47. CHOKE DIAPHRAGM ROD
23. DISCHARGE CHECK VALVE	48. LINK HOLDER
24. OUTLET VALVE PLUG	
25. POWER VALVE JET (EXCEPT CALIFORNIA AND HIGH ALTITUDE)	

49. PUMP CONNECTING ROD	71. AUTOMATIC CHOKE DIAPHRAGM COVER
50. PUMP ARM	72. WASHER
51. PUMP LINK	73. IDLE NOZZLE
52. ACCELERATOR PUMP PLUNGER	74. IDLE NOZZLE PLUG
53. PUMP ARM RETURN SPRING	75. COASTING RICHER SOLENOID (EXCEPT 49 STATE, A/T)
54. THROTTLE ADJUST SCREW SPRING	76. DASH POT ARM (ONLY FOR M/T)
55. CHOKE PISTON LINK	77. THROTTLE VALVE CABLE LEVER (ONLY FOR A/T)
56. THROTTLE RETURN SPRING	78. THROTTLE VALVE CABLE STOPPER (ONLY FOR A/T)
57. TAPE BEARING BUSHING	79. LEAD WIRE CLIP
58. TEFLON BUSHING	80. FAST IDLE LEVER ASSEMBLY
59. THROTTLE RETURN SPRING HANGER	81. FAST IDLE LEVER
60. DIAPHRAGM	82. SPEED UP CONTROLLER ADJUSTING SCREW
61. DIAPHRAGM SPRING	83. LINK HOLDER
62. AUTOMATIC CHOKE VACUUM HOSE	84. THROTTLE ADJUST ARM ASSEMBLY
63. CLAMP	85. PRIMARY THROTTLE ARM
64. THERMOSTAT CASE COVER	86. SPEED UP CONTROLLER ADJUSTING SPRING
65. THERMOSTAT CASE	87. SCREW & WASHER
66. AUTOMATIC CHOKE DIAPHRAGM	88. CLIP SET
67. SELECT ARM LEVER	89. GASKET SET
68. AUTOMATIC CHOKE DIAPHRAGM SPRING	90. SCREW & WASHER SET
69. LEAD WIRE CLIP	
70. SELECT ARM	

Fig. 13-25. Exploded view of two-barrel carburetor. Study part names and locations. This type of illustration would be useful when disassembling and reassembling carburetor.

278

Carburetor Types

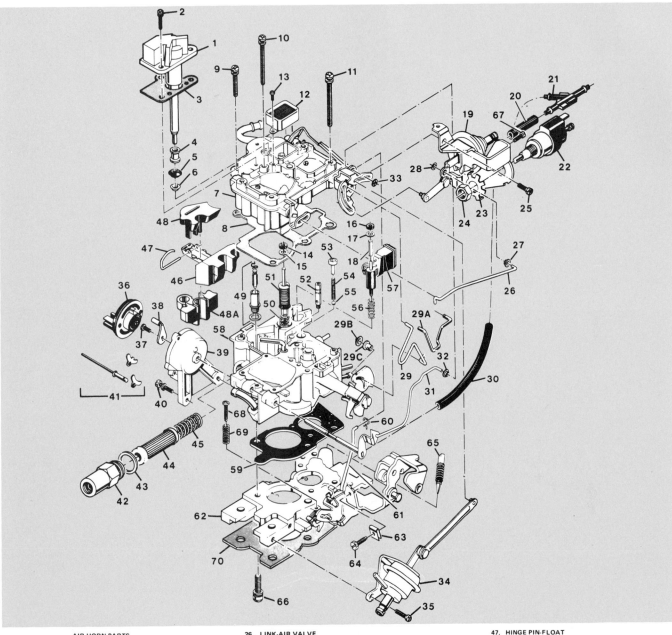

AIR HORN PARTS

1. MIXTURE CONTROL (M/C) SOLENOID
2. SCREW ASSEMBLY-SOLENOID ATTACHING
3. GASKET-M/C SOLENOID TO AIR HORN
4. SPACER-M/C SOLENOID
5. SEAL-M/C SOLENOID TO FLOAT BOWL
6. RETAINER-M/C SOLENOID SEAL
7. AIR HORN ASSEMBLY
8. GASKET-AIR HORN TO FLOAT BOWL
9. SCREW-AIR HORN TO FLOAT BOWL (SHORT)
10. SCREW-AIR HORN TO FLOAT BOWL (LONG)
11. SCREW-AIR HORN TO FLOAT BOWL (LARGE)
12. VENT STACK AND SCREEN ASSEMBLY
13. SCREW-VENT STACK ATTACHING
14. SEAL-PUMP STEM
15. RETAINER-PUMP STEM SEAL
16. SEAL-T.P.S. PLUNGER
17. RETAINER-T.P.S. PLUNGER SEAL
18. PLUNGER-T.P.S. ACTUATOR

CHOKE PARTS

19. VACUUM BREAK AND BRACKET ASSEMBLY-PRIMARY
20. HOSE-VACUUM BREAK PRIMARY
21. TEE-VACUUM BREAK
22. SOLENOID-IDLE SPEED
23. RETAINER-IDLE SPEED SOLENOID
24. NUT-IDLE SPEED SOLENOID ATTACHING
25. SCREW-VACUUM BREAK BRACKET ATTACHING

26. LINK-AIR VALVE
27. BUSHING-AIR VALVE LINK
28. RETAINER-AIR VALVE LINK
29. LINK-FAST IDLE CAM
29A LINK-FAST IDLE CAM
29B RETAINER-LINK
29C BUSHING-LINK
30. HOSE-VACUUM BREAK
31. INTERMEDIATE CHOKE SHAFT/LEVER/LINK ASSEMBLY
32. BUSHING-INTERMEDIATE CHOKE LINK
33. RETAINER-INTERMEDIATE CHOKE LINK
34. VACUUM BREAK AND LINK ASSEMBLY-SECONDARY
35. SCREW-VACUUM BREAK ATTACHING
36. ELECTRIC CHOKE-COVER AND COIL ASSEMBLY
37. SCREW-CHOKE LEVER ATTACHING
38. CHOKE COIL LEVER ASSEMBLY
39. CHOKE HOUSING
40. SCREW-CHOKE HOUSING ATTACHING
41. CHOKE COVER RETAINER KIT
67. SCREW-VACUUM BREAK BRACKET ATTACHING

FLOAT BOWL PARTS

42. NUT-FUEL INLET
43. GASKET-FUEL INLET NUT
44. FILTER-FUEL INLET
45. SPRING-FUEL FILTER
46. FLOAT AND LEVER ASSEMBLY

47. HINGE PIN-FLOAT
48. UPPER INSERT-FLOAT BOWL
48A LOWER INSERT-FLOAT BOWL
49. NEEDLE AND SEAT ASSEMBLY
50. SPRING-PUMP RETURN
51. PUMP PLUNGER ASSEMBLY
52. PRIMARY METERING JET ASSEMBLY
53. RETAINER-PUMP DISCHARGE BALL
54. SPRING-PUMP DISCHARGE
55. BALL-PUMP DISCHARGE
56. SPRING-T.P.S. ADJUSTING
57. SENSOR-THROTTLE POSITION (TPS)
58. FLOAT BOWL ASSEMBLY
59. GASKET-FLOAT BOWL

THROTTLE BODY PARTS

60. RETAINER-PUMP LINK
61. LINK-PUMP
62. THROTTLE BODY ASSEMBLY
63. CLIP-CAM SCREW
64. SCREW-FAST IDLE CAM
65. IDLE NEEDLE AND SPRING ASSEMBLY
66. SCREW-THROTTLE BODY TO FLOAT BOWL
68. SCREW-IDLE STOP
69. SPRING-IDLE STOP SCREW
70. GASKET-INSULATOR FLANGE

Fig. 13-26. Exploded view of more modern computer controlled carburetor. Locate the mixture control solenoid, throttle body, float, and idle speed solenoid. (Buick)

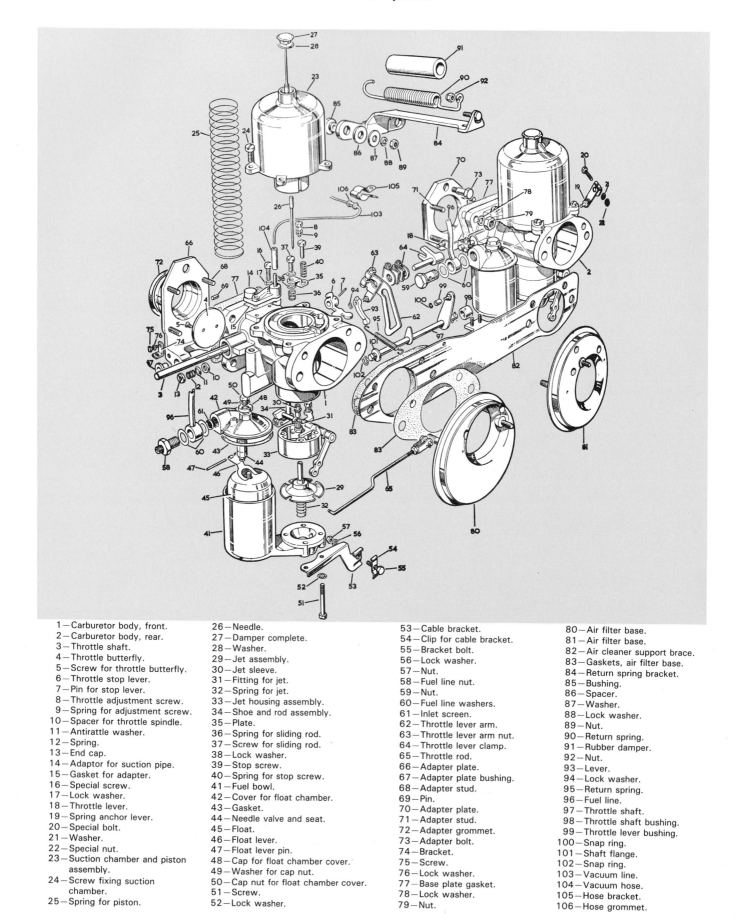

1—Carburetor body, front.
2—Carburetor body, rear.
3—Throttle shaft.
4—Throttle butterfly.
5—Screw for throttle butterfly.
6—Throttle stop lever.
7—Pin for stop lever.
8—Throttle adjustment screw.
9—Spring for adjustment screw.
10—Spacer for throttle spindle.
11—Antirattle washer.
12—Spring.
13—End cap.
14—Adaptor for suction pipe.
15—Gasket for adapter.
16—Special screw.
17—Lock washer.
18—Throttle lever.
19—Spring anchor lever.
20—Special bolt.
21—Washer.
22—Special nut.
23—Suction chamber and piston assembly.
24—Screw fixing suction chamber.
25—Spring for piston.

26—Needle.
27—Damper complete.
28—Washer.
29—Jet assembly.
30—Jet sleeve.
31—Fitting for jet.
32—Spring for jet.
33—Jet housing assembly.
34—Shoe and rod assembly.
35—Plate.
36—Spring for sliding rod.
37—Screw for sliding rod.
38—Lock washer.
39—Stop screw.
40—Spring for stop screw.
41—Fuel bowl.
42—Cover for float chamber.
43—Gasket.
44—Needle valve and seat.
45—Float.
46—Float lever.
47—Float lever pin.
48—Cap for float chamber cover.
49—Washer for cap nut.
50—Cap nut for float chamber cover.
51—Screw.
52—Lock washer.

53—Cable bracket.
54—Clip for cable bracket.
55—Bracket bolt.
56—Lock washer.
57—Nut.
58—Fuel line nut.
59—Nut.
60—Fuel line washers.
61—Inlet screen.
62—Throttle lever arm.
63—Throttle lever arm nut.
64—Throttle lever clamp.
65—Throttle rod.
66—Adapter plate.
67—Adapter plate bushing.
68—Adapter stud.
69—Pin.
70—Adapter plate.
71—Adapter stud.
72—Adapter grommet.
73—Adapter bolt.
74—Bracket.
75—Screw.
76—Lock washer.
77—Base plate gasket.
78—Lock washer.
79—Nut.

80—Air filter base.
81—Air filter base.
82—Air cleaner support brace.
83—Gaskets, air filter base.
84—Return spring bracket.
85—Bushing.
86—Spacer.
87—Washer.
88—Lock washer.
89—Nut.
90—Return spring.
91—Rubber damper.
92—Nut.
93—Lever.
94—Lock washer.
95—Return spring.
96—Fuel line.
97—Throttle shaft.
98—Throttle shaft bushing.
99—Throttle lever bushing.
100—Snap ring.
101—Shaft flange.
102—Snap ring.
103—Vacuum line.
104—Vacuum hose.
105—Hose bracket.
106—Hose grommet.

Fig. 13-27. This two-carburetor arrangement can be found on some foreign cars. Multiple carbs are used to increase engine performance.

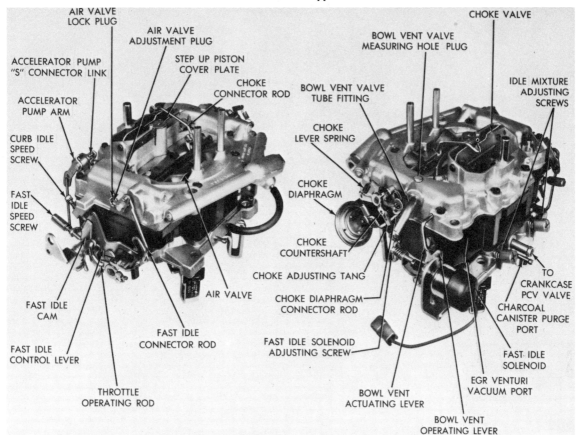

Fig. 13-28. This carburetor, called a "thermo quad," is a high performance four-barrel. Thermo is used because the carburetor main body is made of heat resistant plastic that stays cool. Quad is used to mean four-barrel. (Carter)

at exactly the same time. They are often fastened to the same throttle shaft and lever. With a two-stage carburetor, the primary butterflies open before the secondary butterflies. The secondaries can be operated either by a vacuum diaphragm or by mechanical linkage. Vacuum secondaries are more common.

An auxiliary air valve is used to smooth primary to secondary transition (secondary barrel "kick-in"). The valve plate fits over the top of the secondary barrels and opens after air demand is high enough.

Venturi classification ranges from a single venturi to a triple or three-stage venturi. Vaned venturis are used in high economy carburetors. Variable venturis are also designed for high efficiency. They maintain rapid air speed and fuel mixing action, even at low engine speeds.

Fuel bowls can be integral (built into carburetor body) or removable, integral being more common.

Carburetor size is given in cfm (cubic feet of airflow per minute). It is an indicator of carburetor capacity.

KNOW THESE TERMS

Barrel, Downdraft carburetor, Updraft carburetor, Sidedraft Carburetor, Primary throttles, Secondary throttles, Single stage carburetor, Two-stage carburetor, Vacuum secondaries, Mechanical secondaries, Progressive throttle linkage, Auxiliary air valve, Single venturi, Double-venturi, Triple-venturi, Vaned venturi, Variable venturi, Integral fuel bowl, Removable fuel bowl, CFM rating, Multiple carburetors, Air horn body, Main body, Throttle body.

REVIEW QUESTIONS—CHAPTER 13

1. List six methods of classifying carburetors.
2. A _____ carburetor is normally bolted to the top of the intake manifold, above the engine.
3. A _____-_____ carburetor has two air horns each equipped with its own idle circuit, choke plate, and other parts.
4. Which of the following statements does NOT apply to a four-barrel carburetor?
   a. Has a single air horn.
   b. Often used on large displacement V-8 engines.
   c. Has been partly phased out in favor of fuel injection and smaller, more efficient carburetors.
   d. Has primary and secondary throttles.
5. Discuss the difference between primary and secondary throttles and indicate where they are found.
6. When a carburetor's throttle plates open at the same time, it is known as _____ stage; when they do NOT open at exactly the same time, they are known as _____-stage or _____-stage.
7. Explain the purpose of delaying the opening of secondary throttles when the gas pedal is pressed to the floor.
8. A _____ _____ valve, similar in appearance to a choke valve, keeps secondary barrel air velocity high enough to assure complete fuel mixing and atomization.
9. List five different venturi designs.
10. Integral fuel bowls are found only on single-barrel

carburetors. True or false?

11. Give the basic formula for calculating carburetor size.

12. Compute the general size carburetor suited to a 302 CID engine operating at highway speeds.

13. List the types of materials used in construction of today's carburetors and where they might be used.

14. Which of the following are main sections or castings of the carburetor?

a. Throttle body.
b. Bowl.
c. Air horn body.
d. Main body.
e. Float mechanism.
f. Throttle valves.

15. A bigger carburetor will always produce more engine power because it will allow the engine to "breathe" more easily. True or false?

How many parts can you identify from top view of this carburetor? (Rochester Carburetors)

14

CARBURETOR DIAGNOSIS, SERVICE, REPAIR

After studying this chapter, you will be able to:
- *Describe problems that can develop in a carburetor.*
- *Diagnose common carburetor problems.*
- *Remove and replace a carburetor.*
- *Rebuild a carburetor following the directions in a service manual.*
- *Perform basic carburetor adjustments.*
- *Test and repair a computerized carburetor system.*
- *List the safety rules for carburetor service.*

After extended service, carburetor passages can become clogged with dirt, rust, and other debris. Gaskets rupture. Seals dry out and crack. Air bleeds become filled with dust and dirt. Engine sensors or the computer control system may malfunction.

Any of these problems can cause a carburetor malfunction. The carburetor's air-fuel mixture could become too rich or lean, upsetting engine efficiency, performance, and fuel economy. You, the mechanic, must locate and correct these kinds of troubles to save our natural resources and to keep the customer satisfied.

DIAGNOSING CARBURETOR PROBLEMS

Before tearing down a carburetor, always make sure the carburetor is causing the performance problem. Ignition and engine problems can cause similar symptoms (missing, high fuel consumption, stalling, etc.). Fouled spark plugs, cracked distributor caps, poor plug wire insulation, incorrect ignition timing, defective electronic ignition amplifier or pickup coil, faulty sensors or computer, and inoperative emission control devices may appear to be a malfunction of the carburetor. Only after you have checked these or other possible sources of the problem should you consider the possibility of a carburetor malfunction.

Finding faulty carburetor circuit

Determining which carburetor circuit is bad begins by using the trouble symptoms as a clue. For example, suppose that the engine misses at idle, only acts up when hot, and runs smooth at high speed. Knowing this, you can usually eliminate the high-speed circuit, accelerator pump circuit, and the choke as sources of the problem. However, several other circuits could produce these low-speed problems.

It is more logical to suspect that the idle circuit might be clogged or that the float level is too high. If further investigation shows these circuits to be working properly, check less likely causes (leaking power valve, inoperative temperature compensating valve, etc.). By going from most common to least common problems, you will be using a logical approach to troubleshooting.

Visual carburetor inspection

Visual inspection of the carburetor often provides clues to the problem. Remove the air cleaner housing. Look for evidence of fuel leakage, sticking choke, binding linkage, missing or disconnected vacuum hoses, or any other obvious problems, Fig. 14-1. A thick coating of road dirt usually means the carburetor has been in service for a long time.

As part of the inspection, check the rest of the engine compartment. Be on the lookout for disconnected wires and hoses. Listen for vacuum leaks. Examine the distributor cap for cracks or carbon tracking. Look for anything that could upset engine operation.

Air-fuel mixture problems

Often, internal carburetor problems show up as an air-fuel mixture either too rich or too lean.

An extremely *lean mixture* is the result of any condition allowing too much air and too little fuel to enter the intake manifold. Lean mixtures will cause an engine to MISS ERRATICALLY (now and then). Several conditions could cause a lean mixture: a vacuum leak, incorrect mixture screw adjustment, clogged fuel passage, low float level, or other problems.

An *overly rich mixture* is the result of too much fuel and too little air entering the engine. Such a mixture causes the engine to roll, lope, and give off a black exhaust (smoke). A number of carburetor problems could cause this condition. Among them are a clogged air bleed, high float level, and incorrect choke setting.

Exhaust gas analysis

A mechanic can use exhaust gas analysis to assist in diagnosing carburetor problems. An *exhaust gas analyzer* is a testing device that measures the chemical content of the engine exhaust gases. The analyzer draws a sam-

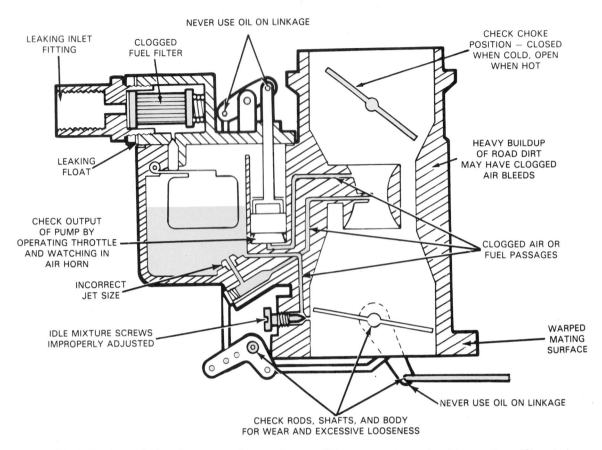

LEAKING INLET FITTING

CLOGGED FUEL FILTER

NEVER USE OIL ON LINKAGE

CHECK CHOKE POSITION – CLOSED WHEN COLD, OPEN WHEN HOT

LEAKING FLOAT

HEAVY BUILDUP OF ROAD DIRT MAY HAVE CLOGGED AIR BLEEDS

CHECK OUTPUT OF PUMP BY OPERATING THROTTLE AND WATCHING IN AIR HORN

CLOGGED AIR OR FUEL PASSAGES

INCORRECT JET SIZE

IDLE MIXTURE SCREWS IMPROPERLY ADJUSTED

WARPED MATING SURFACE

NEVER USE OIL ON LINKAGE

CHECK RODS, SHAFTS, AND BODY FOR WEAR AND EXCESSIVE LOOSENESS

Fig. 14-1. Study typical carburetor problems. Some will be apparent on visual inspection. (Chrysler)

ple of the exhaust gas out of the car's tailpipe. Modern types measure the amount of carbon monoxide (CO), hydrocarbons (HC), carbon dioxide (CO_2), and oxygen (O_2) in the exhaust. This information indicates the air-fuel ratio entering the engine.

The exhaust gas analyzer will quickly tell you whether the engine's air-fuel mixture is too lean or rich.

Reading spark plugs

Reading spark plugs is done by inspecting the color and texture of the deposits on the plug tip. Generally, a white, clean tip indicates a lean mixture. A black, soot covered tip indicates a rich mixture. This provides another easy way of checking carburetor operation.

Float system problems

Symptoms pointing to carburetor *float system* problems include: flooding, rich fuel mixture, lean fuel mixture, fuel starvation (no fuel), stalling, low speed engine miss, and high speed engine miss. See Fig. 14-2. When float operation is at fault, engine operation is affected at all speeds. The float system provides fuel for all of the other carburetor circuits.

High float setting

A *high float level* setting will allow excess fuel to enter the fuel bowl. If only slightly high, excessive fuel consumption, spark plug fouling, crankcase dilution, and poor engine performance will result. Less venturi vacuum

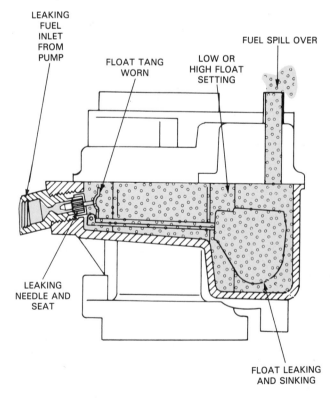

LEAKING FUEL INLET FROM PUMP

FLOAT TANG WORN

LOW OR HIGH FLOAT SETTING

FUEL SPILL OVER

LEAKING NEEDLE AND SEAT

FLOAT LEAKING AND SINKING

Fig. 14-2. Float troubles can affect engine and carburetor operation under all conditions. (Dodge)

will be needed to pull fuel out of the bowl and into the engine. An extremely high float setting can even let fuel leak out the top of the carburetor.

Low float setting

A *low float setting* will not let enough gasoline enter the fuel bowl. The engine may die or stumble during cornering, fast accelerating, or hard stops. The small amount of fuel can splash to one side of the bowl and no longer cover the circuit openings. Engine stalling, backfiring, spitting, and power loss may occur.

Sticking float

A *stuck float* can cause either carburetor flooding or fuel starvation. If stuck open, a continuous stream of fuel will pour into the bowl. As shown in Fig. 14-2, fuel pump pressure can finally push fuel out the top of the carburetor bowl vents.

This is a dangerous condition. Gasoline may leak onto the engine, starting a fire.

A float stuck open is usually caused by dirt inside the needle valve, by an incorrect float drop setting, or by a ruptured hollow brass float. If light tapping on the bowl temporarily corrects the flooding, check the float drop adjustment and action of the needle valve. Normally, a carburetor rebuild is required to permanently correct a carburetor flooding problem.

When a carburetor sits for extended periods, gum can cause the float needle to stick shut. This will not allow fuel to enter the bowl. The engine may not start or will not continue to run after initial starting. Part cleaning and usually a rebuild, is needed.

Idle system problems

Symptoms of carburetor *idle system problems* normally include rough idle, stalling at low speeds, or incorrect idle speed. The engine may run well at high speeds, however. Fig. 14-3 lists some common idle problems.

With *clogged idle passages,* fuel flow to the idle ports is reduced. There is less fuel to mix with the incoming air; thus, a lean mixture produces poor idle. This is a common problem because jets orifices are so small. Even a tiny bit of dirt can affect fuel flow.

By obstructing the premixing of air, a *clogged idle air bleed,* produces a rich air-fuel mixture. It can be the source of an idle problem. Most of the air bleed jets are at the top of the carburetor. Inspect them visually and clean them as needed.

On older cars, you can check whether you have an idle system problem by adjusting the idle mixture screws. If turning the screws has little or no effect on idle speed and smoothness, the idle circuit is NOT functioning properly. The carburetor probably needs to be overhauled.

WARNING! Late model carburetors have sealed idle mixture screws. Federal law prohibits adjusting them unless you are following manufacturer's instructions. Check the service manual if needed!

Acceleration system problems

When the engine is called upon for rapid acceleration, the carburetor must supply a much richer air-fuel mixture. As you learned earlier, this function is performed by the *acceleration system.* An accelerator pump directs a strong stream of fuel droplets into the air horn. Failure of this system will usually cause the engine to hesitate, stall, pop, backfire, or stumble. Look at Fig. 14-4.

To check accelerator pump operation, shut down the engine and remove the air cleaner cover. While looking directly down on the carburetor air horn, open and close the throttle several times. Each time the throttle opens, the accelerator pump should produce a heavy stream of fuel. The spray should begin as soon as the throttle begins to open and should continue for a short time after full throttle has reached maximum opening. If the stream is weak, either adjust the linkage or repair the pump.

High speed system problems

The *high speed system* is simple and dependable. When problems do occur, they are usually the result of a lean air-fuel mixture (miss at cruising speeds) or a rich mixture (poor fuel economy). Refer to Fig. 14-5.

Engine surge is a condition in which engine power at

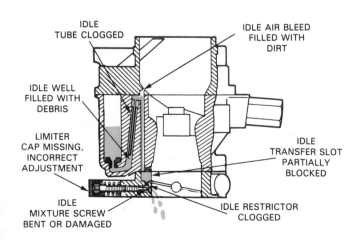

Fig. 14-3. Idle circuit problems are very common. The passages are so small they can fill with debris. (Chrysler)

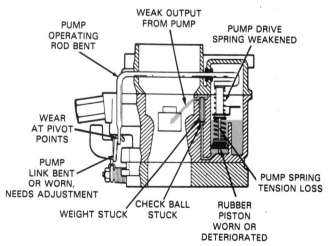

Fig. 14-4. A bad accelerator pump will cause an engine to hesitate or stumble upon acceleration. Always check adjustment of linkage first. (Chrysler)

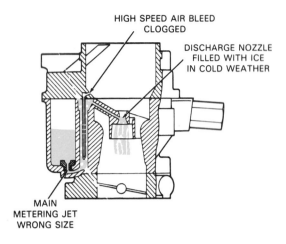

Fig. 14-5. High speed system is very dependable. There are not too many things that can go wrong with this system.

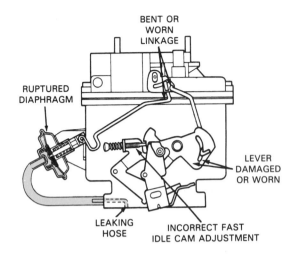

Fig. 14-6. Choke is common problem area. Note potential troubles.

cruising speeds seems to vary automatically. Its cause is frequently an extremely lean air-fuel mixture. This is encountered frequently with the lean, high-efficiency mixtures produced by today's carburetors.

A faulty mixture control solenoid or computer control circuit can upset carburetor air-fuel mixtures at all speeds. As you know from a previous chapter, the mixture control solenoid opens and closes to meter the correct amount of fuel into the engine on signal from the computer. If the solenoid or computer fails, fuel system operation will be upset. We will discuss this shortly.

Full power system problems

The proper technique for finding and repairing full power system problems depends on the type of carburetor system. As was mentioned in an earlier chapter, the types include: power valve, mechanical metering rod, vacuum metering rod, and solenoid metering valve.

A ruptured diaphragm in a power valve will allow engine vacuum to pull fuel through the diaphragm into the air horn. This richening of the air-fuel mixture causes rough idle and poor gas mileage. High speed engine performance can also be affected.

A metering rod set too far into the jet will cause a high-speed mixture to be too lean. Adjusting it too far out of the jet will cause a rich mixture.

A vacuum operated metering rod may malfunction should a diaphragm rupture or a vacuum piston stick. Either a rich or lean fuel mixture could result.

Choke system problems

Carburetor *choke system problems* usually occur during or right after starting the engine. The choke should operate only when cold. Fig. 14-6 shows some choke-related problems.

If the *choke sticks shut,* an extra-rich air-fuel mixture will pour into the engine. Engine operation will be extremely rough and the exhaust will be black. The engine will lack power and may soon stall.

Should the *choke stick open,* a cold engine may not start. Pumping the gas pedal several times may get it started. However, it will accelerate poorly and may stall several times.

A slightly *rich choke setting* can cause a warm engine to run poorly. By remaining partially closed, the choke allows too little air to mix with the fuel entering the engine. A slight engine miss or roll and a small amount of black exhaust may be noted.

A *lean choke setting* will act something like a choke stuck open. The engine will be hard to start in cold weather. No problems may appear in warm weather. A lean choke adjustment can also cause a cold engine to stall during acceleration.

Fast idle cam problems

An idle cam out of adjustment can create idling problems. If adjusted too high, a cold engine will race for long periods. If adjusted too low, the cold engine may stall until it warms up. Fast idle problems may also stem from faulty choke operation or incorrect adjustment.

FACTORY DIAGNOSIS CHARTS

Factory or service manual diagnosis charts are helpful should you have difficulty pinpointing carburetor problems. These charts are designed for servicing each type of carburetor.

CARBURETOR REMOVAL

When diagnosis indicates internal problems such as dirt-clogged passages, leaking gaskets or seals, or worn parts, it is usually necessary to remove the carburetor for repairs. First remove the air cleaner and all the wires, hoses, and linkages connected to the carburetor.

DANGER! To avoid the possibility of fire, wrap a shop rag around the fitting when removing a fuel line. This will absorb fuel that might leak onto a hot engine.

Labeling wires and hoses

Avoid confusion during reinstallation by labeling hoses and wires connecting to the carburetor. See Fig. 14-7. Late model cars have many such connections and proper routing during reassembly may become confusing.

Fig. 14-7. Always label hoses and wires to help speed reassembly. (Toyota)

Lifting carburetor from manifold

Remove the nuts or studs securing the carburetor to the engine's intake manifold. Make sure there are no loose parts which could fall into the intake manifold.

Carefully lift the carburetor off the engine. Keep it level to avoid stirring up sediment in the fuel bowl and getting it into the passages.

Mount the carburetor on a carburetor stand, Fig. 14-8. This prevents accidental damage to the throttle plates. The unit is now ready for disassembly.

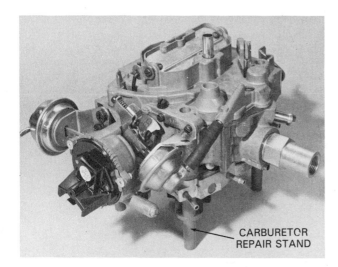

Fig. 14-8. Repair stand will protect throttle plates from damage during service. (Rochester)

CARBURETOR REBUILD

A carburetor should be *rebuilt* or *overhauled* when passages become clogged, when gaskets or seals leak, when rubber parts deteriorate, or when other components wear or fail.

Usually, a carburetor overhaul involves:
1. Disassembling and cleaning parts.

2. Inspecting parts for wear or damage.
3. Installing a carburetor rebuild kit. The kit typically includes new gaskets, seals, needle valve and seat, pump diaphragm or cup, as well as other nonmetal parts. See Figs. 14-9 and 14-10.
4. Adjusting of major parts.

Carburetor disassembly

To disassemble a carburetor, follow the instructions in the service manual. In general, remove all PLASTIC

Fig. 14-9. When ordering parts, you may need identification numbers on carburetor or just make, model, and engine size for car. (Ford)

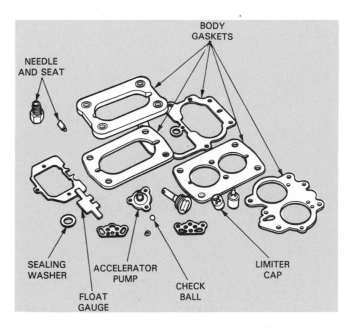

Fig. 14-10. A carburetor rebuild kit includes most of the parts needed, especially plastic and rubber parts and gaskets, to restore carburetor to good condition. (Chrysler)

and RUBBER parts so that metal parts can be soaked in cleaning solution. Also, remove any part that will prevent thorough cleaning of passages.

These general rules should be followed during carburetor disassembly:

1. Lift off the air horn body carefully. Forcing it may cause part damage. See Fig. 14-11.

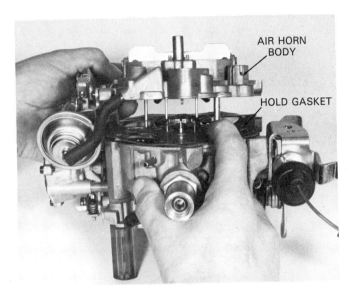

Fig. 14-11. When removing air horn body, be careful not to damage anything and note how linkage rods connect to other components.

2. Keep track of the location of all check balls and check weights. They must be reinstalled in exactly the same location, Fig. 14-12.

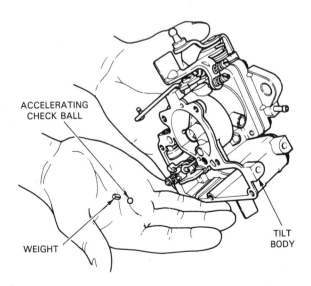

Fig. 14-12. Be careful not to drop small components, like check balls or weights, when disassembling carburetor. Also, note which holes they fit into in body. (Ford)

3. Jets must also be located in the same position in the carburetor. Although outside dimensions may be the same, the orifice (hole) or jet sizes may be different. This is true for main jets, air bleed jets, and any other jets. See Fig. 14-13.

4. Do not let C-clips fly off during removal. They are small and easily lost. Wear safety glasses when removing C-clips, Fig. 14-14.

5. Use special carburetor service tools when needed. Various special tools are available. For example, a common tool is the needle seat or jet tool shown in Fig. 14-15. It will unscrew a seat or jet without damage to the part. If this tool is not available, use a large screwdriver that fits snugly in the screw slot.

6. Remove plugs carefully. Many new carburetors have press-in plugs covering idle mixture screws. Be careful not to damage the carburetor body when removing these plugs. As shown in Fig. 14-16, the service manual will give directions for proper removal. There is usually a special location that

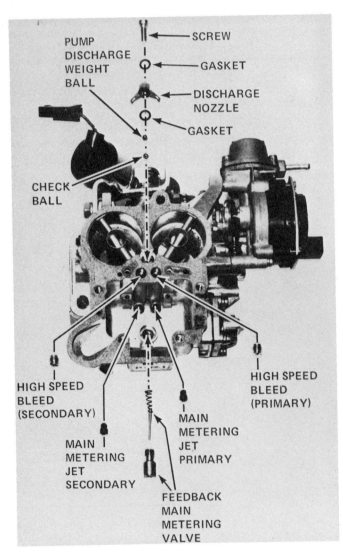

Fig. 14-13. This service manual illustration shows location of small components. Refer to manual if needed when servicing carburetor. (Oldsmobile)

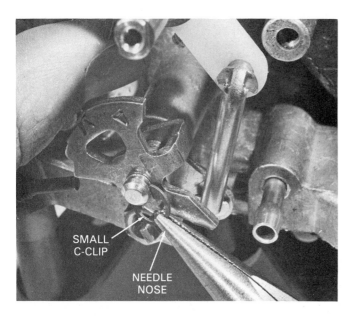

Fig. 14-14. Small C-clips frequently hold linkage rods and levers. Use pliers or special clip tool to remove them. Wear safety glasses and watch where clips go if they fly out of pliers. (American Motors)

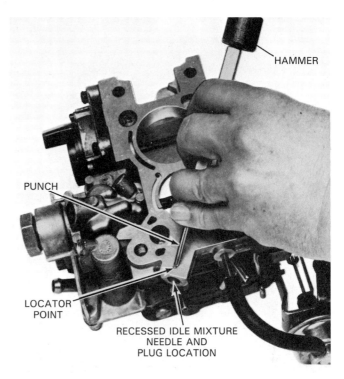

Fig. 14-16. Modern carburetors normally have small plugs that cover mixture adjustment screws. Refer to service manual to find recommended procedure for removing them. This mechanic is using punch to drive out plug for major carburetor service.

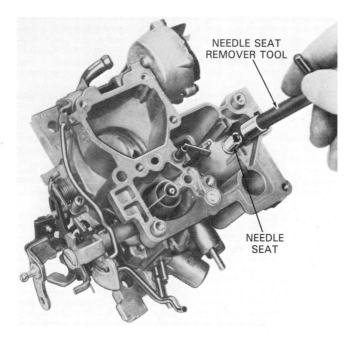

Fig. 14-15. Needle seat tool is best way to unscrew seat. If not available, use large screwdriver. (General Motors)

allows for punching out the plug.

7. Many modern carburetors also have parts, especially choke housings, that are riveted in place. You will have to drill out the old rivets to disassemble the choke. Use the correct size drill bit. Screws are commonly used to reinstall the choke. See Fig. 14-17.

8. Only if damaged or if the shaft or bushings are worn should you remove the throttle and choke plates.

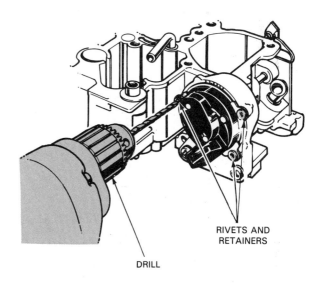

Fig. 14-17. Today's choke housings are commonly riveted in place to prevent misadjustment. Drill out old rivets to service choke. Screws are used to reinstall choke housing. Follow manual procedures for particular carburetor. (Buick)

Scribe mark each plate and the carburetor housing before removal. This will let you reinstall the plates in the same location, assuring a proper fit. Look at Fig. 14-18.

9. Keep all parts organized. You may want to keep

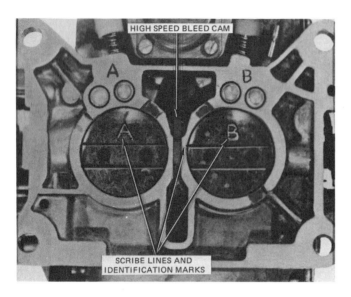

Fig. 14-18. If throttle plates must be removed to service throttle shaft, scribe mark them. You should reinstall throttle plates in same locations to make sure they are centered in throttle bore and do not rub on air horn. (Ford)

parts from each major assembly (float, choke, air horn body, etc.) separated. This will help you find the correct screws and parts during reassembly, Fig. 14-19.

10. Inspect each part carefully as it is removed. Look for damage, corrosion, and other troubles.
11. Use a service manual if in doubt about any task. The manual will give detailed directions for the particular carburetor.
12. Check in the service manual for an exploded view of the carburetor. This will be very helpful in making sure all parts are removed. It will also help during reassembly. An exploded view of one type carburetor is shown in Fig. 14-20.

Cleaning carburetor parts

Carburetor cleaner, also known as decarbonizing cleaner, is a very powerful cleaning agent. It will remove gum, carbon, oil, grease, and other deposits which may have collected in air and fuel passages and on external parts. Look at Fig. 14-21A.

Never allow carburetor cleaner to contact plastic and rubber parts. It will ruin them.

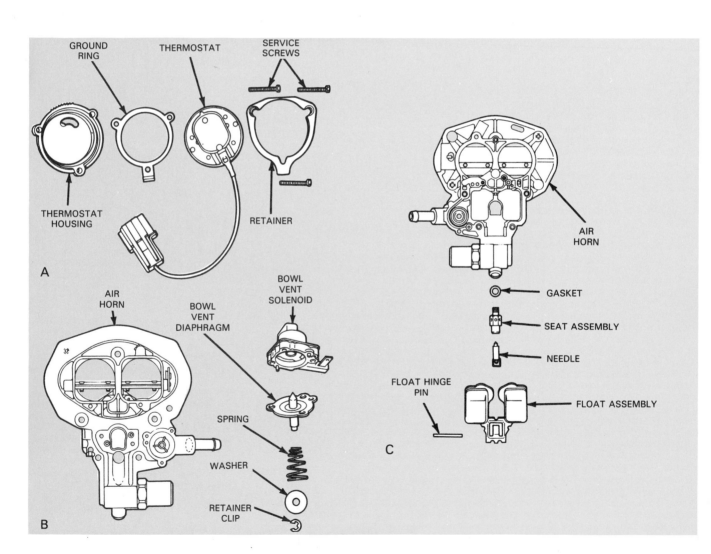

Fig. 14-19. This service manual illustration shows how major assemblies can be kept together to reduce confusion. A—Choke components. B—Float components. C—Air horn components. (Ford)

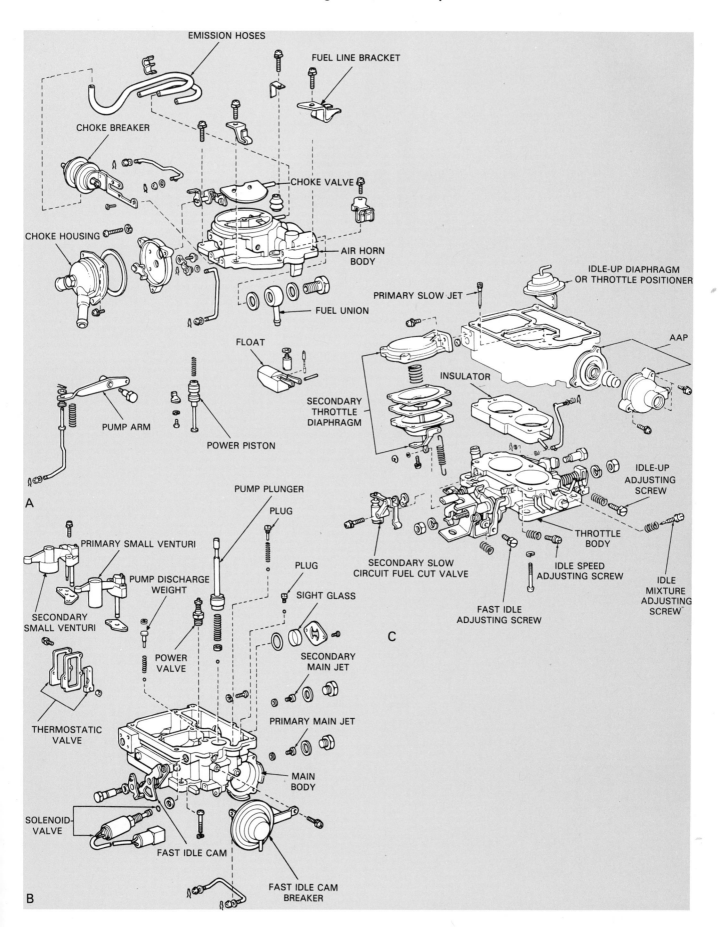

Fig. 14-20. Exploded views of carburetor can be helpful. Service manual will usually have several for particular carburetor you are rebuilding. This aids reassembly. (Mazda)

Place metal components in a tray and lower them into the cleaner. Allow the parts to soak for the recommended amount of time.

> **WARNING!** Carburetor cleaner can cause serious chemical burns to the skin and eyes. Wear rubber gloves and eye protection when working with carburetor type cleaning solution.

After soaking, rinse carburetor parts with water, clean kerosene, or clean cold-soak solution, Fig. 14-21B. Dry parts with compressed air. Force air through all passages to make sure they are clear.

NOTE! Do NOT use wire or a drill to clean out carburetor passages. This could scratch and enlarge the passages, upsetting carburetor operation.

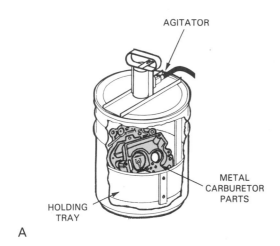

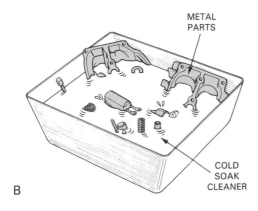

Fig. 14-21. After completely disassembled, all metal parts of carburetor should be cleaned. A—Submerse metal parts in decarbonizing cleaner, often called carburetor cleaner. It is very powerful so do not let cleaner splash into your eyes or get on your hands. Soak parts for a period recommended by solution manufacturer. B—After soaking in carburetor cleaner, wash parts in cold soak cleaner and blow dry. (Tomco and Toyota)

Inspecting carburetor parts

Inspect each cleaned carburetor part closely for:
1. Wear between the throttle shaft and throttle body.
2. Binding of the choke plate and linkage.
3. Warpage, cracks, and other problems with carburetor bodies.
4. Leaking float or bent hinge arm.

5. Nicks, burrs, and dirt on gasket mating surfaces.
6. Damaged or weakened springs and stripped fasteners.
7. Damage to tips of idle mixture screws.

Replace parts showing wear or damage. Discard all of the old parts that will be replaced from your casrburetor rebuild kit.

Carburetor reassembly

Unless you are an experienced mechanic, follow the detailed directions in a service manual to reassemble a

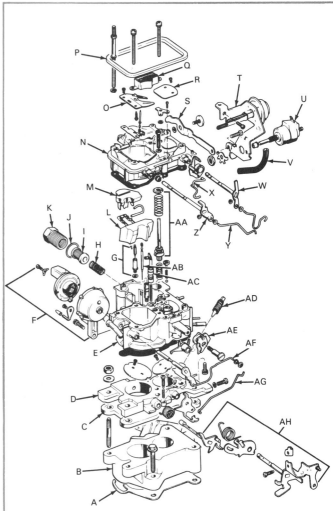

A. BASE PLATE GASKET	R. CHOKE VALVE
B. INTAKE ADAPTER	S. PUMP LEVER
C. INSULATOR	T. VACUUM BREAK AND BRACKET
D. THROTTLE BODY	U. IDLE STOP SOLENOID
E. MAIN BODY	V. VACUUM HOSE
F. ELECTRIC CHOKE ASSEMBLY	W. VACUUM BREAK LEVER
G. NEEDLE SEAT ASSEMBLY	X. CHOKE LINK
H. SPRING	Y. AIR VALVE ROD
I. FUEL INLET FILTER	Z. AIR VALVE LEVER
J. FITTING GASKET	AA. ACCELERATOR PUMP
K. FUEL INLET FITTING	AB. METERING ROD
L. FLOAT ASSEMBLY	AC. POWER PISTON
M. FLOAT BAFFLE	AD. IDLE NEEDLE AND SPRING
N. AIR HORN	AE. FAST IDLE CAM
O. SECONDARY AIR VALVE	AF. INTERMEDIATE CHOKE ROD
P. AIR HORN GASKET	AG. PUMP ROD
Q. VENT SCREEN	AH. THROTTLE LEVER ASSEMBLY

Fig. 14-22. This exploded view shows major carburetor bodies. Note how major parts fit into body sections. (AMC)

modern carburetor. Some car makers have as many as 20 different carburetors for a single model year. There can be thousands of different procedures and adjustments. Each is critical to carburetor performance.

The manual will give procedures for installing each part. An exploded view of the carburetor, like the one in Fig. 14-22, shows how parts fit together.

There are some fundamental rules for carburetor reassembly:

1. Typically, start by assembling the throttle body onto the main body. Then you can mount the carburetor on a repair stand as shown earlier.
2. Tweezers or needle nose pliers will hold small parts during reassembly. As in Fig. 14-23, the mechanic can use a tweezer-like tool to insert a lever into the carburetor body.

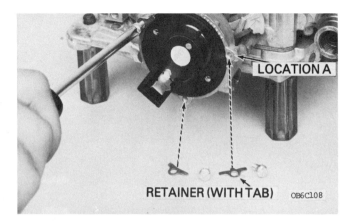

Fig. 14-24. Linkage rods may have to be in place as other parts are assembled. Use common sense and a manual to determine assembly sequence. (Buick)

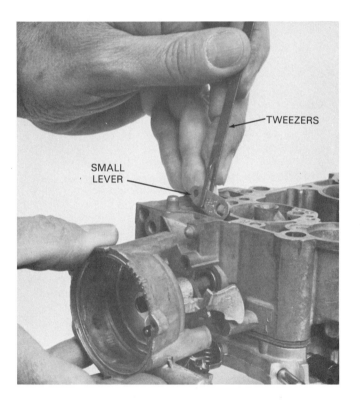

Fig. 14-23. Use tweezers or needle nose pliers when parts are too small to hold with fingers. (Rochester)

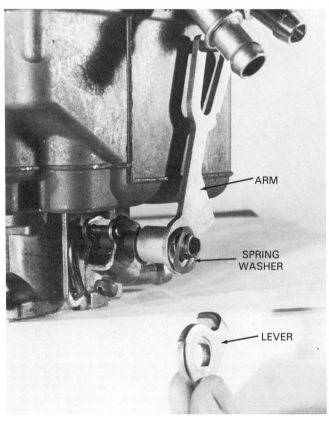

Fig. 14-25. Note how spring washer is used between vent arm and outside lever. It keeps parts from rattling. (Carter)

3. Assemble the parts in proper sequence. It is very easy to install one part and then find out another part should have been installed first. Usually, components install on the main body before the air horn body is attached. Double-check that all linkage rods are correctly placed. Again, refer to the manual for exact directions, Figs. 14-24 and 14-25.
4. Many carburetor parts are delicate. Do not force them into place. Work slowly and gently. For example, slowly lower the venturi assembly into place. It is possible to damage emulsion tubes and other parts by forcing them. See Fig. 14-26.
5. Use new gaskets and match all gasket holes with passages. Sometimes more than one gasket is pro-

vided for the same component. This allows the rebuild kit to be used on more than one make of carburetor. Make sure the gasket is correct for the specific carburetor or the rebuild will have to be done over, Fig. 14-27.
6. Normally, do not use gasket cement or sealer on a carburetor. Bits of sealer could break off and enter carburetor passages.

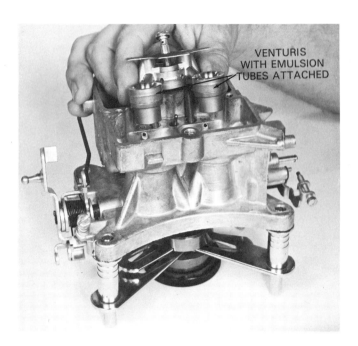

Fig. 14-26. Be careful not to damage delicate parts. Emulsion tubes and float assembly are fragile and can be easily ruined. (AMC)

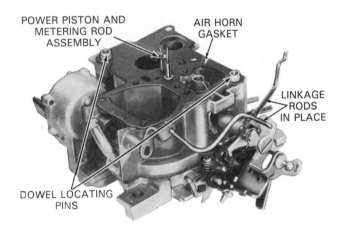

Fig. 14-27. Make sure all gasket holes and any holes in the carburetor body align. If even one hole is not lined up, carburetor operation can be affected. (General Motors)

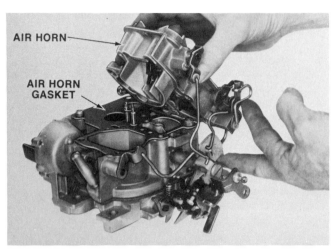

Fig. 14-28. Double-check gasket, check weights, check balls, and linkage rods as you close up carburetor. (Oldsmobile)

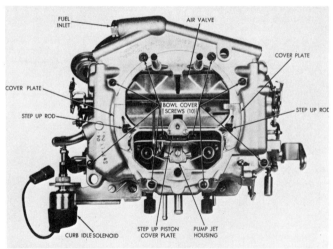

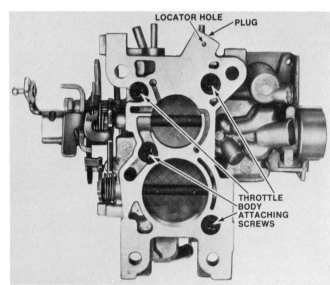

Fig. 14-29. Check screw lengths and tighten screws in a crisscross pattern. Undertightening and overtightening screws can both cause problems. (Carter Carb. Div. and Oldsmobile)

7. Double-check all linkage rods as their air horn is fitted. It is sometimes possible to mix up the linkage rods or fit them on backwards. Look at Fig. 14-28.

8. Refer to an exploded view of your carburetor for proper location of all parts. Also, check closely for extra parts as each major assembly is installed on the carburetor. If even one check ball is left out, the carburetor and engine will not function properly.

9. Make sure screw lengths are correct when attaching the air horn and throttle body, Fig. 14-29. Tighten the screws in a crisscross pattern. This will clamp the parts together evenly and help prevent

part warpage due to engine heat.

10. Complete the assembly by installing any external carburetor parts. For instance, the accelerator pump may be attached after the internal parts. On a diaphragm type pump, do not overtighten the screws but tighten them evenly to prevent part warpage. Look at Fig. 14-30.

11. Operate the throttle lever while watching linkage operation. Make sure the accelerator pump, choke, and other components work properly as the throttle swings open and closed.

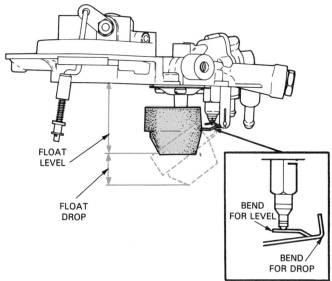

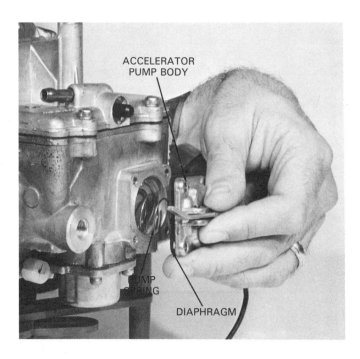

Fig. 14-30. Complete carburetor assembly by installing all external parts. Mechanic is installing accelerator pump in this carburetor. (AMC)

Fig. 14-31. Float drop and float level adjustments are critical. Note basic distance that is measured when making level and drop adjustments. (Renault)

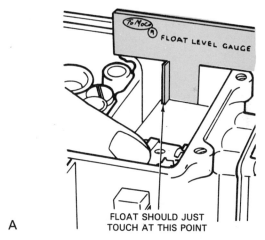

A

FLOAT ADJUSTMENT

Float level settings are extremely important to carburetor operation. A float setting can affect engine operation at almost any engine speed or load condition. The two basic float adjustments are: float level and float drop. Refer to Fig. 14-31.

Carburetor float level can be set either wet or dry.

Dry float level adjustment

A *dry float setting* is made by measuring from some point on the float to a point on the carburetor bowl or air horn. See Fig. 14-31 again. Quite often, the carburetor kit will provide a cardboard gauge to simplify float level measurement.

Changing float level and drop

There are two basic methods for altering float level and float drop. One is to bend the float tangs (tabs) and the other is to turn an adjusting screw.

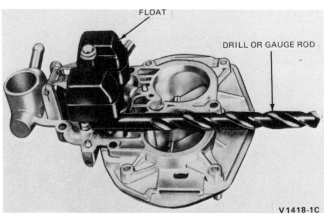

B

Fig. 14-32. Two methods of checking float adjustment. A—Using special float gauge. B—Using recommended drill bit size. Bend float tang or arm as suggested in service manual until float touches measuring device. (Ford)

295

Depending upon design, the float drop tang will have to be bent until drop is correct. Needle nose pliers or a small screwdriver can be used to bend the tabs on the float, Fig. 14-31. Bending the tang towards the needle valve will lower the float level. Bending it away from the valve will raise the float level.

When float tangs are not provided, float adjustment can also be made by bending the float arms. This should be avoided when tabs are present.

Manufacturers may give measurements and describe how a ruler or drill bit may also be used to check dry-float level. See Fig. 14-32.

Wet-float level adjustment

A *wet-float level setting* is made with fuel in the carburetor bowl. It is more accurate than a dry-float adjustment. The bowl can be filled with fuel by running either the engine or the electric fuel pump. A slight plug or glass bowl can be utilized to visually inspect actual fuel level. Also, the top of the carburetor can be removed to check the wet-fuel level. The float is simply adjusted until the fuel level is within specs. Fig. 14-33 shows two methods of setting the wet-float level.

Screw type float adjustment is ideal for wet-float settings. The adjustment screw is turned until the desired fuel level is obtained. Then, a lock nut is tightened to maintain the setting.

Float drop adjustment

A float drop setting is just the opposite of a float level setting. Refer to Fig. 14-31 again. The float must be in the full release position (needle valve completely open).

If a float is allowed to drop too much, it can hit the bottom of the fuel bowl and stick. The engine could flood.

Fig. 14-34 shows one way to measure float drop. A ruler is used to measure from the upper end of the float to the gasket surface on the air horn. Procedures may vary slightly from model to model.

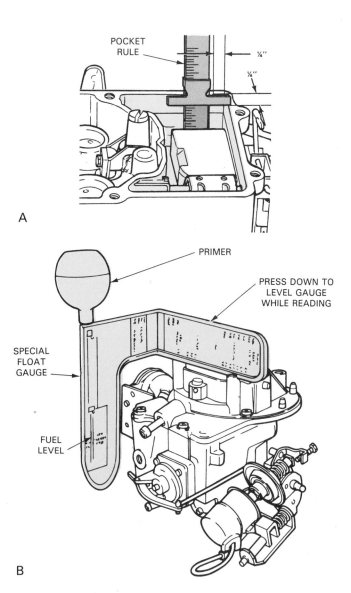

Fig. 14-33. Wet-float level adjustment is done with gasoline in carburetor. A—Using ruler to check level. B—Using special gauge to check float level. Follow service manual recommendations. (Ford)

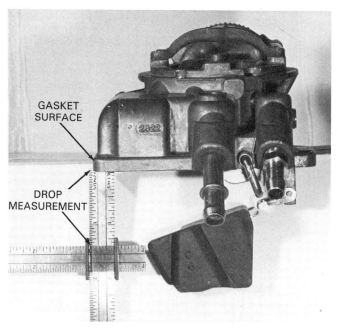

Fig. 14-34. Photo shows one way of checking float drop. Too much drop would let float hit bottom of fuel bowl. This could ruin float or let it stick open. (AMC)

ACCELERATOR PUMP ADJUSTMENT

Differences in carburetor design require different methods of adjusting an accelerator pump. Fig. 14-35 illustrates several methods of adjusting an accelerator pump. Study them carefully.

Usually, engine hesitation and stumbling are caused by a lean accelerator pump setting. Therefore, in most cases, pump stroke must be increased to correct this performance problem. Linkage wear tends to reduce pump stroke and output.

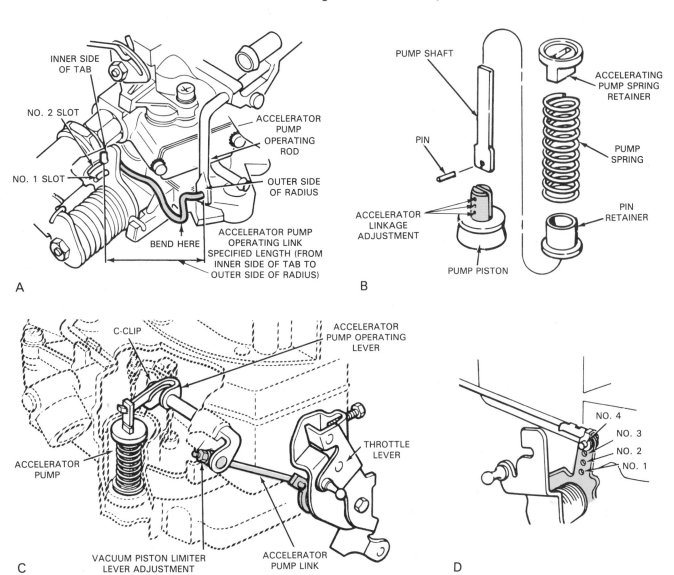

Fig. 14-35. Accelerator pump adjustments. A—Bend linkage rod to adjust. B—Change pin location in pump piston to adjust. C—Turn nut to adjust linkage rod. D—Change linkage rod position in lever to check adjustment. (Ford)

BOWL VENT ADJUSTMENT

An inoperative or misadjusted external bowl vent can increase air pollution. It may not let fuel vapors enter the vapor storage canister. It may also not let engine vacuum purge (clean) fumes out of the canister for combustion in the engine.

In general, use a technique similar to that shown in Fig. 14-36 when adjusting a bowl vent. Measure the amount of vent opening with the throttle plates at curb idle. The measurement should meet factory specs.

CHOKE ADJUSTMENT

Exact choke setting procedures vary with choke type and carburetor model. With an integral choke, Fig. 14-37, manufacturers usually recommend specific alignment of the choke index marks. A mark on the plastic choke cover must be lined up with a mark on the metal choke housing or body. This initial choke setting will

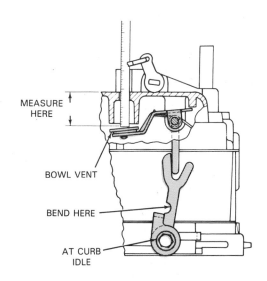

Fig. 14-36. Typical method of adjusting vent or other component is to bend arm. Bend where slot is formed in arm. (Carter)

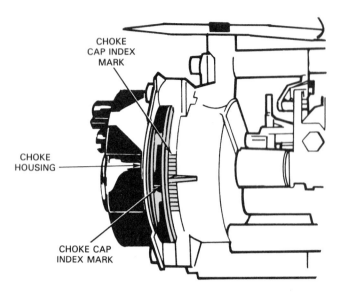

Fig. 14-37. Many choke housings have small notches that allow for easy adjustment. If choke is too far open when cold, rotate housing so mark points to richer position. (Ford)

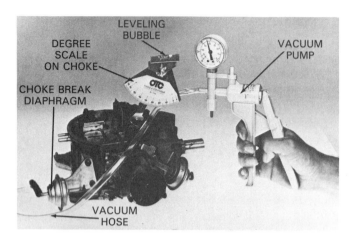

Fig. 14-38. Vacuum pump and degree scale are being used to check adjustment of choke break diaphragm. Choke should open specific amount with vacuum applied. This prevents engine flooding. (OTC)

usually provide adequate choke operation.

With a remote choke, adjustments may be made in two basic ways: by rotating the thermostatic spring or by bending the choke linkage. Bending and shortening the choke linkage rod usually closes the choke and richens the mixture. Straightening or lengthening the rod opens the choke to lean starting mixtures. Always check instructions for exact procedures.

Because of part wear, fatigue, or other changes, factory settings may not always provide satisfactory choke operation. Generally, the choke should almost close when cold. It should open fully as the engine reaches full operating temperature. If factory settings do not result in efficient operation, refer to a service manual, replace faulty parts, and alter settings as needed.

Choke piston adjustment

Like a diaphragm choke break, a vacuum piston cracks open the choke valve for cold engine operation. To adjust it, use the following procedure.

With the thermostatic spring housing removed, fit a short length of wire gauge between the choke piston and its cylinder. (Wire gauge size will be given in a service manual or carburetor rebuild kit.) Rotate the choke linkage to the closed position. Fit a specified size drill bit into the air horn. With the wire gauge and drill bit in place, bend or adjust the choke linkage so the choke opening is the same as the drill bit diameter. The choke plate should just touch the bit. Refer to a manual on how and where to bend or adjust the linkage.

VACUUM CHOKE BREAK ADJUSTMENT

A vacuum choke unloader should open the choke plate the correct amount to prevent an over-rich air-fuel mixture upon initial engine starting. Several methods are used to measure and adjust choke break action.

Vacuum must be applied to the break diaphragm as in Fig. 14-38. Then, a specified size drill bit is fitted

A

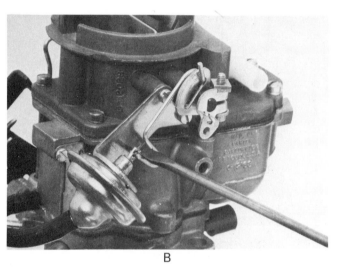

B

Fig. 14-39. Common method of adjusting choke break. A—Shorten rod to make diaphragm open choke more. B—Lengthening rod will richen cold starting mixture. (AMC)

between the choke butterfly and the air horn. The drill bit size measures choke opening. The linkage must be lengthened or shortened until the drill bit fits snugly.

Needle nose pliers can be used to shorten some choke break rods, Fig. 14-39A. In 14-39B, a screwdriver is being used to lengthen the rod. Bend the choke linkage only in the recommended location or choke operation could be upset.

Some auto manufacturers give choke break specs in degrees, not drill bit size or inches. A *choke angle gauge* is shown in Fig. 14-38. It has a degree scale, magnetic foot or stand, and a leveling bubble. The gauge is attached to the choke plate and the bubble is centered. Then, exact choke opening can be measured in degrees.

When dual vacuum breaks are utilized on one carburetor, adjust one and then the other. Again, a hand-vacuum pump activates the break and a degree tool or drill bit measures choke opening.

INSTALLING CARBURETORS

Place a new base plate gasket on the intake manifold. Then fit the carburetor over the mounting studs, if used. Install and properly torque the fasteners that secure the carburetor to the intake manifold.

WARNING! Do NOT overtighten the carburetor hold-down nuts or bolts. Excessive tightening can easily snap off the carburetor base flange or warp the throttle body.

Connect hoses, lines, wires, and linkage rods or cables. See that all vacuum hoses have been installed in their correct positions. Operate the throttle by hand to make sure the throttle plates are not binding or hitting on the base plate gasket.

You are now ready to make the final idle speed and idle mixture adjustments on the carburetor.

Cold idle speed (fast idle cam) adjustment

To set the cold fast idle or fast idle cam, warm the engine to operating temperature. Set the emergency brake, block the wheels of the car, and connect a tachometer to the engine.

While following service manual instructions, set the fast idle cam to hold the throttle open for a fast idle, Fig. 14-40. The fast idle screw must contact a specified step on the fast idle cam.

Turn the fast idle screw until the tachometer reads within specifications, Fig. 14-41. Cold fast idle speeds vary from approximately 750 to 950 rpm. Some specs are given with the automatic transmission in drive. This will reduce the cold fast idle value.

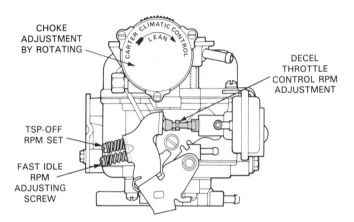

Fig. 14-41. Note locations of speed adjusting screws. This is typical. (Ford)

Hot idle speed adjustment

Look at Fig. 14-42. To set hot idle speed, turn the solenoid adjusting screw until the tachometer reads correctly. Sometimes the adjustment is on the solenoid mounting bracket or it may be on the solenoid plunger.

Typically, hot idle speed is from about 650 to 850 rpm. It is LOWER than cold idle speed but HIGHER than curb idle speed.

CAUTION! When adjusting an idle speed solenoid, make sure it is NOT an air conditioning solenoid.

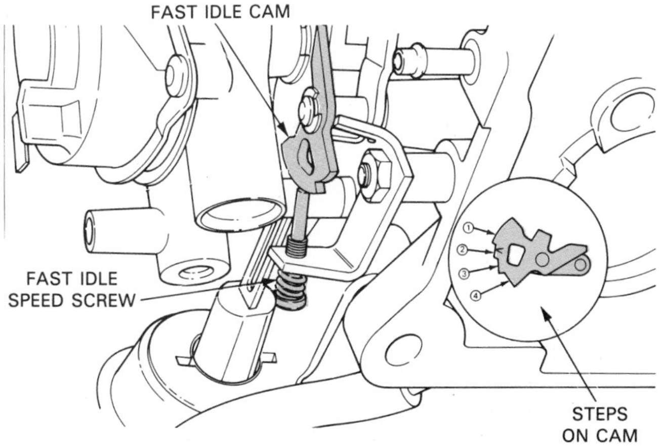

Fig. 14-40. Fast idle adjustment is usually done by setting adjustment screw on specific step of cam. Then, use tachometer to measure engine rpm while turning fast idle screw. (Ford)

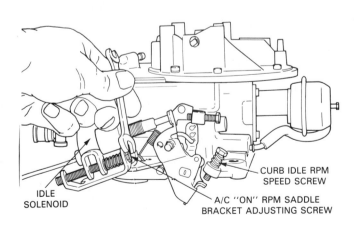

Fig. 14-42. Sometimes, hot idle speed is adjusted with solenoid screw. Solenoid holds throttle open to increase idle speed with engine running but lets idle drop back to curb or low idle when the engine is shut off. This prevents dieseling or run-on. (Ford)

Curb idle speed adjustment

Curb idle speed is the lowest idle speed setting. On older carburetors, it is the idle speed setting that controls engine speed under normal warm-engine conditions.

On late model carburetors using a fast idle solenoid, curb idle can also be termed *idle drop* or *low idle speed adjustment.* In this case, it is a very low idle speed that only occurs as the ignition is turned off. When the solenoid is deenergized, the throttle plates drop to the curb idle setting to keep the engine from dieseling (engine running with ignition key OFF).

To adjust curb idle, disconnect the wires going to the idle speed solenoid. Turn the curb idle speed screw until the tachometer reads within specifications, Fig. 14-42.

Curb idle speed is usually very low, approximately 500 to 750 rpm.

IDLE MIXTURE ADJUSTMENT

After setting idle speed, you may also need to adjust the carburetor idle mixture. There are several methods for doing this.

Pre-emission idle mixture adjustment

On older, pre-emission carburetors, the idle mixture screw is turned in and out until the smoothest idle mixture is obtained, Fig. 14-43.

Basically, the mixture screw is turned in until the engine misses (lean miss). Then, it is turned out until the engine rolls (rich miss). The mixture screw is then set halfway between the lean miss and rich roll settings. This procedure is repeated on the second mixture screw if needed (two or four-barrel carburetor).

Next check and adjust the idle speed as needed. If you must reset the idle speed, then readjust the idle mixture.

Propane idle mixture adjustment

Propane idle mixture adjustment is required on some later model carburetors. As shown in Fig. 14-44, connect a bottle of propane to the engine (air cleaner or intake manifold). The propane provides a richer fuel mixture during carburetor adjustment. Following service manual procedures, you must meter a certain amount of propane into the engine while adjusting the mixture screws. This will provide a leaner (lower exhaust emissions) setting after the propane mixture is disconnected. See Fig. 14-45.

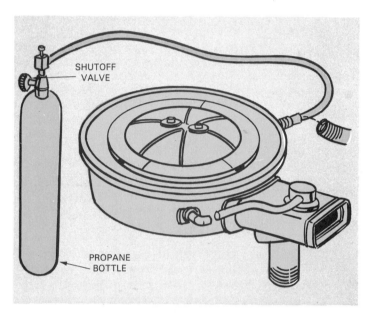

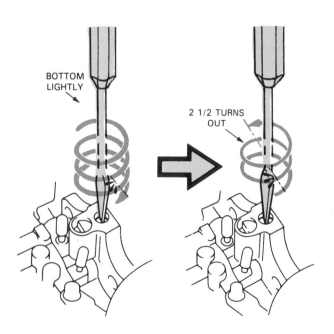

Fig. 14-43. Basic idle mixture screw adjustment is done by bottoming screw lightly. Then, turn it out about two and one-half revolutions. This is typical for most carburetors. Further adjustment is needed once the engine is running. (Toyota)

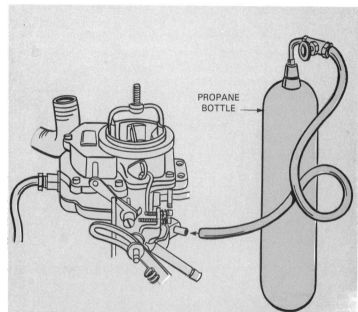

Fig. 14-44. Propane idle mixture adjustment is done on some late model cars requiring lean carburetor setting. Propane bottle is connected to air cleaner or intake manifold fitting. Then, small amount of propane is metered into engine. This will supplement lean mixture and allow for more precise adjustment of mixture screws. Follow manufacturer instructions. (Champion Spark Plugs)

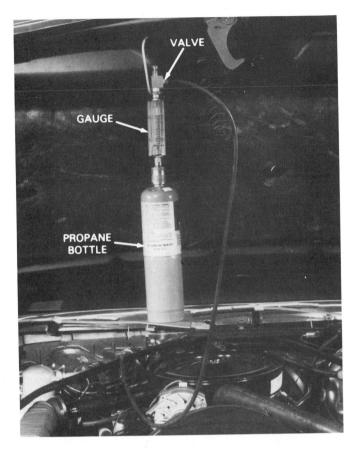

Fig. 14-45. When adjusting idle mixture with propane, hang bottle where it cannot fall into engine fan. Do not leave open bottle unattended. Make sure hose is also away from fan and hot engine parts that could rupture hose and cause gas leak. (Chevrolet)

Late model idle mixture adjustment

WARNING! Most late model carburetors have *sealed idle mixture screws* preset at the factory. Unless major carburetor repairs have been made, do NOT remove and tamper with these screws. To avoid a violation of federal law, follow manufacturer's prescribed procedures when making carburetor adjustments.

An exhaust gas analyzer may be used to adjust the idle mixture on late model cars. Adjust mixture screws until exhaust sampling is within specified limits.

Dead idle circuit

When an idle mixture screw is turned in or out, it should affect how the engine idles. Idle should decrease and roughen as the screw is set too rich or lean. If not, then the idle circuit is said to be *dead.*

Something is wrong inside the carburetor. When the screw must be turned too far OUT for a good idle, dirt may be clogging one of the idle passages. When an idle screw must be turned almost all the way IN to get a smooth idle, a power valve may be ruptured and leaking fuel into the carburetor, the float may be set too high.

Servicing computer controlled carburetors

A computer controlled carburetor system requires specialized service techniques, Fig. 14-46. Most systems have a self-diagnostic mode that allows easy location of system problems. One computer system uses a flashing code light to indicate the trouble. The mechanic can use the code chart in the service manual to pinpoint the trouble.

Computer controlled carburetor problems

Computer controlled carburetors can suffer the same

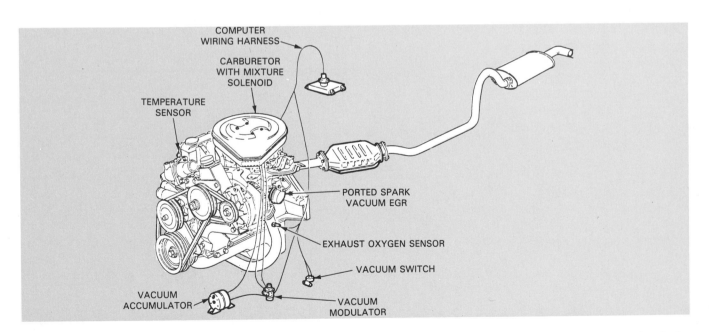

Fig. 14-46. Computer controlled carburetor system uses an electronic control unit and various sensors to operate metering device in carburetor. Oxygen sensor and temperature sensors are common trouble sources. Also check for loose vacuum hoses, disconnected wires, and other more typical causes of performance problems before testing computer system. (General Motors)

problems as a conventional carburetor. Gaskets can split and leak. Rubber seals can harden and split. Passages become clogged. It is helpful to remember this when checking possible problems with the computer, sensors, or mixture control solenoid.

Before suspecting the computer system or any of its components, check all other possible causes. This includes the ignition system (ignition timing, spark plug wires, spark plugs, pick-up coil, etc.). Also check the operation of the air cleaner door and all emission controls (EGR system, evaporative system, PCV system, EFE system, etc.) as well as engine compression (piston rings and valves).

Also, inspect for vacuum leaks at the intake manifold gasket, vacuum hoses, and carburetor. A malfunction of any one of these parts could upset engine operation and appear to be a computer system problem.

If none of these checks reveal the cause of the performance problem, the computer control system may be defective and should be tested. The following symptoms could indicate a possible computer system problem:

1. Detonation.
2. Poor idling.
3. Missing.
4. Hesitation.
5. Surging.
6. Poor gas mileage.
7. Sluggish engine.
8. Hard starting.
9. Exhaust odor or smoke.
10. Engine cutout.

Computer controlled carburetor system testing

In most cases, a built-in diagnostic warning system catches and indicates computer system problems. When a fault exists in the computer network, a "check engine" light may glow in the instrument panel. This warns the driver and mechanic of a possible system problem.

Reading trouble codes

To isolate which part in the computer system is bad, ground the trouble code test lead on the computer (control module) or in the wiring. See Fig. 14-47. This will cause the "check engine" light to flash a code. On many systems, an analog (needle type) voltmeter may have to be connected to a specified point on the system. Each pulse (sweep) of the voltmeter needle would signal the numbers of the code.

For example, the light or voltmeter may flash or sweep two times and then pause, indicating the number two. Then, after a few seconds, it may flash or sweep three times to indicate the number three. This would mean a trouble code of 23 (2, pause, 3) as in Fig. 14-47.

The mechanic can then pinpoint the trouble by referring to a code chart in the service manual. For detailed information on codes and repair procedures, refer to a service manual for the computer system being checked. Systems codes and test procedures vary.

Other tests

A system performance test is used to find problems not included in the system code or diagnostic warning system. The check is needed when the warning light

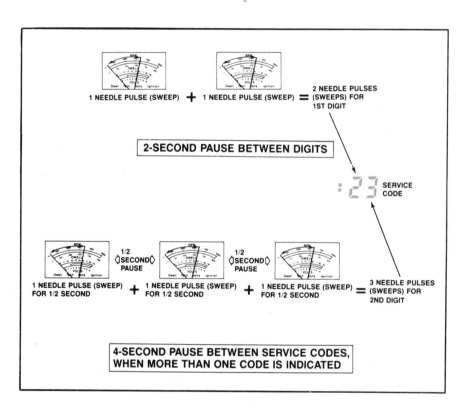

Fig. 14-47. Analog or needle type voltmeter is needed to read the self-diagnosis code of many computer controlled carburetor systems. When activated by jumper wire, computer will send voltage pulses to test socket and meter. Two meter deflections, a short pause, and three more meter deflections would represent a twenty-three code. Your manual will tell what problem twenty-three represents. (Ford)

does not indicate a system problem and no cause for a problem can be found. It is the last test to be made when an engine performance problem exists. The service manual will give step-by-step directions for a performance check of the system.

Measuring dwell signal to carburetor

Several computer controlled carburetor systems require a dwell meter and tachometer to check the computer output. The dwell meter measures the duration of the electrical pulses to the carburetor mixture control solenoid. The tach measures engine rpm or speed.

Generally, connect the dwell meter and tach as shown in Fig. 14-48. The dwell meter connects to a specified electrical connector in the computer wiring harness. The tach connects normally to the ignition system.

Fig. 14-49 shows the relationship between the dwell meter reading and the carburetor's air-fuel mixture. As

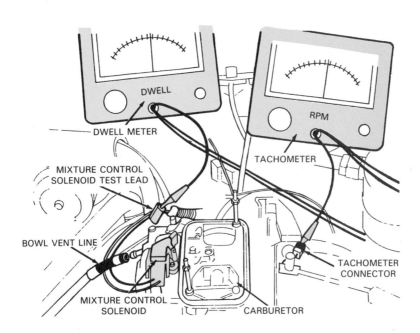

Fig. 14-48. Note basic connections for using a dwell meter to check computer signals to carburetor. Dwell meter will measure pulse width going to mixture control solenoid. This will help you determine whether there is a problem in the control system, sensor system, or carburetor itself. (AMC)

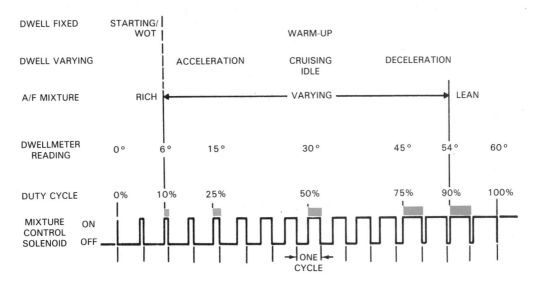

Fig. 14-49. Chart shows relationship between dwell and mixture control solenoid operation. Note how low dwell richens mixture and high dwell leans mixture. (Buick)

dwell decreases, the mixture becomes richer and vice versa.

When testing the system, compare your dwell meter readings to factory specs. If the dwell is too high or low for engine operating conditions, there is a problem in the computer system. Normally, the dwell reading will rise and fall within a certain range (around 35 degrees plus or minus 5 degrees) with engine at full operating temperature and at a constant speed.

Fig. 14-50 is a chart showing dwell meter readings for one system when operating under different conditions. The service manual will give a similar chart for the particular system and car model being tested.

Interpreting dwell signal from computer

If the computer dwell signal is not within specs, use common sense, basic testing methods, and a service manual to find the problem.

For example, a low dwell, rich mixture problem would point to several possible sources. You would need to think of all the system parts or troubles that could richen the mixture. For example, a faulty oxygen sensor, bad engine temperature sensor, defective throttle position sensor, or leaking mixture control solenoid are all possible causes.

Your service manual will normally give a logical step-by-step procedure for isolating specific problems using a digital VOM or special testers.

Note! An analog (needle type) meter should NOT be used to check many electronic devices in a computer system, the oxygen sensor for example. It could damage system components. It can draw too much current through the system and ''burn'' components.

Defective mixture control solenoid

A defective mixture control solenoid can upset car-

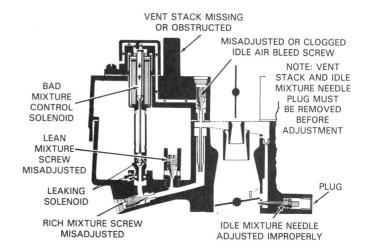

Fig. 14-51. Note some of the troubles that can occur with computer controlled carburetor. (AMC)

FUEL CONTROL SYSTEM OPERATION			
ENGINE CONDITION	**INPUTS TO ECM**	**M/C SOLENOID OPERATION**	**DWELL METER READING**
Starting (Cranking) Warm-Up	• Tach signal less than 200 RPM • Tach above 200 RPM (Engine Running) • O₂ Sensor less than 600°F (315°C) • Coolant less than 66°F (19°C) • Less than 10 seconds elapsed since starting	M/C Solenoid off (Rich Mixture) Fixed Command from ECM to M/C solenoid	0°dwell Fixed Reading between 10° and 50° dwell
Normal Warm Operation Idle and Cruising (''Constant'' Engine Speed)	• O₂ Sensor above 600°F (315°C) • Coolant above 66°F (19°C) • Map sensor (high vacuum)	M/C solenoid signal determined by oxygen sensor information to ECM	Varying anywhere between 10° and 50° (nominal 35°) dwell (faster than higher RPM)
Acceleration and Deceleration (''Changing'' Engine Speeds)	• Throttle position sensor (TPS) • Map sensor • O₂ sensor	Momentary programmed signal from ECM during period after throttle change until oxygen sensor resumes control of M/C solenoid	Momentary change, can't be read on dwell meter. Will vary, but high or low on scale depending upon operating conditions.
Wide-Open Throttle	• TPS fully open • Map sensor (low vacuum)	Very rich command to M/C solenoid	

Fig. 14-50. Study how dwell readings affect solenoid operation, inputs to computer, and engine operation. (Oldsmobile)

buretor operation at all engine speeds. It controls the fuel mixture ratio at both idle and cruising speeds, Fig. 14-51.

When poor gas mileage and/or a rough idle exist, first check the more common causes of engine performance problems (vacuum leaks, ignition timing, spark plugs, float level, etc.). If nothing is found to be wrong, check and test the mixture control solenoid.

Note! Always check the electrical connection to the mixture control solenoid, Fig. 14-52. It is a common source of system trouble.

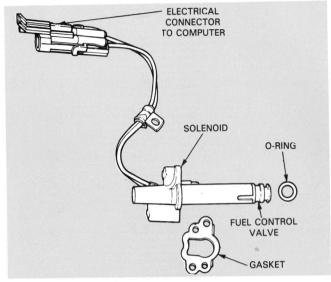

Fig. 14-54. O-rings are common problem area for mixture control solenoids. They can begin to leak and upset carburetor operation. Replace O-rings when solenoid is removed. Also check wires to solenoid for internal breaks that could cause erratic problems. (Oldsmobile)

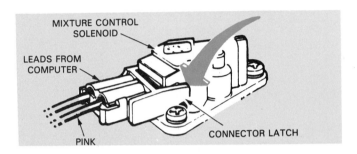

Fig. 14-52. Inspect all electrical connections, especially connection to mixture control solenoid, when you suspect problem with carburetor or computer system. (Pontiac)

Testing mixture control solenoid

For testing, remove the mixture control solenoid from the carburetor, Fig. 14-53. Check the solenoid for sticking, binding, and other troubles, Fig. 14-54.

Connect a 12-volt power source (battery) to the solenoid terminals. Remove the rubber seal and retainer from the end of the solenoid stem, Fig. 14-54. Attach a hand vacuum pump to the end of the solenoid stem, as in Fig. 14-55.

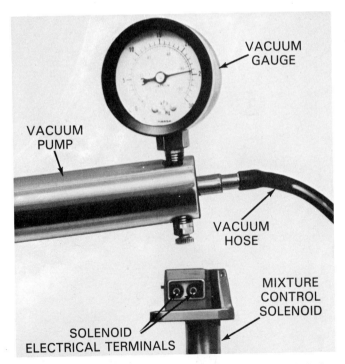

Fig. 14-55. This carburetor mixture control solenoid is being tested with a vacuum pump. Vacuum leakage must be within specs when solenoid is energized with voltage and when it is deenergized. (American Motors Corp.)

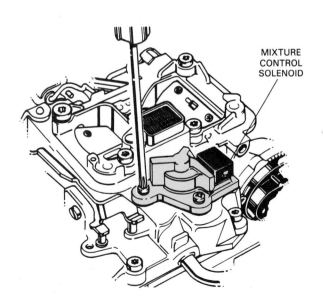

Fig. 14-53. This mixture control solenoid is removed easily from top of carburetor. Always refer to a service manual for specific carburetor when in doubt. (General Motors)

With the solenoid energized in the lean position (battery connected), apply vacuum (at least 25 in./Hg) to the stem. Time the leak-down rate. It generally should NOT exceed 5 in./Hg of leak-down in five seconds. If faster than this, replace the solenoid.

To check the solenoid for sticking in the down (rich) position, remove the voltage source (battery connections). The vacuum pump reading should drop to zero in less than one second. If not, the valve is sticking and should be replaced.

Also inspect the metering jet on the solenoid. Look for loose parts, damage, or clogging with dirt.

Note! Never attempt to adjust the lean mixture screws inside the solenoid circuit. The screws are factory set during a carburetor flow test. Any adjustment will usually result in a misadjustment and an improper carburetor fuel mixture calibration.

Other carburetor control system problems

When the carburetor itself is not at fault, check other system components using factory testing methods. Fig. 14-56 illustrates a few other problems. Loose electrical connections, broken wires, and faulty sensors can all upset engine operation.

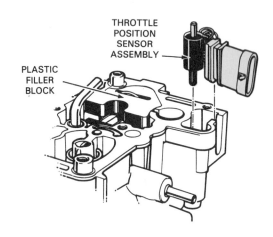

Fig. 14-57. Throttle position sensor can change resistance after prolonged service. This would upset signals going to computer. Use an ohmmeter to make sure sensor resistance values are within specs at different throttle openings. (Pontiac)

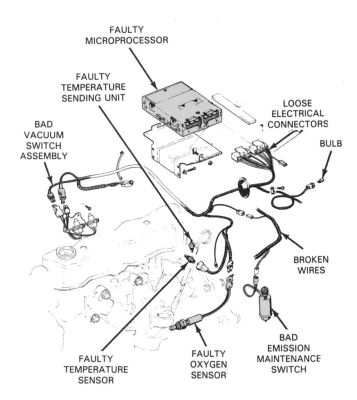

Fig. 14-56. Study types of problems that can occur in a computer controlled carburetor system. (AMC)

For example, the throttle position sensor sends information (electrical signals) to the computer about how wide the throttle plates in the carburetor are opened. If the sensor shorts or opens, the computer will receive false information and will adjust the carburetor fuel mixture incorrectly. You need to test the sensor and replace it if its resistance value (in ohms) is not within specs for different throttle positions, Fig. 14-57.

If defective, the idle speed motor could make the engine idle too high or too low. It positions the throttle plates as it receives signals from the computer. Fig.

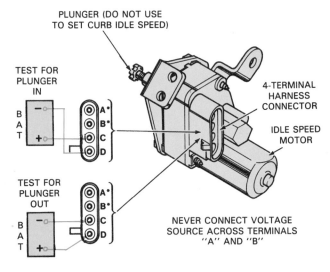

Fig. 14-58. Idle speed control motor can prevent normal control of engine idle rpm if it malfunctions. Check for proper voltage going to each terminal of the motor and compare it to motor operation. Note how battery voltage is connected to this motor to check its operation. (Oldsmobile)

14-58 shows testing methods for one type idle speed control motor.

Added information

Much of the information given in the next two chapters relates to computer controlled carburetors. An electronic gasoline injection system uses many of the same principles as a computerized carburetor system.

SUMMARY

After extended service, a carburetor may develop problems which cause poor engine operation. However, never tear down a carburetor without checking engine and ignition problems which could cause similar symptoms. When these sources are definitely eliminated, begin diagnosis of the carburetor by examining the clues to find out which carburetor system is causing the trouble.

A quick visual inspection of the carburetor will often provide additional clues. Look for evidence of leaking, sticking choke, binding linkage, and missing or disconnected vacuum hoses.

Extremely lean air-fuel mixtures will cause an erratic engine miss. Look for vacuum leaks, maladjusted fuel mixture screw, clogged fuel passages, and low float level. Overly rich fuel mixtures cause engine to roll, lope, and emit black exhaust. Among sources are clogged air bleed, high float level, and rich choke setting.

An exhaust gas analyzer, by sampling the exhaust gases, will quickly show whether air-fuel mixture is too rich or too lean. Inspection of spark plugs will also indicate whether fuel mixture is at fault.

Idle system problems are indicated when the engine has a rough idle, stalls at low speeds, or has incorrect idle speed. However, when this condition exists, the engine will run well at higher speeds. Causes could include a clogged idle tube, dirt-filled idle air bleed, dirt clogged idle well, partial blockage of idle transfer slot, and bent or damaged idle mixture screw.

Failure of the acceleration system will usually cause the engine to hesitate, stall, pop, backfire, or stumble. To check, look directly down the carburetor throat as the throttle is opened and closed several times. The pump should produce a heavy stream or fuel each time the throttle opens. If it does not, adjustment or repair is indicated.

Troubles in the high speed system, though rare, are usually caused by a too-lean or too-rich air-fuel mixture. Engine surge is an indication of too-lean mixture. Usually, a faulty mixture control solenoid or computer control circuit are to blame.

Depending upon the type of system, full power system problems are usually the result of a ruptured diaphragm in the power valve, a poorly adjusted metering rod, or a sticking vacuum piston.

Choke problems usually occur during or right after starting the engine. Common problems are a sticking choke valve, or an improper setting which causes too-lean or too-rich a mixture.

Before removing a carburetor for service, label all hoses and wires connected to the carburetor. Use care not to drop any loose parts into the intake manifold. Lift carburetor off carefully so as not to get sediment from the bowl into passages.

Disassemble the carburetor according to instructions in a service manual, clean parts, inspect for wear, install new parts and reassemble. Adjust float drop, float level, accelerator pump, bowl vent, choke, and choke break following the service manual procedures.

Never install a new or rebuilt carburetor without also installing a new base plate gasket on the intake manifold. Torque the mounting nuts to specifications. After connecting lines, wires, and linkage or cable, make cold fast idle cam adjustment and hot idle speed adjustment. On some vehicles, idle mixture adjustment is made using propane.

Computer controlled carburetor systems require specialized service techniques. Most have a self-diagnostic mode allowing easy location of problems. However, also check for leaking gaskets or seals, and clogged passages.

Before checking computer controls, make certain that other engine systems are working properly. Computer system problems may cause detonation, poor idling, hesitation, surging and a number of other symptoms. Usually the self-diagnostic mode will indicate where the trouble lies. Follow service manual procedures for diagnosis and repair.

KNOW THESE TERMS

Exhaust gas analyzer, Carburetor flooding, Engine flooding, Engine surge, Air horn body, Main body, Throttle body, Carburetor rebuild, Carburetor kit, Carburetor cleaner, Self-diagnosis, Dwell signal.

REVIEW QUESTIONS—CHAPTER 14

1. Before tearing down a carburetor, make sure it is the cause of the problem. _____ and _____ problems can cause similar symptoms.
2. What should you look for when visually inspecting a carburetor?
3. What are the symptoms of a lean air-fuel mixture?
4. Which of the following symptoms indicate(s) a too-rich air-fuel mixture?
   a. Engine misses now and then.
   b. Engine rolls, lopes, and exhaust smoke is black.
   c. Engine backfires.
5. List causes for lean and rich fuel mixtures.
6. Explain how an exhaust gas analyzer can help in troubleshooting carburetor problems.
7. A (high, low) float setting may cause engine stalling, backfiring, spitting, and loss of power.
8. A clogged idle passage will cause air-fuel mixture to be _____ while a clogged idle air bleed will cause the air-fuel mixture to be _____.
9. A customer's car hestiates or loses power upon acceleration from a stop. The problem occurs whether the engine is warm or cold.
   Mechanic A says that the choke setting is too rich and that an adjustment may be needed.
   Mechanic B says that the float level may be set too low or the accelerator pump may need adjustment or replacement. Who is right?
   a. Mechanic A only.
   b. Mechanic B only.
   c. Both Mechanic A and B.
   d. Neither Mechanic A nor B.
10. Engine _____ is a condition in which engine power seams to alternately increase and decrease independent of the driver's actions.
11. A faulty _____ _____ solenoid or the computer control circuit can upset high, midrange, or low speed carburetor operation.
12. What are the symptoms of a leaking power valve?
13. Black exhaust smoke indicates a lean fuel mixture that could be caused by the choke valve being stuck open. True or false?
14. What is the purpose of labeling wires and hoses.

before removing a carburetor?

15. What are the five major steps in a carburetor rebuild?

16. All jets in the carburetor are interchangeable so it is not important to reinstall them in the same location. True or false?

17. Explain why a wire or drill should not be used to clean out carburetor passages.

18. List seven things to look for while inspecting carburetor parts.

19. Always use new gaskets on a rebuild. Make sure that gasket holes match up with _____ in the carburetor castings.

20. A _____ float setting is made by measuring from some point on the float to a point on the carburetor bowl or horn.

21. Indicate the two basic methods for adjusting float level and float drop.

22. A _____ float setting is made with fuel in the carburetor bowl.

23. If a float is allowed to drop too much it will touch the bottom. Explain what could occur in such a case.

24. Describe how to adjust linkage for an accelerator pump.

25. Suggest two basic ways to adjust a remote choke.

26. Immediately upon starting, a _____ piston cracks open the choke valve slightly to prevent flooding.

27. Describe the two basic procedures for adjusting linkage for the vacuum choke break.

28. Cold fast idle speeds vary from about _____ to _____ rpm.

29. Explain how to adjust the hot idle speed solenoid.

30. Propane is used for idle mixture adjustment on some later model carburetors because:
    a. Propane is a more volatile mixture than air-fuel.
    b. It is easier to control than the normal air-fuel mixture.
    c. Propane provides a richer fuel mixture for making the setting and will assure a leaner setting for normal idle operation.

31. Many computer controlled carburetor systems have a _____-_____ mode that can be used to quickly locate malfunctions.

32. List 10 symptoms that might indicate a possible computer system problem.

33. Checking out the computer output on a computer controlled carburetor system usually requires a _____ _____ and a _____.

This engine has electronic, multi-point injection, tuned-runner intake manifold, and produces high horsepower and good gas mileage.
(Ford Motor Co.)

15 GASOLINE INJECTION SYSTEM OPERATION

After studying this chapter, you will be able to:
- *Explain the basic operation of an electronic fuel injection (EFI) system.*
- *List the major parts of an EFI system.*
- *Describe the construction of each major component of an EFI system.*
- *List some of the advantages of an EFI system.*
- *Explain the action of an electronically controlled fuel injector.*
- *Describe the four subsystems of an EFI system: fuel delivery, air induction, sensor, and computer control.*
- *Compare open loop and closed loop modes.*
- *List and describe the operation of typical system sensors.*
- *Compare analog and digital sensor signals.*

This chapter introduces the operating principles of an electronic fuel injection (EFI) system, also called a computer controlled gasoline injection system. It discusses multi-point or port injection because it is the most modern and efficient type system. The next chapter covers other injection system types, including single-point or throttle body injection. Many of the parts (computer, sensors, solenoid type injectors, etc.) explained in this chapter are common to most of the other system types. Study carefully!

WORKING PRINCIPLE

A modern *gasoline injection system,* using pressure from an electric fuel pump, sprays fuel into the engine intake manifold. See Fig. 15-1. Like a carburetor, it provides the correct air-fuel mixture for specific engine operating conditions. However, PRESSURE, not engine vacuum (suction), feeds the fuel into the engine cylinders. This makes gasoline injection very efficient.

Gasoline injection advantages
Gasoline injection has several advantages over a carburetor. The most important of these are:
1. Improved atomization. Fuel is forced into the intake manifold under pressure. This helps break the fuel into a fine mist.
2. Better fuel distribution. Flow of fuel vapors to each cylinder is more uniform.

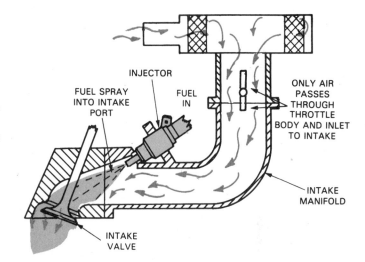

Fig. 15-1. Simplified cutaway shows action of most common fuel injection system for a gasoline engine.

3. Smoother idle. A leaner fuel mixture can be used without rough idle because of better fuel distribution and low-speed atomization.
4. Improved fuel economy. Efficiency is high because of more precise fuel metering, atomization, and distribution.
5. Lower emissions. Lean, efficient air-fuel mixture reduces exhaust pollution.
6. Better cold weather operation. Injection gives better control of mixture enrichment than a carburetor choke.
7. More engine power. Precise metering of fuel to each cylinder and increased airflow can produce more horsepower.
8. Simpler. Late-model, electronic fuel injection systems have fewer parts than modern computer controlled carburetor systems.

EFI SUBSYSTEMS

A modern electronic fuel injection system has four subsystems. As illustrated in Fig. 15-2, these include:

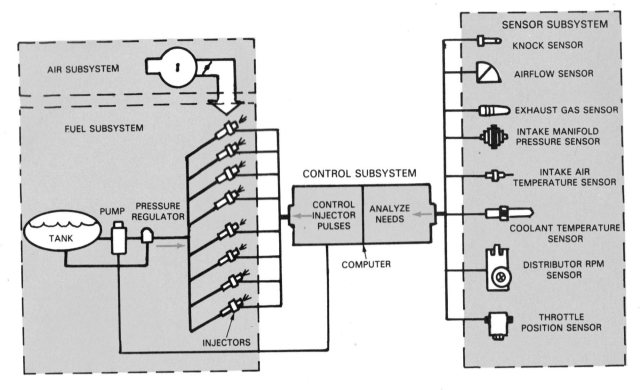

Fig. 15-2. Gasoline injection systems are made up of four subsystems. Boxes (solid and dashed lines) separate and depict various subsystems.

1. Fuel delivery system.
2. Air induction system.
3. Sensor system.
4. Computer control system.

Fuel delivery system

The *fuel delivery system* cleans and meters the right amount of fuel for varied driving conditions. It is made up of an electric fuel pump, fuel filter, fuel rail, pressure regulator, injectors, and connecting lines and hoses. These components are shown in Fig. 15-3.

The electric pump forces fuel out of the tank, through the lines, and into the pressure regulator. When the computer energizes the injectors, fuel sprays into the engine. Extra fuel bleeds out of the pressure regulator and back to the tank.

Note! Many of the parts (fuel pump, filters, lines, etc.) were explained in previous chapters. Refer to the index as needed for more information.

Air induction system

An *air induction system* for EFI provides clean air and delivers it to the engine cylinders. This sytem, typically, consists of an air filter, throttle body, throttle valve, intake manifold, and connecting air ducts. Fig. 15-3 shows the air induction system for one type of EFI-equipped engine.

Design of air induction systems will vary depending upon engine and fuel system design. However, they all have an air filter for removing airborne impurities, a throttle valve for controlling airflow into the engine, and other related parts.

Sensor system

The EFI *sensor system* keeps track of engine operating conditions and passes this information to the computer. See Fig. 15-3. A typical EFI sensor system includes an oxygen (exhaust gas) sensor, engine coolant temperature sensor, air inlet temperature sensor, throttle position sensor, intake manifold pressure (vacuum) sensor, and other sensors.

Just as your body can sense a change in its surroundings, (touching a hot stove for example), the sensor system can sense a change in the operation of many vehicle systems. It enables the modern fuel injection system to CHECK and CORRECT itself.

Computer control system

The *computer control system* processes information and controls operation of the EFI system and certain other automobile systems. Refer to Fig. 15-4.

A *computer wiring harness* carries current to and from the computer. Sensors from the engine, transmission, air conditioning system, and other systems send electrical information to the computer through the harness. Then, the computer can operate the injectors, transmission, ignition system, emission control systems, and other components for maximum efficiency.

ELECTRONIC CONTROL UNIT (COMPUTER)

The *computer,* or *electronic control unit* (ECU), is the ''brain'' of the fuel injection system. The sensors and wiring harness serve as the ''nervous system'': checking temperatures, positions, and other considerations for

Gasoline Injection System Operation

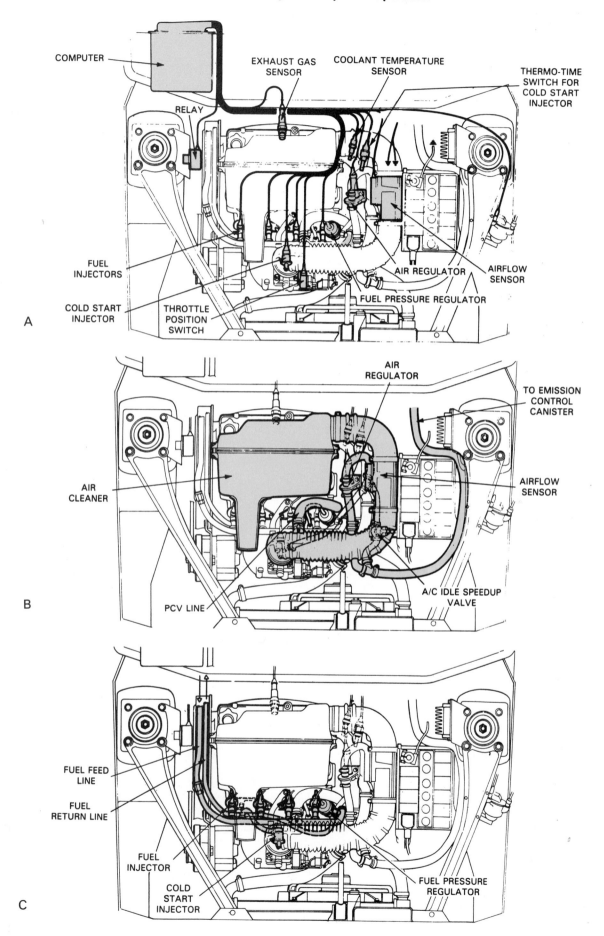

COMPUTER

EXHAUST GAS
SENSOR

COOLANT TEMPERATURE
SENSOR

THERMO-TIME
SWITCH FOR
COLD START
INJECTOR

RELAY

FUEL
INJECTORS

COLD START
INJECTOR

THROTTLE
POSITION
SWITCH

AIR REGULATOR

AIRFLOW
SENSOR

FUEL PRESSURE REGULATOR

A

AIR
REGULATOR

TO EMISSION
CONTROL
CANISTER

AIR
CLEANER

AIRFLOW
SENSOR

PCV LINE

A/C IDLE SPEEDUP
VALVE

B

FUEL FEED
LINE

FUEL
RETURN LINE

FUEL
INJECTOR

COLD
START
INJECTOR

FUEL PRESSURE
REGULATOR

C

Fig. 15-3. Note parts of EFI subsystems. A—Sensor system. B—Air induction system. C—Fuel delivery system.

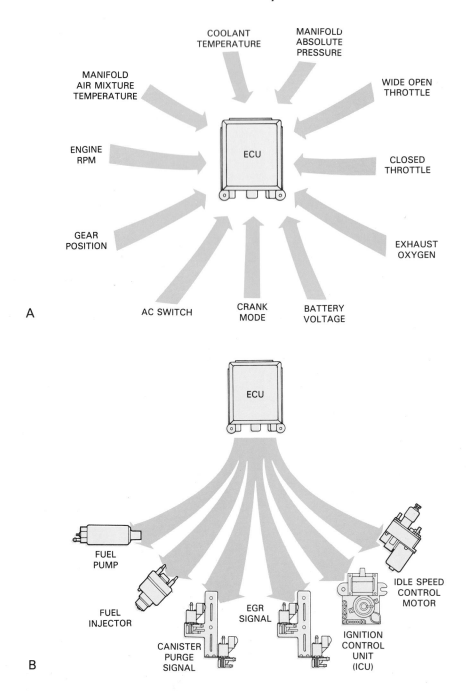

Fig. 15-4. These diagrams summarize computer control system. A—Computer or ECU inputs. B—ECU (electronic control unit) outputs of various system parts. (Renault)

proper injection system operation. Fig. 15-5 shows how electric current is fed to the computer from various sensors and how the computer feeds current to the injectors.

A car's computer is actually a preprogrammed microcomputer (preset, miniaturized electronic circuit). It has microscopic electronic circuits which are formed inside integrated circuits (ICs).

An *integrated circuit* is an electronic chip or circuit manufactured by photographically reducing a circuit and placing it on a special semiconductor (transistor type) material. This enables the computer to have literally hundreds or thousands of transistors, resistors,

capacitors, and similar components in a very small space. Different circuits are provided in the computer for performing different functions.

Fig. 15-6 shows a photo of the inside of an automobile computer. Note the very small components, especially the integrated circuits.

The on-board computer is about the size of a car radio and is often placed in the passenger compartment. Since computers are sensitive to vibration, extreme temperature change, and moisture, they are sometimes located behind or under the dash panel, Fig. 15-7. This places them away from engine heat, moisture, and the

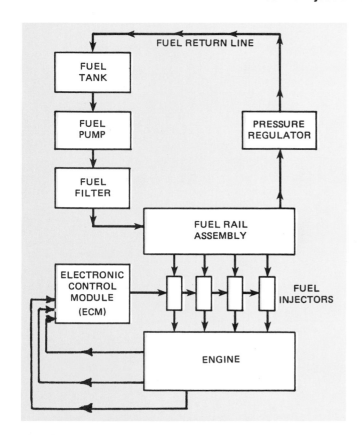

Fig. 15-5. Engine sensors send flow of information to electronic control module (on-board computer) in form of small electric currents. The module, acting on signal received, feeds current that operates injectors.

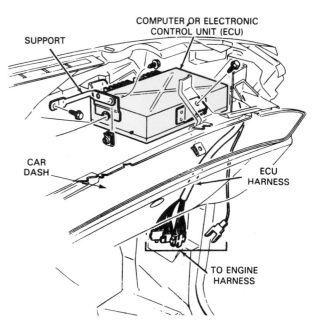

Fig. 15-7. On-board computer is usually behind instrument panel (dash). In this location, it is shielded from damaging engine heat and vibration. Some computers are mounted on air cleaner or elsewhere in engine or passenger compartment. (Cadillac)

Fig. 15-6. On-board computer contains thousands of miniaturized circuits.

elements in the engine compartment.

There are four basic parts or sections to a car's computer: input/output devices, central processing unit, power supply, and memories.

Input/output devices

The *input/output devices* are electronic circuits that convert signals from sensors into digital (on/off or computer) signals for use in the central processing unit (brain or calculator section) of the computer. The devices (circuits) can also change computer language into electrical signals to operate system components.

Central processing unit

The *central processing unit* performs mathematical functions or logic functions to deliver the correct air-fuel ratio and to operate other system devices. It uses digital signals from the input devices to determine what is going on during vehicle operation and what should be done to increase efficiency.

Power supply

The *power supply* in a car's computer prevents voltage fluctuations that could affect computer operation. A computer relies on very smooth dc current, mainly from the car battery. The power supply simply regulates input voltage to other parts of the computer.

Computer memories

Most computers have three basic types of memory circuits: read only memory, random access memory, and programmable read only memory.

The *read only memory* (ROM) is programmed data that can only be analyzed by the computer itself. It is information used by the computer in performing the various functions. The ROM program cannot be changed. If the battery or voltage supply is disconnected from the computer, the data in the ROM will remain in the computer.

The *random access memory* (RAM) is temporary information held in the computer. It is like a "note pad" of inputs and outputs. Data such as self-diagnosis codes

313

can be pulled out of RAM. If battery voltage is removed, all information is erased from RAM.

The *programmable read only memory* (PROM) has information on the particular make and model car. It has data about engine size, vehicle weight, transmission type, rear axle ratio, etc. As you will learn in the chapter on fuel injection service, the PROM is normally removed and reused when replacing the central processing unit (computer). If voltage is disconnected from the PROM, it will retain its information.

GASOLINE INJECTOR

An *injector* for a gasoline injection system is simply an electrically operated fuel valve. When energized by the computer, it must open and produce a uniform fuel spray pattern in the intake manifold. When deenergized, it must close quickly, without leakage.

Injector construction

Most modern injectors are made of metal and plastic. Rubber O-rings seal joints where parts fit together. Usually, the injector fits into a hole machined into the intake manifold. However, as will be discussed in the next chapter, some systems have the injector in the throttle body assembly.

Fig. 15-8 is a cutaway view of a typical gasoline injector. Refer to this illustration.

1. ELECTRIC TERMINALS (electrical connection for circuit between injector coil and computer).
2. INJECTOR SOLENOID (armature and coil that opens and closes valve).
3. INJECTOR SCREEN (screen filter for trapping debris before it can enter injector nozzle).
4. NEEDLE VALVE (end of armature shaped to seal against needle seat).
5. NEEDLE SEAT (machined surface around the hole in end of injector against which the needle valve tip presses to form a seal).
6. INJECTOR SPRING (small spring that returns needle valve to closed position).
7. O-RING SEAL (rubber seal that fits around outside of injector body and seals in intake manifold).
8. INJECTOR NOZZLE (injector outlet that produces fuel-spray pattern).

Gasoline injector operation

In a simplified way, Fig. 15-9 illustrates the operation of a gasoline injector. When the computer sends current to the injector coil, the coil develops a magnetic field. Like an electromagnet, the field attracts and pulls on the injector armature. The armature moves up into the coil's field. The needle is then lifted off its seat and lets fuel spray out the nozzle into the intake manifold.

When the computer shuts off current to the injector coil, the magnetic field collapses. This lets the injector spring push down on the armature forcing the needle against its seat. This blocks fuel flow.

Injector pulse width

Injector pulse width refers to the length of time or duration that the injector is open. The computer controls injector pulse width. A *long pulse width* richens the fuel

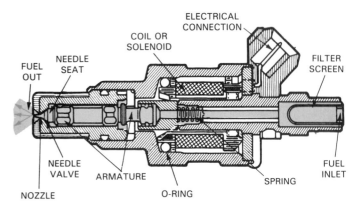

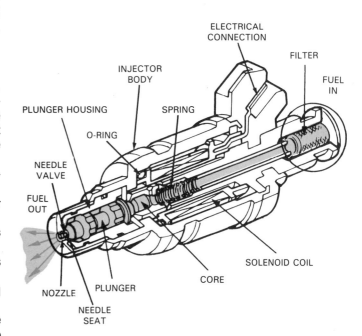

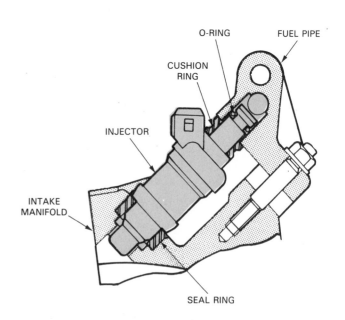

Fig. 15-8. Cutaways show important components of an injector. (Chrysler)

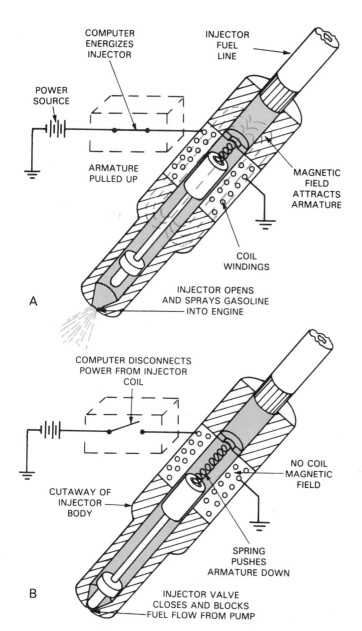

Fig. 15-9. EFI injector operation. A—Current through injector coil builds magnetic field. Magnetism attracts and pulls up on armature to draw injector needle off its seat. Gasoline sprays out. B—Current flow stops when computer breaks circuit. Injector valve closes stopping fuel spray.

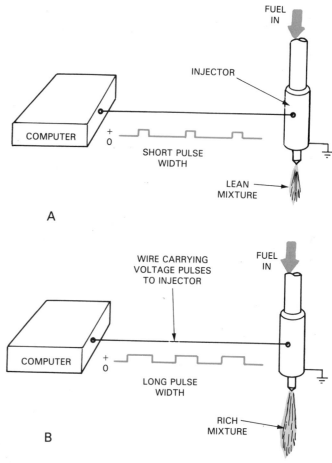

Fig. 15-10. Injector pulse width means the amount of time that computer sends current to injector to keep valve open. A—Short pulse causes less fuel spray because injector valve is not open long percentage of time. Mixture is leaner. B—Long pulse keeps valve open more of the time. Mixture is richer.

mixture because more fuel would spray into the intake manifold on each cycle. A *short pulse width* would lean the mixture because the injector would be kept closed longer between pulses.

Fig. 15-10 illustrates short and long injector pulse widths. Note that the *square sine wave* (sine representing voltage change) denotes the pulse width. When the wave moves up from zero, indicating voltage supply to the injector, the injector is open. When the wave moves back down to zero (base line), the injector is closed because there is no voltage and current flow.

When the square wave is shorter, Fig. 15-10A, the injector pulse width is shorter and the fuel mixture is leaner. When the square wave is longer, Fig. 15-10B,

the pulse width is longer and the mixture is richer.

With many systems, the computer cycles the injectors open and closed several times a second. By changing the percentage of ON and OFF times, it can control the air-fuel mixture ratio.

FUEL RAIL ASSEMBLY

The *fuel rail assembly,* also called a *fuel log,* feeds fuel to all of the injectors, Fig. 15-11. The fuel pressure regulator and sometimes the cold-start injector attach to the fuel rail. The fuel rail can be a length of steel tubing or a cast metal block.

As shown in Fig. 15-12, fuel enters the fuel rail from the electric fuel pump. Equal fuel pressure forms inside the rail and at the inlet to each injector. The pressure regulator bleeds off excess fuel to maintain the proper pressure in the system. This allows cool fuel to constantly circulate between the fuel rail and the fuel tank.

A *service fitting* is a threaded orifice for bleeding off pressure and for installing a pressure gauge. One is normally provided on the fuel rail. It is usually covered with a metal cap that keeps out dust and dirt.

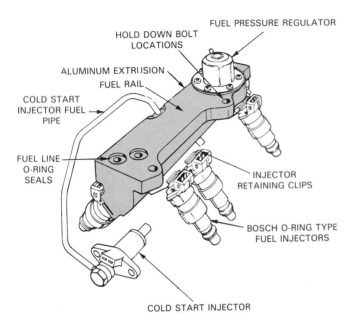

Fig. 15-11. Fuel rail assembly. Fuel pressure regulator and sometimes cold-start injector are considered part of this assembly. (Buick)

Fig. 15-13. Cold-start injector is extra injector which supplies additional fuel for richer starting mixture when engine is cold. (Saab)

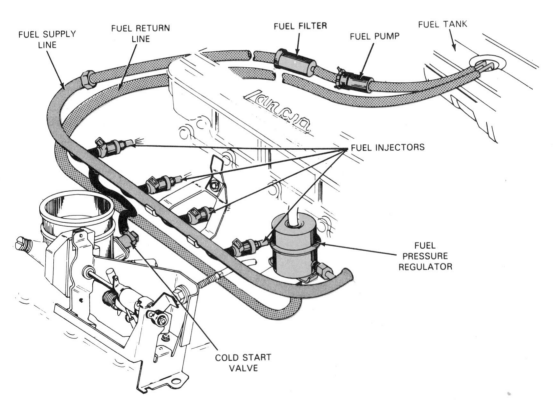

Fig. 15-12. Electric fuel pump supplies fuel to fuel rail. Note that this rail consists of a length of tubing rather than a casting. (Fiat)

Cold-start injector

A *cold-start injector* is an additional fuel injector valve which supplies extra gasoline for cold engine starting. Refer to Figs. 15-13 and 15-14. Either a thermo-time switch or the system computer can activate the cold start injector.

Cold-start injector construction

In general, a cold-start injector is constructed like a main fuel injector. It, too, has a coil that produces a magnetic field when energized. The field pulls the injector open. A spring closes the valve when deenergized. The cold-start injector can be placed in various loca-

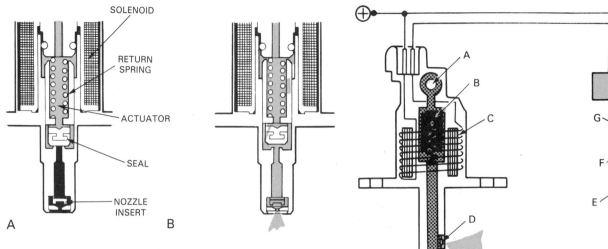

Fig. 15-14. Cutaway of cold-start injector. Note part names.
A—When coil of solenoid is not energized valve is closed and
there is no fuel spray. B—Energized coil opens valve.

A. FUEL INLET   E. CONTACT POINT
B. VALVE   F. BI-METAL ELEMENT
C. COIL   G. HEATER
D. SWIRL TYPE NOZZLE

tions in the intake manifold. It usually sprays fuel near
the entry of the intake so fuel will be carried to all
cylinders.

Cold start injector operation

Fig. 15-15 shows a basic cold start injector thermo-
time switch circuit. It illustrates cold-start valve action.
When the sensor detects a cold engine, the switch

Fig. 15-15. Thermo-time switch for basic cold start injector.
Switch energizes injector when engine temperature is low
enough. Activated injector will spray enough extra fuel into the
cold engine to keep it running smoothly. After engine is warm,
thermo-time switch opens to shut off injector. Heat element
in switch will shut off injector after a short period of time even
if engine is very cold.

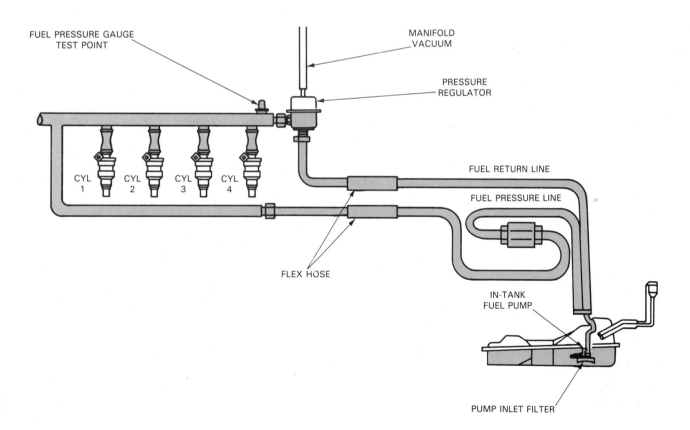

Fig. 15-16. Pressure regulator maintains uniform pressure in rail by allowing fuel to bypass and return to the tank when pressure
limit is reached.

closes to energize the cold-start injector. This injector sprays an additional charge of fuel into the intake manifold to provide a rich fuel mixture.

FUEL PRESSURE REGULATOR

The *fuel pressure regulator* maintains a constant, preset pressure at the injectors. It is usually mounted on the fuel rail. After the fuel flows through the rail, it enters the pressure regulator. Extra fuel flows through an orifice in the regulator, into a return fuel line, and back to the tank. See Fig. 15-16.

Engine intake manifold vacuum is connected to the fuel pressure regulator. This allows the regulator to change fuel pressure with changes in engine load.

Fuel pressure regulator construction

Fig. 15-17 shows a cutaway view of a typical fuel pressure regulator. Study its construction.

The basic parts of a fuel pressure regulator are:

1. FUEL INLET FITTING (allows fuel to enter pressure regulator from fuel rail).

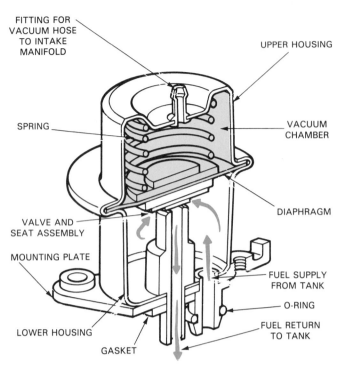

Fig. 15-18. Vacuum chamber shown at top is sealed by diaphragm. It receives vacuum from intake manifold. When intake manifold vacuum is low (engine accelerating or under load) spring keeps bypass valve closed so more fuel is delivered to injectors. (Ford Motor Co.)

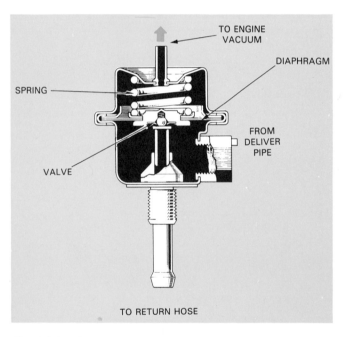

Fig. 15-17. Study construction of pressure regulator from this cutaway. Pressure on diaphragm opens valve to return hose.

2. FUEL RETURN FITTING (allows excess fuel to flow out of rail and regulator and return to fuel tank).
3. CHECK VALVE (opens and closes to control fuel flow through regulator).
4. DIAPHRAGM (flexible disc that can move with changes in fuel pressure).
5. DIAPHRAGM SPRING (coil spring that pushes diaphragm toward fuel and closes check valve).
6. VALVE SEAT (attached to diaphragm, works with check valve to open and close fuel return).
7. VACUUM CHAMBER (allows engine vacuum to act on backside of vacuum diaphragm), Fig. 15-18.

8. VACUUM FITTING (allows vacuum hose from intake manifold to connect to vacuum chamber).

Fuel pressure regulator operation

Whenever the electric fuel pump is operating, fuel flows into the regulator's pressure chamber from the fuel rail. The fuel, being under pressure, pushes on the regulator diaphragm. However, there is still not enough pressure to cause the return valve to open. Additional force must be supplied by engine vacuum.

When the engine is running, vacuum enters the vacuum chamber of the regulator and exerts a "pull" that, together with the force of the fuel in the opposite chamber, causes the diaphragm to flex and open the return valve. Excess fuel pressure is bled from the system to lean the fuel mixture. The excess fuel returns to the fuel tank. See Fig. 15-19.

Under rapid acceleration, the engine requires a richer mixture. The fuel pressure regulator is designed to help richen the mixture. This is what happens:

1. As the engine begins to accelerate, engine vacuum drops.
2. Since fuel pressure alone cannot keep the diaphragm flexed, it returns to its former position, closing the return valve.
3. This causes fuel pressure to build up higher to richen the mixture for more power.

Keep in mind that the computer and sensors are monitoring fuel mixture and other variables. They work with the pressure regulator to maintain the most efficient air-fuel ratio for the needs of the engine.

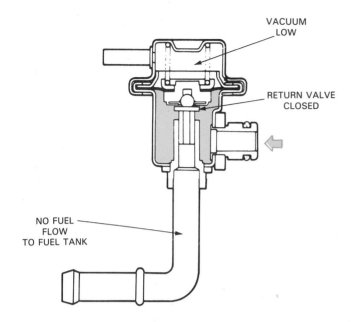

VACUUM LOW

RETURN VALVE CLOSED

NO FUEL FLOW TO FUEL TANK

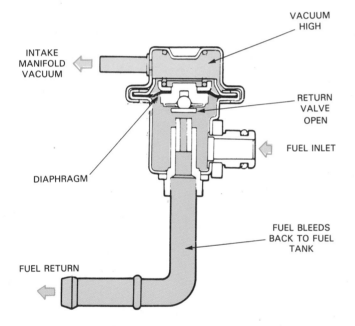

VACUUM HIGH

INTAKE MANIFOLD VACUUM

RETURN VALVE OPEN

FUEL INLET

DIAPHRAGM

FUEL BLEEDS BACK TO FUEL TANK

FUEL RETURN

Fig. 15-19. Cutaway shows how vacuum affects regulator action. When engine vacuum is high (engine at low speed or idle) diaphragm flexes in direction of vacuum. This opens valve and allows fuel to bypass and return to fuel tank. At low vacuum (engine under load) diaphragm flexes down to close bypass valve and increase fuel pressure. (Honda Motor Co.)

Fig. 15-20 shows a cutaway view of a fuel rail and its fuel pressure regulator. Note how the regulator acts to maintain fuel pressure in the rail and to the injectors.

FUEL RETURN CIRCUIT

The *fuel return circuit* is made up of all the components that carry fuel from the outlet of the pressure regulator to the fuel tank, Fig. 15-21. This normally includes rubber fuel hoses, a steel fuel line, and necessary fittings.

The fuel return circuit allows circulation of fuel through

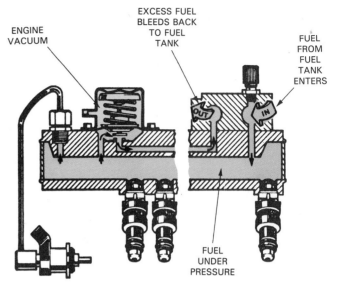

ENGINE VACUUM

EXCESS FUEL BLEEDS BACK TO FUEL TANK

FUEL FROM FUEL TANK ENTERS

OUT

IN

FUEL UNDER PRESSURE

Fig. 15-20. Cutaway of fuel rail and its pressure regulator. Note arrangement of passages for incoming fuel and fuel returning to fuel tank. (Oldsmobile)

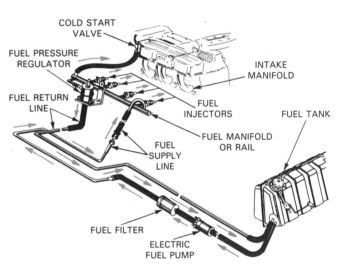

COLD START VALVE

FUEL PRESSURE REGULATOR

INTAKE MANIFOLD

FUEL RETURN LINE

FUEL INJECTORS

FUEL TANK

FUEL MANIFOLD OR RAIL

FUEL SUPPLY LINE

FUEL FILTER

ELECTRIC FUEL PUMP

Fig. 15-21. Above parts are typical of fuel return systems. One end connects to fitting on fuel pressure regulator, other end to fuel tank fitting. (Fiat)

the main fuel line, fuel rail, past the inlet to the injectors, through the regulator, return line, and back to the tank. This circulation prevents a heat buildup in the fuel. Heat could cause bubbles in fuel, upsetting fuel mixture.

Fuel accumulator

A *fuel accumulator* dampens fuel pressure pulses and helps maintain fuel pressure when the engine is shut off. Fuel pressure fluctuations result from the action of the fuel pump and from the injectors constantly opening and closing. Too much pressure fluctuation could upset the operation of the pressure regulator.

The fuel accumulator acts like a "shock absorber." It can increase component life and help quiet system operation. Look at Fig. 15-22.

The fuel accumulator is simply an enclosed container with a spring-loaded diaphragm. Fuel pressure pushes the diaphragm down and compresses the diaphragm spring. This provides energy to maintain pressure or to counteract a sudden pressure drop.

The *throttle body,* itself, is usually made of cast aluminum. It mounts in the induction system just ahead of the intake manifold. As was shown in Fig. 15-23, it sometimes bolts to the inlet of the intake manifold.

There are many variations in throttle body design. These will be discussed in the next chapter. The most

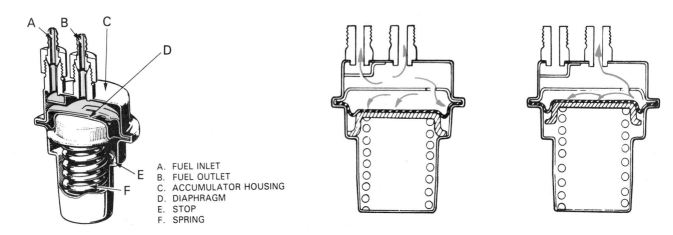

A. FUEL INLET
B. FUEL OUTLET
C. ACCUMULATOR HOUSING
D. DIAPHRAGM
E. STOP
F. SPRING

Fig. 15-22. Fuel accumulator is simply a small canister fitted with fuel inlet and outlet. Canister contains a spring and a diaphragm. The spring and diaphragm absorb pressure pulsations like a cushion. Left. Note names of parts. Center. When engine is running, fuel pressure compresses diaphragm spring. Right. When engine is stopped, spring presses up on diaphragm to maintain system pressure. (Volvo)

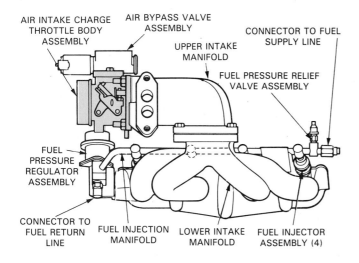

Fig. 15-23. Throttle body assembly only controls airflow to engine with most modern fuel injection systems. (Ford)

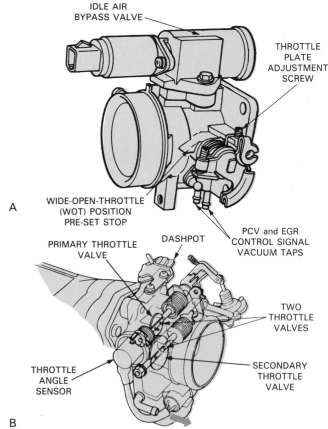

Fig. 15-24. There are many throttle body designs for fuel injected systems. Two variations are shown here. (Ford and Honda)

THROTTLE BODY ASSEMBLY

The *throttle body assembly* for most modern fuel injection systems is used only to control airflow into the engine. See Fig. 15-23.

A *throttle valve,* like the butterfly valve in a carburetor, is attached to the throttle cable and gas pedal. When the driver presses the gas pedal, the shaft rotates. This swings the throttle valve open to admit more air. This increases engine power for acceleration or pulling a load.

modern design is simply used to control airflow, as shown in Fig. 15-24.

Air bypass valve (idle speed control valve)

An *air bypass valve,* also known as an *idle speed control valve,* is frequently used to regulate engine idle speed, Fig. 15-25. The air bypass valve normally mounts on the throttle body assembly. It can be controlled by either a temperature-sensitive device or by the computer.

Fig. 15-26 shows the action of the temperature *(bimetal strip)* operated idle air control valve. When the control valve and engine are cold, the bimetal strip holds

the air bypass open. This increases engine speed. When the engine warms, the bimetal strip bends and moves to close the idle air valve. This drops engine idle speed to normal.

Fig. 15-27 illustrates the basic action of a *computer controlled* idle air valve. When the engine is cold, the engine temperature sensor signals the computer. The computer knows that engine rpm should be increased to prevent engine stalling and stumbling. It then sends current to the idle air control valve. This opens the valve and allows air to bypass the throttle valve. The rest of the injection systems reacts to this increased airflow and engine idle speed increases, just as a carburetor fast idle cam increases cold engine idle rpm.

As the engine warms, the computer operates the idle air control valve to close off the bypass. Then, no extra air flows around the throttle plate and idle speed returns to normal again.

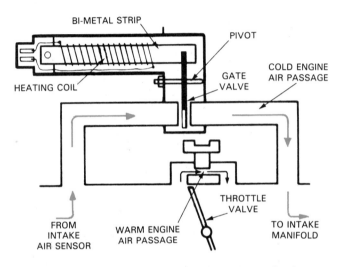

Fig. 15-25. Air bypass valve acts like fast idle mechanism on carbureted fuel system. It speeds up engine idle rpm when it is cold. (Volkswagen)

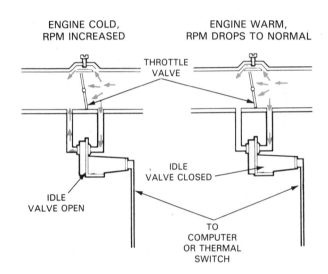

Fig. 15-27. Computer operates this air bypass valve. (AMC)

Fig. 15-28 shows a cutaway view of one computer controlled type of idle speed valve. When the engine is cold, the computer sends current to rotate the small dc motor in one direction. This turns the rotor and screw to pull the air bypass valve open. When the engine warms, the computer reverses the polarity to the motor so it turns the screw in the opposite direction, closing the valve.

AIR INLET DUCTS

Air inlet ducts on most fuel injected engines carry air from the air cleaner to the throttle body, Fig. 15-29. Sometimes, the airflow sensor is mounted in this duct. The duct is usually made of plastic to reduce weight.

Clamps at duct fittings prevent leakage and keep airborne dirt out of the engine. Where ducts are placed between the airflow sensor and the intake manifold leakage could upset the air-fuel ratio.

Fig. 15-29 also shows where to look for components

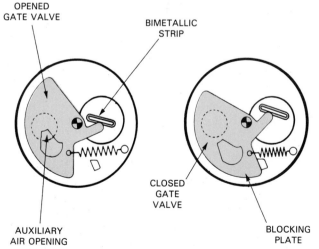

Fig. 15-26. Expansion and contraction of metals in bimetallic strip open and close air door in this air bypass valve. When engine is cold, metal contracts and opens door admitting more air. As engine warms, warm air, assisted by electric heat element, closes blocking plate to lower idle speed. (Renault and AMC)

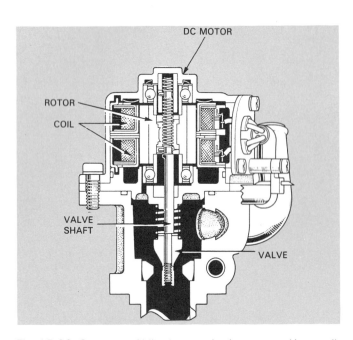

Fig. 15-28. Cutaway of idle air control valve operated by small dc motor and computer. Current from computer energizes motor which turns threaded shaft to open and close air valve. (Toyota)

just discussed in the chapter. Note the locations of the fuel pressure regulator, fuel rail or manifold, throttle cable, throttle body, air bypass valve, cold start valve, and airflow sensor.

ENGINE SENSORS

An *engine sensor* is a device that changes resistance or voltage output with a change in a condition such as temperature, position, movement, etc. A modern electronic fuel injection system uses numerous sensors to improve efficiency. An EFI system might use many of the sensors shown in Fig. 15-30.

Fig. 15-31 shows some of the conditions sensed by one computer system. Note that it checks everything from charging voltage to air conditioning operation.

Throttle position sensor

A *throttle position sensor* is a variable resistor or multiposition switch connected to the throttle valve shaft. When the driver presses on the gas pedal for more power, the throttle shaft and sensor are rotated. This changes the internal resistance of the sensor. The change in current signals the computer and the computer can alter the air-fuel ratio as needed.

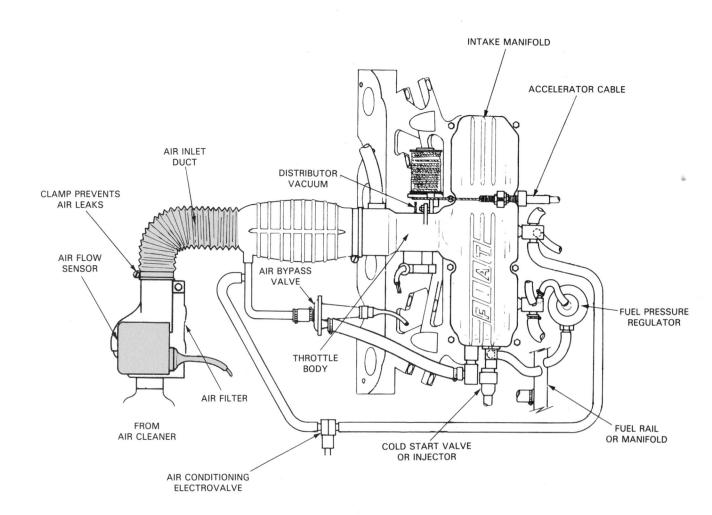

Fig. 15-29. This air intake uses flexible duct to move air from air cleaner into intake manifold. Note location of airflow sensor. Leaks in duct could upset air-fuel ratio. (Fiat)

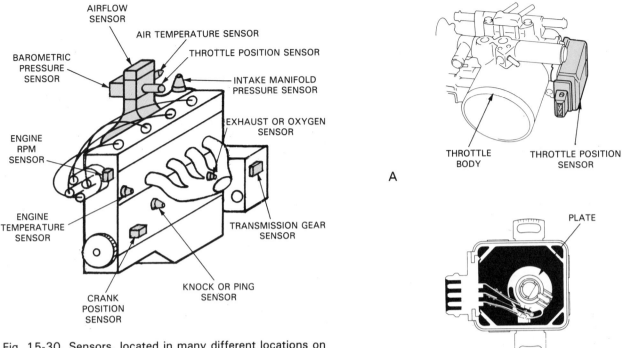

Fig. 15-30. Sensors, located in many different locations on engine, may be used to feed information to computer.

One type of throttle position sensor is illustrated in Fig. 15-32. Note how it mounts on the throttle body and that it has several contacts to change output resistance.

Fig. 15-32. Throttle position sensor. A—Sensor is mounted on throttle body over throttle shaft. B—View of inside of sensor. As throttle shaft rotates, it turns plate which alters internal circuit connections and resistance, sending electric current signals to computer. (Toyota)

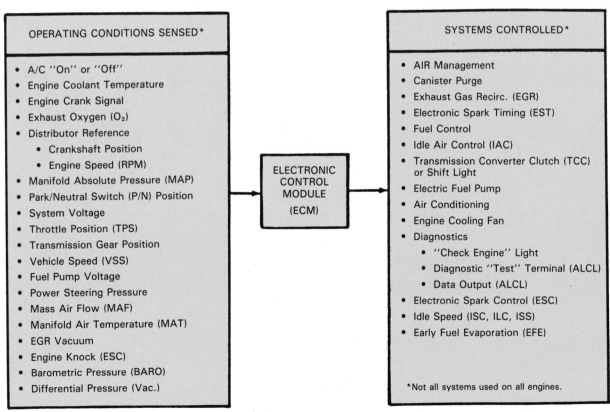

Fig. 15-31. Note typical conditions sensed and controlled by electronic means. (Buick)

Engine coolant temperature sensor

An *engine coolant sensor* monitors the operating temperature of the engine, Fig. 15-33. It is mounted so that it is exposed to the engine coolant.

When the engine is cold, the sensor might provide a high current flow (low resistance). The computer would adjust for a richer air-fuel mixture for cold engine operation. When the engine warms, the sensor would supply information (high resistance for example) so that the computer could make the mixture leaner.

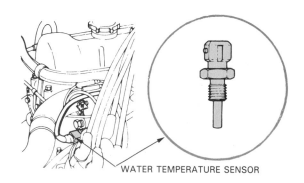

Fig. 15-33. Water temperature sensor usually is threaded into opening in block or head where it will be in contact with coolant. (Subaru)

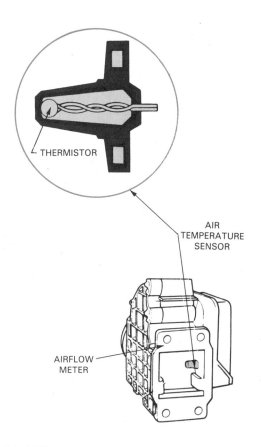

Fig. 15-34. Airflow meter shows air temperature sensor. Sensor uses thermistor which is extremely sensitive to temperature changes. As temperature rises, resistance of thermistor decreases. (Subaru)

Inlet air temperature sensor

An *inlet air temperature sensor* measures the temperature of the air entering the engine, Fig. 15-34.

Cold air is more dense than warm air, requiring a little more fuel. Warm air is NOT as dense as cold air, requiring a little less fuel. The air temperature sensor helps the computer compensate for changes in outside air temperature and maintain an almost perfect air-fuel ratio.

Charge temperature sensor

A *charge temperature sensor,* similar to an inlet air temperature sensor, measures the temperature of the air-fuel mixture. See Fig. 15-35. It is installed in the intake port, in front of the engine intake valve.

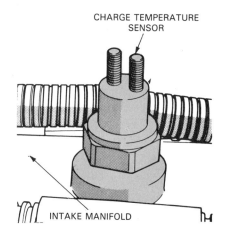

Fig. 15-35. Unlike air temperature sensor which senses only coldness of air, charge temperature sensor checks temperature of the air-fuel mixture. It is located in intake port just before intake valve. (Chrysler)

Knock sensor

A *knock sensor* is installed in the engine and detects abnormal combustion (ping, preignition, or detonation). Many new cars, especially those with turbochargers, can suffer from engine knock. The sensor actually "listens" for the knocking of abnormal combustion. When it detects knock, an electrical signal allows the computer to retard the ignition timing or lower turbocharger boost enough to prevent engine damage.

Crankshaft position sensor

A *crankshaft position sensor* is used to detect engine speed. It allows the computer to change injector openings as engine speed changes. Higher engine speed generally requires more fuel.

This sensor can be located on the front, rear, or center of engine. Its tip is close to the crank so that it can sense the teeth or notches as they rotate past the sensor. The magnetic field around the sensor, and current flow through the sensor, change as the crank rotates, allowing the computer to measure engine rpm.

Manifold pressure sensor

A *manifold pressure sensor,* usually referred to as MAP

(for manifold absolute pressure), measures vacuum inside the engine intake manifold. See Fig. 15-36.

Engine manifold pressure, as discussed in earlier chapter, is a good way to indicate engine load. High pressure (low intake vacuum) occurs with a heavy load. The engine needs a rich fuel-air mixture. When manifold pressure is low (high intake vacuum), there is very little load. A lean mixture is sufficient.

The MAP sensor changes resistance with changes in engine load. This data is used by the computer to alter the fuel mixture.

Flap airflow sensor

A *flap airflow sensor* measures the airflow into the engine. This helps the computer determine how much fuel should be injected into the intake manifold.

The airflow sensor usually mounts ahead of the throttle body assembly in the air inlet duct system. Refer again to Fig. 15-29.

Fig. 15-37 shows how a typical airflow sensor operates. At idle, the sensor flap is nearly closed. Sensor resistance stays high. This tells the computer that the engine is idling and needs very little fuel.

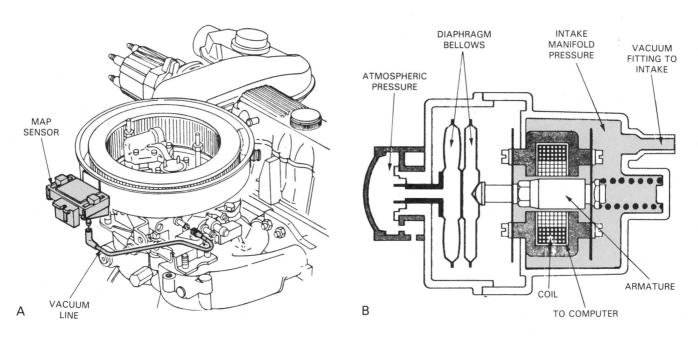

Fig. 15-36. MAP (manifold absolute pressure) sensor tells computer whether engine is under light or heavy load. A—MAP unit with connecting vacuum line to intake manifold. B—Cutaway of MAP sensor. Bellows expand or contract, as affected by engine intake manifold vacuum. This causes motion transfer to coil. Coil sends signal to computer. (Buick and Robert Bosch)

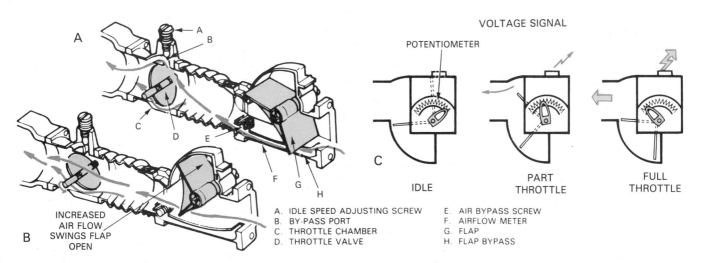

Fig. 15-37. Flap airflow sensor tells computer whether engine is idling or at higher speed by measuring amount of air moving into engine. A—At idle speed, air flap is nearly closed. Responding to small current, computer produces short injection pulse width for small amount of fuel. B—Throttle open for more power, air flap swings open and sends strong signal to computer. Computer then sends wide pulse width to injectors for richer fuel mixture. C—Potentiometer is attached to sensor shaft. Wiping arm moves across potentiometer as flap moves. Depending on flap position, potentiometer will send weak or strong signal to electronic control unit.

As engine speed and airflow increase, air forces the flap to swing open. This moves the variable resistor to the low resistance position. The increased current flow now tells the computer that more air is flowing into the engine. The computer then increases injector pulse width as needed.

Fig. 15-38 shows how the flap type airflow sensor and potentiometer (variable resistor) are connected. Note that a weak spring is used to return the flap to the closed, idle position.

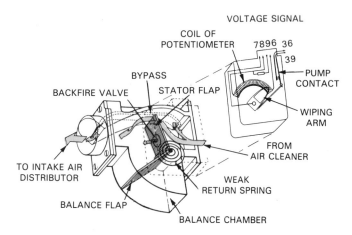

Fig. 15-38. Cutaway shows how air sensor flap connects to and controls potentiometer. Strength of signal to computer depends on where wiping arm rests on resistor of potentiometer. (Volkswagen)

Mass airflow sensor

A *mass airflow sensor* performs about the same function as a flap type sensor, but it sends more precise information to the computer. Fig. 15-39 is a newer type sensor found on some late model cars.

Basically, the mass airflow sensor uses a small electrically energized, resistance wire to detect airflow, Fig. 15-40. The wire's temperature drops as air flows over it. The greater the airflow, the lower its temperature. The drop in temperature changes the wire's resistance, signalling the computer of more air intake. The opposite is true for low airflow.

A mass airflow sensor, sometimes called a "hot wire" sensor, is desirable because it automatically compensates for changes in air temperature and atmospheric pressure. It eliminates the need for an air temperature sensor and air pressure sensor.

Oxygen sensor

The *oxygen sensor* or *exhaust gas sensor* measures the oxygen content in the engine's exhaust gases, Fig. 15-41. The oxygen content is an excellent indicator of whether the air-fuel mixture is too rich or too lean. The oxygen sensor is one of the most important sensors in modern electronic fuel injection systems. It actually checks the efficiency of the fuel system with the engine running.

The oxygen sensor screws into a fitting in the engine exhaust manifold or pipe, as in Fig. 15-42. Its inner tip is exposed to hot exhaust gases. Its outside surface is exposed to outside air. An electrical lead connects to the computer wiring harness.

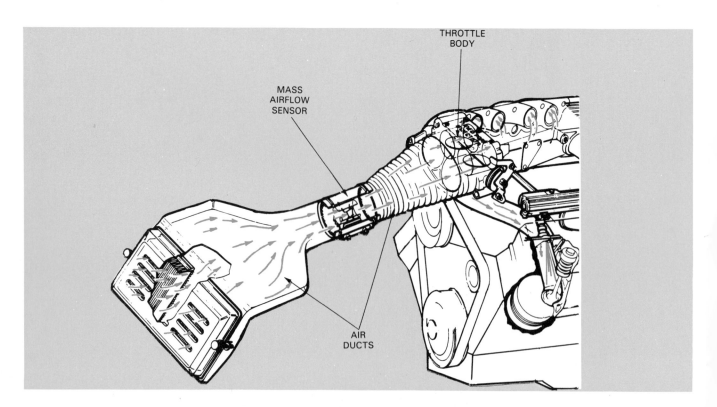

Fig. 15-39. Modern electronic, multi-point fuel injection system. Mass airflow sensor is located in duct ahead of engine. It detects airflow volume with fine resistance wire. (Chevrolet)

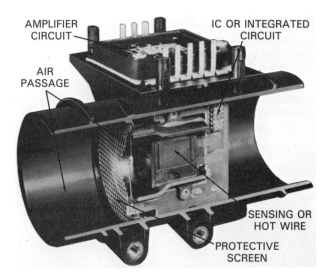

Fig. 15-40. Mass airflow sensor with resistance wire. Wire is heated by electric current and is very sensitive to temperature. The greater the airflow past it, the lower its temperature becomes. The lower its temperature, the more the signal to computer changes. (Chevrolet)

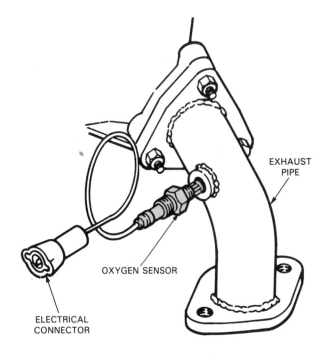

Fig. 15-42. Oxygen sensor is placed in exhaust system either at exhaust manifold or in pipe leading from exhaust manifold, as shown here. (Renault and AMC)

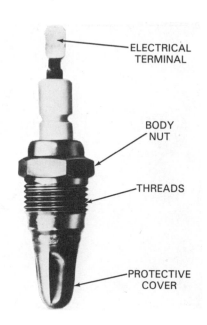

Fig. 15-41. Oxygen sensor detects amount of oxygen in engine's exhaust. (Oldsmobile)

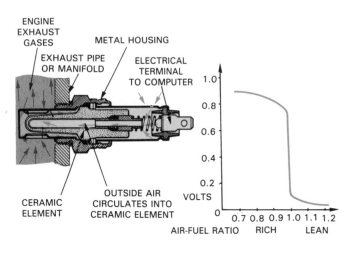

Fig. 15-43. Cutaway of exhaust sensor. Note how it operates. Graph at right shows how oxygen content changes voltage of sensor's signal to computer. (Fiat)

Oxygen sensor construction

The oxygen sensor has a special ceramic core made of zirconium dioxide, Fig. 15-43. The surface of the ceramic core is coated with platinum. The coated ceramic core has the ability to produce a voltage output when exposed to heat and a difference in oxygen levels on each side of the ceramic element. The ceramic voltage-producing device is enclosed inside a metal housing. Terminals are provided for connecting the sensor to the computer wiring harness.

Oxygen sensor operation

When the sensor is cold, it produces no voltage. The system then operates on data preprogrammed into the computer. When the oxygen sensor is heated above about 300°F (149°C), it begins to produce a voltage signal. For example, look at Figs. 15-43 and 15-44.

When the fuel mixture is too rich, there is a small amount of oxygen in the engine exhaust gases. This produces a large difference in the oxygen levels on each side of the ceramic sensing device. Negative oxygen ions flow through the ceramic device and a voltage output is produced for the computer, Fig. 15-44A. About a ONE

VOLT signal is fed to the computer. The computer can then shorten injector pulse width to lean the air-fuel mixture slightly.

When the engine's fuel mixture is too lean, there is an excess amount of oxygen in the engine exhaust. This reduces the difference in the oxygen levels on each side of the sensor's ceramic element. Very few oxygen ions flow through the sensor and the sensor's voltage output drops to a fraction of a volt, Fig. 15-44B. This signals the computer to increase injector pulse width. This helps maintain an almost perfect air-fuel mixture.

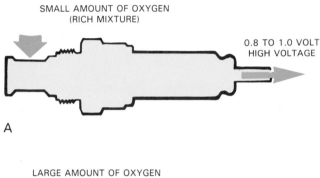

SMALL AMOUNT OF OXYGEN
(RICH MIXTURE)

0.8 TO 1.0 VOLT
HIGH VOLTAGE

A

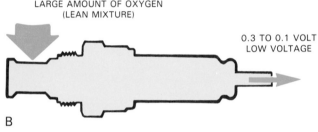

LARGE AMOUNT OF OXYGEN
(LEAN MIXTURE)

0.3 TO 0.1 VOLT
LOW VOLTAGE

B

Fig. 15-44. A—Small amount of oxygen causes high voltage signal from sensor. B—Large amount of oxygen reduces signal strength. (Volkswagen)

Other sensors

Other sensors besides those just covered can be used in a computer control system. Some of them include sensors checking the operation of the transmission, air conditioning system, brake system, and emission control systems. When information is required on any of these sensors, refer to a service manual. It will give the exact details of the specific system. It will explain its operation and how the sensor should be tested or serviced.

OPEN AND CLOSED LOOP

The terms open loop and closed loop refer to the operating mode of a computerized fuel system.

When an engine is cold, the computer operates *open loop* (no feedback from sensors).

After the engine warms to operating temperature, the system "kicks in" to *closed loop* (uses feedback from sensors) to control system operation.

Open loop operation

As was mentioned earlier, an oxygen sensor must heat up to several hundred degrees before it will function

properly. This is the main reason computer systems have an open loop mode. The computer has preprogrammed information (injector pulse width, injector timing, air bypass valve position) that will keep the engine running satisfactorily while the oxygen sensor is warming up.

Look at Fig. 15-45A. When the engine and oxygen sensor are cold, no information flows to the computer. The computer ignores any signals from the sensors. The "loop of information" is open.

Closed loop operation

After the sensor and engine are warm, the oxygen sensor, and other sensors, begin to feed data to the computer, Fig. 15-45B. This forms an "imaginary loop" (closed loop) as electrical data flow from the engine exhaust, to the oxygen sensor, to the computer, to the injector, and back to the oxygen sensor. Normally, the computer system functions closed loop to analyze the fuel mixture provided to the engine. This lets the computer "double-check" itself.

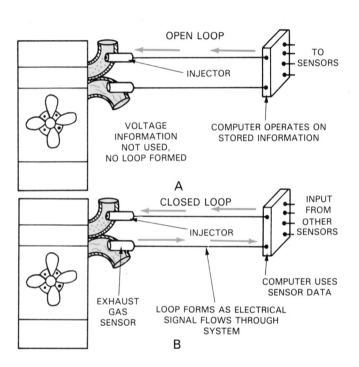

Fig. 15-45. Drawings show difference between closed loop and open loop mode of computer operation.

Analog and digital signals

The signal from the engine sensors can be either a digital or analog type output. See Fig. 15-46. The output from the computer can also be analog or digital.

Digital signals are instant on-off signals. An example of a sensor providing a digital signal is the crankshaft position sensor which shows engine rpm. Voltage output or resistance goes from maximum to minimum, like a switch, to report rpm in number form.

An *analog signal* progressively changes in strength. For example, sensor internal resistance may smoothly increase or decrease with temperature, pressure, or part position. The sensor acts as a variable resistor.

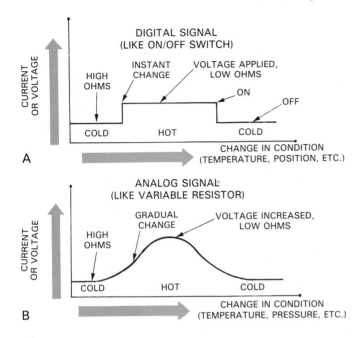

Fig. 15-46. A—Digital signal is ON-OFF signal, like from wall light switch. B—Analog signal steadily increases or decreases voltage signal. It is not an instant ON-OFF signal.

EFI SYSTEM OPERATION

Now that you understand the basics of the major components of a modern fuel injection system, a summary of system operation may be beneficial. Refer to Fig. 15-47 as system operation is explained.

When the engine is first started cold, the computer operates open loop. Data stored in the computer allows the engine to run. The computer ignores any signals from the various sensors. Injector pulse width is held constant, at a wider or longer value than normal. The cold start injector may also function. This richens the mixture to aid cold engine operation.

When the engine and oxygen sensor warm up, the system goes into closed loop. All of the sensors, especially the oxygen sensor, send data about system operation to the computer. This allows the computer to adjust injector pulse width. The computer then attempts to keep the fuel mixture at a stoichiometric ratio (theoretically perfect air-fuel ratio of 14.5 parts of air to 1 part fuel by weight).

As conditions change, driver presses gas pedal for example, the sensors (throttle position sensor and mass airflow sensor in this example), report this change of con-

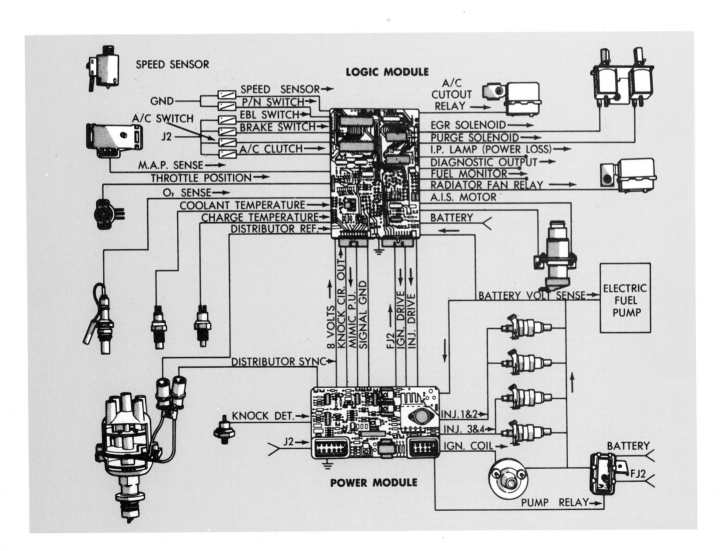

Fig. 15-47. Diagram summarizes operation of EFI system. (Chrysler)

dition. The computer can slightly richen the mixture for more power. It can also lean the mixture if the car begins to decelerate for a stop.

The fuel pressure regulator is a vital part of the system. It must maintain a constant fuel pressure across all of the injectors, including the cold start injector. With fuel pressure constant, the main variable for modifying the fuel mixture is the injector pulse width.

SUMMARY

A gasoline injection system uses pressure to feed fuel into the engine. Air is drawn into the engine separately. Gasoline injection has several advantages over a carburetor system. It provides better atomization of fuel, better fuel distribution, smoother idle, greater fuel economy, lower emissions, better cold-weather operation, more engine power, and it is a simpler system.

Modern electronic fuel injection (EFI) is made up of four subsystems: fuel delivery system, air induction system, sensor system, and computer control system. The fuel delivery system includes the electric fuel pump, fuel filter, fuel rail, pressure regulator, and the connecting lines and hoses. It cleans and meters the proper amount of fuel for every engine condition.

An air induction system includes an air filter, throttle valve, and connecting ducts. It cleans, delivers, and controls air going to the engine's cylinders.

The sensor system keeps track of various engine operating conditions. The sensors change physical conditions to electrical signals which the computer can understand and to which it can react.

The computer control system receives electrical signals from the sensors, interprets them, and uses this information to control fuel and air delivery for best engine performance for varying conditions. The computer, itself, has four separate components: input/output devices, central processing unit, power supply, and memories.

Injectors are simply electrically operated fuel valves which are controlled by the on-board computer. The injector opens when it receives an electrical impulse from the computer. This allows fuel spray into the manifold.

A fuel rail assembly, or fuel log, supplies fuel to all the injectors. A fuel pressure regulator is usually mounted on the fuel rail assembly. Its purpose is to maintain a constant fuel pressure at all the injectors. To maintain pressure, it sometimes allows some fuel to recirculate through the fuel return circuit. This is a fuel line that carries excess fuel back to the fuel supply tank.

A fuel accumulator dampens the pulses of a fuel pump to help maintain even fuel pressure at the injectors.

The throttle body assembly of a fuel injection system is very similar to that of a carburetor system. However its only function is to control airflow to the engine. It has a throttle valve controlled by the gas pedal. An air bypass valve is usually attached to the throttle body to aid in control of engine idle speed.

The engine sensors are the "nerves" of the electronic fuel injection system. The throttle position sensor signals the computer when the driver has pressed down on the accelerator and more power is needed. The coolant temperature sensor lets the computer know whether the engine is cold or warmed up for normal operating temperature. An inlet air temperature measures the temperature of air entering the engine. A charge temperature sensor measures the temperature of the air-fuel mixture. A knock sensor "listens" for the sound of abnormal combustion. A crankshaft position sensor measures engine speed. A manifold pressure measures vacuum inside the intake manifold. A flap airflow sensor and the mass airflow sensor detect the amount of air entering the engine. An oxygen sensor analyzes the exhaust to detect the presence of oxygen after combustion. Other sensors may check the operation of the transmission, air conditioning system, brake system, and emission control system. The computer receives the signals from the sensors and adjusts the system or systems for most efficient operation.

KNOW THESE TERMS

Gasoline injection system, Fuel rail, Sensor system, Air induction system, Computer wiring harness, Integrated circuit, Input-output devices, Power supply, Read Only Memory, Random Access Memory, Programmable Read Only Memory, Injector solenoid, Injector nozzle, Injector pulse width, Cold-start injector, Fuel pressure regulator, Fuel return circuit, Fuel accumulator, Air bypass valve, Throttle position sensor, Engine coolant sensor, Inlet air temperature sensor, Charge temperature sensor, Crankshaft position sensor, Manifold pressure sensor, Flap airflow sensor, Mass airflow sensor, Oxygen sensor, Open loop, Closed loop, Analog signals, Digital signals.

REVIEW QUESTIONS—CHAPTER 15

1. The basic difference between a carbureted engine and a fuel-injected engine is that _____, not engine vacuum (suction), feeds the fuel into the engine cylinders.
2. List the four subsystems that make up the modern electronic fuel injection system.
3. Indicate which of the following is (are) NOT a part of the fuel delivery system:
   a. Choke valve.
   b. Fuel pump.
   c. Fuel filter.
   d. Fuel rail.
   e. Fuel lines and hoses.
   f. Pressure regulator.
   g. Fuel injectors.
   h. Throttle valve.
4. List the essential parts of the air induction system and explain the purpose of each.
5. An EFI sensor system (select ALL CORRECT answers):
   a. Keeps track of engine operating conditions and passes this information to an on-board computer.
   b. May be used with or without an on-board computer.
   c. Converts an engine condition it senses into a small electrical current which travels to the computer.
   d. Sends an electrical current to solenoids or

motors which adjust fuel system operation.

6. The _____ _____ system processes information about and controls operation of an electronic fuel injection system and certain other automobile systems.

7. List the four basic parts of an on-board computer and tell what each does.

8. A _____ for a gasoline injection system is simply an electrically operated fuel valve.

9. List and describe the eight parts of the typical fuel injector.

10. Describe how a gasoline fuel injector operates.

11. Injector pulse width refers to the length of time that the injector is open. True or false?

12. A _____ pulse width richens the fuel mixture; a _____ pulse width leans the mixture.

13. The purpose of fuel rail is to:
   a. Provide a place to attach the fuel pressure regulator.
   b. Regulate pressure from the fuel pump.
   c. Bleed off fuel to keep fuel pressure uniform.
   d. Feed fuel to all of the injectors.

14. The _____ _____ _____ maintains a constant, preset fuel pressure at the injectors.

MATCHING TEST: On a separate sheet of paper, match the terms with the appropriate statements.

15. ____ All hoses, fuel lines, and fittings used to carry fuel from pressure regulator back to fuel tank.

16. ____ Dampens fuel pressure pulses and helps maintain fuel pressure when engine is stopped.

17. ____ Admits air into intake manifold.

18. ____ Frequently used to control engine idle speed.

19. ____ Holds throttle valve and sometimes a fuel injector.

20. ____ A variable resistor which reacts to movement of throttle valve.

21. ____ Measures temperature of air-fuel mixture.

22. ____ Measures temperature of air entering the engine.

a. Throttle position sensor.
b. Inlet air temperature sensor.
c. Air bypass valve.
d. Charge temperature sensor.
e. Throttle valve.
f. Accumulator.
g. Fuel return circuit.
h. Throttle body.

23. The terms _____ _____ and _____ _____ refer to the operating mode of a computerized fuel system.

24. What is the difference between digital signals and analog signals?

This modern V-6 engine is typical of the power plants installed in today's automobiles. It has multi-point electronic fuel injection, which is controllewd by an on-board ECU. To be a successful fuel system mechanic today, you must understand the construction and operation of such systems. (GMC)

16 GASOLINE INJECTION SYSTEM CONSTRUCTION, TYPES

After studying this chapter, you will be able to:
- *List and describe the many classifications of gasoline injection.*
- *Compare multi-point and single-point injection systems.*
- *Discuss the different means of controlling a gasoline injection system.*
- *Explain the difference between timed, intermittent, and continuous injection.*
- *Compare simultaneous, sequential, and group injection.*
- *Describe the parts and operation of TBI (throttle body injection).*
- *Explain the operation of an idle air control valve.*
- *Describe the construction and operation of other specific types of injection systems.*

The previous chapter explained the fundamentals of a modern electronic, multi-point fuel injection system. This system was discussed because it is the most modern fuel injection system. This chapter will expand and build upon this information by discussing specific types of injection systems and how they are constructed.

This is an important chapter because it will help prepare you to work on any make and model of fuel injection. It will give you a strong background for better understanding the next chapter on service and repair.

GASOLINE INJECTION CLASSIFICATIONS

There are many types of gasoline injection systems. Before studying the most common ones in detail, you should have a basic knowledge of the different classifications. This will help you compare similarities and differences between systems.

Both a gasoline injection system and a diesel injection system are commonly called a *fuel injection system.* The two are quite different, however. To avoid confusion, diesel injection is discussed in Chapters 18 and 19.

Single-point and multi-point injection

Gasoline injection systems may be classified in several ways. One way is by the location of the injectors.

A *single-point* or *throttle body injection (TBI) system* has injector nozzles located in the throttle body assembly on top of the engine. Fuel is sprayed into the throttle body where vacuum draws it into the center of the intake manifold. See Fig. 16-1.

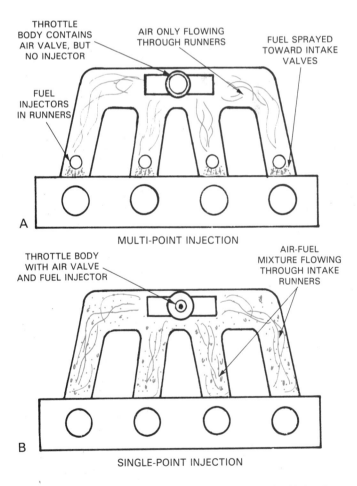

Fig. 16-1. Two main classifications of gasoline fuel injection systems include these. A—Multi-point injection delivers fuel at each valve. B—Single-point injection sprays fuel into air horn.

333

A *multi-point* or *port injection system,* as its name suggests, has a fuel injector nozzle in the runner (air-fuel passage) to each cylinder. See Fig. 16-1B. Gasoline is sprayed into each intake port leading to the intake valve.

Both systems are used on late-model cars. Multi-point is more common and is replacing TBI, Fig. 16-2.

Indirect and direct injection

Fuel injection systems may be either direct or indirect. An *indirect* system sprays the fuel into the engine intake manifold. Most gasoline injection systems are this type. Fuel and air are already mixed before they enter the cylinder.

A *direct* system sprays the fuel directly into the engine's combustion chambers. All diesel injection systems are this type. Both systems are in Fig. 16-3.

GASOLINE INJECTION CONTROL

Control refers to two functions in gasoline injection. One is concerned with the amount of fuel or air-fuel mix-

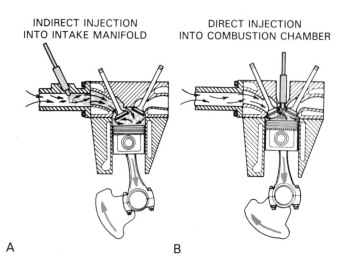

Fig. 16-3. Cutaway view of types of injection. A—Indirect injection sprays fuel into intake manifold. B—Direct injection sprays fuel into combustion chamber. Gasoline injection systems are usually indirect. Diesels are direct.

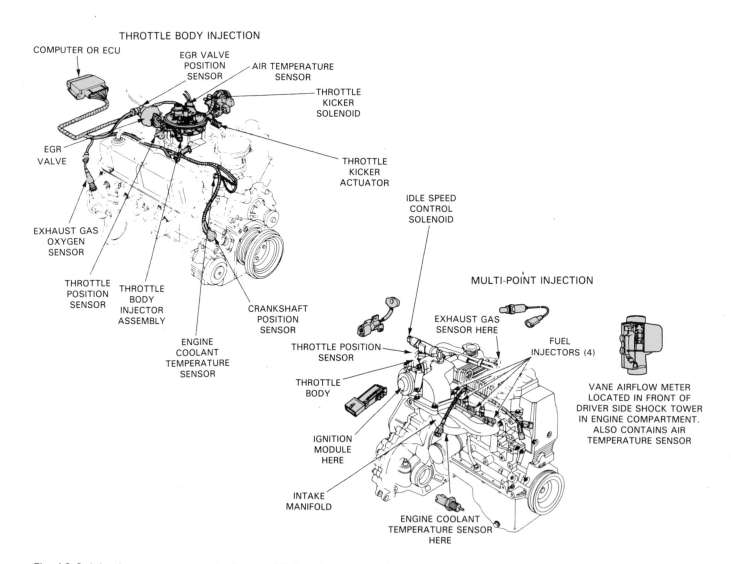

Fig. 16-2. Injection systems may also be classified as throttle body or port. Throttle body injection is indirect because it delivers fuel into intake manifold like older carbureted system. Port injection sprays fuel toward intake valve and is more efficient system.
(Ford)

ture delivered, the other with the timing of its delivery. Three methods are used to control the amount of gasoline injected into the engine: *electronic, hydraulic,* and *mechanical.* Older gasoline injection systems combined all three.

Electronic fuel injection provides control of fuel supply through various engine sensors and a computer. From information sent by the sensors, the computer "decides" when to open and close the injector valves. The most modern and common of all gasoline injection systems, it will be covered in greater detail later. See Fig. 16-4.

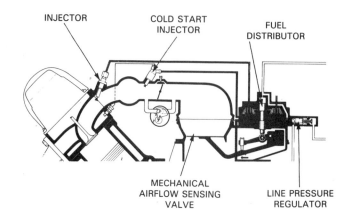

Fig. 16-5. Basic setup for hydraulically controlled fuel injection system. Hydraulic airflow sensor controls fuel distributor pressure control device. (Volvo)

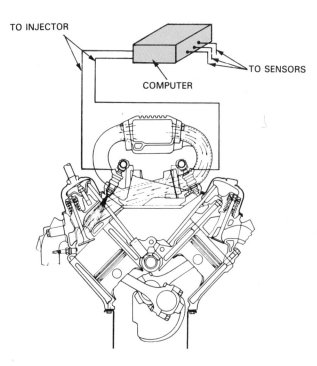

Fig. 16-4. Control of electronic fuel injection is provided by combination of computer, sensors, and solenoids. Sensors monitor conditions of combustion for computer. Computer "examines" data and sends electronic "orders" to solenoids to adjust air-fuel mixture. (Chevrolet)

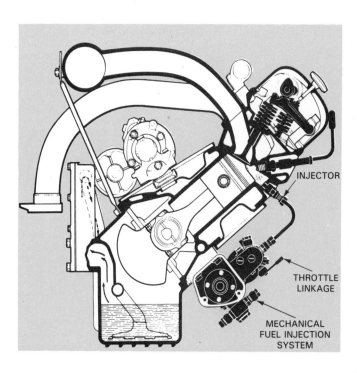

Fig. 16-6. Cutaway is view of mechanical fuel injection system used on gasoline engine. (Mercedes-Benz)

Hydraulic fuel injection provides control by having air or fuel pressure move control devices. Hydraulic control, which will also be discussed later, requires an airflow sensor and a fuel distributor (hydraulic valve mechanism) to meter the gasoline. It is used on several foreign cars. See Fig. 16-5.

Mechanical fuel injection meters fuel through throttle linkage, a mechanical pump, and a governor (speed control device). This is a very old system designed for high performance or racing engines, Fig. 16-6. It is seldom used except for diesel injection.

Gasoline injection timing

Injection timing refers to when fuel is sprayed into the intake manifold. It may or may not be coordinated with the opening of the engine's intake valve. Basically, there are three types of gasoline injection timing: intermittent, timed, and continuous.

An *intermittent* gasoline injection system opens and closes the injector valves independently of the engine intake valves. It may spray fuel into the intake manifold when the valves are open or when they are closed. This type is sometimes called a *modulated injection system.*

A *timed* or *sequential* injection system delivers fuel into the intake port just before or while the intake valve is opening. It is timed to work with intake valve action. Many late model cars now come equipped with sequential port injection.

Diesel fuel injection is the best example of this system. However, many modern gasoline injection systems are also timed to get a slight increase in power and fuel economy.

A *continuous* gasoline injection system sprays fuel into

the intake manifold nonstop. Whenever the engine is running, some fuel is being forced out of the injector nozzles into the intake manifold.

The air-fuel ratio is controlled by an increase or decrease of pressure at the injectors. This variation in pressure increases or decreases fuel flow out of the injectors. This type is used on several foreign automobiles and a few American makes.

Several other terms are used to describe the way injectors deliver fuel. These are: simultaneous injection, sequential injection, and group injection. Refer to Fig. 16-7 for simplified sketches of each.

In *simultaneous injection,* all injectors are pulsed on and off (opened and closed) together. See Fig. 16-7A.

Sequential injection, discussed earlier as timed injection, has the injectors opening one after the other. This type is common in today's electronic fuel injection systems which spray the fuel into the intake port as the engine intake valve begins to open. This improves fuel atomization and combustion since the fuel spray or mist is immediatley pulled into the engine. Fig. 16-7B shows this arrangement.

Several injectors open at the same time in a *group injection* system. For example, on a V-8 engine, four injectors may open at once and then the other four. A six cylinder would have three injectors opening at once. Look at Fig. 16-7C.

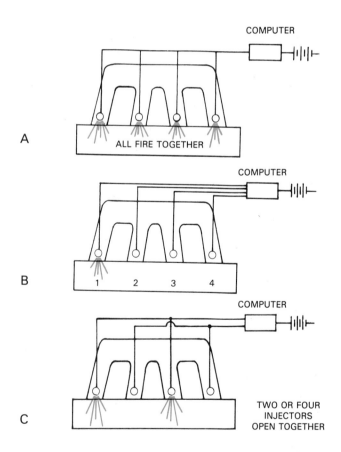

Fig. 16-7. How injectors deliver fuel. A—All injectors open and close together. B—Injectors open one after another. C—Group injection system (several injectors open at once).

GASOLINE INJECTION TYPES

As you have just learned, there are several ways to classify gasoline injection systems. Today's cars normally use one of the following types.

1. *Electronic, sequential, multi-point injection.* An injector is provided for each engine cylinder. It sprays fuel on the intake valve as it opens.
2. *Electronic, modulated, single-point injection.* The solenoid-operated injector is mounted at the inlet of the intake manifold. It is pulsed open and closed independently of the engine valves.
3. *Electronic, group, multi-point injection.* There is an injector for each cylinder. More than one injector is opened at one time, but the intake valve can either be open or closed.
4. *Hydraulic-mechanical, continuous, multi-point injection.* Each cylinder has an injector which sprays fuel anytime the engine is running. Fuel pressure controls the air-fuel mixture.
5. *Electronic, continuous, single-point injection.* A central injector mechanism, mounted at the inlet to the engine intake manifold, sprays fuel anytime the engine is running.

There are many variations of gasoline injection. These five, however, represent the most common types. They will be discussed in more detail as you study the rest of the chapter.

ELECTRONIC MULTI-POINT INJECTION

Electronic, multi-point injection is most common on modern cars. It has the capability of providing more engine power, higher fuel economy, and lower exhaust emissions than other types.

Since fuel is injected directly into the intake port, there is no problem with fuel distribution (fuel mixture flow from inlet of intake manifold to each cylinder). Also, being electronic, the system monitors its own performance using an oxygen (exhaust gas) sensor and other sensors. This helps keep the fuel mixture close to stoichiometric (ideal).

Chapter 15 discussed fundamentals of an electronic multi-point system. We will now expand upon this information by explaining the construction and variations in this type.

EFI multi-point injector

An injector for an EFI system is simply an electrically-operated fuel valve, Fig. 16-8. Fuel, under pressure, is constantly supplied by an electric fuel pump. When the computer feeds current to the injector coil, the injector opens and fuel sprays into the intake manifold.

Note! Chapter 15 listed and explained the basic parts of a multi-point, electronic fuel injector. Refer to this chapter for a review, if needed.

EFI multi-point injector mounting

Most multi-point injectors mount in the engine intake manifold. They may be secured by a press fit, threads, or a bolt and bracket clamping device. Fig. 16-9 shows one type of injector mounting.

A rubber insulator or seal usually fits between the in-

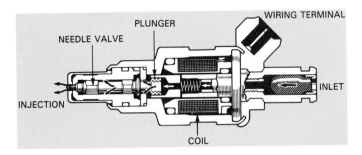

Fig. 16-8. Cutaway shows inside of electronic fuel injector. Note part names. (Toyota)

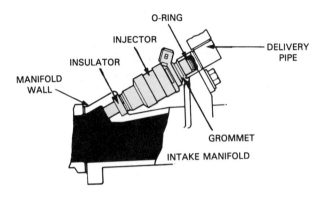

Fig. 16-9. One type of injector is press fitted into manifold. O-ring provides seal between injector and manifold wall. (Toyota)

jector and the intake manifold. It prevents vacuum leakage. As shown in Fig. 16-10, an injector housing may bolt to the engine intake manifold. The injectors fit into this housing.

Various means are used to connect the fuel rail to the injectors. Fig. 16-10 shows special clips locking the fuel rail to each injector. Conventional fittings are also common.

Fuel rail construction

The fuel rail can be made from steel tubing, Fig. 16-11, or cast metal, Fig. 16-12. Both types are common. Various components are installed on the fuel rail. It usually holds the pressure regulator, a service fitting, fittings for attaching to the injectors, and brackets for mounting to the engine.

An inlet fitting on the fuel rail allows fuel to enter from the fuel pump. An outlet fitting allows excess fuel to bleed back to the tank through the pressure regulator. A fitting for feeding fuel to the cold start injector may also be found on the fuel rail.

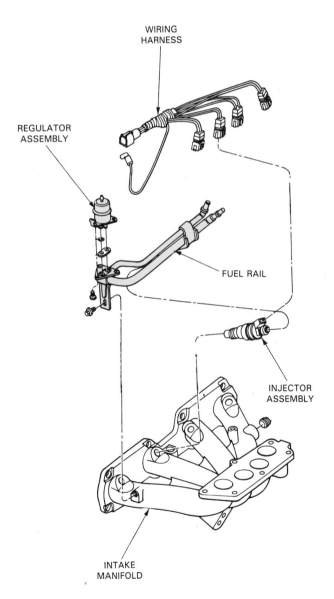

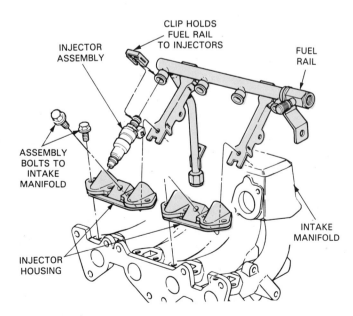

Fig. 16-10. Exploded view shows mounting arrangement of multi-port fuel injectors. Note method of attaching injectors to fuel rail. (Buick)

Fig. 16-11. Some fuel injection assemblies use steel tubing for fuel rail. (Ford)

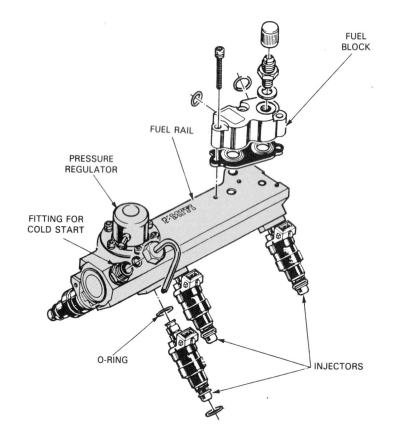

FUEL
BLOCK

FUEL RAIL

PRESSURE
REGULATOR

FITTING FOR
COLD START

O-RING

INJECTORS

A

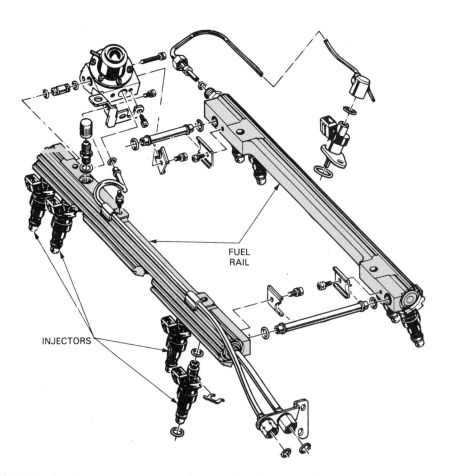

FUEL
RAIL

INJECTORS

B

Fig. 16-12. Fuel rails may also be cast of metal like these. A—Single rail. B—Dual rail. (Pontiac)

Multi-point throttle body construction

A throttle body for a multi-point injection system only controls airflow into the engine. See Figs. 16-13 and 16-14. The throttle body mounts at the inlet to the intake manifold.

Butterfly valves or throttle valves are linked to the driver's gas pedal. When the driver presses down on the gas pedal, the throttle body valves swing open to increase airflow and engine power.

A throttle body for a multi-point injection system is constructed something like a carburetor. However, it does not contain or control fuel. The body is usually cast from aluminum. The steel throttle plates mount on a steel shaft. Gaskets seal the various sections, Fig. 16-14.

Most throttle body assemblies contain an idle air control valve and a throttle position sensor, Fig. 16-14. These devices were explained in the previous chapter.

Various vacuum fittings may be added to the throttle body. See Fig. 16-15. Vacuum fittings supply suction for the PCV system, fuel storage canister purge, and other systems.

Fig. 16-16 shows an exploded view of a typical multi-point throttle body assembly. Note the relationship between the parts. In particular, note the throttle valve, throttle shaft, throttle lever, and main body.

Fig. 16-17 shows an external view of a throttle body

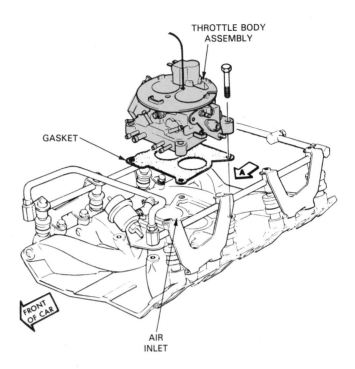

Fig. 16-13. Throttle body for multi-port injection system is similar to carburetor without its fuel systems. (Cadillac)

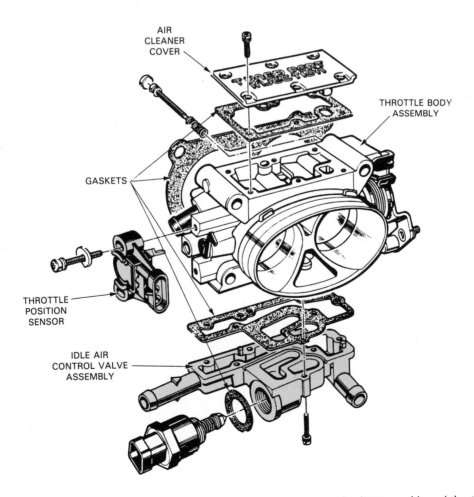

Fig. 16-14. Study this exploded view of throttle body. Note especially idle air control valve assembly and throttle position sensor. (Oldsmobile)

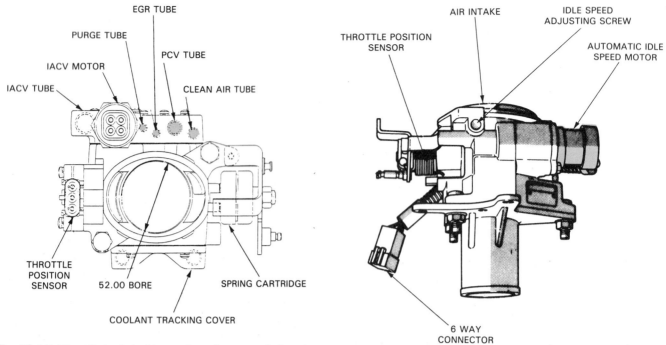

Fig. 16-15. Throttle body holds number of vacuum fittings including those for EGR, PCV, and clean air tubes. (Buick)

Fig. 16-17. External view of throttle body design. (Chrysler)

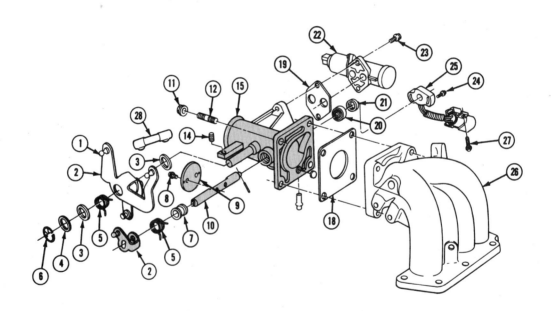

ITEM	PART NUMBER	PART NAME	ITEM	PART NUMBER	PART NAME
1	9E551	BALL – THROTTLE LEVER	15	9E927	BODY – AIR INTAKE CHARGE THROTTLE
2	9583	LEVER – THROTTLE – PRIMARY	18	9E936	GASKET – AIR CHARGE CONTROL TO INTAKE MANIFOLD
3	9W591	SPACER – THROTTLE SHAFT	19	9F670	GASKET – AIR BYPASS VALVE
4	9B853	SPACER – THROTTLE CONTROL TORSION SPRING	20	9C753	SEAL – THROTTLE CONTROL SHAFT
5	9B569	SPRING – THROTTLE RETURN	21	9B508	BUSHING – CARBURETOR THROTTLE SHAFT
6	NG63109	CLIP – THROTTLE SHAFT	22	9F715	VALVE ASSEMBLY – THROTTLE AIR BYPASS
7	9F569	BUSHING	23	N-605773-S100	BOLT – M6x1.0x20 HEX HEAD FLANGE
8	N-603076-S100	SCREW M4x.7x8	24	N-800885-S52	SCREW AND WASHER ASSEMBLY M4x22
9	9E950	PLATE – AIR INTAKE CHARGE THROTTLE	25	9B989	POTENTIOMETER THROTTLE POSITION
10	9E951	SHAFT – AIR INTAKE CHARGE THROTTLE	26	9425	MANIFOLD – INTAKE UPPER
11	N-800103-S52	NUT – M8	27	N-611133-S51	SCREW – M4x0.7x14.0 HEX WASHER TAP
12	N-802411-S100	STUD – M8x42.5	28	9815	LINK – ROD ASSEMBLY THROTTLE CONTROL
14	N800545-S100	SCREW – M5x.8x16.25 SLOT HEAD	29	9A794	LEVER – THROTTLE SECONDARY

Fig. 16-16. Typical multi-point throttle body assembly. Note throttle valve, throttle shaft, throttle lever, and main body. (Ford)

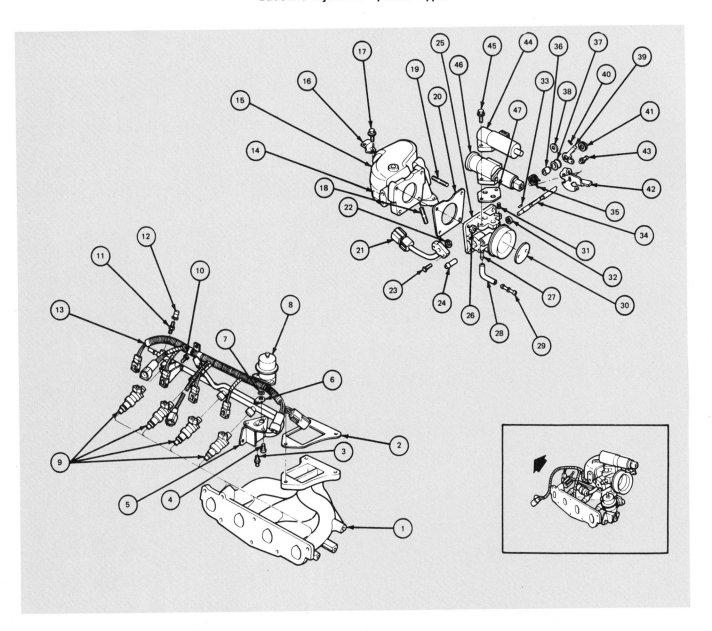

	PART NAME		PART NAME
1	MANIFOLD—INTAKE LOWER	24	TUBE—EMISSION INLET
2	GASKET—INTAKE MANIFOLD UPPER	25	BODY—AIR INTAKE CHARGE THROTTLE
3	CONNECTOR—1/4 FLARELESS X 1/8	26	NUT—M8 X 1.25
	EXTERNAL PIPE	27	TUBE
4	SCREW—M5 X .8 X 10 SOCKET HEAD	28	HOSE—VACUUM
5	MANIFOLD ASSEMBLY—FUEL INJECTION	29	CONNECTOR
	FUEL SUPPLY	30	PLATE—AIR INTAKE CHARGE THROTTLE
6	GASKET—FUEL PRESSURE REGULATOR	31	SCREW—M4 X .7 X 8
7	SEAL—5/16 X .070 "O" RING	32	SEAL—THROTTLE CONTROL SHAFT
8	REGULATOR ASSEMBLY—FUEL PRESSURE	33	PIN—SPRING COILED 1/16 X .42
9	INJECTOR ASSEMBLY—FUEL	34	SHAFT
10	BOLT—M8 X 1.25 X 20 HEX FLANGE HEAD	35	SPRING—THROTTLE RETURN
11	VALVE ASSEMBLY—FUEL PRESSURE	36	BUSHING—ACCELERATOR PUMP
	RELIEF		OVERTRAVEL SPRING
12	CAP—FUEL PRESSURE RELIEF	37	BEARING—THROTTLE CONTROL LINKAGE
13	WIRING HARNESS—FUEL CHARGING	38	SPACER—THROTTLE CONTROL TORSION
14	DECAL—CARBURETOR IDENTIFICATION		SPRING (MTX ONLY)
15	MANIFOLD—INTAKE UPPER	39	LEVER—CARBURETOR TRANSMISSION LINKAGE
16	RETAINER—WIRING HARNESS	40	SCREW—M5 X .8 X 16.25 SLOT HEAD
17	BOLT—M8 X 1.25 X 30 HEX FLANGE HEAD	41	SPACER—CARBURETOR THROTTLE SHAFT
18	STUD—M6 X 1.0 X 40	42	LEVER—CARBURETOR THROTTLE
19	STUD—M8 X 1.25 X 1.25 X 47.5	43	BALL—CARBURETOR THROTTLE LEVER
20	GASKET—AIR INTAKE CHARGE TO	44	VALVE ASSEMBLY—THROTTLE AIR BYPASS (ALT)
	INTAKE MANIFOLD	45	BOLT—M6 X 1.0 X 20 HEX FLANGE HEAD
21	POTENTIOMETER—THROTTLE POSITION	46	VALVE ASSEMBLY—THROTTLE AIR BYPASS
22	BUSHING—CARBURETOR THROTTLE	47	GASKET—AIR BYPASS VALVE
	SHAFT		
23	SCREW AND WASHER ASSEMBLY M4 X 22		

Fig. 16-18. Familiarize yourself with parts of charging assembly. (Ford)

assembly. Note the throttle position sensor and automatic idle speed control motor.

Fig. 16-18 is an exploded view of a complete fuel charging assembly. This includes the throttle body, the injectors, fuel rail, pressure regulator, and related components.

Sensing devices

Multi-point fuel injection systems use different types of sensing devices to feed electrical data or information to the on-board computer.

Airflow sensor operation

As shown in Fig. 16-19, an *airflow sensor* is a flap-operated variable resistor located at the inlet to the intake manifold.

Air flowing through the sensor, causes the *flap* (also called an *air door*) to swing to one side. A variable resistor connected to the flap converts the motion to an electrical signal which travels to the on-board computer. The variable resistor alters current.

When the throttle valve is closed (engine idling), the air door is nearly closed. Refer again to 16-19. The elec-

trical signal is weak. The computer produces only a short pulse width for the injectors. Only a small amount of fuel is injected into the intake ports.

During acceleration, when the throttle plate is swung open, increased airflow pushes the air door open, changing sensor resistance. The computer responds to the signal by increasing the injector pulse width.

Pressure sensor operation

A pressure sensing multi-point injection system uses intake manifold pressure (vacuum) to signal the computer of fuel needs. Look at Fig. 16-20.

A pressure sensor is connected to a passage going into the intake manifold. The pressure sensor converts changes in manifold vacuum to electrical signals. As with the airflow system, the computer uses this electrical data (signal) to calculate engine load and air-fuel ratio requirements.

Fig. 16-21 shows the general location of all of the parts of a pressure sensing, multi-point, electronic fuel injection system. This system uses group injection. Half of the injectors are cycled on together, then the other half is activated.

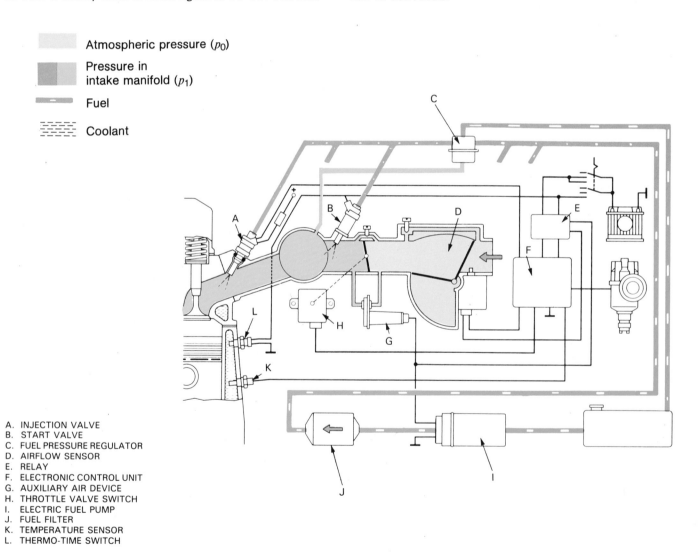

A. INJECTION VALVE
B. START VALVE
C. FUEL PRESSURE REGULATOR
D. AIRFLOW SENSOR
E. RELAY
F. ELECTRONIC CONTROL UNIT
G. AUXILIARY AIR DEVICE
H. THROTTLE VALVE SWITCH
I. ELECTRIC FUEL PUMP
J. FUEL FILTER
K. TEMPERATURE SENSOR
L. THERMO-TIME SWITCH

Fig. 16-19. Diagram simplifying operation of airflow sensor. Note how parts are related to each other. Sensor (marked D) uses a flap to vary resistance to computer. Note regular injector at A and cold start injector at B. (Robert Bosch)

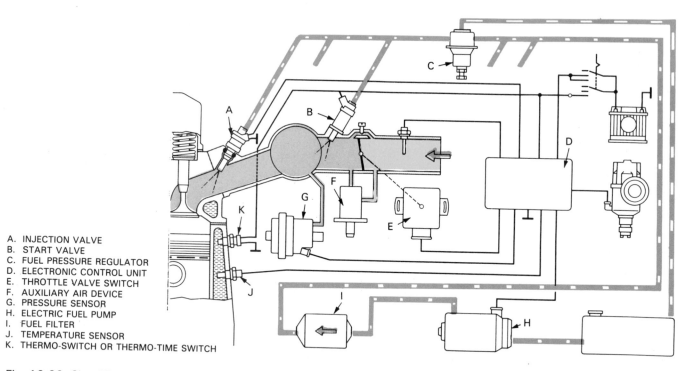

A. INJECTION VALVE
B. START VALVE
C. FUEL PRESSURE REGULATOR
D. ELECTRONIC CONTROL UNIT
E. THROTTLE VALVE SWITCH
F. AUXILIARY AIR DEVICE
G. PRESSURE SENSOR
H. ELECTRIC FUEL PUMP
I. FUEL FILTER
J. TEMPERATURE SENSOR
K. THERMO-SWITCH OR THERMO-TIME SWITCH

Fig. 16-20. Simplified diagram of multi-point injection system using pressure sensor to generate signal to computer. Changes in manifold pressure move sensor so it sends computer signals when engine fuel needs change. (Robert Bosch)

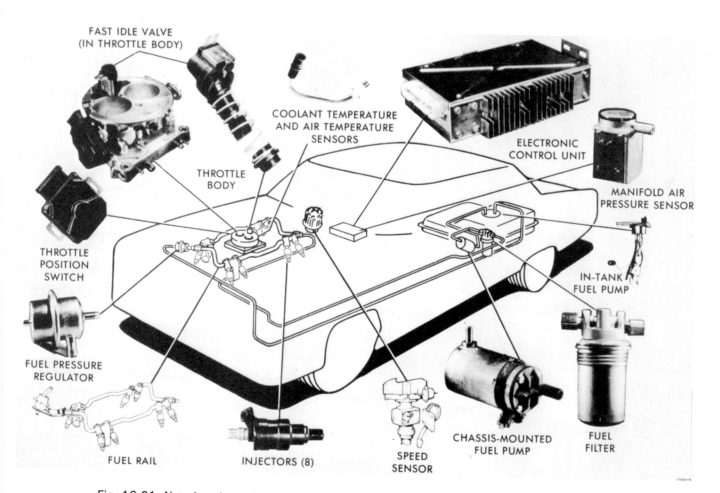

Fig. 16-21. Note locations of all parts of a pressure-sensing EFI multi-point system. (Cadillac)

THROTTLE BODY INJECTION

A *throttle body injection* (TBI) system mounts one or two injector valves in the throttle body. The injectors, when triggered electronically, spray fuel into the top of the throttle body air horn, Fig. 16-22. The fuel spray mixes with the air flowing through the air horn. As with the carbureted system, the mixture is then pulled through the engine intake manifold runners by manifold vacuum.

TBI was very common when auto makers first changed from carburetion to fuel injection. An inexpensive system, it can utilize existing engine components (intake manifolds, air cleaners, etc.) with few design changes.

TBI is slightly *more efficient* than carburetion because electronic controls can more quickly alter the air-fuel mixture with changes in engine demands.

TBI injection, however, is NOT as efficient as multipoint injection. It suffers from the same fuel distribution problems as does a carburetor system. Since both fuel and air are pulled through the intake manifold, one cylinder may receive a richer or leaner fuel charge than another. This reduces engine efficiency slightly.

TBI assembly

Generally, the TBI assembly, Fig. 16-23, includes the:

1. *Throttle body housing.* A metal casting, it holds the injectors, a fuel pressure regulator, throttle valves, and other parts.

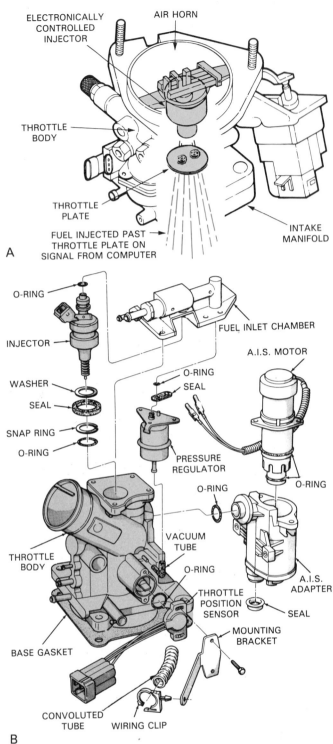

A

B

Fig. 16-23. A—Simple representation of TBI. B—TBI assembly. Refer to this illustration as it is explained in text. (Chrysler)

2. *Fuel injectors.* These are solenoid-operated fuel valves that mount in the upper section of the throttle body assembly.

3. *Fuel pressure regulator.* A spring-loaded bypass valve, this part maintains constant fuel pressure at the injectors.

4. *Throttle positioner.* This motor assembly, acting on commands from the computer, opens and closes the throttle plates to control engine idle speed.

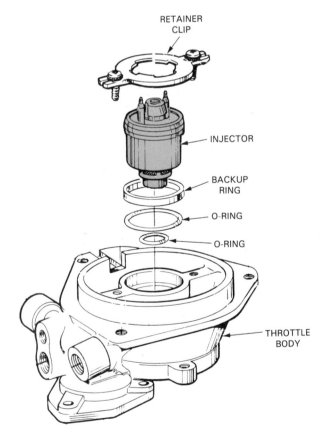

Fig. 16-22. Exploded view shows arrangement of single injector in air horn of throttle body injection system. (Renault/AMC)

5. *Throttle position sensor.* A variable resistor, this component senses changes in the position of the throttle plates.

6. *Throttle plates.* These are flat plates called butterfly valves. They control airflow through the throttle body.

Remember, vacuum is used to meter fuel into the throttle bore with a carburetor. Since fuel is sprayed into the throttle body, atomization is very good. The fuel is quickly broken into tiny droplets as it enters the intake manifold.

TBI throttle body

Like a carburetor body, the *TBI throttle body* bolts to the inlet pad on the intake manifold, Figs. 16-22 and 16-23. Throttle plates are mounted in the lower section of the throttle body. Rods or a cable link the throttle plates with the gas pedal. An inlet fuel line attaches to one fitting on the throttle body. Another fitting accepts a return fuel line to the tank.

Throttle body injector

A *throttle body injector* is made up of an electric solenoid coil, armature or plunger, ball or needle valve, ball or needle seat, and injector spring. These parts are pictured in Fig. 16-24.

Wires from the computer (electronic control unit) connect to the injector terminals. When the computer energizes the injectors, a magnetic field is produced in the injector coil. The magnetic field pulls the plunger up and the valve opens. Fuel can then squirt through the injector nozzle and into the engine.

TBI pressure regulator

The throttle body *pressure regulator* consists of a fuel valve, diaphragm, and spring, Fig. 16-25. When fuel pressure is low (as when engine is first started), the spring prevents the fuel valve from opening. Pressure is allowed to build up as fuel flows into the regulator from

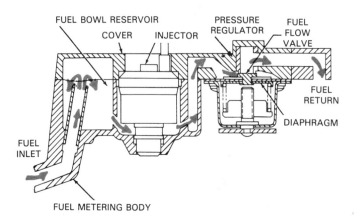

Fig. 16-25. Cutaway of pressure regulator and throttle body. Fuel (note arrows) moves through fuel metering body, around injector and to pressure regulator. If valve of regulator is opened by excess pressure, fuel will be let through so it can return to fuel storage tank. (Cadillac)

the electric fuel pump.

Look at Fig. 16-26. It shows how the injector and pressure regulator work together in one type of throttle body assembly.

When a preset pressure is reached, excess pressure moves the diaphragm. The diaphragm compresses the spring behind it and opens the fuel valve. Fuel moves through the valve, into the return line, and back to the fuel tank. This happens any time the pressure begins to exceed a preset pressure at the injectors.

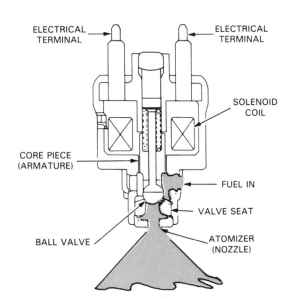

Fig. 16-24. Some TBI injectors, like this one, use ball type valve instead of pointed needle valve. Note part names. (Cadillac)

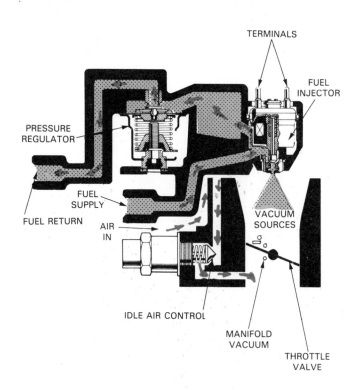

Fig. 16-26. Cutaway shows path of fuel through inlet to injector. Excess fuel, as in Fig. 16-25, bypasses regulator and returns to fuel tank. (Pontiac)

Idle air control valve

An *idle air control valve* in a TBI system, usually fitted onto the throttle body, helps control engine idle speed. It is opened and closed by a solenoid which responds to commands from the system computer or a built-in thermostat. Refer again to Fig. 16-26.

An open idle air control valve allows more air to enter the intake manifold. This tends to increase the idle speed of the engine. When the air valve is closed, no additional air can pass into the intake manifold; idle speed decreases. Air valve action is something like a carburetor fast idle cam.

TBI idle control motor

An *idle control motor* performs the same function as an idle air control valve. It allows the computer to increase or decrease engine idle speed with changes in load and other conditions. One is pictured in Fig. 16-27.

The idle control motor is a small dc motor that turns a screw device. The screw device is used to push the throttle lever and throttle plate(s) open or closed as needed. In this way, the computer can control engine idle rpm.

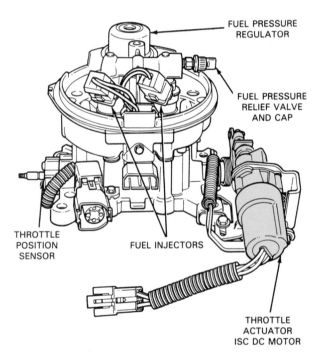

3.8L CFI CHARGING ASSEMBLY
WITHOUT 10 PIN CONNECTOR

Fig. 16-27. Throttle control motor turns idle screw to adjust amount of air entering air horn. (Ford)

For example, if the engine is idling and the driver turns on the air conditioning system, engine load increases. This can cause a drop in engine idle speed. The engine speed sensor would signal the computer causing it to send current to the idle control motor. The motor would screw the plunger out, forcing the throttle plates to open

slightly. See Fig. 16-28. This would let more airflow into the engine to maintain the correct idle rpm.

To retract the idle control motor plunger and reduce idle rpm, the computer reverses the direction of current flow through the small dc motor. Extra wires are usually provided to the motor so that polarity can be reversed.

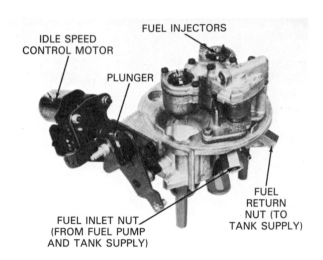

Fig. 16-28. This idle speed control motor moves plunger in and out to move a lever attached to air throttle plates.

TBI variations

There are many TBI assembly design variations. Most, however, use the same basic components.

Fig. 16-29 shows the external components of a typical TBI assembly. Study the locations of the parts.

Fig. 16-30 illustrates a dual throttle body injection setup. This is a high performance application for increasing power output of a small V-8 engine.

Fig. 16-31 shows an exploded view of another design for a TBI unit. Study the relationship of the components.

TBI operation

Under varying driving conditions, a computer controlled TBI system will use various sensor inputs to control the computer output to the injectors and other components. Figs. 16-32 through 16-38 illustrate and explain the various modes of operation for electronic TBI with intermittent injection.

The seven common modes for TBI operation are:
1. Key-on mode.
2. Crank mode.
3. Warm-up mode.
4. Cruise mode.
5. Deceleration mode.
6. Wide-open throttle mode.
7. Key-off mode.

It is important that you be able to visualize injection system operation under different driving conditions. For example, if an engine runs poorly under acceleration, you would be able to visualize system operation. This would help you pinpoint which components would be most likely to cause the symptom or problem.

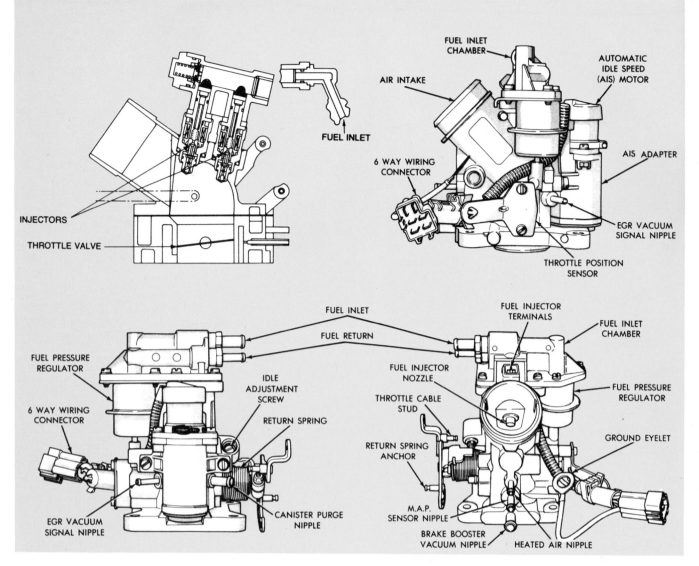

Fig. 16-29. Four views of typical variation of throttle body assembly. Do you recognize any of the parts? (Plymouth)

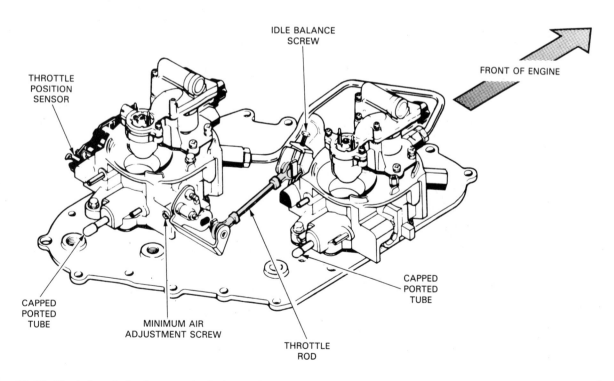

Fig. 16-30. Dual throttle body setup is designed for high performance engine. Note mechanical linkage. (Corvette)

347

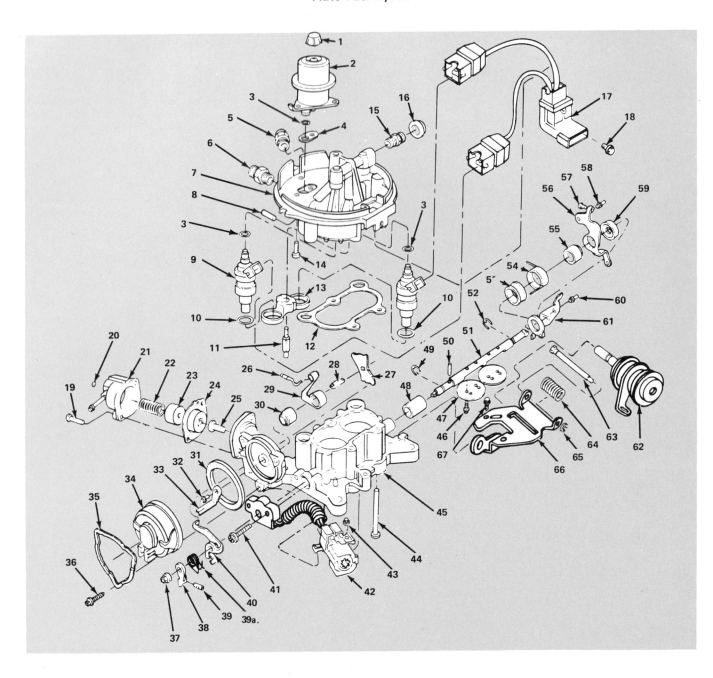

1. PLUG – FUEL PRESSURE REGULATOR ADJUSTING SCREW
2. REGULATOR ASSEMBLY – FUEL PRESSURE
3. SEAL – 5/16 x .070 "O" RING
4. GASKET – FUEL PRESSURE REGULATOR
5. CONNECTOR – 1/4 PIPE TO 1/2-20
6. CONNECTOR – 1/8 PIPE TO 9/16-16
7. BODY – FUEL CHARGING MAIN
8. PLUG – 1/16 27 HEADLESS HEX
9. INJECTOR ASSEMBLY – FUEL
10. SEAL – 5/8 x .103 "O" RING
11. SCREW – FUEL INJECTOR RETAINING
12. GASKET – FUEL CHARGING BODY
13. RETAINER – FUEL INJECTOR
14. SCREW M5.0 x 20.0 PAN HEAD
15. VALVE ASSEMBLY – DIAGNOSTIC VALVE
16. CAP – FUEL PRESSURE RELIEF VALVE
17. WIRING ASSEMBLY – FUEL CHARGING
18. SCREW – M3.5 x 1.27 x 12.7 PAN HEAD
19. SCREW & WASHER – M4 x 7.0 20.00
20. BALL – LEAD SHOT .26 - .24 DIA.
21. COVER ASSEMBLY – CONTROL DIAPHRAGM
22. SPRING – CONTROL MODULATOR
23. RETAINER – PULLDOWN DIAPHRAGM

24. DIAPHRAGM – PULLDOWN CONTROL
25. ADJUSTER – PULLDOWN CONTROL
26. ROD – FAST IDLE CONTROL
27. CAM – FAST IDLE
28. SHAFT – CHOKE HOUSING
29. POSITIONER – FAST IDLE CONTROL ROD
30. BUSHING – CHOKE HOUSING
31. GASKET – THERMOSTAT HOUSING
32. SCREW & WASHER – M3.5 x 0.6 x 6 PAN HEAD
33. LEVER – CHOKE THERMOSTAT
34. HOUSING ASSEMBLY – THERMOSTAT
35. RETAINER – HOUSING ASSEMBLY
36. SCREW
37. NUT & WASHER ASSEMBLY – .7-6H HEX
38. LEVER – FAST IDLE CAM ADJUSTER
39. SCREW – NO. 10 - 32 x .50 SET SLOTTED HEAD
39a. FAST IDLE PICK-UP LEVER RETURN SPRING
40. LEVER – FAST IDLE
41. SCREW & WASHER – M4.07 x 22.0 PAN HEAD
42. THROTTLE POSITION SENSOR
43. SCREW – M4 x .7 x 14.0 HEX WASHER TAP
44. SCREW – M5 x .7 x 55.0
45. BODY – FUEL CHARGING – THROTTLE

46. SCREW – M3 x 0.5 x 7.4 HEX WASHER HEAD
47. PLATE – THROTTLE
48. BEARING – THROTTLE CONTROL LINKAGE
49. "E" RING – 7/32 RETAINING
50. PIN – SPRING COILED
51. SHAFT – THROTTLE
52. "C" RING – THROTTLE SHAFT BUSHING
53. BEARING – THROTTLE CONTROL LINKAGE
54. SPRING – THROTTLE RETURN
55. BUSHING – ACCELERATOR PUMP OVER
        TRAVEL SPRING
56. LEVER – TRANSMISSION LINKAGE
57. SCREW – M4 x 0.7 x 7.6
58. PIN – TRANSMISSION LINKAGE LEVER
59. SPACER – THROTTLE SHAFT
60. BALL – THROTTLE LEVER
61. LEVER – THROTTLE
62. POSITIONER ASSEMBLY – THROTTLE
63. SCREW – 1/4 - 28 x 2.53 HEX HEAD ADJUSTING
64. SPRING – THROTTLE POSITIONER RETAINING
65. "E" RING – RETAINING
66. BRACKET – THROTTLE POSITIONER
67. SCREW – M5 x 8 x 14.0 HEX WASHER TAP

Fig. 16-31. Study parts of typical TBI unit and how parts are assembled. (Ford)

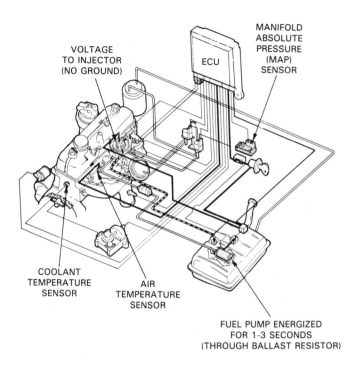

VOLTAGE
TO INJECTOR
(NO GROUND)

MANIFOLD
ABSOLUTE
PRESSURE
(MAP)
SENSOR

ECU

COOLANT
TEMPERATURE
SENSOR

AIR
TEMPERATURE
SENSOR

FUEL PUMP ENERGIZED
FOR 1-3 SECONDS
(THROUGH BALLAST RESISTOR)

Fig. 16-32. Operation of fuel injection system in "key on" mode. When ignition key is turned "on", ECU gets input from coolant, air temperature and MAP (manifold absolute pressure) sensors. Fuel pump relay is energized for several seconds. (AMC/Renault)

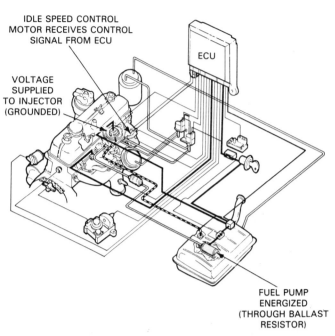

IDLE SPEED CONTROL
MOTOR RECEIVES CONTROL
SIGNAL FROM ECU

ECU

VOLTAGE
SUPPLIED
TO INJECTOR
(GROUNDED)

FUEL PUMP
ENERGIZED
(THROUGH BALLAST
RESISTOR)

Fig. 16-34. EFI operation under warm-up mode. Inputs received from same sources as in crank mode except starter solenoid signal. ECU supplies voltage to injector and controls length of fuel injection through ground circuit. ECU also determines idle speed required and sends control signal to idle speed control motor. (AMC/Renault)

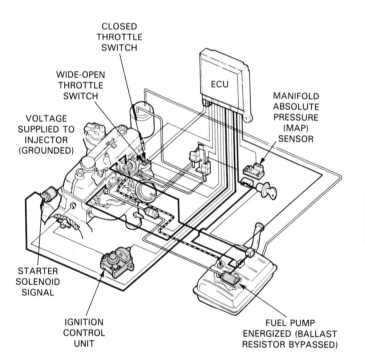

CLOSED
THROTTLE
SWITCH

WIDE-OPEN
THROTTLE
SWITCH

ECU

VOLTAGE
SUPPLIED TO
INJECTOR
(GROUNDED)

MANIFOLD
ABSOLUTE
PRESSURE
(MAP)
SENSOR

STARTER
SOLENOID
SIGNAL

IGNITION
CONTROL
UNIT

FUEL PUMP
ENERGIZED (BALLAST
RESISTOR BYPASSED)

Fig. 16-33. During cranking mode, ECU gets input from: coolant temperature sensor, air temperature sensor, MAP sensor, starter solenoid signal, ignition control unit engine rpm signal, wide-open throttle switch, and closed throttle switch. Fuel pump is energized and ECU completes pump ground circuit. Voltage is supplied to injector and ECU controls length of injection time by controlling injector's ground circuit. If wide-open throttle switch is activated, ECU assumes flooding condition and breaks injector ground circuit. (AMC/Renault)

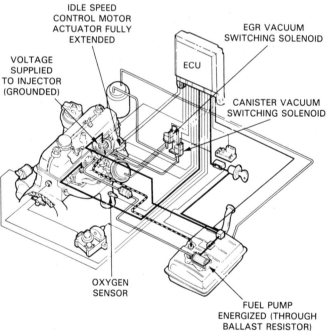

IDLE SPEED
CONTROL MOTOR
ACTUATOR FULLY
EXTENDED

EGR VACUUM
SWITCHING SOLENOID

VOLTAGE
SUPPLIED
TO INJECTOR
(GROUNDED)

ECU

CANISTER VACUUM
SWITCHING SOLENOID

OXYGEN
SENSOR

FUEL PUMP
ENERGIZED (THROUGH
BALLAST RESISTOR)

Fig. 16-35. EFI cruise mode. ECU receives input from oxygen sensor as well as from same sensors and components as in warm-up mode. ECU supplies voltage to injector and controls length of injection period through injector's ground circuit. ECU breaks ground circuit for EGR and charcoal canister switching solenoids. This allows vacuum to pass to EGR valve and charcoal canister valve. Gases circulate through the combustion chambers. ECU signals idle speed control motor to fully extend its actuator (plunger). (AMC/Renault)

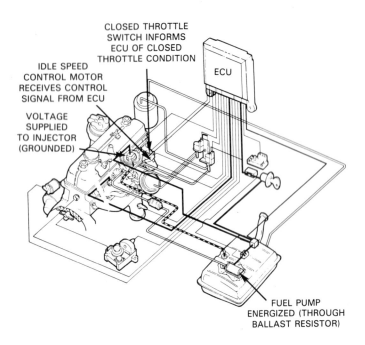

Fig. 16-36. EFI deceleration mode. ECU receives input from same sensors and components noted in warm-up mode plus oxygen sensor. Closed throttle switch tells computer that throttle is shut down. ECU then completes ground to EGR and charcoal canister switching solenoids. The solenoids interrupt vacuum signals and these functions stop. Voltage still goes to injector but ECU controls duration of injection cycle through injector ground circuit. ECU determines correct idle speed and sends signal to control idle speed control motor. (AMC/Renault)

Fig. 16-38. Key off mode. ECU breaks injector ground circuit and injection stops. Then ECU signals idle speed control motor to fully extend its actuator for next start-up. ECU also stores MAP sensor input for next start-up and then shuts itself off. (AMC/Renault)

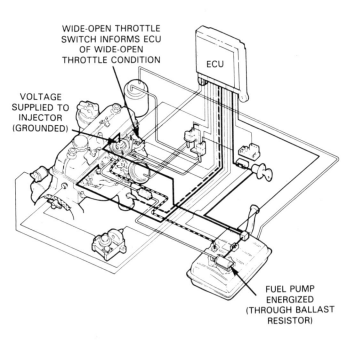

Fig. 16-37. Wide-open throttle mode. ECU receives input from coolant temperature sensor, air temperature sensor, MAP sensor, ignition control unit engine rpm signal, wide-open throttle switch, and closed throttle switch. Wide-open throttle switch tells ECU that throttle is wide open. At this signal, ECU completes ground to EGR and charcoal canister switching solenoids and these functions stop. Voltage is sent to injector and ECU controls the length of the fuel injection through the injector ground circuit. ECU ignores oxygen sensor input signal.

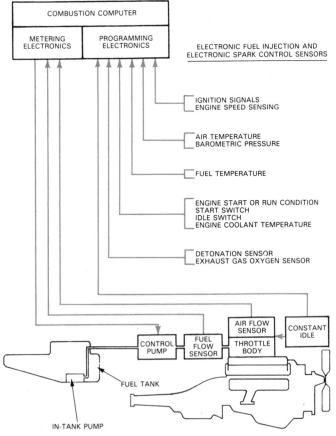

Fig. 16-39. Schematic for continuous flow throttle body injection (CTBI) system. Combustion computer electronically adjusts air-fuel mixture and spark advance. (Chrysler)

CONTINUOUS THROTTLE BODY INJECTION

As you learned earlier in this chapter, *continuous throttle body injection* (CTBI) systems spray a solid stream of fuel into the air horn, Fig. 16-39. Unlike the more common modulated systems, it never pulses the injectors on and off to control the air-fuel mixture.

Increased or decreased fuel flow is accomplished through increasing or decreasing the pressure applied to the nozzles in the throttle body. The air cleaner and throttle body assembly contain the computer, airflow sensor, fuel control motor (a pump), spray bar (injector nozzles), and other parts. See Fig. 16-40.

Fig. 16-41 presents a diagram showing how data is transferred in this system. The system's sensors measure fuel flow, airflow, and other engine conditions that affect fuel mixture. The computer acts on the information to increase or decrease the speed of the control pump (fuel pump).

HYDRAULIC-MECHANICAL, CONTINUOUS INJECTION

In *hydraulic-mechanical, continuous injection* systems (CIS), a *mixture control unit* (airflow sensor-fuel distributor assembly) operates the injectors. Fig. 16-42 illustrates this system. It is not controlled by electronics. The airflow sensor is pictured in Fig. 16-43.

CIS fuel distributor

A *fuel distributor* is a hydraulically-operated valve that controls fuel flow (pressure) to each of the CIS injectors.

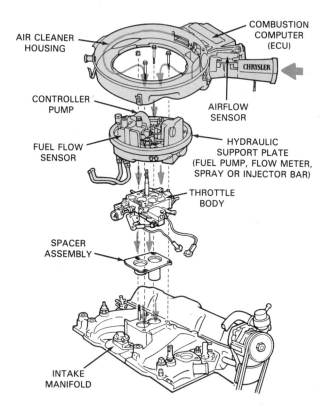

Fig. 16-40. Typical CTBI in exploded view. Combustion computer is mounted on air cleaner and determines most efficient air-fuel mixture based on engine operating conditions. Fuel is delivered through flow meter to simple air control body. After mixing with air, the mixture is delivered to engine. (Chrysler)

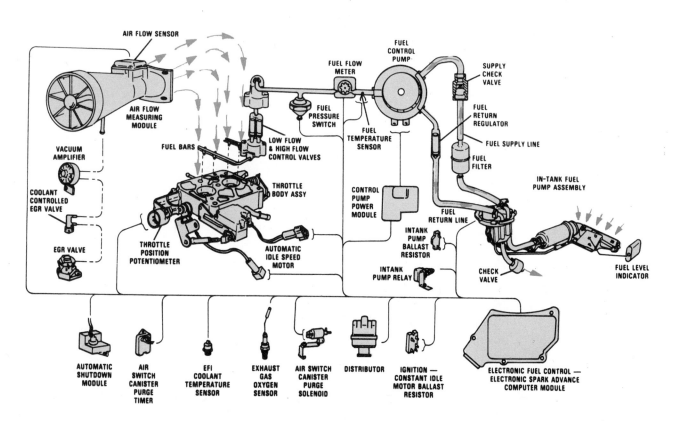

Fig. 16-41. Typical CTBI and relationship of parts. Fuel pump's speed and pressure are changed to alter air-fuel ratio. Increased pump speeds and pressure richens mixture and vice versa. (Chrysler)

351

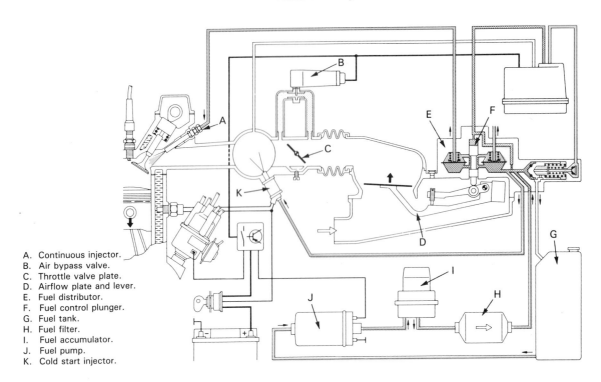

A. Continuous injector.
B. Air bypass valve.
C. Throttle valve plate.
D. Airflow plate and lever.
E. Fuel distributor.
F. Fuel control plunger.
G. Fuel tank.
H. Fuel filter.
I. Fuel accumulator.
J. Fuel pump.
K. Cold start injector.

Fig. 16-42. Hydraulic-mechanical injection system uses mechanical airflow sensor (D) to operate hydraulic fuel distributor (E) assembly. Continuous injector is used to spray fuel into engine whenever it is running. (Robert Bosch)

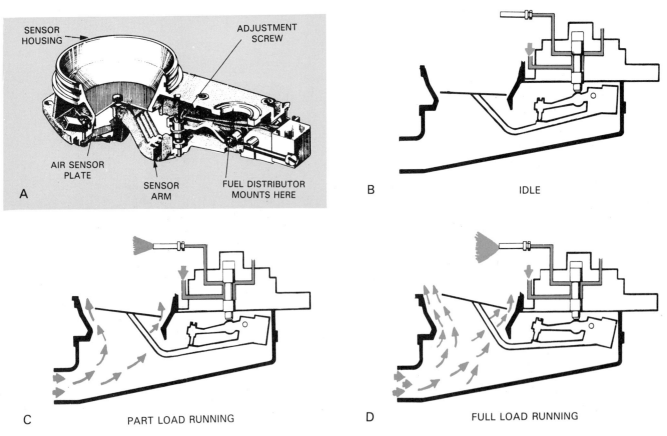

Fig. 16-43. A—Airflow sensor consists of large disc-shaped flap located in air horn. Flap operates lever which controls fuel control plunger. B—At idle, low airflow moves sensor plate only slightly. Lever pushes up lightly on fuel control plunger to inject very small amount of fuel. C—At part throttle, more airflow moves sensor plate and lever farther. More fuel is sprayed from injector. D—Full load condition creates larger airflow pushing sensor plate higher. Fuel control plunger opens fully for maximum injection pressure and flow. (Robert Bosch)

See Fig. 16-44. The fuel control plunger is in the center of the distributor. Fuel fed from the plunger flows outward to spring-loaded diaphragms. The diaphragms make up for pressure differences in each injection line. Their purpose is to see that the same amount of fuel is sent to each injector.

Only one type of CIS system uses a fuel distributor. This system is common on foreign automobiles.

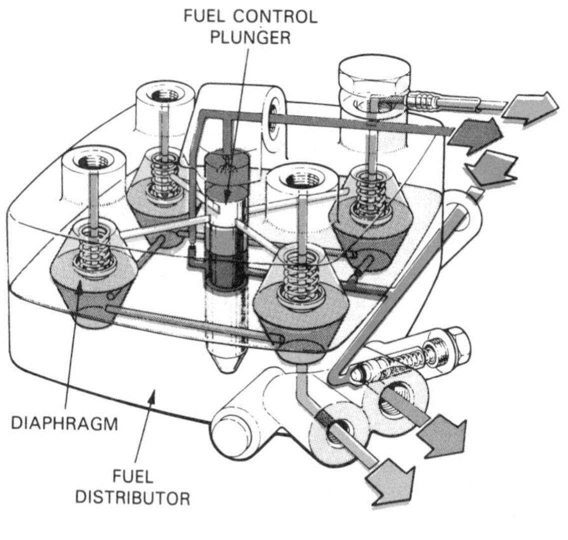

Continuous fuel injector

A *continuous fuel injector* is a spring-loaded valve that injects fuel nonstop, as long as the engine is running. See Fig. 16-45.

The spring normally keeps the valve closed. A filter, located inside the injector, traps dirt.

When the engine is running, and as it is started, fuel pressure is strong enough to push the injector valves open.

With CIS injectors, the amount of fuel sprayed from

FUEL CONTROL
PLUNGER

DIAPHRAGM

FUEL
DISTRIBUTOR

LINE PRESSURE
UPPER CHAMBER PRESSURE
LOWER CHAMBER PRESSURE
INJECTION PRESSURE
CONTROL PRESSURE
RETURN, NO PRESSURE

Fig. 16-44. Fuel distributor is set of four pressure differentiating diaphragm valves. They assure that same amount of fuel is sent to each injector. (Saab)

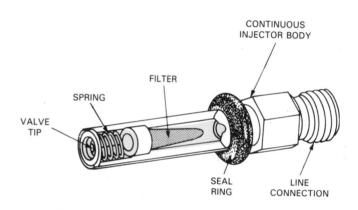

CONTINUOUS
INJECTOR BODY

FILTER

SPRING

VALVE
TIP

SEAL
RING

LINE
CONNECTION

Fig. 16-45. Continuous type injector is simply spring-loaded valve. At correct pressure injector valve opens and fuel sprays into intake port of engine.

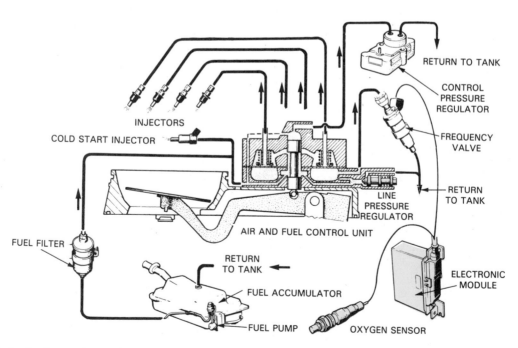

RETURN TO TANK

CONTROL
PRESSURE
REGULATOR

FREQUENCY
VALVE

RETURN
TO TANK

INJECTORS

COLD START INJECTOR

LINE
PRESSURE
REGULATOR

AIR AND FUEL CONTROL UNIT

ELECTRONIC
MODULE

FUEL FILTER

RETURN
TO TANK

FUEL ACCUMULATOR

FUEL PUMP

OXYGEN SENSOR

Fig. 16-46. This hydraulic-mechanical system incorporates computer, oxygen sensor, and line pressure regulator to make it more efficient. (Volvo)

the injectors (air-fuel ratio) is controlled by an increase or decrease of fuel pressure to the injectors.

When the engine is shut down, fuel pressure drops and the spring is able to close the injector valve. This prevents fuel from leaking into the engine.

Fig. 16-46 shows a hydraulic-mechanical system that also has a computer, oxygen sensor, and line pressure regulator. This increases system efficiency. The oxygen sensor can determine if the engine's air-fuel mixture is too lean or too rich. Then the computer can increase or decrease fuel pressure as needed.

AFTERMARKET INJECTION SYSTEMS

Several companies make aftermarket fuel injection systems. They can be used to convert an older car with a carburetor to electronic fuel injection.

Fig. 16-47 shows the components needed to convert a carbureted engine to electronic fuel injection. Fig. 16-48 shows this system installed on an engine.

Fig. 16-47. Aftermarket kit converts carbureted engine to electronic fuel injection. (Edelbrock)

Fig. 16-48. Aftermarket kit shown in Fig. 16-47 is shown installed.

KNOW THESE TERMS

TBI, Indirect injection, Direct injection, Electronic injection controls, Hydraulic injection controls, Mechanical injection controls, Intermittent gasoline injection system, Modulated injection system, Timed (or sequential) gasoline injection system, Continuous gasoline injection system, Simultaneous injection, Group injection, Electronic sequential multi-point injection, Electronic group multi-point injection, Electronic modulated single-point injection, Electronic continuous single-point injection, Hydraulic-mechanical continuous multi-point injection, TBI pressure regulator, Fuel distributor.

SUMMARY

There are many different ways to classify gasoline injection systems. The first and most general of classifications is single-point and multi-point systems.

Injectors for single-point injection are located in the throttle body on top of the engine. Multi-point systems have injectors in the runners at each intake valve. Also, fuel injection systems may be direct or indirect. When the fuel is sprayed into the intake manifold, it is called indirect. When the fuel is delivered right into the combustion chamber, it is known as direct injection.

Fuel injection must be controlled for proper fuel delivery and mixture of fuel and air. This can be done by electronic, hydraulic, or mechanical means.

Delivery of the fuel to the intake manifold or cylinders may or may not be timed with the action of the intake valves. Of the three basic types, one is timed; the others, known as intermittent and continuous, are not. In timed systems, the injectors spray fuel just before or while the intake valves are open; intermittent systems have injectors that spray off and on without regard to when the valves are open; continuous systems spray fuel nonstop whenever the engine is running.

Sensors are needed at different points on an engine or fuel system to monitor operation of the engine. On multi-point systems, there are two different types. One is operated by airflow, the other by pressure.

Throttle body injection or TBI, is a name given systems which use one or two injectors in the throttle body, part of the air induction system. Like the carburetor jet, the injector adds fuel to the airstream being drawn in through the air horn.

REVIEW QUESTIONS—CHAPTER 16

1. A _____-_____ injection system has an injector or injectors located in the throttle body.
2. In a _____-_____ injection system, the injectors are located in the intake manifold runners near the intake valves.
3. To what two functions does ''control'' of fuel injection refer?
4. Name the three types of injection timing.
5. A(n) _____ type of gasoline injection opens and closes the injector valves independently of the engine intake valves.
6. In simultaneous injection, several sets of injectors alternate in spraying fuel together. True or false?

MATCHING TEST: On a separate sheet of paper, match the terms with the appropriate statements.

7. _____ Injector at each cylinder sprays fuel at intake valve as it opens.

8. _____ Solenoid-operated injector mounted at inlet of intake manifold is pulsed open and closed independently of intake valves.

9. _____ Injectors are located at each cylinder. More than one is open at a time. Intake valves may be open or closed.

10. _____ Individual cylinder injectors spray fuel whenever engine is running. Fuel pressure controls opening and closing and air-fuel mixture.

11. _____ Central injector mechanism, mounted at inlet to engine intake manifold and electronically controlled, sprays fuel anytime engine is running.

a. Electronic, group, multi-point injection.

b. Hydraulic-mechanical, continuous, multi-point injection.

c. Electronic, continuous, single-point injection.

d. Electronic, sequential, multi-point injection.

e. Electronic, modulated, single-point injection.

12. In a multi-point gasoline injection system, the injectors are usually mounted:
   a. In the throttle body.
   b. In the intake manifold near the throttle body.
   c. In the intake manifold near the intake valve for each cylinder.
   d. Where fuel can be injected directly onto the exhaust valves.

13. What is the purpose of the throttle body in a multi-point injection system and where is it located?

14. Where is the airflow sensor located in a multi-point injection system and what is its purpose?

15. A throttle body injection system (select correct statements):
   a. Mounts one or two injector valves in the throttle body.
   b. Is more common than a port injection system.
   c. Sprays fuel into the top of the throttle body air horn.
   d. Sprays fuel onto intake valves.
   e. Generally uses a fuel pressure regulator (spring-loaded bypass valve) to keep constant fuel pressure at the injectors.

16. List the main parts of a TBI injector.

17. Describe the function of an idle control motor in throttle body injection.

18. In continuous throttle body injection (CTBI), fuel flow is regulated by increasing or decreasing the _____ applied to the _____ in the throttle body.

19. Define a fuel distributor and explain its purpose.

20. A hydraulic-mechanical, continuous injection system (select all correct answers):
   a. Operates injectors through a mixture control unit.
   b. Controls injectors electronically.
   c. Always uses a fuel distributor.
   d. Generally does not use a fuel distributor.

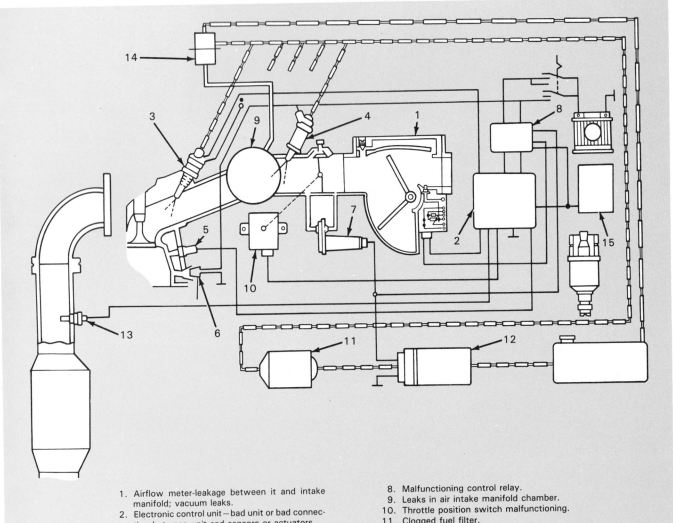

1. Airflow meter-leakage between it and intake manifold; vacuum leaks.
2. Electronic control unit—bad unit or bad connection between unit and sensors or actuators.
3. Fuel injector leaking, dirty, or inoperative.
4. Cold start injector—bad therm-time switch, bad connection, faulty circuit, poor spray pattern, inoperative.
5. Inoperative coolant temperature sensor.
6. Inoperative temperature thermo-time switch.
7. Bad supplementary air control valve.
8. Malfunctioning control relay.
9. Leaks in air intake manifold chamber.
10. Throttle position switch malfunctioning.
11. Clogged fuel filter.
12. Electrical fuel pump—burned out, receiving no power.
13. Oxygen sensor contaminated, or has poor electrical connections.
14. Fuel pressure regulator malfunctioning, poor connection, etc.
15. Ignition control module not functioning.

Fig. 17-1. Note possible malfunctions in EFI systems. (Renault/AMC)

17 GASOLINE INJECTION DIAGNOSIS AND REPAIR

After studying this chapter, you will be able to:
- Locate common gasoline injection system problems.
- Check fuel pressure regulator output.
- Diagnose problems common to electronically controlled and continuous types of fuel injectors.
- Interpret fuel injector spray patterns.
- Explain the self-diagnostic mode incorporated into modern EFI systems.
- Use a service manual when making basic adjustments on gasoline injection systems.

As you have learned from previous chapters, several types of gasoline systems are used on modern automobiles. Each design has its variations and each is unique. However, components such as injectors, pressure regulators, computers, and sensors are very similar. This chapter will explain common symptoms and show you how to test and adjust major parts of these systems. By studying these typical steps for service and repair, you will be able to follow the more detailed service manual instructions.

GASOLINE INJECTION PROBLEM DIAGNOSIS

To find the cause of gasoline injection system problems, you must call upon:
1. Your knowledge of system operation.
2. Basic troubleshooting skills.
3. A service manual for the system being checked.

As you look for the trouble, picture the operation of the four subsystems (air, fuel, sensor, and control). Try to relate the function of each component to the problem. You can then logically eliminate several possible problem sources and concentrate on others. See Fig. 17-1.

There are some similarities between a carburetor fuel system and a gasoline injection system. Several parts in each system perform the same function. Fig. 17-2 compares carburetor and EFI systems.

CARBURETOR	GASOLINE INJECTION SYSTEM
THIS PART SERVES SAME PURPOSE............	AS THIS PART
1. Choke	1. Engine temperature sensor or cold-start injector
2. Float	2. Fuel pressure regulator
3. Idle circuit	3. Air bypass valve
4. Accelerator pump	4. Throttle position switch, manifold pressure sensor
5. Fast idle cam	5. Throttle positioner or air bypass valve
6. Power valve or metering rod	6. Manifold pressure sensor, throttle position switch
7. Fuel metering jets or mixture control solenoid	7. Injector valves, fuel pressure regulator
8. Throttle valves	8. Throttle valves
9. Carburetor body	9. Throttle body
10. Venturi and main discharge	10. Airflow sensor
11. System computer (if used)	11. System computer
12. System sensors (if used)	12. System sensors

Fig. 17-2. Gasoline injection systems have parts which perform same function as their counterparts in carburetors.

Verify the problem

Before testing, verify the complaint. Make sure the customer has accurately described the symptoms. Never conclude that a gasoline injection system is faulty until you have checked that all other systems are functioning properly. The ignition system, for example, normally causes more problems than an injection system.

EFI self-diagnosis

Usually, modern EFI systems can diagnose or test themselves and indicate where the problem lies. If the system has a bad sensor or connection, the computer will produce a number code. The service manual will explain the number code so that you can quickly locate the faulty part. Systems vary from one make to another. However, the most common computer self-diagnoses work as follows:

1. Computer generates and displays DIGITAL (number) CODE, indicating problem location.
2. Computer produces ON/OFF VOLTAGE PULSE as a numerical code to help pinpoint bad component.
3. Computer activates LIGHT EMITTING DIODE(S) (small light bulbs) on the computer itself which indicate specific number code and problem.
4. A special EFI ANALYZER can be used in conjunction with COMPUTER OUTPUTS, to check the condition of the system.

Preparing for self-diagnosis

Before activating the self-diagnosis mode of the computer, you must normally warm the engine to full operating temperature. This will assure that the system has gone into the closed loop mode, using sensor feedback information. Then, you must usually turn the engine off and place the ignition key in the ON or RUN position. With some cars you may also have to place the transmission shift lever in neutral and switch off the air conditioning. ALWAYS check in a manual for specific instructions.

Digital self-diagnosis code

A digital output, self-diagnosing type of EFI system usually produces the number code on the climate control temperature (number) readout. Look at Fig. 17-3.

When the computer detects trouble, a number (88 in this specific system) will flash in the climate control or temperature digital display. To make the computer produce the trouble code, the mechanic must activate the self-diagnosis mode. Sometimes, two buttons on the climate control must be activated simultaneously.

To find out what the number means, use a service manual TROUBLE CODE CHART. It will list problems for each number code, as shown in Fig. 17-3. A ''14'' may mean a bad coolant sensor. A ''21'' may mean a shorted throttle position sensor. This lets you test the specific component or section of the computer system. You can then more quickly locate the problem.

On/off voltage self-diagnosis code

Some EFI systems produce an on/off voltage pulse that represents a trouble code. A flashing CHECK ENGINE light may be displayed. The mechanic, as before, will activate the self-diagnosis mode to find the exact problem.

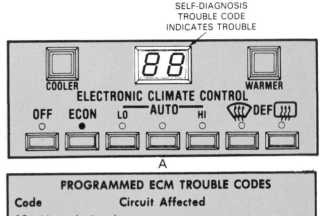

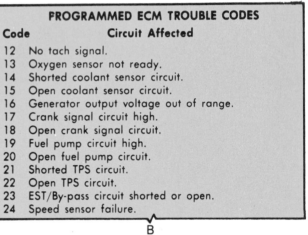

Fig. 17-3. EFI systems have self-diagnostic mode. Special code will come up either as a digital read-out, printed message, or a flashing code. A—Dash number indicates trouble. B—Partial list of trouble code from service manual for one automobile. (Cadillac)

To energize the computer self-diagnosis mode, a jumper wire is commonly used to short across two terminals in the computer wiring harness. This is shown in Fig. 17-4. Normally, during this step, the engine must be OFF and the ignition key turned ON.

To read on/off voltage signals, watch the engine light flash or analog voltmeter needle movement. Connect voltmeter to recommended test point. A digital meter will not show voltage pulses as easily as an analog meter. Service manual will give specific directions, Fig. 17-5.

Warning! An analog or needle voltmeter is commonly recommended for reading trouble codes. However, a digital meter should be used when testing components and circuits. The digital meter has higher internal resistance and will not damage some components like an analog type meter.

On/off voltage signals are easy to read. Each engine light flash or voltage pulse equals ONE. Normally, there will be several quick signals (flashes or voltage pulses) for the first number code; then a short pause and rapid signals for the second number of the code. One flash or needle sweep, a pause, and then two flashes or sweeps would equal code 12. Another example: if the first signal has three voltage pulses, it equals a three. After a short pause, a second set of two on/off pulses equals a two. A three (3 needle movements) followed by a two (2

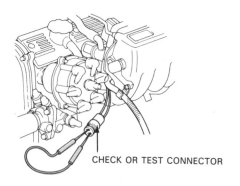

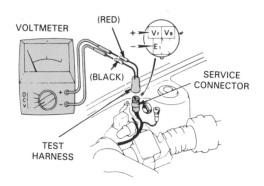

Fig. 17-4. To activate the self-diagnosis mode of some ECUs, it is necessary to connect jumper wire across two terminals in wiring harness of computer. (Toyota)

Fig. 17-5. Service manual will always give directions for connecting voltmeter to system. (Toyota)

Code Number	Component	Check
1 (Needle Sweep)	O₂ Sensor	▪ Wire harness and connectors. ▪ O₂ Sensor. ▪ ECU Computer.
2 (Needle Sweeps)	Ignition Signal	▪ Wire harness and connectors. ▪ ESC Igniter. ▪ ECU Computer.
3 (Needle Sweeps)	Airflow Sensor	▪ Wire harness and connectors. ▪ Airflow Sensor. ▪ ECU Computer.
4 (Needle Sweeps)	Pressure Sensor	▪ Wire harness and connectors. ▪ Pressure Sensor. ▪ ECU Computer.
5 (Needle Sweeps)	Throttle Position Sensor	▪ Wire harness and connector. ▪ Throttle Position Sensor. ▪ ECU Computer.
6 (Needle Sweeps)	ISC Motor Position Switch	▪ Wire harness and connectors. ▪ ISC Servo. ▪ ECU Computer.
7 (Needle Sweeps)	Coolant Temperature Sensor	▪ Wire harness and connectors. ▪ Coolant temperature sensor. ▪ ECU Computer.
8 (Needle Sweeps)	Vehicle Speed Sensor	▪ Wire harness and connectors. ▪ Vehicle Speed Sensor. ▪ ECU Computer.

VOLTMETER CODE NUMBER DISPLAY

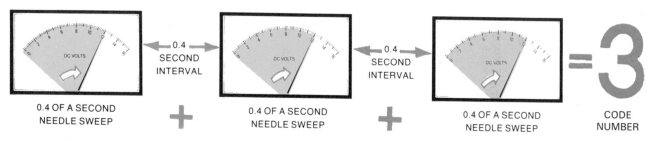

Fig. 17-6. Top. Service manual will explain code signals flashed by ECU. Bottom. How to recognize and count needle sweep of voltmeter. (Ford)

meter deflections) would represent a "32" number code. This is illustrated in Fig. 17-6.

Fig. 17-7 shows a similar, but slightly different, on/off code. Note how the voltage fluctuates from 2.5 volts to 5.0 volts. (Each meter deflection from 2.5 to 5.0 equals one.) The system previously discussed pulsed from zero to 12 volts.

indicate there is still a problem in the system. Further tests would be needed.

EFI tester diagnosis

When the fuel injection system does not have a complete self-diagnosis function, an EFI TESTER can be used to locate system troubles. Look at Fig. 17-9. The tester,

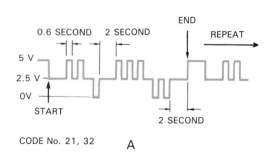

CODE No. 21, 32   A

Code No.	Voltage Pattern	Diagnosis System	Diagnostic light "CHECK ENGINE".
21		O₂ sensor signal	ON
22		Coolant temperature signal	ON
23		Intake air temperature signal	No indication
31		Air flow meter signal	ON

B

Fig. 17-7. Typical code indication. A—Graphic explanation of how malfunction code is read. First, there will be a 5 volt indication for two seconds followed by 2.5 volt indication for two seconds. Then, number of times needle deflects between 2.5 and 5 volts is first digit of two-digit code. Second digit is given in same way. Usually more than one code number will be given with 2.5 volt needle indication of 2 seconds duration between the numbers. B—Chart explains various code numbers. (Toyota)

LED self-diagnosis code

Either single or multiple LEDs (light emitting diodes) may be used to produce a number code. The LEDs are normally located right on the computer, Fig. 17-8A.

If one LED is used, it will flash a number code similar to the one which uses a VOM. Several quick flashes show the first number of the code. The second set of flashes represent the second number of the code.

Fig. 17-8B shows how to read one car maker's LED when more than one light is used to form the code. Since there are variations from one system to another, always refer to a manual. It will give the specific details needed to properly read the self-diagnosis codes.

Clearing trouble codes

To CLEAR or remove the trouble codes from an EFI computer system, you must usually disconnect the battery or the fuse to the computer. Disconnect the battery negative cable or the fuse for about 10 seconds.

After clearing the codes, check to see if the engine light comes back on after engine operation. This would

also called an EFI analyzer, is connected to the wiring harness of the system.

An EFI tester may use indicator lights and sometimes a digital meter (volt-ohm-milliammeter) to check system operation. Refer to the instructions with the tester, and use the indicator light action to make the various tests.

The digital meter readings can be compared to specifications. If a voltage, resistance, or current is NOT within specs, it indicates what repairs are needed.

Fig. 17-10 shows another computer system tester. It will check dozens of components in an electronic fuel injection system.

Oscilloscope tests

An *oscilloscope* can sometimes be used to test or view the electrical waveforms (voltage values) at the EFI injectors. It is a quick, easy way of diagnosing injector, wiring harness, and computer problems.

Inspecting injection systems

A general inspection of the engine and related com-

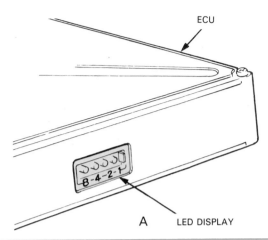

LED Display	Possible Cause	Symptom
○ ✴ ✴ ○ 　4　2	• Disconnected coolant temperature sensor coupler • Open circuit in coolant temperature sensor wire • Faulty coolant temperature sensor (thermostat housing)	• High idle speed during warm-up • High idle speed • Hard starting at low temp
○ ✴ ✴ ✴ 　4　2　1	• Disconnected throttle angle sensor coupler • Open or short circuit in throttle angle sensor wire • Faulty throttle angle sensor	• Poor engine response to opening throttle rapidly • High idle speed • Engine does not rev up when cold
✴ ○ ○ ○ 8	• Short or open circuit in crank angle sensor wire • Crank angle sensor wire interfering with high tension cord • Crank angle sensor at fault	• Engine does not rev up • High idle speed • Erratic idling
✴ ○ ○ ✴ 8　　　1	Same as above	Same as above

B

Fig. 17-8. Light emitting diodes are sometimes used to flash trouble codes. A—LED display mounted in ECU. B—Portion of service manual chart explaining trouble code. (Honda)

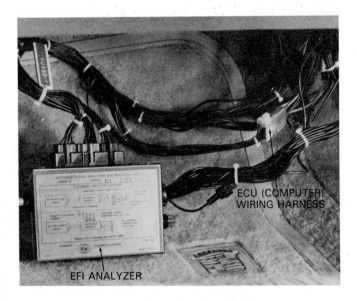

Fig. 17-9. EFI analyzer is used when system does not have self-diagnosis function. Analyzer connects to main wiring harness. Tester comes with special instruction for finding sources of problems. (Kent-Moore Tools)

ponents will sometimes locate gasoline injection troubles. Check the condition of all hoses, wires, and other parts, Fig. 17-11. Look for fuel leaks, vacuum leaks, kinked lines, loose electrical connections, and other troubles. Spend more time checking components most likely to cause the particular symptoms.

With an EFI system, you may need to check the terminals of the wiring harness. Look at Fig. 17-12. Disconnect and inspect them for rust, corrosion, and burning. High resistance at terminal connections is a frequent cause of problems, Fig. 17-13.

CAUTION! Never disconnect an EFI harness terminal with the ignition switch ON. This could damage the computer. Refer to a service manual for details. You may be told to disconnect the negative battery terminal during EFI wiring service.

VOM troubleshooting of EFI

A VOM or digital multimeter can be used to do many tests on an electronic fuel injection system. Normally, the service manual will give voltage, resistance, and current specifications for many circuits and components (sensors and actuators). Measure actual values and

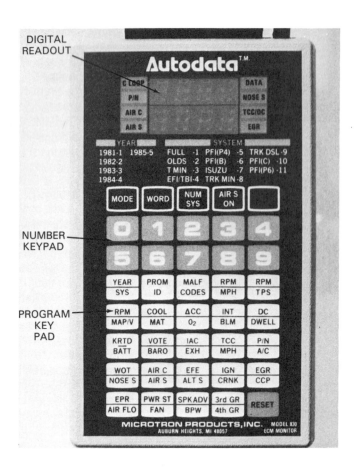

Fig. 17-10. This EFI tester is designed to be used with several makes and years of cars. Information is provided in a digital readout (top). (Microtron Products, Inc.)

Fig. 17-11. Visual inspection often turns up source of gasoline injection problems. (Saab)

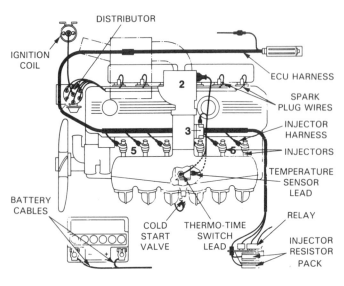

Fig. 17-12. Disconnect various electrical connections. Inspect for corrosion, rust, and burning.

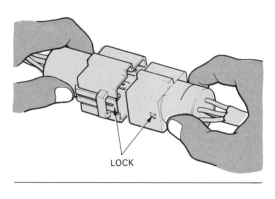

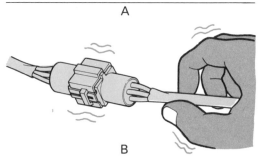

Fig. 17-13. Frequently, bad contact in wiring connectors cause problems. A—Check for bent terminals and that connector is pushed in completley and locked. B-Tap or wiggle connector and observe if signal changes or if motion produces trouble code. (Toyota)

compare them to specs.

Fig. 17-14 shows service manual illustration for a test terminal on a multi-point fuel injection system. Note that it gives voltage values that should be present when the engine is idling in the closed loop mode. It also gives a few values for engine off and key on conditions. If any of the voltages are not correct, you will need to test further in that section of the circuit.

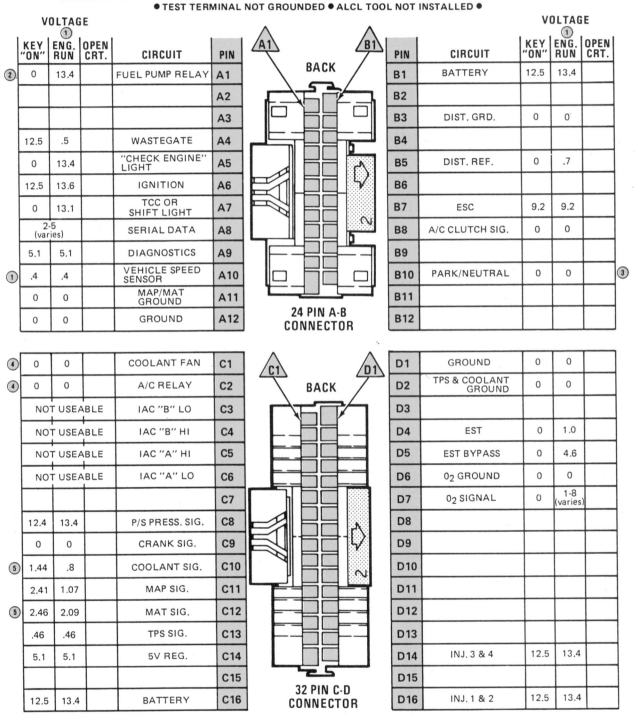

PORT FUEL INJECTION ECM CONNECTOR IDENTIFICATION

THIS ECM VOLTAGE CHART IS FOR USE WITH A DIGITAL VOLTMETER TO FURTHER AID IN DIAGNOSIS. THE VOLTAGES YOU GET MAY VARY DUE TO LOW BATTERY CHARGE OR OTHER REASONS, BUT THEY SHOULD BE VERY CLOSE.
THE FOLLOWING CONDITIONS MUST BE MET BEFORE TESTING:
● ENGINE AT OPERATING TEMPERATURE ● ENGINE IDLING IN CLOSED LOOP (FOR "ENGINE RUN" COLUMN) ●
● TEST TERMINAL NOT GROUNDED ● ALCL TOOL NOT INSTALLED ●

24 PIN A-B CONNECTOR

	KEY "ON"	ENG. RUN ①	OPEN CRT.	CIRCUIT	PIN
②	0	13.4		FUEL PUMP RELAY	A1
					A2
					A3
	12.5	.5		WASTEGATE	A4
	0	13.4		"CHECK ENGINE" LIGHT	A5
	12.5	13.6		IGNITION	A6
	0	13.1		TCC OR SHIFT LIGHT	A7
	2-5 (varies)			SERIAL DATA	A8
	5.1	5.1		DIAGNOSTICS	A9
①	.4	.4		VEHICLE SPEED SENSOR	A10
	0	0		MAP/MAT GROUND	A11
	0	0		GROUND	A12

PIN	CIRCUIT	KEY "ON"	ENG. RUN ①	OPEN CRT.
B1	BATTERY	12.5	13.4	
B2				
B3	DIST. GRD.	0	0	
B4				
B5	DIST. REF.	0	.7	
B6				
B7	ESC	9.2	9.2	
B8	A/C CLUTCH SIG.	0	0	
B9				
B10	PARK/NEUTRAL	0	0	③
B11				
B12				

32 PIN C-D CONNECTOR

	KEY "ON"	ENG. RUN	OPEN CRT.	CIRCUIT	PIN
④	0	0		COOLANT FAN	C1
④	0	0		A/C RELAY	C2
	NOT USEABLE			IAC "B" LO	C3
	NOT USEABLE			IAC "B" HI	C4
	NOT USEABLE			IAC "A" HI	C5
	NOT USEABLE			IAC "A" LO	C6
					C7
	12.4	13.4		P/S PRESS. SIG.	C8
	0	0		CRANK SIG.	C9
⑤	1.44	.8		COOLANT SIG.	C10
	2.41	1.07		MAP SIG.	C11
⑤	2.46	2.09		MAT SIG.	C12
	.46	.46		TPS SIG.	C13
	5.1	5.1		5V REG.	C14
					C15
	12.5	13.4		BATTERY	C16

PIN	CIRCUIT	KEY "ON"	ENG. RUN	OPEN CRT.
D1	GROUND	0	0	
D2	TPS & COOLANT GROUND	0	0	
D3				
D4	EST	0	1.0	
D5	EST BYPASS	0	4.6	
D6	O_2 GROUND	0	0	
D7	O_2 SIGNAL	0	1-8 (varies)	
D8				
D9				
D10				
D11				
D12				
D13				
D14	INJ. 3 & 4	12.5	13.4	
D15				
D16	INJ. 1 & 2	12.5	13.4	

① ENGINE RUNNING IN PARK, AT IDLE, NORMAL OPERATING TEMPERATURE, IN CLOSED LOOP.
② ALL VOLTAGES SHOWN AS "0" SHOULD READ LESS THAN 1.0 VOLT WITH A DIGITAL VOLT-OHM METER SUCH AS J-29125.
③ READS BATTERY VOLTAGE IN GEAR.   ④ A.C. OFF   ⑤ VOLTAGE DEPENDS ON ENGINE TEMPERATURE

ENGINE 1.8 TURBO
CARLINE J

Fig. 17-14. Service manual will identify ECM connector terminals and specify proper voltages for each. (General Motors)

Using EFI wiring diagrams

It helps to study electronic fuel injection wiring diagrams when you are trying to find problems. Look at Fig. 17-15. They show all possible circuit connections and components that may be faulty.

Loose electrical connections cause many troubles in EFI systems. High resistance, resulting from the poor connection, can prevent sensor input to the computer or proper output from the computer. Fig. 17-16 shows most of the electrical connections that could cause trouble in one multi-point injection system.

EFI air leaks

As with a carbureted engine, *air* or *vacuum leaks* can upset the operation of an injected engine. When an air leak occurs between the airflow sensor and the intake manifold, the mixture can become too lean. The sensor will show an air inlet quantity that is too small. The computer will shorten injector pulse width even though a large volume of air is being drawn into the engine through the leak.

Always inspect air ducts carefully. They should be tight and all seals should be intact, Fig. 17-17A. Also

check vacuum hoses to the engine intake manifold. They can harden and leak or fall off their fittings, Fig. 17-17B.

EFI fuel leaks

Fuel leaks in an electronic fuel injection system can be extremely dangerous. Fuel pressures are normally higher in an EFI system than in a carburetor type system. If a leak develops, a large quantity of fuel can spray out. Also, the fuel pump is electric and can keep forcing fuel out a ruptured line even though the engine is not getting fuel.

As shown in Fig. 17-17C and D, make sure all lines and connections are secure. All fuel line supports should be tight. Check all hose clamps and fittings for leakage.

RELIEVING EFI SYSTEM PRESSURE

DANGER! Always relieve fuel pressure before disconnecting an EFI fuel line. Many gasoline injection systems retain fuel pressure (as high as 50 psi or 345 kPa), even when the engine is NOT running. At this pressure, fuel spraying out could cause eye injury or a fire!

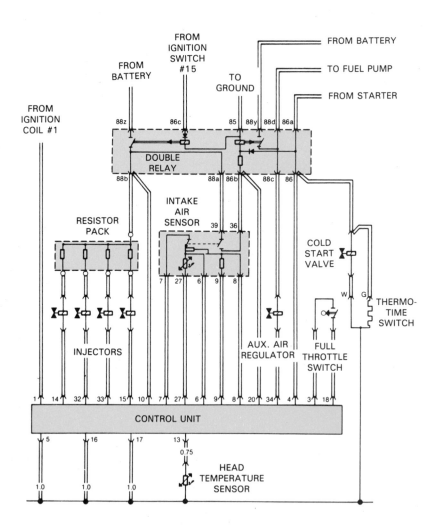

Fig. 17-15. Wiring diagrams such as this will indicate where connectors are located in EFI system. On this diagram every connector position is marked by a number. (Volkswagen)

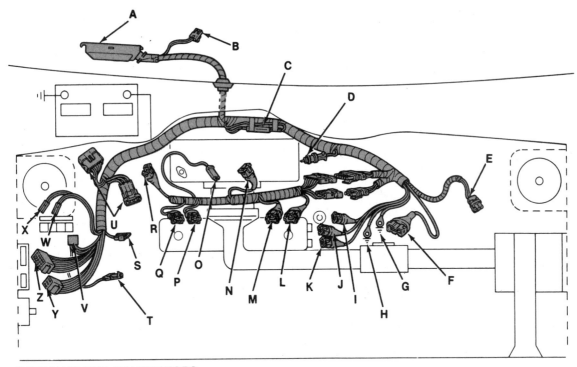

WIRE HARNESS CONNECTORS

A - ECU	J - Thermo-time coolant	R - Cold start injector
B - IP	switch	S - Ballast resistor bypass
C - Diodes	K - Coolant temp. sensor	T - Tach voltage
D - Oxygen sensor	L - Injector	U - Engine wire harness
E - Power steering switch	M - Injector	V - Control relay
F - Airflow meter	N - Throttle position sensor	W - Ballast resistor
G - Signal ground	O - Fast idle solenoid	X - Ballast resistor
H - Power ground	P - Injector	Y - Diagnostic connector D1
I - Supplementary air valve	Q - Injector	Z - Diagnostic connector D2

Fig. 17-16. Any of the connections on this wire harness could be the cause of malfunctions in EFI operation. (Renault)

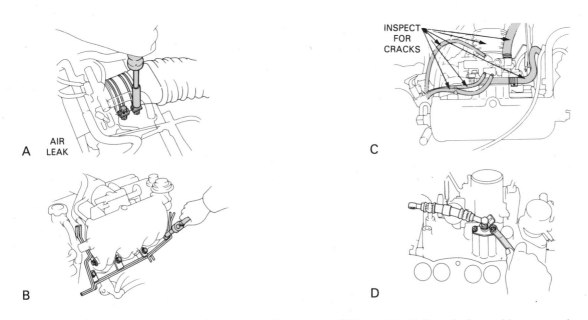

Fig. 17-17. Fittings at any point between air sensor and intake manifold must be tight or leaks could upset engine operation. Likewise, fuel fittings must not leak. A—Make certain clamps on air ducts are tight. B—Inspect hoses for cracks and deterioration. C—Fasteners holding fuel lines should be kept tight. D—Fittings at cold start injector must be kept tight also or it will leak air or fuel. (Toyota)

Most EFI systems have a special relief valve for bleeding the pressure back to the fuel tank. On some, the pressure regulator may allow pressure relief. Look at Fig. 17-18.

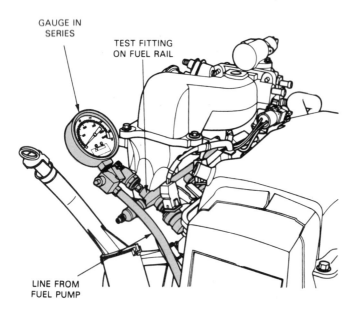

Fig. 17-18. To check pressure regulator, fuel pump, and filters, connect gauge ahead of injectors. If pressure is not within specs check filters and pump before replacing regulator. (Ford)

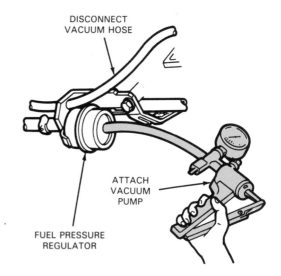

Fig. 17-19. Pressure in regulator can be removed by applying vacuum with hand vacuum pump. (Ford)

When a relief valve or a pressure regulator relief are NOT provided, remove the fuse for the fuel pump or disconnect its wires. Start and idle the engine. When the engine stalls from lack of fuel, the pressure has been removed from the system. You can also relieve pressure by applying vacuum to the pressure regulator, Fig. 17-19. It is then safe to work.

FUEL PRESSURE REGULATOR SERVICE

A *faulty fuel pressure regulator* can cause an extremely rich or lean mixture. If system pressure is high, too much fuel will spray out of the injectors, causing a rich mixture. If the regulator bypasses too much fuel to the tank (low pressure), fuel spray from each injector will be too light. This causes a lean mixture.

Testing fuel pressure regulator

You can test a fuel pressure regulator with a pressure gauge. Although exact procedures vary, you should connect the gauge into the line to the pressure regulator. Usually this will be at a test fitting on the fuel rail. Start the engine and note the pressure gauge reading.

LOW FUEL PRESSURE could be caused by a bad fuel pressure regulator, clogged fuel filter, weak fuel pump, or injector(s) sticking open. To isolate the problem, disconnect the vacuum line to the pressure regulator, Fig. 17-20A. This should make the pressure rise slightly. If not, use pliers to pinch the fuel return hose closed, Fig. 17-20B. This should also make the fuel pressure rise.

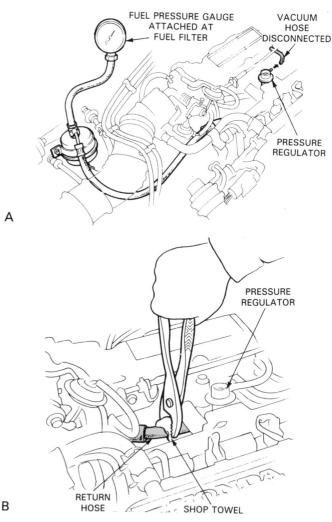

Fig. 17-20. Checking low fuel pressure. A—Measure fuel pressure. Disconnecting vacuum hose from pressure regulator should cause fuel pressure to rise. B—If fuel pressure does not rise, pinch return hose with pliers. If fuel pressure then increases, replace pressure regulator. (Honda)

If pinching the return hose and disconnecting the vacuum hose does not increase pressure, a fuel filter could be clogged or the fuel pump weak. If the pressure rises when pinching the return hose closed, the pressure regulator is probably bleeding off too much fuel. It should be replaced.

If there is HIGH FUEL PRESSURE, the pressure regulator may be stuck closed or the fuel return line may be restricted. Disconnect the fuel return line and route it into a container. Start the engine and note fuel pressure. If pressure remains too high, the pressure regulator is bad. If pressure drops, check for a clogged or smashed fuel return line.

For more information on testing fuel pumps, filters, and other fuel supply system parts, refer again to Chapter 9.

Fuel pressure regulator replacement

A single-point injection system's fuel pressure regulator is normally located on the throttle body assembly. With multi-point injection, it is generally found on a fitting or line on the fuel rail. It may also be mounted elsewhere in the engine compartment, as in Fig. 17-21.

To replace the regulator, bleed off system pressure using the service fitting. Disconnect the lines from the regulator. Unbolt the brackets and install the new unit. Reattach the lines. Start the engine and check for leaks. You may want to measure fuel pressure to double-check your work.

Fig. 17-22 shows the fuel flow through a multi-point fuel injection system. Visualize how a leaking pressure regulator could lower fuel pressure. Also visualize how a blocked regulator or restricted return line could cause high fuel pressure.

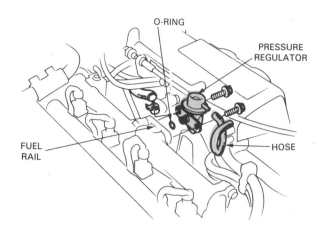

Fig. 17-21. In multi-point fuel injection systems, fuel pressure regulator is usually located at fuel rail. (Honda)

INJECTOR PROBLEMS

A *bad injector* may cause a wide range of problems including rough idle, hard starting, poor fuel economy, and engine miss. Each fuel injector must provide the correct fuel spray pattern. Fig. 17-23 shows possible EFI injector troubles.

A *leaking injector* richens the fuel mixture by letting extra fuel through the closed injector valve. The injector valve may become worn or dirty. The return spring may be weak or broken.

A *dirty injector* can restrict fuel flow and cause a poor fuel spray pattern. The air-fuel ratio would be too lean

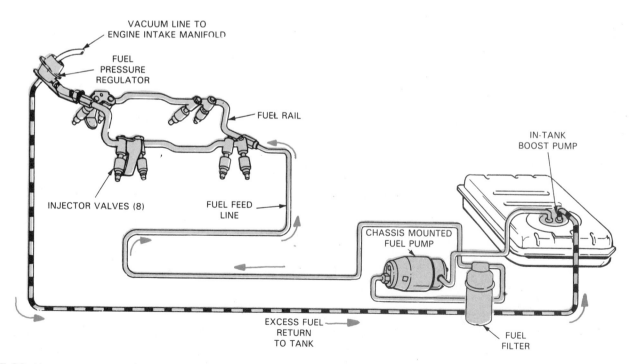

Fig. 17-22. How pressure regulator functions on multi-point fuel injection system. Fuel pump provides pressure as it pumps fuel continuously into fuel rail, injectors, and pressure regulator. When pressure is high enough, pressure regulator allows excess fuel to bleed off and return to fuel tank. Engine vacuum to pressure regulator allows regulator to vary fuel pressure according to engine load. (Cadillac)

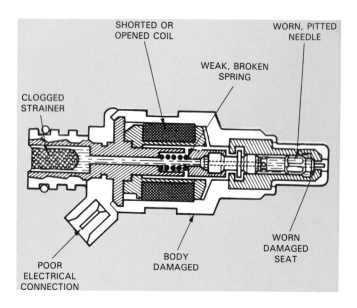

Fig. 17-23. Above problems can affect operation of fuel injector. (Renault/AMC)

for proper engine operation.

An *inoperative EFI injector* normally has shorted or opened coil windings. Current is reaching the injector, but since the coil is bad, a magnetic field cannot form and open the injector valve.

A continuous injector that does not use a solenoid will usually operate, but may have other problems. Among these are poor spray pattern, or weak spring (incorrect opening pressure or leakage).

Checking EFI multi-point injector operation

To check opening and closing of injectors, place a STETHOSCOPE against each injector, Fig. 17-24. Absence of clicking sound indicates the injector is not opening and closing. The injector solenoid, wiring harness, or computer control circuit may be bad.

With the engine off, check the condition of the coils on inoperative injectors. Use an ohmmeter, Fig. 17-25. Measure the resistance across the injector coil and check for shorts to ground. If the coil is open (infinite resistance) or shorted (zero resistance to ground), you must replace the injector.

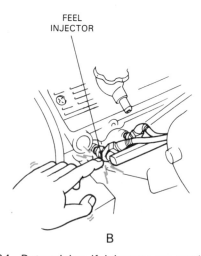

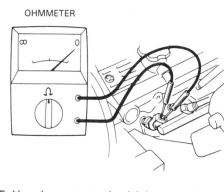

Fig. 17-25. Use ohmmeter to check injector resistance. Unplug connector from injector. With probes of ohmmeter, test continuity of both terminals. Resistance should be about 1.5 to 3 ohms. Check service manual for specifications. (Honda)

If the injector is good, you may need to check the wiring going to that injector. Following the service manual, check supply voltage to the inoperative injector. You may also need to measure the resistance in the circuit between the injector solenoid and the computer. A high resistance would indicate a frayed wire, broken wire, or poor electrical connection.

Refer to a wiring diagram when solving complex fuel injection electrical problems. The diagram will show all electrical connections and components that can upset the function of the injection system.

CAUTION! Some EFI multi-point systems use dropping resistors before the injectors. The resistors lower the supply voltage to the injectors. Do NOT connect direct battery voltage to this type of injector or coil damage may occur.

EFI multi-point injector balance test

An *injector balance test* quickly measures the fuel out-

Fig. 17-24. Determining if injectors are working. A—Use stethoscope, touching it to each injector you should hear clicking sound if they are opening and closing. B—Feeling for vibration as injector opens and closes. (Honda)

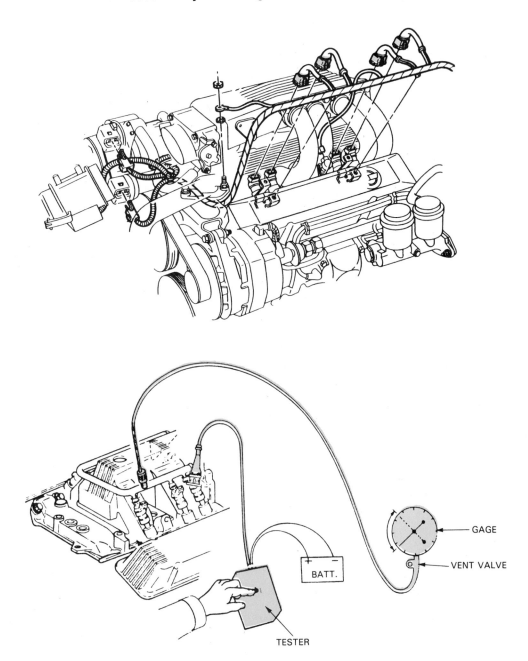

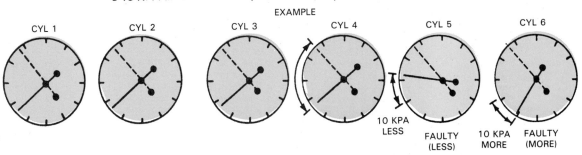

Fig. 17-26. Performing balance test on injectors. A—Disconnect wiring harness from all injectors. B—Hook up special test tool. Connect pressure gauge to fuel rail. Activate fuel pump momentarily by turning ignition on briefly. Read pressure gauge. Energize tester and note whether injector opens (sprays fuel and pressure decreases). Pressure in fuel rail should drop if injector opens.
C—Each injector should cause equal fuel pressure drop within about 1.5 psi (10 kPa) of each other. (Chevrolet)

put from each injector for a specific time span. Poor output indicates a leaking or clogged injector, weak spring, damaged needle and seat, and other problems.

To do an injector balance test, connect a pressure gauge to the fuel rail. Disconnect wires to all injectors, Fig. 17-26A. Connect a special injector balance test tool to one of the injectors, Fig. 17-26B. Turn the ignition off and then on for about 10 seconds. This will turn the fuel pump on and pressurize the system.

Energize the tester. This will provide voltage to the injector for about 5 seconds. With the injector open and spraying fuel, measure the drop in fuel pressure at the gauge. Note the pressure as soon as the injector closes or the pressure stops falling.

If an injector does NOT cause enough pressure drop, that injector is not allowing enough fuel into the engine. It could be clogged, be stuck closed, or have other troubles restricting fuel flow through the injector. A lean mixture would result in that cylinder.

If an injector causes TOO MUCH fuel drop, it is permitting too much fuel flow. It could be leaking, have a weak return spring, damaged needle and seat, or other problems. That cylinder would have a rich mixture with the engine running.

Each injector should cause an equal drop in fuel pressure, as shown in Fig. 17-26C. Generally, make sure all of the injectors have a pressure drop within 1.5 psi (10 kPa) of each other. If one causes too much or too little pressure drop, replace it.

Replacing EFI multi-point injectors

An EFI multi-point injector is easy to replace. First, bleed off fuel pressure. Then simply remove the hose from the injector and fuel manifold, or disconnect the injector from the fuel rail, Fig. 17-27. Unplug the electrical

connection and remove any fasteners holding the injector. Pull the injector out of the engine. See Fig. 17-28.

Inspect the boot and other rubber parts closely. Refer to Fig. 17-28 again. Some manufacturers suggest that you replace the boot, seals, and hose if the injector is removed for service.

Install the new or serviced injector in reverse order. Follow the directions in the shop manual.

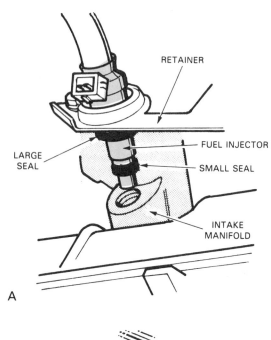

A

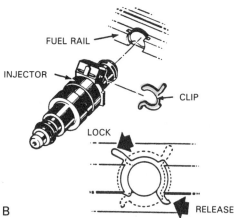

B

Fig. 17-28. One type of fastener for holding injector in manifold. A—Upon removal, inspect seals for cracks and other damage. Some manufacturers recommend replacement rather than reuse. B—Some injectors are fastened to fuel rail with retaining clips. (Fiat and Buick)

TESTING COLD START INJECTOR

If the fuel injected engine is hard to start when cold, check the operation of the cold start injector. Remember from earlier study that this is a solenoid-operated fuel valve that richens the mixture for cold starting.

To test the injector, remove it from the engine, Fig. 17-29. Aim the injector into a container. Crank the cold engine and observe the fuel spray pattern. The injector

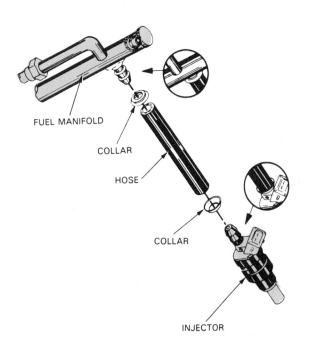

Fig. 17-27. Removing multi-point fuel injector is simple. Disconnect fuel hose and electrical plug first. Injector is either press-fitted into manifold or held with bolts. (Fiat)

Fig. 17-29. Checking cold start injector spray pattern. Test is made with engine temperature sensor below specified temperature. (Saab)

should spray fuel whenever the engine temperature sensor is below a set temperature.

If the cold start injector does NOT function, test the thermo-time switch, circuit, and the injector.

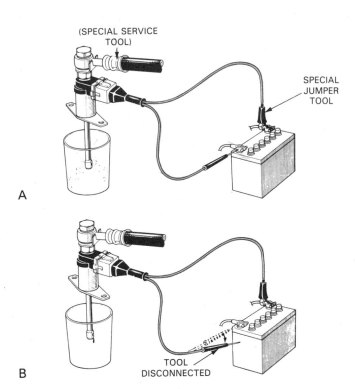

Fig. 17-30. Method of battery test for cold start injector. A—With probe making contact with other battery terminal, switch on, and fuel pump energized cold start injector should spray fuel. B—With no battery current, injector should close. Check for leaking. (Honda)

One manufacturer suggests connecting the cold start injector to a battery, Fig. 17-30. If the injector begins to function (did not work connected to system), the supply circuit (thermo-time switch, electronic control unit, etc.) is faulty. Further circuit tests are needed to find the circuit trouble. If the cold start injector does not function properly (not open, leakage, etc.) on the battery test, it should be replaced.

SERVICING CONTINUOUS INJECTORS

From earlier chapters you know that continuous injectors are simply spring-loaded fuel valves. They do not need a solenoid to open and close the injector.

To check continuous injectors, remove them from the intake manifold, Fig. 17-31.

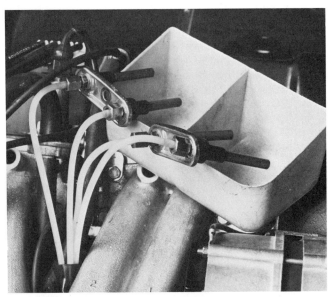

Fig. 17-31. Continuous injectors have been removed for testing. Spray should be directed into unbreakable tray. Do not spill fuel or allow it to spray over engine. (Saab)

Testing continuous injectors

To test continuous injector *spray patterns,* place the injectors in a suitable, graduated container. See Fig. 17-32. Activate the electric fuel pump and move the airflow sensor plate as described in a service manual. This will make fuel spray out the injectors. Check the fuel spray patterns, Fig. 17-33. Also check for leakage when the fuel pump is deactivated.

DANGER! Injectors can spray out enough fuel to start a large fire. Make sure the fuel container and injectors are secured during testing.

Testing fuel volume

One CIS manufacturer suggests that you measure the *injector fuel volume output* (amount of fuel output for specific time period). For efficient engine operation, it is important that each engine cylinder receive the same

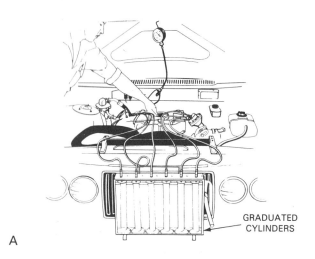
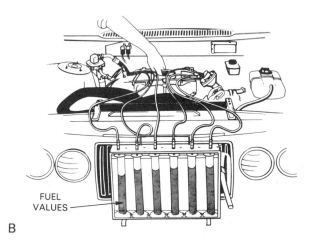

A

B

Fig. 17-32. Another test using graduated containers compares fuel output of continuous injectors. A—Test arrangement with injectors in graduated cylinders. B—Test in progress showing fuel values for each injector. (Volvo)

Fig. 17-33. Gasoline injector spray patterns. A—Normal. Pattern shows spray is even and partially atomized. B—Bad. Solid stream of fuel, poor spray pattern. C—Bad. Dirty nozzle affecting spray pattern. D—Bad. Uneven spray pattern. (Saab)

amount of fuel. For this test, place the injectors in graduated containers. Look again at Fig. 17-32. Activate the fuel system or tester and measure the amount of fuel spraying out of each injector.

If the volume of fuel coming out of the injectors is NOT equal, there are injector or fuel distributor problems.

To pinpoint the trouble, switch the problem injector (injector with incorrect output) with a good one. If the same injector has an incorrect output, that injector is faulty. If the same fuel line produces an improper output, the injector is good, but the fuel line or fuel distributor is defective.

FUEL DISTRIBUTOR SERVICE

If the continuous injector tests find the problem to be the fuel distributor, replace the unit. If the valves in the unit do not supply the same amount of fuel to each in-

jector, the unit should be removed from the flow sensor and a new one installed. Fig. 17-34 shows a fuel distributor being serviced.

Refer to a shop manual when servicing a fuel distributor assembly. It is a very precise fuel metering device. The slightest change can upset its operation.

SINGLE-POINT INJECTOR SERVICE

You can quickly check the operation of a single-point (throttle body) injection system by watching the fuel spray pattern in the throttle body. Remove the air cleaner. With the engine cranking (no-start problem) or running, each injector should form a rapidly pulsing spray of fuel. Refer to Fig. 17-35.

Check TBI wiring

If the throttle body injectors do not work, check out

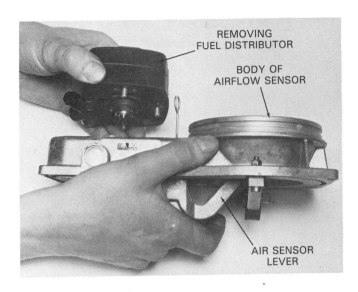

Fig. 17-34. Replace fuel distributor if injectors of CIS system prove to be all right. Here distributor is being lifted from body of airflow sensor. (Saab)

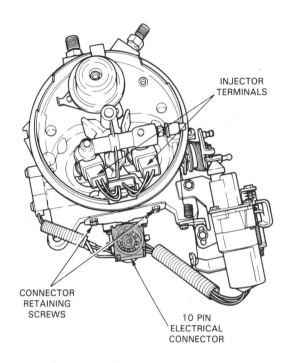

Fig. 17-36. Check electrical connector to TBI. It may be loose and not making good contact. (Ford)

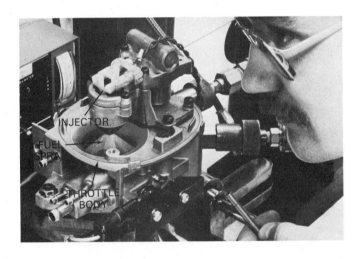

Fig. 17-35. TBI unit may have one or two injectors mounted on air horn. To check operation, simply remove air cleaner and observe spray pattern. (Pontiac)

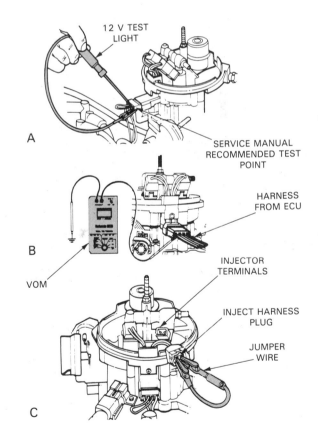

Fig. 17-37. Follow auto makers service manual for correct procedure when checking out EFI system voltages and resistance. A wrong meter setting or connection could damage parts. A—One manufacturer's method of checking supply voltage to injector from computer. B—Measuring supply voltage with digital VOM. C—Use of jumper wire to check circuit resistance. Ohmmeter can be used at other end of harness. (Ford Motor Co.)

the wiring to the TBI unit first. A loose connection could be preventing good electrical contact. Wiggle and reconnect wiring at the terminals to the injectors, Fig. 17-36.

Testing TBI injectors

If the system is pressurized and terminals are tight, follow service manual procedures to test for power to the injector solenoid. Fig. 17-37 illustrates some simple test procedures recommended by one auto maker.

NO POWER (current) to the injector indicates a problem with the wiring harness or computer (ECU). If you have power, but NO FUEL SPRAY PATTERN, then the injector may be bad. Make sure you have adequate fuel pressure before condemning a fuel injector.

Rebuilding TBI unit

While exact methods vary with design, there are a few

general rules to follow when servicing any TBI unit:

1. Relieve system fuel pressure before removing throttle body components.
2. Disconnect the negative battery cable.
3. Label hoses and wires before removal.
4. Remove TBI unit, Fig. 17-38.
5. Mount TBI unit on repair legs, Fig. 17-39.
6. Read number to identify TBI unit. Use this information to order new parts.
7. Avoid damage to the throttle body housing when pulling or prying out the injector, Fig. 17-40.
8. Remove all rubber and plastic parts before cleaning in decarbonizing solvent, Fig. 17-41.
9. Install new rubber O-rings. You may need to

10. lubricate them with approved lubricant to aid new injector installation, Fig. 17-42.
11. Double-check that you have installed all injector washers and O-rings correctly.
12. Push injector into the throttle body by hand. Refer once more to Fig. 17-40D. Avoid using a hammer or pliers. They may damage the new injector.
13. Make sure the injector is fully seated in the throttle body. Normally, a tab must fit into a notch, assuring correct alignment.
14. Reinstall the unit. Make any adjustments described in the service manual. Check for leaks after TBI unit installation. Torque all fittings to specs.

Air throttle body rebuild

An *air throttle body rebuild,* like a carburetor or throttle body injection unit rebuild, involves replacing all gaskets, seals, and other worn parts. You must remove and disassemble the throttle body, making sure all plastic, rubber, and electrical parts are removed.

The metal parts are soaked in carburetor cleaner, washed in cold soak cleaner, and blown dry.

Inspect each part carefully for signs of wear or damage. Then, reassemble and install the TBI unit following the instructions in a manual. Refer to Fig. 17-43.

> **DANGER!** Carburetor cleaner is a very powerful solvent. Wear eye protection and rubber gloves when working with this type of cleaner. Follow the directions on the solvent label in case of an accident.

Chapter 14, Carburetor Diagnosis, Service, and Repair, covers subjects related to rebuilding a throttle body assembly. It covers solvent use, mounting the unit on a holding fixture, and other procedures.

Idle air valve service

A *bad idle air valve* can make engine idle too low or too high. The engine may race at idle or stall when cold. Whenever you have idle problems with a fuel injected engine, check the idle air valve. This valve normally mounts in the throttle body assembly, as shown in Fig. 17-44. It may also mount in a remote location in the engine compartment. Use service manual information for additional service details.

Idle speed motor service

An *inoperative idle speed motor* will usually fail to hold constant idle speed regardless of engine load. Engine speed may also be too high or too low. The small dc motor or the screw mechanism that moves the throttle lever may be bad. Refer to Fig. 17-43 again.

First, check for current to the idle speed motor (engine running). This will indicate that the computer and wiring are not at fault. As you move the throttle by hand, the motor should try to compensate and correct engine idle speed. The stem should move in and out.

To check the motor, you can also connect battery voltage to it using jumper wires. The motor stem should extend when connected with one polarity. It should retract with the polarity reversed.

If the idle speed motor is defective, install a new unit. Follow service manual directions to mount and adjust the motor assembly properly.

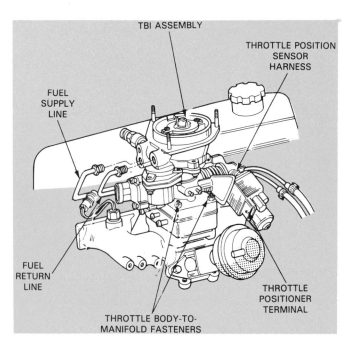

Fig. 17-38. Getting ready to remove TBI unit. Remove four fasteners holding throttle body to manifold. Be sure to disconnect and label all hoses and wires. (Renault-AMC)

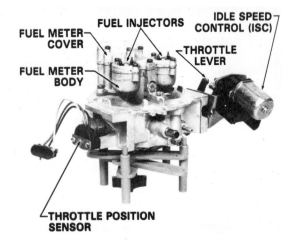

Fig. 17-39. Mount TBI unit on repair legs before attempting repair or rebuild. (Cadillac)

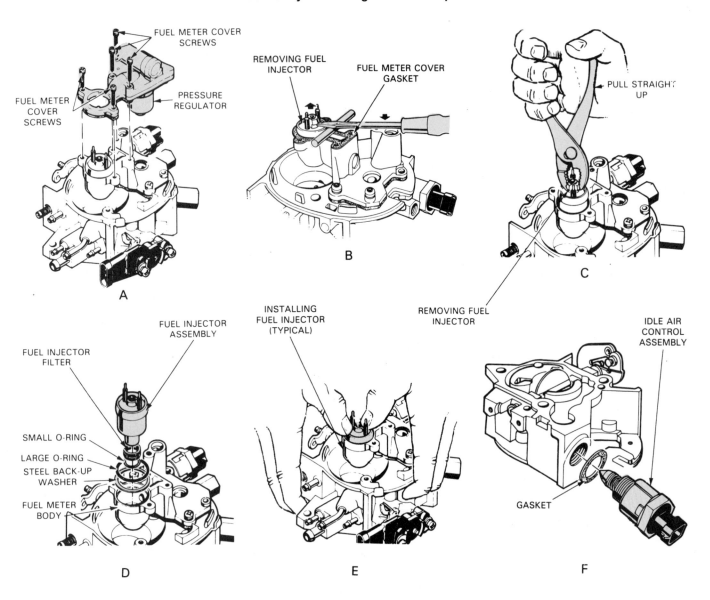

Fig. 17-40. Removing TBI injector. A—Remove fasteners holding fuel metering body. Lift off while keeping screw lengths organized. B—Use screwdriver and pin to pry injector loose from throttle body. C—Pliers may be used to pull out old injector. D—As necessary, replace O-rings and washers. E—Align new injector and press it into throttle body with thumbs. F—Install idle air control assembly. (Pontiac)

ENGINE SENSOR SERVICE

Most EFI engine sensors can be tested with a special tester (analyzer), as covered earlier in the chapter. Some engine sensors, however, should be checked with a digital meter or a test light. Remember to always follow manufacturer's directions.

CAUTION! When a service manual directs you to measure voltage or resistance in an EFI sensor, use a high impedence meter or a digital meter. An inexpensive conventional needle type meter can damage some sensors by drawing too much current.

Fig. 17-45 is a sensor location illustration from a service manual. It is helpful when trying to find and test various engine sensors.

Throttle position sensor service

A *throttle position sensor* should produce a given amount of electrical resistance for different throttle openings. For example, the sensor might have high resistance when the throttle plates are closed and a lower resistance when they are open. As shown in Fig. 17-46A. compare ohmmeter readings to specs to determine the condition of a throttle position sensor.

To replace a throttle position sensor, you must usually unstake the attaching screws. They may be soldered in place to prevent tampering. Use a file to grind off the stake and remove the sensor, Fig. 17-46B.

After installing the new throttle position sensor, adjust the sensor. Refer to your manual for details. For most systems, the sensors should read a prescribed resistance or voltage with the throttle plates in specified positions.

MAP sensor service

A *faulty MAP sensor* can make the air-fuel mixture too rich or lean. The engine may miss and ping, indicating

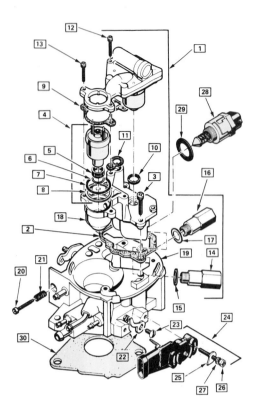

1—FUEL METER ASSEMBLY
2—GASKET—FUEL METER BODY
3—SCREW AND WASHER ASSEMBLY ATTACH. (3)
4—FUEL INJECTOR KIT
5—FILTER—FUEL INJECTOR NOZZLE
6—SEAL—SMALL O-RING
7—SEAL—LARGE O-RING
8—BACK-UP WASHER—FUEL INJECTOR
9—GASKET—FUEL METER COVER
10—DUST SEAL—PRESS, REGULATOR
11—GASKET—FUEL METER OUTLET
12—SCREW AND WASHER ASSEMBLY—LONG (3)
13—SCREW AND WASHER ASSEMBLY—SHORT (2)
14—NUT—FUEL INLET
15—GASKET—FUEL INLET NUT
16—NUT—FUEL OUTLET
17—GASKET—FUEL OUTLET NUT
18—FUEL METER BODY ASSEMBLY
19—THROTTLE BODY ASSEMBLY
20—SCREW—IDLE STOP
21—SPRING—IDLE STOP SCREW
22—LEVER—TPS
23—SCREW—TPS LEVER ATTACHING
24—SENSOR—THROTTLE POSITION KIT
25—RETAINER—TPS (2)
26—SCREW—TPS ATTACHING (2)
27—WASHER—TPS SCREW (2)
28—IDLE AIR CONTROL VALVE
29—GASKET—CONTROL VALVE TO T.B.
30—GASKET—FLANGE MOUNTING

Fig. 17-41. Certain TBI parts (marked in color) should never be placed in decarbonizing solvent. Injectors, electrical parts, plastics, and rubber materials will be ruined by the cleaner. O-rings and gaskets are replaced during a rebuild. (Oldsmobile)

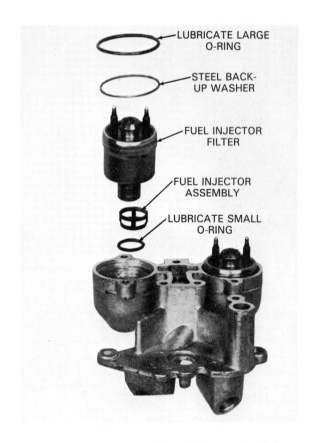

Fig. 17-42. Install new rubber O-rings on fuel injector. (Cadillac)

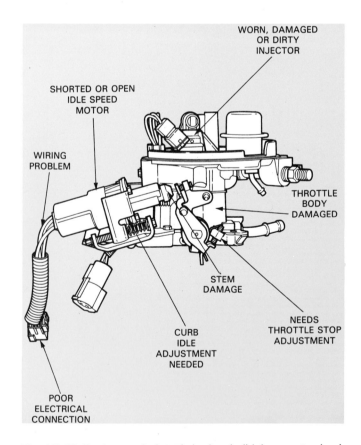

Fig. 17-43. During an air throttle body rebuild, be sure to check for defective or damaged parts. (Ford)

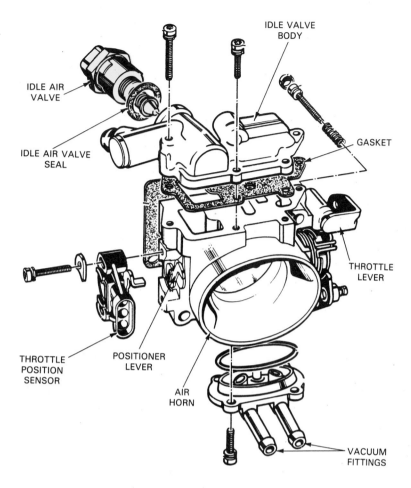

IDLE VALVE
BODY

IDLE AIR
VALVE

GASKET

IDLE AIR VALVE
SEAL

THROTTLE
LEVER

THROTTLE
POSITION
SENSOR

POSITIONER
LEVER

AIR
HORN

VACUUM
FITTINGS

Fig. 17-44. Exploded view of air throttle body. Note location of air valve.
Sometimes it is located elsewhere in engine compartment. (Chevrolet)

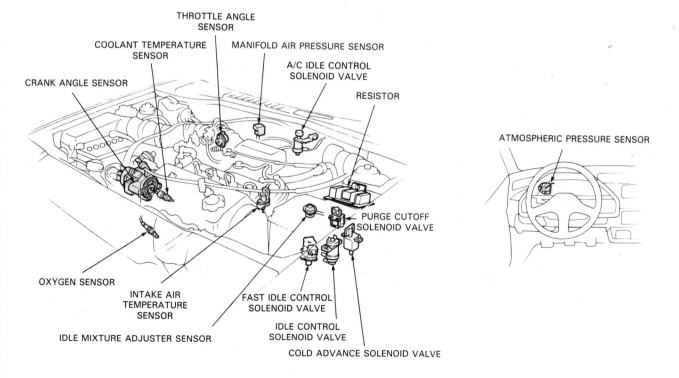

THROTTLE ANGLE
SENSOR

COOLANT TEMPERATURE
SENSOR

MANIFOLD AIR PRESSURE SENSOR

A/C IDLE CONTROL
SOLENOID VALVE

CRANK ANGLE SENSOR

RESISTOR

ATMOSPHERIC PRESSURE SENSOR

PURGE CUTOFF
SOLENOID VALVE

OXYGEN SENSOR

INTAKE AIR
TEMPERATURE
SENSOR

FAST IDLE CONTROL
SOLENOID VALVE

IDLE MIXTURE ADJUSTER SENSOR

IDLE CONTROL
SOLENOID VALVE

COLD ADVANCE SOLENOID VALVE

Fig. 17-45. Service manual will show location of engine's sensors. (Honda)

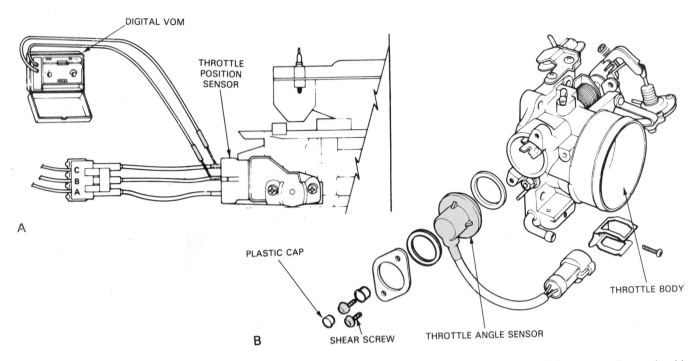

Fig. 17-46. Servicing the throttle position sensor. A—Check resistance or voltage output of throttle position sensor. Specs should be checked for unit being serviced. B—Exploded view of throttle position sensor. Usually, several screws hold it in place. Some must be filed off for removal of the sensor. (Buick and Honda)

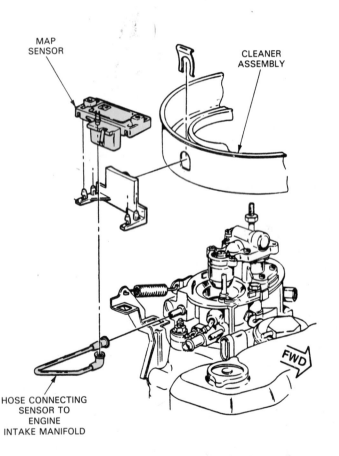

Fig. 17-47. MAP sensor measures manifold pressure changes which result from engine load and speed changes. It converts this to a voltage output to computer. When it fails, check for loose hoses and electrical connections. If sensor is found to be faulty, it must be replaced. (General Motors)

a lean mixture. It may also be sluggish and get poor gas mileage, indicating a rich mixture. One MAP sensor is shown in Fig. 17-47. Note the vacuum hose connecting the sensor to the throttle body.

If resistance tests show the MAP sensor to be bad, replace the unit. Recheck system operation after installing a new sensor.

Thermo-time switch service

A *bad thermo-time switch* can keep the cold starting injector from functioning or it may cause it to remain open all the time. When a car suffers starting troubles, check the thermo-time switch.

As shown in Fig. 17-48, an ohmmeter can normally be used to check the switch. When cold, the switch

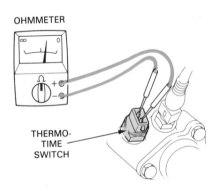

Fig. 17-48. Using ohmmeter to check thermo-time switch. Resistance between terminals is specified in service manuals. (Toyota)

should have a specified resistance. When warm, its resistance should normally increase to shut the cold start injector off. Replace the thermo-time switch if its resistance values are not within specifications.

Engine temperature sensor service

Most EFI systems use both a coolant temperature sensor and an inlet air temperature sensor. If these sensors are bad, they will make the engine run either rich or lean.

A digital ohmmeter is generally recommended for testing a temperature sensor. As shown in Fig. 17-49, the service manual will give resistance values for various temperatures. If ohmmeter test readings are not within specs for each temperature, the sensor is bad.

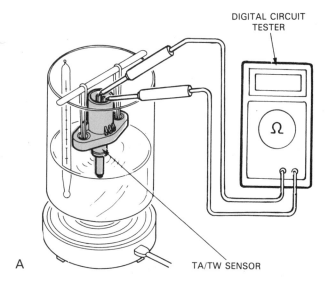

A

DIGITAL CIRCUIT TESTER

TA/TW SENSOR

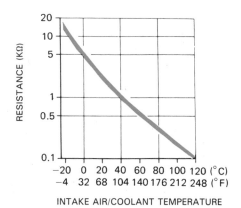

B

INTAKE AIR/COOLANT TEMPERATURE

Fig. 17-49. Testing air/coolant temperature sensor. A—Suspend unit in cold water and heat slowly. Measure resistance between terminals as temperature rises. B—Chart gives acceptable resistance readings for each temperature change. (Honda)

Intake air temperature sensor

A *faulty intake air temperature sensor* can also affect the fuel system's air-fuel ratio. Acting on a bad signal from the sensor, the computer can produce a wrong in-

jector pulse width for the air temperature. Cooler air requires a slightly richer mixture. Warmer air needs a leaner mixture because the air contains less oxygen.

If symptoms or the self-diagnosis indicate malfunction of the air temperature sensor, check its resistance with an ohmmeter, Fig. 17-50. If resistance is not within specs, replace the sensor.

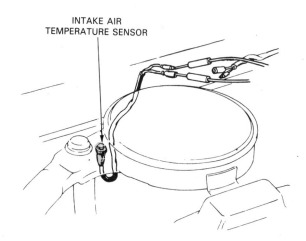

INTAKE AIR TEMPERATURE SENSOR

Fig. 17-50. Check resistance of suspected air temperature sensor with ohmmeter. (Honda)

Airflow sensor service

A *bad airflow sensor* will cause more obvious problems than most other sensors. There are two basic types: flap sensor and mass (hot wire) sensor, Fig. 17-51.

Use service manual procedures for testing airflow sensors to prevent damage. With a flap type, the sensor should produce a specific ohms value for specific flap positions. The mass airflow sensor is usually installed in a system with self-diagnosis. The computer will indicate problems with the sensor. Check wiring before condemning the sensor, however.

Oxygen sensor problems

An *inoperative oxygen sensor* can also upset the air-fuel ratio and cause very poor engine performance. Normally, oxygen sensors are designed to last about 50,000 miles (81 000 km). However, its life can be shortened by contamination, blocked outside air, and poor electrical connections, Fig. 17-52.

Oxygen sensor contamination can result from:
1. USE OF LEADED FUEL. Leaded fuel is the most common cause of oxygen sensor contamination. Lead coats the ceramic element and the sensor cannot produce voltage output for computer.
2. SILICONE. Sources are anti-freeze, RTV silicone sealers, waterproofing sprays, and gasoline additives. Silicone forms a glassy coating.
3. CARBON. Carbon contamination results from rich fuel mixtures. Carbon in fuel coats the sensor.

Carbon and moderate lead contamination can sometimes be reversed. Run the engine at high speeds with a large vacuum hose removed and with only unleaded

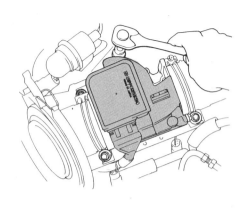

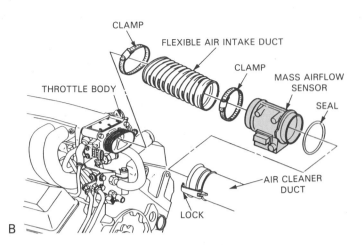

A

B

Fig. 17-51. Faulty airflow sensors cause more obvious problems than most other sensors. A—Removing airflow sensor from air duct system. B—Exploded view of mass airflow sensor which uses a thermister to convert temperature to an electrical resistance.

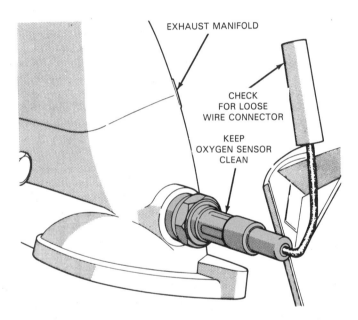

Fig. 17-52. Foreign matter or loose electrical connections may cause malfunction of oxygen sensor. (Chrysler)

fuel in the tank. This will sometimes burn off light lead and most carbon deposits. The sensor may start working normally again.

Also, check that the outside of the sensor and its electrical connection are free of oil, dirt, undercoating, and other deposits. If outside air cannot circulate through the oxygen sensor, the sensor will not function.

An oxygen sensor only generates a tiny voltage (an average of about .5 volts). A poor electrical connection can prevent this small voltage from reaching the computer. Always check the sensor's electrical connections.

Oxygen sensor testing

The self-diagnosis mode will usually indicate problems with the oxygen sensor by flashing its number code. To test the sensor, you can usually measure voltage out-

put with a high impedence (10 megohm or more) multimeter (voltmeter).

Warm the engine to full operating temperature and let it idle. As the throttle is OPENED and CLOSED, the oxygen sensor should produce a voltage that varies around .5 volts. Normal voltage output is between .2 and .9 volts. If voltage does not change with the opening the sensor may be bad.

If you pull off an engine vacuum hose (creating a lean mixture), the oxygen sensor voltage should DECREASE (go down to about .1 to .3 volts). The sensor should try to signal the computer of the lean condition.

If you block the air inlet at the air cleaner (creating a rich mixture), the oxygen sensor voltage should INCREASE (go up to about .8 to 1.0 volts). It should try to signal the computer that too much fuel, or not enough air, is entering the combustion chambers.

If the oxygen sensor voltage does not change properly as you simulate rich and lean air-fuel ratios, the oxygen sensor is faulty. You might try running the engine at high speeds, with a large vacuum hose removed, to clean off light lead or carbon contamination.

Oxygen sensor replacement

Disconnect the negative battery cable. Then, separate the sensor from the wiring harness by unplugging the connector. Never pull on the wires themselves as damage may result.

The oxygen sensor may have a permanently attached pigtail. Never attempt to remove it. Unscrew the sensor and inspect its condition. Some sensors may be difficult to remove at temperatures below 120 °F. Use care to avoid thread damage.

Follow these rules when replacing an oxygen sensor:
1. Do NOT touch the sensor element with anything (water, solvents, etc.)
2. Coat sensor threads with anti-seize compound.
3. Do NOT use silicone based sealers on or around exhaust system components. Use sealers sparingly on engine components. PCV system can draw silicone fumes into engine intake manifold.
4. Hand start sensor to prevent cross threading.

5. Do NOT overtighten sensor. It could be damaged.
6. Make sure outside vents are clear so that air can circulate through the sensor.
7. Make sure wiring is reconnected securely to sensor.
8. If sensor checks out good, Fig. 17-53, check continuity of wiring between sensor and computer. See Fig. 17-54.
9. Check oxygen sensor output on fuel system operation after installing sensor.

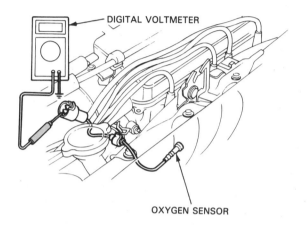

Fig. 17-53. Checking out oxygen sensor output. Engine should be warmed up for several minutes before this test is made. Note VOM hookup. (Honda)

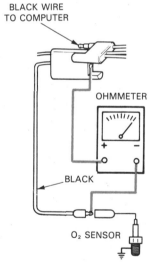

Fig. 17-54. If oxygen sensor passes voltage check, check out continuity of wiring with ohmmeter. (Ford)

Reading oxygen sensor

To *read an oxygen sensor,* inspect the color of the sensor's tip:

1. LIGHT GREY TIP. This is a normal color for an oxygen sensor.
2. WHITE SENSOR TIP. This indicates silicone contamination. Sensor MUST usually be replaced.
3. TAN SENSOR TIP. This could be lead contamina-

tion. It can sometimes be cleaned away by briefly running the engine lean on unleaded fuel.
4. BLACK SENSOR TIP. Normally, this indicates carbon contamination which can usually be cleaned after correcting the cause.

Servicing other EFI sensors

Other sensors in an electronic fuel injection system are tested using the same general procedures just discussed. You would use the self-diagnosis mode, a special analyzer, or a digital meter to check each sensor. Refer to a service manual for exact procedures.

COMPUTER SERVICE

A *malfunctioning computer* may have a wide range of symptoms. The engine may run poorly or not at all. Several types of testers are available for checking the computer condition. You could also use a test computer. Plug a computer known to be good into the system. If this corrects the trouble, install a new computer.

Most modern on-board computers are located under the car dash, away from engine heat, moisture, and vibration. To remove the unit, remove the bolts holding the computer under the dash, Fig. 17-55. Unplug the wiring harness carefully. Then, remove the computer from the car.

Many computers have a PROM (programmable read only memory) that calibrates the computer for the specific vehicle. It is normally removed from the old com-

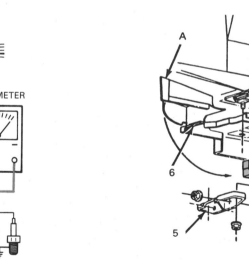

1	DASH UPPER PANEL
2	ADAPTER
3	HOUSING
4	ECM
5	BRACKET
6	HINGE AND HOUSING COVER
	A FLANGE (FLAP)
	B FLANGE

Fig. 17-55. Electronic control modules on most modern cars are located in passenger compartment. To remove, unbolt them from brackets. (Pontiac)

puter and installed in the new unit. Fig. 17-56 shows the basic steps for PROM installation.

After installing the PROM, install the new computer. Make sure the wiring and brackets are secure. Check engine and fuel system operation after installation.

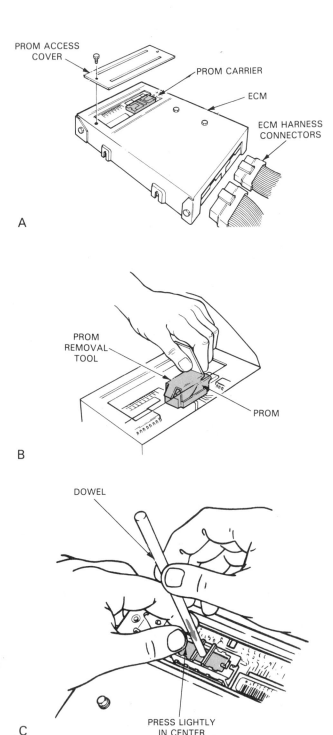

A

B

C

Fig. 17-56. Steps for installing PROM (programmable read only memory). A—Disconnect connectors from ECM. Remove mounting hardware and take ECM from passenger compartment. Remove PROM access cover. B—Use special tool and alternately rock each end of PROM carrier upward until it is free of socket. C—Make sure you have correct PROM. Carefully press it into socket. Use dowel if needed. (Buick)

FUEL INJECTION SYSTEM WIRING DIAGRAMS

When electrical problems in a fuel injection system are difficult to locate, refer to service manual wiring diagrams. Two examples are given in Figs. 17-57 and 17-58. Wiring diagrams will let you trace wires from one component to another. You can then more easily find loose connections, shorts, opens, and other problems.

GASOLINE INJECTION ADJUSTMENTS

Several tune-up adjustments are needed on a gasoline injection system. These include:

1. Curb idle speed adjustment, Fig. 17-59.
2. Fast idle speed adjustment.
3. Throttle plate stop adjustment, Fig. 17-59.
4. Idle air-fuel mixture adjustment, Fig. 17-59.
5. Throttle cable adjustment, Fig. 17-60.

Since there are so many types of systems, refer to a service manual for exact procedures.

NOTE! Many auto makers do not provide or warn against making some adjustments on a gasoline injection system. Only make adjustments when major problems exist or an exhaust analyzer shows high emission levels. If you must make adjustments, there are usually problems in the system.

SUMMARY

Most modern EFI systems are designed to test themselves and indicate where the problem lies by displaying some type of code. The computer's self-diagnosis mode is normally activated after the engine is warmed to operating temperature. In every situation, the service manual procedures must be followed.

Self-diagnosis output codes include several different types:

1. Digital readout display that flashes a number corresponding to the trouble.
2. On-off voltage pulse that must be read by a voltmeter.
3. LED display which flashes a number code.

EFI systems without self-diagnostic capabilities require you to use an EFI tester. Various meter readings must be compared with specifications. Oscilloscope tests are sometimes also used to find wiring, harness, and computer problems.

An EFI harness terminal should never be disconnected with the ignition switch turned ON. This could damage the computer. In some instances, the negative battery terminal may need to be disconnected.

Studying wiring diagrams in the shop manual will help in finding problem sources. You will be able to locate service connections and components that must be checked out for such problems as loose connections and faulty operation.

There are problems beyond the electrical/electronic components of the EFI. Air and fuel leaks can cause malfunctions and must also be checked. Inspect both subsystems carefully, making certain all fittings are tight and leakfree.

Fuel pressure that is either too high or too low will cause improper fuel mixtures. A faulty pressure regulator

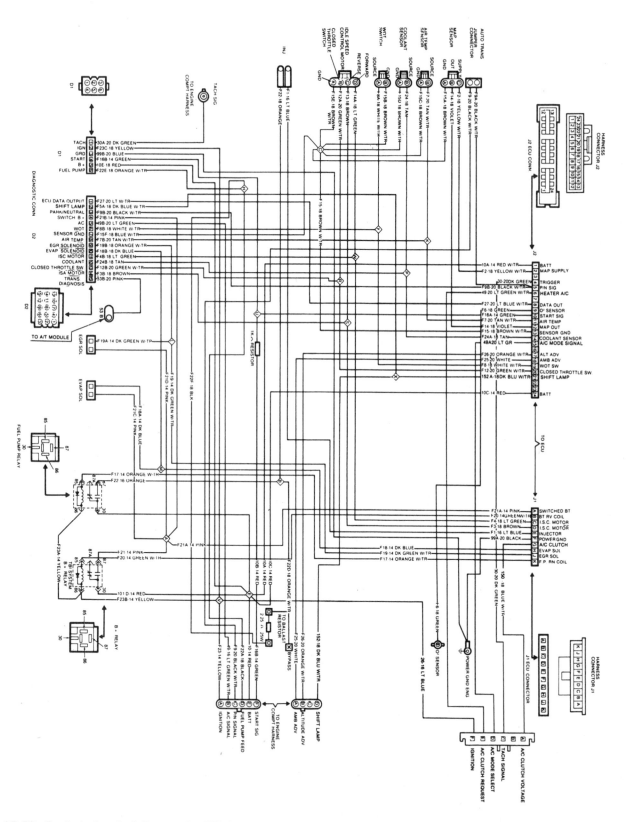

Fig. 17-57. Typical electrical diagram for TBI fuel system. Every service manual includes such illustrations. (AMC/Renault)

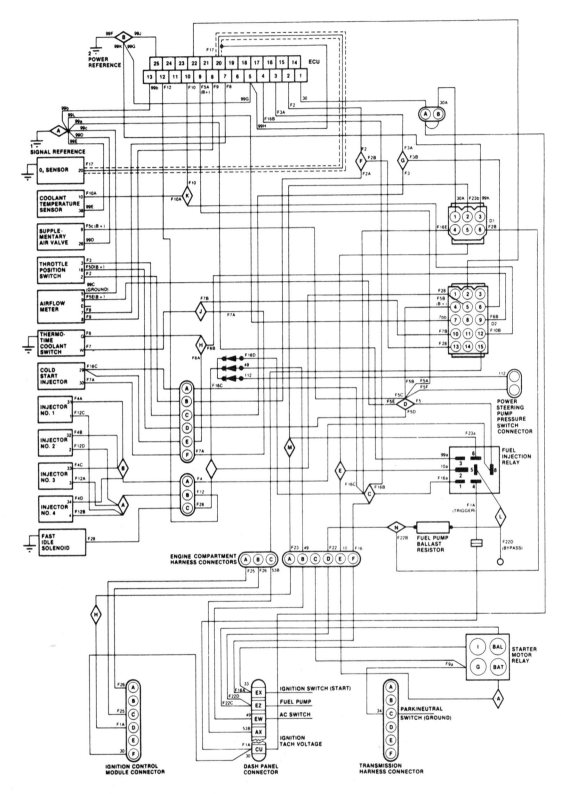

Fig. 17-58. Diagram for electrical system for multi-point fuel injection system. Note ECU at top of illustration.

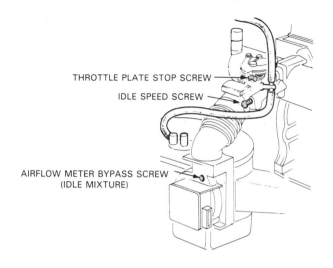

Fig. 17-59. Adjustments and their access locations will vary with each model and gasoline injection type. These are typical of one system. (Renault)

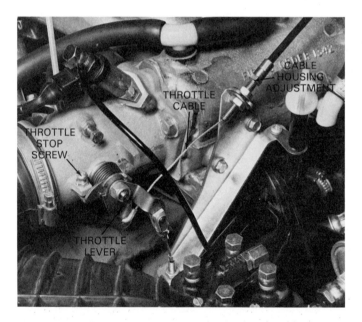

Fig. 17-60. Making throttle cable adjustment. Note names of parts and check with the shop manual for specifications. (Saab)

may be the cause. You will need to test the regulator and replace it if necessary.

Bad fuel injectors will also affect fuel delivery and cause problems such as hard starting, rough idle, missing, and poor fuel economy. Leaking injectors richen the mixture. Dirty injectors restrict fuel flow, causing a too-lean mixture and a poor spray pattern. A shorted or open coil will prevent the injector from working. Injectors can be checked by listening to them and by checking them with an ohmmeter.

A balance test measures the output of each injector of a multi-point fuel injection system. This involves using a pressure gauge on the fuel rail to check the drop as

each injector is opened. Fuel pressure drop should be about equal or within 1.5 psi of each other.

Be sure to bleed off fuel pressure before replacing an injector. At the same time, inspect the boot and any other rubber parts.

To inspect a cold start injector, remove it from the engine. Aim it into a container and crank the cold engine. It should spray a good pattern. If operation is faulty, check the thermo-time switch and circuit.

Special graduated containers are used for testing continuous injectors, and for testing fuel volume of individual injectors. Problem injectors should be switched to other positions to see if the problem is in the injector or elsewhere in the system.

Checking the operation of single-point injectors is simple. Remove the air cleaner or air cleaner cover and observe the fuel spray pattern in the throttle body.

Idle air valves and idle speed motors should be inspected when engine idle problems arise. Speeds may be erratic, too high, or too low.

Most EFI engine sensors are tested with a special analyzer. However, some require use of a digital meter. Check sensor resistances and compare with manufacturer's specs.

Testers are available for checking a malfunctioning ECU. Or, as an alternative, you can check its condition by plugging in a test computer. Many ECUs have a PROM (programmable read only memory) which calibrates the computer for the automobile in which it is being used. Normally, it will be removed from an old ECU and reinstalled in the new one.

Tune-up procedures may require service or adjustment of some parts of the EFI system. These include curb idle speed, fast idle, throttle plate stop, idle air-fuel mixture, and throttle cable. Refer to a manual for procedures.

KNOW THESE TERMS

EFI self-diagnosis, EFI trouble code, EFI tester, Leaking injector, Improper injector spray pattern, Open injector coil, Shorted injector coil, TBI rebuild, Injector output volume, PROM.

REVIEW QUESTIONS—CHAPTER 17

1. Describe the four common ways a self-diagnostic EFI mode might indicate the location of a malfunction.
2. To assure that the system has gone into the _____ _____ mode, the engine being tested should be warmed to operating temperature.
3. Why should a digital meter be used when testing circuits and components?
4. An _____ can sometimes be used to test the electrical waveforms (voltage values) at the EFI injectors.
5. Indicate which of the following statements are correct:
   a. Always turn the ignition switch to "on" before disconnecting an EFI terminal.
   b. It may be necessary to disconnect the negative battery terminal before disconnecting any EFI harness terminal.

c. Disconnecting an EFI harness terminal with the ignition "on" could damage the computer.

6. Where would you find instructions and specifications for checking out an EFI system with an EFI analyzer?

7. Look at the specification chart in Fig. 17-14. If you were testing the EFI system of a port fuel injected engine (ignition "on") and got a 12.5 volt reading when testing the voltage at pin B1, what would you conclude about the condition of that terminal?

8. Studying EFI wiring diagrams will help you find locations of _____, _____, and _____ that may be faulty.

9. When an air leak occurs between the airflow sensor and the intake manifold, the air-fuel mixture can become too rich. True or false?

10. What symptoms might a faulty fuel pressure regulator create?

11. Describe how to check a fuel pressure regulator.

12. Which of the following conditions could cause low fuel pressure at the fuel pressure regulator:
   a. Fuel return line restricted.
   b. Bad pressure regulator.
   c. Weak fuel pump.
   d. Clogged fuel filter.
   e. Injector(s) not opening.

13. A fuel pressure regulator pressure test shows high fuel pressure at the regulator.
   Mechanic A says that the fuel pressure regulator is bad and should be replaced.
   Mechanic B says that it could be either a bad pressure regulator or a clogged/smashed fuel return line.
   Who is right?
   a. Mechanic A only.

   b. Mechanic B only.
   c. Both Mechanic A and B.
   d. Neither Mechanic A nor B.

14. Indicate which of the following symptoms could be caused by a faulty fuel injector:
   a. Engine miss.
   b. Poor fuel economy.
   c. Hard starting.
   d. Rough idle.
   e. None of the above.

15. Explain what check you might make to determine if injectors are opening and closing.

16. What is an injector balance test and how is it done?

17. If a cold-start injector fails to operate you should (check all correct answers):
   a. Replace the injector. It is defective.
   b. Test the thermo-time switch before condemning the injector.
   c. Test the electronic circuit before condemning the injector.
   d. Test the injector before condemning it.

18. Continuous injector spray patterns are tested by placing the injectors in a _____ _____ while activating the _____ _____ and moving the _____ _____ _____ as described in a service manual.

19. Explain one method of performing a fuel volume test on continuous injectors.

20. In checking out a throttle body injector, what would you do first?

21. List the first two steps in removing TBI injectors for rebuilding or service.

22. Explain how you would check whether an inoperative idle speed motor is defective or not.

23. Give the basic procedure for typical PROM installation in an ECU.

18

DIESEL INJECTION SYSTEM OPERATION

After studying this chapter, you will be able to:
- Explain the fundamentals of a diesel injection system.
- List the basic parts of a diesel injection system.
- Compare gasoline and diesel fuel system differences.
- Describe the construction of injector nozzles.
- Explain the construction and operation of glow plugs.
- Describe the operation of a glow plug control circuit.
- Explain the function of a water detector, fuel heater, and block heater.
- Compare the construction and operation of typical diesel injection pumps.
- Describe the operation of a fuel shutoff solenoid, cold start solenoid, and other injection pump mounted devices.
- Explain the operation of electronically controlled diesel injection systems.

A *diesel injection system* is a high pressure fuel system that delivers fuel oil directly into the engine combustion chambers, Fig. 18-1. It is unlike a lower pressure gasoline system that meters fuel into the engine intake manifold. Diesel fuel injection is relatively simple. It uses a mechanical pump as the main control of the engine's air-fuel mixture. On modern systems, electronic devices can be used to help control injection pump operation and fuel metering.

Note! Several earlier chapters discuss information essential to this chapter. If needed, use the index to locate and review coverage of diesel fuel, fuel pumps, fuel filters, fuel lines, and diesel combustion.

DIESEL INJECTION SYSTEM

Fig. 18-2 illustrates the essential parts of a diesel injection system. Refer to it as each component is explained.

1. SUPPLY PUMP—Feeds fuel from tank to injection pump.

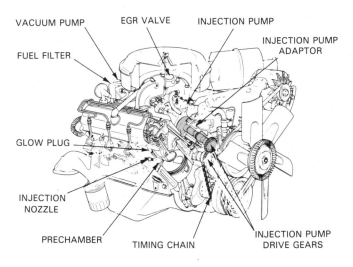
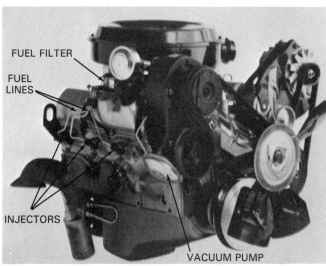

Fig. 18-1. Diesel high pressure fuel injection system. A—Cutaway of system. Note names of parts. B—Modern smaller displacement automotive diesel engine is a V-6. (Oldsmobile)

387

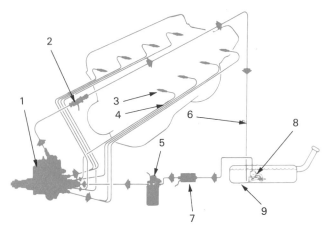

1. INJECTION PUMP
2. HOUSING PRESSURE ALTITUDE ADVANCE SOLENOID
3. NOZZLE
4. INJECTION LINE
5. FUEL FILTER
6. FUEL RETURN
7. FUEL PUMP
8. FILTER
9. TANK

Fig. 18-2. Schematic of diesel system. Note that there are two fuel pumps. One delivers fuel at low pressure to injection pump. Second pump pressurizes fuel before delivery to injectors.

2. INJECTION PUMP—High pressure, mechanical pump that meters the correct amount of fuel and delivers it to each injector nozzle at the right time.
3. INJECTION LINES—High strength, steel tubing that carries fuel to each injector nozzle.
4. INJECTOR NOZZLES—Spring-loaded valves that

spray fuel into each combustion chamber.
5. GLOW PLUGS—Electric heating elements that warm air in precombustion chambers to aid cold engine starting.

The fuel supply system, using either a mechanical or an electric pump, feeds fuel to the injection pump. The engine-driven injection pump then controls when and how much fuel is sent to the injector nozzles. High pressure injection lines carry fuel to the injectors. The spring-loaded nozzles are normally closed. However, when the injection pump produces enough pressure, each nozzle opens and squirts a fuel charge into the engine to start combustion. A return line carries excess fuel back to the fuel tank, Fig. 18-3.

DIESEL AND GASOLINE ENGINE DIFFERENCES

Before discussing the parts of a diesel injection system, you should review the major differences between gasoline and diesel engines. This will help you understand diesel injection. Refer to Fig. 18-4 for a review of the diesel cycle.

Compression ignition engine

A diesel engine is a *compression ignition engine* because it uses the heat from highly compressed air to ignite and burn the fuel.

Diesel engines use a very high compression ratio (approximately 17:1 to 23:1.) A gasoline engine's compression ratio is only 8:1 or 9:1. Gasoline engines are classified as *spark ignition engines* because they use an electric arc (spark plug) to ignite the fuel.

No control of airflow

Both carbureted and gasoline injected engines use a

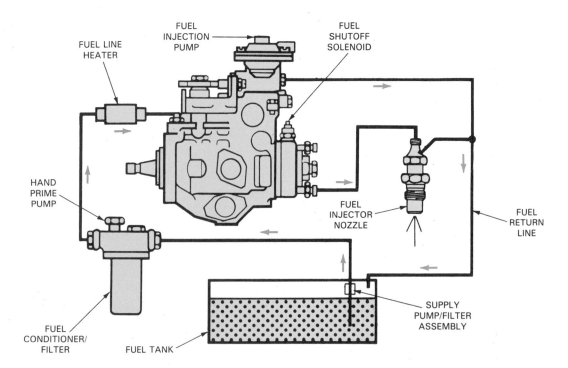

Fig. 18-3. Schematic of fuel injection system. In-tank pump supplies fuel to injection pump. Injection pump meters high-pressure fuel to each injector as needed.

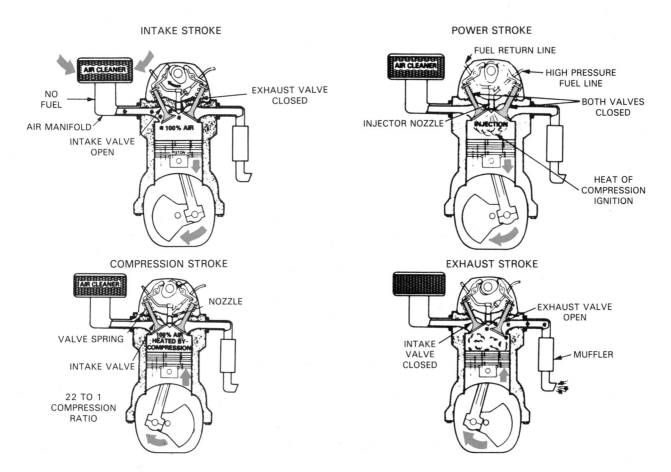

Fig. 18-4. Basic operation of diesel engine. Air intake stroke. Intake valve opens as piston travels downward. Clean, fresh air is drawn into combustion chamber. Compression stroke. Valve closes on piston's upward stroke and air is compressed which causes it to heat up. Power stroke. Injector sprays fuel charge into combustion chamber. It mixes with heated air and ignites. Piston moves downward with tremendous force. Exhaust stroke. Exhaust valve opens. Piston, moving upward, drives out spent fuel to complete the cycle.

throttle valve to control airflow and engine power. A diesel engine has NO throttle valve.

Compresses only air

A diesel engine compresses only air on its compression stroke. A gasoline engine compresses an air and fuel mixture.

Direct fuel injection

A diesel engine forces fuel directly into the combustion chambers. A gasoline engine meters fuel into the intake manifold.

Fuel controls engine speed and power

A diesel controls engine speed and power safely by controlling the amount of fuel injected into the engine. More fuel produces more power. A gasoline engine controls engine power by regulating both air and fuel flow.

DIESEL INJECTORS

Diesel injector nozzles, like injectors on CIS gasoline engines, are basically spring-loaded valves. They are located where they can spray fuel directly into the engine precombustion chambers. See Fig. 18-5.

1. INJECTION NOZZLE
2. THREADED RING
3. PRECHAMBER
4. SEAL RING
5. NOZZLE SEAL
6. PIN-TYPE GLOW PLUG
7. CYLINDER HEAD GASKET

Fig. 18-5. Cutaway of diesel injection nozzle. Fuel is sprayed directly into prechamber where it is ignited by heated or compressed air.

389

The injectors, one for each cylinder, are threaded into the cylinder head. Like a spark plug tip in a gasoline engine, the inner tip of the nozzle is exposed to the heat of combustion.

Diesel injector parts

Parts of a diesel injector are pictured in Figs. 18-6 and 18-7. Refer to these illustrations as the parts are described:

1. A diesel injector *heat shield* or *nozzle holder* helps protect the injector from engine combustion heat. It threads into the cylinder head and helps make a good seal between the injector and the cylinder head.

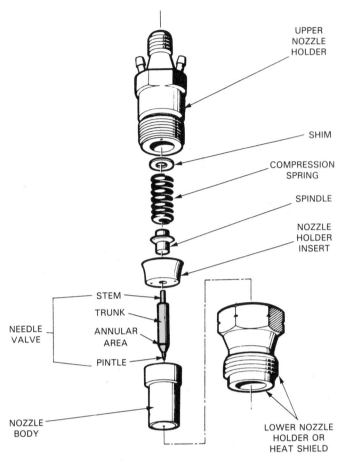

Fig. 18-7. Exploded view of injector. Note parts of needle valve.

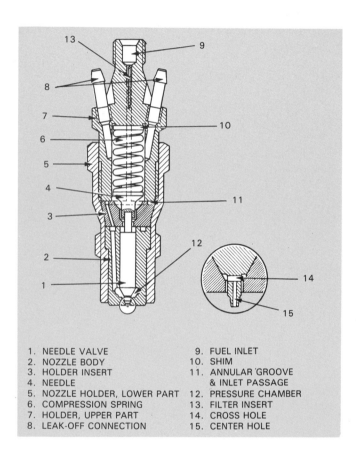

1. NEEDLE VALVE	9. FUEL INLET
2. NOZZLE BODY	10. SHIM
3. HOLDER INSERT	11. ANNULAR GROOVE
4. NEEDLE	& INLET PASSAGE
5. NOZZLE HOLDER, LOWER PART	12. PRESSURE CHAMBER
6. COMPRESSION SPRING	13. FILTER INSERT
7. HOLDER, UPPER PART	14. CROSS HOLE
8. LEAK-OFF CONNECTION	15. CENTER HOLE

Fig. 18-6. Study cutaway of diesel injector. Note names and relationship of parts.

2. The *injector body,* which threads into the heat shield, is the main section of the injector that holds the other parts, Fig. 18-7. Fuel passages are provided in the injector body. A needle seat is formed by the lower opening in the injector body.

3. The *needle valve* opens and closes the nozzle (fuel opening), Fig. 18-8. It is a precisely machined rod with a specially shaped tip. When the valve is closed, the tip of the needle seals against the injector body.

4. The *injector spring* holds the injector needle in a normally closed position. It fits around the needle

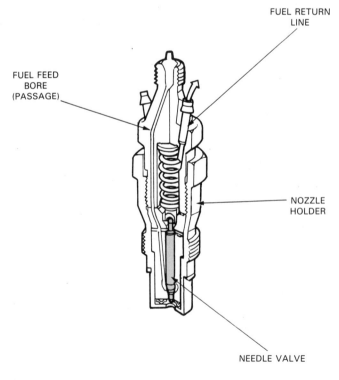

Fig. 18-8. Needle valve (in color) controls fuel entry into engine's combustion chamber.

and against the injector body. Spring tension helps control injector opening pressure.

5. An *injector pressure chamber* is a cavity in the injector body formed around the tip of the injector needle. Injection pump pressure forces fuel into this chamber. Its upward pressure pushes the needle valve open. Fig. 18-9 shows two types of nozzle tips. Compare the two pressure chambers.

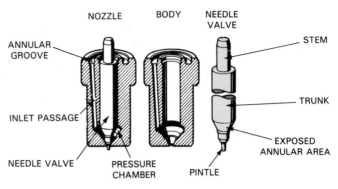

Fig. 18-10. Pintle type nozzle. Pintle keeps spray hole free of carbon and influences spray pattern.

Diesel injector nozzle operation

When the injection pump produces high pressure, fuel flows through the injection line and into the inlet of the injector nozzle. The fuel continues down through the fuel passage in the injector body and into the pressure chamber.

The high pressure of the fuel in the pressure chamber forces the needle upward, compressing the injector spring, Fig. 18-11. This allows diesel fuel to spray out in a CONE-SHAPED pattern, Fig. 18-12. Excess fuel leaks past the injector needle and is sent back to the fuel tank through the return lines.

Diesel injector nozzle types

Several types of injector nozzles are found in diesel engines, Fig. 18-12. The most common of these are the:
1. Inward opening injector nozzle.
2. Outward opening injector nozzle.
3. Pintle injector nozzle.
4. Hole injector nozzle.

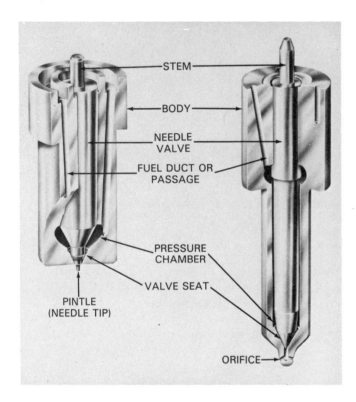

Fig. 18-9. Cutaway of inward opening injector nozzles. Left. Pintle type injector in which needle projects through spray hole is common on automotive diesel injection system. Right. Hole type nozzle is not as popular. (American Bosch)

Nozzle body and seat construction

The nozzle body and seat are very important to diesel engine operation. Note in Fig. 18-10 how the needle valve fits in its body.

An *annual groove* in the body allows fuel pressure to flow into the pressure chamber at the base of the needle valve.

The *pressure chamber* is a small cavity formed around the annular area on the needle valve and the inside of the nozzle body.

The *needle valve,* Fig. 18-10, typically has four parts:
1. *Stem* (upper end of needle valve that is machined smaller in diameter).
2. *Trunk* (main section that supports needle valve as it slides up and down in nozzle body).
3. *Annular area* (tapered surface on needle valve exposed to high fuel pressure for nozzle opening).
4. *Pintle* (small tip on bottom end of needle valve to help form cone-shaped spray pattern).

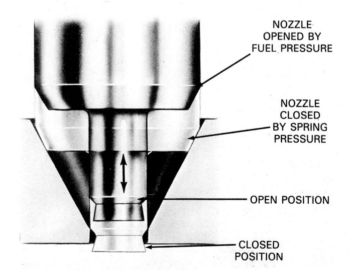

Fig. 18-11. Opening and closing of nozzle valve is controlled by fuel pressure. (American Bosch)

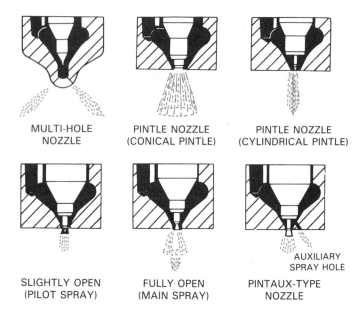

MULTI-HOLE
NOZZLE

PINTLE NOZZLE
(CONICAL PINTLE)

PINTLE NOZZLE
(CYLINDRICAL PINTLE)

SLIGHTLY OPEN
(PILOT SPRAY)

FULLY OPEN
(MAIN SPRAY)

AUXILIARY
SPRAY HOLE

PINTAUX-TYPE
NOZZLE

Fig. 18-12. Different types of injector nozzles produce vastly different spray patterns. (Chrysler)

Most automotive diesels use inward opening pintle injector nozzles. Figs. 18-13 and 18-14 show two more injector variations.

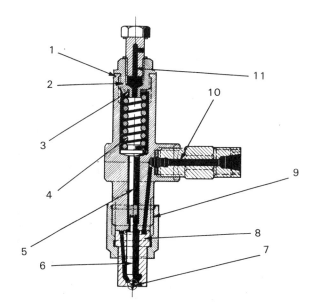

1. INJECTOR HOLDER
2. CAP NUT
3. ADJUSTING SHIMS (CALIBRATION)
4. SPRING
5. PUSH ROD
6. INJECTOR NEEDLE VALVE
7. NIPPLE
8. INJECTOR NOZZLE
9. NOZZLE RETAINING NUT
10. FEED LINE
11. RETURN LINE

Fig. 18-14. Another variation of pintle type nozzle. This style can be calibrated using adjusting shims between cap nut and spring. (Peugeot)

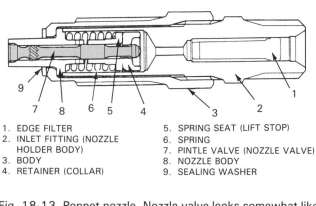

1. EDGE FILTER
2. INLET FITTING (NOZZLE HOLDER BODY)
3. BODY
4. RETAINER (COLLAR)
5. SPRING SEAT (LIFT STOP)
6. SPRING
7. PINTLE VALVE (NOZZLE VALVE)
8. NOZZLE BODY
9. SEALING WASHER

Fig. 18-13. Poppet nozzle. Nozzle valve looks somewhat like poppet valves in engine.

DIESEL PRECHAMBER CUP

A *diesel prechamber cup* is a steel insert pressed into the face of the cylinder head. One is provided for each engine cylinder on most automotive engines.

The cup encloses the tips of the injector nozzle and the glow plug. Fig. 18-15 shows the parts of a prechamber cup. Fig. 18-16 shows a prechamber cup installed in its cylinder head.

GLOW PLUGS

Glow plugs are small heating elements that warm the air in the precombustion chambers to help start a cold

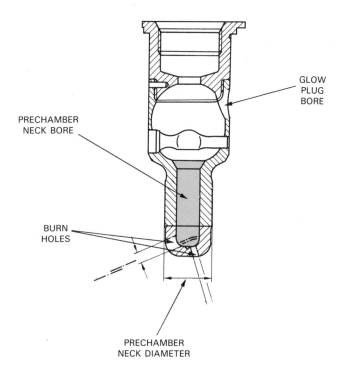

GLOW
PLUG
BORE

PRECHAMBER
NECK BORE

BURN
HOLES

PRECHAMBER
NECK DIAMETER

Fig. 18-15. Prechamber cup protects glow plug and injector. Glow plug heats air in prechamber to start combustion in cold engine.

392

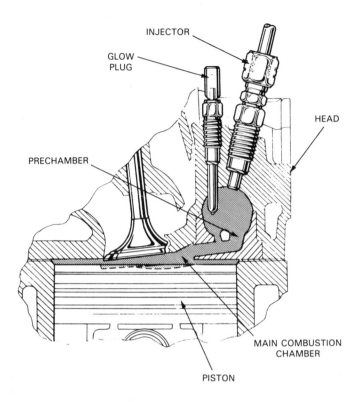

Fig. 18-16. Venturi-shaped prechamber moves flame into main combustion chamber at greater velocity and with swirling action. Burn is more complete. (Oldsmobile)

diesel engine. Refer to Fig. 18-16 again.

The glow plugs are threaded into holes in the cylinder head. The inner tip of the plug extends into the precombustion chamber.

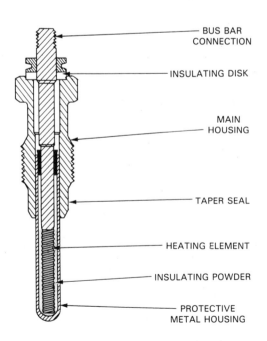

Fig. 18-17. Glow plug in cutaway shows basic construction. Resistance (electric) heating element gets red hot when current is supplied to plug terminal.

Glow plug construction

Fig. 18-17 is a cutaway view of a typical glow plug. Note how the electric heating element is enclosed inside a metal housing. An electrical connection allows current to enter the heating element.

Glow plug wiring harness

The glow plug *wiring harness* feeds high current to each of the glow plugs when the engine is cold. See Fig. 18-18. The wiring harness on modern cars connects to the glow plug control circuit. Look at Fig. 18-19.

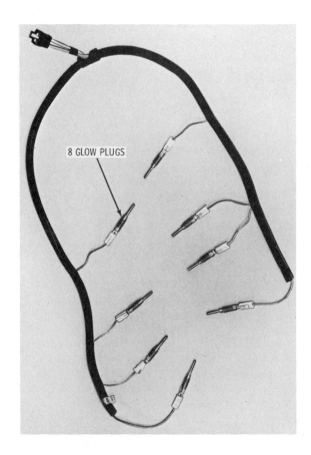

Fig. 18-18. Wiring harness conducts electrical (battery) current to each glow plug. (Oldsmobile)

A *glow plug control circuit* automatically turns the glow plugs OFF after a few seconds of operation. Fig. 18-20 shows a typical glow plug circuit.

A sensor checks the temperature of the engine coolant. It feeds this electrical data to a control unit. Thus, if the engine is already warm, the control unit will not turn on the glow plugs.

Indicator lights, also operated by the control unit, inform the driver when the engine is ready to start. Glow plugs normally need but a few seconds to heat up precombustion chamber air.

Glow plug operation

When the engine is cold and the ignition switch is turned to RUN, a large current flows from the battery

393

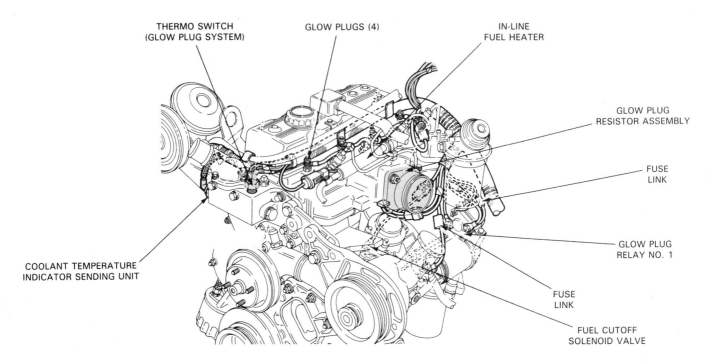

Fig. 18-19. Glow plug circuit feeds out of engine's electrical wiring harness. A thermo-switch reacts to temperature and activates glow plugs when engine is cold.

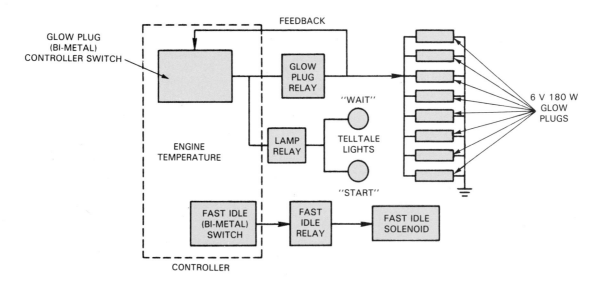

Fig. 18-20. Typical schematic of glow plug circuit. Through a feedback system, controller (at left) will automatically break electrical circuit to plugs after few seconds so that voltage is sent to plugs in short pulses.

to the glow plugs. In a few seconds, the glow plug tips will heat to a dull red glow.

When the glow plug indicator light goes out, the driver can start the engine. The compression stroke pressure and heat, along with the heat from the glow plugs, causes the fuel to ignite easily.

WATER DETECTOR

A *water detector* may be used to warn the driver of moisture in the diesel fuel. Such contamination is very harmful to a diesel fuel system. The water mixes with the fuel and may cause corrosion of precision parts in the injection pump and injectors. Fig. 18-21 shows a circuit using an in-tank water detector.

FUEL HEATER

A *fuel heater* is sometimes used to warm the diesel fuel, preventing the fuel from JELLING (turning into a semisolid). This optional device, is needed in very cold climates. The heater is simply an electric heating element placed in the fuel line ahead of the injection pump. See Fig. 18-21 again.

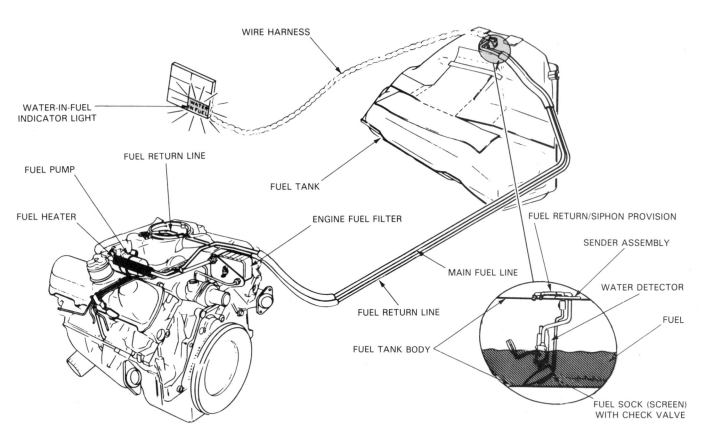

Fig. 18-21. Water-in-fuel circuit. Sensing unit and separating system are in tank. Separating unit will turn on dash signal light if large amounts of water are detected in fuel tank.

BLOCK HEATER

A *block heater* may be used to warm the engine block in cold weather. It is a heating device that plugs into a 120 V electrical outlet. It keeps the engine warm overnight to make the diesel engine easier to start in cold weather.

INJECTION LINES

Diesel *injection lines,* designed for high pressure, are made of double-wall steel. They connect the injection pump with the injector nozzles. Usually, injection lines have flare or special fittings on each end. They are all the same length regardless of distance from the injector the pump. This is necessary so that injection timing and duration remain the same for every injector nozzle, as shown in Fig. 18-22.

Fuel return or *leak-off lines* are lower pressure hoses or tubing that let the diesel fuel circulate out of the nozzles and back to the tank. This cools and lubricates the moving parts (especially the needle in the injector nozzle), to prevent wear. See Fig. 18-22 again.

DIESEL INJECTION PUMPS

A *diesel injection pump* has several important functions. It:

1. Carefully meters the correct amount of fuel to each injector.

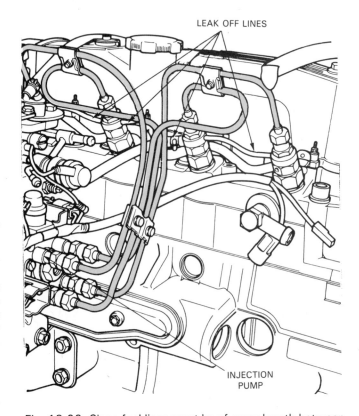

Fig. 18-22. Since fuel lines must be of same length between injection pump and injectors, they are bent in different configurations and held by various brackets to reduce variation.

2. Circulates fuel through fuel lines and nozzles.
3. Produces extremely high fuel pressure.
4. Times fuel injection to meet speed and load of engine.
5. Provides a means for the driver to control engine power output.
6. Controls engine idle speed and maximum engine speed.
7. Helps close injector nozzles after injection.
8. Provides a means of shutting OFF engine.
9. Sometimes has sensors (on more modern types) to supply injection system data to the on-board computer.
10. May also use computer controlled activators to allow the computer to alter injection timing and quantity.

Injection pump action

Fig. 18-23 illustrates diesel injection pump action. Basically, the injection pump is mechanical. An eccentric or cam lobe acts on a pumping plunger. The plunger pressurizes the fuel and pushes it out of the injector nozzle.

The amount of fuel injected into the engine is controlled by changing the size of the eccentric or cam lobe. When the lobe is larger, more fuel is injected into the combustion chamber. To reduce engine speed, the injection pump reduces the size of the lobe and the pumping plunger does not slide out as far. Less fuel is pumped.

2. Chain drive, Fig. 18-24B.
3. Belt drive, Fig. 18-24C.

Injection pump locations

Normally the diesel injection pump is mounted on the engine. Then engine power can be used to power the pump. Generally, an injection pump can be located on the:
1. Top, front of the engine, Fig. 18-24A.
2. Side, front of the engine, Fig. 18-24B and 18-24C.
3. Side, rear of the engine.

DIESEL INJECTION PUMP TYPES/OPERATION

There are two broad categories of automotive diesel injection pumps: in-line and distributor. See Fig. 18-25. Within these two classifications, there are slight variations.

IN-LINE DIESEL INJECTION PUMPS

An *in-line diesel injection pump* has one pumping plunger (piston) for each engine cylinder. The pumping plungers are aligned, like pistons of an in-line engine.

The major parts of an in-line injection pump are shown in Figs. 18-26 and 18-27. Refer to these illustrations as the parts are described:
1. The in-line injection pump *camshaft* operates the pumping plungers. It has lobes like an engine cam-

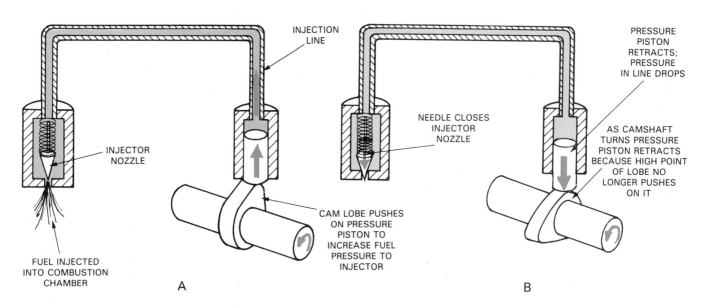

Fig. 18-23. Cam action in injector pump times opening of injector. A—When cam is at highest point and pressing hardest on pressure piston, fuel pressure in lines is highest and injector needle will retract, releasing fuel into combustion chamber. B—As cam moves off high point pressure will drop and injector needle will close stopping fuel spray.

Injection pump drives

An *injection pump drive* transfers engine rotating power to the injection pump shaft. It usually turns the pump at camshaft speed (one-half engine rpm). The three types of injection pump drives are the:
1. Gear drive, Fig. 18-24A.

shaft. When the engine turns the pump camshaft, the lobes push on the roller tappets to move them up and down.
2. The injection pump *roller tappets* transfer camshaft action to the pumping plungers. Like roller lifters in an engine, the rollers reduce friction and wear on

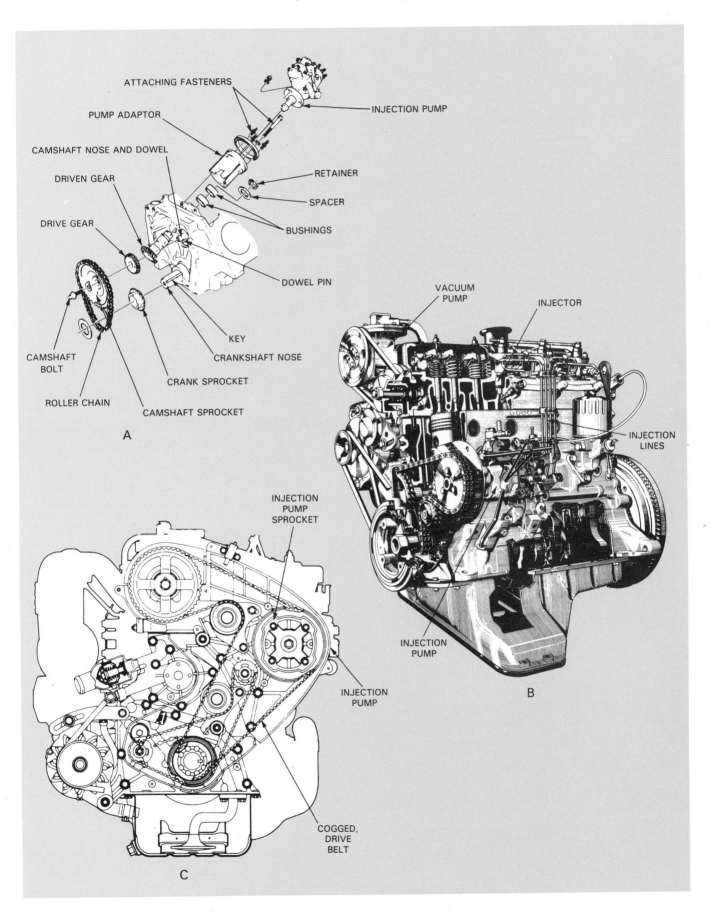

ATTACHING FASTENERS

PUMP ADAPTOR

CAMSHAFT NOSE AND DOWEL

DRIVEN GEAR

DRIVE GEAR

CAMSHAFT BOLT

ROLLER CHAIN

CAMSHAFT SPROCKET

CRANK SPROCKET

CRANKSHAFT NOSE

KEY

DOWEL PIN

BUSHINGS

SPACER

RETAINER

INJECTION PUMP

A

VACUUM PUMP

INJECTOR

INJECTION LINES

INJECTION PUMP

B

INJECTION PUMP SPROCKET

INJECTION PUMP

COGGED, DRIVE BELT

C

Fig. 18-24. Injector pumps can be driven by different methods and are located at different points on engine. A—Exploded view of drive system located on top of engine and driven by gears. B—Cutaway of chain driven system mounted on side of engine. C—Belt driven system which is also side mounted.

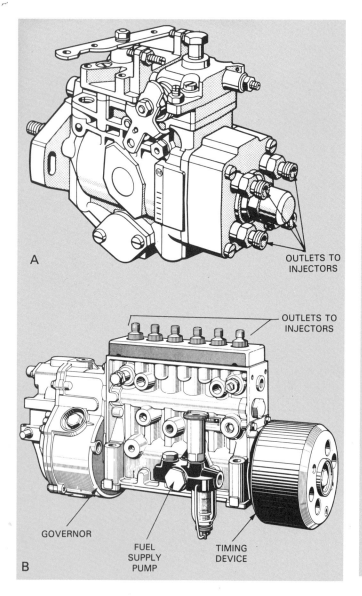

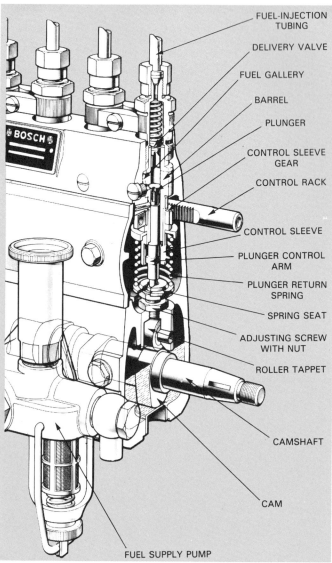

Fig. 18-25. There are two basic types of diesel injection pumps. A—Distributor type. B—In-line type. (American Bosch)

Fig. 18-26. In-line diesel injection pump has one end in cutaway to show parts of pump which feeds fuel to one injector. (Robert Bosch)

the cam lobes, Fig. 18-26.

3. In-line pump *plungers* are small pistons that push on and pressurize the diesel fuel. When the cam lobe acts on the roller tappet, both the tappet and the plunger are pushed upward.

4. The *barrels* are small cylinders that hold the pumping plungers. When the plunger slides upward in its barrel, extremely high fuel pressure is developed.

5. The *plunger return springs* keep a downward pressure on the pumping plungers and roller tappets. This action holds the tappets against the camshaft when the lobes rotate away from the rollers.

6. *Control sleeve* turns on the pumping plungers to alter fuel volume to each injector nozzle. See Fig. 18-26 and locate the control sleeves.

7. A *rack* is a toothed shaft that acts as a throttle to control diesel engine speed and power. It rotates the control sleeves to increase or reduce injection pump output and engine power.

In-line injection pump fuel metering

To control the amount of fuel injected into the engine, the control rack moves across the control sleeves. The teeth on the rack engage those on the sleeves and rotate them.

A *helix-shaped* (spiraled) groove and a slot are cut into the side of the plunger, Fig. 18-28. When the helix is aligned with the lower port (hole) in the side of the barrel, the plunger CANNOT develop pressure. Fuel will flow down the slot, through the helix groove, and out the port.

The *effective plunger stroke* is the amount of plunger travel that pressurizes fuel. Refer again to Fig. 18-28. It controls the amount of fuel delivered to the injectors.

When the plunger moves up and the helix is NOT aligned with the barrel port, fuel is trapped and pressurized. In this way, rotation of the sleeve can be used to regulate how much fuel is injected into the engine's combustion chambers.

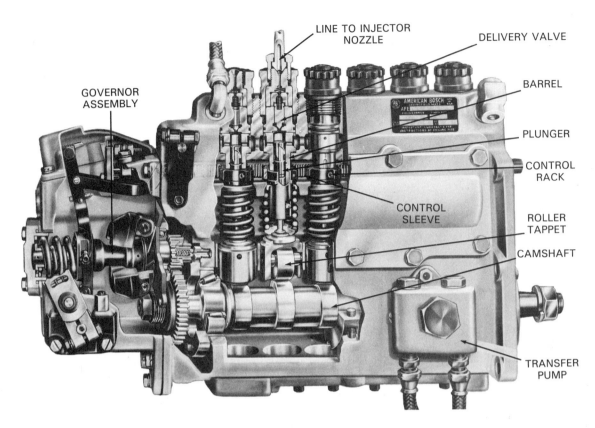

Fig. 18-27. Another cutaway view of in-line diesel injection pump. Study part names and note, especially, location of control sleeves and control rack. (American Bosch)

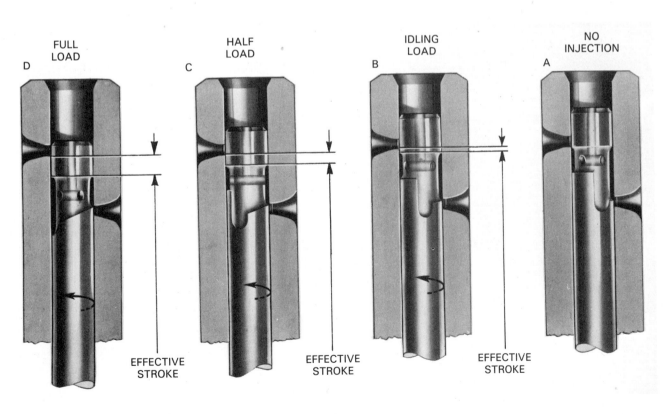

Fig. 18-28. Cutaway showing effect of rotation on effective stroke. As plunger is turned by control rack, amount of travel which creates pressure increases.

Fig. 18-29 shows how the control rack turns the sleeve to change effective plunger stroke.

Delivery valves

Delivery valves are small check valves located at the high pressure outlets of the diesel injection pump. They assure quick, leak-free closing of the injector nozzles. During injection, the delivery valve is pushed open. After injection, the valve snaps shut. This is shown in Fig. 18-30.

In-line injection pump governor

A *governor* on an in-line injection pump controls engine idle speed and limits maximum engine speed. A diesel engine can be DAMAGED if allowed to run too fast. Look at Fig. 18-31.

Notice how the governor uses centrifugal (spinning) weights, springs, and levers, Figs. 18-31 and 18-32. The levers are connected to the control rack or rod. If engine speed increases too much, the governor weights are thrown outward. This moves the levers and control

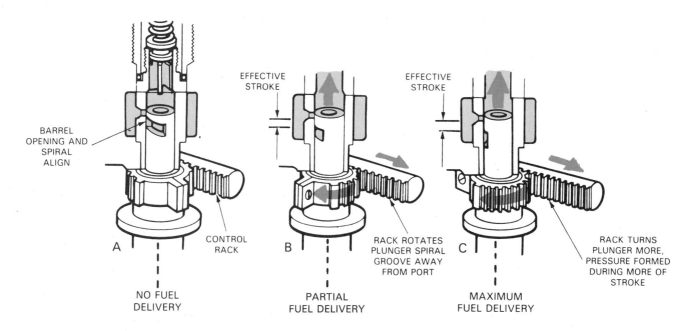

Fig. 18-29. Control rack is connected to gas pedal and governor. A—Control rack has turned plunger so groove is always open to port. This prevents pressure buildup. B—Control rack has moved turning plunger spiral groove away from port. Fuel is delivered. C—More movement of rack increases plunger's effective stroke. (Plunger can move farther before spiral groove aligns with a port.) (Chrysler)

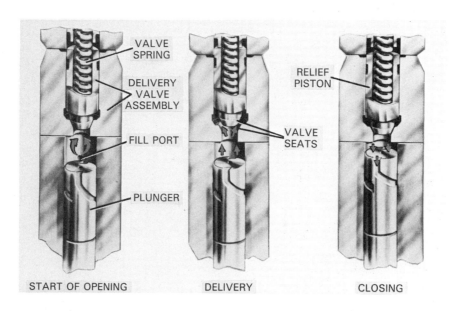

Fig. 18-30. Delivery valves prevent back flow of pressurized fuel after fuel has been delivered to injector nozzle. It opens and closes automatically from plunger action. (American Bosch)

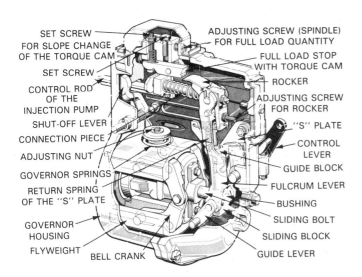

SET SCREW
FOR SLOPE CHANGE
OF THE TORQUE CAM
SET SCREW
CONTROL ROD
OF THE
INJECTION PUMP
SHUT-OFF LEVER
CONNECTION PIECE
ADJUSTING NUT
GOVERNOR SPRINGS
RETURN SPRING
OF THE "S" PLATE
GOVERNOR
HOUSING
FLYWEIGHT
BELL CRANK

ADJUSTING SCREW (SPINDLE)
FOR FULL LOAD QUANTITY
FULL LOAD STOP
WITH TORQUE CAM
ROCKER
ADJUSTING SCREW
FOR ROCKER
"S" PLATE
CONTROL
LEVER
GUIDE BLOCK
FULCRUM LEVER
BUSHING
SLIDING BOLT
SLIDING BLOCK
GUIDE LEVER

Fig. 18-31. Cutaway of mechanical governor for diesel engine. It is linked to control rod (rack) of injection pump. Note adjusting points. (Robert Bosch)

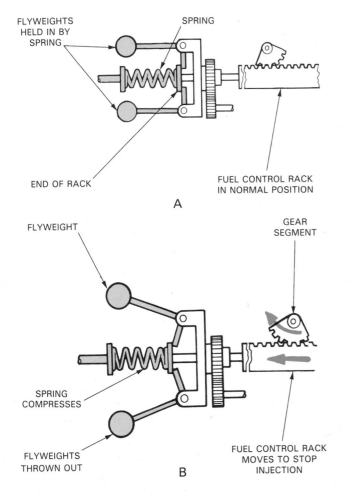

FLYWEIGHTS
HELD IN BY
SPRING
SPRING
END OF RACK
FUEL CONTROL RACK
IN NORMAL POSITION

A

FLYWEIGHT
GEAR
SEGMENT
SPRING
COMPRESSES
FLYWEIGHTS
THROWN OUT
FUEL CONTROL RACK
MOVES TO STOP
INJECTION

B

Fig. 18-32. How diesel injection pump governor operates. A—Before maximum engine speed is reached, spring tension holds flyweights. Control rack is held in normal position for that throttle lever position. B—At maximum speed, flyweights are spinning so fast that centrifugal force compresses spring. Then lever action moves control rack into "no injection" position. Power is shut off to engine until rpm drops. (Chrysler)

rack to reduce the effective stroke of the plungers. Engine speed and power output are limited.

When the driver depresses the gas pedal for more power, the pedal moves a control or throttle lever on the side of the injection pump governor. See Fig. 18-33. This causes throttle lever spring pressure to overcome governor spring pressure. The control rack is moved to increase fuel delivery and engine power. Only when engine rpm reaches a preset level does the governor overcome the full throttle lever position.

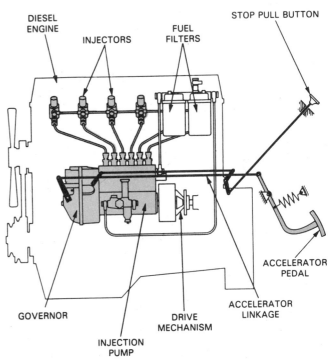

DIESEL
ENGINE
INJECTORS
FUEL
FILTERS
STOP PULL BUTTON
ACCELERATOR
PEDAL
GOVERNOR
INJECTION
PUMP
DRIVE
MECHANISM
ACCELERATOR
LINKAGE

Fig. 18-33. Basic linkage between accelerator pedal and diesel injection pump. Its action overrides governor spring pressure. (Robert Bosch)

Injection timing

Injection timing refers to when fuel is injected into the combustion chambers in relation to the engine piston position. It is similar to spark timing in a gasoline engine.

Injection timing in an in-line injection pump is usually controlled by spring-loaded weights. As engine speed increases, the weights fly outward to advance injection timing. This gives the diesel fuel enough time to ignite and burn properly.

DISTRIBUTOR INJECTION PUMPS

A *distributor injection pump* normally uses only one or two pumping plungers to supply fuel for ALL of the engine's cylinders. It is the most common type used on passenger cars, Fig. 18-34.

In many ways, the operation of a distributor type pump is similar to the action of an in-line injection pump. Both use small pumping plungers to trap and pressurize fuel. Both align and misalign grooves with fuel ports to control

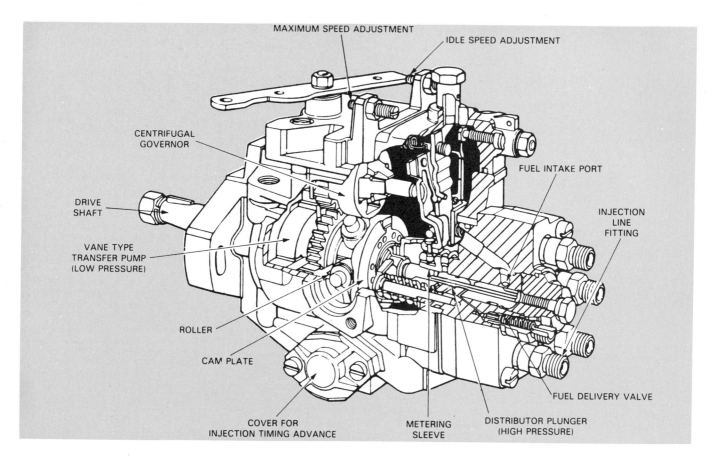

MAXIMUM SPEED ADJUSTMENT

IDLE SPEED ADJUSTMENT

CENTRIFUGAL
GOVERNOR

FUEL INTAKE PORT

DRIVE
SHAFT

INJECTION
LINE
FITTING

VANE TYPE
TRANSFER PUMP
(LOW PRESSURE)

ROLLER

CAM PLATE

FUEL DELIVERY VALVE

COVER FOR
INJECTION TIMING ADVANCE

METERING
SLEEVE

DISTRIBUTOR PLUNGER
(HIGH PRESSURE)

Fig. 18-34. Cutaway of single plunger distributor injection pump. This type is one of most common in automotive use. (Chrysler)

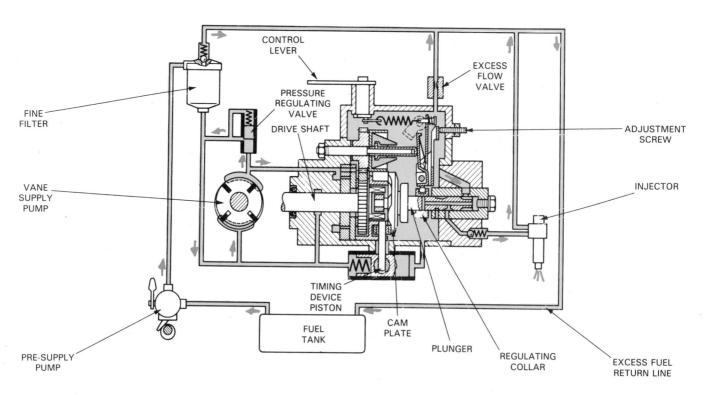

CONTROL
LEVER

EXCESS
FLOW
VALVE

PRESSURE
REGULATING
VALVE

DRIVE SHAFT

ADJUSTMENT
SCREW

FINE
FILTER

VANE
SUPPLY
PUMP

INJECTOR

TIMING
DEVICE
PISTON

FUEL
TANK

CAM
PLATE

PLUNGER

REGULATING
COLLAR

EXCESS FUEL
RETURN LINE

PRE-SUPPLY
PUMP

Fig. 18-35. Schematic of fuel injection system using distributor injection pump.
Arrows trace flow of fuel through lines, pump, injector, and back to tank.

fuel flow to the injector nozzles. Buth use delivery valves, governors, and other similar parts. Refer to Fig. 18-34 again.

There are also important differences that you must understand.

There are design variations of the distributor injection pump, the most general being the single-plunger and two-plunger types. Both will be explained.

Single-plunger distributor injection pump

The major parts of a single-plunger distributor injection pump are shown in Figs. 18-34 and 18-35. Refer to these illustrations as the parts are discussed:

1. The *drive shaft* uses engine power to operate the parts in the injection pump. The outer end of the shaft holds either a gear, chain sprocket, or a belt sprocket. This provides a drive mechanism for the pump, Fig. 18-34.
2. A *transfer pump,* also called supply pump, is a small pump that forces diesel fuel into and through the injection pump, Fig. 18-35. This lubricates the pump and fills the pumping chamber. Most transfer pumps for distributor pumps are a VANE TYPE.
3. A *pumping plunger* for a distributor type injection pump is a small piston that produces high fuel pressure, Fig. 18-36. It is comparable to an in-line plunger.
4. A *cam plate* is a rotating, lobed disc that operates the pumping plunger. Like an in-line pump camshaft, it forces the pumping plunger to move and develop injection pressure.
5. A *fuel metering sleeve* can be slid sideways on the pumping plunger to change the effective plunger stroke (plunger movement that causes fuel pressure). As shown in Fig. 18-37, it surrounds the pumping plunger. The fuel metering sleeve performs

the same function as the sleeves and control rack in an in-line pump. The sleeve controls injection quantity, engine speed, and power output.

6. The *hydraulic head* is the housing around the pumping plunger. It contains passages for filling the plunger barrel with fuel and for allowing fuel to be injected into the delivery valves.
7. A *centrifugal governor* helps control the amount of fuel injected and engine speed. Flyweight action moves the metering sleeve to limit top rpm.

Single-plunger distributor pump operation

As the injection pump shaft rotates, the fill port in the hydraulic head lines up with the port in the plunger. At this point, the transfer pump can force fuel into the high pressure chamber in front of the plunger. Refer to Fig. 18-36 again.

With more shaft and plunger rotation, the fill port moves out of alignment and an injection port lines up. At this instant, the cam plate lobe pushes the plunger sideways. Fuel is forced out the injection port to the correct injector nozzle.

This process is repeated several times during each rotation of the injection pump drive shaft. Fuel injection must be timed to occur at each nozzle as that engine piston nears TDC on its compression stroke.

Single-plunger distributor pump fuel metering

In a single-plunger distributor injection pump, the amount of fuel injected is controlled by movement of the sleeve on the pumping plunger. This is illustrated in Fig. 18-37. The sleeve slides one way to increase fuel delivery (by covering the spill port). The sleeve moves the other way to reduce delivery (by uncovering the spill port).

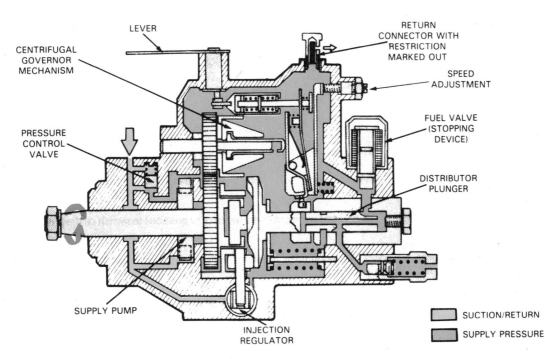

Fig. 18-36. Cutaway shows makeup and parts of single plunger distributor pump. (Volvo)

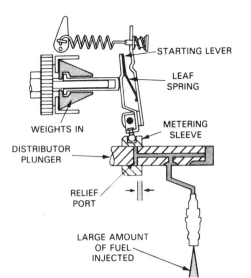

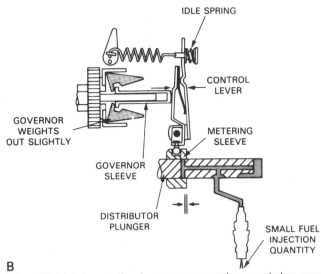

A

Starting — Leaf spring presses starting lever to left so metering sleeve moves to right. Distributor plunger moves further before relief port is exposed. Injection lasts longer.

B

Idle — Weights in centrifugal governor are partly expanded so governor sleeve moves to right. Metering sleeve moves to left. Distributor plunger now moves a short distance before relief port is uncovered.

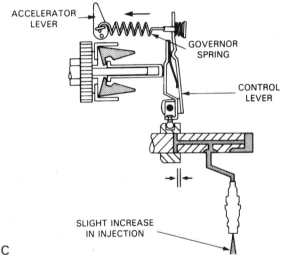

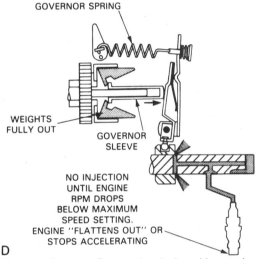

C

Acceleration — Control lever is pulled to left by linkage from accelerator pedal. Metering sleeve is moved to right. Engine speed increases until governor "neutralizes" effect of pedal linkage.

D

Maximum Speed — Governor is spinning with enough centrifugal force for governor sleeve to stretch governor spring. Metering sleeve uncovers relief port at beginning of distributor plunger stroke.

Fig. 18-37. Study each illustration to learn how single plunger distributor injection pump works.

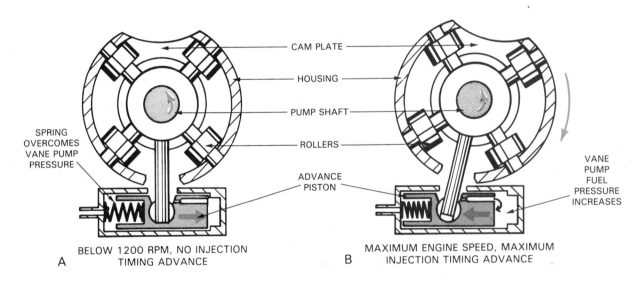

Fig. 18-38. Vane pump controls injection timing. A—Low engine speed, lower pump pressure cannot compress spring. Timing stays retarded. B—Higher speeds, pump develops more pressure. Piston can rotate roller housing. Lobes engage rollers sooner and injection begins sooner. (VW)

Single-plunger distributor pump injection timing

At the end of the engine compression stroke, diesel fuel must be injected directly into the precombustion chamber. Injection must continue past TDC to make sure all of the fuel burns and adequate power is developed.

As engine speed increases, injection must occur sooner to ensure peak combustion pressure right after TDC. Fig. 18-38 shows how one type of injection pump advances injection timing with an increase in engine speed.

Increased engine rpm causes the transfer (vane) pump to spin faster. This increases the pressure output of the transfer pump. The pressure is used to move an injection advance piston. The piston, in turn, causes the cam plate ramps (lobes) to engage the pumping plunger sooner, advancing the injection timing.

Two-plunger distributor pump

A *two-plunger distributor injection pump* is pictured in Fig. 18-39. Some parts (transfer pump, hydraulic head, and delivery valve) are almost the same as those in a single-plunger type pump. Besides these basic parts, a two-plunger injection pump consists of:

1. Two pumping plungers—two small pistons that move in and out to pressurize fuel to each injector nozzle.
2. Distributor rotor—slotted shaft that controls fuel flow to each injector nozzle.
3. Internal cam ring—lobed collar that acts as a cam to force plungers inward for injection of fuel.
4. Fuel metering valve—rotary valve that regulates fuel injection quantity by controlling how far two pumping plungers move apart on fill stroke.

Two-plunger distributor pump operation

When the engine is running, the drive shaft turns the transfer (vane) pump. This pulls fuel into the injection pump. When the charging ports line up, fuel fills the high pressure pumping chamber. Refer to Figs. 18-39 and 18-40.

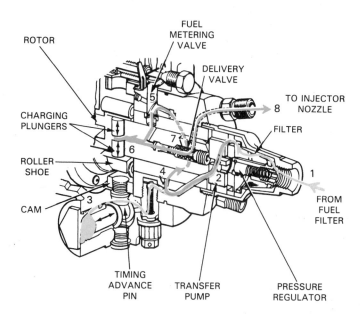

Fig. 18-40. Fuel movement through injection pump. Incoming fuel is filtered again at 1. At 2, transfer pump increases fuel pressure and volume. Then pressurized fuel moves to timing advance chamber (3) and through annulus (4). It continues around rotor to fuel metering valve at 5. This valve regulates flow of fuel into charging chamber at 6. It moves through delivery valve at 7 and to injector nozzle (8).

As the shaft continues to turn, the charging ports move out of alignment and the discharge ports line up. At this instant, the plungers are forced inward by the cam ring. This pressurizes the fuel and pushes it out the hydraulic head to the injector nozzles.

Vane pump operation

The *vane pump,* located in the diesel injection pump, feeds fuel to the high pressure pumping plunger(s) and operates the injection advance mechanism. It is a sliding vane pump driven by injection pump shaft rotation.

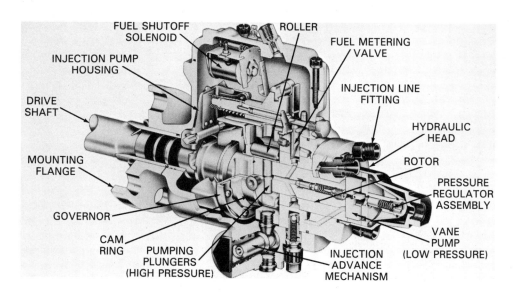

Fig. 18-39. Two-plunger distributor pump is similar to single-plunger type. Do you see major differences as well as similarities? (Oldsmobile)

Injection timing advance

In a distributor type injection pump, injection timing is advanced by the action of a *hydraulic piston.* Refer again to Figs. 18-38 and 18-40. As engine and injection pump speed increase, the fuel pressure output of the vane pump increases. This causes the advance piston to slide further in its cylinder. This rotates the control sleeve or roller housing to make the pumping plunger(s) operate sooner. As a result, injection timing is advanced as engine rpm increases. Injection timing returns to its normal, initial setting when engine rpm drops back to idle.

Fig. 18-41 is a diagram showing the fuel flow and major components of one type of distributor injection pump. Electronic control of diesel injection timing will be discussed later.

Fuel cutoff solenoid

The *fuel cutoff solenoid* is used to stop a diesel engine by preventing fuel flow to the pumping plungers and injectors. Remember, diesels do not have an electric/electronic ignition system to stop the cylinders from firing and engine from running. A fuel cutoff, also called *shut-off solenoid,* is needed.

A simple cutaway drawing of a fuel cutoff solenoid is shown in Fig. 18-42. It mounts on the injection pump and opens and closes a fuel passage. Current is supplied to the solenoid windings whenever the engine is running. This opens the fuel passage. When ignition key is turned off, the field collapses in the solenoid and the plunger moves down to block fuel flow and engine operation.

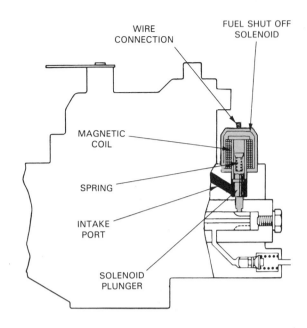

Fig. 18-42. Fuel cutoff solenoid closes fuel passage to stop engine when ignition is off.

Vacuum regulator valve

A *vacuum regulator valve* is sometimes used to control the amount of vacuum pump suction reaching the car's accessory units. As shown in Fig. 18-43, it normally mounts on the injection pump.

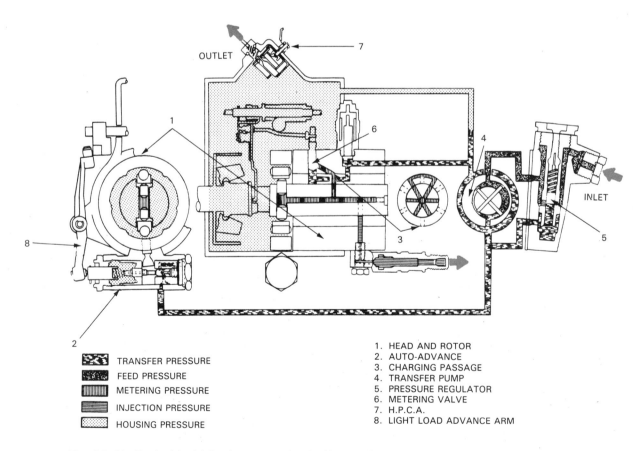

1. HEAD AND ROTOR
2. AUTO-ADVANCE
3. CHARGING PASSAGE
4. TRANSFER PUMP
5. PRESSURE REGULATOR
6. METERING VALVE
7. H.P.C.A.
8. LIGHT LOAD ADVANCE ARM

TRANSFER PRESSURE
FEED PRESSURE
METERING PRESSURE
INJECTION PRESSURE
HOUSING PRESSURE

Fig. 18-41. Typical fuel injection pump circuit. Note various ports and pressures in the circuit. (Buick)

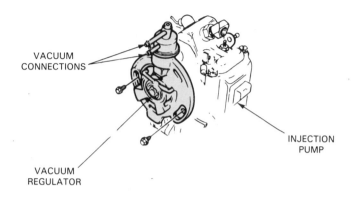

Fig. 18-43. Vacuum regulator valve controls vacuum to accessories. (Buick)

Vent wire

Excess fuel is allowed to bypass through a vent and return to the fuel tank. The vent is nothing more than a screw with a hole machined through it. A fine wire, called a *vent wire,* is fitted through the hole. As fuel flows through the vent, the wire agitates to prevent clogging from debris in the fuel.

COMPUTER CONTROLLED DIESEL INJECTION

Computer controlled diesel injection, also called *electronic diesel injection,* uses engine sensors, injection pump mounted actuators, and an electronic control unit (computer), to increase efficiency. Fig. 18-44 shows one type system.

Basically, the computer helps control injection timing and the amount of fuel delivered to each injector nozzle. This increases fuel economy and reduces emissions (HC and CO). It also improves engine power, acceleration, and cuts down engine smoke (particulates), especially after system parts begin to wear from high mileage.

Diesel engine sensors

A typical computer controlled diesel system has *sensors* that monitor engine speed, manifold pressure, crankshaft position, nozzle needle position, throttle lever position, and EGR operation. Diesel sensors function the same way as sensors in a gasoline injection system operate. They monitor a condition and convert it into an electrical signal that the computer can measure. This lets the computer alter injection pump operation for maximum efficiency.

Fig. 18-45 shows an *engine speed sensor.* Note how it mounts in the vacuum pump. As the engine runs, the sensor produces a pulsing ac signal representing engine rpm.

A *throttle position sensor* for a diesel is illustrated in Fig. 18-46. Note how it bolts to the side of the injection pump. Movement of the throttle lever causes equal movement of the sensor and an equal change in sensor internal resistance. This electrical data is used by the computer to increase fuel injection with increased throttle opening.

A *diesel coolant temperature sensor* is a thermistor

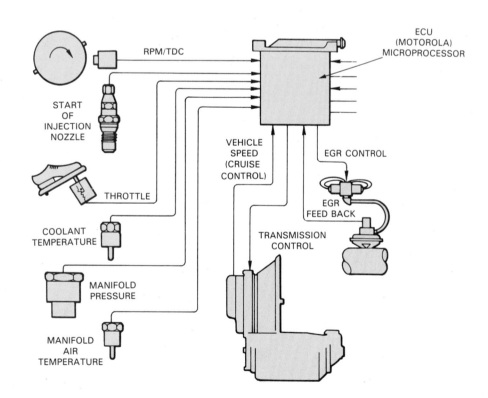

Fig. 18-44. Schematic of electronically controlled diesel injection system. Engine mounted sensors analyze engine speed, top dead center, start of injection, coolant temperature, manifold pressure, and exhaust gas recirculation valve position. (Stanadyne Inc.)

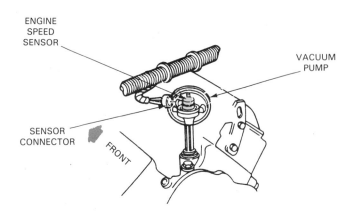

Fig. 18-45. Speed sensor is mounted somewhere in engine. Unit shown attaches to vacuum pump.

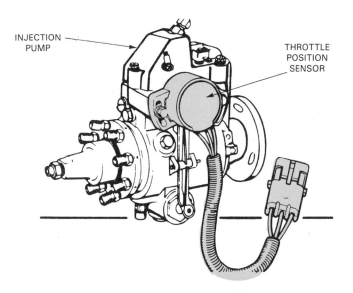

Fig. 18-46. Typical throttle position sensor mounts to injection pump throttle valve where it can signal throttle valve angle to computer.

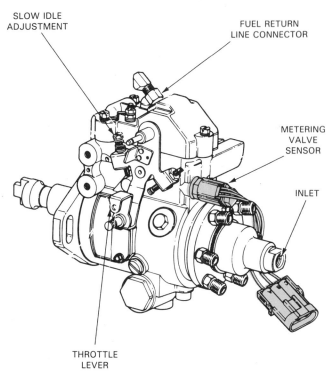

Fig. 18-47. Metering valve sensor also mounts on injection pump. It senses amount of fuel delivered to each injector by monitoring rotation of control sleeves.

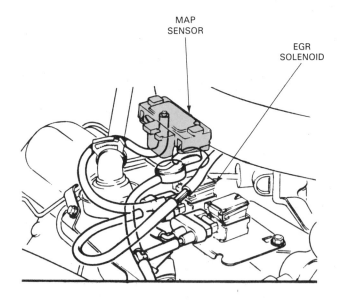

Fig. 18-48. MAP sensor measures vacuum in the exhaust gas recirculation system.

that works the same as a coolent sensor in a gasoline engine. When the sensor is cold, its electrical output tells the computer to change injection timing and injection duration for improved cold engine drivability.

The *metering valve sensor,* mounted on the injection pump, is used to detect injection quantity. It is a variable resistor connected to the fuel metering valve. As shown in Figs. 18-41 and 18-47, it provides fuel feedback data for the computer. It measures rack or control sleeve rotation. It checks the operation of the fuel injection control actuators which will be discussed later.

The *injection nozzle sensor* is usually a hall-effect type sensor that measures needle lift or nozzle opening. The sensor is incorporated into one or more of the nozzles. This provides data on the start of injection and duration of injection. It allows the computer to compensate for wear in the injection pump, engine components, or nozzle itself.

The MAP or *absolute pressure sensor* monitors

vacuum in the EGR (exhaust gas recirculation) vacuum circuit of a diesel. It may also send the computer a barometric pressure or altitude signal. This allows the computer to adjust injection timing and quantity for exact engine needs. See Fig. 18-48.

A *vehicle speed sensor* can also be used in a computer controlled diesel injection system. It reports information on vehicle road speed, allowing the computer to control fuel injection more precisely.

Cold advance solenoid

A *cold advance solenoid* advances diesel injection timing when engine temperature is below a preset temperature. It is used to reduce emissions, white smoke, and diesel knock when starting a diesel engine.

When the temperature sensor signals a cold engine, the computer energizes the cold advance solenoid. Usually, the solenoid plunger extends and pushes the injection pump pressure regulator check ball off its seat, as shown in Fig. 18-49. This reduces injection pump housing pressure and causes the injection timing to advance for better engine performance.

Altitude advance solenoid

The *altitude advance solenoid* is located in the fuel return line from the injection pump and can be used to change injection timing. It is dependent upon the map sensor signal to the computer. When energized by the computer, it alters fuel flow back to the tank and the fuel pressure in the injection pump to alter advance action. Refer again to Fig. 18-49.

Fast idle solenoid

The *fast idle solenoid* is also computer controlled and is normally deenergized or retracted, Fig. 18-50. When the engine is cold, the computer sends current to the solenoid. The solenoid plunger extends to act on the throttle mechanism and to maintain engine rpm.

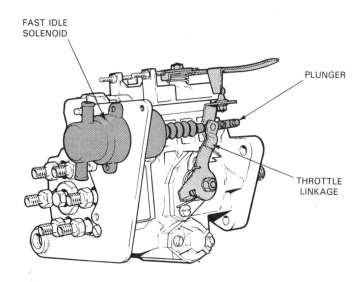

Fig. 18-50. Typical diesel fast idle solenoid. When energized, the solenoid extends the plunger to move throttle.

Computer controlled injection pump

A *computer controlled injection pump* is similar to a conventional all-mechanical pump. However, electric control motors are used to improve fuel metering in the injection pump. The actuators or motors serve as an interface or communication link between the electronics of the computer and the mechanical parts in the injection pump. The computer can send an electrical signal (voltage) to an actuator and the actuator converts it into

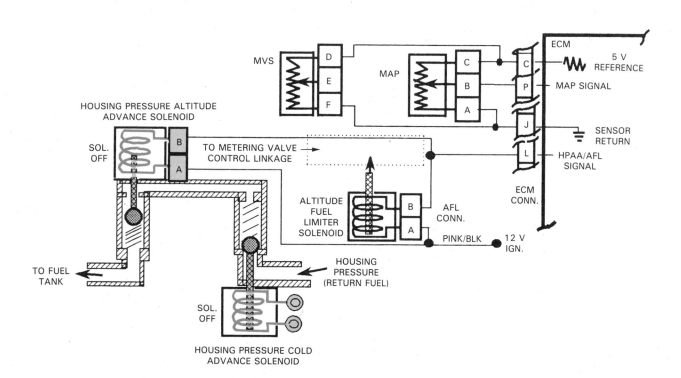

Fig. 18-49. Schematic view of two computer controlled solenoids to advance injection timing. Housing pressure altitude advance solenoid opens when MAP sensor indicates a low pressure manifold. Cold advance solenoid must open at same time to reduce fuel pressure so injection timing will advance properly.

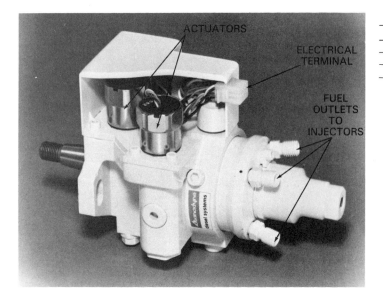

Fig. 18-51. Electronically controlled injection pump. Actuators (motors) receive electrical impulse from computer and adjust fuel output for efficient engine operation. (Stanadyne, Inc.)

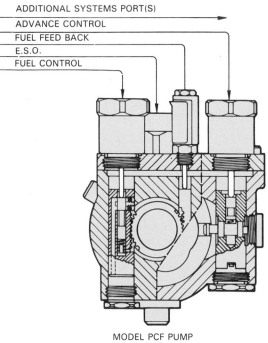

MODEL PCF PUMP

Fig. 18-52. Cutaway of typical electronically controlled diesel injection pump. Actuators (DC motors) control timing advance and fuel volume. (Stanadyne, Inc.)

movement to change the output of the pump.

An electronic-mechanical injection pump is shown in Fig. 18-51. Note the actuators on top of the pump.

The *injection pump actuators* are small reversible, DC MOTORS that operate a screw mechanism. In this way, the computer can make the motors turn a rod outward or inward to change injection volume and timing. The computer does this by changing the electrical polarity to the motors. With one polarity, the small motor spins in one direction to lengthen its output rod and with the other polarity, it reverses direction to shorten its output rod.

The *fuel control motor* alters the effective plunger stroke and amount of fuel injection, Fig. 18-52. The computer, after almost instantaneous gathering of data from the sensors, can energize the fuel control motor to move the fuel metering sleeve to increase or decrease plunger stroke.

For example, if the throttle position sensor detects wide open throttle, the computer would energize the fuel control motor to move the control sleeve for a longer effective plunger stroke. This would cause more fuel to spray out the injection nozzles to increase engine power.

The *injection advance control motor* is used to alter injection timing with changes in engine speed, load, temperature, and other variables. Refer again to Fig. 18-52. It works just like the fuel control motor, but is used to bleed off pressure to the hydraulic piston. The computer rapidly cycles the motor to change pressure in the hydraulic piston. Injection timing is then more closely matched to exact engine needs.

Electronic governor

An *electronic governor* uses the engine speed sensor, computer, and fuel control devices in the pump to limit maximum engine rpm. Unlike an all-mechanical injection pump that uses flyweight action, electronic pulse con-

trols fuel metering at high speed. This reduces the number of components in the injection pump and protects the engine from high rpm damage. An electronic governor system can also check idle speed and change it as needed.

Diesel system computer

The *diesel system computer* is similar in construction to the computer for an electronic gasoline injection system. Tiny electronic chips, called *integrated circuits*, process the large amount of information from all of the sensors and send electrical signals to the various actuators (motors or solenoids).

As the components in the injection pump wear, the computer, by monitoring nozzle opening and fuel metering valve action, can compensate for part wear and keep injection timing and volume set for maximum efficiency. This keeps the engine "in-tune" for longer periods of service.

The computer sends a reference voltage to most of the engine sensors. This reference voltage is usually between 5 and 12 volts. When the resistance in a sensor changes, a different current level returns to the computer. The computer can then alter injection pump operation for the changed condition.

If needed, review Chapter 15, Gasoline Injection System Operation. It has more information on computer control system operation.

Limited and full authority

Limited authority means that the computer only has partial control over injection pump operation. *Full authority* means that the computer has total control over injection pump fuel metering, injection timing, and other variables. A full authority system would even use the computer to link the driver's gas pedal to the injection pump's throttle lever.

Fail safe system

The *fail safe system* allows the diesel injection system to provide engine operation with failure of the computer or engine sensors. This will let the car be driven to a service center for repairs.

Diesel self-diagnosis

Some electronic diesel injection systems have *self-diagnosis* so that the computer will produce a number code indicating specific problems. Again, it is similar to the self-diagnosis found in electronic gasoline injected engines. Refer to Chapter 15 for details.

OTHER DIESEL SYSTEM INJECTION FEATURES

There are numerous design variations of the diesel injection systems. For details of each design, refer to the manufacturer's service manual. It will explain the construction and operation of the particular system.

SUMMARY

A diesel injection system delivers fuel at high pressure directly into the engine's combustion chambers. A mechanical injection pump is the main control of the air-fuel mixture. Other important components are a supply pump, high pressure fuel lines, glow plugs, and injectors.

There are important differences between gasoline injection engines and diesel injection engines. Diesels operate at very high compression ratios (17:1) and ignite the fuel mixture by heat of compression. Gasoline engines compression ratios are low (8:1 or 9:1) and depend on spark ignition.

Further, the diesel engine has no throttle valve to control airflow into the combustion chambers. Rather, the diesel injection pump output is controlled to vary engine power output.

Diesel engines compress only air on the compression stroke and deliver the fuel directly into the combustion chamber. Gasoline injection systems deliver fuel to the intake manifold and the engine compresses the air-fuel mixture.

Injectors, because the system works on pressure, and because fuel is delivered directly into the combustion chamber, use spring loaded valves and are protected from combustion temperatures by a heat shield. The nozzle of the diesel injector is so constructed that high fuel pressure lifts a needle valve off its seat and allows fuel to enter the combustion chamber.

A prechamber cup in the combustion chamber encloses the tips of the injectors and glow plugs heat the chamber on a cold diesel engine to improve fuel combustion.

Diesel injection pumps perform many important functions. Their chief purpose is to meter, time, and deliver fuel at high pressure to the injectors. They may be gear, chain, or belt driven. There are two broad classifications of diesel injection pumps: in-line and distributor. In-line

pumps have a pumping plunger (type of piston) for each cylinder; distributor-type pumps normally use only one or two pumping plungers to supply fuel to all of the engine's cylinders.

As in gasoline injection systems, computers and sensing devices may be used to increase fuel delivery efficiency of diesel injection systems. Small electric motors, called actuators, placed in the injection pump, receive signals from the computer and adjust fuel delivery for maximum efficiency.

KNOW THESE TERMS

Diesel injection system, Heat shield, Pressure chamber, Annular groove, Stem, Needle valve, Trunk, Pintle, Inward opening injection nozzle, Outward opening injection nozzle, Pintle injection nozzle, Hole injection nozzle, Diesel prechamber cup, Water detector, Fuel heater, Block heater, In-line diesel injection pump, Distributor injection pump, Pumping plunger, Transfer pump, Fuel metering sleeve, Centrifugal governor, Distributor rotor, Internal cam ring, Fuel metering valve, Fuel cutoff solenoid, Vacuum regulator valve, Vent wire, Actuators, Limited authority, Full authority, Fail safe system.

REVIEW QUESTIONS—CHAPTER 18

1. The main control of the air-fuel mixture in a diesel injection system is a _____ _____; however, on modern diesel systems _____ devices help control this operation.
2. List the essential components of a diesel injection system and explain the function of each.

Indicate whether each characteristic listed belongs to a diesel or a gasoline injection fuel system.

3. _____ _____ Has compression ratio of about 17:1 to 23:1.
4. _____ _____ Compression ratio is in the range of 8:1 and 9:1.
5. _____ _____ Uses glow plug to aid ignition.
6. _____ _____ Has no way to control airflow into the engine.
7. _____ _____ Injects fuel into intake manifold.
8. _____ _____ Controls engine speed and power solely by controlling the amount of fuel injected into the engine.
9. Name the five principle parts of a diesel injector and explain the function of each.
10. The tapered surface on the injector needle valve which is exposed to high fuel pressure to open the nozzle is called the _____ _____.
11. List the five most common diesel injector nozzle types.
12. A prechamber cup is (select ALL correct answers):
    a. Partially enclosed area in the intake manifold to protect injectors.
    b. A steel insert pressed into the face of the diesel cylinder head.
    c. A cup enclosing the tips of the injector nozzle

and the glow plug.

   d. A chamber in the combustion chamber where the glow plug can heat up the air and help start combustion when the nozzle opens.

13. Glow plugs are small _____ elements.

14. Name two devices (other than glow plugs) designed to aid cold starting of diesel engines.
diesel engines.

15. Which, if any, of the following are NOT a function of the diesel injection pump?

   a. Controls the volume of air entering the combustion chamber.

   b. Controls amount of fuel and delivers it, at high pressure, to the injectors.

   c. Helps close injector nozzles after injection.

   d. Times injection for speed and load.

   e. Controls operation of the glow plugs.

16. Diesel injection pumps can be divided into two broad categories or types: _____ and _____.

17. Compare the design of the two main categories of diesel injection pumps.

18. Explain how a fuel cutoff solenoid operates to stop a diesel engine.

19. Computer controlled diesel injection uses engine sensors, injection pump mounted _____ and an electronic _____ _____ to control injection of fuel into the engine.

20. What is an injection pump actuator and what does it control?

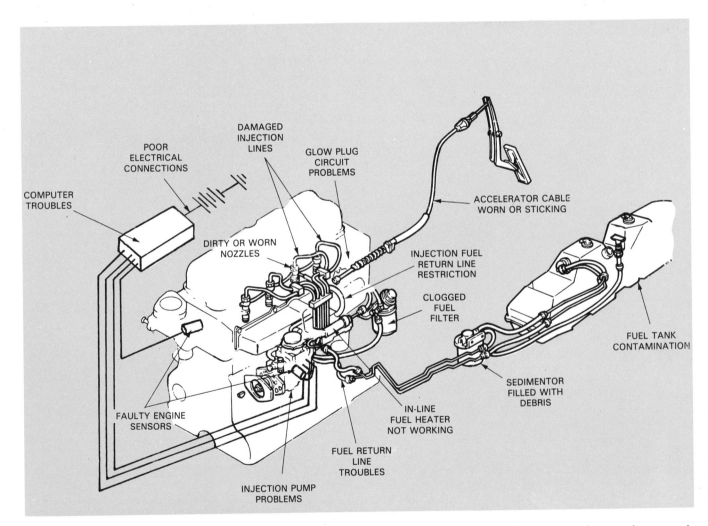

Fig. 19-1. Summary of possible causes for malfunction or failure of diesel injection system. Check appropriate service manual for troubleshooting procedures. (Ford Motor Co.)

19

DIESEL INJECTION DIAGNOSIS, SERVICE, REPAIR

After studying this chapter, you will be able to:
- *List typical diesel injection system problems and problem symptoms.*
- *Diagnose common diesel injection system troubles.*
- *Service diesel injection nozzles.*
- *Test and replace glow plugs.*
- *Perform a cylinder balance test, glow plug resistance balance test, injection pressure test, and digital pyrometer balance test.*
- *Bleed and flush a diesel injection system.*
- *Service diesel injection pumps.*
- *Perform basic adjustments common to diesel injection systems.*

Nearly all auto manufacturers—domestic and foreign—now offer diesel engines. It is therefore important that you understand how to test, service, and repair a diesel injection system. While it is true that diesel injection components are different from gasoline injection components and require different procedures, the methods are quite simple to master. At the same time, you will find that testing of the modern electronic controls for diesel systems requires the same methods used on electronic gasoline injection systems.

DIESEL INJECTION DIAGNOSIS

As you begin your study of diesel diagnostic procedures, you will be able to draw upon your knowledge of engines and injection system operation, learned in previous chapters. Likewise, the basic troubleshooting skills you have developed will be useful.

Diagnosis should begin by a check of engine operation. Any of the following could provide clues to the problem:
1. Observe the exhaust. White, black, or blue smoke may indicate trouble.
2. Listen for excessive knocking.
3. Also note if the engine misses (not firing on one or more cylinders).
4. No-start condition.
5. Lack of engine power.
6. Fuel economy poor.

For difficult problems, always consult the troubleshooting chart in the car's service manual. It will indicate where to look for the cause of most symptoms. Further, there are many variations of basic diesel injection systems and the service manual is designed for the system you are testing. It is bound to be more accurate than general charts.

Some of the diesel injection system problems are shown in Fig. 19-1. Note, especially, the potential for malfunctions in the computer control system.

Abnormal diesel exhaust

Heavy *white diesel smoke* is just about always an indication of INCOMPLETE COMBUSTION. This condition is normal during cold starts due to the condensation of unburned fuel particles. Its cause is condensation of the fuel droplets on cold engine parts. Liquid fuel will not ignite and, therefore, is mixed with the exhaust. If the condition continues after the engine is warmed up, there is a problem. See Fig. 19-2. The most common causes are: glow plugs not working, engine compression low, thermostat stuck open, injector spray pattern bad, injection timing late or, possibly, a problem with the injection pump.

NOTE! White exhaust can also be caused by a coolant leak into the combustion chamber. In such cases, the engine may have a bad head gasket, cracked cylinder head, or a cracked block.

Excessive *black smoke* is caused by an overly RICH MIXTURE. Too much fuel or too little air in the mixture produces a heavy concentration of carbon (ash) or "coked fuel" (partially burned fuel) in the exhaust. It generally indicates problems with the injection pump, injection timing, air cleaner, injectors, fuel, or the engine itself.

Blue smoke in a diesel engine exhaust indicates OIL consumption. The oil may be getting past worn piston rings, scored cylinder walls, or worn valve guides and seals. However, white-blue smoke is normally the result of incomplete combustion and injection system problems.

Smoke meter

A *smoke meter,* Fig. 19-3, is a test device that measures the amount of ash or soot in diesel exhaust. It senses the amount of light coming through a sample of exhaust gases. A low reading indicates a need of repairs or adjustment to the injection system.

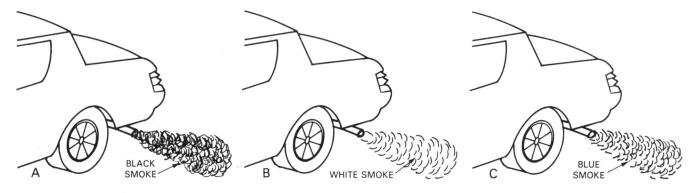

Fig. 19-2. Condition of exhaust is often a clue of source of trouble in fuel system. A—Black smoke indicates too rich fuel mixture. B—White smoke indicates presence of unburned fuel—as when engine is cold. C—Blue smoke means engine oil is entering combustion chamber.

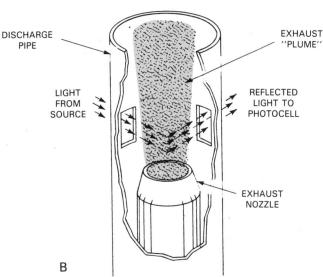

Fig. 19-3. Smoke meter "reads" diesel exhaust to detect unburned fuel. A—Smoke meter test unit. B—How light dispersion unit works. (Robert Bosch)

Diesel knock

A knocking sound is normal to all diesel engines. It is the natural result of rapidly burning fuel ignited at high compression. The high pressure produces a familiar rumble or dull clattering sound inside the engine.

However, several conditions will cause the knock to become very pronounced. One is *ignition lag.* (This is time interval between diesel fuel being injected into the combustion chamber and ignition of fuel.) Some lag is normal. If lag is EXCESSIVE, too much fuel ignites at once and a mild explosion occurs, creating the knocking sound.

Other common causes of *diesel knock* are: low engine operating temperature (thermostat stuck open), early injection timing, low engine compression, contamination of fuel, and oil consumption.

Diesel engine miss

A *diesel engine miss* is the failure of one or more cylinders to fire (burn fuel) properly. Several conditions can be responsible for this problem: a faulty injector, a clogged fuel filter, incorrect injection timing, low cylinder compression, an injection system leak, air leak, or a faulty injection pump.

No-start condition

Failure of a diesel engine to start may be caused by inoperative glow plugs, restrictions in air or fuel flow, malfunctioning fuel shutoff solenoid, fuel contamination, or problems in the injection pump.

Note! One common cause of a no-start condition is slow cranking speed. Since the engine depends on compression to ignite the fuel, it must be cranked fast enough to produce sufficient combustion heat.

Engine lacks power

When a diesel engine runs, but *lacks power,* the problem could be one of several conditions. Check the throttle cable for proper adjustment, governor setting, air and fuel filters (for clogging), engine compression, and any other condition that could affect combustion.

Remember that a diesel engine does NOT develop as much power as a gasoline engine of equal size.

Poor fuel mileage

Fuel leaks, clogged air filter, incorrect injection timing, or leaking injectors may cause *poor fuel economy*. Usually, a diesel engine will have better fuel mileage than a gasoline engine. Before making repairs, check EPA economy values for the make and size engine.

DIESEL INJECTION SERVICE SAFETY

1. While diesel fuel is not as flammable as gasoline, it is still a serious fire hazard. Follow the same safety precautions.
2. NEVER attempt to remove any part of the injection system with the engine running. Pressures of 6000 to 8000 psi (41 340 to 55 120 kPa) are developed by the injection pump. Spurting fuel can strike the skin with enough force to puncture it. BLOOD POISONING or DEATH could follow!
3. Do not attempt to stop a diesel engine by covering the air inlet opening. The suction is strong enough to cause INJURY or draw objects into the engine intake manifold.

DIESEL INJECTION MAINTENANCE

Like gasoline injection systems, diesel injection systems need periodic maintenance. Check the service manual for details. In addition to changing or cleaning filters periodically, you will need to inspect the system for signs of trouble. If inspection turns up signs of fuel leakage, check around all fittings using a piece of cardboard, Fig. 19-4. If there is a serious leak, it will show up on the cardboard. Injection lines or hoses not in good condition should be replaced, Fig. 19-5.

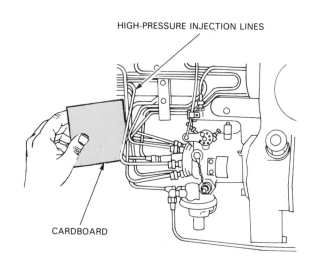

HIGH-PRESSURE INJECTION LINES

CARDBOARD

Fig. 19-4. Use a piece of cardboard, not your hand, to check fuel injection fittings for leaks. Never forget that fuel injection fuel pressure is strong enough to cause injury. (Chrysler)

For good performance, clean filters are important. Water contaminated diesel fuel causes rapid corrosion and pitting of injection system components. Thus, periodic draining is important for proper operation and long life of the system. The main fuel filter may have a drain for bleeding off trapped water.

As you learned in earlier chapters, fuel systems may have several filters. One, usually a "sock" filter, will be located in the fuel line. Sometimes, the injector assembly will have a final filter screen.

The schematic in Fig. 19-6 shows the typical location of filters. Study where each is positioned. Also study the

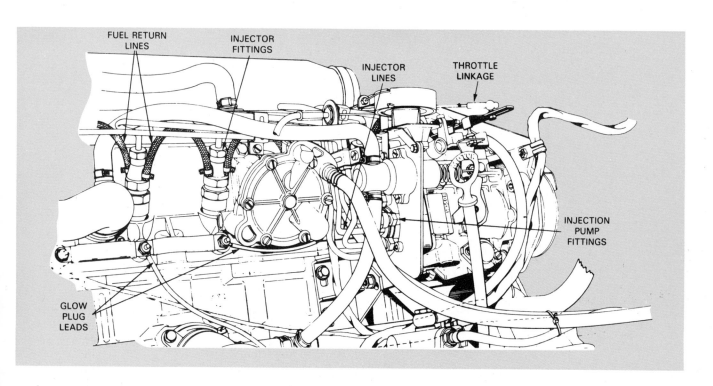

Fig. 19-5. Sometimes visual inspection of hoses, lines, filter, and linkages will reveal sources of problems. (Volvo)

COMPONENT IDENTIFICATION

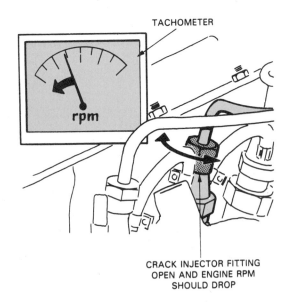

COMPONENT IDENTIFICATION

(See illustration for component location)

COMPONENT	FUNCTION	COMMENTS
1. Sedimentor (On frame under vehicle)	Separates water from fuel and traps it inside; alarm system warns of high water level; drain valve has remote release in cab behind driver's seat	• Must be drained every 8 046 kilometers or 5,000 miles or when water-in-fuel lamp is lit
2. Cold Start Knob	Manually advances injection pump timing and provides fast idle setting for starting	• Up to 8° advance
3. Neutral Switch	Causes control module to cut off after-glow system operation	• Neutral switch closed in all gears except neutral
4. No. 1 Glow Plug Relay	Supplies current to glow plugs for starting below 30°C (86°F) when key is in ON or START position	• Part of quick-start system (Q.S.S.)
5. Clutch Switch	Causes control module to cut off after-glow system operation	• Clutch switch closed when clutch pedal is released
6. Fuel Cutoff Solenoid Valve	Cuts off fuel supply to injection pump when key is in OFF position	• Mounted on injection pump • Valve opens with key ON • Valve closes with key OFF
7. Glow Plug Resistor	Cuts after-glow plug voltage to 4.2–5.3 volts	• Located between intake manifold and air inlet tube flange
8. Fuel Filter	Filters fuel before it enters injection pump and injection nozzles	• Filter change interval 48 280 kilometers or 30,000 miles
9. Injection Pump	Distributor-type pump with built-in governor, altitude compensator and metering controls	• Serviced by replacement only • Do not disassemble or break seals • Only adjustments are pump timing and idle speed
10. Coolant Thermo Switch	Senses coolant temperature for after-glow system control	• Closed below 30°C (86°F)
11. Vacuum Pump	Supplies vacuum for power brakes, speed control and heater A/C controls	• Belt driven • Vacuum reservoir with speed control system
12. Low-Vacuum Warning Switch	Turns on BRAKE warning signal when system vacuum drops too low	• Closes with low vacuum • Has 7-second time delay to prevent signal flickering
13. In-Line Fuel Heater	Warms fuel before it enters filter to prevent wax formation clogging at low fuel temperatures	• Internal thermostatic control • Cuts in below −1°C (30°F) • Cuts off above 8°C (46°F)
14. Injection Nozzles	Combustion chamber fuel spray devices with calibrated, spring-loaded discharge valves	• Nozzle valve opens at 135–140 kg/cm² (1 920–1 990 psi)
15. Glow Plugs	Ignites fuel when starting and during initial warm-up (after-glow)	• Ready to start in 3 seconds
16. No. 2 Glow Plug Relay	Continues to supply current to glow plugs (after-glow) if engine is cold	• Operates below 30°C (86°F) when vehicle is not in motion
17. Glow Plug System Control Module	Controls current to glow plug relays	• Quick-start system for starting aid • After-glow system for warm-up heating

Fig. 19-6. Service manual part locations in typical system.

chart to become familiar with the maintenance aspects of a diesel fuel system.

TESTING DIESEL INJECTION OPERATION

There are numerous ways of checking the operation of a diesel injection system and the procedure will vary from one system to another. However, we will explain the most common testing methods. When testing ANY system, refer to the instructions for that system. Testing procedures vary with the design of the injection pump, return lines, and injection nozzles.

Cylinder balance test

A diesel *cylinder balance test* is a systematic disabling of cylinders, one after another, to determine if all cylinders are "firing" (fuel igniting). In a gasoline engine, this is done by removing spark plug wires. In a diesel engine, it is done by loosening the injector line to leak fuel and, thus, "starve" the cylinder so it cannot fire.

To perform this test, first wrap a shop rag around the injector. Then loosen the injector fitting. See Fig. 19-7.

NOTE! This operation can be dangerous. Loosen the injection line only enough to allow fuel to drip. Wear safety glasses, leather gloves, and get approval

TACHOMETER

CRACK INJECTOR FITTING
OPEN AND ENGINE RPM
SHOULD DROP

Fig. 19-7. Cylinder balance test. To check for "dead" cylinders crack each injector line at fitting (with engine running) until fuel drips. This prevents injector from opening. If rpm drops, injector and cylinder are operating normally. If engine rpm is not affected when injector is disabled the cylinder is not "firing." (Volvo)

from your instructor before starting the test. Refer to the service manual, as well, and follow the prescribed testing method.

When the fitting has been loosened, fuel should slowly leak from it. It should prevent the injector from opening to provide fuel for the combustion chamber.

With fuel leaking from the fitting, engine SPEED SHOULD DROP and the engine idle should become rough. If engine operation is NOT affected, the cylinder has NOT been firing. The injector may be bad, compression low, or the injection pump not functioning properly. Further tests are needed. Check all cylinders in the same way to see which ones are "dead."

Diesel engine compression test

Good compression is critical to the operation of a diesel engine since it depends on extremely high compression to ignite the fuel. If compression is too low, the fuel cannot ignite. A compression test will determine if compression is up to specs.

While a diesel engine compression test and a gasoline engine compression test are much alike, never use a compression gauge intended for a gasoline engine. The high diesel engine compression can damage the gauge. A diesel compression gauge must read up to about 600 psi (4 134 kPa).

Making the test

To begin a diesel compression test, remove either the injectors or the glow plugs, depending on instructions given in the engine's service manual. Usually, you must remove the glow plugs. Install the compression gauge. The heat shield may be needed to seal the gauge when installed in the injector hole.

Disconnect the fuel shutoff solenoid to disable the injection pump. Crank the engine several revolutions and note the highest reading on the gauge.

Typical compression pressure for an automotive diesel engine will be in the 400 to 500 psi (2 756 to 3 445 kPa) range. Readings from cylinder-to-cylinder should be within 50 to 75 psi (344 to 517 kPa) of each other. Otherwise, an engine repair is needed.

WARNING! Some auto makers discourage the use of wet compression testing of diesel engines. Too much oil introduced into the cylinder may cause hydraulic lock and engine damage. The piston cannot compress excess oil in cylinder.

Compression test procedures will be covered in greater detail later.

Removing injectors and glow plugs

Following service manual instructions, disconnect the battery to prevent engine cranking. Remove the injection line. See Fig. 19-8. Be careful not to bend or kink the line.

WARNING! Wrap a shop rag around the fitting before cracking the fitting open, and wear safety glasses.

Most automotive injectors are threaded into the cylinder head. On larger truck engines, you may find that they are press-fitted into the head and secured with

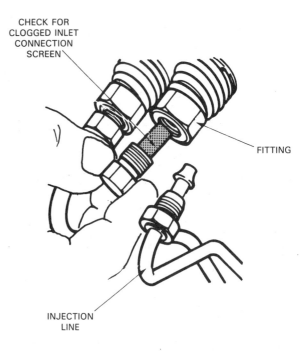

Fig. 19-8. Screens and filters should be checked for possible clogging when lines are removed. (Chrysler)

bolts. You may need a special impact tool to remove them.

Before removing glow plugs, disconnect the wires going to them. Then, using a deep well socket and a ratchet, carefully unscrew the glow plugs.

DANGER! Temperatures of glow plugs can soar to more than 1000 °F (538 °C). Use extreme caution to avoid serious burns on your hands.

Reinstalling plugs and injectors

Coat threads with antisieze compound and start them by hand. Then tighten to specifications. Use care not to overtighten. This will damage the unit. Reconnect wires or lines and check for proper operation.

Glow plug resistance-balance test

A heated glow plug will have greater resistance to current than a cold one. This provides a safe, simple method of finding out if every cylinder is firing.

To run a *glow plug resistance-balance test,* disconnect all wires to the glow plugs. Connect one lead of a digital ohmmeter to a glow plug and the other lead to ground. Note the ohms (resistance) reading. Repeat the test with each of the other glow plugs. See Fig. 19-9.

Next, start the engine and let it run for a few minutes. With the engine off, repeat the test, noting the change in resistance of each glow plug. Amount of ohms increase should be relatively even among all the plugs. If any are lower, the cylinder is NOT firing.

Digital pyrometer balance test

A *digital pyrometer* is an electronic device which makes very accurate measurement of temperature. It is sometimes used to check whether diesel cylinders are all firing.

417

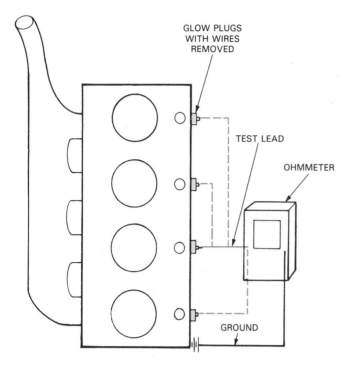

Fig. 19-9. Glow plug resistance-balance test compares resistance of glow plugs to determine if all cylinders are firing. Resistance is measured at every glow plug.

With the engine running, simply touch the probe of the pyrometer to the exhaust manifold at each exhaust port, Fig. 19-10. Temperatures at each port should be about the same. A cooler reading at any exhaust port would

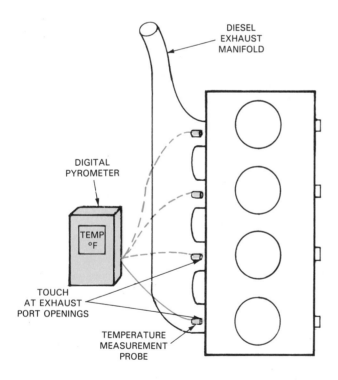

Fig. 19-10. Digital pyrometer test is made by measuring temperature of exhaust manifold at every port.

be a fairly reliable indication that the cylinder is NOT FIRING.

Measuring injection pressure

Using special valves and high pressure gauges, an *injection pressure tester* measures the amount of pressure in the injection lines between the injection pump and the injectors, Fig. 19-11. Connect the tester as shown in Fig. 19-12 and follow instructions supplied with the tester.

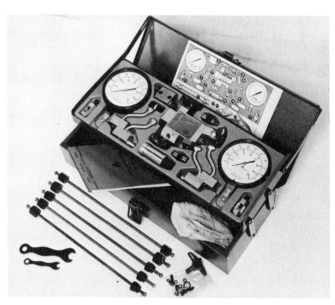

Fig. 19-11. Test unit checks injection pressure without disconnecting major parts from system. (Hartridge)

In some cases, an on-car injection pressure tester will provide a check of the following:
1. Injector opening pressure.
2. Leaking at injector nozzle.
3. Injection line pressure balance.
4. Injection pump condition.

An injection pressure test can provide much useful information about the condition of the injection system. It pinpoints problems without removing major parts of the system. The test will locate bad injector nozzles, clogged injector filters, or a bad injection pump.

Fig. 19-13 shows still another tester for measuring system pressure. Note how it connects to the system. Also look at Fig. 19-14. It illustrates several special diesel engine service tools.

DIESEL INJECTION CLEANLINESS

Maintaining cleanliness is important in diesel fuel system service. Even the smallest speck of foreign matter can damage or upset the operation of an injection pump or an injector. Parts have been machined to tolerances that are measured in *microns* (millionths of a meter).

To help maintain this degree of cleanliness:

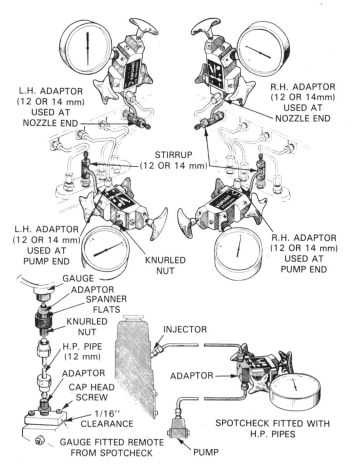

L.H. ADAPTOR
(12 OR 14 mm)
USED AT
NOZZLE END

R.H. ADAPTOR
(12 OR 14mm)
USED AT
NOZZLE END

STIRRUP
(12 OR 14 mm)

L.H. ADAPTOR
(12 OR 14 mm)
USED AT
PUMP END

KNURLED
NUT

R.H. ADAPTOR
(12 OR 14 mm)
USED AT
PUMP END

GAUGE

ADAPTOR
SPANNER
FLATS

KNURLED
NUT

H.P. PIPE
(12 mm)

INJECTOR

ADAPTOR

ADAPTOR

CAP HEAD
SCREW

1/16''
CLEARANCE

GAUGE FITTED REMOTE
FROM SPOTCHECK

SPOTCHECK FITTED WITH
H.P. PIPES

PUMP

Fig. 19-12. Instructions for hooking up test unit for measuring injection pressure. These are supplied with the test kit.

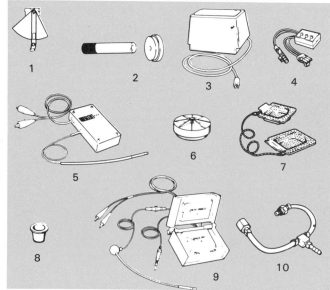

1. ANGLE GAUGE
2. NOZZLE ASSEMBLY TOOL
3. ULTRASONIC NOZZLE CLEANER
4. METERING SENSOR ADJUSTING HARNESS
5. TACHOMETER
6. AIR CROSSOVER COVER
7. MANIFOLD SCREEN COVERS
8. PLASTIC PLUGS, PUMP LINES, AND NOZZLES
9. TIMING METER
10. HOUSING PRESSURE ADAPTER

Fig. 19-14. Special diesel service tools. Note their names. (General Motors)

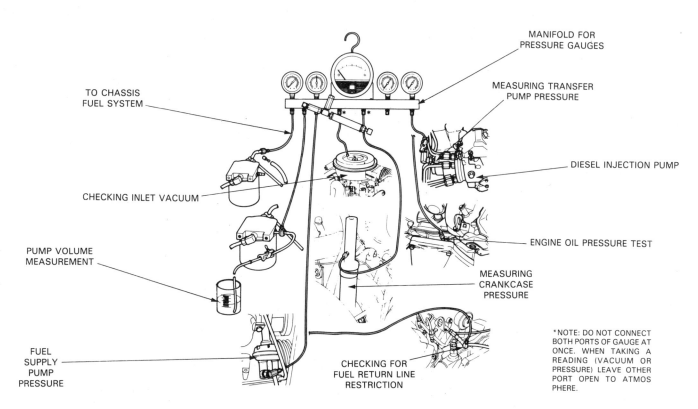

MANIFOLD FOR
PRESSURE GAUGES

TO CHASSIS
FUEL SYSTEM

MEASURING TRANSFER
PUMP PRESSURE

DIESEL INJECTION PUMP

CHECKING INLET VACUUM

ENGINE OIL PRESSURE TEST

PUMP VOLUME
MEASUREMENT

MEASURING
CRANKCASE
PRESSURE

FUEL
SUPPLY
PUMP
PRESSURE

CHECKING FOR
FUEL RETURN LINE
RESTRICTION

*NOTE: DO NOT CONNECT
BOTH PORTS OF GAUGE AT
ONCE. WHEN TAKING A
READING (VACUUM OR
PRESSURE) LEAVE OTHER
PORT OPEN TO ATMOS
PHERE.

Fig. 19-13. Another type of diesel injection pressure tester. Note various hookups.

1. Always cap and seal any injection system fitting that is disconnected, Fig. 19-15. This will prevent dust from entering the system.
2. Use only clean, lint-free shop rags when wiping off components. Even a single fiber can upset the system.
3. Use compressed air and clean shop rags to remove dirt around fittings.
4. Never spray water on a hot diesel engine. Cracking or warping of the injection pump could cause serious injection problems.

Performing an injector pop test

While the injector is removed from the engine, it may be *pop tested*, Fig. 19-16. This will determine its opening pressure, its spray pattern, and whether or not it is leaking.

To make a pop test, fill the tester reservoir with the recommended CALIBRATION FLUID. Never use diesel oil for this test. It is too flammable and results may be too unreliable.

With the tester valve open, pump the tester handle. Watch the spray coming out of the tester. When there are no more air bubbles in the spray, close the valve.

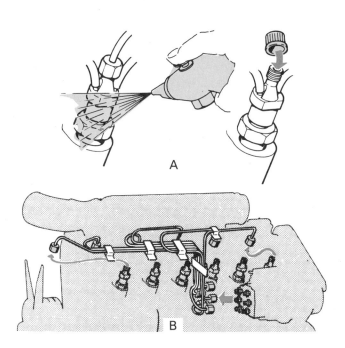

Fig. 19-16. Pop tester is device for checking injector for opening pressure, spray pattern, and leakage.

Fig. 19-15. A—Blow away dirt with compressed air before disconnecting fuel lines. B—Cap all lines to prevent dirt from entering system. (Volvo)

NOZZLE SERVICE

Being exposed to the direct heat of combustion and the by-products of combustion, injector nozzles can erode, become clogged with carbon, or become damaged in other ways.

A *bad injector* will usually cause an engine to miss. Other symptoms may be reduced engine power, smoking, or knocking. These effects are the result of an incorrect opening pressure, incorrect spray pattern, fuel leakage, and other problems.

Injector substitution

One way to confirm that an injector is faulty is to replace it with one known to be good. Install the good injector in the "dead cylinder." If this solves the problem, you will know that the injector was bad. If the problem remains, then other fuel system or engine problems exist.

Clogged filters and screens can also upset injector operation. While you are removing the injection lines, check the screens as well.

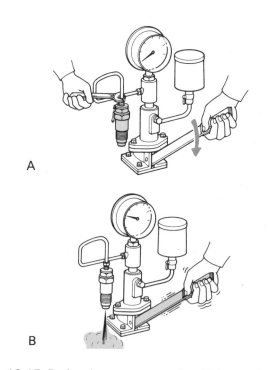

Fig. 19-17. Performing pop test on diesel injector. A—Install injector to tester fitting and bleed air from fitting. Pump handle to discharge carbon from injection hole. B—Pump handle slowly and observe at what pressure injector opens. (NOTE: Spray should always be directed into transparent container, but is omitted here for clarity). (Toyota)

Connect the injector and perform the spray test as shown in Fig. 19-17.

DANGER! Extremely high pressures are developed during pop testing. Wear eye protection and keep your hands away from the fuel spray coming out of the injector nozzle. Aim the nozzle into a clean, approved container. Wrap a clean shop rag around the opening of the container.

Measuring injector opening pressure

Before attempting to measure the injector's opening pressure, purge all air from the nozzle by pumping the tester lever up and down several times. Then, move the handle slowly while observing the pressure reading. Note the pressure when the nozzle opens. Repeat this procedure until you are satisfied that you have an accurate reading.

Opening pressure is usually about 1700 to 2200 psi (11 713 to 15 158 kPa). If an injector opens at pressures not within the service manual specs, rebuild or replace it.

Checking injector spray pattern

Operate the pop tester handle while watching spray coming out of the nozzle, Fig. 19-18. The pattern should be a narrow cone-shaped mist. Replacement or service is needed if the spray pattern varies.

Checking injector leakage

Check injector leakage by slowly operating the tester handle to maintain a pressure of about 300 psi (2 060 kPa) lower than the nozzle opening pressure. The diesel injector should NOT leak or drip during a 10 second test.

Leakage is normally caused by dirt, carbon buildup, or worn components. Refer to Fig. 19-19.

Note! Certain diesel injectors make a *chattering sound* during operation. All nozzles should make a *swishing* or *pinging sound* when spraying fuel.

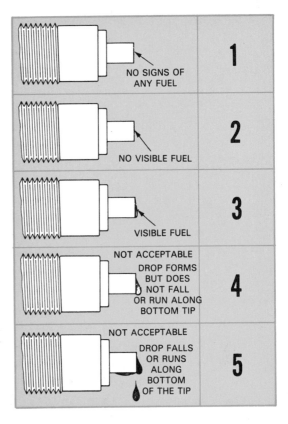

Fig. 19-19. Check injector nozzles for leakage with pressure about 150 psi (1 033 kPa) below opening pressure. Maintain pressure for 5 seconds. At that time, you should be able to observe whether injector is leaking. Patterns 1, 2, and 3 are acceptable but 4 and 5 are not.

Fig. 19-20 shows what kinds of trouble to look for when inspecting a nozzle. Make sure the needle and seat are in perfect condition. With the needle and housing

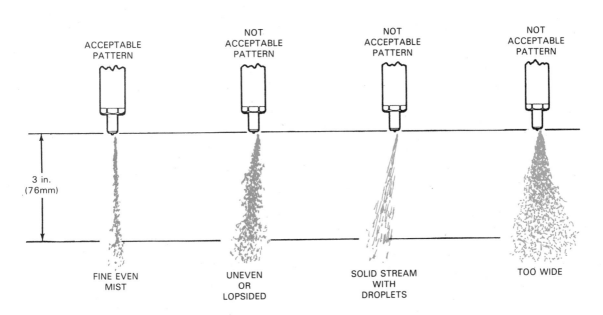

Fig. 19-18. Typical spray patterns. Only narrow, even mist is acceptable. (Chrysler)

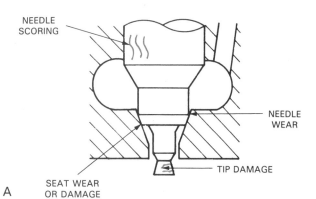

NEEDLE
SCORING

NEEDLE
WEAR

TIP DAMAGE

SEAT WEAR
OR DAMAGE

A

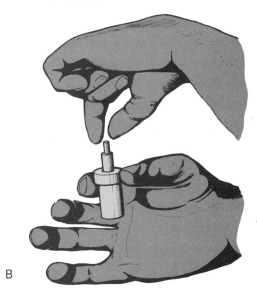

B

Fig. 19-20. A—Inspect nozzle needle and seat for damage and wear. B—Cleaned, dry needle should slide or drop freely into housing. (Peugeot and Toyota)

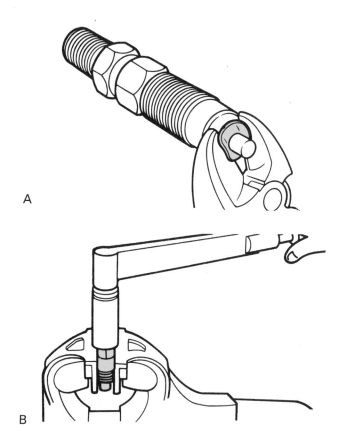

A

B

Fig. 19-21. Worn or damaged injectors can be rebuilt by mechanic. A—Old sealing washer is being removed with side cutters. B—Removing nozzle holder. Injector is clamped in vise. Use soft jaw on vise to avoid injector damage. (GM and Ford)

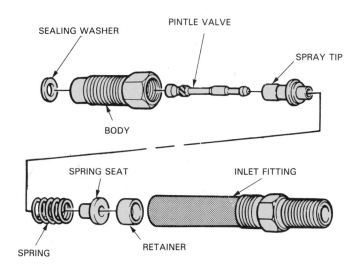

SEALING WASHER

PINTLE VALVE

SPRAY TIP

BODY

SPRING SEAT

INLET FITTING

SPRING

RETAINER

Fig. 19-22. Exploded view of injector nozzle assembly. Note arrangement and names of parts. (Buick)

clean and dry, the needle should fall freely into the housing.

Rebuilding diesel injector nozzles

Injector designs vary. Always follow the instructions in the service manual. It will include specific instructions for disassembly, cleaning, inspecting, assembly methods, torque values, and other information critical to a rebuild. See Fig. 19-21A.

Begin the disassembly by unscrewing the body from the nozzle assembly using a six-point wrench or socket, Fig. 19-21B. Remove and inspect all the injector parts, Figs. 19-22 and 19-23.

Use solvent to clean the nozzle parts. You may need special cleaning tools such as brass scrapers and brass brushes to remove hardened carbon deposits. See Fig. 19-24. Be very careful not to mar the needle and nozzle. Even the smallest scratch can upset the injector's operation. Parts showing wear or damage should be replaced.

If several injectors are disassembled at once, use care NOT TO MIX PARTS. Components may have been select fitted at the factory and will not fit any other injector. Use an organizing tray to separate parts, Fig. 19-25.

Inspecting nozzle components

During disassembly and reassembly, inspect all nozzle parts closely for signs of wear or damage. Diesel injector nozzles are machined to tolerances measured in micro dimensions. The slightest surface irregularity can affect nozzle operation.

422

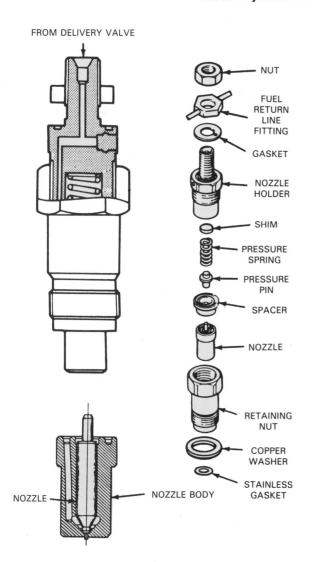

FROM DELIVERY VALVE

NUT

FUEL RETURN LINE FITTING

GASKET

NOZZLE HOLDER

SHIM

PRESSURE SPRING

PRESSURE PIN

SPACER

NOZZLE

RETAINING NUT

COPPER WASHER

STAINLESS GASKET

NOZZLE   NOZZLE BODY

Fig. 19-23. When disassembling injectors, carefully note position of each part. (Ford Motor Co.)

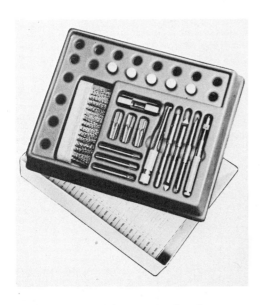

Fig. 19-24. Special service kit contains brushes, scrapers, and gauges for rebuilding injectors. (Hartridge)

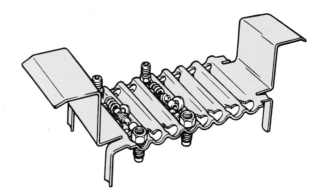

Fig. 19-25. Tray type organizer keeps small parts in proper order for reassembly. (Buick)

Shimming injectors

An *injector shim* is a thin metal spacer used to adjust spring tension and valve opening pressure. Low pressure can be remedied by installing a thicker shim, high pressure by installing a thinner one. A manufacturer's instruction sheet will give more details on how thickness affects pressure. Refer to Fig. 19-26.

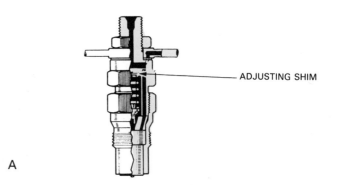

ADJUSTING SHIM

A

QUICK REFERENCE NOZZLE SHIM ADJUSTMENT CHART

Part No.	Thickness (mm)	Resulting Pressure Change (kPa)
E3TZ-9M577-A	0.50	Base
E3TZ-9M557-B	0.54	480
E3TZ-9M557-C	0.58	960
E3TZ-9M557-D	0.62	1440

B

Fig. 19-26. A—Cutaway shows thicker washer used to adjust spring tension on injector. B—Part of service manual chart showing washer thicknesses and pressure changes produced by each.

Injector assembly

Reassemble the injector according to steps given in a service manual. Lubricate parts with diesel oil before beginning. Make sure that all parts are assembled in proper order. Torque the injector body to specifications. Pop test the injector again, Fig. 19-27.

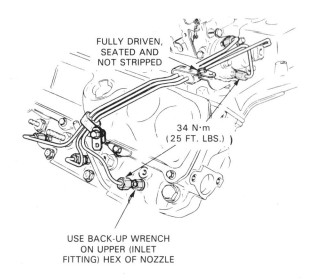

Fig. 19-29. Injectors must be torqued to manufacturer's specs. (Oldsmobile)

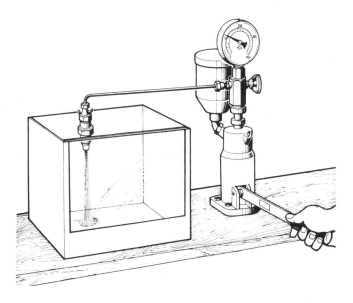

Fig. 19-27. Pop testing a rebuilt injector. Spray pattern should be a narrow spray of fine mist. (Peugeot)

Installing diesel injectors

Before installing injectors, coat them with antisieze compound. You may need to install a new heat shield or seal to prevent compression leak. Turn the injector into the cylinder head by hand until it is snug, Fig. 19-28. Torque it to recommended specs, Fig. 19-29.

Reconnect the injector fuel line without bending it. Tighten it with a torque wrench if possible.

Reconnect the wire coming from the fuel shutoff solenoid and start the engine. Check for leaks and determine if the injectors are working properly. You may need to bleed air from some systems. Check a manual for proper procedures.

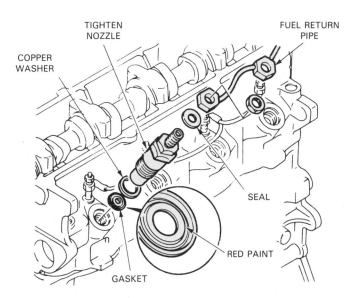

Fig. 19-28. Proper injector assembly order. Turn injector in by hand. (Ford)

INOPERATIVE GLOW PLUGS

Inoperative glow plugs will cause cold starting problems. If none of them are working, the engine may not start. There will be too little heat of compression to ignite the fuel. If one or two of the glow plugs are not working, the engine may miss until it is up to operating temperature.

Testing glow plugs

If glow plug problems are suspected, you can use a test light or VOM to check voltage. Touch the tester to the feed wires going to the glow plugs. Voltage should be up to specs.

Clip-on ammeters are also made that check the current applied through the feed wires. An incorrect reading at or on the glow plugs indicates trouble in the supply circuit or in the glow plugs themselves.

WARNING! Some 6-volt glow plugs may be damaged by connecting them to direct battery voltage. Refer to the service manual before beginning the test. Fig. 19-30 shows a typical glow plug circuit.

Glow plug tests can also be made with an ohmmeter, as shown in Fig. 19-31. With supply wires to the plugs removed, Fig. 19-32, connect the ohmmeter across the glow plug terminal and ground. Resistance should be about 0.5 ohms cold or 1.0 ohms hot. If above or below specs, replace the glow plug.

Note! Weak batteries and slow cranking rpm will prevent diesel starting.

Injection line service

When servicing diesel injection lines, there are several rules to remember:

1. Injection lines are usually factory bent to form equal length lines. Use a factory line when replacing a damaged line to make sure lengths are equal. Unequal lines could affect injection quantity, injection timing, and engine performance.

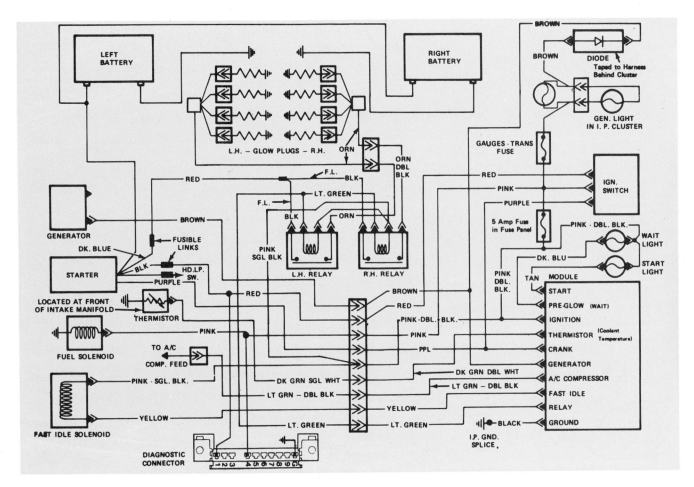

Fig. 19-30. When you must test glow plug circuit, it is helpful to study circuit diagrams in service manual to locate wires and terminals. (Oldsmobile)

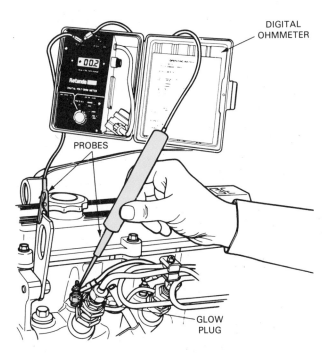

Fig. 19-31. Using an ohmmeter to check glow plug resistance. Check service manual for resistance values. Ground one probe to engine and touch other probe to each glow plug terminal. If there is no continuity, replace glow plug. (Ford)

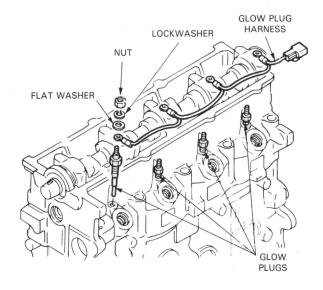

Fig. 19-32. Disconnect glow plug from wiring harness and remove it for inspection.

2. Cap all lines and fittings when they are disconnected. The slightest bit of dirt could upset the operation of the nozzles.

3. When installing a new injection line, wash it in solvent and blow it clean. Direct a blast of compressed air through the line to remove any debris.
4. Do NOT bend an injection line when removing it. This could kink the line and restrict fuel flow. It could also weaken the line and cause it to rupture and leak in service.
5. Torque injection line fittings to specifications. Under or overtightening a line could cause leakage or fitting failure.
6. Remember the danger of the high pressure inside a diesel injection line. Fuel pressure is high enough to cause the fuel to puncture the skin if leaking out of the system. This could cause blood poisoning or blindness.

reinstalled correctly.

Make sure you note how the initial timing marks (usually indentations) on the pump housing and engine line up. Then, you can remove the bolts securing the pump to the engine.

The parts typically requiring removal for diesel injection pump service are shown in Fig. 19-34.

Make sure you cap all fittings to prevent debris from entering the pump or nozzle lines. See Fig. 19-35.

Replacement injection pumps

Diesel injection pumps are precision mechanisms. Specialized tools and equipment are needed to repair them. Most garages install a NEW or FACTORY REBUILT unit when internal parts are faulty.

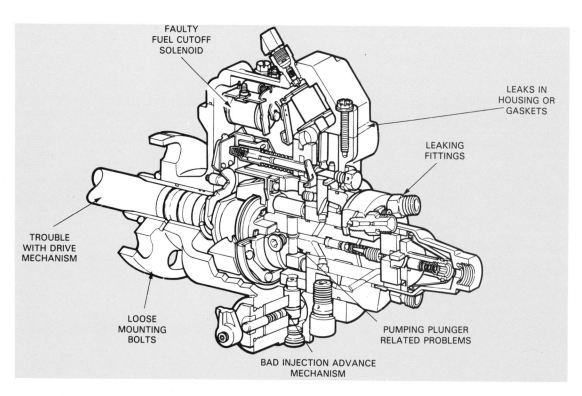

Fig. 19-33. In checking malfunctioning injection pumps, look for these problems.

INJECTION PUMP SERVICE

A *malfunctioning injector pump* can cause a number of operating problems: engine misses, smoking, won't run. Though usually trouble-free, the pump may be damaged by water contamination, accidents, wear from long service, or leaking seals. Adjustment, repair, or replacement may be needed. Fig. 19-33 shows common problems.

Removing the pump

To begin injection pump removal, disconnect the cables, linkages, lines, and wires from the pump. Be careful not to damage or bend the fuel lines while removing them. If needed, mark the wires so that they can be

Injection pump test stand

An *injection pump test stand* is needed to analyze the performance of an injection pump after major repairs. One is pictured in Fig. 19-36. It has a large electric motor that spins the injection pump drive shaft, simulating engine operation.

The output lines from the pump are connected to injector nozzles. Pressure sensing devices can then be used to monitor injection pressure, injection timing advance, pump rpm, etc.

Since an injection pump test stand is very expensive and complicated to use, it is usually found in only specialized shops that rebuild diesel injection pumps. This is another reason why most automotive garages do NOT rebuild injection pumps. In-shop repairs are usually

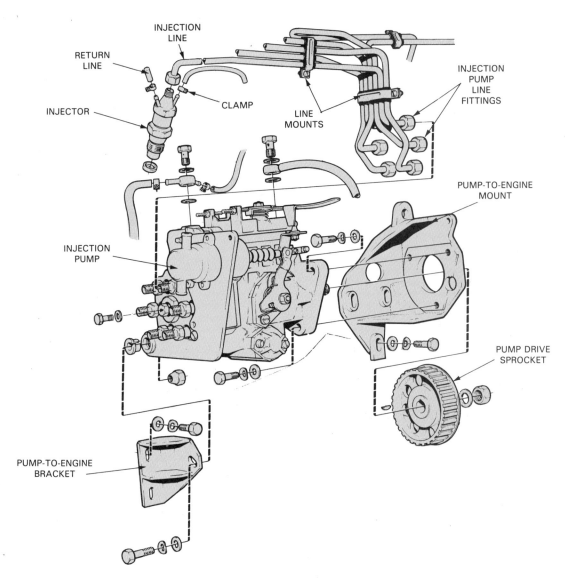

Fig. 19-34. Exploded view of typical injection pump. These parts should be inspected during service of unit. (Volvo)

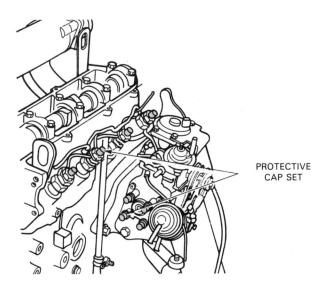

Fig. 19-35. Cap all fittings during pump service to keep out dust and debris. (Ford)

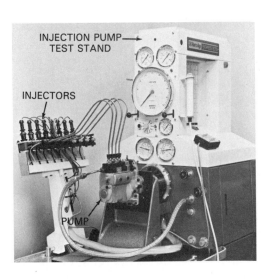

Fig. 19-36. Shops which specialize in injection pump service will use a test stand for checking pumps' operation. (Hartridge)

limited to replacing housing gaskets, fuel shutoff solenoids, idle stop solenoids, vacuum valves, cold advance solenoids, metering valve sensors, advance sensors, and control actuators.

When doing any injection pump service, always refer to a factory service manual. It will give the essential details for doing competent work.

Installing an injection pump

Transfer parts such as solenoids, vacuum valves, and the drive sprocket from the old pump to the new one. Align timing marks on the pump and the engine. Install the pump. Torque fasteners and lines to specs. Adjust timing before starting the engine.

WARNING! Do not hammer or pry on the injection pump housing or drop it. It is easily damaged and expensive to replace.

If the pump is belt driven, make sure the belt is properly adjusted. See Fig. 19-37.

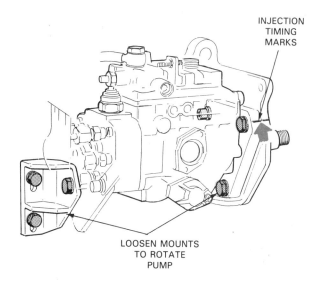

Fig. 19-38. In static timing, mark on pump is lined up with mark on engine.

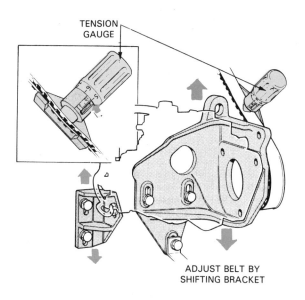

Fig. 19-37. Special tool is recommended by some auto manufacturers to adjust belt tension on belt-driven injection pumps. (Volvo)

Adjusting injection timing

Timing adjustment is made by rotating the injection pump on its mount after loosening the mounting bolts. To advance timing, turn the housing in the opposite direction as the shaft rotation. To retard, turn it in the same direction as shaft rotation.

Timing can also be adjusted by several other methods. Procedures will vary by manufacturer, type of engine, and injection system.

In one method, the engine is NOT running. This is called *static timing.* The timing marks on the engine and the injection pump are aligned, Fig. 19-38. This is only an approximate (rough) setting but is used by some auto manufacturers.

More accurate timing requires that the injection pump components be moving. This is called *dynamic timing.*

There are several methods for adjusting dynamic injection timing:
1. With a dial indicator to measure the stroke of the injection pump plunger in relation to the position of the piston (when it fires). See Fig. 19-39.
2. With a luminosity (light) instrument which detects combustion in the No. 1 cylinder. See Figs. 19-40 and 19-41.
3. With a fuel injection pressure detector.
4. With a diagnostic connector and a timing meter.

Always consult a factory shop manual for instructions on the approved method for checking and adjusting timing and the timing device to use, Fig. 19-42. Follow the instructions carefully; injection timing is critical to engine performance.

Regardless of the method used to adjust injection timing, check it at IDLE and also at a HIGHER RPM. Make the initial setting at idle. The higher rpm reading will tell you whether the advance mechanism in the pump is working properly.

Adjusting throttle cable linkage

A check of the *throttle cable* or *linkage adjustment* will determine if the pump's throttle lever is operating freely. Wear makes periodic adjustment necessary. A check should be made for binding at the same time. Lubricate the linkage, if needed, Fig. 19-43.

SPEED ADJUSTMENTS

Injection pumps can usually be adjusted for curb idle speed, maximum speed, and fast (cold) idle speed. Look at Fig. 19-44.

Diesel idle speed adjustment

Usually, diesel engine *idle speed* is set with a special diesel tachometer. An idle stop screw on the injection pump is rotated left or right to vary the engine speed.

The *diesel tachometer* is similar to or may be part of the injection timing tool. Sometimes an instrument like

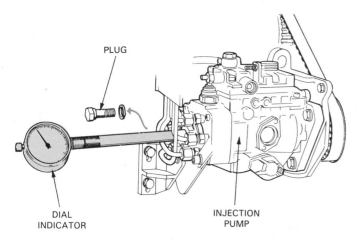

Install dial indicator or injection pump.

A

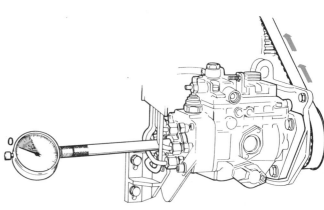

Turn engine counter-clockwise until dial indicator is at minimum. Set indicator to zero.

B

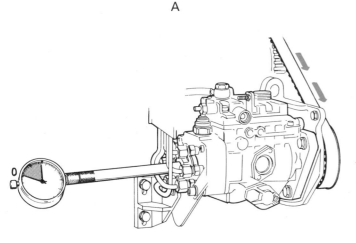

To check setting of injection pump, turn engine clockwise until engine timing mark is at 0 or TDC. Dial indicator should read according to manufacturer's specs. Make certain engine has not been turned past mark.

C

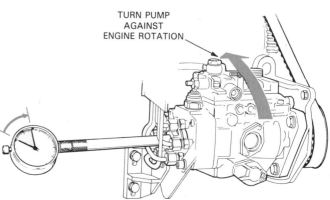

If dial indicator is reading below specs, loosen retaining bolts and turn pump against engine rotation to read within specs. Retighten retaining bolts.

D

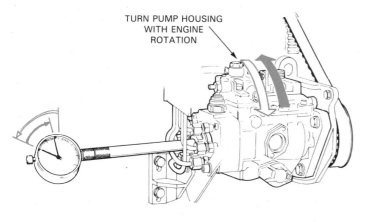

If dial indicator is reading higher than specs, loosen retaining bolts and turn pump housing with engine rotation until within specs. Retighten retaining bolts.

E

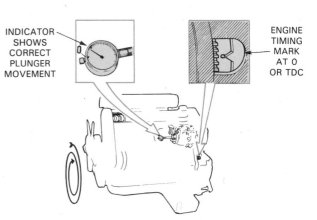

Turn engine twice and check setting. Repeat adjustment steps if necessary.

F

Fig. 19-39. Typical steps for checking and adjusting injection pump timing. After test, remove dial indicator and replace plug.

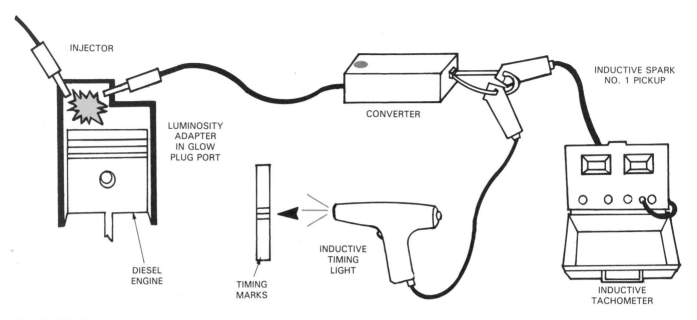

Fig. 19-40. Magnetic timing device sometimes called "luminosity" probe can be used. It senses firing of cylinder to indicate if timing is correct. (Hennessy Industries Inc.)

Fig. 19-41. Luminosity tester is being used to check diesel engine timing. (OTC Tools)

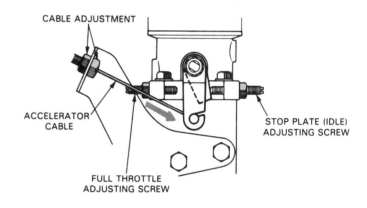

Fig. 19-43. There are several typical basic adjustments on diesel injection pump: throttle cable, idle speed, and maximum speed. (Toyota)

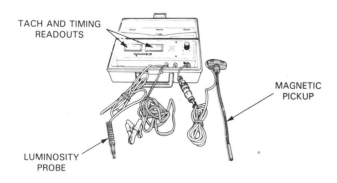

Fig. 19-42. Dynamic timing meter has luminosity probe to sense when cylinder (usually No. 1) fires while a magnetic pickup determines crankshaft position when cylinder combustion takes place. (Rotunda)

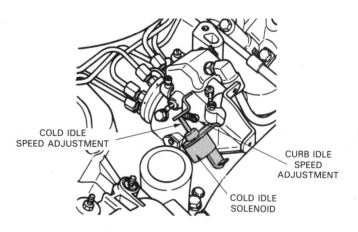

Fig. 19-44. Adjusting cold idle speed solenoid is not unlike adjusting carburetor cold idle speed solenoid. Solenoid will be located on injection pump. (Ford)

the one in Fig. 19-45 is used. Regardless of design, the diesel tach will provide a readout of engine speed in rpm.

Diesel curb idle speed

Curb idle speed adjustment is simple. Place the transmission in neutral or park. Bring the engine up to operating temperature. Connect the tachometer to the engine and compare the readings with specs. These should be printed on the engine compartment emission control decal or given in the service manual.

To raise or lower curb idle speed, you must usually loosen a locknut before you can turn the adjusting screw. Hold the screw as you retighten the locknut. See Fig. 19-42, again.

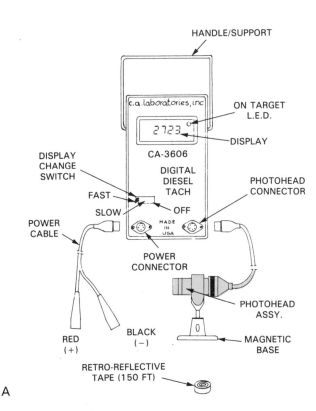

Cold idle speed adjustment

To reset the cold idle speed adjustment, you will need a jumper wire to activate the fast idle solenoid. To do this, connect the jumper wire between the positive battery terminal and the solenoid terminal. Momentarily raise engine speed to release the solenoid plunger. Take tachometer readings and compare them with specifications. Adjust the setting, Fig. 19-44.

On some engines, a *cold start lever* replaces the cold idle solenoid. Refer to the shop manual for details on adjustment. Methods are similar.

Maximum speed adjustment

A *governor* in the injection pump limits diesel engine rpm, as you will recall from Chapter 18. To check it, place your tachometer so it can be read from the engine compartment. Place the transmission in neutral or park and start the engine. Hold your left foot on the brake as you press the accelerator slowly to the floor, Fig. 19-46. Observe the tachometer to see if maximum speed of the engine is within specs.

CAUTION! On a maximum speed test, release the accelerator immediately if engine rpm goes above specs. Continued high rpm could damage the engine.

To bring speed to specification, turn the maximum speed adjusting screw on the injection pump. Retighten the locknut and retest.

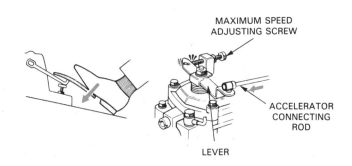

Fig. 19-46. Testing and adjusting maximum engine speed. Left. Press accelerator to floor and observe rpm on tachometer. Right. Injection pump will have adjusting screw which can be turned to reset rpm to specs. (Toyota)

VACUUM VALVE ADJUSTMENT

Vacuum valve adjustment is needed to assure the correct vacuum (suction) output at different throttle openings. Frequently, a carburetor angle gauge is used to adjust the vacuum valve on the side of the diesel injection pump. The valve has elongated slots that allow adjustment of valve opening. Follow service manual directions for the particular unit.

FUEL CUTOFF SOLENOID SERVICE

A *bad fuel cutoff solenoid* will usually keep the diesel engine from running. Windings may fail so it can no

Fig. 19-45. Digital diesel tachometer. Photo head takes "readings" off flywheel dampener to which strips of reflective tape have been attached. A—Digital "Magtac." B—Details on components of digital tachometer.

longer open to allow fuel entry into the pumping chambers in the diesel injection pump. It is also possible for the solenoid to fail in the open position, keeping the diesel engine from shutting off.

Always check the supply circuit to the fuel cutoff solenoid. Wires feeding current to the solenoid could be opened or shorted, causing faulty operation of the solenoid.

Testing fuel cutoff solenoid

Check for power to the solenoid first. Use a test light as shown in Fig. 19-47. The test light should glow with the ignition key on or the engine cranking. If not, disconnect the wire from the solenoid. Check for power on the incoming wire alone. If you do not have voltage, the circuit is bad. If you do have voltage, the solenoid is probably faulty.

An ohmmeter can also be used to measure the resistance of the windings in the solenoid. If open or shorted, replace the unit.

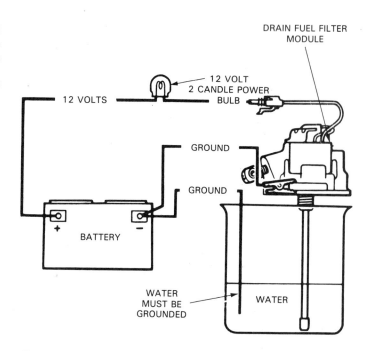

Fig. 19-48. This basic arrangement is typical of methods for testing water detector. Note that water supply in container must be grounded to battery's negative terminal. (Buick)

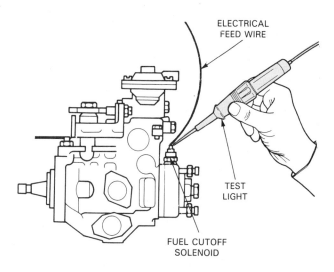

Fig. 19-47. Testing fuel cutoff solenoid. Touch test light to solenoid. With ignition on and wire connected it should light. Also test wire for voltage. (Ford)

Replacing fuel cutoff solenoid

To replace the fuel cutoff solenoid, simply unscrew the old unit. Be careful not to let dirt or other debris fall into the opening in the injection pump. After making sure it is clean, screw the new solenoid into the pump. Torque it to specs. Connect its feed wires and check engine operation.

WATER DETECTOR SERVICE

A *faulty water detector* can keep the water-in-fuel indicator light from working when there is water in the fuel. Fig. 19-48 shows how a battery and test light can be used to check the sensor for this circuit. Submerge the sensor in a can of water. Connect a battery and test

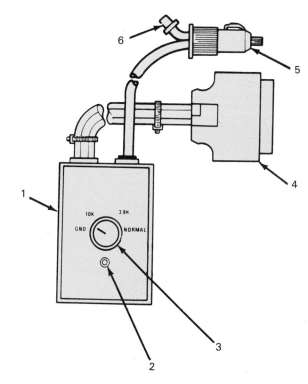

1—DIESEL DIAGNOSTIC CHECK TOOL
  OR EQUIVALENT
2—CHECK ENGINE LIGHT
3—DIAGNOSTIC MODE SELECTOR SWITCH
4—CONNECTOR
5—CIGAR LIGHTER PLUG-IN
6—BAT CONNECTOR

Fig. 19-49. Diesel diagnostic test tool. This type can draw power either from battery lead or from terminal that can be plugged into cigarette lighter. (GM)

light. Ground the housing. If working properly, the water sensor should turn on the test light. If it does not, replace the unit.

COMPUTER CONTROLLED DIESEL INJECTION SERVICE

Most computer controlled diesel injection systems have self-diagnosis. The computer will produce a number code indicating possible troubles with components in the system. This is similar to the self-diagnosis discussed for gasoline injection.

For example, a dash-mounted engine warning light may glow. This tells the driver and mechanic that something is wrong. Then the mechanic can activate self-diagnosis by jumping across two terminals in the wiring harness or by connecting a special tester to the system, Fig. 19-49. The computer will then produce a coded number.

The code can be used, along with a service manual, to find the problem. The service manual will have a chart for all numbers. By comparing the service manual code number with the number produced by the computer, you can usually find the trouble with the system.

Always refer to a factory shop manual for details of activating and testing a computer controlled diesel injection system. Procedures vary and the slightest mistake could cause part damage.

Sensor and actuator problems

Like a gasoline injection system, engine sensors and actuators on diesel injection systems can fail and upset engine operation. See Fig. 19-50.

You must use the same basic techniques to check the internal resistance of components, voltage output at various wires, and current levels. If a voltage, resistance, or current value is not within specs, corrective action must be taken.

Fig. 19-51 gives an example of a service manual illustration for checking voltage to the metering valve sensor or an injection pump. Voltage is checked at a specified point on the wiring harness. If too high or too low, something is wrong with the current. The computer, wiring, wire connections, or sensor itself may be at fault.

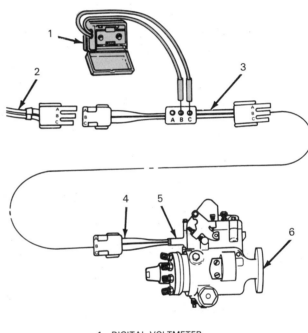

1—DIGITAL VOLTMETER
2—ENGINE HARNESS
3—HARNESS
4—HARNESS
5—METERING VALVE SENSOR
6—INJECTION PUMP

Fig. 19-51. Service manuals carry illustrations which show how and where to make voltage checks of electronic components for diesel injection system. (General Motors)

Fig. 19-52 shows one procedure for adjusting the throttle position sensor on a diesel injection pump. As with sensors on gasoline injection systems, simply measure sensor resistance at different throttle openings. If not within specs, rotate the sensor on its mount until resistance readings are correct. Replace the unit if it will not adjust within specifications.

A wiring diagram for a computer controlled diesel injection system is given in Fig. 19-53. Note how wires feed to and from the diesel injection pump. This kind of diagram can be used to trace wires between components

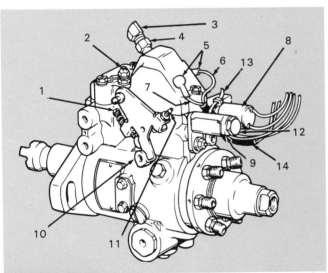

1—SLOW IDLE ADJUSTMENT SCREW
2—FUEL SHUT-OFF SOLENOID (HIDDEN)
3—TO FUEL RETURN LINE
4—FUEL RETURN LINE CONNECTOR
5—HOUSING PRESSURE COLD ADVANCE LEADS
6—PINK-BLACK
7—GREEN-BLACK
8—ALTITUDE FUEL LIMITER
9—METERING VALVE SENSOR (MVS)
10—THROTTLE LEVER
11—DO NOT ADJUST
12—MVS ADJ. HOLE PLUG
13—STRAP
14—PROTECTOR SLEEVE

Fig. 19-50. Note components on this computer controlled diesel injection pump. (Oldsmobile)

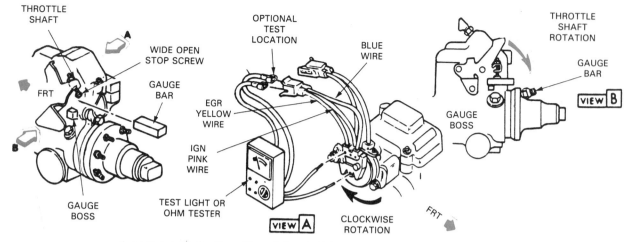

1. Loose assemble throttle position switch to fuel injection pump with throttle lever in closed position.

2. Attach a continuity meter across the IGN (pink) and EGR (yellow) terminals or wires.

3. Insert the proper "switch-closed" gage block as shown on Emission Control Label between the gage boss on the injection pump and the wide open stop screw on the throttle shaft.

4. Rotate and hold the throttle lever against the gage block.

5. Rotate the throttle switch clockwise (facing throttle switch) until continuity pivot occurs (high meter reading) across the IGN and EGR terminals or wires. Hold switch body at this position and tighten mounting screws to 5-7 N•m (4-5 ft. lbs.)

   NOTE: Switch point must be set only while rotating switch body in clockwise direction.

6. Release throttle lever and allow it to return to idle position. Remove the "switch-closed" gage bar and insert the "switch-open" gage bar. Rotate throttle lever against "switch-open" gage bar. There should be no continuity (meter reads ∞) across the IGN and EGR terminals or wires. If no continuity exists, switch is set properly. However, if there is continuity, then the switch must be reset by returning to step 1 and repeating the entire procedure.

Fig. 19-52. This is one procedure for adjusting throttle position sensor on diesel injection pump.

when troubles are difficult to locate. A service manual will have a wiring diagram for the particular system being tested.

Note! If necessary, review Chapter 17, Gasoline Injection System Diagnosis, Service, Repair. It covers testing information relating to electronic diesel injection systems.

DIESEL INJECTION SERVICE RULES

Remember these basic rules when servicing diesel injection systems:

1. Cap lines and fittings to prevent foreign matter from entering the fuel system.
2. Treat the diesel injector or injection pump gently. Do not pry on them or drop them.
3. High pressure inside a diesel injection system can cause serious injury. Use care.
4. Remember, some diesel injection systems must be bled (air removed) after repairs.
5. Clean dirt around fittings before disconnecting them.
6. Follow all torque specifications. This is extremely important on a diesel engine.
7. Replace bent, kinked, or damaged fuel injection lines.
8. Place a piece of screen mesh over the air inlet when the engine is to be operated without the air cleaner. This will keep other objects from being sucked into the engine. Also, do not cover

the air inlet with your hand while the engine is running. Injury may result.
9. Check fuel filters and water separators periodically. Water can cause expensive corrosion of injection system parts.
10. Observe safe practices concerning ventilation and fire hazards when operating a diesel engine.
11. Wear safety glasses when working on a diesel injection system.
12. When in doubt, refer to instructions in a service manual for the make and model of vehicle being serviced. Avoid mistakes which could upset engine performance or cause engine damage.
13. Use service manual procedures when working on an electronic or computer controlled diesel injection system. The slightest mistake could ruin expensive components (computer, sensors, actuators, etc.).

SUMMARY

Diagnosis of diesel injection system problems should begin by observing how the engine operates. This will provide your best clues to the source of the problem. Consulting the troubleshooting section of the car's service manual is also recommended.

Indications of problems might include white, black, or blue exhaust; lack of power; missing; knocking; and

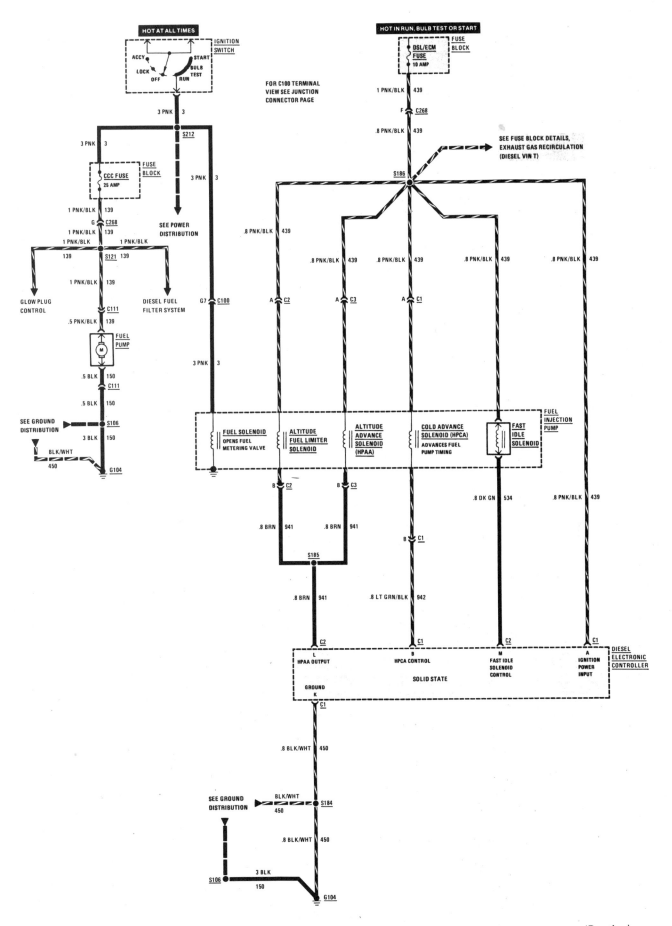

Fig. 19-53. This wiring diagram shows wiring harness for diesel fuel system controlled by computer. (Pontiac)

failure of the engine to start. Good diagnosis techniques will help you determine the causes of these conditions.

Because of high pressures in the diesel fuel injection system, use extreme care when removing any part of the injection system. Never attempt to stop a diesel engine by covering the air inlet. Suction could draw objects or your hand into the intake manifold.

Periodic maintenance should include changing or cleaning of filters, checking for fuel leaks, replacement of deteriorating hoses and lines, and bleeding off trapped water in the fuel.

An important aspect of troubleshooting is testing various parts of the fuel injection system to locate the source of trouble. A cylinder balance test will indicate which cylinder is not firing when there is a miss in the engine. This is performed by disabling one injector after another to create a deliberate miss.

A diesel compression test requires a gauge which reads up to 600 psi. This test should be done when there is doubt concerning engine mechanical condition.

Injectors must be tested to determine if they are opening and closing properly and that they are not leaking. Glow plugs and the glow plug wiring harness need to be tested if they are not functioning properly.

Various instruments are used in testing diesel injection systems. A digital ohmmeter is used to test the resistance of glow plugs and other electrical components. Higher than normal resistance indicates a problem.

A digital pyrometer is an electronic device which makes extremely accurate measurement of temperature. It is useful in determining whether a cylinder is firing or not. An injection pressure tester measures the amount of pressure in the injection lines between the injection pump and the injectors.

Cleanliness is all-important in fuel injection service. A tiny piece of dirt can cause a problem because of the close tolerances in the system. Cap all fittings as soon as they are opened and use only clean, lint-free shop rags when wiping components.

An injector suspected of being bad can be checked by substituting a good injector. If the problem disappears, the old injector was the problem. A pop test, done with injectors removed from the engine, determines if their opening pressures and spray patterns are alright and if they are leaking. Parts that are not performing properly must be replaced.

An injection pump can cause engine misses, smoking, no-run conditions. Since pumps are precision instruments, special tools and instruments are needed to repair them. It is best to install new or factory rebuilt units when the old unit is faulty. The mechanic will need to adjust timing, idle curb speed, cold idle speed, and maximum engine speed (governor).

Other service includes vacuum valve adjustment and testing of the fuel cutoff solenoid. The latter can malfunction, keeping the diesel engine from shutting off or preventing it from starting. A water detector can be tested with a battery and a special test light.

Most computer controlled diesel systems have self-diagnosis. A service manual will detail the proper procedures for activating this system to find the source of the problem.

KNOW THESE TERMS

Smoke meter, Diesel knock, Ignition lag, Cylinder balance test, Diesel compression test, Glow plug resistance-balance test, Digital pyrometer, Injection pressure tester, Injector substitution, Pop tester, Injector spray pattern, Injector leakage, Injector shim, Injector rebuild, Injection test stand, Injection pump timing, Diesel tachometer, Diesel maximum speed adjustment.

REVIEW QUESTIONS—CHAPTER 19

1. List five engine conditions that might provide a clue to the diesel fuel injection problems.
2. Heavy _____ diesel smoke is just about always an indication of incomplete combustion.
3. An overly rich fuel mixture will cause excessive _____ smoke due to a heavy concentration of _____ in the exhaust.
4. What is a smoke meter and what does a low smoke meter reading indicate?
5. Explain why ignition lag may cause pronounced diesel knock.
6. List five other conditions that may cause diesel knock.
7. Which of the following fuel system problems are likely to prevent the diesel engine from starting?
   a. Oil consumption.
   b. Slow cranking speed.
   c. Fuel leaks.
   d. Fuel shutoff valve not functioning.
8. Filters or filter screens may be placed at several locations in a diesel fuel system. List the likely locations.
9. In a cylinder balance test, the engine speed dropped and the engine idle speed became rough when one of the injectors was disabled. This indicates that:
   a. The cylinder served by the injector is not firing.
   b. The cylinder served by the injector was firing until the injector was disabled.
10. A diesel compression test can be taken either through the glow plug port or the fuel injector port. True or false?
11. Explain how to use a digital pyrometer in determining which cylinders of a diesel engine are not firing.
12. A pop test is conducted for all the following reasons EXCEPT:
   a. Determine whether injection pump pressure is up to specs.
   b. Determine opening pressure of injector.
   c. Observe spray pattern of injector.
   d. Determine if injector is leaking.
13. What is an injector shim and why would you use it?
14. A resistance test on a diesel engine with a cold starting problem indicates that all of the glow plugs are at about 0.5 ohms. Indicate which of the following would be your likely next step:
   a. Remove all glow plugs for a visual inspection.
   b. Check current applied through the feed wires to see if each is up to specs.
   c. Do a voltage test on the glow plug.
   d. Replace the glow plugs as they are defective.
15. Injection pumps are usually not repaired by the

average mechanic. Service is usually limited to replacement of actuators, sensors, vacuum valve, and solenoids. True or false?

16. List four types of instruments used for testing the injection pump for proper timing.

17. To set idle speed, a(n) _____ _____ _____ on the injection pump is rotated left or right.

18. Explain how to adjust maximum engine idle speed on a diesel engine.

19. A fuel cutoff solenoid is found to be inoperative. A resistance check indicates infinite resistance in the solenoid. Which is the probable cause?
   a. The wiring supplying current to the solenoid is grounded or open.
   b. The windings in the solenoid are open.
   c. The windings in the solenoid are shorted.

20. Describe how to check a water detector for proper operation.

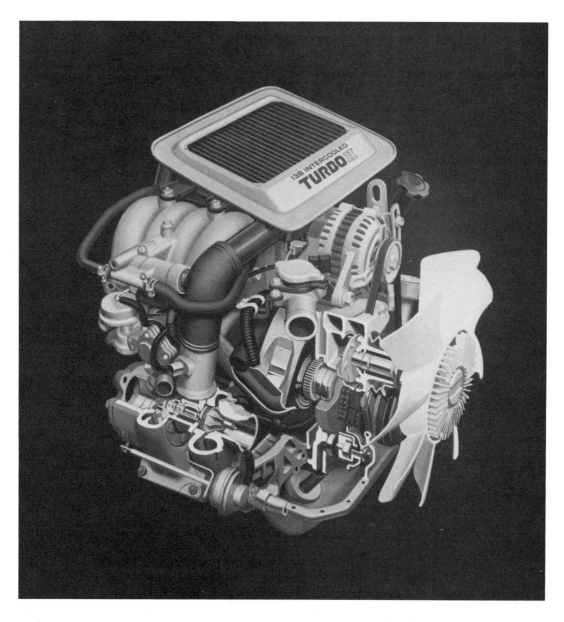

A modern Wankel rotary engine with turbocharger and intercooler is capable of great power despite its small size. Turbocharger is shown in cutaway at lower left on engine. (Mazda)

TURBOCHARGER CONSTRUCTION AND OPERATION

After studying this chapter, you will be able to:
- *Explain the principles of supercharging and turbocharging.*
- *List the basic parts of a typical turbocharger system.*
- *Define the function of each turbocharging system component.*
- *Summarize the advantages and disadvantages of turbocharging.*
- *Describe the engine modifications needed with turbocharging.*
- *Describe the function and construction of an intercooler.*
- *Explain the operation of a waste gate.*
- *Summarize the operation of a detonation sensor in a computer controlled turbo system.*

A *turbocharger* is a device which increases engine power by increasing the amount of air and fuel that enters the engine's combustion chambers. It is a fan driven by the exhaust gas flow out of the engine.

Turbochargers have been associated with high performance and racing for many years. Now, with the increased demand for performance and fuel efficiency, they are popular on passenger cars.

Note! *Supercharging* originally referred to any method of increasing intake manifold pressure and engine power. Now, the term generally means a belt, gear, or chain-driven blower while turbocharging refers to an EXHAUST DRIVEN blower.

NORMAL ASPIRATION

An engine with *normal aspiration* (taking in air) uses only outside air pressure or atmospheric pressure to cause airflow into the combustion chambers. At sea level, this would be 14.7 psi (101 kPa).

Normal aspiration means the engine is NOT supercharged or turbocharged. Refer to Fig. 20-1. With only outside air pressure to carry oxygen into the engine cylinders, engine power is limited by the engine's low volumetric efficiency.

Volumetric efficiency is a measure of how much air or air-fuel mixture the engine can draw in on its intake

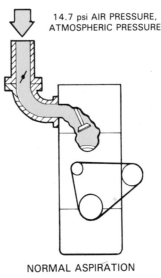

A — NORMAL ASPIRATION

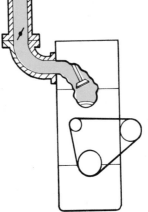

B — TURBO OR SUPERCHARGING

Fig. 20-1. A—Engine with normal aspiration uses air at atmospheric pressure (less dense). B—In turbocharging, engine's efficiency is increased by being able to take in air at higher pressure.

strokes. A high volumetric efficiency means the engine "breathes" easily because of good intake port, valve, combustion chamber, and camshaft design.

Because of *pumping losses* (restriction to airflow through intake manifold and cylinder head), most engines do not have high volumetric efficiency. As a result, they do not produce much horsepower for their size. To improve efficiency and power output, turbocharging is frequently used on today's gasoline and diesel engines.

Passenger car supercharging

A supercharger for a passenger car is a small unit mounted near the front of the engine. It is usually belt driven off the engine crankshaft damper, as in Fig. 20-2. The supercharger has an electromagnetic clutch, similar to the clutch on an air conditioning system compressor. This allows the supercharger to be turned on and off.

For example, one system uses a wide open throttle switch to activate the supercharger, Fig. 20-2. When the gas pedal is pushed to the floor for passing or getting up to highway speed, the switch sends power to the electromagnetic clutch. This engages the supercharger and engine power is instantly increased for rapid acceleration. When the gas pedal is released, the clutch disengages and the engine returns to normal aspiration. A computer and various sensors can also be used to operate the supercharger.

Supercharging has the advantage of instant air pressure increase on demand. With a turbocharger, there is a slight delay in pressure and power increase as the unit has to build up rpm (speed). This will be discussed in detail shortly.

Note! Superchargers (mechanically driven blowers) are not commonly used on passenger cars. They are being experimented with by some auto makers presently and show promise for the future.

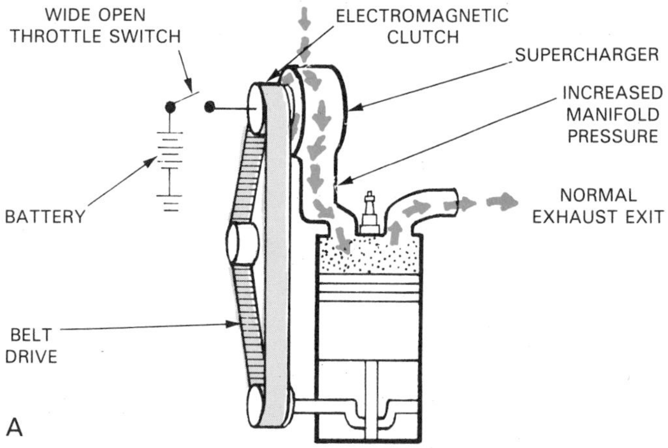

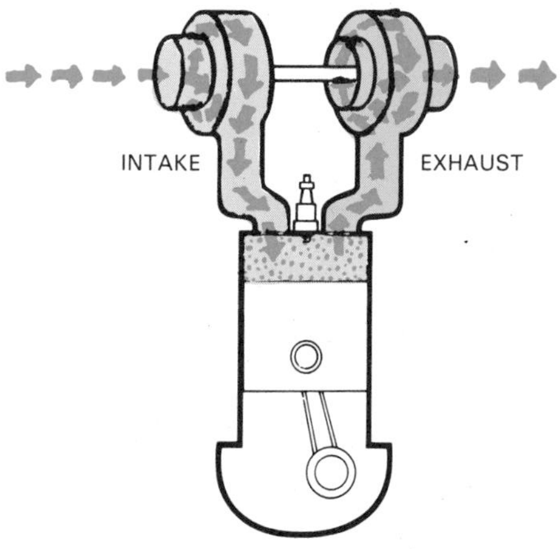

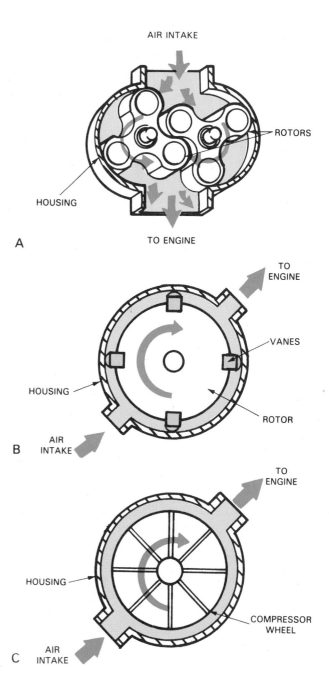

Fig. 20-2. Two types of devices for increasing volumetric efficiency (ability of engine to take in more air). A—Belt-driven supercharger takes its power from engine crankshaft. B—Turbocharger uses energy of exhaust gases to pump more air into engine. (Ford)

Fig. 20-3. There are three basic types of supercharger. A—Rotor or Rootes type. B—Vane type. C—Centrifugal supercharger, commonly called a turbocharger. (Chrysler)

Supercharger types

Basically, there are three classifications of superchargers. These classes relate to how they produce pressure and are known as: rotor type, vane type, and centrifugal type.

The *rotor* or *Rootes supercharger* is a positive displacement type blower that uses specially shaped rotors to produce a pumping action. As shown in Fig. 20-3A, two rotors are enclosed in a housing. Gears are used to synchronize rotor rotation. As the two rotors turn, they trap air or air-fuel mixture, between their lobes. This carries the charge around inside the housing and compresses it to produce pressure. This supercharger is used on large diesel truck engines and some drag racing engines.

The *vane supercharger* uses small, straight blades mounted on a round rotor, Fig. 20-3B. Like a power steering pump or air injection pump, the rotor is offset to one side of its housing. This causes the volume between the rotor and housing to change as the air is carried around in the blower. It is also a positive displacement type supercharger because it displaces a specific volume on each revolution.

The *centrifugal supercharger* has a fanlike wheel with fins that blow air into the engine, Fig. 20-3C. It is NOT a positive displacement type pump because the blades do NOT seal tightly against the inside of the housing. As the compressor wheel spins, centrifugal force throws the air or air-fuel mixture outward to increase its pressure. This type supercharging method is commonly incorporated in a turbocharger.

TURBOCHARGERS

As we mentioned earlier, a *turbocharger,* frequently called simply a "turbo," is an exhaust-driven supercharger (fan or blower) that sends air or air-fuel mixture into the engine under higher than atmospheric pressure. See Figs. 20-4 and 20-5. Such devices are frequently used on small gasoline and diesel passenger car engines to increase their horsepower. Through HARNESSING normally wasted energy in the EXHAUST GASES, the turbocharger also improves engine efficiency through greater fuel economy and lower emission levels. This is especially true with a diesel engine.

Turbocharger components

As Fig. 20-6 shows, the parts of a turbocharger are:
1. TURBINE WHEEL—exhaust-driven fan that turns the turbo shaft and the compressor wheel.
2. TURBINE HOUSING—outer enclosure that routes exhaust gases around a turbine wheel, Fig. 20-7.
3. TURBO SHAFT—steel shaft that connects turbine and compressor wheels. It passes through the center of the turbo housing.
4. COMPRESSOR WHEEL—driven fan that forces air into the engine intake manifold under pressure.
5. COMPRESSOR HOUSING—part of turbo housing that surrounds the compressor wheel. Its shape helps pump air into the engine.
6. BEARING HOUSING—enclosure around turbo shaft that contains bearings, seals, and oil passages.

Turbocharger operation

Look at Fig. 20-8. The exhaust manifold and connecting tubing route hot exhaust gases into the turbo's housing. While passing through the turbine housing, the hot gases strike the fins or blades of the turbine wheel. When engine load is high enough, the exhaust gas flow is strong enough to spin the turbine wheel at high speed.

Being connected to the turbine wheel by the turbo

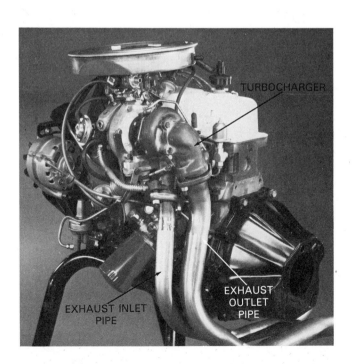

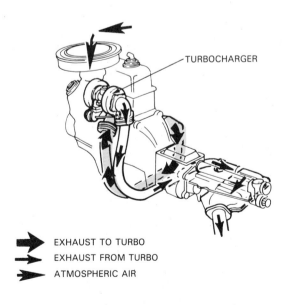

EXHAUST TO TURBO
EXHAUST FROM TURBO
ATMOSPHERIC AIR

Fig. 20-4. Typical turbocharger. A—Pipes route exhaust gases through turbine housing. Compressed air from turbo enters intake manifold and then engine combustion chamber. B—Arrows in drawing show path of exhaust gases to and from turbocharger. (Ford)

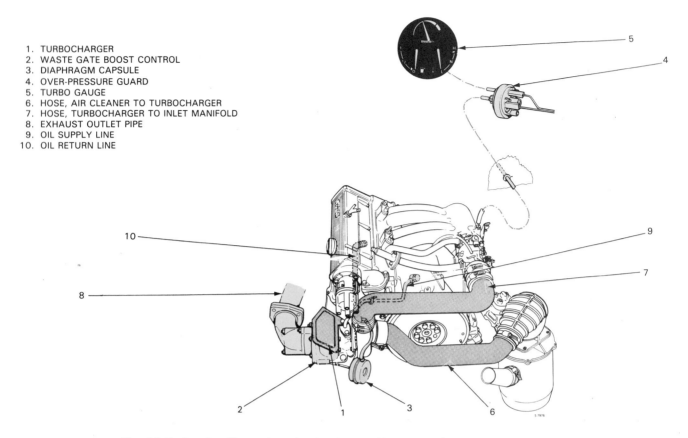

1. TURBOCHARGER
2. WASTE GATE BOOST CONTROL
3. DIAPHRAGM CAPSULE
4. OVER-PRESSURE GUARD
5. TURBO GAUGE
6. HOSE, AIR CLEANER TO TURBOCHARGER
7. HOSE, TURBOCHARGER TO INLET MANIFOLD
8. EXHAUST OUTLET PIPE
9. OIL SUPPLY LINE
10. OIL RETURN LINE

Fig. 20-5. Another illustration of turbocharger. Note names of its various parts. (Saab)

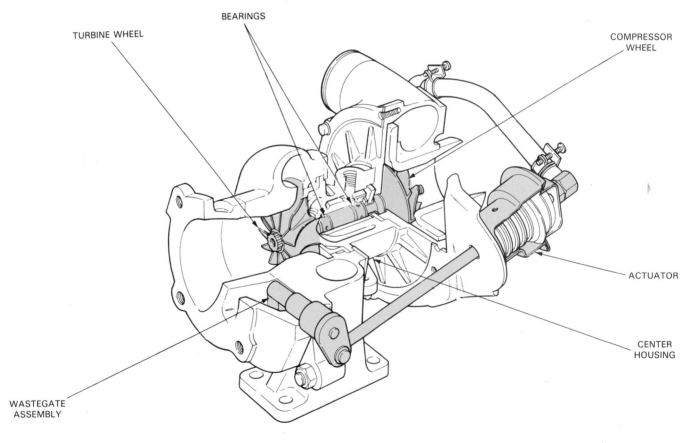

Fig. 20-6. Cutaway shows typical internal parts. Study them carefully.

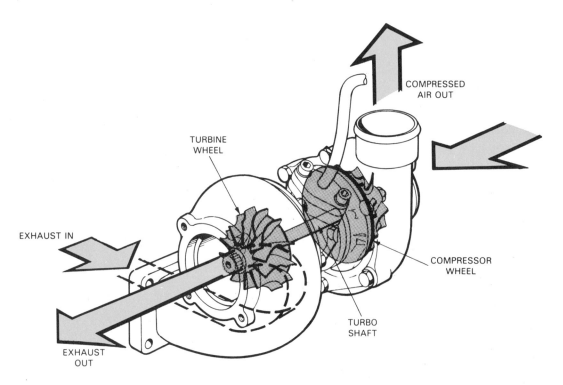

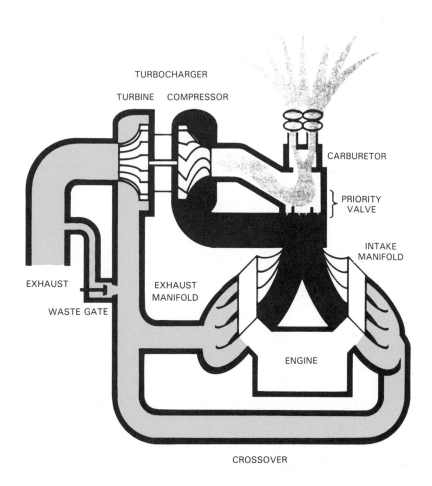

Fig. 20-7. Phantom view of turbocharger. Housing supports turbine wheel and compressor wheel and routes exhaust gases and compressed air. (Saab)

Fig. 20-8. Schematic of turbo system. Note how exhaust gases are routed to turbine wheel. When engine is under heavy load, exhaust pressure is high enough to drive turbine wheel and compress air entering intake manifold. (Turbosonic)

shaft, the compressor wheel rotates just as fast and pulls air into the compressor housing. Centrifugal force throws the air outward and into the engine.

Turbocharger location

Usually, a turbocharger is located to one side of the engine. An exhaust pipe runs between the engine exhaust manifold and the turbine housing to carry exhaust flow to the turbine wheel. As shown in Fig. 20-9, one pipe connects the compressor housing intake to a throttle body or carburetor. Another pipe conducts compressed air to the intake manifold.

In a *blow-through turbo system,* the turbocharger is located ahead of the carburetor or throttle body so that the air is compressed before it enters either the carburetor or throttle body. Fuel mixes with the incoming air after the air leaves the compressor wheel.

A *draw-through turbo system's* turbocharger is placed after the carburetor or throttle body assembly. Thus, both air and fuel (gasoline engine) pass through the compressor housing. Fig. 20-9 is this type.

In theory, the turbocharger should be located as close to the engine exhaust manifold as possible. This should allow the greatest amount of exhaust heat to enter the turbine housing. As the hot gases move across the spinning turbine wheel, they are still expanding. Greater force is created to help spin the turbine.

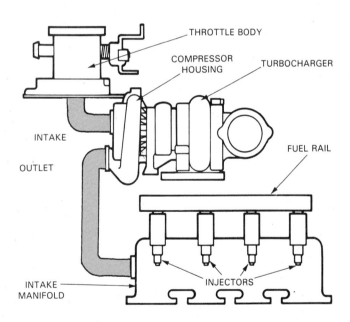

Fig. 20-9. Schematic of turbocharger of gasoline injected engine. This type is called a draw-through system since it is located between throttle body and intake manifold. Air and fuel are compressed together. (Chrysler)

Carbureted engine turbocharging

When a carbureted engine is turbocharged, both air and air-fuel mixture pass through the turbo housing. Since hot exhaust gases are passing through the other side of the turbo, this tends to increase the temperature of the fuel charge. However, it also helps mix the fuel

and air into a fine mist for good combustion.

Fig. 20-8 shows a turbocharger on a carbureted engine. Note the use of the priority valve. It is used to bypass the turbo when needed so that the air-fuel mixture can flow into the engine normally. The valve remains open for normal aspiration until the turbo has built enough pressure to operate properly. Then, the priority valve closes to increase intake manifold pressure.

Gasoline injected engine turbocharging

On a port fuel injection engine, only air passes through the turbo housing. Fuel is injected afterward at the intake valves. However, throttle body injection, like carburetion, passes both air and fuel through the turbo.

Fig. 20-9 shows a basic layout for a turbocharged engine with port fuel injection. With turbocharging, the fuel system must be able to compensate for the increased pressure inside the intake manifold. Fuel pressure must be higher to counteract the high air pressure in the intake manifold. If not, too little fuel would spray out the injectors.

Diesel engine turbocharging

Since a diesel engine's fuel is delivered into the combustion chambers, only air passes through the diesel turbocharger. Because of their much higher compression, diesel engines also benefit more from turbocharging than gasoline engines. Both power and fuel economy are improved. See Fig. 20-10.

Exhaust and intake sides of turbo

The *turbo exhaust side* includes everything from the engine exhaust valve to the end of the exhaust system. See Fig. 20-10B. These would be the engine exhaust valve, cylinder head exhaust port, exhaust manifold, turbine wheel, turbine housing, exhaust pipes, muffler, and catalytic converter, if one is used.

The *turbo intake side* takes in everything from the air cleaner to the engine intake valve. Included are the air cleaner, connecting air ducts, compressor wheel, compressor housing, intake manifold, cylinder head intake port, and intake valve. Look at Fig. 20-10B.

TURBOCHARGED ENGINE MODIFICATIONS

A turbocharged engine needs several modifications so it can withstand the increased combustion heat, cylinder pressure, and horsepower. A few of these changes shown in Fig. 20-11 include:
1. Lower engine compression ratio.
2. Stronger rods, pistons, and crankshaft.
3. Higher volume oil pump and an oil cooler.
4. Larger cooling system radiator.
5. O-ring type head gasket.
6. Heat resistant valves.
7. Knock sensor (ignition retard system).

A turbocharged engine can produce up to 50 percent more horsepower than a naturally aspirated engine of the same size. This puts greater strain on many critical engine components. Theoretically, turbocharging is like increasing the compression ratio of the engine. Under a load, more pressure is developed during the engine's compression strokes. Combustion of the air-fuel charge

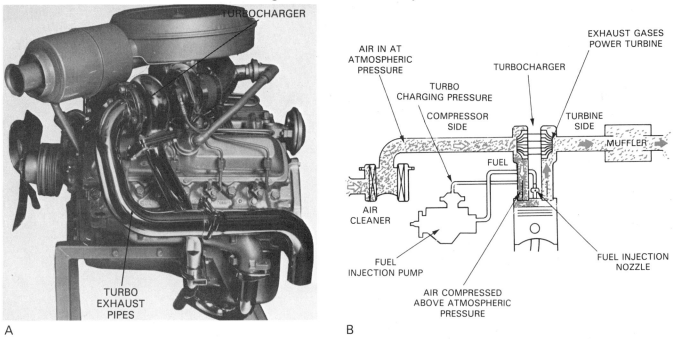

Fig. 20-10. Diesel turbocharger. A—Modern automotive diesel engine fitted with turbocharger to improve mileage and power. B—Schematic of basic diesel turbocharger. Exhaust enters one end of turbo and powers compressor fan. (Roto-Master and Chrysler)

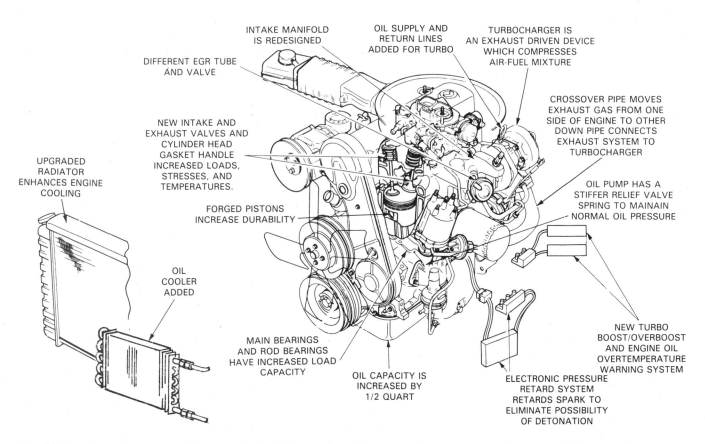

Fig. 20-11. Many changes must be made in turbocharged engines to make them stronger. Note that oil lines run from engine to supply turbo with lubrication. (Ford)

is more violent and more pressure is placed on the pistons, rods, and crankshaft. If not designed properly, an engine could destroy itself from the added heat and load created by turbocharging.

Turbocharger lubrication

Turbochargers may operate at speeds above 100,000 rpm. To help protect the turbo shaft and bearings from overheating and damage, constant lubrication is required. This is supplied by the engine's lubrication system, Fig. 20-12.

The turbocharger has a system of passages in its housing and bearings for lubrication. An oil supply line brings the oil, under pressure, from the engine to the turbo.

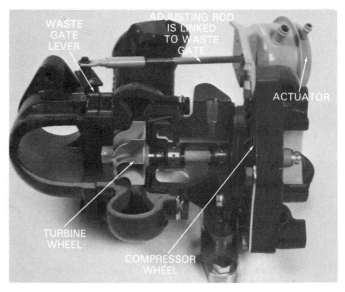

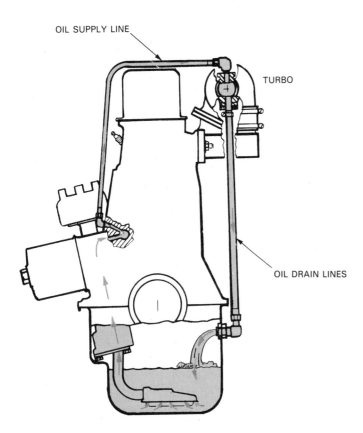

Fig. 20-12. Oil under pressure enters turbo from supply line to lubricate bearings and journals. Then gravity flow returns oil to engine sump via oil drain line. (Roto-Master)

WASTE GATE

A *waste gate* is like a relief valve that limits the amount of boost developed by the turbocharger. *Boost pressure* is the amount the turbocharger raises the air pressure before it enters the intake manifold.

The *gate* is a butterfly or poppet valve that allows exhaust to bypass the turbine wheel and enter the exhaust system. See Fig. 20-13.

Without some method of controlling the amount of exhaust energy hitting the turbine wheel of the turbocharger, too much pressure could develop in the engine combustion chambers. This could cause detonation (spontaneous, premature combustion) resulting in engine damage. The waste gate controls the amount of exhaust pressure reaching the turbine.

The waste gate is controlled by an actuator (diaphragm) shown in Fig. 20-14. The diaphragm acts on manifold pressure. When the manifold pressure reaches a certain level, the diaphragm moves and acts on linkage to the waste.

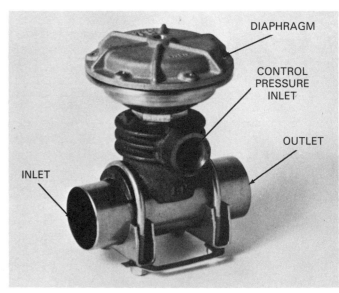

Fig. 20-13. Top. Cutaway of turbocharger. Waste gate at left bypasses exhaust back to exhaust system to relieve pressure on turbine wheel. Bottom. Unit, called activator, controls opening of waste gate. (Ford)

Waste gate operation

When the engine is under partial load, all of the exhaust gases are routed through the turbine housing, Fig. 20-15A. The strength of the actuator diaphragm spring keeps the waste gate closed. Thus, there is adequate boost to increase engine power.

But, when the engine is under full load, boost may become so high that manifold pressure becomes high enough to overcome the diaphragm spring pressure. The spring compresses and pulls on the linkage to open the waste gate valve, Fig. 20-15B. Now, some of the exhaust gases flow through the waste gate passage into the exhaust system. Pressure of exhaust on the turbine is less; the turbine slows. *Boost pressure* is limited to a preset value.

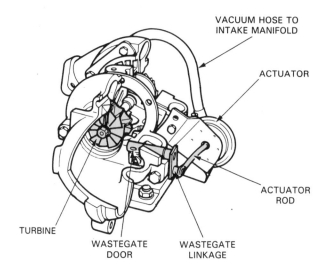

Fig. 20-14. Waste gate is linked to a vacuum-controlled diaphragm called an actuator. (Ford)

Fig. 20-15C is an over-pressure or over-boost protection system. Should the waste gate fail to open normally (malfunction), a pressure switch in the intake manifold opens, disconnecting electrical power to the fuel pump. The engine shuts down and is, thus, protected from severe damage.

Turbo lag

When the accelerator of a turbocharged automobile is depressed, the engine may not respond for a few seconds (engine lacks power). This delay, called *turbo lag,* occurs because the turbo has NOT developed sufficient BOOST. It is the result of the compressor and turbine wheels not yet spinning fast enough. Time is needed for the gases to bring the turbo up to speed.

Modern turbo systems have overcome most of this lag. The two wheels are made very light so that they can be brought up to speed quickly.

A *twin-scroll turbo,* a relatively new design, uses two exhaust gas inlet runners to prevent turbo lag. At lower

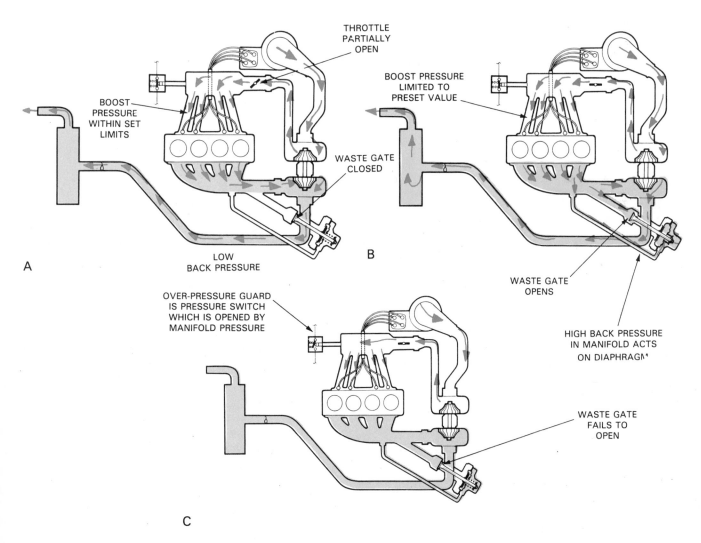

Fig. 20-15. How turbo system operates. A—Under part throttle and normal boost conditions, waste gate stays closed. Total exhaust output flows over turbine wheel. B—With engine under full load, boost pressure may become too high. Intake manifold pressure will act on diaphragm causing waste gate to open. Some exhaust bleeds off into exhaust system. C—If waste gate fails to open normally, pressure switch in intake manifold will open, shutting off electrical power to fuel pump. This prevents damage to engine. (Saab)

speed and load conditions, a flap or valve blocks one of the exhaust inlet scrolls (runners). This increases gas flow velocity to help get the turbo wheel up to speed faster. At higher engine speeds and loads, since there is enough exhaust flow to maintain velocity, the second scroll passage opens for more power.

Turbocharger intercooler

A *turbocharger intercooler* is usually an air-to-air heat exchanger that cools combustion air entering the engine. Looking a little like a radiator, it mounts at the pressure outlet of the turbocharger, Fig. 20-16. Some intercoolers use water, not air, as a cooling medium, Fig. 20-17.

As water or air flows through the intercooler and engine inlet air enters the cooler, heat is removed from the air. The cooler inlet air increases engine power because cool air is denser than warm air and contains more oxygen by volume. Cooling also reduces likelihood of engine detonation. Refer to Fig. 20-17.

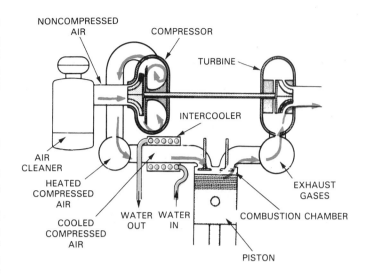

Fig. 20-17. Schematic of water-cooled intercooler. Heated compressed air is cooled by water flow through coil before it enters combustion chamber. (Waukesha)

Fig. 20-16. Intercooler is like a radiator. It cools air entering turbocharger for more efficient engine operation. (General Motors)

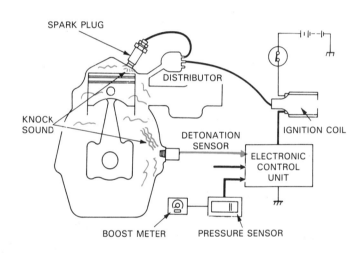

Fig. 20-18. Schematic of knock sensor operation. Sensor detects knock and sends signal to igniter. Igniter can then retard spark to spark plug. (Chrysler)

KNOCK SENSOR

A *knock sensor* is a device that allows the computer to retard ignition timing when the engine begins to knock (detonate or ping). From its mounting on the engine, Fig. 20-18, it acts like a microphone. A knock will cause it to send an electrical signal to the on-board computer which retards timing until the knock stops.

Use of the knock sensor allows the ignition timing to be advanced as much as possible for every load condition. This improves engine power and gas mileage. It protects the engine from detonation damage, as well.

COMPUTER CONTROLLED TURBOCHARGING

A *computer controlled turbocharging system* uses the electronic control unit to operate the waste gate control valve through sensor signals. Fig. 20-19 shows a diagram of one system.

The computer output is conducted (sends current) to a vacuum control solenoid. The vacuum solenoid controls engine vacuum going to the waste gate diaphragm. Several sensors, especially the knock sensor, speed sensor, and oxygen sensor, provide data to the computer.

The computer can then control vacuum to the waste gate diaphragm as needed. For example, if the knock sensor detects preignition, the computer can shut off current to the vacuum solenoid. This will let the solenoid open the waste gate to prevent detonation and damage.

The computer normally energizes the solenoid all of the time. When it does not, the waste gate opens. Other systems vary, however.

Fig. 20-20 shows still another computer controlled system. This engine has a distributorless ignition system.

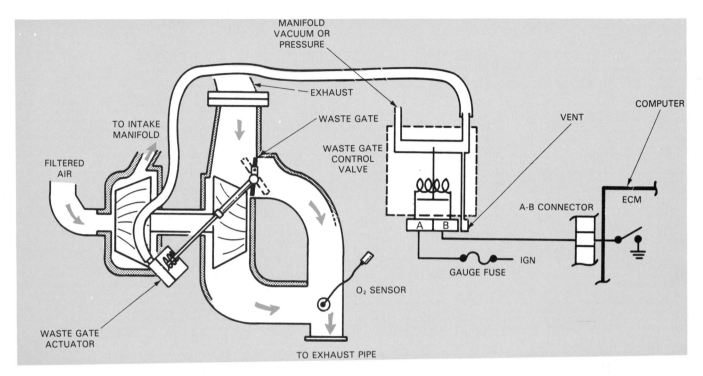

Fig. 20-19. Computer controlled turbocharging system. ECM is fed information by oxygen sensor located in exhaust system. (Buick)

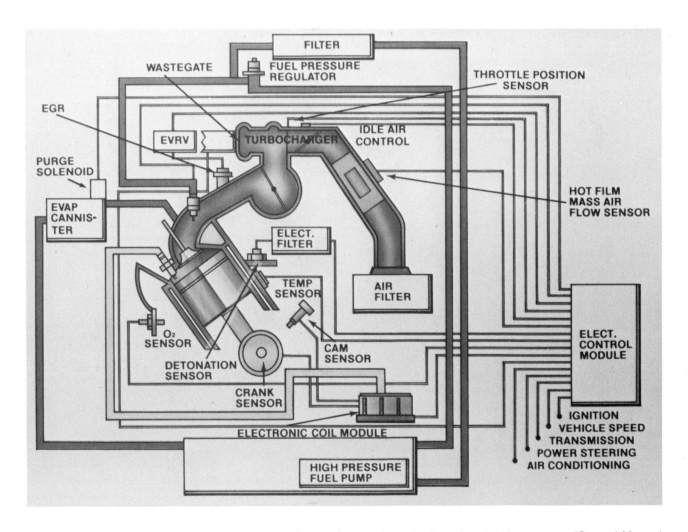

Fig. 20-20. This computer controlled turbocharger system is used on distributorless ignition system. (General Motors)

SUMMARY

Turbochargers and superchargers are devices which mount on the engine and increase the amount of air and fuel which enters the engine's combustion chambers. This is accomplished by a blower which compresses air going into the intake manifold.

A supercharger is powered by a belt, gear, or chain. A turbocharger is powered by engine exhaust.

There are three basic types of superchargers based on how they produce pressure. The rotor or Rootes type uses two rotors which trap air between them. The vane type has a rotor which is offset in its housing so air is squeezed and compressed as the rotor spins. The more common centrifugal type uses a compressor wheel which spins fast enough to throw the air or air-fuel mixture out of the housing at a higher pressure.

The basic turbocharger consists of a pair of wheels, a connecting shaft, and a housing. The turbine wheel is driven by the exhaust from the engine. It, in turn, drives the compressor wheel which compresses air entering the engine. The housing is an enclosure for routing exhaust gases and air, and keeping the two separated. A shaft connects the two wheels. The bearing housing supports bearings, seals, and contains oil passages.

Because more horsepower is developed by turbocharging, the engine and engine parts must be built stronger. If not properly designed, the engine can be destroyed by the added heat and load.

Since they operate at high rpm, turbochargers must have constant, adequate lubrication. This is supplied by the engine oiling system.

Too much pressure (boost) from the engine's turbine could damage the engine. Therefore, a bypass, called a waste gate, opens to relieve pressure when it reaches set limits.

Turbo lag is a delay in delivery of power by the engine after the accelerator has been depressed and the turbocharger goes into operation. It is caused by the inability of the turbocharger to deliver compressed air to the engine until it is brought up to speed by exhaust gases. Modern systems overcome this by using lighter turbine and compressor wheels that come up to speed faster. Some systems use twin scrolls (exhaust inlets) to overcome turbo lag. At low speeds, only one scroll operates. When more power is needed, the second begins to operate.

Modern turbochargers are fitted with intercoolers that lower the temperature of the inlet air and sensors which tell a computer how the turbocharged engine is operating. Several sensors, provide data on which the computer acts to improve engine efficiency.

KNOW THESE TERMS

Turbocharger, Supercharger, Normal aspiration, Pumping losses, Rootes supercharger, Vane supercharger, Centrifugal supercharger, Turbine wheel, Turbo shaft, Compressor wheel, Blow-through turbo system, Draw-through turbo system, Waste gate, Waste gate actuator, Boost, Turbo lag, Twin scroll turbo, Intercooler.

REVIEW QUESTIONS — CHAPTER 20

1. Explain the difference between a supercharger and a turbocharger. Explain how they are the same.
2. An engine with _____ _____ uses only air at normal atmospheric pressure in its combustion chambers.
3. Give the main advantage of a supercharger over a turbocharger.
4. Name the three main types of supercharger.
5. List the six main components of a turbocharger.
6. When the turbocharger is located ahead of the carburetor or throttle body it is known as a _____ _____ turbo system.
7. Indicate which of the following DO NOT belong to the intake side of the turbocharger:
   a. Turbine wheel.
   b. Compressor wheel.
   c. Air cleaner.
   d. Cylinder head intake port.
   e. Intake valve.
   f. Intake manifold.
   g. Air ducts connecting compressor housing and intake manifold.
8. A turbocharged engine can produce up to 50 percent more horsepower than a naturally aspirated engine of the same size. True or false?
9. List five modifications necessary on a turbocharged engine to improve its durability.
10. A _____ _____ is a device for limiting the amount of boost developed by a turbocharged engine.
11. A _____-_____ turbo uses two exhaust gas inlet runners to prevent turbo lag.
12. An intercooler can be either an air-to-air heat exchanger or a water-to-air heat exchanger whose purpose is to cool the turbocharged air going into the intake manifold. True or false?
13. What can be done to control knocking in a turbocharged engine?
14. In computer controlled turbocharging, a current from the computer controls the _____ _____ which opens and closes the waste gate.

21 TURBOCHARGER DIAGNOSIS, SERVICE, REPAIR

After studying this chapter, you will be able to:
- *Diagnose common turbocharger problems.*
- *Inspect for turbocharger system troubles.*
- *Measure boost pressure as a means of system diagnosis.*
- *Remove and replace a turbocharger.*
- *Measure turbocharger shaft and bushing wear.*
- *Disassemble and reassemble a turbocharger.*
- *Service a turbo waste gate.*

Turbochargers operate at over 100,000 rpm in "metal melting" temperatures above 1000 °F (538 °C). Even the slightest driver abuse or service neglect can lead to rapid turbocharger failure. As a fuel system technician, it is important for you to be able to quickly and efficiently diagnose and fix turbocharging systems.

INCREASING TURBOCHARGER LIFE

The average service life of a turbocharger is approximately 50,000 miles (about 80 500 km). However, this figure can be considerably lower if certain rules are not followed. Some turbos last over 100,000 miles when properly maintained. With the high replacement cost of a turbo, it is wise to pass on to your customers some basic tips on driving and servicing a turbocharged engine:
1. Change the engine oil and filter at regular intervals. Many automakers recommend an oil change TWICE AS OFTEN when the engine has a turbocharger. With the turbo shaft spinning at high rpm, impurities in the oil will cause greater shaft and bearing wear and possible failure.
2. Allow the engine and engine oil to reach full operating temperature before "kicking in" or "boosting" the supercharger. An oil temperature gauge may help.
3. If oil temperature is too high, it can also cause turbocharger failure. An oil cooler may be installed to help protect the turbocharger.
4. After driving the car at highway speeds, allow the engine to idle for a few moments before shutting the engine off. This will let the turbo cool down from its potential internal temperature of about 1000 °F (538 °C). At idle, exhaust gases are much cooler and

will help prevent heat damage as the unit cools. Idling may also possibly prevent oil starvation damage. It allows the turbo to slow down before the engine and oil supply stop.

SERVICING TURBOCHARGING SYSTEMS

Common turbocharging system PROBLEMS include inadequate boost pressure (which shows up as lack of engine power) and leaking shaft seals which cause engine oil consumption. Damaged turbine or compressor wheels may cause vibration and noise. An inoperative waste gate may cause excess boost (detonation).

For specific troubleshooting information, refer to a factory service manual. It will list common troubles for the turbo system being serviced. Fig. 21-1 shows typical problems.

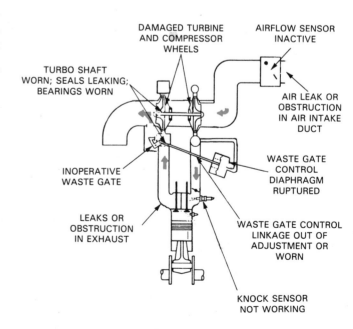

Fig. 21-1. Any of these problems with turbocharger system will affect operation of turbocharged engine. (Ford)

Checking the turbocharging system

Frequent *checks* should be made of the turbocharger's condition. An inspection should include:

1. Check for deterioration of vacuum lines to waste gate control. Make certain connections do not leak.
2. Check of condition of oil lines to the turbocharger. At turbine speeds, interruption of lubrication for even a few seconds could cause severe bearing damage.
3. Using a regulated low-pressure hose check for waste gate control diaphragm leakage and determine if the diaphragm is working properly.
4. Checking the turbo boost pressure (pressure developed by turbo under load) by observing the dash gauge or use a test gauge. It may be necessary to connect the gauge to an intake manifold fitting to conduct this test. See Fig. 21-2.

Checking the turbocharger

Checking out the condition of the inside of the turbocharger requires removing it from the engine. Disconnect the oil lines, and unbolt the unit from its mountings, Fig. 21-3. Move the unit to a workbench.

Before disassembling the unit, scribe an alignment mark on the housing.

Open the housing and inspect the interior for oil con-

Fig. 21-2. If there is no turbo pressure dash gauge, use small pressure gauge with long section of hose for on-road testing of boost pressure. (Saab)

Removal

1. Remove air cleaner.
2. Remove turbo heat shield (not shown).
3. Remove exhaust outlet pipe.
4. Remove exhaust down pipe.
5. Remove boost control tube.
6. Remove oil supply line.
7. Remove throttle bracket.
8. Remove rear wastegate actuator vacuum line.
9. Remove EGR tube.
10. Remove dipstick and tube if necessary.
11. Remove necessary vacuum hoses.
12. Remove 4 bolts attaching turbocharger to intake manifold.
13. Remove 4 bolts from turbocharger rear brace and remove turbocharger from engine.
14. For installation of turbocharger, replace the 3 O-rings, grease the compressor outlet O-ring and reverse the above procedure.

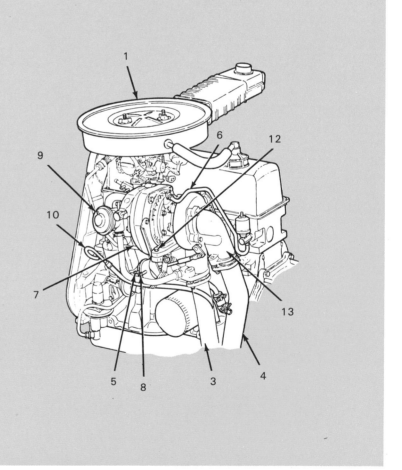

Fig. 21-3. Note typical procedure for removing turbocharger from vehicle. (Ford Motor Co.)

tamination, Fig. 21-4. Also check the turbine and compressor wheels. They should be clean and free of damage. Even the slightest imperfection could throw the wheels out of balance and cause severe vibration or disintegration. See Fig. 21-5. Fig. 21-6 illustrates how bearing and shaft wear are measured. Make sure the turbine assembly spins freely and does not rub on the housing when rotated by hand.

WARNING! Never use any hard, metal tool or sandpaper to remove carbon deposits from the turbine wheel. Gouges or abrasions can remove enough metal to cause severe vibration or even destruction of the turbo. Cleaning can be accomplished with a soft wire brush and a solvent.

Turbocharger repairs

Most problems with a turbo cannot be repaired in the field. If the wheels, shaft, bearings, or center section of the turbo fail, a new or factory rebuilt unit is commonly installed. We will discuss this procedure later.

Minor problems, however, such as a bad waste gate control diaphragm, leaking seals or hoses, damaged housings, etc., can be corrected in-shop. Always refer to a service manual for the exact make and model engine or turbo being serviced.

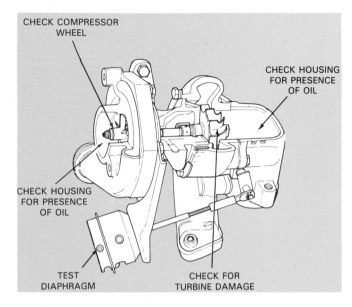

Fig. 21-4. During inspection of opened turbocharger housing, look for oil in housing which would indicate leaking seals. Also check turbine and compressor wheels. They should be free of debris, knicks, wear, or defects of any kind. Rotate them to check for binding or rubbing. (Ford)

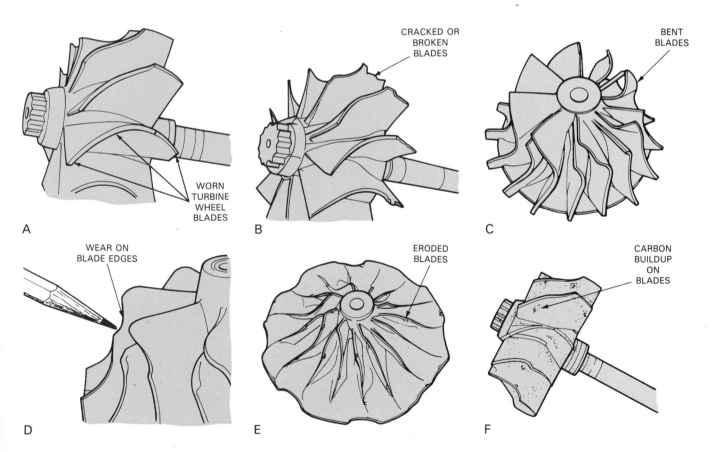

Fig. 21-5. Typical turbine and compressor wheel damage and wear. A—This wear was caused by heavy blade contact against housing after bearing failure. B—Blades damaged by ingestion of foreign objects. Note that damage is about same to all blades. C—Compressor wheel damage resulting from ingestion of soft material such as shop rags or rubber booting. D—This compressor wheel shows wear caused by abrasive material such as dirt. Blades have rounded edges as though sanded. E—Extreme erosion from sand striking compressor wheel. Ingestion was caused by loose duct between turbocharger inlet and air cleaner. F—Turbocharger seal leak allowed oil into turbine where it burned. Hardened ash stuck to turbine blades. (Ford)

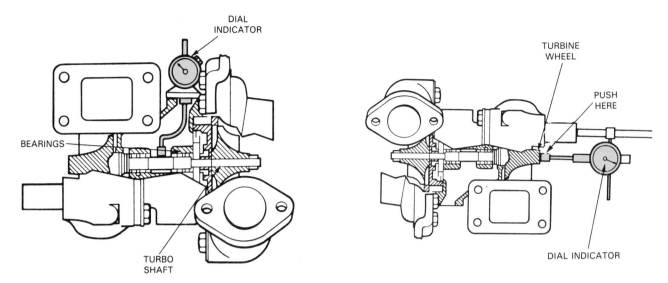

Fig. 21-6. Dial indicator should be used to measure bearing and shaft condition. Left. Measuring radial clearance between shaft and bearings. Right. Measuring shaft end play.

If the main components of the turbo are in good shape, reassemble the unit, Fig. 21-7. Make sure there are no deposits or debris in the housings. Realign scribe marks. Install new gaskets and seals and make sure they are properly positioned. Torque fasteners to specs in recommended sequence. Pre-oil the unit by squirting motor oil into the oil inlet opening. Spin the impeller-turbine to check for binding before installing unit to engine.

Installing a new turbocharger

When a new or rebuilt turbocharger must be installed, you should:

1. Double-check that the new unit is the correct one for the engine. Compare the part numbers.
2. Install new gaskets and seals.
3. Torque fasteners to specs.
4. Change engine oil and flush oil lines, if needed, before

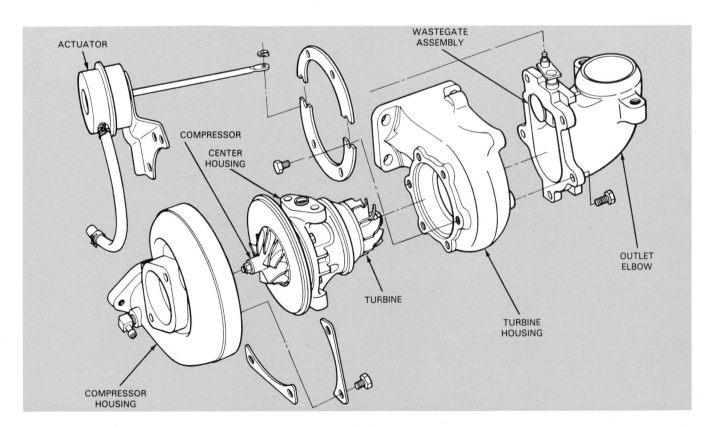

Fig. 21-7. Exploded view of typical turbocharger. Refer to similar illustrations in shop manual when reassembling unit. (Ford)

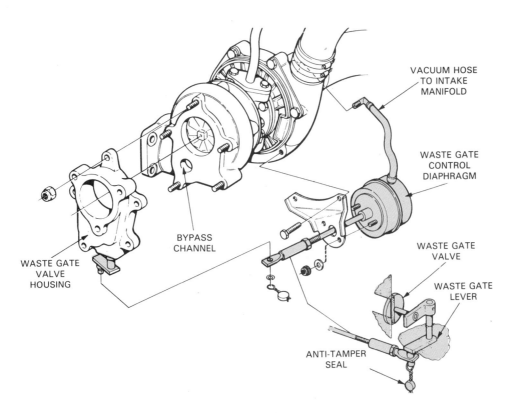

Fig. 21-8. Waste gate control, sometimes called an actuator, bolts to side of turbo housing. Linkage rod connects control to waste gate mechanism. Vacuum from intake manifold controls movement of diaphragm in waste gate control. Seal prevents unauthorized tampering with boost setting. Overboost, while increasing power, can cause severe engine and turbocharger damage. (Saab)

starting the engine.
5. If oil problems were the cause of the failure of the old unit, check oil supply pressure in the feed line to the turbo.

Waste gate service

Too little or too much boost pressure could be caused by an INOPERATIVE WASTE GATE. If the gate is sticking open, exhaust will bypass the turbine and the unit will not produce sufficient boost pressure. The engine will lack power. If the waste gate remains closed, regardless of engine load, there will be too much boost. Detonation and engine damage can result.

Several conditions may cause a waste gate malfunction. Carbon buildup can cause binding or failure of the gate to close. A defective waste gate control (bad diaphragm or leaking vacuum hose) can also prevent waste gate from operating.

Before condemning the waste gate or the waste gate control, always search for and try to eliminate other problems. Check the knock sensor (spark retard system) if there is one. Also check the ignition timing. Make sure the vacuum pressure lines are all properly connected. If the engine has a computer control unit, check its operation. A computer self-diagnosis test might find a faulty sensor or control device.

Follow suggested procedures in the appropriate service manual for testing and/or replacing the waste gate or waste gate control. As shown in Fig. 21-8, removal

of the waste gate control is simple. Unbolt the unit, disconnect the vacuum lines and lift the unit away. Most manuals recommend replacement of the control unit rather than attempt in-shop repairs.

SUMMARY

Because turbochargers operate at high speed and in temperatures over 1000 °F, they require proper care and maintenance.

Turbo life can be increased by: frequent oil and filter changes, refraining from using turbo until engine and engine oil are at operating temperature, keeping oil temperature as low as possible (installing an oil cooler is often advisable), allowing the engine to idle for a few minutes after coming off a high-speed trip.

Common problems, when turbos do require service include poor boost pressure, leaking shaft seals, damaged turbine or compressor wheels, and too much boost. To determine the general condition of the unit, the mechanic should:
1. Check vacuum lines and connections to waste gate and oil lines to the turbo.
2. Check the wate gate diaphragm for proper operation.
3. Measure boost pressure.
4. Listen for bad turbocharger bearings.

An internal inspection and cleaning is sometimes necessary. The unit should be removed and the work done at a bench.

Turbo wheels can be cleaned with a soft wire brush and solvent. If there is damage or wear, the entire unit should be replaced by a new or rebuilt unit.

Minor repairs, such as replacement of seals, hoses, waste gate diaphragm, or damaged housing can be done in the shop.

Before disassembly, scribe a mark on the housing. When reassembling, align the scribe marks. Spin the impeller/turbine unit to make sure it is not binding or rubbing inside the housing.

KNOW THESE TERMS

Oil starvation, Leaking shaft seals, Ruptured control diaphragm, Inoperative waste gate, Impeller or turbine damage.

REVIEW QUESTIONS—CHAPTER 21

1. Turbochargers operate in temperatures over:
   a. 500 °F.
   b. 800 °F.
   c. 1000 °F.
   d. 1500 °F.
2. Give four general rules for prolonging turbocharger service life.
3. Inadequate boost pressure in a turbocharger would show up as a lack of _____ _____.
4. List four basic checks that can be made to determine general condition of a turbocharger.

5. Which of the following would NOT be used to clean turbine/compressor wheels?
   a. Solvent.
   b. Soft wire brush.
   c. Fine sandpaper.
6. It is normal for the turbo shaft and wheels to turn hard after the turbocharger has been reassembled. True or false?
7. How can you check that a replacement turbo is the correct one for the vehicle?
8. If a waste gate is stuck open, which of the following conditions could result?
   a. It would have no effect on engine operation.
   b. The engine would lack power.
   c. Detonation, excessive boost, and possible engine damage.
9. A customer complains of severe ping or knock. The turbocharged engine has good power.
   Mechanic A says the waste gate must be stuck closed, allowing too much boost.
   Mechanic B says that the ignition timing and boost pressure under load should be measured.
   Who is correct?
   a. Mechanic A.
   b. Mechanic B.
   c. Both Mechanic A and B.
   d. Neither Mechanic A nor B.
10. A bad turbocharger cannot cause engine oil consumption. True or false?

22 EXHAUST SYSTEM OPERATION

After studying this chapter, you will be able to:
- *Identify the basic parts of a modern exhaust system.*
- *Explain the operation of the exhaust system.*
- *Compare exhaust system design differences.*
- *Explain the operation of a heat riser.*
- *Compare differences in exhaust manifold designs.*
- *Describe how mufflers and catalytic converters are constructed.*

An exhaust system is an important part of a fuel system. It must treat the poisonous by-products of combustion and carry them away. See Fig. 22-1.

A properly maintained and designed exhaust system adds to the power and efficiency of the engine. An unmaintained exhaust system can adversely affect engine power and fuel economy. It is important, therefore, for you to understand exhaust system operation and basic service methods used to repair exhaust system troubles.

EXHAUST SYSTEM FUNDAMENTALS

An *exhaust system* quiets engine operation and carries exhaust fumes to the rear of the car. The parts of a typical system are shown in Fig. 22-2:

1. EXHAUST MANIFOLD—Connects cylinder head exhaust ports to header pipe.
2. HEADER PIPE—Specially shaped steel tubing that carries exhaust gases from exhaust manifold to catalytic converter or muffler.
3. CATALYTIC CONVERTER—Canister-like device for removing pollutants from engine exhaust.
4. INTERMEDIATE PIPE—Connecting pipe sometimes used between header pipe and muffler or catalytic converter and muffler.
5. MUFFLER—Metal chamber for damping pressure pulsations and reducing exhaust noise.
6. TAILPIPE—Length of pipe that carries exhaust from

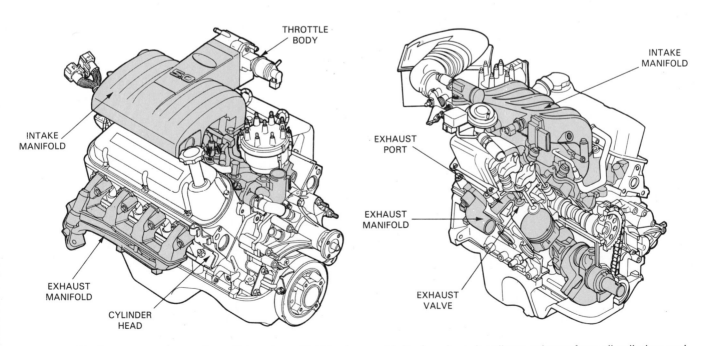

Fig. 22-1. Modern automotive engines. Exhaust manifold is shown at left of engines. It collects exhaust from all cylinders and routes it to header pipe. (Ford)

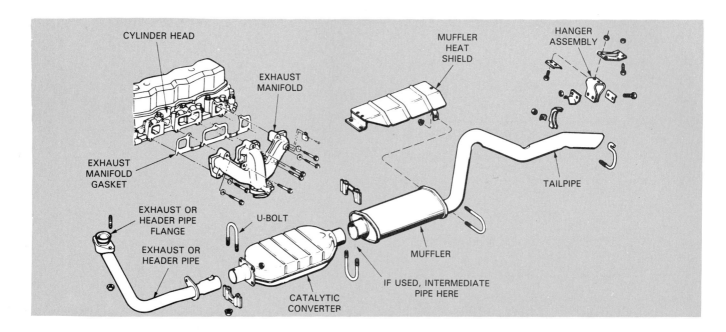

Fig. 22-2. Exploded view of typical exhaust system. Exhaust exits cylinder head, enters exhaust manifold, travels through header pipe to catalytic converter, intermediate pipes, through muffler, and out the tailpipe. (AMC)

muffler to rear of car body.

7. HANGERS—Devices for securing exhaust system to underside of car body or frame.

8. HEAT SHIELDS—Metal plates that prevent transfer of exhaust heat to flammable materials.

9. MUFFLER CLAMPS—U-bolts for holding together sections and parts of exhaust system.

Exhaust system operation

A running engine blows extremely hot gases out of the cylinder head exhaust ports. The gases enter the exhaust manifold where they expand and start to cool. They flow through the header pipe, catalytic converter, intermediate pipe, muffler, and out the tailpipe.

Exhaust back pressure

Exhaust back pressure is the amount of pressure above atmospheric pressure developed in the exhaust system while the engine is running. High back pressure reduces engine power because spent gases cannot leave the combustion chamber as easily as they should. A well designed system should have LOW back pressure.

Every part of the exhaust system—pipes, catalytic converter, muffler, and tailpipe—contributes to exhaust back pressure. Larger pipes and a free-flowing muffler, for example, would help reduce back pressure.

EXHAUST MANIFOLD

An *exhaust manifold* bolts to the cylinder head to enclose the exhaust port openings, Fig. 22-3. The cylinder head mating surface is machined carefully so it is absolutely flat. An exhaust manifold gasket is commonly used between the cylinder head and manifold to help prevent exhaust leakage.

The outlet of the exhaust manifold, a single round

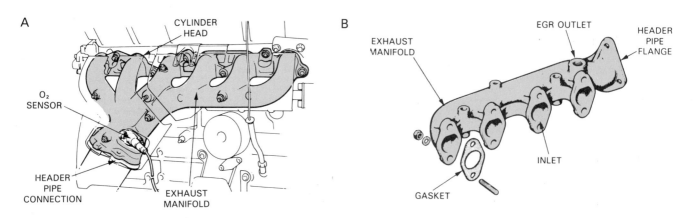

Fig. 22-3. A—Surface bolted to cylinder head is machined for close fit. B—A typical four-cylinder exhaust manifold. Gaskets prevent exhaust leakage. (Toyota and AMC)

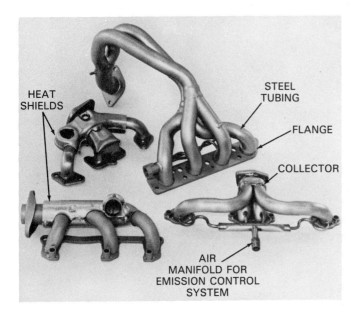

Exhaust manifold construction

Exhaust manifolds can be made from cast iron, cast aluminum, or steel tubing. Cast iron manifolds are heavy but last a long time under normal service. Cast aluminum manifolds are not commonly used but may be found on a few specialty sports cars.

Exhaust manifolds of thin-wall, steel tubing are becoming popular for several reasons, Fig. 22-4. They are lightweight and can be bent into complex configurations to make each exhaust tube equal in length. Having tubes equal in length increases engine power and efficiency since the pressure pulsations in the exhaust manifold are equalized. This helps make each engine cylinder do the same amount of work (same amount of exhaust scavenge and reversion). Look at Fig. 22-5.

HEAT SHIELDS

Parts of the exhaust system, especially catalytic converters and mufflers, become extremely hot during engine operation. *Heat shields* provide protection when these components are close to the auto body. They absorb and deflect heat that could damage the body or start fires under the car. Refer to Figs. 22-2 and 22-6.

> DANGER! Always reinstall exhaust system heat shields. Without a heat shield, car undercoating, carpeting, dry leaves on the ground, and other flammable materials could catch on fire!

Exhaust manifold heat valve

An *exhaust manifold heat valve* diverts hot exhaust

Fig. 22-4. Thin-wall, lightweight steel tubing is becoming popular for fabricating exhaust manifolds. Aside from being lighter, they can be shaped to make individual tubes equal in length which balances out pressure pulsations.

opening, is fitted with stud bolts or cap screws to hold the header pipe. A gasket or O-ring (doughnut) seals the joint between the exhaust manifold and header pipe. It can be made of fiber or steel.

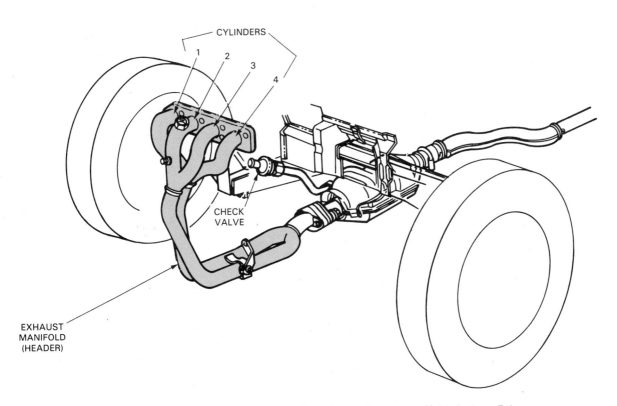

Fig. 22-5. Simplified drawing of modern exhaust manifold design. Exhaust manifold and header pipes are designed so all four cylinders have same back pressure. (Ford)

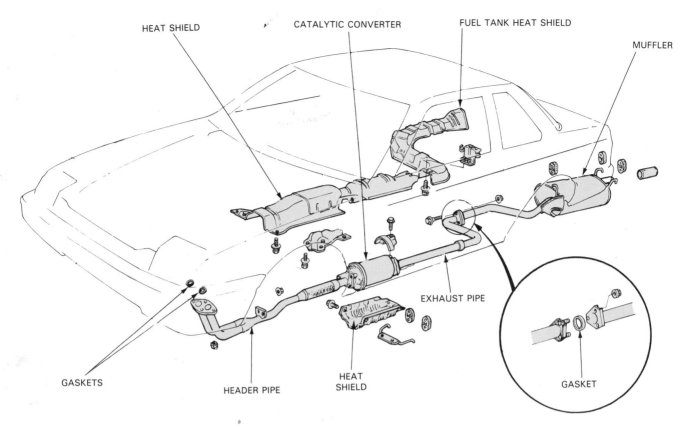

HEAT SHIELD

CATALYTIC CONVERTER

FUEL TANK HEAT SHIELD

MUFFLER

EXHAUST PIPE

GASKETS

HEADER PIPE

HEAT SHIELD

GASKET

Fig. 22-6. Complete exhaust system is shown with appropriate shielding in areas where extreme heat is developed. Note heat shielding around catalytic converter. (Honda)

gases into the intake manifold for better cold weather starting. Look at Figs. 22-7 and 22-8.

It is simply a butterfly valve, usually located in the

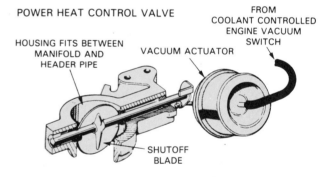

POWER HEAT CONTROL VALVE

HOUSING FITS BETWEEN MANIFOLD AND HEADER PIPE

VACUUM ACTUATOR

FROM COOLANT CONTROLLED ENGINE VACUUM SWITCH

SHUTOFF BLADE

Fig. 22-8. Cutaway shows heat control valve, sometimes called heat riser. When closed, it redirects hot exhaust gases into intake manifold. It helps engine run smoothly during warmup. (Chrysler)

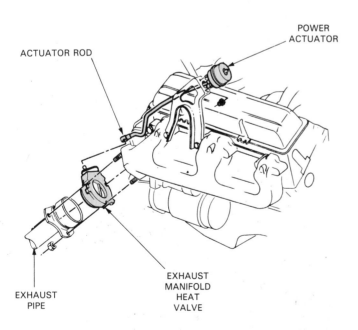

ACTUATOR ROD

POWER ACTUATOR

EXHAUST MANIFOLD HEAT VALVE

EXHAUST PIPE

Fig. 22-7. Typical exhaust manifold heat valve arrangement. Note how actuator (vacuum activated control mechanism) is linked to valve to open and close it according to engine temperature. (Pontiac)

outlet of the exhaust manifold. A heat-sensitive spring or a vacuum diaphragm and temperature sensing vacuum switch may control the valve.

On a cold engine, the heat valve blocks the exhaust outlet. This increases exhaust pressure and hot gases are forced into an exhaust passage in the intake manifold, Fig. 22-9. This warms the floor of the intake manifold improving fuel vaporization. Once the engine is warmed up, the heat valve opens.

Heat tubes

Heat tubes are sometimes used to carry exhaust gases

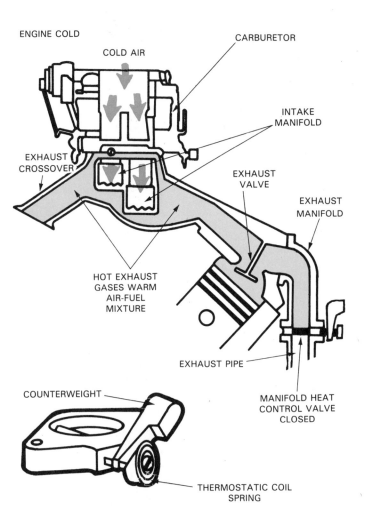

Fig. 22-9. When heat control valve is closed, back pressure in exhaust forces large amounts of hot exhaust into chamber in bottom of intake manifold. This warms and helps vaporize fuel. (Florida Dept. of Vocational Education)

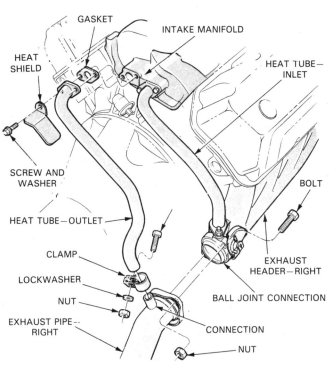

Fig. 22-10. Heat tubes on some types of engines route hot exhaust through intake manifold. (Chrysler)

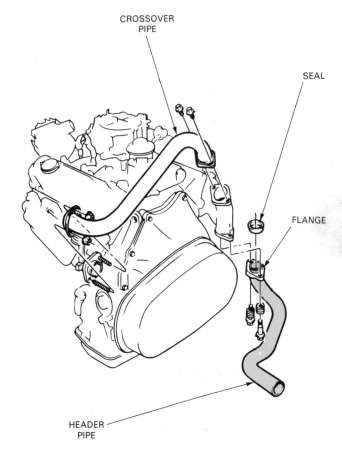

Fig. 22-11. Header pipe connects to end of manifold. Flange and seal, sometimes called "doughnut," assure no-leak connection. (Pontiac)

from the header pipe to the intake manifold. This is shown in Fig. 22-10. The tubes are made of thin, steel tubing and have clamps or flanges on each end. They serve the same basic function as a heat riser valve—to help warm the plenum of the intake manifold for better cold weather drivability.

EXHAUST SYSTEM CONSTRUCTION

Exhaust pipes carry the burned gases from the exhaust manifold, to the catalytic converter (if used), to the muffler, and, then, to the rear of the car. They are usually made of rust resistant steel tubing or stainless steel.

The *header pipe* is a pipe section that bolts directly to the exhaust manifold, as shown in Fig. 22-11. A FLANGE and O-ring gasket connect it to the manifold.

The *header pipe "doughnut"* or *O-ring* gasket is made of heat resistance fibers or metal. It fits between a recess in the exhaust manifold and a lip formed on the header pipe. It prevents exhaust leakage between the two. As shown in Fig. 22-12, stud bolts extend through the header pipe flange and into the exhaust manifolds. Nuts

461

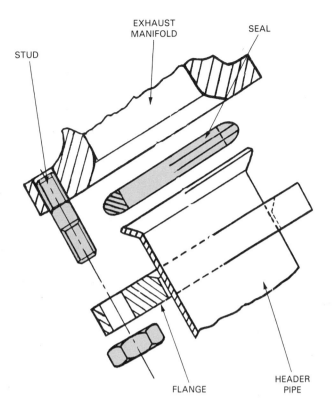

Fig. 22-12. Cross section shows stud and seal arrangement at joint formed by exhaust manifold and header pipe. (Saab)

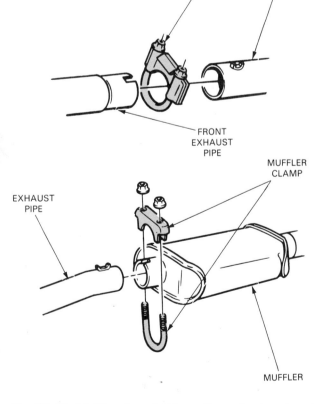

Fig. 22-13. Muffler clamp holds sections of exhaust system securely together. One is used at either end of muffler. (Buick)

clamp the doughnut between the assembly.

Muffler clamps are large U-bolts that clamp exhaust pipes together. The clamps squeeze the outer pipe around the inner pipe to hold them securely together. They can also be used to hold the pipes to their hangers. See Fig. 22-13.

Hangers hold the exhaust system in place under the car. They fasten to the underbody of the car and to the exhaust pipes. Fig. 22-14 shows common designs.

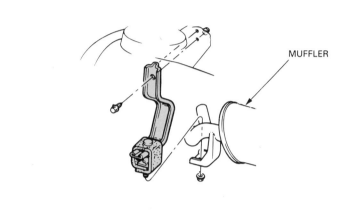

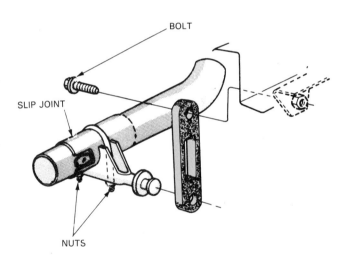

Fig. 22-14. Hangers fasten exhaust system to underbody of car. (Oldsmobile and Chrysler)

MUFFLERS

Fig. 22-15 shows the inside of the muffler. Note how various chambers, tubes, holes, and baffles are arranged to cancel out the pressure pulsations which occur each time an exhaust valve opens.

A *reverse-flow muffler* changes the direction of exhaust gas flow through the inside of the muffler. This is the most common type of muffler on passenger cars. Look at Fig. 22-16.

A *straight-through muffler* allows exhaust gases to flow straight through a single tube. The tube has perforations that help break up the pressure pulsations. A straight-through muffler does not quiet the exhaust as much as a reverse flow muffler.

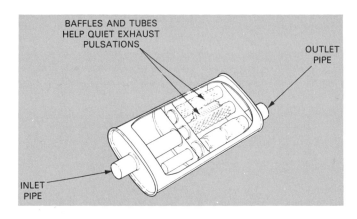

Fig. 22-15. Note interior construction of muffler. It is designed to reduce back pressure and quiet sound. (Walker)

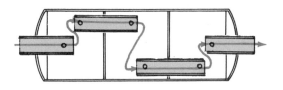

STRAIGHT-THROUGH MUFFLER

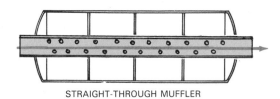

REVERSE—FLOW MUFFLER

Fig. 22-16. Two types of mufflers are used on passenger cars. (Deere & Co.)

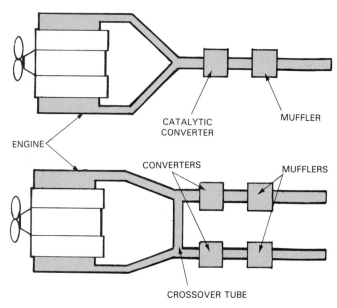

Fig. 22-17. Simplified drawing shows two types of muffler systems. Top. Single exhaust system. Bottom. Dual exhaust is used on high performance cars.

CATALYTIC CONVERTER

A *catalytic converter* is used to help burn any fuel that enters the exhaust system. Extreme heat in the converter oxidizes (combusts) the emissions (unburned and partially burned fuel) that flow out of the engine.

The catalytic converter is located ahead of the muffler in the system, Fig. 22-18. Thus, more engine heat enters the converter to help it burn off emissions.

There are several types of catalytic converters: mini catalytic converter, monolithic catalytic converter, and

A *resonator* is sometimes added to the exhaust system after the muffler to reduce exhaust noise even more. It is constructed like a straight-through muffler.

Single and dual exhaust systems

A *single exhaust system* has one path for exhaust flow through the system, Fig. 22-17, top. Typically, it has only one header pipe, main catalytic converter, muffler, and tailpipe. The most common, it is used from the smallest 4-cylinder engines, up to large V-8 engines.

A *dual exhaust system* has two separate exhaust paths to reduce back pressure, Fig. 22-17, bottom. It is two single exhaust systems combined into one. A dual exhaust system is sometimes used on high performance cars with large V-6 or V-8 engines. It lets the engine "breathe" better at high rpm.

A *crossover pipe* connects the right and left sides of the exhaust system on V-type engines. On a single exhaust system, it simply allows both exhaust manifolds to empty into the same system. However, on dual exhaust systems, the crossover pipe is used to balance back pressure in each side and increase engine power and efficiency.

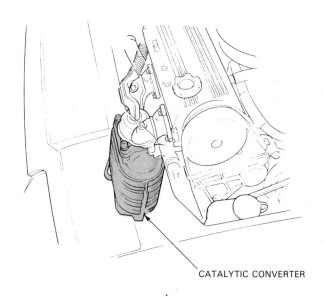

Fig. 22-18. Catalytic converter is added to modern cars to burn off polluting emissions. Engine exhaust heat assists in their action. This type is known as a mini converter. (Honda)

463

pellet catalytic converter. There are other designs within these categories.

A *mini catalytic converter* is a very small unit located right next to the engine exhaust manifold, Fig. 22-18. It is used to quickly heat up and start burning the exhaust pollutant quickly. A main catalytic converter is normally located right after the mini converter.

A *monolithic catalytic converter* is made of a large ceramic block shaped like a honeycomb or bee hive. It is coated with special chemicals that help the converter act on the exhaust gases. Refer to Fig. 22-19A.

A *pellet catalytic converter* has hundreds of small beads that act as the catalyst agent. They do the same thing as the honeycomb shaped block. However, the pellets can be removed and replaced to restore the converter after extended service. Look at Fig. 22-19B.

PARTICULATE OXIDIZER

A *particulate oxidizer* is a special catalytic converter designed to burn diesel engine particulates. *Particulates* are solid particles of carbon-like soot that blow out the diesel engine as black smoke.

A particulate oxidizer, also called a *trap oxidizer,* is located between the exhaust manifold and turbocharger. It has a ceramic monolithic element housed inside a steel cylinder, Fig. 22-20A.

The surface of the monolithic element is electroplated with a silver alloy. The ceramic element has small holes or pores in it. Since the passages through the element are alternately closed, this forces any particulates to flow through the small pores in the element, Fig. 22-20B. The particulates stick in these pores and are burned before

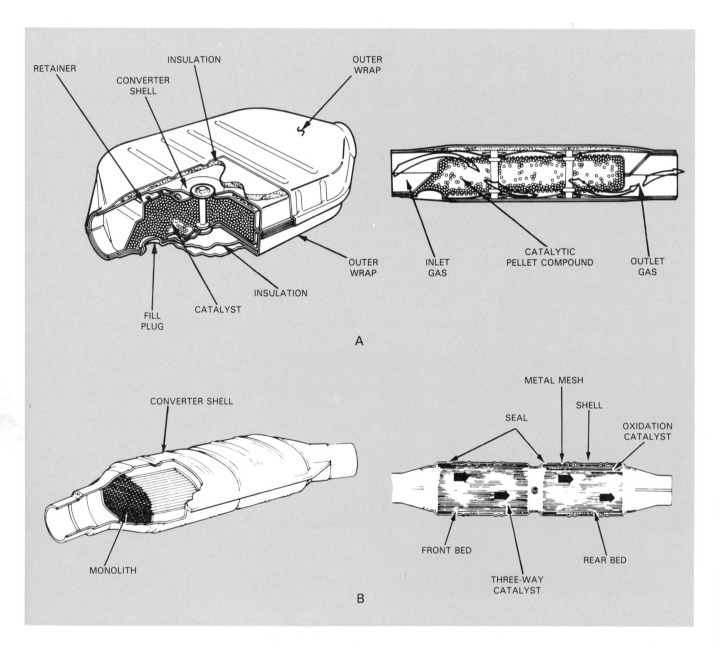

Fig. 22-19. Two popular types of catalytic converter. A—Pellet type. B—Monolithic type. Note names of parts. (Ford and General Motors)

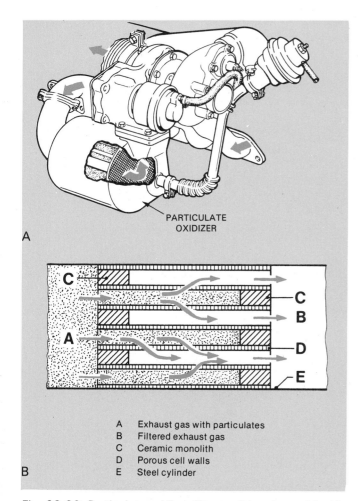

A

B

A Exhaust gas with particulates
B Filtered exhaust gas
C Ceramic monolith
D Porous cell walls
E Steel cylinder

Fig. 22-20. Particulate oxidizer filters solid, unburned fuel in diesel exhaust. A—Routing of exhaust through oxidizer. B—How oxidizer traps particulates. (Mercedes-Benz)

they can flow out of the exhaust system. The burning or oxidation of the particulates produces carbon dioxide.

SUMMARY

The exhaust system is an important part of the fuel system. It treats and carries the byproducts of combustion to the back of the car before discharging them.

The system is made up of several parts. These include: manifold, header pipe, catalytic converter, intermediate pipe, muffler, tailpipe, hangers, heat shields, and muffler clamps.

While the exhaust system is treating exhaust and muffling the combustion sounds, it must not create too much back pressure.

The exhaust manifold, which bolts to the cylinder head, collects the exhaust from the cylinders after combustion and delivers it to the header pipe. The manifold may be cast from iron or aluminum or fabricated from steel tubing.

Heat shields are located around portions of the exhaust system that get very hot. They must always be reinstalled after the system is repaired to protect the car's undercoating, carpeting, etc. from fire.

An exhaust manifold heat valve diverts hot exhaust

gases into the intake manifold to aid cold engine starting. As the engine warms, a temperature-sensing switch or spring will open the valve.

Exhaust pipes are of several types, each serving a special purpose. The header pipe connects to the exhaust manifold and carries the exhaust to the catalytic converter. An intermediate pipe carries the exhaust from the catalytic converter to the muffler. It is also used between the header pipe and catalytic converter.

The muffler is an arrangement of tubes within a canister that muffle the pressure pulsations as each exhaust valve opens. A reverse-flow muffler changes the flow of exhaust gases before they exit into the tail pipe. A straight-through muffler has a single perforated tube which allows the gases to continue in the same direction. The perforations muffle the sound.

Sometimes a resonator is added to the system after the muffler. Built like a straight-through muffler, its job is to reduce the exhaust noise still more.

A single exhaust system provides one path for exhaust through the system. A dual exhaust is actually two separate systems. It is sometimes used to reduce the back pressure on high performance cars.

On V-type engines, a crossover pipe connects the right and left sides of the exhaust system together. On a single system, it simply allows both exhaust manifolds to empty into the same system. On dual systems, it balances back pressure between the two systems.

A catalytic converter oxidizes unburned and partially burned fuels before releasing them to the atmosphere. A mini catalytic converter is located next to the exhaust manifold where it heats up quickly and burns off exhaust pollutants quickly. A monolithic converter is a larger ceramic block, shaped like a honeycomb. A coating of special chemicals helps it act on the exhaust gases. A pellet catalytic converter has hundreds of small beads that act as the catalytic agent.

A particulate oxidizer is a special catalytic converter used on diesel engines. It burns the soot that would otherwise exit the diesel exhaust as black smoke.

KNOW THESE TERMS

Exhaust system, Exhaust fumes, Exhaust manifold, Header pipe, Catalytic converter, Intermediate pipe, Muffler, Tailpipe, Hangers, Heat shields, Muffler clamps, Exhaust back pressure, Exhaust ports, Stud bolts, Exhaust manifold heat valve, Heat tubes, Baffles, Reverse-flow muffler, Straight-through muffler, Resonator, Header pipe donut, Single exhaust system, Dual exhaust system, Crossover pipe, Mini-catalytic converter, Monlithic catalytic converter, Pellet catalytic converter, Particulate oxidizer, Trap oxidizer, Ceramic element.

REVIEW QUESTIONS—CHAPTER 22

Match the terms with their proper definitions.

1. ____ Connects cylinder head exhaust ports to header pipe.
2. ____ Steel tube or pipe that carries exhaust gases from exhaust manifold

a. Catalytic converter.
b. Muffler clamps.
c. Exhaust manifold.

to catalytic converter.

3. ____ Canisterlike device for removing pollutants from exhaust.

4. ____ Connecting pipe sometimes used between header pipe or catalytic converter and muffler.

5. ____ Metal chamber for damping pressure pulsations and and reducing exhaust noise.

6. ____ Length of pipe which carries exhaust from muffler to rear of car.

7. ____ Devices for securing exhaust system to auto underbody.

8. ____ Metal plates that pre-

d. Heat shields.
e. Header pipe.
f. Hangers.
g. Intermediate pipe.
h. Muffler.
i. Tailpipe.

vent transfer of exhaust heat to combustible parts of the body or to flammable materials on the ground.

9. ____ Connectors that hold parts of the exhaust together.

10. A (n) _____ _____ _____ valve diverts hot exhaust gases into the intake manifold to aid cold weather starting.

11. Name the two basic types of muffler and describe their construction.

12. Name three types of catalytic converter.

13. A _____ catalytic converter is made of a large ceramic block which is coated with special chemicals to help oxidize the exhaust gases.

14. What is a particulate oxidizer?

This modern rear-engine sports car has four-valves per cylinder, twin overhead cams, and electronic fuel injection. It produces tremendous power for its size. (Toyota)

23 EXHAUST SYSTEM SERVICE

After studying this chapter, you will be able to:
- Locate exhaust system troubles.
- Remove exhaust system components.
- Torque an exhaust manifold to specs.
- Shape the end of an exhaust pipe.
- Properly install an exhaust system.
- Check for high exhaust system back pressure.

Exhaust system repairs are usually needed when one of the components rusts out and starts to leak exhaust gases. Furthermore, engine combustion produces water and acids that attack the exhaust system. This corrosion can cause parts to fail in a relatively short period of time.

This chapter will outline the most common testing and repair operations done on a modern exhaust system. Refer to other text chapters for related information—turbocharger service, for example.

EXHAUST SYSTEM WARNING

The running engine produces deadly gases that can cause death. Always use an EXHAUST VENT HOSE to prevent a buildup of this gas in an enclosed shop. See Fig. 23-1.

Another warning! Parts of the exhaust system, especially the catalytic converter, can be VERY HOT after engine operation. Be careful not to get burned on hot components. Allow them to cool before working.

EXHAUST SYSTEM INSPECTION

Visual inspection of the exhaust system will find most problems. You must check for leakage, missing clamps or hangers, cracked exhaust manifolds, high back pressure, stuck heat riser, and similar types of problems.

Exhaust system leaks

Exhaust system inspection usually starts by raising the car on a hoist. With the transmission in park or neutral and the engine running, listen for exhaust leakage from the system. A leak will make a hissing or rumbling type of sound.

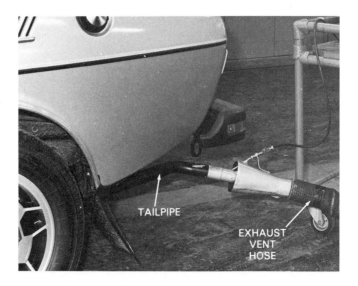

Fig. 23-1. Never run an engine in an enclosed shop without using exhaust vent to route exhaust gases outdoors. (Saab)

Check closely for leakage around the O-ring gasket, exhaust manifold gaskets, pipe connections, muffler, and tailpipe. These are the most common points of leakage. Look for rust, holes, smashed pipes, and muffler rot.

A section of vacuum hose can be used as a listening device to find small leaks. Place one end of the hose next to your ear. Move the other end of the hose around potential points of leakage. This will make a small leak very loud and easy to find.

Fig. 23-2 illustrates some common problems that can occur in the exhaust system. Study each closely.

DANGER! Since engine exhaust is poisonous, a leaky exhaust can allow toxic gases to flow through any opening in the auto body and into the passenger compartment. This could endanger people's health!

Hanger problems

Always check the condition of hangers when inspecting an exhaust system. If any are broken, missing, or damaged, replace them. Without them, the exhaust

A

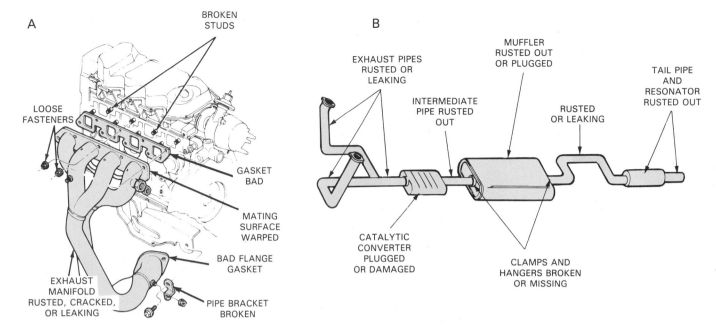

B

Fig. 23-2. When inspecting exhaust systems, look for these problems. A—Exhaust manifold problems. B—Problems with rest of exhaust system.

system will drop down from its normal position. Then, parts, especially the muffler, can strike the ground and be torn away or severely damaged.

High exhaust back pressure

High exhaust back pressure can be caused by a collapsed pipe, clogged catalytic converter, or any other trouble that limits gaseous flow through the exhaust system. Such problems will greatly reduce engine performance because burned exhaust gases cannot exit the combustion chambers normally.

Fig. 23-3 shows an easy way of checking for high exhaust system back pressure. Install a pressure gauge into the air injection system manifold fitting. Run the engine at about 2500 rpm and check back pressure. A pressure reading of more than about 2 1/2 to 3 1/2 psi (17-24 kPa) indicates a restriction and high back pressure. You will need to disconnect individual components until you find the source of the problem.

A vacuum gauge can also be used to check for high back pressure when air injection is not used on the engine. Connect the vacuum gauge to the engine intake manifold. Check vacuum at about 1000 rpm. Then, increase engine speed to about 2500 rpm. If VACUUM DROPS at higher speeds, a restriction might be present in the exhaust system.

Exhaust system repairs

When exhaust system components are faulty, they must be replaced. Usually, a muffler is the first part of the system to rust out. After prolonged service, however, several parts or the ENTIRE SYSTEM may need replacement.

To make repair easier, remember the following:
1. Use RUST PENETRANT on all threaded fasteners that will be reused, Fig. 23-4A. This is most impor-

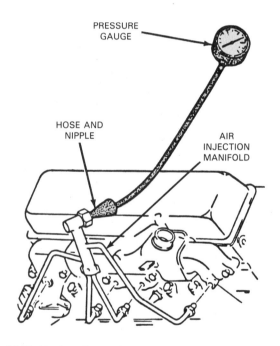

Fig. 23-3. Hookup for testing high back pressure with pressure gauge. Nipple is inserted into exhaust manifold A.I.R. pipe. If back pressure exceeds about 2.5 to 3.5 psi at 2500 rpm, there is probably a restriction. (GM)

tant on the exhaust manifold flange nuts and studs.
2. Use an air chisel, cutoff tool, cutting torch, or a hacksaw when removing other faulty parts. Be sure not to damage any part that is to be reused. See Figs. 23-4 and 23-5.
3. Use a SIX-POINT socket and ratchet or an AIR IMPACT WRENCH to remove fasteners quickly

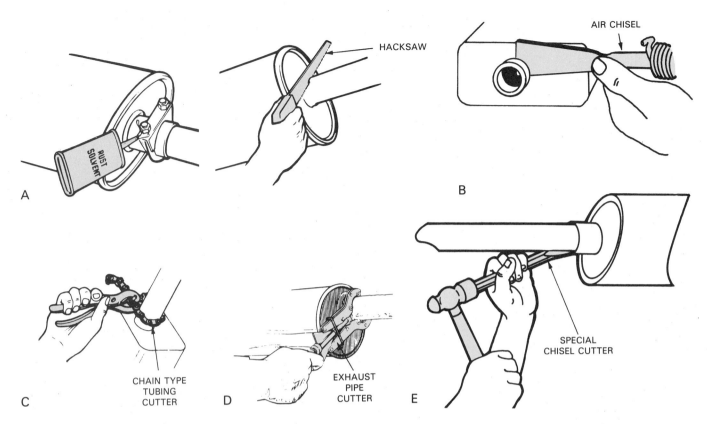

Fig. 23-4. Parts of exhaust to be reused should be carefully removed. Parts being replaced should be removed by cutting or sawing. A—Using penetrant to dissolve rust on threaded parts. B—Using hacksaw or air chisel to remove parts. C—Chain type cutting tool. D—Exhaust pipe cutter. E—Specially shaped hand-type cutter or chisel. (AP Parts, Lisle, and Florida Dept. of Voc. Ed.)

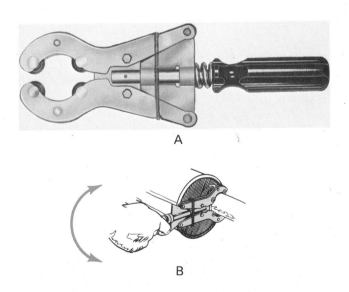

Fig. 23-5. Using muffler tailpipe cutoff tool. A—Tool. B—Tool works like copper tubing cutter. Twist handle to increase "bite" while rotating tool. (Lisle)

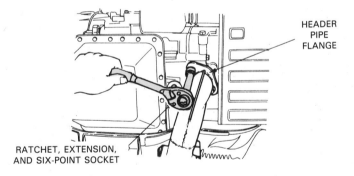

Fig. 23-6. Removing header pipe fasteners can be difficult. Work with plenty of rust penetrant. Use six-point socket, extension, and ratchet. (Subaru)

without rounding the corners of the nuts or heads. Look at Fig. 23-6.

4. Wear SAFETY GLASSES or GOGGLES to keep rust and dirt out of your eyes.
5. Get the correct replacement parts.

6. A PIPE EXPANDER is a handy tool for enlarging pipe ends as needed. Refer to Fig. 23-7. A PIPE SHAPER straightens dented pipe ends, Fig. 23-8.
7. Use care that all pipes are fully inserted into mating parts. Position the clamps correctly, Fig. 23-9.
8. Double-check for correct routing of the system. There must be adequate clearance between exhaust parts, the car body and frame. See Figs. 23-10 and 23-11.
9. Clamps and hangers must be tightened evenly. Torque them only enough to hold the parts securely. Overtightening will crush and deform the pipe,

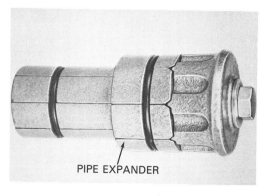

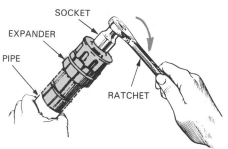

Fig. 23-7. Pipe expander is designed to enlarge ID (inside diameter) of exhaust pipes so one pipe will fit over another. Turning bolt clockwise with ratchet and socket will enlarge expander diameter. (Lisle)

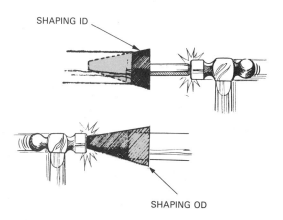

Fig. 23-8. Dented, out-of-round pipe ends will not seal properly. Pipe shaper will correct this problem. (Lisle)

leading to leakage and early failure, Fig. 23-11.
10. Reinstall all heat shields, Fig. 23-13.
11. If an exhaust manifold must be replaced, check the sealing surface for flatness and use an exhaust manifold gasket, Fig. 23-12. If warped, the manifold must be machined. Torque bolts to shop manual specs and install related parts, Fig. 23-14.
12. Never reuse old gaskets and O-rings.
13. Make sure the heat riser is operating properly. Follow procedures outlined in the shop manual.
14. Complete the repair by testing the system for leaks and rattles.

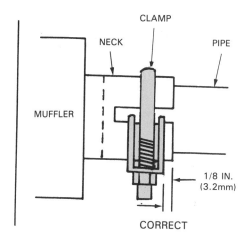

Fig. 23-9. Carefully insert one pipe into another and place clamp so it surrounds both pipes. Otherwise, one pipe will pull out of the other. (Florida Dept. of Voc. Ed.)

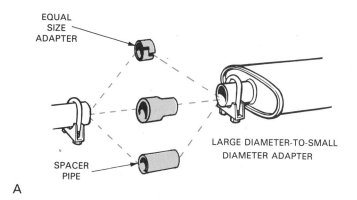

A

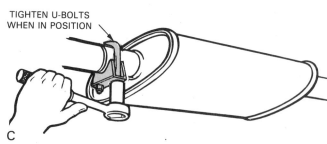

B

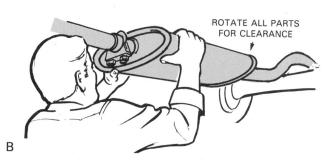

C

Fig. 23-10. Using adapters and attaching new muffler. A—Various adapters can be used to fit new muffler to old exhaust pipes. B—Check clearances carefully and adjust parts for best fit. C—When satisfied with clearances, tighten clamps evenly. Avoid overtightening which could destroy parts or cause leakage. (AP Parts)

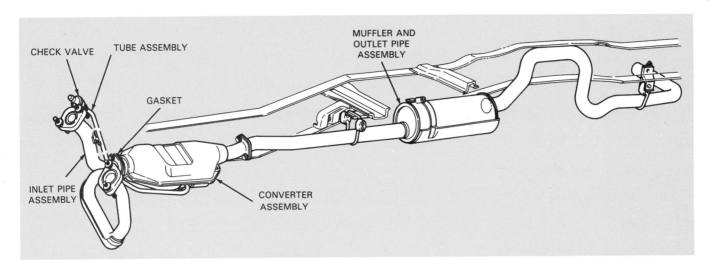

Fig. 23-11. You may need to check a shop manual for proper routing and proper torque values. (Pontiac)

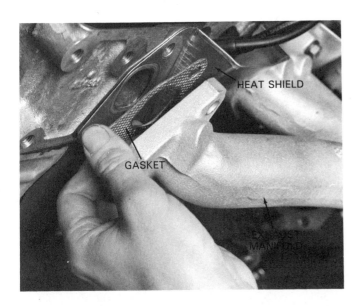

Fig. 23-12. Replacing exhaust manifold. Use gaskets at mating surfaces where manifold bolts to cylinder head. Tighten bolts to torque values specified, using a crisscross pattern. (Saab)

SUMMARY

Exhaust systems fall prey to rust and the corrosive action of engine-produced water and acids. As a result, exhaust system parts fail and must be replaced.

Because exhaust gases are toxic, it is important to keep the exhaust system in good repair to protect occupants of the car. Likewise, when the system is being tested in an enclosed shop, use an exhaust vent hose to carry exhaust gases outdoors.

Exhaust inspection includes checking for leaking and rusted or rotted components, holes or crushed pipes, etc. Look also for broken or disconnected hangers and check for evidence of high back pressure from clogged or crushed exhaust parts.

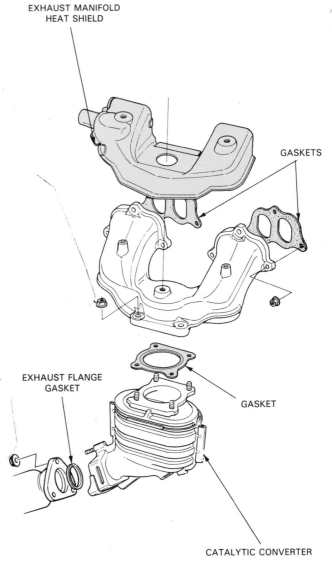

Fig. 23-13. Be certain to replace all heat shields after servicing exhaust system. (Honda)

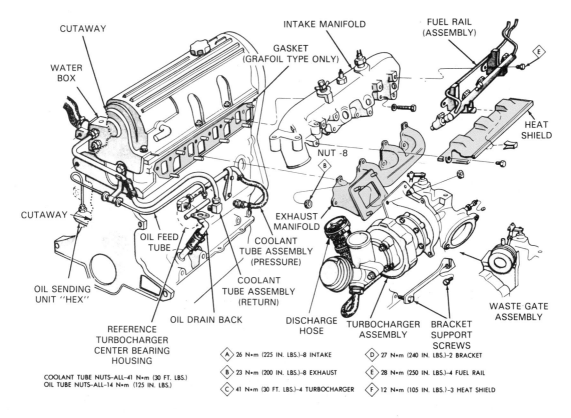

CUTAWAY

WATER BOX

INTAKE MANIFOLD

GASKET (GRAFOIL TYPE ONLY)

FUEL RAIL (ASSEMBLY)

E

HEAT SHIELD

NUT -8

B

CUTAWAY

OIL FEED TUBE

EXHAUST MANIFOLD

COOLANT TUBE ASSEMBLY (PRESSURE)

OIL SENDING UNIT "HEX"

COOLANT TUBE ASSEMBLY (RETURN)

OIL DRAIN BACK

DISCHARGE HOSE

TURBOCHARGER ASSEMBLY

BRACKET SUPPORT SCREWS

WASTE GATE ASSEMBLY

REFERENCE TURBOCHARGER CENTER BEARING HOUSING

COOLANT TUBE NUTS—ALL-41 N•m (30 FT. LBS.)
OIL TUBE NUTS—ALL-14 N•m (125 IN. LBS.)

A 26 N•m (225 IN. LBS.)-8 INTAKE

B 23 N•m (200 IN. LBS.)-8 EXHAUST

C 41 N•m (30 FT. LBS.)-4 TURBOCHARGER

D 27 N•m (240 IN. LBS.)-2 BRACKET

E 28 N•m (250 IN. LBS.)-4 FUEL RAIL

F 12 N•m (105 IN. LBS.)-3 HEAT SHIELD

Fig. 23-14. Exploded view of engine components from your shop manual will help you reassemble manifolds and heat shields. Note torque values for different parts. (Chrysler)

Replacement of parts or all of a system is not a difficult task if a few simple rules are followed:

1. Use a rust penetrant on threaded fasteners and other parts that are to be reused.
2. On parts to be discarded, use an air chisel, torch, cutoff tool, or a hacksaw for removal.
3. Use a pipe expander to enlarge pipe ends. It will speed up the assembly work.
4. Check assembled system for correct routing and that there is adequate clearance between the system and the auto body or frame.
5. Use care in torquing clamps and hangers; parts can be damaged or crushed by too much tightening.
6. In the case of manifold leaks, check for flatness at mating surfaces.
7. Check the heat riser for proper operation.
8. Test the repaired system for leaks and rattles.

KNOW THESE TERMS

Exhaust vent hose, Leaking exhaust system, Toxic gases, High exhaust back pressure, Clogged catalytic converter, Collapsed pipe, Rust penetrant, Pipe expander, Sealing surface flatness.

REVIEW QUESTIONS—CHAPTER 23

1. You are testing the exhaust system of a car in an enclosed shop. Which of the following is preferred to avoid danger from exhaust buildup?
   a. Open a window.
   b. Only run the engine for short periods.
   c. There is no danger.
   d. Attach an exhaust vent hose to the exhaust.
2. Exhaust inspection usually starts with the car raised on a _____, the transmission in _____ or _____, and the engine running.
3. What are the most common points of leakage in an exhaust system?
4. Even small leaks are easier to find if you:
   a. Hold your hand over the end of the tailpipe and look for signs of exhaust all along the exhaust system.
   b. Use a section of vacuum hose as a listening device.
   c. Pour water over the various connections in the system.
5. Why should faulty hangers be repaired immediately?
6. Explain how to use a pressure gauge to detect high exhaust back pressure.
7. All threaded fasteners that are to be reused in an exhaust system repair should be treated with _____ _____ so they will not be damaged during removal.
8. Neither a cutting torch nor a hacksaw should ever be used in removing parts of the exhaust system. True of false?
9. Explain the use of a pipe expander.
10. What repair procedure is indicated when an exhaust leak is found between the cylinder head and the exhaust manifold?

24 EMISSION CONTROL SYSTEMS

After studying this chapter, you will be able to:
* *Define the basic terms that relate to automotive emission control systems.*
* *Discuss the automotive-related sources of air pollution.*
* *Explain the operating principles of emission control systems.*
* *Compare differences in the design of emission control systems.*
* *Explain how a computer controls the operation of emission control systems.*

Manufacturers must equip cars with *emission control systems* that reduce the amount of harmful combustion gases released into the atmosphere. See Fig. 24-1. Federal regulations require these systems to help keep our air cleaner. This chapter introduces emission system terms, system parts, and their operation. This important chapter prepares you to service and repair such systems.

AIR POLLUTION SOURCES AND FEDERAL LAW

Air pollution comes from a number of sources. Some natural causes are volcanoes, forest fires, wind-blown dust, pollen, and decay of vegetation. Factories, home furnaces, fireplaces, and automobiles also pollute our air. Look at Fig. 24-2.

The federal government has passed strict laws to reduce automotive air pollution. These laws, enforced by the EPA (Environmental Protection Agency), limit the amount of emissions from cars. Auto manufacturers have designed and added a number of components to meet EPA emissions standards.

Smog

Smog is a new name coined for a combination of visible air pollutants. The name comes from the words "smoke" and "fog." Smog is common in large cities.

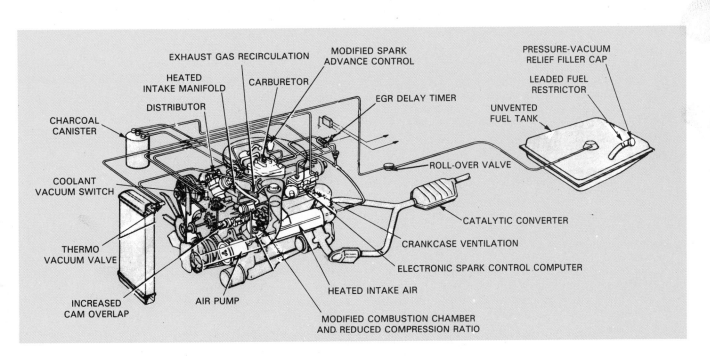

Fig. 24-1. Modern automobiles use numerous designs of emission control systems. Components are designed to work together and reduce air pollution from auto exhaust. (Chrysler)

Fig. 24-2. Smog is the result of a concentration of pollutants from factories, homes, and cars. (Shell Oil Co.)

When dense, it can be harmful to humans, animals, and vegetation. It also attacks paint, rubber, and other materials.

In more technical language, smog is photochemical fog. Photochemical means that airborne chemicals react with sunlight, We see it as a yellow-gray cloud of pollution.

VEHICLE EMISSIONS

Vehicle emissions are the byproducts of burning automotive fuels. There are four basic types of vehicle emissions: hydrocarbons, carbon monoxide, oxides of nitrogen, and particulates. These are important terms that should be understood by the auto technician.

Hydrocarbons (HC)

Hydrocarbons, shortened to HC, are a form of emission resulting from the release of *unburned fuel* into the atmosphere. As you learned in earlier chapters, petroleum (crude oil) products are made of hydrocarbons (hydrogen and carbon compounds). This includes gasoline, diesel fuel, LP-gas, and motor oil. Hydrocarbon emission can be caused by incomplete combustion or by fuel evaporization. For example, HC is produced when unburned fuel leaves an untuned engine's exhaust system. Another cause is the escape of fuel vapors from a car's fuel system. Hydrocarbon emissions are a health hazard. They contribute to eye, throat, and lung irritation, other illness, and, possibly, cancer.

Carbon monoxide (CO)

Carbon monoxide, also called *CO*, is an extremely toxic gaseous emission that is found in PARTIALLY BURNED FUEL. It can occur from the incomplete combustion of a petroleum-based fuel.

Carbon monoxide is colorless, odorless, and deadly. It causes headaches, nausea, and respiratory (breathing) problems. Inhaled in large quantities, it can cause death.

CO prevents human blood cells from carrying oxygen to body tissues. Any factor that reduces the amount of oxygen present during combustion increases carbon monoxide emissions. A rich air-fuel mixture (high ratio of fuel to air) would increase CO, for example. As the mixture becomes more lean (more air, less fuel), CO emissions drop. See Fig. 24-3.

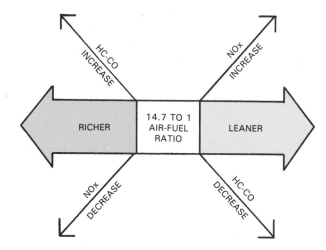

Fig. 24-3. It pays to study the relationships between air-fuel ratios and emissions. Maintaining tight control of an engine's fuel mixture will help keep emissions minimal. (Fiat)

Oxides of nitrogen (NOx)

NOx (oxides of nitrogen) are emissions produced by extremely *high temperature combustion*. Air is about 80 percent nitrogen and 20 percent oxygen. When heated above approximately 2500°F (1 370°C), nitrogen and oxygen in the air-fuel mixture form NOx emissions.

Oxides of nitrogen account for smog's dirty yellow/gray color. They also produce ozone which has an unpleasant smell and irritates eyes and lungs. NOx is also harmful to many types of plants and rubber products.

Basically, an engine with a high compression ratio, lean fuel mixture, high temperature thermostat, and resulting high combustion heat, emits high levels of NOx. This poses a problem. These same factors tend to improve gas mileage and reduce HC and CO exhaust emissions. Emission control systems must interact to lower each of these forms of pollution.

Particulates

Particulates are solid particles of carbon soot and fuel additives from a car's exhaust. Carbon particles make up the largest percentage of these emissions. The rest of the particulate consists of lead and other additives sometimes used in gasoline and diesel fuel.

Particulate emissions are a serious problem with diesel engines. You have probably seen a diesel truck or car with BLACK SMOKE (particulate) coming out of it's exhaust. Diesel particulate is normally caused by an extremely rich, high-powered fuel mixture or a mechanical problem in the injection system. About 30 percent of all particulate emissions are heavy enough to settle out of the air. The other 70 percent, however, can float in the air for extended periods, contributing to health hazards.

Sources of vehicle emissions

Vehicle emissions, Fig. 24-4, have three basic sources:

1. ENGINE CRANKCASE BLOWBY FUMES (20 percent).
2. FUEL VAPORS (20 percent).
3. ENGINE EXHAUST GASES (60 percent).

Various engine modifications and emission control systems are used to reduce air pollution from these sources.

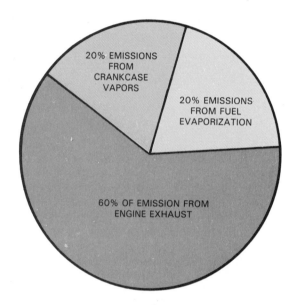

Fig. 24-4. Auto air pollution comes from three general sources.

ENGINE MODIFICATIONS FOR EMISSION CONTROL

Generally, auto makers agree that the best way to reduce exhaust emissions is to burn all of the fuel inside the engine. For this reason, several engine modifications have been introduced to improve combustion efficiency. Today's engines generally have the following modifications to lower emissions:

1. LOWER COMPRESSION RATIOS that allow the use of unleaded gasoline. Unleaded fuel burns quickly to lower HC emissions. It has no lead additives to cause particulate emissions. Lower compression stroke pressure also reduces combustion temperatures and NOx emissions.
2. LEANER AIR-FUEL MIXTURES that lower HC and CO emissions. Leaner mixtures have more air to improve fuel burn.
3. HEATED INTAKE MANIFOLDS that speed warm-up and permit the use of leaner mixtures during start-up.
4. SMALLER COMBUSTION CHAMBER SURFACE that reduces HC emissions. A smaller surface area in the combustion chamber increases combustion

efficiency by lowering the amount of heat removed from the fuel mixture. Less combustion heat escapes to the cylinder head and more heat is retained to burn the fuel. A hemispherical (hemi) type of combustion chamber typically has the smallest surface volume (SV).

5. INCREASED VALVE OVERLAP which cuts NOx emissions. A camshaft with more overlap dilutes the incoming air-fuel mixture with inert (burned) exhaust gases. This modification reduces peak combustion temperatures and NOx.
6. HARDENED VALVES AND SEATS that can withstand unleaded fuel. When lead additives were used in fuel, they served as a high-temperature lubricant. They reduced wear at valve faces and seats. Hardened valves and seats are needed with unleaded fuel to prevent excessive wear.
7. WIDER SPARK PLUG GAPS that properly ignite "super-lean," clean burning air-fuel mixtures. Wider gaps produce hotter sparks which can ignite hard-to-burn, lean air-fuel mixtures.
8. REDUCED QUENCH AREAS in the combustion chambers lower HC and CO emissions. When the engine pistons move too close to the cylinder head, they tend to quench (stop) combustion and increase emissions of unburned fuel. Cylinder heads and pistons in modern engines are redesigned to prevent high quench areas.
9. HIGHER OPERATING TEMPERATURES to reduce HC and CO emissions. Thermostats in today's engines have higher temperature ratings. If the metal parts in an engine are hotter, less combustion heat will leave the burning fuel. More heat will remain in the burning mixture to produce gas expansion, piston movement, and complete combustion. Various other methods are used to reduce engine emissions. Many of them are covered later in this chapter.

VEHICLE EMISSION CONTROL SYSTEMS

Different systems are used to reduce the amount of air pollution produced by the automobile. The major ones are:

1. PCV SYSTEM (positive crankcase ventilation system). It keeps engine crankcase fumes out of atmosphere.
2. HEATED AIR INLET SYSTEM (thermostatic controlled air cleaner maintains incoming air at a constant temperature for improved combustion).
3. FUEL EVAPORATION CONTROL SYSTEM. This closed vent system prevents fuel vapors from entering the atmosphere.
4. EGR SYSTEM (exhaust gas recirculation) returns burned exhaust gases to the combustion chamber to lower combustion temperature and prevent NOx pollution.
5. AIR INJECTION SYSTEM. An air pump forces outside air into exhaust system to help burn unburned fuel.
6. CATALYTIC CONVERTER. (This is a thermal reactor for burning and chemically changing exhaust byproducts into harmless substances).

Variations of these systems and computer control systems are all used to make the modern car very efficient.

POSITIVE CRANKCASE VENTILATION (PCV)

Modern gasoline and diesel engines normally use a positive crankcase ventilation *PCV system*. The system employs engine vacuum to draw toxic blowby gases into the intake manifold for reburning in the combustion chambers. Look at Fig. 24-5.

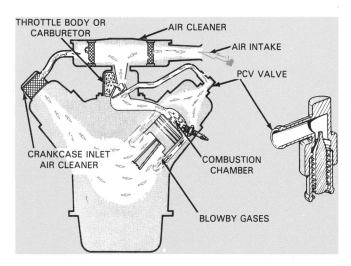

Fig. 24-5. A PCV system draws toxic vapors from the crankcase and returns them to the intake manifold, allowing them to be burned by the engine. (Chrysler)

As noted in earlier chapters, *engine blowby* is caused by pressure leakage past the piston rings on the power strokes. A small percentage of combustion gases can escape past the ring end gaps, back of the piston ring grooves, and into the crankcase.

Engine blowby gases contain unburned fuel *(HC)*, partially burned fuel (CO), particulate, and small amounts of water, sulfur, and acid. Blowby gases, if not removed from the engine crankcase, can cause:

1. Air pollution (HC and CO).
2. Release of sulfur and acid can corrode engine parts.
3. Dilution of engine oil (HC, water, sulfur, and acid).
4. Sludge (chocolate-pudding-like substance that can clog oil passages).

A PCV system is designed to prevent these problems. It helps keep the inside of the engine clean and also reduces air pollution.

Note! Years ago, cars vented crankcase fumes to the atmosphere. A road draft tube allowed crankcase fumes to escape out the back of the engine. This caused air pollution. Another engine ventilation system, called an open PCV system, was NOT sealed and fumes could leak out when the engine was shut off. These systems have been totally replaced by the closed PCV system.

A closed PCV system utilizes a sealed oil filler cap, sealed oil dipstick, ventilation hoses, and either a PCV valve or flow restrictor. The fumes are drawn into the

engine and burned. The system stores the fumes when the engine is not running.

PCV system operation

While designs vary, all PCV systems operate basically the same. Look at Fig. 24-5. A hose usually connects the intake manifold to the PCV valve in one of the engine valve covers. With the engine running, vacuum acts on the crankcase area of the engine. Air is drawn in through the engine air cleaner, through a vent hose into the other valve cover, and down into the crankcase. After the fresh air mixes with the crankcase fumes, the mixture is pulled up past the PCV valve, through the hose, and into the engine intake manifold. The crankcase gases are then drawn into the combustion chambers for burning. Fig. 24-6 shows a cutaway of an air cleaner assembly for a fuel injected engine. Notice the location of parts.

PCV valve

A PCV valve controls the flow of air through the crankcase ventilation system. It may be located in a rubber grommet in a valve cover or in a breather opening in the intake manifold. The PCV valve adjusts airflow for idle, cruise, acceleration, wide open throttle, and engine-off conditions. Refer to Fig. 24-7. It shows the action of a PCV valve under various operating conditions.

At idle, the PCV valve is pulled toward the intake manifold by high vacuum. This restricts flow and prevents a too lean air-fuel mixture. When cruising, lower intake manifold vacuum allows the spring to open the PCV valve. More air can flow through the system to clean out crankcase fumes. At wide open throttle or with the engine off (low or no intake manifold vacuum), the PCV valve closes completely. In case of an engine backfire (air-fuel mixture in intake manifold ignites), the PCV valve plunger is seated against the body of the valve. This keeps the backfire (burning) from entering and igniting the fumes in the engine crankcase.

Diesel engine PCV system variations

A PCV system for a diesel engine is illustrated in Fig. 24-8. Note how outside air is drawn in through the oil filter cap. The breather cap contains a check valve that prevents crankcase fumes from leaking out of the engine. This maintains a closed PCV system.

HEATED AIR INLET SYSTEM

The heated air inlet system reduces engine warm-up time and holds the temperature of the air entering the engine at about 70°F (21°C). Late-model gasoline engines commonly use this system. By regulating inlet air temperature, the carburetor or fuel injection system can be calibrated leaner to reduce emissions. The heated air inlet system is also known as the thermostatic air cleaner system. It also helps prevent carburetor icing during warm-up. Typically, the system consists of the parts shown in Fig. 24-9.

A thermal valve is normally located in the air cleaner where it controls the vacuum motor and heat control door. Refer to Fig. 24-10A. The thermal valve is supplied with vacuum from the engine. Another vacuum hose runs from the thermal valve to the vacuum motor

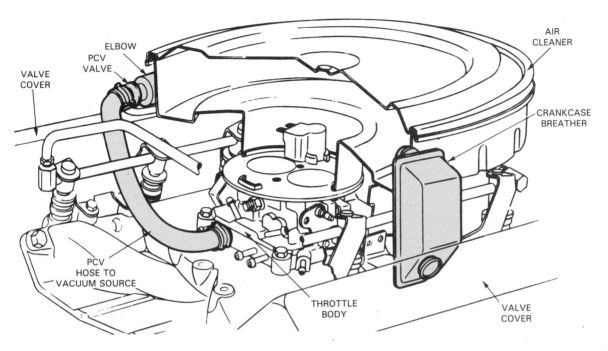

Fig. 24-6. PCV valves are usually located in a valve cover. Breather allows fresh air to enter the crankcase through the other valve cover. Hose connects PCV valve to a source of engine vacuum.   (Cadillac)

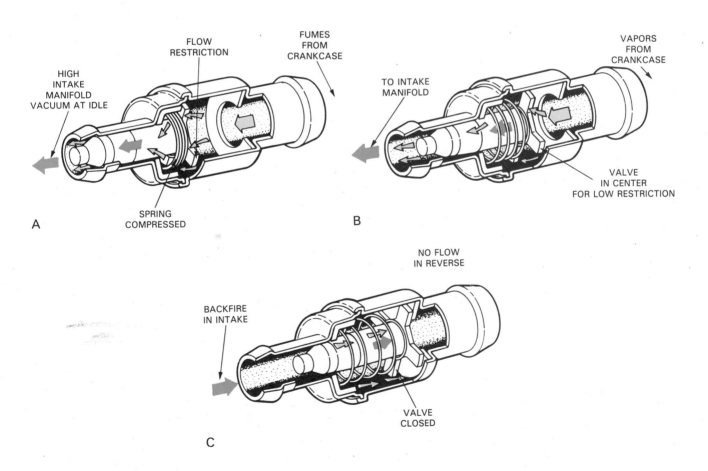

Fig. 24-7. PCV operation adjusts to various operating conditions. A—At idle, high manifold vacuum pulls on plunger, allowing minimal flow. Little air is drawn through valve. Idle fuel mixture is not upset. B—Under acceleration, intake manifold vacuum is extremely low; PCV valve moves to a center position and maximum flow. C—With engine stopped, spring presses valve against its seat, closing the valve. During a backfire, valve is also in this position.   (AC Spark Plug)

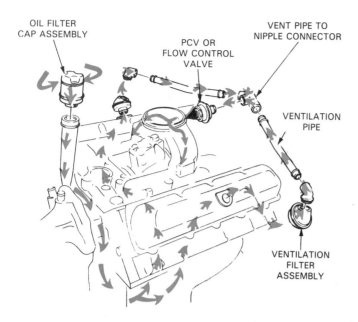

Fig. 24-8. Arrows indicate PCV system flow through a diesel engine. (Cadillac)

(diaphragm). The vacuum motor, or vacuum diaphragm, as it is often called, operates the air control door or flap in the air cleaner inlet. Look at Fig. 24-1OB. The vacuum motor has a flexible diaphragm, a spring, a rod, and a diaphragm chamber. When vacuum is supplied to the unit, the diaphragm and rod are pulled upward, moving the air cleaner door. When the air cleaner door is closed,

hot air from the exhaust manifold shroud enters the air cleaner and the engine. When the air cleaner door is open, cooler, outside air enters the engine.

Heated air inlet system operation

When the engine is cold, the thermal valve is closed, allowing full vacuum to the vacuum motor, Fig. 24-1OB. The vacuum acts on the diaphragm, compressing the spring, and closing the door. Warm air from the exhaust manifold shroud is drawn into the engine. This provides faster warm-up. As the inlet air temperature rises, the thermal valve closes. Spring tension can then begin to open the inlet door, Fig. 24-1OC. Then both heated air and cool air mix and flow into the engine. In this way, a constant inlet air temperature is maintained. When both the engine and outside air are hot (above inlet system operating temperature), the thermal valve closes. Without vacuum, spring tension in the diaphragm chamber completely opens the vacuum door. This blocks the warm air inlet and opens the cool air inlet, Fig. 24-1OD.

AIR CLEANER CARBON ELEMENT

An air cleaner carbon element may be used to absorb fuel vapors when the engine is shut off. Refer to Fig. 24-11. When a hot engine is turned off, engine heat may cause excessive formation of fuel vapors. This condition is known as hot soak. Vapor can collect in the carburetor or throttle body air horn. The carbon element attracts

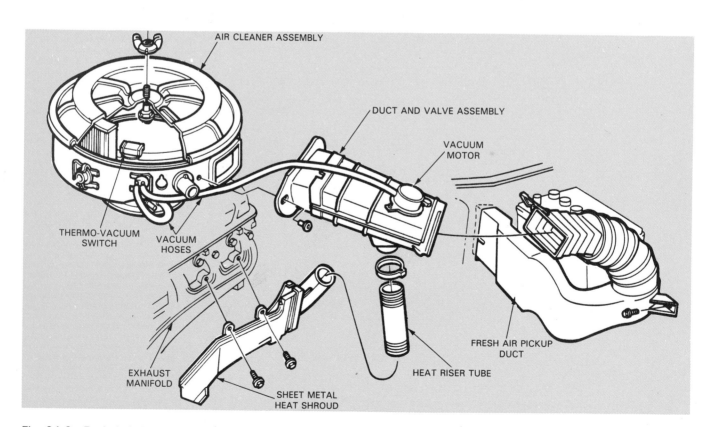

Fig. 24-9. Exploded view shows basic parts of a thermostatic air cleaner assembly. Heat shroud fits around exhaust manifold to warm air charge. (Ford)

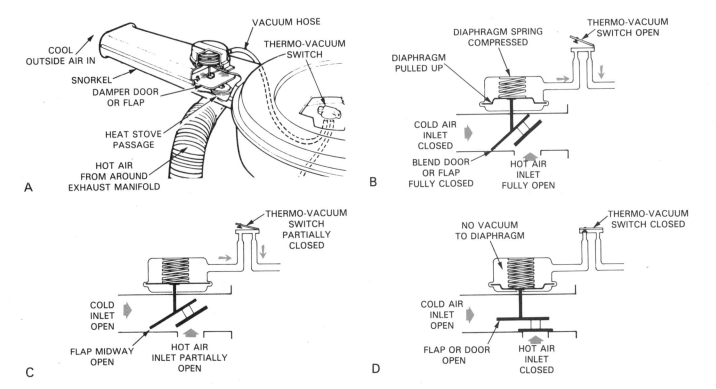

Fig. 24-10. Thermostatic air cleaner construction and operation. A—Hinged damper or flap is located inside snorkel. Thermo-vacuum switch activates vacuum motor that controls flap. B—On a cold engine, thermo-vacuum switch is open. Vacuum deflects diaphragm to pull flap shut. Warm air from the exhaust manifold shroud is allowed to enter air cleaner. C—With engine at operating temperature, thermo-vacuum switch is partly open. This allows vacuum motor to partly close. This allows a mixture of hot and cold air to enter the air cleaner. D—With engine and outside air both hot, thermo-vacuum switch closes. Vacuum motor receives no vacuum. Spring closes door to block warm air from shroud. (Buick and Fiat)

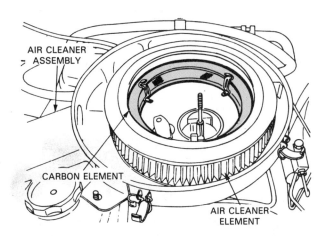

Fig. 24-11. Some late-model air cleaners use a carbon element in addition to a filter element. It holds fuel vapors until the engine is started. (Chrysler)

and stores these fumes. When the engine is started, airflow through the element will pull the fumes into the engine combustion chamber.

FUEL EVAPORIZATION CONTROL SYSTEM

The *fuel evaporization control system* contains vapors from the fuel tank and carburetor preventing the vapors from entering the atmosphere. See Fig. 24-12. Older, pre-emission control cars used vented gas tank caps and carburetor bowls. Fuel admissions were considerable. A fuel evaporization control system prevents this source of air pollution.

A sealed fuel tank cap keeps fuel vapors from entering the atmosphere through the tank filler neck. As the fuel expands (from warming), tank pressure forces fuel vapors out a vent line or lines at the top of the fuel tank, not out the tank cap. The cap may also have pressure and vacuum valves for safety. These valves will open in case of extreme pressure or vacuum.

Some vehicles have a fuel tank air dome. This is a hump built into the top of the fuel tank for fuel expansion. It provides about 10 percent more space to allow for fuel expansion. Refer to Fig. 24-12A.

LIQUID-VAPOR SEPARATOR

A *liquid-vapor separator* is frequently used to keep liquid fuel out of the evaporization control system. It is nothing more than a metal tank located above the main fuel tank, Fig. 24-12A. Liquid fuel condenses on the walls of the liquid-vapor separator. The separated liquid fuel returns to the fuel tank.

Some manufacturers place a *roll-over valve* in the vent line from the fuel tank. It keeps liquid fuel from entering the vent line should the car roll over in an accident. Refer to Fig. 24-12B. A metal ball or plunger valve in the device blocks the vent line when the valve is turned over.

A charcoal canister, located in the engine compartment, stores fuel vapors that originate in the gas tank and

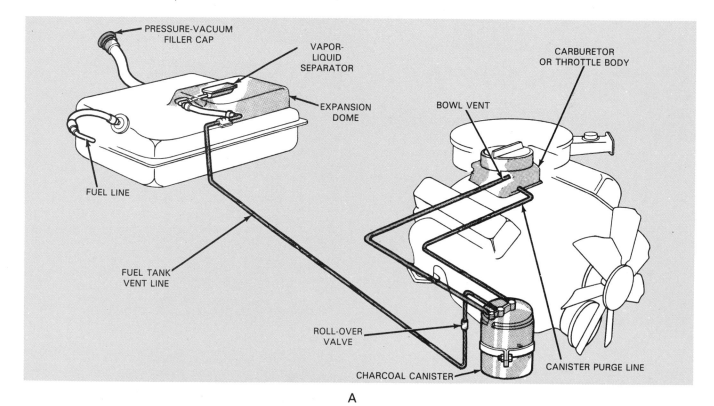

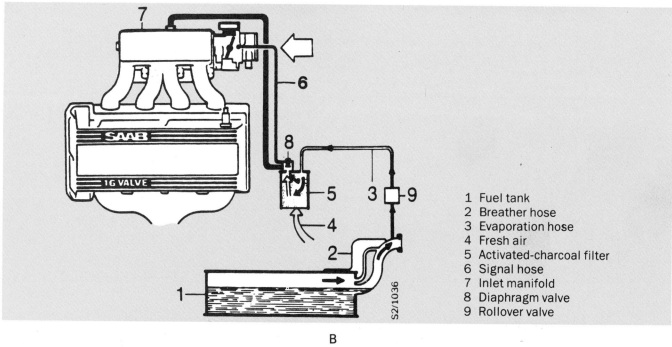

1 Fuel tank
2 Breather hose
3 Evaporation hose
4 Fresh air
5 Activated-charcoal filter
6 Signal hose
7 Inlet manifold
8 Diaphragm valve
9 Rollover valve

Fig. 24-12. Evaporation control system draws fuel vapors into the engine for burning. A—Parts and fittings at the fuel tank. (Plymouth) B—Diagram of a complete system. Note names of important parts. (Saab)

the carburetor bowl. A fuel tank vent line carries fuel tank vapors up to the *charcoal canister.* The metal or plastic canister is filled with activated charcoal granules. See Fig. 24-12B. The charcoal absorbs fuel vapors and stores them while the engine is not running. The top of the canister has fittings for the fuel tank vent line, a carburetor vent line, and the *purge* (cleaning) *line.* The bot-

tom of the canister has an air filter that cleans outside air down into the canister.

When the engine is running, a control signal from the throttle housing acts on the diaphragm valve, opening the part to the inlet manifold. Manifold vacuum draws the full vapors from the activated charcoal.

Fig. 24-13 is a cutaway view of a charcoal canister.

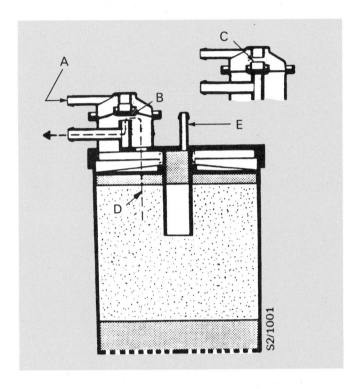

Fig. 24-13. Cutaway view of a charcoal canister. Carbon granules store gasoline vapors when the engine is not running. When engine starts, manifold vacuum draws air through the bottom of the canister. A—Signal input from computer opens diaphragm valve to engine. B—Passage to inlet manifold. C—Diaphragm valve closed. D—Diaphragm valve open. E—Inlet from fuel tank. (Saab)

Note that the canister utilizes a vapor vent valve. This valve opens to allow carburetor bowl venting when the engine is off. When the engine is running it is closed by vacuum to seal the bowl vent line.

The carburetor vent line connects the carburetor fuel bowl with the charcoal canister. Bowl vapors flow through this line and into the canister.

A purge line carries the stored vapors to the engine intake manifold when the engine is running. Engine vacuum draws the vapors out of the canister and through the purge line.

Fig. 24-14 illustrates operation of the basic fuel evaporization system. With the engine running, intake manifold vacuum acts on the charcoal canister purge line. This action pulls fresh air through the filter in the bottom of the canister. The incoming fresh air picks up the stored fuel vapors and moves them through the purge line. The vapors enter the intake manifold and then the combustion chambers where they are burned. When the engine stops, engine heat produces more vapors.

EXHAUST GAS RECIRCULATION SYSTEM

The exhaust gas recirculation (EGR) system allows burned exhaust gases to enter the engine intake manifold to help reduce NOx emissions. Adding exhaust gases to the air-fuel mixture decreases peak combustion temperatures. This lowers the amount of NOx in the engine exhaust.

A basic EGR system consists of a vacuum-operated *EGR valve* controlled by a vacuum line from the carburetor. This is pictured in Figs. 24-15 and 24-16. The

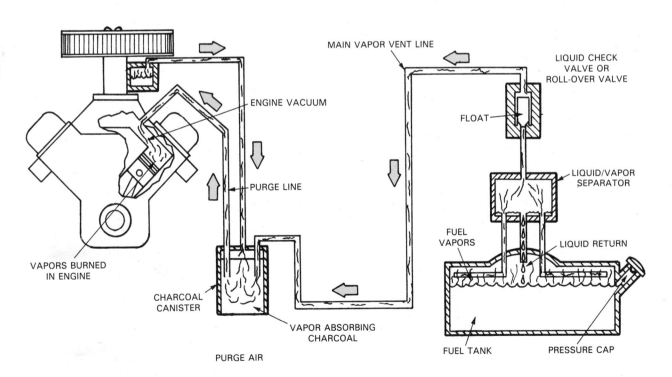

Fig. 24-14. Trace the operation of an evaporization control system through its related components. Vapors from fuel tank and carburetor are captured by the charcoal canister when engine is off. With engine running, vacuum pulls vapors from the canister and burns them. Liquid-vapor separator keeps liquid fuel out of the vent line. A roll-over valve closes if car rolls over. This prevents fires.

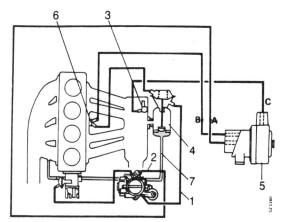

1 Signal outlet upstream of throttle butterfly
2 Signal outlet at throttle butterfly
3 Signal outlet at inlet manifold
4 EGR valve
5 Signal converter
6 Thermostatic valve
7 EGR pipe

Hose Identification
A = From ignition advance outlet (1) at throttle housing
B = From EGR valve (spring-loaded side) via the thermostatic valve
C = From inlet manifold (brake servo outlet)

A

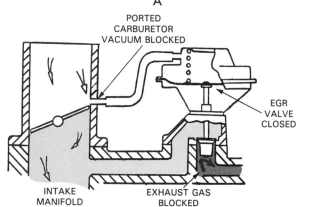

B

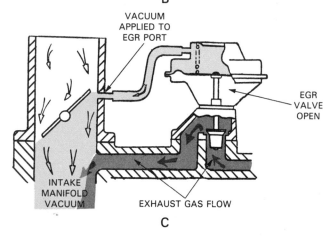

C

Fig. 24-15. A—Diagram of a typical linkup of an EGR valve with exhaust manifold and throttle butterfly. Note also signal linkages between carburetor and EGR valve. When engine is warm, and throttle butterfly is open the EGR valve opens allowing exhaust gases to enter intake manifold. (Saab) EGR valve operation. B—Throttle is closed for engine idle speed. Vacuum to EGR valve blocked and valve remains closed. C—Throttle opens for more engine speed. This opens vacuum port to EGR valve. Diaphragm is pulled up and exhaust gases enter engine intake manifold.

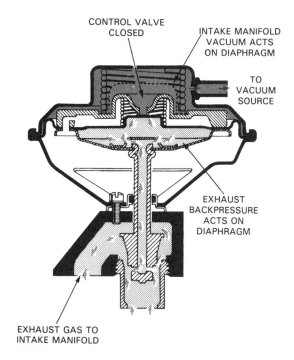

Fig. 24-16. Back pressure EGR valve controls valve opening using engine vacuum and pressure from the exhaust system. (Buick)

EGR valve consists of a vacuum diaphragm, spring, plunger, exhaust gas valve, and a diaphragm housing. It is designed to control exhaust flow into the intake mainfold. See Fig. 24-17. The EGR valve usually bolts to the engine intake manifold or a carburetor plate. Exhaust gases are routed through the cylinder head and exhaust manifold to the EGR valve.

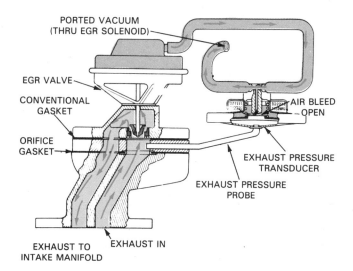

Fig. 24-17. This type of EGR valve uses a back pressure transducer to control vacuum to EGR diaphragm. An increase in engine load can increase back pressure. In turn, the back pressure acts on the transducer diaphragm. Then the transducer causes an increase of vacuum to the EGR valve. This maintains the right amount of exhaust entering the engine for varying operating conditions. (Cadillac)

482

Basic EGR system operation

At idle, the throttle plate in the carburetor or fuel injection throttle body remains closed, Fig. 24-15. Engine vacuum is blocked. It cannot act on the EGR valve. The EGR spring holds the valve shut and exhaust gases CANNOT enter the intake manifold. If the EGR valve were to open at idle, it could upset the air-fuel mixture; the engine could stall.

When the throttle plate opens to increase speed, engine vacuum is applied to the EGR hose. Vacuum pulls the EGR diaphragm up. In turn, the diaphragm pulls the valve open. Engine exhaust can now enter the intake manifold and combustion chambers. At higher engine speeds, there is enough air flowing into the engine so that the air-fuel mixture is not upset.

EGR system variations

There are several variations of the basic EGR system. The function, however, is fundamentally the same for all. Look at Fig. 24-17. It illustrates a back pressure EGR valve. This type utilizes both engine vacuum and exhaust back pressure to control valve action. More accurate control of EGR valve opening is possible. An engine coolant temperature switch may be used to prevent exhaust gas recirculation when the engine is cold. A cold engine does not have extremely high combustion temperatures and does not produce very much NOx. With vacuum to the EGR valve blocked below 100 °F (37.8 °C), the drivability and performance of the cold engine is improved.

A wide open throttle valve (WOT valve) is sometimes placed in the vacuum line to the EGR valve. It opens under full acceleration to provide venturi vacuum to the EGR valve. At wide open throttle, intake manifold vacuum is very low, while venturi vacuum is high. Figs. 24-18 and 24-19 show two other EGR valve designs. Study them!

Fig. 24-20 shows an EGR system that has small jets in the bottom of the intake manifold. They meter a small amount of exhaust gases into the air-fuel mixture. The jets are small enough so that they do not upset the idle air-fuel mixture.

AIR INJECTION SYSTEM

An *air injection system* forces fresh air into the engine's exhaust ports to reduce HC and CO emissions. See Fig. 24-21. Exhaust gases can contain unburned and partially burned fuel. Oxygen from the air injection system allows this fuel to continue to burn.

Fig. 24-22 shows the major parts of an air injection system. The belt-driven air pump forces air at low pressure into the system. As pictured in Fig. 24-22, a rubber hose connects the pump output to a diverter valve.

Fig. 24-22 is a cutaway view of an air injection pump. Note how the spinning vanes (blades) pull in air at one side of the pump. As the vanes rotate, the air is trapped, compressed, and forced out a second opening in the pump.

The *diverter valve* stops air from entering the exhaust system during deceleration. This prevents backfiring in

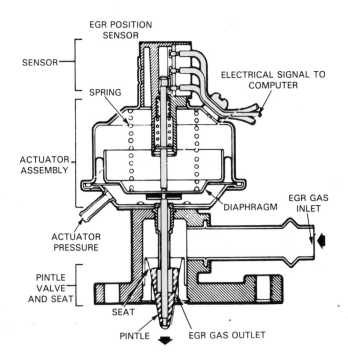

Fig. 24-18. Cutaway of an EGR valve with its sensor connected to an on-board computer. Motion of the EGR's plunger changes amount of current through the sensor. Computer alters operation of engine and other systems according to the signal received from the EGR. (Ford)

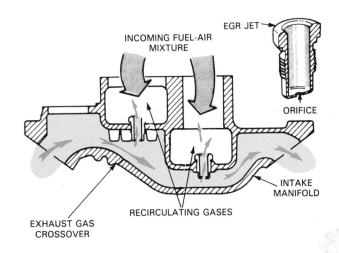

Fig. 24-19. EGR jets are small fittings located in the base of the engine's intake manifold. Jet openings are carefully calibrated to allow metered amounts of exhaust gas to mix with air-fuel mixture. (Chevrolet)

the exhaust system. Refer to Figs. 24-22 through 24-24. The diverter valve also limits maximum system air pressure when needed. Excessive pressure bleeds off through a silencer or muffler.

An air distribution manifold directs a stream of air toward each exhaust valve. Fittings on the air distribution manifold screw into threaded holes in the exhaust manifold or cylinder head. Fig. 24-21 shows a typical air distribution manifold.

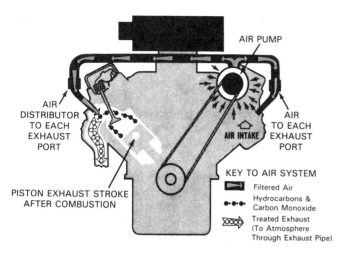

Fig. 24-20. Diagram showing operation of an air injection system. Air pump which runs off a belt from the drive shaft, forces outside air into exhaust manifolds. Introduction of air helps burn hot exhaust gases as they come from the open exhaust valves. (GMC)

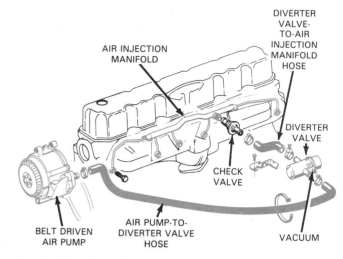

Fig. 24-21. An air injection system listing basic parts. Air pump normally bolts to front of engine. Hose carries air from pump to a diverter valve. Next, air enters exhaust manifold through check valve and distribution manifold. (AMC)

An air check valve is usually found in the line between the diverter valve and the air distribution manifold. It keeps exhaust gases out of the air injection system.

Air injection system operation

When the engine is running, the spinning vanes in the air pump force air into the diverter valve. Look at Fig. 24-24. If the engine is not decelerating, the air moves through the diverter valve, check valve, air injection manifold, into the engine. Fresh air blows on the engine exhaust valves. During deceleration, the diverter valve blocks airflow into the engine exhaust manifold. This prevents any possibility of a backfire that could damage

the car's exhaust system. As needed, the diverter valve's relief valve releases excess pressure.

PULSE AIR SYSTEM

Like an air injection system, a *pulse air system* supplies fresh air into the exhaust part of the engine. Instead of an air pump, it uses natural pressure pulses in the exhaust system. The pulses operate check valves. Fig. 24-25 shows one type. Note how the pulse air system lines and aspirator valves are positioned on the engine. The aspirator valves, also called check valves, gulp valves, or reed valves, block airflow in one direction and

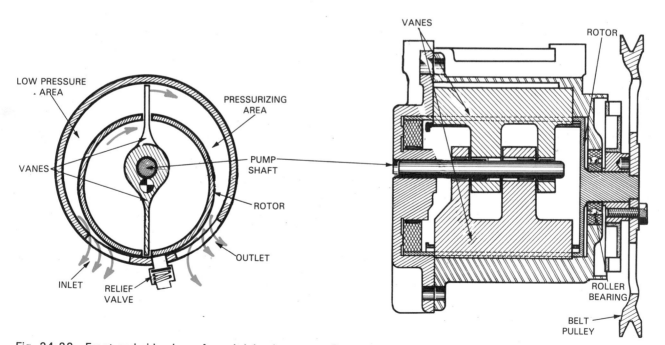

Fig. 24-22. Front and side view of an air injection pump. Cutaway shows action of the pump. Belt turns pump pulley, shaft, and rotor. Vanes trap and pressurize air. (Peugeot)

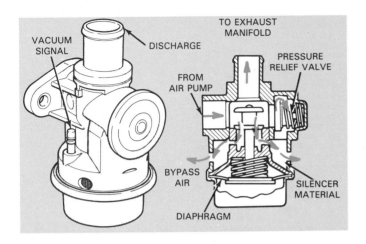

Fig. 24-23. External and cutaway views of a typical diverter valve. It detects sudden increases in intake manifold vacuum and prevents backfire upon sudden deceleration of engine. When it is open, air is prevented from mixing with and igniting rich mixtures caused by deceleration.   (Chrysler)

allow airflow in the other direction. This is illustrated in Fig. 24-25B and 24-25C.

Pulse air system operation

Pressure in the exhaust manifold rises and falls as the engine valves open and close. The aspirator valves are open to allow fresh air to enter exhaust manifold on the low pressure pulses (when exhaust valves first close). They block backflow on the high pressure pulses. This pulse occurs as exhaust valves open.

CATALYTIC CONVERTER

A *catalytic converter* burns off the remaining HC and CO emissions that escape through the exhaust system. Look at Fig. 24-26. Extreme heat (approximately 1400 °F F or 760 °C), ignite the emissions and change them into harmless carbon dioxide (CO_2) and water (H_2O).

The active chemical catalyst in a catalytic converter is usually platinum, palladium, rhodium, or a mixture of two of these substances. A catalyst is any substance that speeds a chemical reaction without being changed itself. The catalyst agent is coated on either a ceramic honeycomb shaped block, or on small ceramic beads. The catalyst is contained in a stainless steel housing that is heat resistant. See Fig. 24-27.

Types of catalytic converters

A catalytic converter using a ceramic block type catalyst is also known as a monolithic type converter. One containing the ceramic beads is called a pellet type catalytic converter. See Fig. 24-28.

A mini (small) catalytic converter is placed close to the engine. It heats up quickly to reduce emissions during engine warmup. A mini catalytic converter is used along with a larger, main converter. See Fig. 24-29.

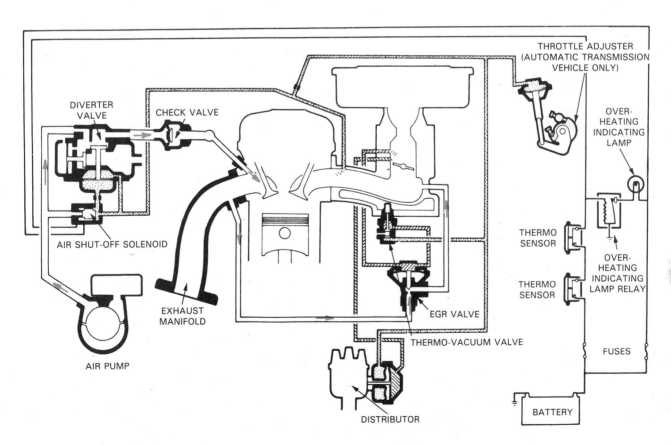

Fig. 24-24. Schematic view shows how components of an air injection system and an EGR system are interconnected. (Chrysler)

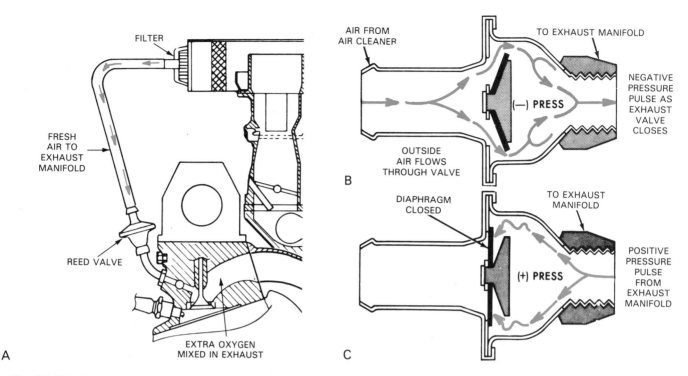

Fig. 24-25. Pulse-type air injection system. A—System uses no air pump. Aspirator valves act as check valves. B—Aspirators allow air to enter engine exhaust manifold when exhaust valves first close and produce vacuum pulse. C—Aspirators cut off airflow when exhaust valves open and produce a pressure pulse.

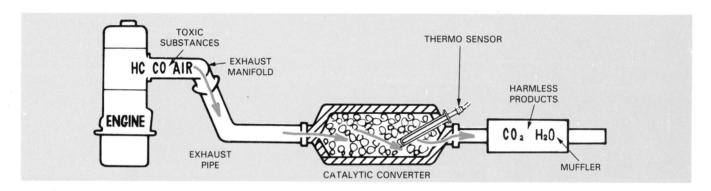

Fig. 24-26. A catalytic converter burns flammable exhaust emissions and changes noncombustibles into harmless carbon dioxide and water. (Toyota)

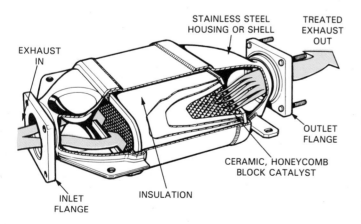

Fig. 24-27. Monolithic catalytic converter consists of a honeycomb block of ceramic covered with chemical substances for treating exhaust gases. (Lancia)

A two-way catalytic converter, sometimes called an OXIDATION type converter, can only reduce two types of exhaust emissions (hydrocarbons and carbon monoxide). A two-way catalyst is normally coated with platinum. A three-way or reduction type catalytic converter, is capable of reducing all three types of exhaust emissions (HC, CO, and NOx). A three-way catalyst is usually coated with rhodium and platinum, Fig. 24-30. A dual-bed catalytic converter contains two separate catalyst units enclosed in a single housing. A dual-bed converter normally has both a three-way (reduction) catalyst and a two-way (oxidation) catalyst. Air is forced into the mixing chamber, located between the two to help burn emissions.

Dual-bed catalytic converter operation

When the engine is cold (below about 128 °F or 53 °C), the air system routes air into the exhaust manifold

486

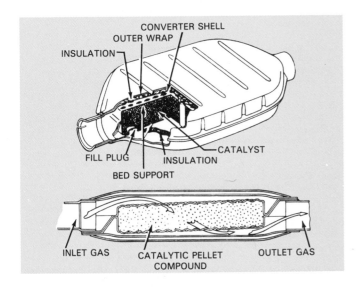

Fig. 24-28. Pellet-type catalytic converter. Beads are coated with a catalytic agent. (GMC)

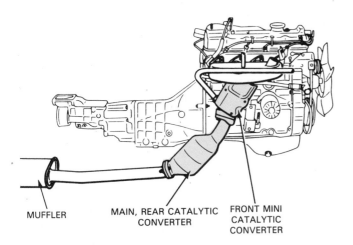

Fig. 24-29. Some engines use a mini catalytic converter in addition to a main converter. In the illustration, the mini converter is placed just outside the exhaust manifold. It heats up quickly and reduces emissions when the engine and main converter are still cold. (Chrysler)

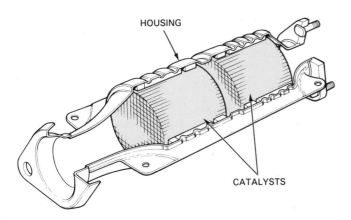

Fig. 24-30. A three-way catalytic converter is used to convert hydrocarbons, carbon monoxide, and oxides of nitrogen in the exhaust gas to carbon dioxide, dinitrogen (N_2), and water vapor. (Honda)

Fig. 24-31. Exhaust heat and the injected air burn the exhaust emissions. When the engine warms, the system forces air into the catalytic converter. First, the exhaust gases pass through the front three-way catalyst that removes HC, CO, and NOx. Then, the exhaust gas flows into the area between the two catalysts. The oxygen in the air flowing into the chamber keeps the gases burning. The exhaust flows into the rear two-way catalyst which removes even more HC and CO. Refer again to Fig. 24-31 and study how the system operates.

COMPUTERIZED EMISSION CONTROL SYSTEMS

A *computerized emission control system* commonly utilizes engine sensors, a three-way catalytic converter, an on-board computer, and either a computer controlled carburetor or fuel injection system. Fig. 24-32 illustrates the operation of a digital fuel injection system using a dual-bed catalytic converter and an exhaust gas sensor. Study how the engine sensors and exhaust sensor feed electrical information to the computer. The computer analyzes this information and adjusts the throttle body air-fuel mixture for maximum efficiency. Also, note how HC, CO, and NOx are lowest at a stoichiometric (theoretically perfect) fuel mixture.

Fig. 24-33 is a schematic of a computer which controls the fuel and emission control systems. The computer receives data from the engine sensors and the exhaust (oxygen) sensor. Responding to sensor signals, it controls the carburetor air-fuel mixture, ignition timing, EGR valve, air injection system, and charcoal canister purge.

Complete details of computer control systems, can be found in the text chapters on carburetion and fuel injection. These chapters explain components such as the engine sensor, exhaust gas sensor, and computer.

SUMMARY

The purpose of emission control systems is to reduce the amount of harmful combustion gases released into the atmosphere. There are four basic types of emission from automobiles: hydrocarbons (HC), carbon monoxide (CO), oxides of nitrogen (NOx), and particulates. Hydrocarbons are unburned fuel; carbon monoxide is a colorless, odorless and deadly gas caused by partially burned fuel; oxides of nitrogen are compounds consisting of 80 percent nitrogen and 20 percent oxygen that result from high combustion temperatures in the car engine; particulates are solid particles of carbon soot and fuel additives.

Automotive engines have been modified in several ways:
1. Lower compression ratios to allow use of unleaded gasoline.
2. Leaner air-fuel mixtures that lower HC and CO emission.
3. Heated intake manifolds that allow quicker warm-ups.
4. Smaller combustion chamber surfaces that reduce HC emission.
5. Greater valve overlap that cuts down on NOx emissions by diluting incoming air-fuel mixture.

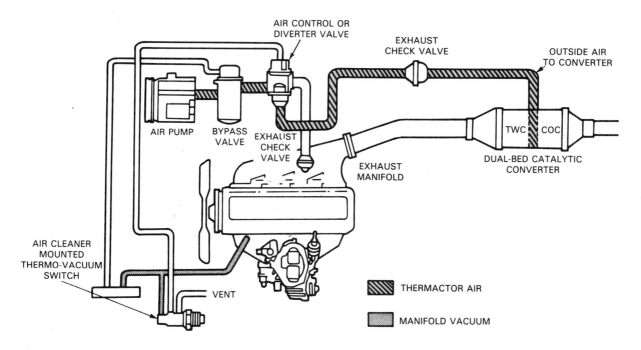

Fig. 24-31. Simplified diagram shows how the air pump forces air into a dual-bed catalytic converter to improve combustion. A thermal-vacuum switch operates air control valve. It opens only when engine coolant is above set temperature. (Ford)

6. Hardened valves and seats that can withstand the lack of lead's lubricating effect.
7. Wider spark plug gapping that produces a hotter spark to ignite hard-to-burn leaner air-fuel mixtures.
8. Reduction of quench areas in the combustion chambers. This helps lower HC and CO emissions by keeping pistons farther away from the cylinder head.
9. Higher operating temperatures that also reduce CO emissions.

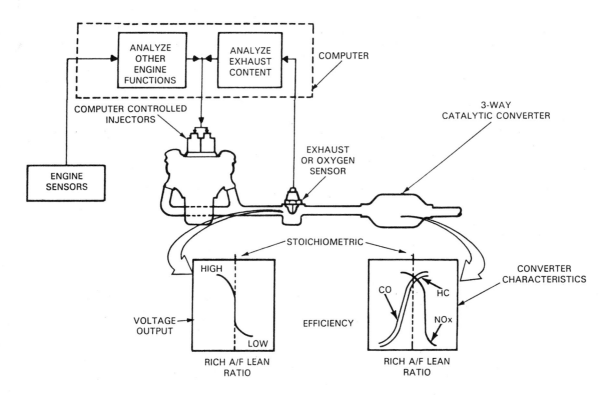

Fig. 24-32. Exhaust gas sensors are also known as oxygen sensors. They monitor oxygen levels in exhaust systems. In general, a high oxygen content indicates complete burning of fuel. When oxygen content is too low, computer will lean the fuel mixture. (Oldsmobile)

To further reduce an engine's emissions, special control systems have been added to the engines. The major ones are: an improved PCV system, an air injection system, and a catalytic converter. Each of these systems must be understood before consideration is given to servicing and repairing them.

KNOW THESE TERMS

Emission control system, Air pollution, Smog, HC, CO, NOx, Particulates, PCV system, Engine blowby, Thermostatic air cleaner, Air cleaner carbon element, Fuel evaporization system, Liquid-vapor separator, Roll-over valve, Charcoal canister, Purge line, EGR valve, Air injection system, Air pump, Diverter valve, Pulse air system, Catalytic converter, Computerized emission control system.

REVIEW QUESTIONS—CHAPTER 24

1. Name some causes of air pollution.
2. Which of the following agencies enforces air pollution standards?
   a. EGR.
   b. SAE.
   c. ASE.
   d. EPA.
3. Define the term "smog."
4. List the four kinds of vehicle emissions and explain what each is.
5. CO emissions are the result of partially burned fuel. True or false?
6. An increase in peak combustion temperature tends to reduce NOx emissions. True or false?

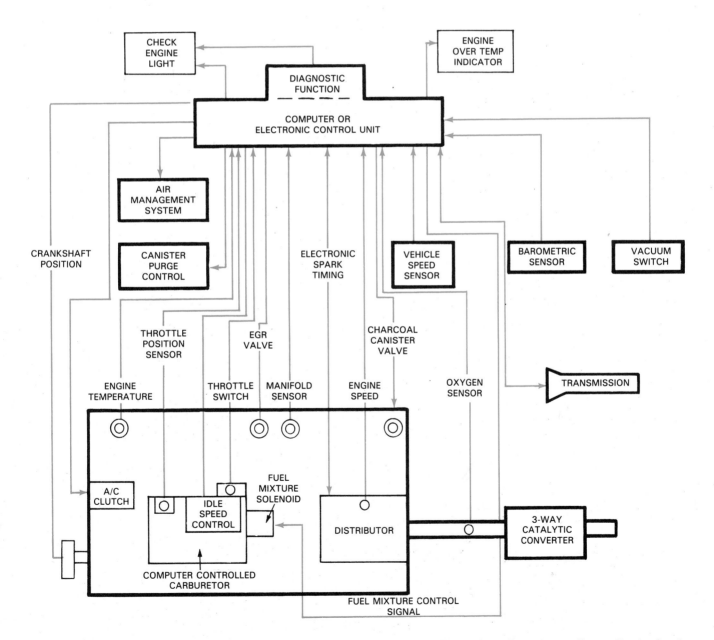

Fig. 24-33. How a computer monitors and controls many engine functions. The on-board computer adjusts all units for the best operating efficiency and minimum air pollution. (Buick)

7. _____ are the solid particles of carbon soot that blow out the tailpipe of a car or truck.

8. List the three basic sources of vehicle emissions.

9. Indicate which of the following is NOT a typical engine modification for reducing emissions.
   a. Lower compression ratios.
   b. Leaner air-fuel ratios.
   c. Decreased valve overlap.
   d. Wider spark plug gaps.

10. Explain the operation of six major emission control systems.

11. A _____ or PCV system utilizes engine vacuum to draw toxic blowby gases into the intake manifold for burning in the combustion chambers.

12. The _____ system speeds engine warmup and keeps the temperature of the air entering the engine at about _____ °F.

13. Explain the operation of the charcoal canister in a fuel evaporization control system.

14. Explain the operation of an EGR system.

15. An air injection system forces fresh air into the _____ of the engine to reduce HC and CO emissions.

16. The diverter valve keeps air from entering the exhaust system during engine deceleration, preventing backfiring. True or false?

17. What is a catalytic converter and what is its purpose?

18. Which of the following is NOT related to operation of catalytic converters?
   a. Monolith.
   b. Pellet.
   c. Stores unburned fuel.
   d. Oxidizes or burns emissions.

19. For what purpose is a mini catalytic converter sometimes used?

20. Briefly explain the operation of a computer controlled emission control system.

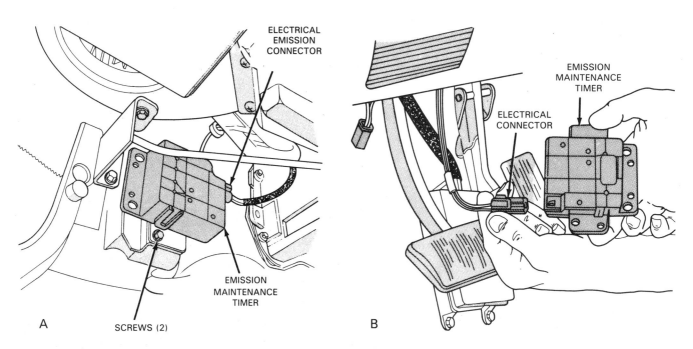

An emission maintenance reminder indicator light. There are many methods used to reset them. In the one shown you have to replace the timer. In others you may need to move a hidden switch, jump across a connector, or perform some other step to turn the reminder light out. A—This timer is mounted under the dash. B—While installing a new timer, make sure electrical connector is fully installed.   (Chrysler)

25 EMISSION CONTROL SYSTEM TESTING, SERVICE, REPAIR

After studying this chapter, you will be able to:
- *Explain and demonstrate the use of two- and four-gas exhaust analyzers.*
- *Inspect and troubleshoot emission control systems.*
- *Perform periodic service operations on emission control systems.*
- *Test various emission control components.*
- *Replace or repair major emission control components.*

Like any automotive system, emission control systems can break down. When this happens, emission levels increase. An auto technician must be able to diagnose and remedy emission control system troubles. Keeping these systems in good working order protects something very important—THE AIR WE BREATHE!

EXHAUST ANALYZER

An *exhaust analyzer* measures the chemical makeup of engine exhaust gases. See Fig. 25-1. When a measurement is wanted, the technican places the analyzer probe (sensor) in the car's tailpipe, Fig. 25-2. With the engine running, the exhaust analyzer will indicate the level of pollutants and other gases in the exhaust. This information helps to determine the condition of the engine and other systems affecting emissions.

An excellent diagnostic tool, the exhaust gas analyzer will indicate problems with the:
1. Carburetor or fuel injection.
2. Engine.
3. Vacuum system.
4. Ignition system.

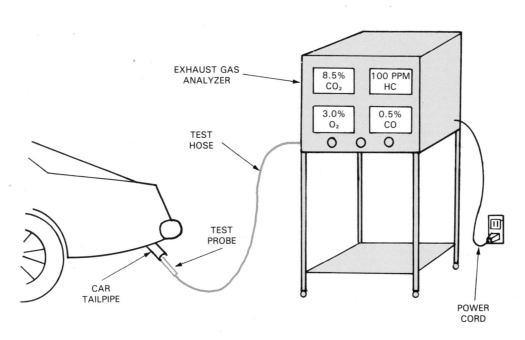

Fig. 25-1. An exhaust gas analyzer is designed to examine and measure the chemical makeup of auto engine exhaust. (Sun Electric)

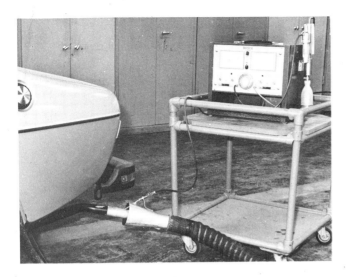

Fig. 25-2. Hose from exhaust analyzer is installed on vehicle's tailpipe. Hose adapter prevents toxic vapors from escaping. Meters should be warmed and calibrated before placing test probe in tailpipe. Compare meter readings with specifications for engine. (Saab)

5. PCV.
6. Air filter.
7. Air injection system.
8. Evaporative control system.
9. Computer control system.
10. Catalytic converter.

Engine exhaust gases

As discussed in the previous chapter, engine exhaust gases contain chemical substances that indicate problems with combustion efficiency of an engine. Some of these substances, such as hydrocarbons (HC), carbon monoxide (CO), and oxides of nitrogen (NOx), are harmful. Other combustion by-products, such as carbon dioxide (CO_2), oxygen(O_2), and water (H_2O) are not harmful. By measuring HC, CO, O_2, and CO_2, we can find out if the engine and its emission systems are working properly.

Two- and four-gas exhaust analyzers

Modern repair shops use two different kinds of exhaust gas analyzers: the two-gas and the four-gas. The *two-gas exhaust analyzer* measures the amount of hydrocarbons (HC) and carbon monoxide (CO). A common type, it has been in use for a number of years. Now, however, the two-gas analyzer is being replaced by the more informative four-gas analyzer.

The *four-gas exhaust analyzer* measures the quantity of hydrocarbons (HC), carbon monoxide (CO), carbon dioxide (CO_2), and oxygen(O_2), in an engine's exhaust. Look at Fig. 25-3. Late model engines are so clean-burning that a four-gas exhaust analyzer is needed to evaluate accurately the makeup of the exhaust gases. It provides extra information for diagnosing problems and making adjustments or repairs.

HC readings

An exhaust gas analyzer measures hydrocarbons (HC) in parts per million (ppm) by volume, Fig. 25-3A. For ex-

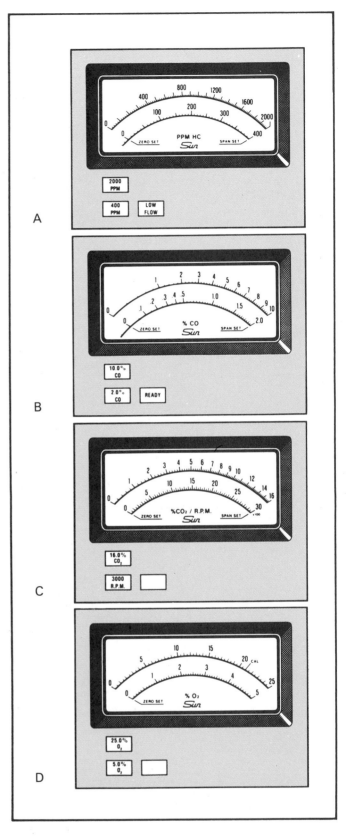

Fig. 25-3. Metering scales for exhaust gas analyzers. Two-gas analyzers have scales for only HC and CO. Four-gas analyzers also measure CO_2 and O_2. A—Hydrocarbon meter shows amount of unburned carbons in parts per million. B—CO meter measures partially burned fuel. C—CO_2 meter shows percentage of harmless carbon dioxide gas. D—O_2 meter reads oxygen content by percentage. (Sun Electric)

ample, an analyzer reading of 100 ppm means there are 100 parts of HC for every million parts of exhaust gas.

A car that is a few years old, for instance, might have an HC specification of 250 ppm. A newer car, having stricter emission requirements, could have a 100 ppm HC specification. If HC is higher, the car's HC emissions, indicating unburned fuel, are too high. Ajustment or repair is needed.

NOTE! Always refer to the emission control sticker in the engine compartment or to a service manual for emission level specs. Values vary year by year.

One or more of the following conditions can cause higher-than-normal HC readings:
1. Either rich or lean air-fuel mixture (carburetor or fuel injection system problem).
2. Improper ignition timing (distributor, computer, or adjustment problem).
3. Engine problems (blowby, worn rings, burned valve, blown head gasket).
4. Faulty emission control system (bad PCV, catalytic converter, evaporative control system).
5. Ignition system troubles (fouled spark plug, cracked distributor cap, open spark plug wire).

CO readings

An exhaust analyzer measures carbon monoxide (CO) in percentage by volume, Fig. 25-3B. For instance, a 1 percent analyzer reading would mean that 1 percent of the engine exhaust is made up of CO. The other 99 percent are other substances. Basically, high CO is caused by incomplete burning and a lack of air (oxygen) during the combustion process.

If the exhaust analyzer reading is higher than specification, the engine is producing too much carbon monoxide. You need to locate and correct the cause.

The exhaust analyzer's CO reading is related to the air-fuel ratio. A HIGH CO reading indicates an over-rich mixture. (This means too much fuel; not enough air.) A LOW CO reading would indicate a lean air-fuel mixture (not enough air compared to fuel).

The following are typical causes of high CO readings:
1. Fuel system: bad injector, high float setting, clogged carburetor air bleed, restricted air cleaner, choke out of adjustment, bad engine sensor, computer troubles.
2. Emission control system: almost any emission control system problem can upset CO.
3. Incorrect ignition timing: timing too far advanced or improper vacuum going to vacuum advance unit.
4. Low idle speed: carburetor or injection system setting wrong.

Four-gas exhaust analyzers measure oxygen (O_2) in percentage by volume, Fig. 25-3D. Typically, O_2 readings should be between 1 and 7 percent. The catalytic converter needs oxygen to burn HC and CO emissions. Without O_2 in the engine exhaust, exhaust emissions can pass unchanged through the converter.

From the previous chapter you learned that there are two systems that add O_2 to the engine exhaust. One is the air injection system; the other an air pulse system. With air added to the exhaust, CO and HC emissions decrease. With air added to the exhaust, CO_2 and HC emissions decrease. Thus, O_2 readings can be used to

check the operation of the carburetor, fuel injection system, air injection system, catalytic converter, and control computer.

Oxygen in the engine exhaust accurately indicates a LEAN air-fuel mixture. In an engine running lean, O_2 increases in proportion to the air-fuel ratio.

When the air-fuel mixture becomes lean enough to cause a LEAN MISFIRE (engine misses), O_2 readings rise swiftly. Thus an O_2 reading accurately measures lean and efficient air-fuel ratios. New car fuel systems must be adjusted almost to the lean misfire point to reduce exhaust emissions and increase fuel economy.

CO_2 readings

Carbon dioxide, a by-product of combustion, is a combination of one carbon molecule with two oxygen molecules. It is not toxic at low levels. Humans and air-breathing animals, for example, exhale carbon dioxide.

The four-gas exhaust analyzer can also measure the percentage of carbon dioxide (CO_2) in exhaust gases, Fig. 25-3C. Typically, CO_2 readings should be above 8 percent. Such readings provide more data for checking and adjusting the air-fuel ratio. For example, if the CO_2 level exceeds O_2 levels, the air-fuel ratio is too rich for a stoichiometric (theoretically perfect) mixture.

How to use an exhaust gas analyzer

To use an exhaust gas analyzer, follow the warmup instructions from the manufacturer. See Fig. 25-3. Next, zero and calibrate the meters. Generally, zeroing is done while sampling clean air (no exhaust gases present).

CAUTION! When using an exhaust analyzer, NEVER allow engine exhaust fumes to escape into an enclosed shop area. Exhaust gases can kill! Use a shop exhaust vent system to trap and remove these toxic fumes.

Exact procedures vary. Follow the operating instructions. This will assure accurate measurements.

Generally, you must measure HC and CO at idle and approximately 2500 rpm. With a four-gas analyzer, you should also measure O_2 and CO_2. That done, compare the analyzer readings with specifications. When testing some electronic (computer) fuel injection systems without a load, only idle readings on the exhaust analyzer will be accurate. A dynamometer must be used to load the engine, simulating actual driving conditions. Several other textbook chapters discuss how exhaust analyzer readings can be used. Refer to the text index for more information.

EMISSION CONTROL SYSTEMS INSPECTION

If your exhaust analyzer readings show problems, you must find the source. As a rule, inspect all engine vacuum hoses and wires first, Fig. 25-4. A leaking vacuum hose or disconnected wire will upset the operation of the engine and the emission control systems.

To detect leaks you can use a section of vacuum hose as a stethoscope (listening device). As in Fig. 25-5, hold one end of the hose next to your ear. Move the other end of the hose among vacuum hoses and connections in the engine compartment. If there is vacuum leak, you will hear a HISSING SOUND through the tubing.

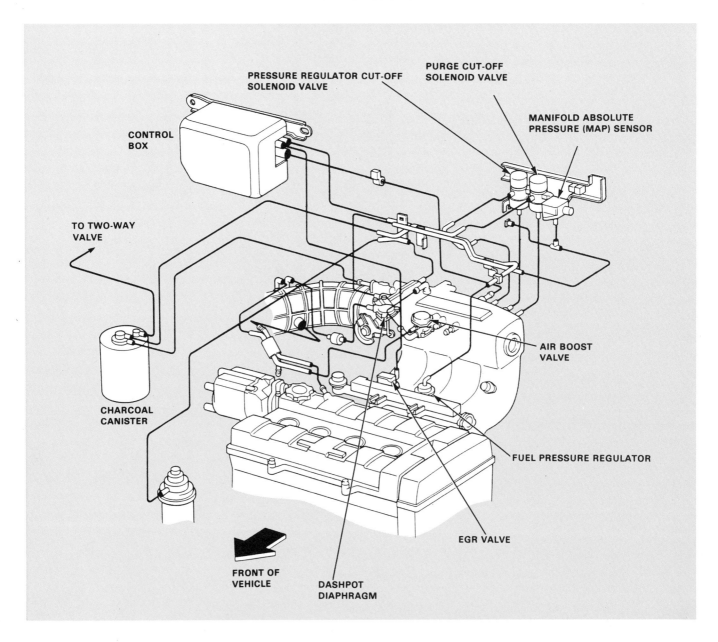

Fig. 25-4. A leaking vacuum hose or a disconnected wire will increase vehicle emissions. A number of parts must be checked. (Honda)

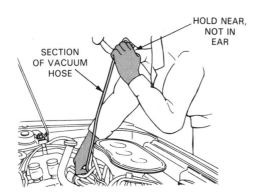

Fig. 25-5. Using a length of small hose as a listening device to find vacuum leaks. Hold one end to ear as you move the other end around the engine compartment. A loud hiss indicates you are near the source of a vacuum leak. (Subaru)

Also inspect the air cleaner for clogging. Check and adjust the air pump belt as necessary. Check, also, for any visual and obvious problems. If no problems turn up during your inspection, check and test each system.

PCV SYSTEM SERVICE

A PCV system that is not functioning not only increases exhaust emissions but causes engine sludge, engine wear, rough engine idle, and other problems. A system leak can cause a vacuum leak producing a lean air-fuel mixture at idle. A restricted PCV system can enrich the fuel mixture. This, again, affects engine idle.

PCV system maintenance

Most automakers recommend periodic maintenance of the PCV system. First, inspect the condition of the PCV

494

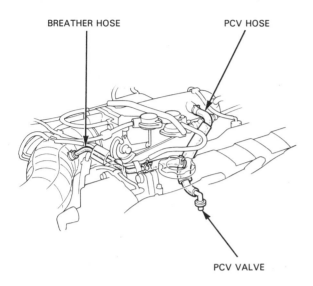

BREATHER HOSE   PCV HOSE

PCV VALVE

Fig. 25-6. Check the crankcase ventilation hoses and connections for leaks and clogging. With the engine at idle, lightly pinching the hose between the PCV valve and the intake manifold should cause a clicking sound in the valve. If it does not, check the PCV valve grommet for cracks or damage.

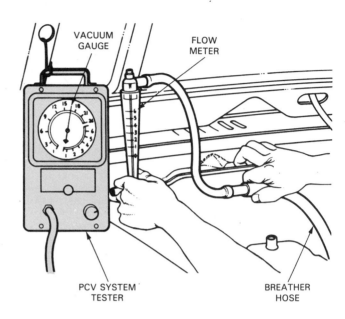

VACUUM GAUGE   FLOW METER

PCV SYSTEM TESTER   BREATHER HOSE

Fig. 25-7. A PCV valve tester measures airflow through the valve. Follow instructions for the tester and compare the reading to specs.   (AMC)

hoses, grommets, fittings, and breather hoses, Fig. 25-6. Replace any hose showing signs of deterioration. Clean or replace the breather filter, if needed. Also, check or replace the PCV valve. Since replacement intervals vary, always refer to the car's service manual.

PCV system testing

To test a PCV system, remove the PCV valve from the engine. With the engine idling, place your finger over the end of the valve. With airflow through the hose blocked, you should feel suction. Also, idle speed should drop about 25-80 rpm. If you cannot feel suction (no vacuum), the PCV valve or hose may be plugged with sludge. If engine rpm drops more than 25-80 rpm and the engine begins to idle smoothly, the PCV valve may be stuck open.

These tests can be made with a PCV valve tester. Look at Fig. 25-7. To use the tester, first make sure that engine intake manifold vacuum is correct. Then connect the tester to the PCV valve following the operating instructions. Start the engine and allow it to idle. Note the airflow rate on the tester. If airflow is not within specified limits, replace the PCV valve.

Paper test

Some auto manufacturers suggest placing a piece of paper over the PCV breather opening to test the PCV system. Seal the dipstick tube with tape, then start and allow the engine to idle. After a few minutes of operation, the crankcase vacuum should pull the paper against the breather opening. If not, there is a leak in the system. There may be a ruptured gasket, a cracked hose, or the system may be plugged.

A four-gas exhaust analyzer can also be used to check the general condition of a PCV system. Measure and note the analyzer readings with the engine idling. Then, pull the PCV valve out of the engine, but not off the hose. Compare the readings after the PCV valve is removed.

A plugged PCV system will show up on the exhaust analyzer when O_2 and CO do NOT CHANGE. Crankcase dilution (excessive blowby or fuel in oil) will usually show up as an excessive (1 percent or more) increase in O_2 or a 1 percent or more decrease in CO.

EVAPORIZATION CONTROL SYSTEM SERVICE

A faulty evaporization control system can cause many problems. Among them are fuel odors, fuel leakage, fuel tank collapse (vacuum buildup), excess pressure in the fuel tank, or a rough engine idle. These problems usually stem from:

1. Defective fuel tank pressure-vacuum cap.
2. Leaking charcoal canister valves.
3. Deteriorated hoses.
4. Incorrect hose routing.

Maintenance and repair on an evaporization control system generally involves cleaning or replacing the filter in the charcoal canister, Fig. 25-8. Service intervals for the canister filter vary. However, if the car is operated in a dusty environment, clean or replace the filter more often.

At the same time, inspect the condition of the fuel tank filler cap. Make sure the cap seals are in good condition. There are testers for checking the opening of the pressure and vacuum valves in the cap. The cap should be tested when excessive pressure or vacuum problems are noticed. Hoses should be inspected in the evaporization system. Look for signs of deterioration (hardening, softening, or cracking). When replacing a hose, make sure you use special fuel-resistant hose. Old vacuum hoses can be ruined quickly by fuel vapors.

HEATED AIR INLET SYSTEM SERVICE

An inoperative heated air inlet system (thermostatic air cleaner) can cause several engine performance prob-

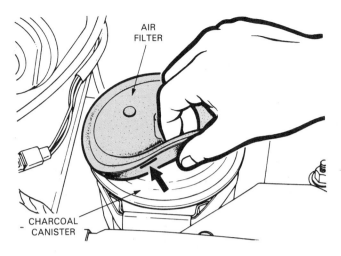

Fig. 25-8. If the charcoal canister has an air filter, remove, clean, or replace it at intervals. (Plymouth)

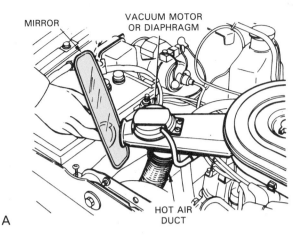

A

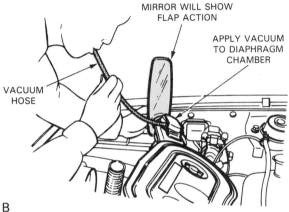

B

Fig. 25-9. Using a mirror and a hose to check thermostatic air cleaner operation. A—Watch reflection in mirror to check action of air blend door or flap inside the snorkel. Flap should be closed during start-up and should close as engine becomes warm. B—When you suck on the hose, applying vacuum to the vacuum motor, the flap should move. (Subaru)

lems. If the air cleaner flap remains OPEN (cold air position), the engine could miss, stumble, stall, and warm up slowly. If the air cleaner flap remains CLOSED (hot air position), the engine could perform poorly at full operating temperature.

Heated air inlet system maintenance

The heated air inlet system requires little maintenance. As a general rule, inspect the condition of the vacuum hoses and hot air hose from the exhaust manifold heat shroud. The hot air tube is frequently made of heat-resistant paper and metal foil. It will tear very easily, Fig. 25-9A. If torn or damaged, replace the hot air tube.

Testing heated air inlet system

A quick method of checking operation of the heated air inlet system, is to watch the action of the air flap in the air cleaner snorkel. Start and idle the engine, as shown in Fig. 25-9A. When the air cleaner temperature sensor is cold, the air flap should be closed. Place an ice cube on the sensor if needed. Then, when the engine and sensor warm to operating temperature, the flap should swing open. If the air cleaner flap does not function, test the vacuum motor and the temperature sensor. To test the vacuum motor, apply vacuum to the motor diaphragm with a hand vacuum pump or your mouth, Fig. 25-9B. With the prescribed amount of suction, the motor should pull the air flap open. Replace the vacuum motor if it leaks or does not open the flap. Recheck heated air inlet system operation to make sure the air temperature sensor is working properly.

To test the thermo-vacuum switch in the air cleaner, place a thermometer next to the unit. With the sensor cooled below its closing temperature, apply vacuum to the thermo-vacuum switch. It should allow passage of vacuum to the vacuum motor. Next, warm the thermo-vacuum switch to its closing temperature. Use a heat gun or hair dryer. When warm, the switch should block vacuum and the vacuum motor should open the flap. Replace the thermo-vacuum switch if it fails to open and close properly.

EGR SYSTEM SERVICE

EGR system malfunctions can cause engine stalling at idle, rough engine idling, detonation (knock), and poor fuel economy. If the EGR valve sticks open, it will act as a large vacuum leak, causing a lean air-fuel mixture. The engine will run rough or stall at idle. If the EGR fails to open or the exhaust passage is clogged, higher combustion temperatures can cause abnormal combustion (detonation) and knocking.

EGR system maintenance

Maintenance intervals for the EGR system vary from one vehicle manufacturer to another. Refer to a service manual for exact mileage intervals. Newer cars have a reminder light in the dash. The light will glow when EGR maintenance is needed. Also, check that the vacuum hoses in the EGR system are in good condition. They can become hardened, causing leakage.

EGR system testing

To test an EGR system, start, idle, and warm the engine. While operating the carburetor or throttle body accelerator linkage by hand, increase engine speed to 2000-3000 rpm very quickly. If visible, observe the movement of the EGR valve stem. The stem should move as the engine is accelerated. If it does not move, the EGR system is not functioning. Sometimes the EGR valve stem is not visible. You then need to test each EGR system component separately. Follow the procedures described in a service manual.

Test the EGR valve while the engine is idling. Connect a hand vacuum pump to the EGR valve, Fig. 25-10. Plug or pinch shut the supply vacuum line to the EGR valve. As vacuum is applied to the EGR valve with the pump, the engine should begin to MISS or STALL. This lets you know that the EGR valve is opening and that exhaust gases are entering the intake manifold. If EGR valve does NOT affect engine idling, remove the valve. It or the exhaust manifold passage could be clogged with carbon. If needed, clean the EGR valve and exhaust passage. If the EGR valve still does not open and close properly, replace the valve.

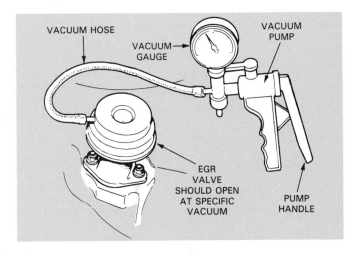

Fig. 25-10. As the engine idles, apply vacuum to EGR valve. Engine should miss or stall if EGR valve is working. (Honda)

Some EGR valves provide electrical data to a computer control system. Such valves require special testing procedures. Refer to a shop manual covering the specific system. Component damage may result from an incorrect testing method.

AIR INJECTION SYSTEM SERVICE

Air injection system problems can cause engine backfiring (loud popping sound), other noises, and increased HC and CO emissions. Remember, air injection is used to help burn any fuel that enters the exhaust manifolds and exhaust system. Without this system, the fuel could ignite all at once (backfire) with a loud bang. Inadequate air from the air injection system could also prevent the catalytic converter from functioning properly.

Air injection system maintenance

Maintenance of an air injection system may require replacing the pump inlet filter (if included in the system), adjusting pump belt tension, and inspecting the condition of hoses and lines. If the pump belt or any of its hoses show signs of deterioration, they should be replaced. Shop manual specs will give maintenance intervals.

Testing air injection system

A four-gas exhaust analyzer tests an air injection system quickly and easily. Run the engine at idle and record the readings. Then disable the air injection system. Remove the air pump belt or use pliers or clamps to close off the hoses to the air distribution manifold. Compare the exhaust analyzer readings before and after disabling the air injection system.

Without air injection, the exhaust analyzer O_2 reading should drop approximately 2-5 percent. *HC and CO readings* should increase. This test would show that the air injection system is working.

If the analyzer readings do NOT change, the air injection system is not working. Test each component until the source of the problem is found. To test the air pump, remove the output line from the pump. Refer to Fig. 25-11.

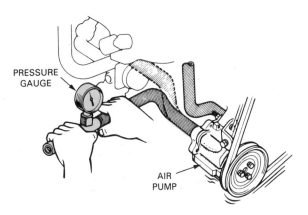

Fig. 25-11. Use a pressure gauge to check air pump output. Replace pump if pressure is not up to specs. (Toyota)

Measure the amount of pressure developed by the pump at idle. Normally, an air pump should produce about 2 to 3 psi (14 to 21 kPa) of pressure. If a low-pressure gauge is not available, place your finger over the line to check for pressure. Replace the pump if faulty.

To test the diverter valve or other air injection system valves, use a service manual. It will explain testing procedures for the specific components.

PULSE AIR SYSTEM SERVICE

Many of the maintenance and testing methods discussed for an air injection system apply to the pulse air system. Inspect all hoses and lines.

Measure O_2 with a four-gas analyzer. Exhaust analyzer oxygen readings should drop when the pulse air system is disabled. If readings DO NOT drop, check the action of the aspirator (Reed) valves. Fig. 25-12 shows a technician testing the aspirator valves. With the engine running, you should be able to feel vacuum pulses on your fingers. However, you should NOT feel exhaust pressure pulses. If you do, it means that exhaust is blowing back through the valves. Replace the valves if they do not function as designed.

Fig. 25-12. Place thumbs over aspirator valves with engine running. You should feel suction pulses. (Saab)

CATALYTIC CONVERTER SERVICE

Catalytic converter problems are usually the result of lead contamination, overheating, and extended service, Fig. 25-13.

A catalytic converter, clogged from lead deposits and overheating, can significantly increase exhaust system backpressure. High backpressure decreases engine performance because gases cannot leave the combustion chamber and flow freely through the converter.

After extended service, the catalyst in the converter becomes coated with deposits. These deposits keep the catalyst from acting on the HC, CO, and NOx emissions. The result: increased air pollution. The inner baffles and shell of the converter can also deteriorate. In a pellet type catalytic converter, exhaust can blow BB-size particles out of the tailpipe. Monolithic (honeycomb) catalytic converters must be replaced when the catalyst becomes damaged (overheated) or contaminated (use of leaded fuel or extended service).

Pellet catalytic converters normally have a plug that

allows replacement of the catalyst agent. The old pellets can be removed and new ones installed. If the converter housing is damaged or corroded, replace the converter.

Testing the catalytic converter

A four-gas exhaust analyzer will check the catalytic converter's condition. Follow the specific directions provided with the analyzer.

Warm up the engine and allow it to idle. With some systems, you may need to disable the air injection or pulse air system. Measure the oxygen and carbon monoxide at the tailpipe.

If O_2 readings are above approximately 5 percent, it is safe to say there is enough oxygen for the catalyst to burn the emissions. However, if the CO readings are still above about .5 percent with other systems operating properly, the catalytic converter is not oxidizing (burning) the emissions from the engine. The converter or catalyst could require replacement.

NOTE! Before replacing a catalytic converter, refer to a factory service manual. It will give added information on checking other systems before converter replacement.

Catalyst replacement

To install new pellets in a catalytic converter, follow service manual instructions. Generally, do as illustrated in Fig. 25-14.

Catalytic converter replacement

Sometimes the catalytic converter is an integral part of the exhaust pipe. With this design, the converter and pipe may have to be replaced together. However, with many cars, the converter can be removed separately. With the catalytic converter removed, make a visual

Fig. 25-13. Cutaway shows badly damaged catalytic converter. This converter would block any exhaust flow and prevent normal engine operation. (Champion Spark Plug Co.)

498

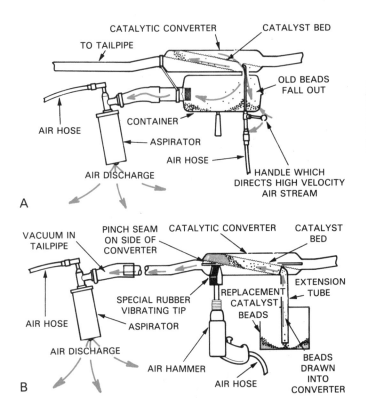

A

B

Fig. 25-14. Changing catalyst in a pellet type catalytic converter. A—Applying air pressure will agitate pellets and cause them to fall out of the opening. B—When replacing pellets, place aspirator over tailpipe. This will draw pellets from the container into the converter. Rubber tip over air hammer will vibrate the converter and help settle the pellets.   (Kent-Moore)

check for plugging by looking through the inside. Inspect the housing for cracks or other damage, Fig. 25-15. When installing the new converter, use new gaskets and reinstall heat shields.

CAUTION! The operating temperature of a catalytic converter can be over 1400°F (760°C). This heat can cause serious burns. Do NOT touch a catalytic converter until you are sure it has cooled.

COMPUTER CONTROLLED EMISSION SYSTEM SERVICE

Computer controlled engine and emission systems are subject to a wide range of problems. The computer may control the carburetor, fuel injection system, EGR valve, evaporization control system, and other components. Any of these parts affect the operation of the total computer system and can increase emissions.

Testing computer controlled emission systems

Since computer systems are so complex, special system analyzers are commonly used to pinpoint specific problems. Fig. 25-16 shows one type of analyzer. Following the manufacturer's operating instructions, plug the analyzer into the wiring of the electronic control unit (computer). Then, use the analyzer to determine the cause for most system troubles.

Many computerized emission control systems have a built-in diagnostic system. If a problem develops, a dash warning light will glow.

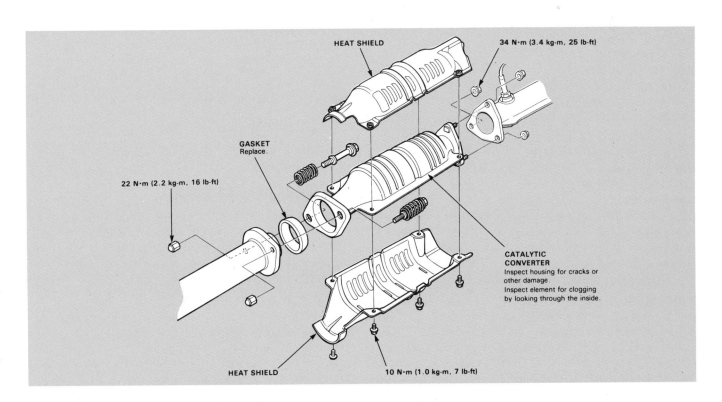

Fig. 25-15. Exploded view of a typical catalytic converter. If an excess of back pressure is suspected, remove and visually inspect the converter. Look through the openings to inspect for clogging. Inspect the housing for cracks or other physical damage.
(Honda)

499

Fig. 25-16. Using a special analyzer to check computer control systems. Manufacturers usually recommend a specific tester for their vehicles. (OTC Tools)

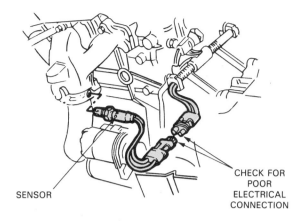

Fig. 25-17. It is common to have loose or poor electrical connections in computer control systems. (Oldsmobile)

To pinpoint the trouble, follow service manual test procedures. For example, one computer system is checked by grounding a special "test terminal." This will cause the computer to flash a special Morse style code on a dash indicating light. The shop manual will tell how to read the code (on-off flashes). This provides a quick and easy way of locating problems.

> WARNING! Do not connect a VOM (volt-ohm-milliammeter) to a computer system unless told to do so by a service manual. Several components, including the computer, can be damaged by incorrect testing procedures.

Replacing parts in computerized emission system

Depending upon what problem the system tester or self-test mode indicated, you would need to replace faulty components. High emission levels could be due to:
1. A faulty oxygen sensor.
2. Coolant temperature sensor.
3. Throttle position sensor.
4. Vehicle speed sensor.
5. Intake manifold pressure sensor.
6. Fuel injector solenoid, a wiring problem.
7. A faulty computer.

Before replacing any computer system part, check all electrical connections until satisfied they are clean and tight. See Fig. 25-17. A corroded connection can affect system operation. When replacing the computer, you may need to remove and reuse the prom. The PROM (programmable read only memory) is a computer chip (miniaturized electronic circuit) designed for use with the specific vehicle. Usually, it is not faulty and can be reused. Only the control section of the computer usually fails.

Fig. 25-18 shows a wiring diagram for a modern computer controlled emission control system. Study the location and relationship between the parts. This type of diagram is useful when tracing or troubleshooting circuit problems.

EMISSION CONTROL INFORMATION STICKER

The emission control information sticker gives important instructions, diagrams, and specs for complying with EPA regulations. One is shown in Fig. 25-19. This sticker is normally located in the engine compartment, on the radiator support, or valve cover.

EMISSION MAINTENANCE REMINDER

The emission maintenance reminder is a circuit that automatically turns on a dash light to indicate the need for emission control system service. The system is designed to automatically illuminate an indicator light so the car can be returned to the shop for service.

After repairs are made, you must turn off the emission maintenance light. There are numerous methods to turn off the light. You might have to move a small lever hidden in the dash, remove the speedometer cluster, jump across a specific connector, etc. Since there are so many variations, refer to the service manual for instructions. It will give the exact location and methods for deactivating the emission maintenance reminder light.

SUMMARY

It is important to maintain emission control systems on automobiles. When they break down or malfunction, their emission of air pollutants increases.

A tool that will measure the chemical makeup of engine exhaust is called an exhaust analyzer. It will indicate the condition of engine and nine components of the engine and emission system controls. Measuring the gases emitted by the engine exhaust indicates the combustion efficiency of the engine. Some substances, like hydrocarbons, carbon monoxide, and oxides of nitrogen are harmful. Other substances like oxygen, carbon dioxide, and water, are not considered harmful. However, analysis of all these substances will reveal the working condition of the engine and emission control system.

There are two types of exhaust analyzers in use; the two-gas analyzer and the four-gas analyzer. The two-

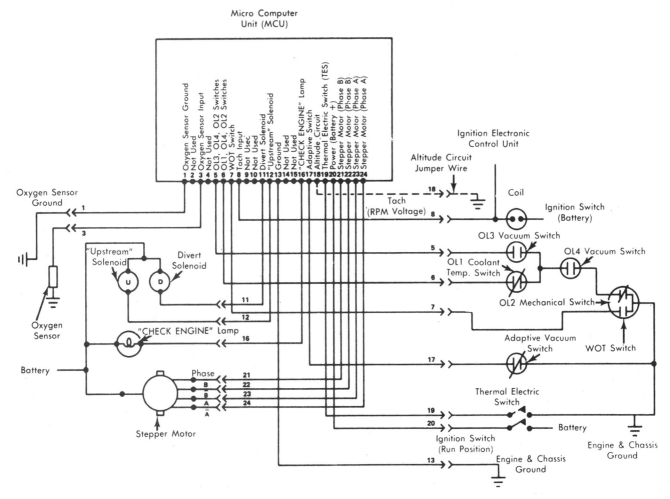

Fig. 25-18. A wiring diagram is useful for tracing hookups for a modern computerized emission control system. (AMC)

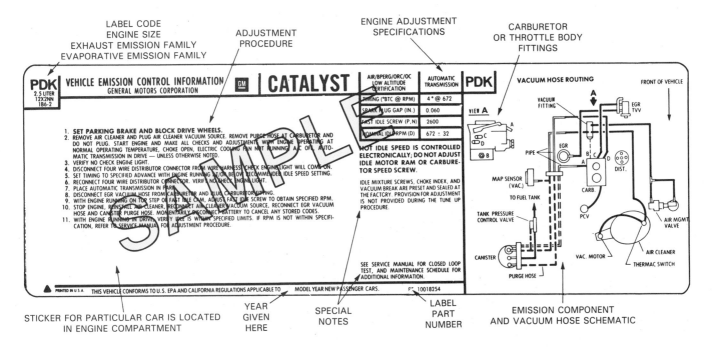

Fig. 25-19. Typical emission control information stickers are found in engine compartment. Refer to it during a tune-up. (Oldsmobile)

gas analyzer measures hydrocarbons and carbon monoxide. It is being replaced by the four-gas analyzer which measures carbon dioxide and oxygen as well as hydrocarbons and carbon monoxide.

Several conditions can cause higher-than-normal hydrocarbon readings: improper air-fuel mixture, ignition timing off, engine problems, faulty emission control system, and ignition system troubles.

A high carbon monoxide reading indicates a too-rich air-fuel mixture. A low carbon monoxide reading indicates that the air-fuel mixture is too lean.

Oxygen readings can be used to check the operating conditions of the carburetor, the fuel injection system, the air injection system, the catalytic converter, and the control computer. Oxygen readings are an accurate measure of lean and efficient air-fuel ratios.

Carbon dioxide readings provide additional data for checking and adjusting the air-fuel ratio. Generally, readings should be above 8 percent.

Emission control systems include the PCV system, evaporization control system, heated air inlet system, EGR system, air injection system, and catalytic converter.

KNOW THESE TERMS

Exhaust analyzer, Two-gas analyzer, Four-gas analyzer, HC readings, O_2 readings, CO_2 readings, Stoichiometric.

REVIEW QUESTIONS — CHAPTER 25

1. An_____ _____is a testing instrument that measures the chemical content of the engine exhaust gases.
2. Indicate the difference between a two- and a four-gas exhaust analyzer.
3. Give five reasons why HC readings may be higher than normal.
4. High levels of CO in the exhaust can be caused by fuel system problems. True or false?

5. With a four-gas exhaust analyzer, oxygen in the engine exhaust is an accurate indicator of an air-fuel mixture. True or false?
6. Typically, if the percent of CO exceeds that of O_2, the air-fuel mixture is:
   a. On the lean side of stoichiometric.
   b. On the rich side of stoichiometric.
   c. Stoichiometric.
   d. None of the above.
7. Describe how to check the general operation of a PCV valve with your finger and with an exhaust gas analyzer.
8. If a vacuum pump is used to activate an EGR valve at idle, the engine is NOT affected. True or false?
9. An air pump in an air injection system should produce about two psi or two kPa. True or false?
10. Describe what results from a clogged catalytic converter.
11. The operating temperature of a catalytic converter may be over _____ or _____.
12. A driver complains that a check engine (emission trouble) light glows in the dash. The car's periodic service switch has not been activated. The car also has a self-diagnostic computer that controls many emission system components.
   Technician A says that, first, an ohmmeter should be used to measure the resistance of the computer system oxygen sensor. It is a common problem with this make and model.
   Technician B says that the computer should be activated so that it can show the possible problem location.
   Which Technician is correct?
   a. Technician A.
   b. Technician B.
   c. Both Technician A and B are correct.
   d. Neither Technician A nor B are correct.
13. What is a computer PROM?
14. Why, in many instances, should the PROM be reused in a new computer?
15. The emission control gives important instructions, diagrams, and specs for complying with EPA regulations. True or false?

26 USING SERVICE MANUALS AND WORK ORDERS

After studying this chapter, you will be able to:
- *Find and use the service manual index and content sections when servicing fuel systems.*
- *Recognize the different types of service manuals available.*
- *Describe the three basic types of troubleshooting charts found in service manuals.*
- *Explain how to use flat rate manuals.*
- *Fill out a work order.*

Earlier chapters of this text have helped you learn the most important terms, tools, and procedures for servicing modern automotive fuel systems. Today's cars, however, are so complex that every good mechanic must sometimes refer to a service manual for additional instructions and specs on a repair job.

Here you will review some basic information on the use of the service manual. You will also learn how to fill out a shop work order. You will need both these skills on the job.

SERVICE (SHOP) MANUALS

Service manuals, frequently called *shop manuals,* are publications having part names and numbers, step-by-step procedures, specifications, diagrams, illustrations of parts and service steps, and other detailed information on each model of car. They are a help to mechanics on difficult repairs and critical adjustments. They are written in concise, technical language.

Types of service manuals

There are several types of service manuals from which to choose. They are published by various sources.

Manufacturer's manuals are published by every auto maker, foreign and domestic. Also called *factory manuals,* each covers every system of every car produced by that company. New manuals are produced with each year's new models.

Specialized service manuals cover repair procedures for only certain parts of the car. One may cover engines and engine systems, another body components or electrical systems. They will usually be issued in several volumes, each covering one section of the automobile.

Publishers include auto makers and after-market (suppliers to automotive field) companies.

General repair manuals are those prepared by companies whose business is publishing (Mitchell Manuals, Motor Manuals, Chilton Manuals). While like manufacturer's manuals, these publications are not as detailed. Moreover, one publication may include data on automobiles manufactured over several years. Others will include only repairs for foreign cars, light trucks, or large trucks.

Because of their cost, a garage may buy one or two general repair manuals rather than a manual for each make and year of car. These manuals will summarize the most important information.

Service manual sections

To make information easier to find, service manuals are divided into various sections. These sections might include: general information, engines, fuel systems, transmissions, etc. See Fig. 26-1.

The *General Information Section* covers basic information which will generally relate to all models of the manufacturer such as identification numbers, belt tension, part identification numbers, general maintenance and lubrication, and fluid capacities.

In the General Information section, an important topic is the *vehicle identification number.* Usually referred to as the ID NUMBER or VIN, it contains a series of code numbers and letters that provide data about the type and style of car and its equipment, Fig. 26-2. This number is commonly used when ordering parts. The manual will explain what each part of the VIN code means, Fig. 26-3.

Repair sections of the service manual cover every major system of the car. By following instructions found here, the mechanic can diagnose problems, make inspections, perform tests, and repair any component. There will be many line drawings showing parts and how they are assembled. One page may describe how to remove a throttle body injector; another page may show how to disassemble the throttle body, itself.

Service manual specifications

Repair sections also give *specifications* needed for various tests and repairs. Servicing of fuel systems may

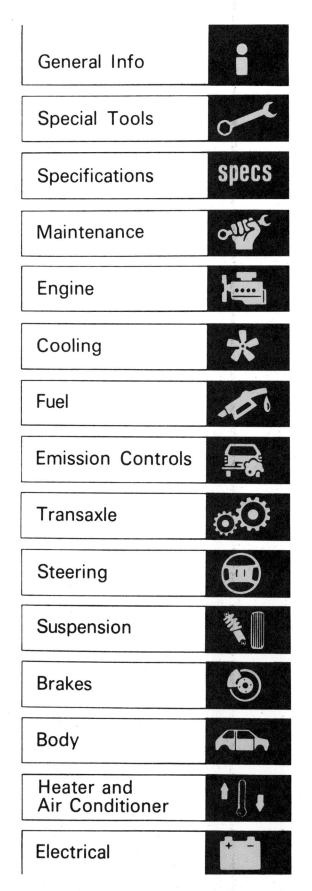

Fig. 26-1. Service manual is divided into sections covering various systems and parts of the automobile. To quickly find section with this type of index, look at edge of manual and locate block corresponding to section desired. (Honda)

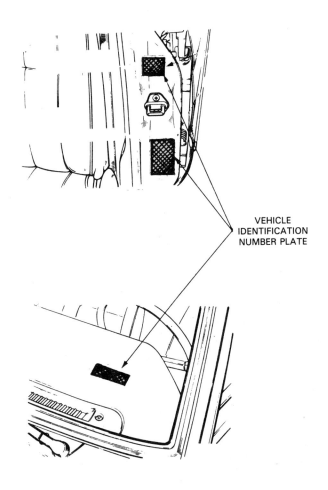

VEHICLE IDENTIFICATION NUMBER PLATE

Fig. 26-2. Vehicle identification number (VIN) may be located on door, dashboard, in engine compartment, or other body section. Manual for specific car will tell where it is located. (Subaru)

require reference to fuel pressure, fuel tank capacity, clearances, operating temperatures, voltages and resistances, and torque values for fasteners.

Special tools

Repair sections will also often refer to *special tools.* These tools will usually be described during explanation of specific repairs or adjustment of certain parts. They may even be pictured, along with their part numbers, at the end of the repair section.

Service manual illustrations

Illustrations of various types supplement the detailed instructions. They are included for many different reasons. Some demonstrate how to measure part wear or how to install a part. Others show an *exploded* (disassembled) view of a component. Still others will illustrate part removal and replacement. See Fig. 26-4.

As you use the service manual, pay particular attention to the illustrations. They are essential to a full understanding of procedures and specifications. Not only do they show what the part looks like, but how parts fit together, where leaks might occur, and how parts should operate.

VEHICLE IDENTIFICATION NUMBER

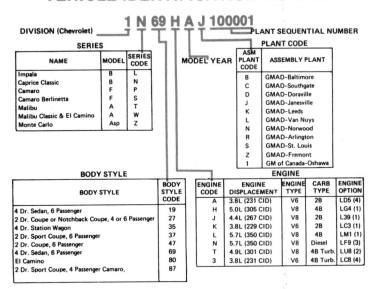

DIVISION (Chevrolet) — 1 N 69 H A J 100001 — PLANT SEQUENTIAL NUMBER

SERIES

NAME	MODEL	SERIES CODE
Impala	B	L
Caprice Classic	B	N
Camaro	F	P
Camaro Berlinetta	F	S
Malibu	A	T
Malibu Classic & El Camino	A	W
Monte Carlo	Asp	Z

PLANT CODE

ASM PLANT CODE	ASSEMBLY PLANT
B	GMAD–Baltimore
C	GMAD–Southgate
D	GMAD–Doraville
J	GMAD–Janesville
K	GMAD–Leeds
L	GMAD–Van Nuys
N	GMAD–Norwood
R	GMAD–Arlington
S	GMAD–St. Louis
Z	GMAD–Fremont
1	GM of Canada–Oshawa

BODY STYLE

BODY STYLE	BODY STYLE CODE
4 Dr. Sedan, 6 Passenger	19
2 Dr. Coupe or Notchback Coupe, 4 or 6 Passenger	27
4 Dr. Station Wagon	35
2 Dr. Sport Coupe, 6 Passenger	37
2 Dr. Coupe, 6 Passenger	47
4 Dr. Sedan, 6 Passenger	69
El Camino	80
2 Dr. Sport Coupe, 4 Passenger Camaro,	87

ENGINE

ENGINE CODE	ENGINE DISPLACEMENT	ENGINE TYPE	CARB TYPE.	ENGINE OPTION
A	3.8L (231 CID)	V6	2B	LD5 (4)
H	5.0L (305 CID)	V8	4B	LG4 (1)
J	4.4L (267 CID)	V8	2B	L39 (1)
K	3.8L (229 CID)	V6	2B	LC3 (1)
L	5.7L (350 CID)	V8	4B	LM1 (1)
N	5.7L (350 CID)	V8	Diesel	LF9 (3)
T	4.9L (301 CID)	V8	4B Turb.	LU8 (2)
3	3.8L (231 CID)	V6	4B Turb.	LC8 (4)

(1) PRODUCED BY GM - CHEVROLET MOTOR DIVISION (3) PRODUCED BY GM - OLDSMOBILE DIVISION
(2) PRODUCED BY GM - PONTIAC MOTOR DIVISION (4) PRODUCED BY GM - BUICK MOTOR DIVISION

Fig. 26-3. VIN is a code. Service manual will explain significance of each set of numbers or letters. (General Motors Corp.)

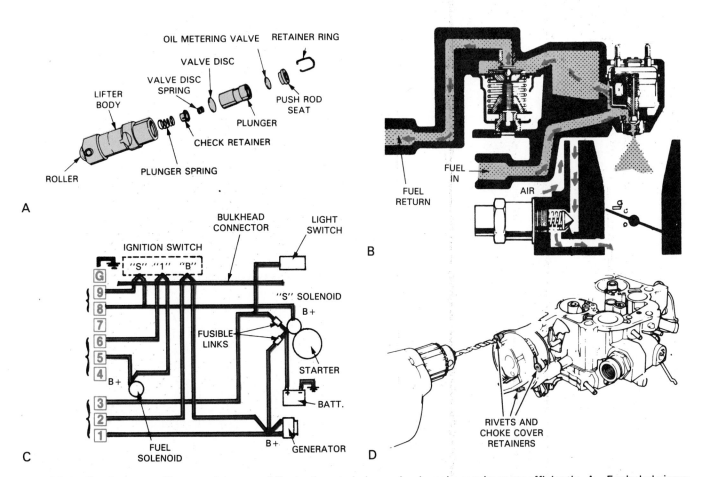

Fig. 26-4. Service manual has special types of illustrations to help mechanic make repairs more efficiently. A—Exploded views show how parts are assembled. B—Cutaways show parts and how they function. C—Wiring diagrams show proper routing and connection of wiring harness to components. D—Line art and photos show how to remove parts or make repairs. (Oldsmobile, Chevrolet, and Buick)

Service manual diagrams

Diagrams are drawings which show routing (paths) of electrical circuits, vacuum hoses, and hydraulic circuits. They show where wires, hoses, and passages go and where connections are made between different parts.

Wiring diagrams, as shown in Fig. 26-4C, show how the automobile's wiring is hooked up from one electrical component to another. Check these diagrams carefully when you are working on electronic components of fuel systems.

Vacuum diagrams are somewhat like wiring diagrams. They help the mechanic trace vacuum hoses between the engine and vacuum-operated devices, Fig. 26-5.

Service manual abbreviations

To save space in their publications, auto manufacturers use *abbreviations* which are shortened words con- sisting of a few letters used in the word. Sometimes these abbreviations are explained the first time they are used. In other cases, a chart will be included in the front or back of the manual which defines them.

Abbreviations will vary from one manufacturer to another. We have, therefore, included in this text only those which are generally used and recognized by all auto manufacturers and auto mechanics. Fig. 26-6 lists abbreviations used by one auto manufacturer.

Troubleshooting charts

Troubleshooting or *diagnosis charts* are step-by-step, logical methods for finding and correcting malfunctions. They include inspections, tests, measurements, and repairs. They are useful guides for correcting the most difficult and common problems.

A *tree diagnosis chart* is a test sequence designed to

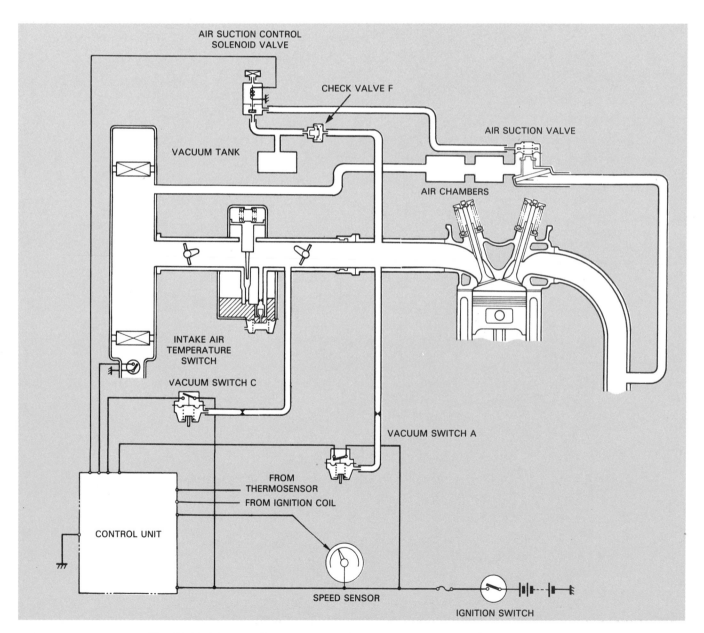

Fig. 26-5. Vacuum diagrams are useful in tracing vacuum hoses.   (Honda)

TYPICAL ABBREVIATIONS	
AI	Air Injection
A/T	Automatic Transmission
BTDC	Before Top Dead Center
EGR	Exhaust Gas Recirculation
EVAP	Evaporative (Emission Control)
EX	Exhaust (manifold, valve)
Ex.	Except
IN	Intake (manifold, valve), Inch
MC	Mixture Control
M/T	Manual Transmission
O/S	Oversized
PCV	Positive Crankcase Ventilation
SC	Spark Control
SST	Special Service Tool
STD	Standard
S/W	Switch
TDC	Top Dead Center
TP	Throttle Positioner
U/S	Undersized
W/O	Without

Fig. 26-6. Manufacturers list abbreviations they use in their repair manuals. (Toyota)

find the malfunctioning part by checking one part at a time until the problem is located. You simply work your way down through the chart or ''tree'' eliminating one part at a time until the bad part is fixed.

Block diagnosis charts list symptoms, probable causes, and method of correcting the problem in columns. In the first column, the symptom would be listed. Additional columns list the diagnostic test for various parts that could be the cause of the trouble, Fig. 26-7.

Illustrated diagnosis charts use photos, symbols, and instructions to guide the mechanic through test sequences. These types of troubleshooting aids lead the mechanic, step-by-step, through the diagnosis and repair. They tell AND show what to do.

USING A SERVICE MANUAL

If you have carefully studied the preceding pages of this chapter, you will be familiar with the various parts of the service manual. To get the most out of the manual, follow these basic steps:
1. Select the correct manual for the repair. Some manuals are published in volumes making up a set. Each volume will cover certain repair areas. Others cover all parts of the automobile and all makes. If you are working on a diesel injection system, look for the manual which covers that repair the best.

NOTE:

● First, confirm that the idle speed is normal by pinching the No. 10 vacuum hose slightly. Then inspect the air cleaner element, ignition timing, spark plug and valve clearance.

● Before going through the Electrical Troubleshooting Charts, make sure that the vacuum hoses are not loose and securely tightened, and that the solenoid valves, throttle body and fast idle mechanism are in good order.

Part / Symptom	Idle control solenoid valve	Fast idle control solenoid valve	A/C idle control solenoid valve	Throttle body	Fast idle mechanism	Starter switch signal	Alternator FR terminal signal	A/C switch signal	ECU	Remarks
Idle speed does not increase after initial start-up.	Valve failure/pinched hose				Adjust screw out of adjustment	Open circuit			Failure in ECU	Is signal available at ECU?
Idle speed too high in neutral	Leaky solenoid valve	Leaky solenoid valve		Valve stuck open	Adjust screw out of adjustment. Leaky fast idle valve.				Failure (signal not stopped)	Pinch idle control solenoid valve hose and re-adjust.
Idle speed changes under electrical load.	Valve failure/pinched vacuum hose			Throttle angle sensor out of adjustment. Valve stuck open.	Adjust screw out of adjustment				Failure (signal not available)	• Is idle control solenoid valve working? • Is fast idle adjust screw adjustment correct?
Idle speed drops when blipping throttle with electrical load.							Open circuit		Failure in ECU	Is there big difference between no-load and loaded conditions?
Idle speed drops when A/C switch is turned ON.			Valve failure Pinched vacuum hose		Opener opening out of adjustment			Open circuit	Failure in ECU	• Is vacuum applied to opener? • Is opener opening adjusted properly?
Idle speed fluctuates when idle control comes into operation.	Valve failure								Failure in ECU	• Is condition improved when solenoid valve is replaced?
Fast idle speed too low at high altitude (above 1,200 m, 4,000 ft).		Valve failure Pinched tube							Failure in ECU	• Is condition improved when solenoid valve is replaced?

Fig. 26-7. Block diagnosis chart lists symptoms and relates them to malfunctions or damage in parts. (Honda)

2. Refer to the table of contents to locate the section covering the system you are repairing or use the index. NEVER thumb through the manual hoping to find your subject.
3. Most manuals have small contents tables at the beginning of each major section. Check these also. They will help you find a topic quickly.
4. Read the procedures carefully. Study them until you are certain you understand what is to be done. A service manual's instructions are highly technical. Overlooking any step could spoil the repair and CAUSE FAILURE.
5. Study the illustrations closely and refer to them as needed during the repair. They make the instructions easier to understand and follow.

Flat rate manual

A *flat rate manual* is a reference tool that will help you calculate how much labor to charge the customer for the repair. It estimates the AMOUNT OF TIME a specific repair should take. This time should be multiplied by the shop's hourly labor rate to find the dollar amount of the labor charge.

A flat rate manual will also list typical prices for parts. For example, it might give the price of an air cleaner element and a TBI unit for a repair job. This would let you estimate the job for the customer before the part is purchased.

When time allows, you should study a flat rate manual. Practice looking up labor times and part costs. This will prepare you for the first time you are required to fill out a work order or give an estimate to a customer.

WORK ORDERS

Work orders serve as records of the repairs and parts needed on a car entering the shop. One is shown in Fig. 26-8. Note the various sections or areas on a work order.

Space for information about the customer and car usually is found at the top of the work order. Space for listing parts may be on the left. The center area of a work order is commonly reserved for filling out information on

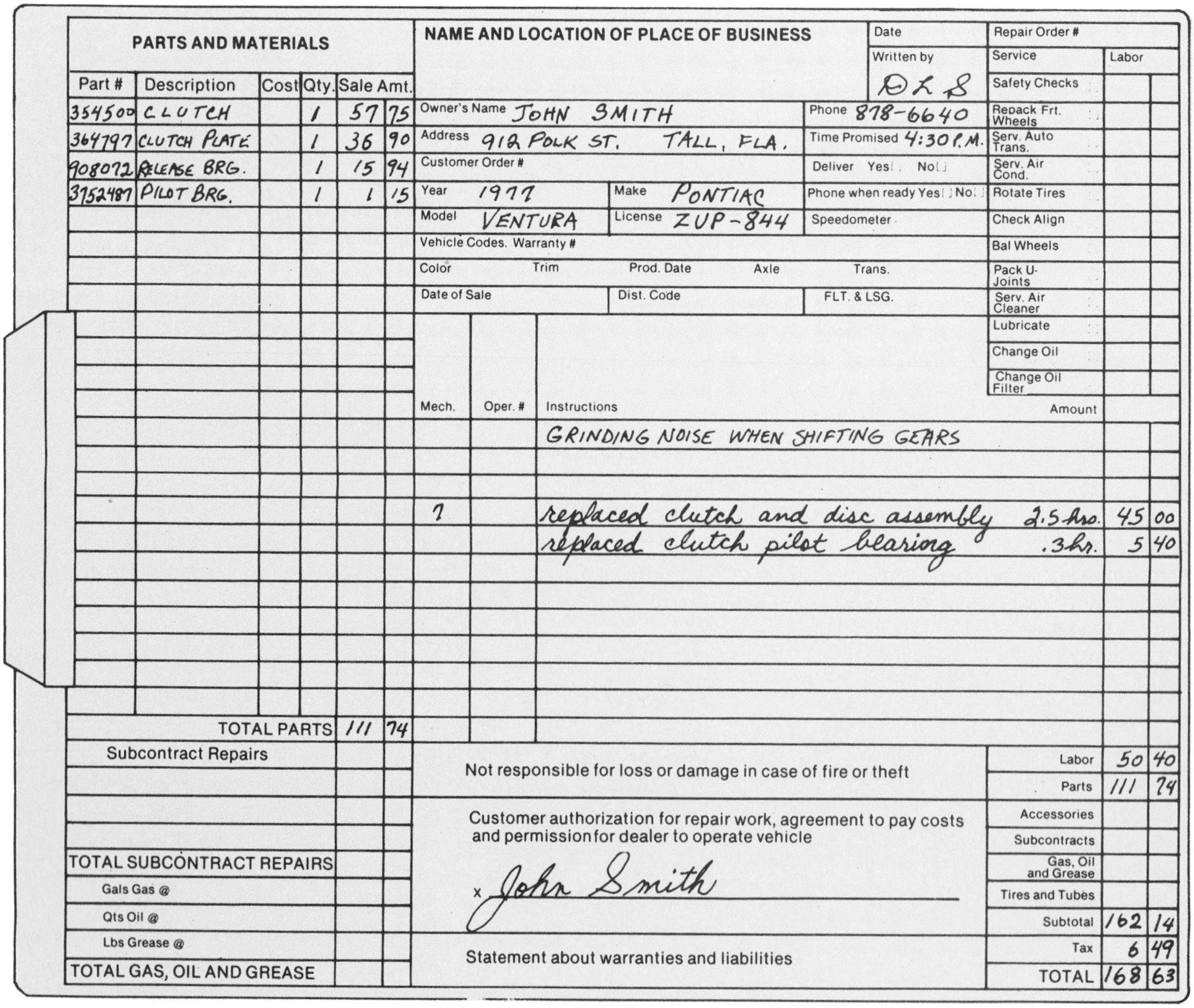
PARTS AND MATERIALS

Part #	Description	Cost	Qty.	Sale Amt.	
354500	CLUTCH		1	57	75
364797	CLUTCH PLATE		1	36	90
908072	RELEASE BRG.		1	15	94
3752487	PILOT BRG.		1	1	15
	TOTAL PARTS			111	74

Subcontract Repairs
TOTAL SUBCONTRACT REPAIRS
Gals Gas @
Qts Oil @
Lbs Grease @
TOTAL GAS, OIL AND GREASE

NAME AND LOCATION OF PLACE OF BUSINESS
Date
Written by DLS
Repair Order #
Owner's Name JOHN SMITH — Phone 878-6640
Address 912 POLK ST. TALL, FLA. — Time Promised 4:30 P.M.
Customer Order # — Deliver Yes No
Year 1977 — Make PONTIAC — Phone when ready Yes No
Model VENTURA — License ZUP-844 — Speedometer
Vehicle Codes. Warranty #
Color — Trim — Prod. Date — Axle — Trans.
Date of Sale — Dist. Code — FLT. & LSG.

Service / Labor: Safety Checks; Repack Frt. Wheels; Serv. Auto Trans.; Serv. Air Cond.; Rotate Tires; Check Align; Bal Wheels; Pack U-Joints; Serv. Air Cleaner; Lubricate; Change Oil; Change Oil Filter

Mech.	Oper. #	Instructions	Amount	
		GRINDING NOISE WHEN SHIFTING GEARS		
7		replaced clutch and disc assembly 2.5 hrs.	45	00
		replaced clutch pilot bearing .3 hr.	5	40

Not responsible for loss or damage in case of fire or theft

Customer authorization for repair work, agreement to pay costs and permission for dealer to operate vehicle

x John Smith

Statement about warranties and liabilities

Labor	50	40
Parts	111	74
Accessories		
Subcontracts		
Gas, Oil and Grease		
Tires and Tubes		
Subtotal	162	14
Tax	6	49
TOTAL	168	63

Fig. 26-8. Work order is record of complaint, parts needed, and work done. (Florida Voc. Ed.)

the needed repairs and what repairs have been done by the mechanic. The right border of the work order is for filling out costs for repairs. The bottom of the work order has space for totaling parts, labor, and tax. It also has a blank for the customer's signature to verify or OK repairs.

Using work orders

The service writer or shop supervisor usually fills out the information on the customer and car. He or she will write down the vehicle make, model, year, and the customer's name, address, phone number, and a description of the problem.

The mechanic will use the description of the problem to double-check and diagnose the source of trouble. (The mechanic knows that the description is a guess and might not be accurate.)

After deciding what must be done to repair the car, the mechanic will use the work order to order parts. The parts, with prices, will be listed on the work order. Any markup or profit on the parts will be the figure shown on the work order.

The mechanic will also list repairs or service operations completed: ''remove and rebuild a throttle body injector'' for example. Then, using the flat rate manual or actual time taken on the job, the mechanic will write in the labor charge for the job.

For example, if the job took two hours (flat rate or actual), the hours would be multiplied by the shop's hourly charge.

If the shop charged $38 dollars an hour, then 2 times $38 would equal a shop labor charge of $76 dollars. This figure is written on the work order.

Next, the mechanic or service writer totals the parts and labor and adds a tax charge to obtain the final total for the job. This is usually written at the lower right corner of the work order, Fig. 26-8.

SUMMARY

Service or shop manuals are carefully written, technical guides which tell how to diagnose and repair all automobile systems. They include specifications, diagrams, illustrations, and step-by-step procedures for maintenance and repair of each model of car.

There are three types of manuals: manufacturer's manuals, specialized service manuals, and general repair manuals. To use them the mechanic should refer to the table of contents or to the index for the specific topic. Special attention should be paid to the illustrations. They are designed to help explain the technical informa-

tion provided in the text.

Other important aids are the troubleshooting charts. They provide concise, logical procedures for finding the source of the problem and how to test and repair it. There are three types of troubleshooting charts: tree diagnosis charts, block diagnosis charts, and illustrated charts.

Flat rate manuals help the mechanic calculate how much labor to charge the customer for a specific repair. The rate is based on the amount of time the job usually takes.

Work orders are the records of the repairs and parts used on a customer's car. There are spaces on the work order for various types of information: customer's address, phone, and a description of the problem, parts used, and cost of the part.

KNOW THESE TERMS

Service manuals, Shop manuals, Manufacturer's manuals, Factory manuals, Specialized service manuals, General repair manuals, General information section, Vehicle identification number, Abbreviations, Tree diagnosis chart, Block diagnosis chart, Flat rate manual, Work orders, Service writer, Shop supervisor.

REVIEW QUESTIONS—CHAPTER 26

1. List and describe the three types of service manual.
2. Because of the cost of manuals, garages may buy one or two _____ _____ _____.
3. If you were planning to test a fuel pump, in what part of the service manual would you look and what kind of information would you need?
4. _____ are drawings which show routing of electrical circuits, vacuum hoses, and hydraulic circuits.
5. A _____ _____ chart is a series of steps in diagnosis of a malfunction which is designed to lead the mechanic through a systematic test sequence that eliminates trouble sources one after another until the cause of the problem is found.
6. Indicate how you would find a specific repair section in a service manual.
7. The following helps you calculate how much labor to charge a customer for a certain automotive repair:
   a. Work order.
   b. Shop manual.
   c. Service manual specification.
   d. Flat rate manual.
8. A record of the repairs and parts needed for a customer's car is called a _____ _____.

Auto Fuel Systems

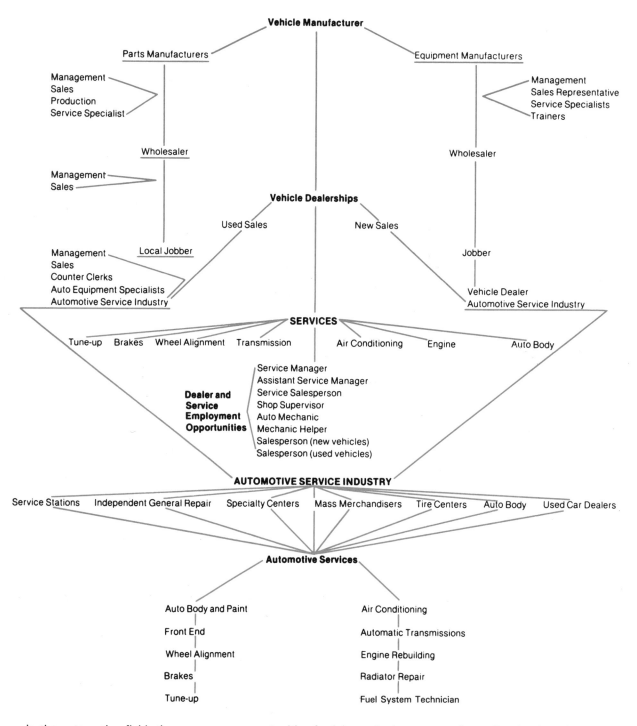

Vehicle Manufacturer

Parts Manufacturers — Equipment Manufacturers

Management
Sales
Production
Service Specialist

Management
Sales Representative
Service Specialists
Trainers

Wholesaler

Management
Sales

Wholesaler

Vehicle Dealerships

Used Sales — New Sales

Local Jobber

Management
Sales
Counter Clerks
Auto Equipment Specialists
Automotive Service Industry

Jobber

Vehicle Dealer
Automotive Service Industry

SERVICES

Tune-up   Brakes   Wheel Alignment   Transmission   Air Conditioning   Engine   Auto Body

Dealer and Service Employment Opportunities

Service Manager
Assistant Service Manager
Service Salesperson
Shop Supervisor
Auto Mechanic
Mechanic Helper
Salesperson (new vehicles)
Salesperson (used vehicles)

AUTOMOTIVE SERVICE INDUSTRY

Service Stations   Independent General Repair   Specialty Centers   Mass Merchandisers   Tire Centers   Auto Body   Used Car Dealers

Automotive Services

Auto Body and Paint

Front End

Wheel Alignment

Brakes

Tune-up

Air Conditioning

Automatic Transmissions

Engine Rebuilding

Radiator Repair

Fuel System Technician

In the automotive field, there are many opportunities for jobs and advancement for well-trained mechanics.
(Florida Dept. of Voc. Ed.)

27 SUCCEEDING IN FUEL AND EMISSION CONTROL SYSTEM SERVICE

After studying this chapter, you will be able to:
- *Explain why quality repairs and work speed are both equally important to fuel system service.*
- *Describe how personal habits on the job can affect advancement.*
- *Explain how tool selection and care affect profits and safety.*
- *Summarize the importance of good communication with other workers.*
- *List basic rules for succeeding on the job.*

Today's auto mechanics are constantly being challenged with new technology. As a result, good mechanics are always striving to improve their knowledge and skills to keep up with new car designs.

This chapter will summarize basic methods for becoming a competent auto technician. It will discuss work habits, communication skills, personal attitude, and other factors that contribute to success as a fuel system technician.

KEEPING UP WITH TECHNOLOGY

The auto mechanic must keep up with the advanced technology being used in new cars. Recent innovations include computers, electronic ignitions, and front-wheel drive. A good mechanic will constantly study service manuals, service bulletins, and other literature to learn how new components work and how to repair them.

Develop the good habit of reading during breaks and lunch hours. Subscribe to some of the automotive magazines. If you work for a large auto dealership, ask the service manager for service bulletins. Auto makers release these regularly to inform mechanics of special methods for correcting problems.

Concentrate on learning about the changes in your specific field. Staying knowledgeable about changing components will make your work more enjoyable, give you a sense of pride and accomplishment, and increase your value to your employer.

AVOID the pitfall of ever thinking you KNOW EVERYTHING ABOUT YOUR JOB. This only leads to trouble. You will soon be left behind by the new technology.

WORK QUALITY

A primary concern of any mechanic, shop owner, or customer is *work quality*. Repairs must be done correctly, following factory specifications.

It is tempting for a mechanic to hurry too much in an attempt to make more profit. When a mechanic works too fast, comebacks (repair must be done over) increase. More customers return unsatisfied with the repairs done on their cars. Then the mechanic must do the repair over again, free-of-charge. He or she loses money on lost work time, the shop owner loses money on wasted stall time, and the owner of the car loses because he or she must return the car to the shop.

WORK SPEED

Work speed is still very important to the mechanic. For those working on a percentage basis, income will depend upon the amount of work completed each week. For those on an hourly basis, employers may not give pay raises because the mechanics are not making the shop enough money.

To increase work speed without affecting work quality, think about ways you can increase work efficiency. Consider a new tool or technique that will save time without lowering the quality of the repair.

This kind of mental attitude will help you improve your skills. As a result, you will use time well and become more productive.

EARNINGS

Shops use three basic methods to pay their mechanics: hourly rate, commission, and commission plus parts.

The *hourly rate* is a stated amount of pay for each hour worked. This pay method is found in every type of shop. It may be desirable when you do not like the pressure of producing a quantity of work each week. The hourly rate is usually slightly lower than the commission method of pay.

When you earn a *commission,* you receive a percentage of the labor charged the customers. For example, with a 50 percent commission at $32 an hour, you would get $16 an hour for work completed.

If you are fairly quick and can match or exceed the *flat rate manual* (book that states how long each repair should take), you can earn a good wage. However, if you are slow or if there is not enough work to keep you busy, your income could be as low or lower than a mechanic paid on an hourly basis.

Commission plus parts is simply a commissioned mechanic that also gets a small percentage for the parts installed on cars. This gives the mechanic more incentive to sell parts to customers. The mechanic will be more inclined to check each automobile for things like worn belts and deteriorated hoses.

TYPES OF SHOPS

The type of shop you work in can affect your success as a mechanic. Some may be more desirable than others.

Large *auto dealerships* are good because they usually pay well and have a large volume of work to be done. However, they also require their mechanics to do warranty work.

Warranty work is repairs done on cars still under warranty. The rate of pay is considerably lower. If you are fairly experienced and fast, a large dealership may still be a good place to seek employment.

A *privately owned garage* can also be an excellent place to work. All of the work done is nonwarranty and the pay is normally good. As long as the shop is reputable and has enough volume to keep you busy, you might want to seek a job in a small garage. The shop owner controls the work atmosphere and conditions for the mechanics.

A *service station* having a repair area may be an excellent place to work, especially when you are a beginner. Much of the work will be *quick service* (repairs that only take an hour or two). You will replace belts, hoses, water pumps, thermostats, tires, and other items. Many service stations offer a parts percentage incentive which can increase your earnings.

INSURANCE

Insurance is meant to protect you from sickness or injury that prevents you from working. The shop may pay for or provide insurance or you may have to pay for insurance out of your paycheck. Always inquire as to benefits when applying for a job. This can make a big difference in your spendable income.

WORK CLOTHES

Many shops provide a standard set of *work clothes* for their mechanics. The mechanics usually must pay a small fee for purchasing or renting the work clothes and having them cleaned.

Work clothes are very important. They should be comfortable, well fitting, attractive in appearance, and clean. To the public, your work clothes reflect your mechanical abilities and work attitudes. If you look dirty and sloppy, they suspect your work may be less than desirable. You must always project a PROFESSIONAL IMAGE!

COOPERATION WITH OTHERS

Cooperation among mechanics can be very beneficial to everyone in a shop. Many repairs are difficult to complete without a helper. You may need someone to increase engine speed while checking engine performance, for example. Always keep a good working relationship with other mechanics. Offer to help them when they need assistance. In return, when you need help, they will be glad to assist.

SHOP SUPERVISOR AND SERVICE WRITER

Always maintain good working relations with the shop supervisor and service writer. In larger shops, they have the authority to assign jobs to mechanics. If you do not treat them with respect, you may be assigned more than your share of less desirable jobs.

PARTS DEPARTMENT

As just mentioned, keep good relations with everyone in the shop. This includes the workers in the parts department. You will have to depend on them to secure parts quickly and correctly for repairs you are making. This also applies to the people working at a parts house. If you do not give them respect, do not expect them to go out of their way to help you obtain a hard-to-get part.

ADVANCEMENT IN AUTO MECHANICS

Keep in mind the many opportunities in auto mechanics. Being a mechanic can be rewarding. However, there are many positions that you can advance to after mastering the skills of a mechanic. A few of these include:
1. Automotive instructor.
2. Service manager.
3. Shop supervisor.
4. Technical representative for an auto maker.
5. Technical representative for an auto aftermarket company.
6. Auto aftermarket sales representative.
7. Automotive engineer.
8. Shop owner.

Some of these positions will require you to take special training. You may have to attend college or take specialized courses. However, a basic knowledge of auto mechanics will give you an edge over others seeking the same kind of position.

The auto repair and manufacturing field is one of the largest employment areas in the nation. Your chances for succeeding in this field depend primarily upon your initiative and WILLINGNESS TO LEARN.

SUMMARY

Being an auto mechanic today offers many challenges and rewards. Your success will depend upon your willingness to develop competent skills, cooperate with employers and fellow workers, and keep abreast of new automotive technological developments.

As a mechanic you will be expected to work efficiently

and rapidly while producing neat, quality work. Depending on the shop or employer, you will be paid an hourly rate, a straight commission on the dollar amount of your work, or a commission plus a percentage of the parts you sell.

The type of shop will have some effect on your success as a mechanic. Large auto dealerships, as a rule, pay well and have a large volume of work. On the other hand, mechanics will be required to do warranty work which is lower paying.

Private garages normally pay well and do no warranty work. Service stations can be a good place for a beginner.

Besides offering rewarding work, auto mechanics also have opportunities to advance into occupations where automotive knowledge will be an asset. Some positions will require you to attend college or take specialized courses.

KNOW THESE TERMS

Service bulletins, Work quality, Work speed, Hourly rate, Commission, Commission plus parts, Warranty work, Quick service, Shop supervisor, Automotive instructor, Service manager, Technical representative, Auto aftermarket sales representative, Automotive engineer, Shop owner.

REVIEW QUESTIONS—CHAPTER 27

1. What is a "comeback"? Explain why mechanics should avoid them.
2. Select statement which is MOST correct:
   a. Work speed is most important to a mechanic's success.
   b. Speed is not important; a mechanic should work slowly so he or she will not make mistakes which could ruin the repair.
   c. A mechanic should work as rapidly as possible without affecting quality of work.
3. List and define the three methods by which mechanics are paid.
4. Discuss the advantages and disadvantages of working for a large auto dealership.
5. List eight possibilities for advancement in auto mechanics.

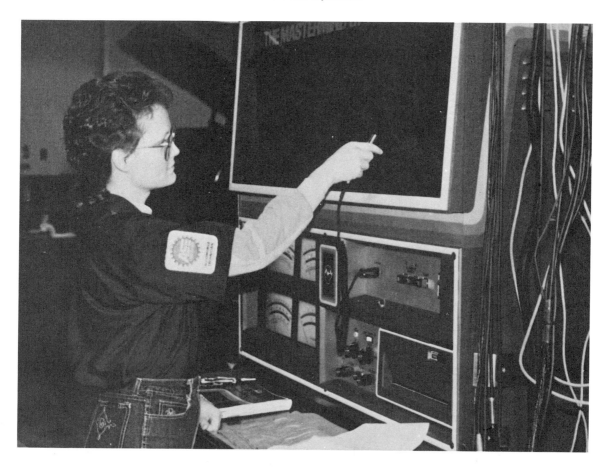

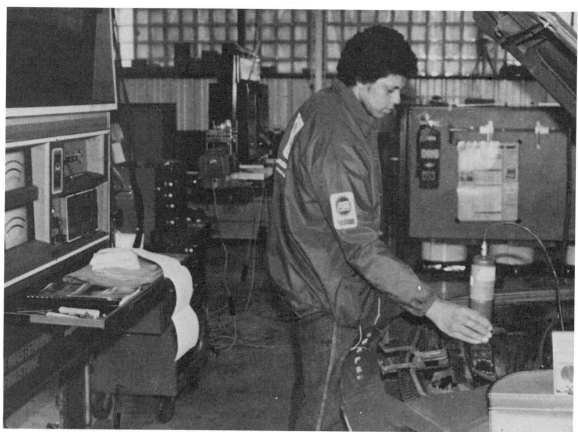

Mechanic who has passed the voluntary certification tests of the Automotive Service Excellence is entitled to wear the arm badge on their work shirt. It tells the customer that the wearer is a qualified mechanic.

28 CERTIFICATION

After studying this chapter, you will be able to:
- *Explain the purpose of the ASE testing program.*
- *Describe ASE testing categories.*
- *Describe why ASE certification can be beneficial to a mechanic and shop owner.*
- *List the certification test categories.*
- *Summarize the types of questions found in an ASE certification test*

Certification tests are multiple-choice quizzes designed to evaluate the technical knowledge of mechanics working in the field. They provide a method of measuring whether a mechanic is competent to service a specific system of a car, Fig. 28-1.

Fig. 28-1. Voluntary certification is a way of testing the competence of the auto mechanic. (ASE)

This text has covered the most important information for testing and servicing modern fuel systems. As a result, you should be prepared to pass several ASE certification tests, especially tests on engine performance and electrical systems.

Remember, always strive to learn more and more about your job! Read, talk to other mechanics, and make every effort to keep up-to-date about recent design changes in the automobile. This will make you a better, more successful automotive technician.

ASE (NIASE)

ASE, formerly NIASE, stands for "Automotive Service Excellence." It was shortened from "National Institute for Automotive Service Excellence." ASE is a nonprofit, nonaffiliated (no ties to industry) organization formed to help assure the highest standards in automotive service.

ASE directs an organized program of self-improvement under the guidance of a 40-member board of directors. These members represent all aspects of the automotive industry—educators, shop owners, consumer groups, government agencies, aftermarket parts companies, and auto manufacturers. This broad group of experts guides the ASE testing program and helps it keep in touch with the needs of the industry.

VOLUNTARY CERTIFICATION

ASE tests are voluntary. They do not have to be taken and they do not license mechanics.

Some countries have made mechanic certification a requirement. In the U.S.A., however, mechanics take the tests for personal benefit and to show their employer and customers that they are fully qualified to work on a system of a car.

ASE statistics show that over 300,000 mechanics have passed certification tests. Thousands of these have been retested and recertified after five years to maintain their credentials.

TEST CATEGORIES

There are eight test categories for auto mechanics: Engine Repair, Automatic Transmission/Transaxle, Manual Drive Train and Axles, Front End, Brakes, Electrical Systems, Heating and Air Conditioning, and Engine

515

Performance. You can take any one or all of these tests. However, only four tests (200 questions maximum) can be taken at one testing session. There are also six tests in heavy-duty truck repair and two tests in auto body.

Fig. 28-2 gives a breakdown of what each automotive service test involves. Study what they cover!

WHO CAN TAKE ASE TESTS?

To take ASE certification tests and receive certification, you must either have two years of on-the-job mechanic experience or one year of approved educational credit and one year of work experience.

However, you may take the tests even if you do NOT have the required two years experience. You will be sent a score for the test but not certification. After you have gained the mandatory experience, you can notify ASE and they will mail you a certificate.

You will be granted credit for formal training by one, or a combination, of the following types of schooling:
1. Three full years of high school training in auto mechanics may be substituted for one year of work experience.
2. Two years of post-high school training in a public or private facility can be substituted for one year of work experience.
3. Two months of short training courses can be substituted for one month of work experience.
4. Three years of an apprenticeship program, where you work under an experienced mechanic as a form of training, can be substituted for both years of work experience.

To substitute schooling for work experience, you must send a copy of your transcript (list of courses taken), a statement of training, or certificate to verify your training or apprenticeship. Each should give your length of training and subject area. This should accompany your registration form and fee payment.

TEST LOCATION AND DATES

ASE administers tests twice a year in over 300 locations across the United States. The test sites are usually community colleges or high schools. Tests are given in May and November. Contact ASE for more specific dates and locations for the tests. See Fig. 28-3.

TEST RESULTS

The results of your test will be mailed to your home. Only you will find out how you did on the tests. You can

Fig. 28-3. A certification test under way. Tests will consist of 40 to 80 multiple-choice questions.

TEST TITLE:		TEST CONTENT:
Engine Repair (80 Questions)	TEST A1	Valve train, cylinder head, and block assemblies; lubricating, cooling, ignition, fuel and carburetion, exhaust, and battery and starting systems
Automatic Transmission/Transaxle (40 Questions)	TEST A2	Controls and linkages; hydraulic and mechanical systems
Manual Drive Train and Axles (40 Questions)	TEST A3	Manual transmissions, clutches, front and rear drive systems
Front End (40 Questions)	TEST A4	Manual and power steering, suspension systems, alignment, and wheels and tires
Brakes (40 Questions)	TEST A5	Drum, disc, combination, and parking brake systems; power assist and hydraulic systems
Electrical Systems (40 Questions)	TEST A6	Batteries; starting, charging, lighting, and signaling systems; electrical instruments and accessories
Heating and Air Conditioning (40 Questions)	TEST A7	Refrigeration, heating and ventilating, A/C controls
Engine Performance (80 Questions)	TEST A8	Oscilloscopes and exhaust analyzers; emission control and charging systems; cooling, ignition, fuel and carburetion, exhaust, and battery and starting systems

Fig. 28-2. To become ASE certified, mechanics must pass one of these tests.   ASE)

Questions 7-9 deal with fuel systems. Practice question 7 deals with carburetion while 8 and 9 are import-type fuel injection questions. The Engine Performance Test (A8) includes (1) questions on carburetion and (2) an equal number of questions on import-type fuel injection systems. Technicians taking the Engine Performance Test will choose between these two groups when taking this test.

7. Which of these could happen when the accelerator pump inlet check ball is left out of a carburetor?

    (A) Flooding during acceleration

    (B) Hard starting and hesitation when the throttle is opened suddenly

    (C) Hesitation or stalling when the throttle is closed suddenly

    (D) Very rich mixtures during low speed driving

8. An engine with an electronic fuel injection (EFI) system stalls at idle.

    Technician A says that a binding air-flow sensor flap could be the cause.

    Technician B says that a loose connection in the fuel injection wiring harness could be the cause.

    Who is right?

    (A) A only   (B) B only   (C) Both A and B   (D) Neither A nor B

9. The primary system pressure is below specs on an engine with a continuous injection system (CIS). The fuel pump delivery volume is OK.

    Technician A says that the primary pressure regulator should be adjusted.

    Technician B says that the warm-up regulator should be checked.

    Who is right?

    (A) A only   (B) B only   (C) Both A and B   (D) Neither A nor B

Fig. 28-4. Sample test questions. Note comment at top of how fuel system is covered in Engine Performance test and how you can choose between carburetion- and fuel injection-related questions.

then inform your employer if you like.

Test scores will be mailed out a few weeks after you have completed the test. If you pass a test, you can consider taking more tests. If you fail, you will know that more study is needed before retaking the test.

TAKING TESTS

Each test, except Engine Performance and Engine Repair, has 40 multiple choice questions. You must read a statement or evaluate the question, and select the most correct answer. Primarily, the test covers diagnosis, service, and repair. You will not be required to recall specific specifications, unless they are general and apply to most makes and models (charging voltage for example).

Remember these tips when taking the tests:

1. Read the statements or questions slowly. You might want to read through them twice to make sure you fully understand the question.

2. Analyze the statement or question. Look for hints that make some of the possible answers wrong.

3. Analyze the question as if you were the mechanic trying to fix the car. Think of all possible situations and use common sense to pick the most correct response.

4. When two mechanics give statements concerning a problem, try to decide if either one is incorrect. If both are valid statements about a situation, mark both mechanics correct. If only one is correct, neither are correct, or just one, mark the answer accordingly. This is one of the most difficult types of questions.

5. If the statement only gives limited information, make sure you do not pick one answer as correct because it may be a more common condition. If the statement does not let you conclude one answer is better than another, both answers are equally correct.

6. Your first thought about which answer is correct is usually the correct response. If you think about a question too much, you will usually read something into the question that is not there. Read the question carefully and make a decision.

7. Do not waste time on any one question. Make sure you have time to answer all questions on the test.

8. Visualize how you would perform a test or repair when trying to answer a question. This will help you solve the problem more accurately.

9. Like the text, the questions refer to generic automotive service tasks and problems. They do not refer to specific makes and models of cars.

Fig. 28-4 gives a few sample questions from an ASE Bulletin. Can you answer these questions? The answers are 7(B), 8(C), and 9(A). Remember that these are the MOST CORRECT answers. Other answers may be correct in only certain situations.

BENEFITS OF ASE CERTIFICATION

When you pass an ASE test, you will be given a shoulder patch for your work uniform. Shown in Fig. 28-5, the patch has the Blue Seal insignia with either the words "Automotive Technician" (passed test in one area or more) or "Master Auto Technician" (passed all tests).

The patch will serve as good public relations, show-

Fig. 28-5. Passing the certification test entitles mechanic to wear shoulder patch indicating mastery of a particular area (or all areas) of automotive service. (ASE)

ing everyone that you are well trained to work on today's complex vehicles. It will also tell employers that you are someone special. You have expended extra effort to prove your value as a mechanic. It should lead to more rapid advancement and more income as customers indicate their preference for a certified mechanic.

MORE INFORMATION

For more information on ASE certification tests, write for a "Bulletin of Information" from: ASE, 1920 Association Drive, Reston, VA 22091. The bulletin will give test locations, testing dates, charges, sample questions, and other useful information.

SUMMARY

A national certification program under guidance of ASE (Automotive Service Excellence) is available to auto mechanics on a voluntary basis. The testing program measures the competence of auto mechanics in eight different categories. A mechanic may take any or all of the tests. Passing the test entitles him or her to wear the special shoulder patch attesting their competence.

To qualify for the tests, the auto mechanic must meet standards of training and education established by ASE's board of directors.

KNOW THESE TERMS

ASE, NIASE, Certification, Certificate, ASE Bulletin.

REVIEW QUESTIONS—CHAPTER 28

1. ASE stands for _____ _____ _____.
2. ASE is (indicate which of the following statements is NOT true):
   a. A federal testing program which licenses auto mechanics.
   b. A nonprofit voluntary program of testing.
   c. Not affiliated with any company or automotive industry.
   d. Under the direction of a board whose members come from all parts of the automotive industry.
3. List the eight categories of tests offered by ASE.
4. In general, to take the ASE tests you must have either two years of on-the-job training or one year of approved educational credit and a year of work experience. True or false?
5. Certification tests are given in _____ and _____ of each year.

INDEX

Index

ACKNOWLEDGEMENTS

The authors would like to thank the companies and individuals that helped make this book possible.

AMERICAN AUTO MANUFACTURERS
American Motors Corp.; Buick Motor Car Division; Cadillac Motor Car Division; Chevrolet Motor Division; Chrysler Motor Corp; Ford Motor Company; GMC Truck & Coach Division; Oldsmobile Division; Pontiac Motor Division.

FOREIGN AUTO MANUFACTURERS
American Honda Motor Co.; Alfa Romeo, Inc.; Aston Martin Lagonda, Inc.; BMW of North America, Inc.; British Leyland Motors, Inc. (Triumph and MG); Nissan Motor Corp., Datsun; Fiat Motors of North America, Inc.; Jaguar; Maserati Automobiles, Inc.; Mazda Motors of America, Inc.; Mercedes-Benz of North America, Inc.; Mitsubishi Motor Sales of America; Peugeot, Inc.; Renault USA, Inc.; Rolls-Royce, Inc.; Saab-Scandia of America, Inc.; Subaru of America, Inc.; S.A. Automobiles Citroen; Toyota Motor Sales, USA, Inc.; Volkswagen of America, Inc.-Porsche; Volvo of America.

DIESEL AND TRUCK RELATED COMPANIES
Airesearch Industrial Div.; Caterpillar Tractor Co.; Cummins Engine Co., Inc.; Detroit Diesel Allison Div.; GMC Truck and Coach Div.; International Harvester; Motor Vehicle Manufacturer's Assn.; White Diesel Div.

AUTOMOTIVE RELATED COMPANIES
AC-Delco; Airtex Automotive Division; Alloy; American Bosch; American Hammered Automotive Replacement Division; Ammco Tools, Inc.; AP Parts Co.; Armstrong Bros. Tool Co.; AP Parts Co.; Applied Power, Inc.; Automotive Control System Group; Beam Products Mfg. Co.; Bear Automotive; Belden Corp.; Bendix; Binks Mfg. Co.; Black & Decker, Inc.; Blackhawk Mfg. Co.; Bonney Tools; Borg-Warner Corp.; Bosch Power Tools; Carter Div. of AFC Ind.; Champion Spark Plug Co.; C.A. Laboratories, Inc.; Clayton Manufacturing Co.; Cleveland Motive Products; Clevite Corp.; Colt Industries; Chicago Rawhide Mfg. Co.; CRC Chemicals; Cy-lent Timing Gears Corp.; D.A.B. Industries, Inc.; Dana Corp.; Dake;

Dayco Corp.; Debcor, Inc.; Deere & Co.; Delco-Remy Div. of GMC; Detroit Art Services, Inc.; The DeVilbiss Co.; Duro-Chrome Hand Tools; The Echlin Mfg. Co.; Edu-Tech-A Division of Commercial Service Co.; H.B. Egan Manufacturing Co.; Ethyl Corp.; Exxon Co. USA; Fairgate Fuel Co., Inc.; Federal Mogul; Fel-Pro Inc.; Firestone Tire and Rubber Co.; Ford Parts and Service Division; Florida Dept. of Vocational Education; FMC Corporation; Fram Corp.; Gates Rubber Co.; General Tire & Rubber Co.; Goodall Manufacturing Co.; The B.F. Goodrich Co.; The Goodyear Tire & Rubber Co.; Gould Inc.; Gunk Chemical Div.; Hartridge Equipment Corp.; Hastings Mfg. Co.; Heli-Coil Products; Helm Inc.; Hennessy Industries; Holley Carburetor Div.; Hunter Engineering Co.; Ingersoll-Rand Co.; International Harvester Co.; Kansas Jack, Inc.; K-D Tools Manufacturing Co.; Keller Crescent Co.; Kem Manufacturing Co., Inc.; Kent-Moore; Killian Corp.; Kline Diesel Acc.; Kwick-Way Mfg. Co.; Lincoln St. Louis, Div. of McNeil Corp.; Lisle Corp.; Lister Diesels, Inc.; Lufkin Instrument Div.-Cooper Industries Inc.; Marquette Corp.; McCord Replacement Products Division; Mac Tools Inc.; Maremont Corp.; Minnesota Curriculum Services Center; Mobile Oil Corp.; Moog Automotive Inc.; Motorola; National Institute for Automotive Service Excellence; NAPA; OTC Tools & Equipment; Owatonna Tool Co.; Parker Hannifin Corp.; Precision Brand Products; Proto Tool Co.; Purolator Filter Division; Quaker State Corp.; Rochester Div. of GM; Roto-Master; Sealed Power Corp-Replacement Products Group; SATCO; Schwitzer Cooling Systems; Sears, Roebuck and Co.; Sellstrom Mfg. Co.; Sem Products, Inc.; Shell Oil Co.; Simpson Electric Co.; Sioux Tools, Inc.; Snap-on Tools Corp.; Speed Clip Sales Co.; Stanadyne, Inc.; The L.S. Starrett Co.; Stewart-Warner; Sun Electric Corp.; Sunnen Product Co.; Test Products Division-The Allen Group, Inc.; Texaco, Inc.; 3-M Company; Tomco (TI) Inc.; TRW Inc.; TWECO Products, Inc.; Uniroyal, Inc.; American Bosch Diesel Products; Vaco Products Co.; Valvoline Oil Co.; Victor Gasket Co.; Waukesha Engine Division, Dresser Industries, Inc.; Weatherhead Co.